Computer-Aided Chemical Engineering, 36

Sustainability of Products, Processes and Supply Chains

Theory and Applications

Computer-Aided Chemical Engineering, 36

Sustainability of Products, Processes and Supply Chains

Theory and Applications

Edited by

Fengqi You
McCormick School of Engineering
Northwestern University
Evanston, IL, USA

ELSEVIER

AMSTERDAM • BOSTON • HEIDELBERG • LONDON
NEW YORK • OXFORD • PARIS • SAN DIEGO
SAN FRANCISCO • SINGAPORE • SYDNEY • TOKYO

Elsevier
Radarweg 29, PO Box 211, 1000 AE Amsterdam, Netherlands
The Boulevard, Langford Lane, Kidlington, Oxford OX5 1GB, UK
225 Wyman Street, Waltham, MA 02451, USA

ISBN: 978-0-444-63472-6
ISSN: 1570-7946

For information on all Elsevier publications
visit our web site at http://store.elsevier.com/

Contents

PART V SUSTAINABILITY ANALYSIS METHODOLOGIES

List of Contributors

Arief Adhitya
Institute of Chemical and Engineering Sciences, Jurong Island, Singapore

Claire S. Adjiman
Department of Chemical Engineering, Centre for Process Systems Engineering, Imperial College London, London, UK

Suttichai Assabumrungrat
Center of Excellence in Catalysis and Catalytic Reaction Engineering, Department of Chemical Engineering, Faculty of Engineering, Chulalongkorn University, Bangkok, Thailand

Sara Badr
Institute for Chemical- and Bioengineering, Swiss Federal Institute of Technology, Zurich, Switzerland

Bhavik R. Bakshi
William G. Lowrie Department of Chemical and Biomolecular Engineering, The Ohio State University, OH, USA

Ana P. Barbosa-Povoa
Centro de Estudos de Gestao, IST, Universidade de Lisboa, Lisboa, Portugal

Fabrizio Bezzo
CAPE-Lab, Computer-Aided Process Engineering Laboratory, Department of Industrial Engineering, University of Padova, Padova, Italy

Lorenz T. Biegler
Department of Chemical Engineering, Carnegie Mellon University, Pittsburgh, PA, USA

Chen Bingzhen
Department of Chemical Engineering, Tsinghua University, Beijing, China

Aaron D. Bojarski
Department of Chemical Engineering, Universitat Politècnica de Catalunya, Barcelona, Spain

Ana Carvalho
Centro de Estudos de Gestao, IST, Universidade de Lisboa, Lisboa, Portugal

Alexandros Chremos
Department of Chemical Engineering, Centre for Process Systems Engineering, Imperial College London, London, UK

A. Kimi Coaldrake
Pan Pacific Technologies, Adelaide, Australia

Theodoros Damartzis
Chemical Process and Energy Resources Institute, Centre for Research and Technology-Hellas, Thermi; and Department of Mechanical Engineering, Aristotle University of Thessaloniki, Thessaloniki, Greece

Prodromos Daoutidis
Department of Chemical Engineering and Materials Science, University of Minnesota, Minneapolis, MN, USA

Alexander W. Dowling
Department of Chemical Engineering, Carnegie Mellon University, Pittsburgh, PA, USA

Eric H. Dunlop
Pan Pacific Technologies, Adelaide, Australia

Thomas F. Edgar
McKetta Department of Chemical Engineering, The University of Texas at Austin, Austin, TX, USA

Mahmoud M. El-Halwagi
Chemical Engineering Department, Texas A&M University, College Station, TX, USA

Antonio Espuña
Department of Chemical Engineering, Universitat Politècnica de Catalunya, Barcelona, Spain

Xiao Feng
Institute of New Energy, China University of Petroleum, Beijing, China

Antonio Flores-Tlacuahuac
Departamento de Ingeniería y Ciencias Químicas, Universidad Iberoamericana, México D. F, México

Esther Forte
Department of Chemical Engineering, Centre for Process Systems Engineering, Imperial College London, London, UK

Rebecca Frauzem
Department of Chemical and Biochemical Engineering, Technical University of Denmark, Kongens Lyngby, Denmark

Amparo Galindo
Department of Chemical Engineering, Centre for Process Systems Engineering, Imperial College London, London, UK

Rafiqul Gani
Department of Chemical and Biochemical Engineering, Technical University of Denmark, Kongens Lyngby, Denmark

Sara Giarola
CPSE, Centre for Process Systems Engineering, Imperial College London, London, UK

H. Giray Resat
Department of Industrial Engineering, Koc University, Istanbul, Turkey

Maria I. Gomes
Centro de Matematica e Aplicacoes, FCT, Universidade Nova de Lisboa, Caparica, Portugal

J. Betzabe González-Campos
Institute of Chemical and Biological Research, Universidad Michoacana de San Nicolás de Hidalgo, Morelia, Michoacán, México

Ignacio E. Grossmann
Department of Chemical Engineering, Carnegie Mellon University, Pittsburgh, PA, USA

Iskandar Halim
Institute of Chemical and Engineering Sciences, Jurong Island, Singapore

Rebecca J. Hanes
William G. Lowrie Department of Chemical and Biomolecular Engineering, The Ohio State University, OH, USA

Wang Hangzhou
Department of Chemical Engineering, Tsinghua University, Beijing, China

Konrad Hungerbühler
Institute for Chemical- and Bioengineering, Swiss Federal Institute of Technology, Zurich, Switzerland

George Jackson
Department of Chemical Engineering, Centre for Process Systems Engineering, Imperial College London, London, UK

Zhao Jinsong
Department of Chemical Engineering, Tsinghua University, Beijing, China

Kriti Kapoor
McKetta Department of Chemical Engineering, The University of Texas at Austin, Austin, TX, USA

Yoshiaki Kawajiri
School of Chemical & Biomolecular Engineering, Georgia Institute of Technology, Atlanta, GA, USA

Pichayapan Kongpanna
Center of Excellence in Catalysis and Catalytic Reaction Engineering, Department of Chemical Engineering, Faculty of Engineering, Chulalongkorn University, Bangkok, Thailand

Andrzej Kraslawski
School of Industrial Engineering and Management, Lappeenranta University of Technology, Lappeenranta, Finland; and Faculty of Process and Environmental Engineering, Technical University of Lodz, Lodz, Poland

Carlos Krieft
Institute for Mathematics, Universität Paderborn, Paderborn, Germany

Ying Labreche
School of Chemical & Biomolecular Engineering, Georgia Institute of Technology, Atlanta, GA, USA

José M. Laínez-Aguirre
Department of Industrial and Systems Engineering, University at Buffalo, Amherst, NY, USA

Jay H. Lee
Department of Chemical and Biomolecular Engineering, Korea Advanced Institute of Science and Technology, Yuseong-gu, Daejeon, Republic of Korea

Luis Fernando Lira-Barragán
Chemical Engineering Department, Universidad Michoacana de San Nicolás de Hidalgo, Morelia, Mich, México

Guilian Liu
Department of Chemical Engineering, School of Chemical Engineering and Technology, Xi'an Jiaotong University, Xi'an, China

Pei Liu
Department of Thermal Engineering, Tsinghua University, Beijing, China

Ryan P. Lively
School of Chemical & Biomolecular Engineering, Georgia Institute of Technology, Atlanta, GA, USA

Davide Manca
PSE-Lab, Process Systems Engineering Laboratory, Dipartimento di Chimica, Materiali e Ingegneria Chimica "Giulio Natta", Milano, Italy

Mariano M. Martín
Department of Chemical Engineering, University of Salamanca, Salamanca, Spain

W. Alex Marvin
Department of Chemical Engineering and Materials Science, University of Minnesota, Minneapolis, MN, USA

Richard Mastragostino
Department of Chemical Engineering, McMaster University, Hamilton, ON, Canada

Bruna Mota
Centro de Estudos de Gestao, IST, Universidade de Lisboa, Lisboa, Portugal

Rajib Mukherjee
Vishwamitra Research Institute, Chicago, IL, USA

Stavros Papadokonstantakis
Institute for Chemical- and Bioengineering, Swiss Federal Institute of Technology, Zurich, Switzerland

Athanasios I. Papadopoulos
Chemical Process and Energy Resources Institute, Centre for Research and Technology-Hellas, Thermi, Greece

Varong Pavarajarn
Center of Excellence in Particle Technology, Department of Chemical Engineering, Faculty of Engineering, Chulalongkorn University, Bangkok, Thailand

Mar Pérez-Fortes
Joint Research Centre, Institute for Energy and Transport, Petten, The Netherlands

Efstratios N. Pistikopoulos
Artie McFerrin Department of Chemical Engineering & Texas A&M Energy Institute, Texas A&M University, College Station, TX, USA

José María Ponce-Ortega
Chemical Engineering Department, Universidad Michoacana de San Nicolás de Hidalgo, Morelia, Mich, México

Tamara Popovic
School of Industrial Engineering and Management, Lappeenranta University of Technology, Lappeenranta, Finland

Luis Puigjaner
Department of Chemical Engineering, Universitat Politècnica de Catalunya, Barcelona, Spain

Matthew J. Realff
School of Chemical & Biomolecular Engineering, Georgia Institute of Technology, Atlanta, GA, USA

Fateme Rezaei
School of Chemical & Biomolecular Engineering, Georgia Institute of Technology, Atlanta, GA, USA

Martín Rivera-Toledo
Departamento de Ingeniería y Ciencias Químicas, Universidad Iberoamericana, México D. F, México

Kosan Roh
Department of Chemical and Biomolecular Engineering, Korea Advanced Institute of Science and Technology, Yuseong-gu, Daejeon, Republic of Korea

José E. Santibañez-Aguilar
Chemical Engineering Department, Universidad Michoacana de San Nicolás de Hidalgo, Morelia, Mich, México

Katherine Searcy
School of Chemical & Biomolecular Engineering, Georgia Institute of Technology, Atlanta, GA, USA

Panos Seferlis
Chemical Process and Energy Resources Institute, Centre for Research and Technology-Hellas, Thermi; and Department of Mechanical Engineering, Aristotle University of Thessaloniki, Thessaloniki, Greece

Warren D. Seider
Department of Chemical and Biomolecular Engineering, University of Pennsylvania, Philadelphia, PA, USA

Debalina Sengupta
Texas A&M University, TX, USA

Medardo Serna-González
Chemical Engineering Department, Universidad Michoacana de San Nicolás de Hidalgo, Morelia, Mich, México

Subhas K. Sikdar
NRMRL, US EPA, Cincinnati, OH, USA

Cory S. Silva
Department of Chemical and Biomolecular Engineering, University of Pennsylvania, Philadelphia, PA, USA

Rajagopalan Srinivasan
Indian Institute of Technology Gandhinagar, Ahmedabad, Gujarat, India

Christopher L.E. Swartz
Department of Chemical Engineering, McMaster University, Hamilton, ON, Canada

Subramanian Swernath
School of Chemical & Biomolecular Engineering, Georgia Institute of Technology, Atlanta, GA, USA

Qiu Tong
Department of Chemical Engineering, Tsinghua University, Beijing, China

Metin Turkay
Department of Industrial Engineering, Koc University, Istanbul, Turkey

Andrea Walther
Institute for Mathematics, Universität Paderborn, Paderborn, Germany

Han Wang
Department of Chemical Engineering, McMaster University, Hamilton, ON, Canada

Yajun Wang
Department of Chemical Engineering, Carnegie Mellon University, Pittsburgh, PA, USA

Craig Whittaker
Research & Development, Weyerhaeuser NR, Federal Way, WA, USA

Linlin Yang
Department of Chemical Engineering, Carnegie Mellon University, Pittsburgh, PA, USA

Kevin Yeh
School of Chemical & Biomolecular Engineering, Georgia Institute of Technology, Atlanta, GA, USA

Qiao Zhang
Department of Chemical Engineering, School of Chemical Engineering and Technology, Xi'an Jiaotong University, Xi'an, China

Chengchuan Zhou
Department of Earth and Environmental Engineering, Columbia University, New York, NY, USA

Edwin Zondervan
Department of Chemistry and Chemical Engineering, Eindhoven University of Technology, Eindhoven, The Netherlands

Preface

Concerns about climate change, environmental pollution, energy security, and resource depletion are driving society to explore more sustainable solutions. Sustainability has become a major challenge and also leads to exciting research opportunities. In the quest for sustainability, process systems engineering plays an increasingly important role due to its holistic systems approach, which is essential for addressing the complex sustainability problems. This book presents recent advances at the interface between sustainability and process systems engineering. The twenty-five chapters, by authors from twenty-one countries, cover a variety of topics on the recent development of systems engineering methodologies and computer-aided tools. Sustainability-related problems in the design and operations of multiscale process systems, ranging from molecular and products systems to energy and chemical processes, and to supply chains, are addressed. These chapters have been grouped into five parts.

Part I covers a number of topics on process design and optimization methodologies for sustainability applications. Yang, Martín, and Grossmann (Chapter 1) provide an overview of systematic methods for the design of water networks based on mathematical programming and targeting approaches and apply these methods toward reducing water consumption in the evolving biofuel industry. Kapoor and Edgar (Chapter 2) address the design of a chiller plant using a model-based approach to maximize energy efficiency. Dunlop, Coaldrake, Silva, and Seider (Chapter 3) describe an integrated systems approach that incorporates sustainability metrics and analysis into computer-aided modeling tools for the design of sustainable chemical processes for biofuels. Chapter 4 by Zondervan, Bojarski, Espuña, and Puigjaner presents a systematic framework for the integrated assessment of economic and environmental impacts in industrial chemical process design and applies the framework to a reactive distillation process for the production of fatty acid esters. A novel pinch analysis method for hydrogen network integration with purification and reuse is proposed by Feng et al. (Chapter 5). Zhao and co-workers (Chapter 6) present a bifurcation analysis method for bioprocess design.

Part II focuses on systems design and optimization for carbon capture, utilization, and storage. Gani and co-workers (Chapter 7) develop a processing network for CO_2 utilization and identify sustainable process design of CO_2 conversion into valuable products using a superstructure optimization approach. Biegler and co-workers (Chapter 8) propose an efficient computational method for the optimal design and operations of pressure swing adsorption process for carbon capture and separation by adopting a superstructure-based approach to allow simultaneous selection of adsorption process cycle steps and optimization of process operating parameters. Zhou, Liu, and Pistikopoulos (Chapter 9) describe superstructure-based modeling and optimization approaches to address the optimal design of a large-scale pipeline network for carbon dioxide transportation. Swernath, Searcy, Rezaei, Labreche, Lively, Realff, and Kawajiri (Chapter 10) develop a dynamic optimization strategy to find the optimal operating conditions of a rapid temperature swing adsorption process using hollow fibers for postcombustion CO_2 capture; they also conduct a detailed technoeconomic analysis based on the optimal operation strategies. Chapter 11 (Hungerbühler, Adjiman, Seferlis, and co-workers) describes an efficient computational approach for screening commercial and novel solvents and process configurations for sustainable chemisorption-based CO_2 capture and also presents detailed analysis results from economic, life cycle, and hazard assessment perspectives. Mukherjee, Sengupta, and Sikdar (Chapter 12) propose a sustainability footprint-based method for the selection of sustainable processes and apply this method to the case of methanol production from CO_2.

Part III includes five contributions addressing sustainability issues in supply chain design, analysis, and optimization. Barbosa-Povoa and co-workers (Chapter 13) present a multiobjective mixed-integer linear programming model for the economic, environmental, and social impact optimization of supply chains and apply it to a real case study of a Portuguese food retail company. Swartz et al. (Chapter 14) address the flexibility and responsiveness of process supply chains using a mathematical programming framework for operability analysis, and they apply these methodologies to a case study on a sustainable forest biorefinery. Chapter 15 by Srinivasan and co-workers proposes a systematic framework that integrates dynamic simulation with life cycle assessment for sustainability assessment of supply chains. Resat and Turkay (Chapter 16) describe a definition of sustainable transportation and propose major performance indexes, which are applied to multiobjective sustainable transportation problems. Chapter 17 (Flores-Tlacuahuac and co-workers) presents a novel dynamic optimization approach for the optimal synthesis of water treatment networks with simultaneous consideration of environmental and sustainability constraints.

Part IV consists of five contributions that cover issues pertaining to the design and optimization of sustainable bioenergy supply chains. Laínez-Aguirre, Pérez-Fortes, and Puigjaner (Chapter 18) address the design and planning of biomass supply chains for cocombustion plants under economic and environmental objectives using a bicriteria mixed-integer linear programming model, which was efficiently solved by a novel Lagrangean decomposition method. Chapter 19 by El-Halwagi and co-workers describes three mathematical programming models that consider multiple key issues for the optimal design and planning of a biorefinery system and proposes solution methods to these multiobjective problems. Marvin and Daoutidis (Chapter 20) develop a bicriteria optimization model for maximizing the net present value and minimizing the greenhouse gas emissions of a biofuel supply chain in the Midwestern United States. Chapter 21 by Yeh, Whittaker, Realff, and Lee addresses the long-term land management and production capacity optimization for a timberlands supply chain, modeled with a cyclic scheduling framework. Giarola and Bezzo (Chapter 22) provide a comprehensive overview of applying mathematical tools to address a variety of issues in biofuel infrastructure optimization, including feasibility, sustainability, market effect, and technology uncertainties.

Part V presents several recent advances on sustainability analysis theory and methodologies, beginning with the presentation of a multiscale modeling framework by Hanes and Bakshi (Chapter 23) that combines detailed sustainable process design models with coarser, aggregated models at the full life cycle value chain and economy scale. Popovic and Kraslawski (Chapter 24) discuss systematic assessment methods for social sustainability using a complex systems approach. Chapter 25 by Manca proposes a systematic approach to economic assessment and sustainable development and applies it to the design and operation of a chemical plant.

I would like to take this opportunity to extend my sincere appreciation to all the contributing authors for their chapters. I also greatly appreciate the time and efforts taken by the reviewers to comment on the manuscripts and improve the quality of the chapters. The support of Elsevier in publishing this book is gratefully acknowledged. Finally, special thanks go to the CACE series editors Professor Rafiqul Gani and Professor Stratos Pistikopoulos for their help and support throughout this project. It is hoped that this book will be a valuable contribution to the evolving field of sustainable process systems engineering.

Fengqi You

PART

I

Process design and optimization for sustainability

CHAPTER

1 Optimization Models for Process Water Networks and Their Application to Biofuel Processes

Linlin Yang*, Mariano M. Martín[§], Ignacio E. Grossmann*,[1]

[§]*Department of Chemical Engineering, University of Salamanca, Salamanca, Spain*

**Department of Chemical Engineering, Carnegie Mellon University, Pittsburgh, PA, USA*

[1]*Corresponding author: E-mail: grossmann@cmu.edu*

1.1 INTRO/MOTIVATION

Water is one of the most important natural resources on earth. In spite of its relative abundance, certain regions of the globe already suffer water stress, see Figure 1.1. Furthermore, the current usage of water and the climatic conditions suggest that by 2025, two-thirds of the world population will suffer from water scarcity (Rosegrant et al., 2002).

In Figure 1.2, the global distribution of the water consumption and water withdrawal is presented. Water withdrawal refers to the water that is diverted from a surface water or groundwater source. Part of it can be reintegrated with or without treatment, depending on its use. Water consumption represents the water that permanently leaves the water cycle either used in industry as a raw material or lost by evaporation. It can be seen that developed countries present an efficient ratio between water withdrawal and water consumption, but that is not the case for developing countries.

Water consumption and energy production (fuels or power), are two linked concepts. Water is consumed during energy production. It is consumed in reactions (i.e., hydrolysis), lost in the exhaust of the cooling towers, or contaminated during the process. On the other hand, energy is consumed in water treatment and purification processes. In particular, biofuels generation present two main water sinks—a large volume of water is required in the cultivation of biomass and additional water is consumed in the fuel production process. Water consumption level in growing biomass, which is the larger share of water requirement in biofuel water use, is highly dependent on the region, e.g., Brazil does not need manual irrigation for sugar cane growth. The second stage, the fuel production process, is the focus of this chapter. For some time ethanol production has been deemed inefficient in terms

Computer Aided Chemical Engineering, Volume 36. ISSN 1570-7946. http://dx.doi.org/10.1016/B978-0-444-63472-6.00001-X

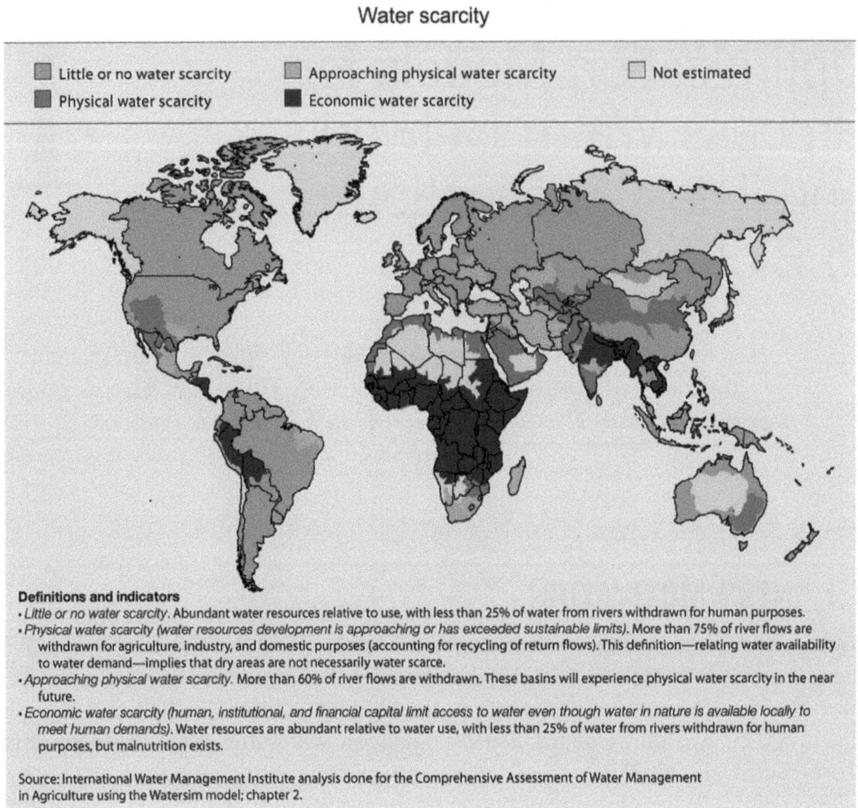

FIGURE 1.1

Water scarcity in different regions in the world.

of water and energy consumption. However, we must bear in mind that efficiency of other processes such as those involving the petrochemical industry has been obtained over a long time and with innovations in process development. Process system engineering approach provides the means for efficient and sustainable designs, and in this case, reduced water consumption.

1.2 WATER NETWORK

1.2.1 REVIEWS ON WATER NETWORK

1.2.1.1 Milestones

Conventional water use in the process industry does not reuse or recycle of water streams of different quality. However, a waste stream can be reused, recycled, or regenerated to lower the overall freshwater consumption. Water network (WN)

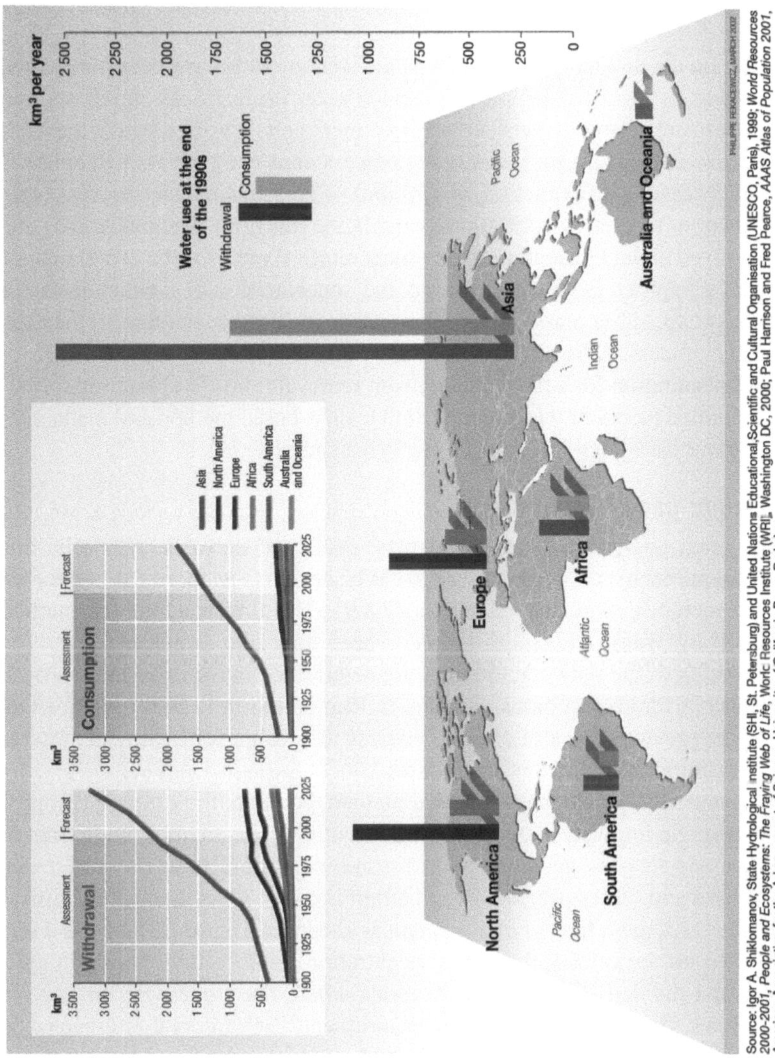

FIGURE 1.2

Water consumption and withdrawal throughout the world (UNEP, 2008).

design has been a major topic in process synthesis for the past 20 years. The major elements in a WN formulation are water-using process units and wastewater treatment units. A set of mixers and splitters are required to distribute the streams. Freshwater is supplied to the process units where contaminants are picked up, then the streams flow to the treatment units where contaminants are removed. The focus of this chapter is on steady state operations, though there are several works on batch processes.

Depending on the architecture of the WN, different models have been considered in the literature (e.g., water-using networks include water-using process units only, and wastewater treatment networks consider only treatment units) as discussed in the work by Bagajewicz and Faria (2009). Water-using process units can be classified either as mass transfer process (e.g., extraction, absorption) or nonmass transfer process (e.g., cooler exchanger or reactor that consumes water in the reaction). Treatment units can be defined by fixed outlet concentration or a fixed removal ratio for the contaminants. The information required is stream flow rate and concentration of contaminants. To define a process unit, either mass loads or flow rate is required depending on the classification of the process. A treatment unit design equation could be defined either for fixed outlet concentration for a fixed contaminant removal ratio. The resulting formulation is a simplified model of the network. Both insight-based method and mathematical programming have been investigated for WN synthesis.

1.2.1.2 Insight-based approaches

Inspired by pinch analysis for heat- and mass-exchange network synthesis, this insight-based approach has been applied to WN designs. A thorough review of insight-based methods is presented in Foo (2009). The first water pinch method was developed by Wang and Smith (1994) where flow rate targeting as well as the detailed design of the network structure are determined in a sequential approach through the use of limiting composite diagram. Kuo and Smith (1997) provided an extension to the previous work in order to address the design for treatment systems in both single and multiple contaminant cases.

To demonstrate the pinch analysis-based method, consider the limiting composite curve for a single impurity water-using process unit WN network without loss or surplus. Processes such as gas absorption assume negligible flow rate difference between the inlet and outlet streams. The diagram is constructed using contaminant concentration versus mass load of contaminant as shown in Figure 1.3. The blue line is plotted based on the inlet and outlet concentration for the contaminant in each process unit and divided into multiple intervals where the slopes are the inverse of the total flows. Based on the flow rate requirement, each interval is assumed to produce the curve. Then a water supply line can be drawn and rotated counterclockwise from the origin until it touches the limiting composite curve, the location where the two curves meet is the pinch location. The water supply predicted the minimum freshwater requirement for the set of process units. In addition, the concentration level indicated by the end point of the water supply line corresponds to the final contaminant concentration of the stream leaving the system. The limiting composite

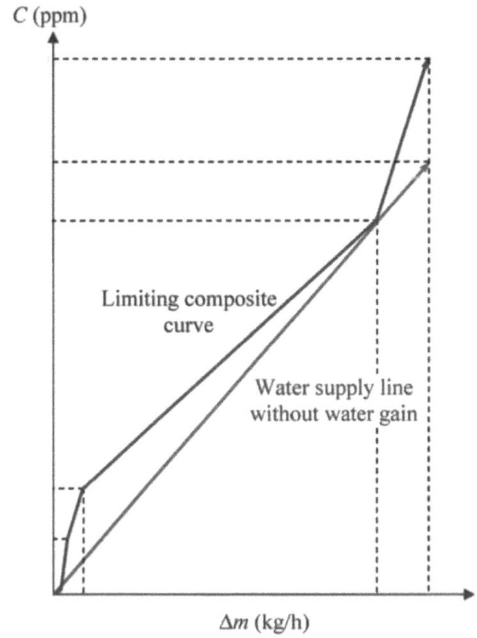

FIGURE 1.3

Composite curve for water targeting.

curve can be modified to reflect gain and loss of water in the process units such as in cooling tower by adjusting the water supply line. In addition, other insight-based procedures have also been developed such as the material recovery pinch diagram. Hallale (2002) presented a graphical approach called water surplus diagram, which provides a rigorous targeting technique for nonmass transfer processes with a single contaminant.

The main difference between pinch analysis in heat exchanger network (HEN) and WN is that whereas only one quality—heat—is being transferred in HEN, WN involve multiple contaminants. Tan et al. (2007) developed a targeting approach to target nonmass transfer water-using process units for two contaminants. However, even though the method is well-studied for WN with a single contaminant, no rigorous algorithm has been developed for a general multicontaminant WN.

1.2.2 MATHEMATICAL PROGRAMMING APPROACH

1.2.2.1 Superstructure with fixed recovery

Mathematical programming generally employs a superstructure that allows for all feasible network structures. Thorough reviews of mathematical optimization techniques can be found in Bagajewicz (2000) and Jezowski (2010). The advantages of this technique are that multiple contaminants can be rigorously taken into account and that targeting and designing can be done simultaneously. The first use of superstructure-based optimization approach for WN appears in the seminal work by Takama et al. (1980). The paper incorporates an NLP model for both water-using processes and treatment units for a multicontaminant system in one framework. Alva-Argaez et al. (1999) developed an MILP model that applies water pinch to wastewater minimization through the use of the transshipment model. Savelski and Bagajewicz developed the necessary conditions of optimality for minimum freshwater requirement for the case of single contaminant (2000) and then later for the case of multiple contaminant (2003). Huang et al. (1999) presented an NLP model to determine the optimal freshwater consumption and wastewater treatment capacity.

Karuppiah and Grossmann (2006) formulated an NLP problem to address both process and treatment units for a multicontaminant system, see Figure 1.4. Due to the effect of mixing, bilinear terms (flow rate multiplied by contaminant concentration) are present to account for individual contaminant flow, causing nonconvexities in the model. As a result, global optimization is required to achieve optimality. The nonlinear programming formulation for a multicontaminant WN with both water-using process units and wastewater treatment units are presented in Eqn (1.1).

$$
\begin{aligned}
\min \quad & \phi = FW + \sum_{\substack{t \in TU \\ i \in t_{out}}} F^i \\
& F^k = \sum_{i \in m_{in}} F^i \quad \forall m \in MU,\ k \in m_{out} \\
& F_k C_j^k = \sum_{i \in m_{in}} F_i C_j^i \quad \forall j, \quad \forall m \in MU,\ k \in m_{out} \\
& F^k = \sum_{i \in s_{out}} F^i \quad \forall s \in SU,\ k \in s_{in} \\
s.t. \quad & C_j^k = C_j^i \quad \forall j, \quad \forall s \in SU, \quad \forall i \in s_{out}, k \in s_{in} \\
& P_{in}^p C_j^k + L_j^p = P_{out}^p C_j^i \quad \forall j, \forall p \in PU, k \in p_{in}, i \in p_{out} \\
& F^k = F^i \quad \forall t \in TU, \quad \forall i \in t_{out}, \quad \forall k \in t_{in} \\
& C_j^i = \beta_j^t C_j^k \quad \forall j, \quad \forall t \in TU, \quad \forall i \in t_{out}, \forall k \in t_{in} \\
& F^{k,\min} \le F^k \le F^{k,\max} \quad \forall k \\
& C_j^{k,\min} \le C_j^k \le C_j^{k,\max} \quad \forall j, k
\end{aligned}
\tag{1.1}
$$

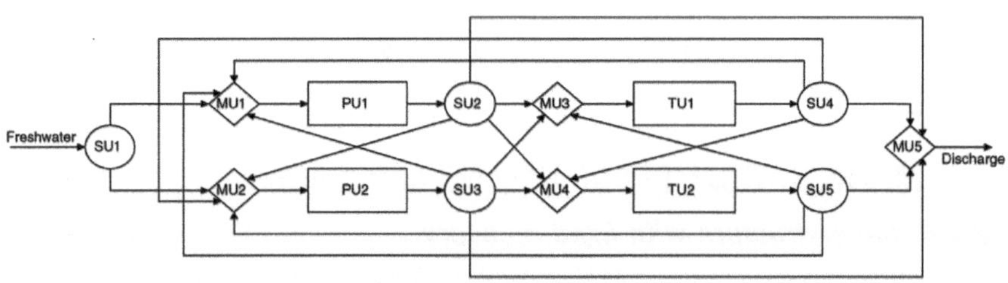

FIGURE 1.4

Integrated water network superstructure.

where F^k and F^i are flow rates (t/h) of streams i and k in the superstructure, respectively; C_j^k and C_j^i are concentrations (ppm) of contaminant j in streams k and I. P_{in}^p and P_{out}^p are parameters indicating the process unit inlet and outlet water flow rates; L_j^p are the mass load of contaminant j in unit p (kg/h); β_j^t indicate the removal fraction of treatment unit t in terms of treating contaminant j. The constraints consist of a set of contaminant mass balances in the mixer units, splitter units, process units, and treatment units. Note that in Eqn (1.1) the objective is simply to minimize freshwater consumption and flow for wastewater treatment. Obviously a more general cost function can be specified. Also, note that bilinearities arise in the second constraint.

A nonredundant-bound strengthening cut (Eqn (1.2)), representing the individual mass balance of each contaminant in the overall network, was proposed by Karuppiah and Grossmann (2008) to significantly improve the quality of the relaxation in the global optimization algorithm. In addition, this work proposes a generalized disjunctive programming formulation that allows the optimizer to select the best among multiple treatment technologies for the removal of each type of contaminant.

$$\sum_{p \in PU} L_j^p \times 10^3 = \sum_{\substack{t \in TU \\ i \in t_{in}}} \left(1 - \beta_j^t\right) + f_j^{out} \quad \forall j \tag{1.2}$$

Ahmetović and Grossmann (2011) expanded model (Eqn (1.1)) to include multiple freshwater sources and pretreatment options while considering all possible interactions among the units. The model is then formulated as an MINLP problem with binary variables to account for the cost of piping. The superstructure of the integrated WN model in Ahmetović and Grossmann (2011) is shown in Figure 1.5. This example considers one freshwater source, five water-using process units, three wastewater treatment units, and one sink. All combinations of reuse, regeneration reuse, local recycling, and mixing of streams are allowed.

The MINLP model for this illustrative example consists of 72 binary variables, 233 continuous variables, and 251 constraints. The optimal allocation of freshwater and recycle rates is obtained by globally optimizing the nonconvex MINLP problem. Using the global optimizer BARON, the problem can be solved to 1% optimality gap in 197.5 CPUs. The optimal network design is shown in Figure 1.6. The network consumes 40 ton/h of freshwater; in comparison, a conventional WN without recycling or reuse requires 300 ton/h of freshwater. In addition, only 13 removable connecting pipes are required to achieve this optimal configuration.

The issue of uncertainty is addressed in several publications. Karuppiah and Grossmann (2008) consider the integrated WN problem where the network is designed to be feasible over a range of operating conditions. Specifically, a multiscenario model is proposed, where each scenario differs from the others in the mass load of contaminant in the process units and the contaminant removal ratio in the treatment units. Since the resulting MINLP model, a deterministic equivalent of a two-stage stochastic programming model, is computationally much more complex, a Lagrangean decomposition method is used to solve the MINLP model to global optimality.

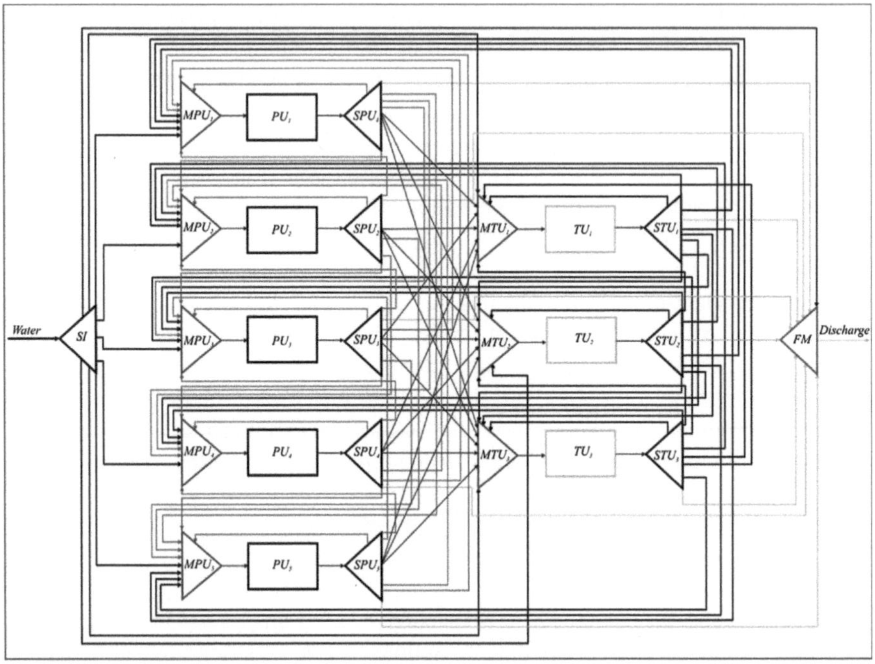

FIGURE 1.5

Integrated water network superstructure for Example 1.

1.2.2.2 Superstructure with wastewater treatment technologies

The standard formulation for a wastewater network design problem greatly simplifies the WN design by using fixed recoveries (i.e., the output concentration of the contaminants is expressed as a linear relation of the inlet concentration). However, this relation does not apply to all treatment technologies. Galan and Grossmann (1998) propose a general structure that can handle linear model with fixed removal ratio, technologies selection, and membrane technologies through nondispersive solvent extraction where a nonlinear model is used to describe the performance of the separation unit. Note that this work makes the first use of the optimization approach to the distributed wastewater treatment network problem. The NLP formulation is solved through an algorithm based on successively solving the relaxed linear lower bound and the original nonlinear nonconvex upper bounding problem is used to yield global or near-globally optimal solutions. The optimal solution for a distributed treatment system of three waste streams with three contaminants is shown in Figure 1.7.

There has been considerable work on wastewater treatment network based on reverse osmosis (RO). El-Halwagi (1992) presented an MINLP formulation using a state-space representation for optimal RO system synthesis. Lu et al. (2006)

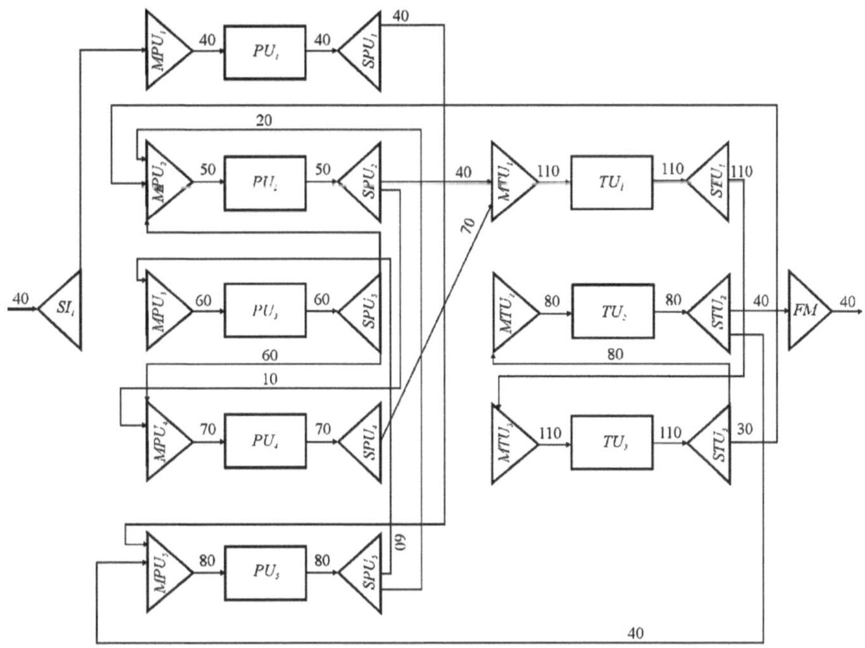

FIGURE 1.6

Optimal design of the integrated water network for Example 1.

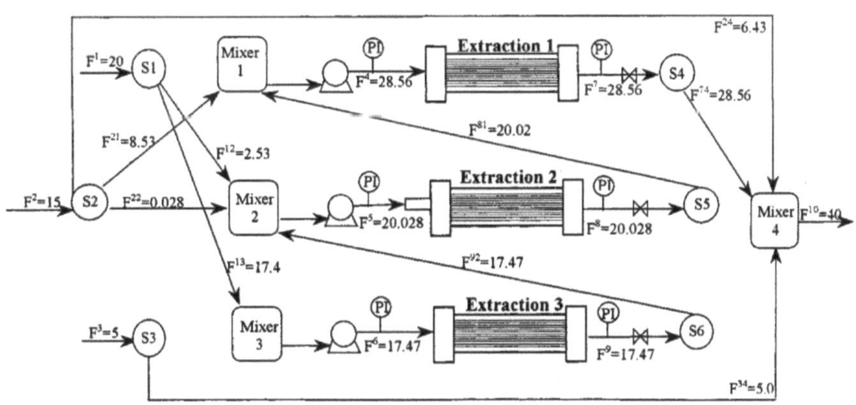

FIGURE 1.7

An optimal solution for a wastewater treatment system.

addressed the optimal cost design of RO desalination system including membrane modules for cleaning and replacement using an MINLP model. The authors extended the mass transfer solution-diffusion model by El-Halwagi (1992) and proposed a fouling model that describes membrane module performance. Khor et al. (2011) proposed an MINLP model for synthesizing water regeneration networks using a rigorous nonlinear RO model. Karuppiah et al. (2012) uses a detailed modeling of spiral wound membrane in a superstructure-based optimization framework to perform RO-based water treatment network synthesis. A two-stage stochastic programming formulation is adopted to incorporate the uncertainty associated with membrane performance.

Based on the superstructure for WN synthesis, Yang et al. (2014) presented an approach to incorporate treatment unit models that are common industrial practice to allow for a more thorough understanding of the trade-offs between unit performance and the economics of the WN design. The authors replaced simplified unit models by short-cut models to describe mass transfer in wastewater treatment units. The treatment units are selected among the best available techniques (Table 1.1). Nonlinear wastewater treatment cost functions are also taken into consideration. In addition, the superstructure allows for selection from multiple wastewater treatment units for the removal of each type of contaminant. Finally, different operating conditions are represented through the use of a three-scenario model, which was demonstrated by Karuppiah and Grossmann (2008), to solve the deterministic equivalent model of a two-stage stochastic programming model with recourse.

An example for this approach is a modified refinery case study previously reported by Ullmer et al. (2005). The freshwater sources, process units, treatment selections, and wastewater sinks are presented in Figure 1.8. The contaminants

Table 1.1 Best Available Techniques

	Suspended Solids (TSS)	Heavy Metals (HM)	Inorganic Salt (TDS)	Organic Unsuitable for Bio. Treat. (ORG)	Organic Suitable for Bio. Treat. (BOD)
Sedimentation	☑	☑			
Flotation	☑	☑			
Filtration	☑	☑			
Ultrafiltration	☑				
Precipitation		☑			
Ion exchange		☑	☑		
Reverse osmosis		☑	☑	☑	
Evaporation			☑	☑	
Biological sulfate treatment			☑		
Air oxidation				☑	
Adsorption				☑	
Anaerobic treatment					☑
Aerobic treatment					☑

Bio. Treat., biological treatment; TSS, total suspended solids; HM, heavy metals; TDS, total dissolved solids; ORG, organics unsuitable for biological treatment

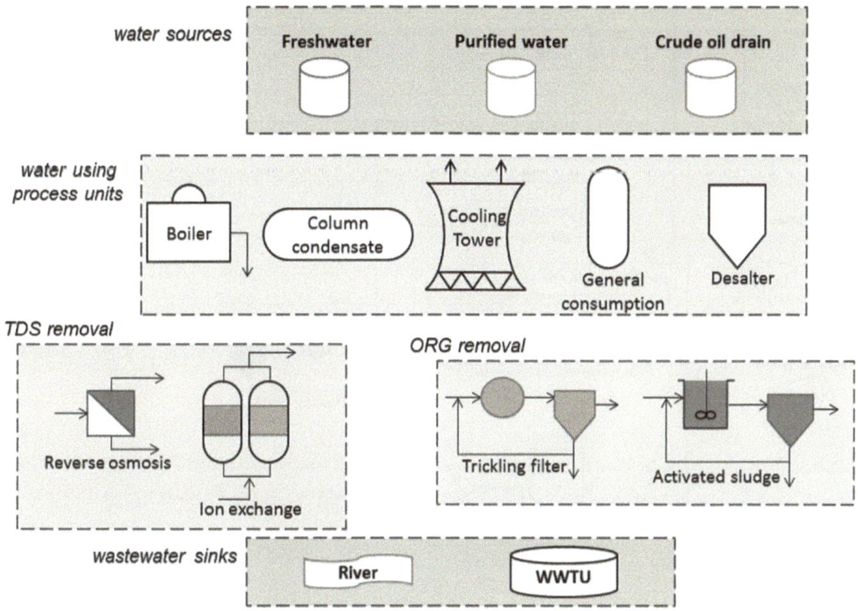

FIGURE 1.8

Refinery process water system for Example 2.

considered in this example are TDS and organics, which can be removed by RO or ion exchange, and trickling filter or activated sludge, respectively.

The resulting MINLP has 1768 constraints, 1331 continuous variables, and 85 binary variables. Note that the problem size is significantly larger than that of Example 1, which has five process units, three treatment units, one freshwater source, one sink, and three contaminants leading to an MINLP with 72 0–1 variables, 233 continuous variables, and 251 constraints. This difference in problem size is mainly due to the three-scenario model to account for uncertainty in the problem. Figure 1.9 shows the optimal network configuration, where trickling filter is selected for organics. The consumption of freshwater and purified water are 360.5 t/h and 47.5 t/h, respectively, which is a 734.5 t/h reduction compared to no water recycling. Note that no salt-removal unit is selected due to a combination of the high-TDS tolerance level at the inlet and the high rate of evaporation loss in the cooling tower. The cooling tower allows for reuse of most of the outlet stream flow, leading to a cooling loop that is more efficient than a once-through design.

1.2.3 TARGETING FOR SIMULTANEOUS HEAT/WATER OPTIMIZATION

An extension to WN synthesis is a simultaneous heat and water integration. Savulescu and Smith (1998) is the first work to consider a heat-integrated single-

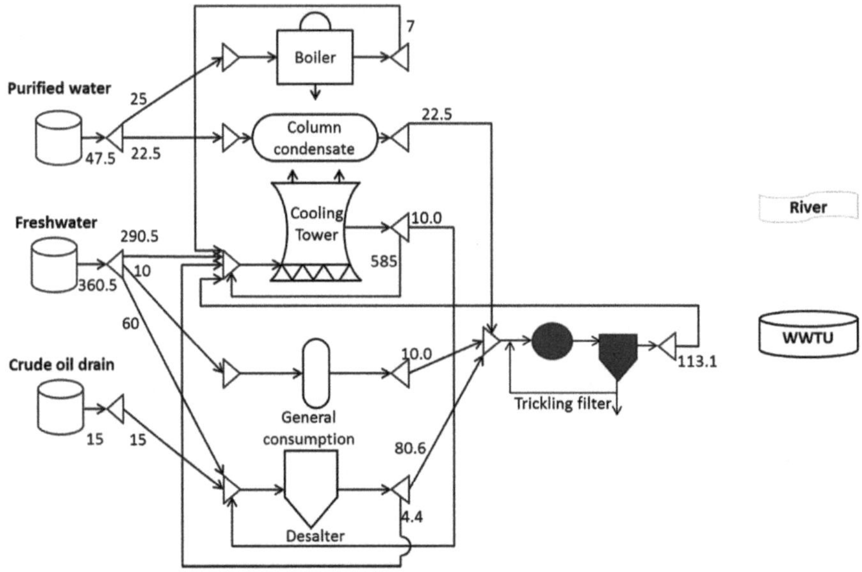

FIGURE 1.9

Optimal network configuration for Example 2.

contaminant WN with only water-using process units. The approach is insight-based sequential procedure, where direct heat transfer is maximized while maintaining the minimum freshwater consumption in the first stage; the HEN is designed in the second stage. A superstructure-based mathematical optimization model is subsequently presented by Bogataj and Bagajewicz (2008) for nonisothermal multicontaminant WN with only water-using process units. This MINLP model considers the typical WN model simultaneously with a modified HEN superstructure presented by Yee and Grossmann (1990). The challenge is handling the nonlinearities and nonconvexities in the model.

Similar to simultaneous optimization of process flowsheets and heat integration as demonstrated in Duran and Grossmann (1986), trade-offs among raw material, investment cost, and utility consumption for simultaneous process flowsheet and water integration can also be captured by allowing variable stream qualities. The challenge is solving the detailed HEN and WN, which are generally formulated as nonconvex MINLP or NLP. To simplify the model, Yang and Grossmann (2013) present an LP targeting model for minimizing the freshwater consumption of a multicontaminant system. All mass balances from formulation (1.2) remain unchanged with the exception of the individual mass balance in the mixer units, which contains bilinear terms. The equality is replaced by the linear inequality in Eqn (1.3) to for the LP targeting model.

$$F_k C_j^{k,\max} \geq \sum_{i \in m_{in}} F_i C_j^{i,\max} \quad \forall j, \quad \forall m \in MU, \; k \in m_{out} \tag{1.3}$$

This water targeting model, along with the available heat targeting model, is then incorporated into flowsheet optimization process to achieve minimum heating utility, cooling utility, and freshwater consumption. The first step of the process is to simultaneously optimize the process flowsheet model, water targeting model, and heat targeting model using the linear programming formulation (1.4).

$$\begin{aligned} \min \quad & \phi = F(x,u,v) + c_H Q_H + c_C Q_C + c_{fw} FW \\ \text{s.t} \quad & h(x,u,v) = 0 \\ & g^P(x,u,v) \leq 0 \\ & g^{HEN}(u, Q_H, Q_C) \leq 0 \\ & g^{WN}(v, FW) \leq 0 \\ & x \in X, u \in U, v \in V \end{aligned} \tag{1.4}$$

where $F(x,u,v)$, $h(x,u,v)$, and $g(x,u,v)$ are the objective function and equality and inequality constraints of the flowsheet in terms of the variables x, u, and v, where the u variables are involved in heat integration constraints, the variables v in the water integration constraints, x are the remaining variables to model the process, Q_H and Q_C are the heating and cooling loads, and FW is the freshwater consumption.

Once the operating conditions for the flowsheet are determined, the second step is to use WN and HEN models to synthesize their respective structures, see Figure 1.10. This approach can be applied to the design of biofuel plants as will be seen in the next section of this chapter.

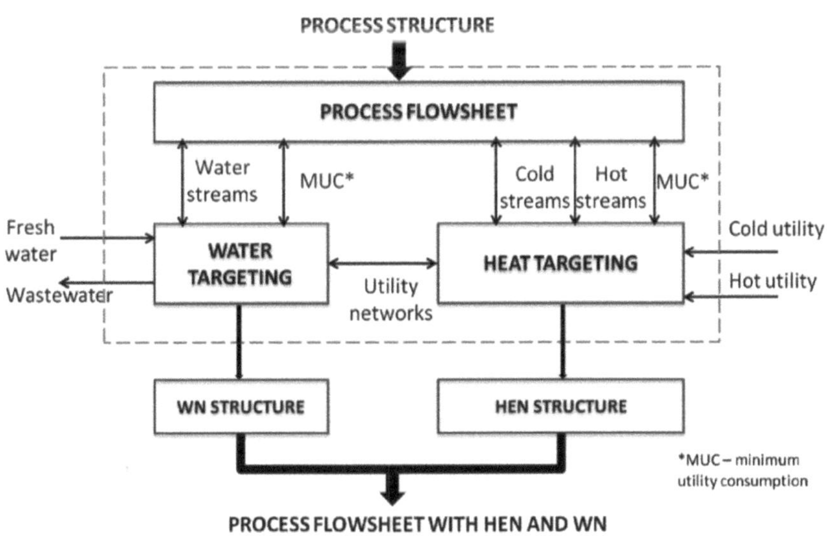

FIGURE 1.10

Simultaneous optimization diagram.

1.3 BIOREFINERY PROCESSES

1.3.1 GASIFICATION

Gasification is a process by which a carbonous mass is partially oxidized to produce raw syngas, consisting mainly of CO, CO_2, hydrogen, hydrocarbons, tar, and char. The specific composition of the gas phase depends on the gasifier design. We can identify mainly two designs, direct gasification and indirect gasification (Bridgewater, 2003). Direct gasification (Dutta and Phillips, 2009) is carried out in a single unit that operates at medium pressure and high temperature using steam and oxygen as fluidification agents. The advantage is that the raw syngas is relatively free from hydrocarbons. However, in order to not dilute the product gas, pure oxygen is required, which increases the production costs. Indirect gasification (Phillips et al., 2007) uses two units, a gasifier and a combustor, both operating at lower pressure. Olivine is used to provide the energy for the gasification. Gas is generated in the first unit using steam and a fluidized bed of olivine. The sand provides the energy and cools down. The solids, sand and char, are separated from the gas phase in a system of cyclones. Burning the char in the combustor provides the heat needed for the sand to be reused in the gasifier. Combustion can be carried out with air. The gas phase from the combustor is sent to an electrostatic precipitator to capture the ash.

Raw syngas is the raw material for fuels and chemicals. Since CO and hydrogen are the main components of interest, raw syngas is treated to remove impurities and to adjust its composition depending on the fuel to be produced, which ranges from hydrogen to synthetic diesel and gasoline as well as methanol or ethanol. In any case, the first treatment stage is reforming the hydrocarbons. In this step, we transform the hydrocarbons into CO and hydrogen through a number of alternative paths. The most typical method is steam reforming, which is an endothermic process that uses steam to break down the hydrocarbons. The advantage is that the hydrogen yield is larger, which is a desirable feature since the typical ratio of H_2 to CO needed is larger than 1. Another method is partial oxidation, which uses oxygen to partially oxidize the hydrocarbons into CO and hydrogen. To avoid complete combustion of the hydrocarbons, oxygen availability is limited. The process is exothermic but it requires pure oxygen, and leads to a lower yield of hydrogen. Autoreforming combines the two methods, and uses oxygen and steam so that the partial oxidation reactions provide the energy required by the steam-using reactions resulting in an adiabatic process. Lately, dry reforming, a process that uses CO_2 to convert hydrocarbons into syngas has been introduced as a way to mitigate CO_2 emissions. For higher hydrocarbons, such as glycerol, aqueous phase reforming has been tested for the production of hydrogen, mainly to avoid phase change costs (Shabaker et al., 2004).

The next stage is gas cleanup, which can be either cold cleaning using a scrubber, or hot cleaning using a ceramic filter. The second option is recommended for operations under high pressure. Pressure swing adsorption (PSA) can be used to remove the final traces of hydrocarbons using the proper bed.

Subsequently, the gas composition is to be adjusted. Ethanol production requires a ratio of H_2 to CO of 1 (Martín and Grossmann, 2001). However, for methanol production we need a ratio of 2. In the case of Fischer–Tropsch fuels, (FT fuels) the ratio depends on the desired product. In this particular case, we obtain a mixture of hydrocarbons. The concentration of each component (e.g., gasoline, diesel, aviation fuels) depends on the operating conditions (Dry, 2002). Apart from adjusting the operating conditions at the gasifier typically water gas shift reaction (WGSR) is used to increase the hydrogen in the syngas.

$$CO_{(g)} + H_2O_{(g)} \leftrightarrow CO_{2(g)} + H_{2(g)} \qquad (1.5)$$

There are two alternatives to perform WGSR depending on the presence of H_2S in the gas phase due to its poisonous effect on the catalyst. Sour water–gas shift is a single stage process that drives the WGSR toward hydrogen and CO_2 below 350 °C by adding steam. Membrane reactors can be used to directly produce hydrogen at this stage (Ji et al., 2009). When there is no H_2S in the gas, a two-stage process is used to improve the conversion toward hydrogen. The first stage is called high temperature WGS and the reaction is allowed to progress adiabatically up to 550 °C. Next, the stream is cooled down to 200 °C and again is allowed to react adiabatically up to 370 °C. Typically, the syngas from biomass contains traces of H_2S and therefore sour WGSR takes place meaning that the catalysts used is not sensitive to the presence of H_2S.

If the stream contains a larger amount of hydrogen than that is required for the synthesis (e.g., the production of ethanol), it may be economically attractive to recover the hydrogen through a membrane system.

The next stage consists of the removal of CO_2 and H_2S from the gas stream. Carbon capture technologies can be used for this purpose. In recent years, a number of technologies have been tested and implemented for the removal of CO_2 in order to reduce its emissions (GPSA, 2004). The absorption using ethanolamines is a widely practiced process. A chemical reaction between the sour gases, CO_2 and H_2S, and a basic solution of ethanolamine removes them from the gas phase. The concentration of the different amines in water is limited by the corrosivity. Next, the amine is regenerated by distillation and reused. The main disadvantage of the process is that a large amount of energy is needed in the recovery of the amine (GPSA, 2004). PSA has also been tested, in which a bed of zeolites operating at low temperature and moderate pressure can be used to adsorb the sour gases. The bed is then regenerated by reducing the pressure (Ko et al., 2003). Until very recently, only CO_2 could be captured this way. Another alternative is the use of selective membranes, where a fluid is needed to transport the permeate. As in the case of the absorption in ethanolamines, much of the cost is due to recovery of the liquid solution. For CO_2 capture, other methods such as the use of dolomite have been developed lately (Abanades et al., 2011). The decomposition of dolomite is equilibrium controlled by the concentration of CO_2 and the temperature, allowing for CO_2 capture and bed regeneration. The process, such as the PSA, consists of two stages. In the first stage, CO_2 in the gas stream is adsorbed at high temperature in a bed of lime, CaO, generating dolomite. Next, $CaCO_3$ is decomposed in a combustor. This method is appropriate when the gas stream is at high temperature, otherwise it is not economically attractive.

$$CO_{2(g)} + CaO \leftrightarrow CaCO_3 \tag{1.6}$$

Once syngas is prepared, the desired fuel can be synthesized.

Ethanol: For ethanol production we can follow two distinct paths, syngas fermentation or mixed alcohols synthesis (Martín and Grossmann, 2011). Syngas fermentation dates back to the 1980s, and has been lately adopted in some of the first second-generation bioethanol processes at pilot and industrial scale plants by Coskata (www.beca.com). The fermentation of syngas occurs at 38 °C and atmospheric pressure, and the process is an exothermic reaction that requires a large volume of cooling water to generate an ethanol—water mixture at below 5%. Apart from the water implications, a large amount of energy is required to dehydrate the ethanol. After a distillation column, a number of alternatives are considered to obtain fuel-grade ethanol. We will discuss the distillation of water and ethanol later. The final stage of dehydration can be a combination of rectification, whose limit is the azeotrope, adsorption in molecular sieves, and pervaporation, although other beds have also been considered. Only the use of molecular sieves and pervaporation allows fuel-grade ethanol production, and the energy consumption of the two is comparable with molecular sieves being slightly favored.

The second alternative is the production of a mixture of alcohols (Phillips et al., 2007; Dutta and Phillips, 2009). In this case, the mechanism of reaction is similar to FT, operating at 68 atm and 300 °C. Methanol, ethanol, propanol, and butanol are the main alcohols produced. A sequence of distillation columns can be used to recover the methanol and increase the yield of ethanol.

Methanol: The conditions for the reactor are affected by the concentration of CO_2 and CO to avoid deposition of carbon on the catalyst. CO_2 from 2% to 8% is recommended (Lee, 2007). Furthermore, in the reactor the ratio of CO to H_2 is around 2 and the ratio $\frac{H_2 - CO_2}{CO + CO_2} \approx 2.1$ (Lee, 2007). In addition to methanol, water and sometimes dimethyl ether are produced. Thus, after recovery of the unreacted gas through condensation of the water and methanol, the liquid phase is distilled.

FT fuels: The operation in the reactor depends on the types and quantities of FT products desired, ranging from either low (200—240 °C) or high (300—350 °C) temperatures, and pressures from 10 to 40 bar using either cobalt or iron-based catalysts. Currently iron-based catalyst is used at high temperature for the production of gasoline and small hydrocarbons, while either cobalt or iron-based catalysts that operate at low temperature (200—240 °C) are used for the production of diesel or heavier products. The iron catalyst provides high selectivity for C_{10}-C_{18}, which leads to a high yield of diesel. Moreover, the reactions with iron catalyst are usually conducted at 30 bar. Furthermore, FT synthesis requires careful control of the H_2 to CO ratio to satisfy the stoichiometry of the synthesis reactions as well as to avoid deposition of carbon on the catalysts (coking). An optimal H_2 to CO ratio from 1:1 to 2:1 for the production of diesel and gasoline is recommended (Dry, 2002). The wide range of products obtained in the FT reaction results in the fact that for enhancing the yield to gasoline and diesel, hydrotreating is recommended to break down the heavier hydrocarbons (Brown et al., 2012).

Hydrogen: The production of hydrogen requires fewer steps since it is completed after composition adjustment. Lately membrane reactors have been implemented to produce hydrogen in situ during the WGSR (Ji et al., 2009).

In Table 1.2, we present the optimal technologies to transform different raw materials including biomass, by-products from other processes, into useful fuels and intermediates based on energy optimization.

1.3.2 BIOCHEMICAL

In this section, we describe different processes that operate at low and medium temperature to produce bioethanol or biodiesel. This section considers the production of biodiesel can be obtained via a biochemical reaction either using inorganic catalysts, or in the absence of catalysts. Biochemical treatments are more specific depending on the raw material since we need to break down different structures and their strength varies. We divide this section into stages that allow biomass processing, and we will discuss the differences among the raw materials within each stage.

1.3.2.1 Pretreatment

Lignocellulosic raw materials: This is a wide range of materials characterized by a structure of lignin and two more components that we will focus our interest for recovery; hemicellulose and cellulose. The amount of lignin defines the pretreatment (Sung and Cheng, 2002; Keshwani and Cheng, 2009). Lignin is a polymer made of different aromatic monomers that provides the shape and structure to the plant. Hemicellulose is mainly composed of pentoses, and cellulose consists of glucose. The aim is to obtain these two sugars for further use. The goal for the pretreatment is to expose the hemicellulose and cellulose before hydrolysis. First, the raw material is broken down into small pieces, using physical means such as grinding, milling, or chipping, for the further treatment to have proper yield. Next, we can distinguish among physical and chemical pretreatment such as steam, CO_2, or ammonia fiber explosion (AFEX). Among these, AFEX is the only one used on the industrial scale. It works efficiently for low lignin biomass and has the advantage of producing little or none of inhibitors such as furans. The disadvantage is that a large amount of energy is needed in the recovery of ammonia (Alizadeh et al., 2005). The second group of pretreatments corresponds to the chemical ones, including the use of ozone, acids, alkali, organic solvents, and peroxides. Among them dilute sulfuric acid operating at 12 bar and 160–180 °C has been the most widely implemented process (Aden and Foust, 2009; Kazi et al., 2010). The yield is high but this method generates furans by the dehydration of the sugars that can inhibit the subsequent fermentation process. The use of enzymes has also been proposed. However, the slow reaction time makes it unattractive on an industrial scale.

Table 1.2 Optimal Selection of Technologies for Transforming Biomass and By-products Into Fuels

Pretreatment	Reforming	Cleanup	Adjustment	CO_2	Synthesis	Purification	Product
Switchgrass							
Direct gasification	Steam	Cold	Recovery of H_2	PSA and MEA	Catalytic alcohols synthesis	Direct distillation	Ethanol
Indirect gasification	Steam	Cold	WGSR	PSA and MEA	FT	Hydrotreating	FT Fuels
Indirect gasification	Steam	Cold	WGSR			Membrane reactor	Hydrogen
Glycerol							
Furnace	Aqueous phase	Cold	WGSR			Membrane reactor	Hydrogen
Furnace	Autoreforming	Cold	Recovery of H_2	PSA	Catalyst	Hydrotreating	FT Fuels
Furnace	Autoreforming	Cold	Recovery of H_2	PSA	Catalyst	Distillation	Methanol

Algae: Algae are an interesting raw material with a growth rate at between 20 and 100 times that of any other biomass in terms of oil production. Algae are microorganisms that can accumulate large amounts of lipids, as well as carbohydrates and protein. Once the algae have reached the proper lipid accumulation, the most energy-demanding step is dewatering. Algae grow in large ponds or flow reactors at low concentrations in seawater or wastewater. Typically, harvesting algae consists of flotation, centrifugation, and drying. Lately, methods based on capillarity are being proposed to reduce the energy consumption. Once dried, the oil must be extracted (Martín and Grossmann, 2012).

Seeds or algae oil extraction: The extraction of oil from the algae or seeds can be carried out following physical and/or chemical methods. Typically a combination of both, a mechanical press and the use of different solvents have provided good recovery of the oil content in the algae or the seeds. The biomass that remains after the extraction of the oil contained in the algae consists of starch and protein.

Grain: Although current efforts are focused on shifting from the use of food-related biomass as raw material for biofuels, it is interesting to note the pretreatment required by first generation bioethanol (Jacques et al., 1999). The pretreatment of the corn grain consists of first grinding to reduce the size so that the cooking, where steam is used to break down the physical structure and expose the starch for further treatment, is more effective.

1.3.2.2 Hydrolysis

Cellulose and Hemicellulose: Hydrolysis of cellulose and hemicellulose aims at the production of mainly glucose and xylose. The reactions take place at 50 °C using cellulases, for about 72 h (Kazi et al, 2010), Eqns (1.7)–(1.8):

$$(C_6H_{10}O_5)_n + nH_2O \rightarrow nC_6H_{12}O_6 \quad \Delta H = 22.1n \quad kJ/mol \tag{1.7}$$

$$(C_5H_8O_4)_m + mH_2O \rightarrow mC_5H_{10}O_5 \quad \Delta H = 79.0m \quad kJ/mol \tag{1.8}$$

Starch: Regardless of the source (algae or grain), starch is processed through a series of steps using moderate temperatures and enzymes to break it down to glucose. The first stage is called liquefaction. The pH range of liquefaction is 6–6.5 and the temperature is kept at 85 °C (Jacques et al., 1999). The chemical reaction involved in this step is the hydrolysis of starch to maltose, which involves the use of the endoenzyme α-amylase described in Eqn (1.9).

$$2(C_6H_{10}O_5)_n + nH_2O \xrightarrow{\alpha - amylase} nC_{12}H_{22}O_{11} \tag{1.9}$$

This step is followed by the use of the exoenzyme glucoamylase to achieve the conversion of maltose to glucose in the saccharification process. Saccharification operates most efficiently at around 65–75 °C described in (Eqn (1.10)).

$$C_{12}H_{22}O_{11} + H_2O \xrightarrow{\text{glucoamylase}} 2C_6H_{12}O_6 \tag{1.10}$$

1.3.2.3 Reaction

Fermentation: Once the sugars are exposed, they can be fermented to produce different species such as ethanol, butanol, acetone, ibutene, or even biodiesel. For the sake of clarity, we present the chemical reactions that produce ethanol from glucose and xylose, typical conversions, as well as the range of additional chemicals obtained in the fermentation. Fermentation is an anaerobic exothermic process that takes place from 32 °C to 38 °C at a pressure slightly above the atmospheric pressure to avoid entrance of oxygen into the system. Large volume of water is used for cooling. *Zymomonas mobilis* allows the fermentation of xylose, while for glucose alone *Saccharomices cerevisiae* is typically used, see Table 1.3.

Transesterification: The production of biodiesel from oil is based on the reaction between the oil and an alcohol, which is typically methanol due to the lower cost and faster reaction. Lately ethanol has also being used (Severson et al., 2013), which has proved an interesting alternative in terms of process integration (Martín and Grossmann, 2013c). The transesterification reaction to produce biodiesel is an equilibrium reaction. The biodiesels, fatty acid ethyl or methyl esters (FAEE and FAME), can be obtained depending on the use of ethanol or methanol, see Eqn (1.11):

$$\begin{array}{lcccccc}
CH_2\text{–}OOC\text{–}R_1 & & & & R_1\text{–}COO\text{–}R' & & CH_2\text{–}OH \\
| & & & \text{Catalyst} & & & | \\
CH\text{–}OOC\text{–}R_2 & + & 3R'OH & \Leftrightarrow & R_2\text{–}COO\text{–}R' & + & CH\text{–}OH \\
| & & & & & & | \\
CH_2\text{–}OOC\text{–}R_3 & & & & R_3\text{–}COO\text{–}R' & & CH_2\text{–}OH
\end{array} \quad (1.11)$$

Table 1.3 Typical Conversions of Sugars in the Production of Ethanol

Reaction	Conversion
Glucose → 2 ethanol + 2 CO_2	Glucose 0.92
Glucose + 1.2NH_3 → 6 *Zymomonas mobilis* + 2.4H_2O + 0.3 O_2	Glucose 0.035
Glucose + 2H_2O → 2 glycerol + O_2	Glucose 0.002
Glucose + 2 CO_2 → 2 succinic acid + O_2	Glucose 0.008
Glucose → 3 acetic acid	Glucose 0.022
Glucose → 2 lactic acid	Glucose 0.013
3 Xylose → 5ethanol + 5 CO_2	Xylose 0.8
Xylose + NH_3 → 5 *Z. mobilis* + 2H_2O + 0.25 O_2	Xylose 0.03
3 Xylose + 5H_2O → 5 glycerol + 2.5 O_2	Xylose 0.02
3 Xylose + 5 CO_2 → 5 succinic acid + 2.5 O_2	Xylose 0.03
2 Xylose → 5 acetic acid	Xylose 0.01
3 Xylose → 5 lactic acid	Xylose 0.01

In order to increase the yield to biofuels we need to shift the equilibrium to products. We can achieve that by adjusting the variables affecting the conversion, such as the catalyst type and load, the operating temperature and pressure (in case supercritical conditions) and the volume of excess of alcohol. The energy spent in heating up the feed, in the recovery of the excess of alcohol, as well as in the purification steps, is to be considered. The purification steps depend on the type of catalyst. Typically, we have homogenous alkali catalysis, i.e., NaOH, KOH, acid catalysis, i.e., H_2SO_4, enzymes, heterogenous acid or basic ones, and with no catalysis under supercritical conditions. If the oil contains water or free fatty acids (FFA), the use of alkali catalysts requires a pretreatment stage to remove the FFA to avoid saponification, which would increase the costs. Acid catalyst typically requires large ratios of alcohol to oil, with the corresponding consumption of energy to recover it. Heterogeneous catalysis has been developed recently, and although there is no industrial experience, it eliminates the purification stages since no biodiesel washing is required. Enzymes can be supported catalysts in which case they can be considered as heterogenous catalysts themselves. Although biodiesel yield is low and the reaction is slow, there have been significant recent improvements. Finally, the use of supercritical conditions benefits from quick reaction times and absence of catalyst; however, it operates at extreme temperatures and pressures with large ratios of alcohol to oil. Among the different catalysts, the decision depends on the type of alcohols and the raw materials (Martín and Grossmann, 2012; Severson et al., 2013). Furthermore, for the selection among the use of ethanol or methanol, some recent optimization studies show the competitiveness of the use of ethanol so long as its price remains below \$0.8/gal (Severson et al., 2013) and also in integrated processes (Martín and Grossmann, 2013c).

1.3.2.4 Purification

Ethanol: Ethanol dehydration go through the same stages as presented for bioethanol production from syngas fermentation.

Biodiesel: Once we have obtained the biodiesel, FAME or FAEE, the mixture is distilled to recover the excess of alcohol. Next, in case homogenous catalyst is used, the stream is neutralized. Subsequently, the glycerol phase, containing the traces of alcohol and water is separated from the biodiesel phase, containing the rest of unconverted oil, and finally, biodiesel is purified by distillation. We have to maintain the operating temperatures of the different distillation columns below the decomposition temperatures of glycerol, 150 °C, biodiesel, 250 °C, and oil, 350 °C. Furthermore, the separation of the liquid phases has higher yield at temperature around 40 °C. Table 1.4 shows the optimal technologies to process different types of biomass into products through a biochemical route.

1.3.3 HYBRID PROCESSES

Both renewable resources and fossil ones fall in this category. Among the fossil fuels coal and natural gas, either conventional or shale gas, can be considered. Actually,

Table 1.4 Optimal Technologies for Biochemical Paths

Raw Material	Initial Stage	Pretreatment	Hydrolysis	Reaction/Catalysis	Purification
Ethanol					
Lignocellulosic	Grinding	Dilute acid	Yes	Fermentation Zymomonas mobilis	Distillation and molecular sieves
Corn	Grinding	Cooling	Saccharification liquefaction	Fermentation Saccharomices cerevisiae	Distillation and molecular sieves
Glycerol				Fermentation S. cerevisiae	Distillation and molecular sieves
Biodiesel					
Algae	Dewatering capillarity	Oil extraction		Transesterification alkali-methanol	Liquid–liquid extraction Biodiesel washing Distillation
Cooking oil				Transesterification heterogenous-methanol	Liquid–liquid extraction Distillation
Biodiesel and bioethanol					
Algae biorefinery complex	Dewatering capillarity	Oil extraction	Yes	Transesterification enzymes-ethanol	Liquid–liquid extraction Distillation

the processes that convert them into fuels have already been described above. The typical approach is a FT technology. If we start with coal, gasification is the first step. In the case of shale gas or natural gas, after purification, the first stage is reforming. Purification of the natural gas, no matter its source, is not only needed to avoid catalyst deactivation, but allows recovering higher hydrocarbons that can be interesting for further processes such as ethane that typically accompanies the shale gas from some wells. The combination of biomass and shale gas (Martín and Grossmann, 2013b), biomass, coal, and natural gas (Baliban et al., 2011) results in flexible processes that allow the production of gasoline and diesel substitutes evaluating the limit prices for the different raw materials to produce competitive products.

1.3.4 HEAT INTEGRATION

1.3.4.1 Multieffect distillation columns (dilute mixtures/debottlenecking)

In the previous sections, we have described some processes that rely on fermentation for the production of different fuels, among them bioethanol. One of the main concerns about first generation bioethanol was the high energy consumption. The most demanding unit within the process is the beer column, representing more than half the energy consumed (Karuppiah et al., 2008). The beer column is the distillation column allocated just after the fermentor. This column not only handles a dilute mixture of ethanol in water, but it has to tackle a solution characterized by an azeotrope that avoids reaching fuel quality in the distillate. In order to reduce the energy consumption, the use of multieffect distillation columns was proposed. Multieffect distillation consists of a system of columns that process a fraction of the initial feed each at different pressures so that the reboiler of a column operating at an immediate lower pressure acts as condenser of the column that operates at an immediate higher pressure. The investment cost of the system is increased but savings in steam of the order of 3 million dollars per year can be obtained by implementing this solution (Karuppiah et al., 2008). By using multieffect distillation columns we not only reduce the energy consumption, but also the cooling water. This fact has an impact in the freshwater consumption since most of the losses of water are due to evaporation in the cooling tower. In fact, the use of multieffect columns saves around two-thirds of cooling needs.

1.3.4.2 Simultaneous optimization with heat/water integration

Duran and Grossmann in 1986 developed a model for pinch location for heat integration. Based on this model, they showed that for processes where the conversion at the reactor depends heavily on the operating conditions, simultaneously flowsheet optimization and heat integration allow for improvement in both the yield and the economics of the process. In the previous section we have presented a few case studies that fall into this type: production of methanol from syngas, FT processes, bioethanol, and transesterification of oil to produce biodiesel. Other reactors, such as second-generation bioethanol via mixed alcohols systems or the FT, can benefit from simultaneous optimization approach as well. A predictive model for the conversion in the reactor is crucial so that we can evaluate the yield as function of the operating conditions such as feedstock composition. If we focus on the biodiesel case, the yield depends on the temperature, catalysis load, ratio alcohol to oil, residence time, and pressure (for supercritical). For some time experimental studies focused on the optimization of the reactor alone reporting optimal operating conditions. However, none of these studies considers the recovery of alcohols after the reaction, nor the adjustment of the feedstock temperature or pressure. Martín and Grossmann (2012) demonstrated that the reported values for the optimal operation of the reactors were actually suboptimal for the case of biodiesel using methanol as transesterification agent since no further evaluation of the separation stages was carried out. Severson et al. (2013) simultaneously optimized and heat integrated the production of biodiesel using ethanol and similarly, the optimal operating conditions at the reactor were different from the ones reported as optimal in the literature due to the fact that previously the authors only studied the reactor and its conversion.

In the case of biodiesel, the only water consumption is the losses in the cooling tower. Thus, energy integration also reduced water consumption, although typically this is not the case.

1.4 WATER OPTIMIZATION IN BIOPROCESSES

1.4.1 TWO-STAGE PROCEDURE FOR WATER OPTIMIZATION IN BIOFUEL PRODUCTION PROCESSES

The complexity of solving the simultaneous optimization problem led us to propose (Grossmann and Martín, 2010) a two-stage solution procedure based on the link between energy and water consumption. We first optimize the flowsheet for the production of a particular fuel for minimum energy consumption or minimum production cost. Next, we develop a HEN using SYNHeat (Yee and Grossmann, 1990) that allows the reduction of cooling needs. Finally, we design the optimal WN (Ahmetović et al., 2010). At this point we identify the units that generate wastewater within the process (sources) those which require water (sinks), those that generate contaminants, and those which use water (demands). Table 1.5 presents the units for several bioethanol production processes of first and second generation, as well as the treatment technologies to be used, such as screens, biological treatment, and RO. Next, we compute the loads of contaminants and flows from the optimal process flowsheet. Finally, we design the detailed WN structure.

Figure 1.11 shows the optimal WN for a corn-based ethanol with a production capacity of 60 Mgal/year. As can be seen, the freshwater in t/h is used in the cooling tower, in the boiler and the fermentor since the purity levels are high to avoid contamination issues. The highly diluted ethanol–water mixture from the fermentor

Table 1.5 Units for Water Network Design in Bioethanol Production

Process	Process Units	Demand Units	Source Units
Corn	P1: Washing P2: Boiler P3: Cooling tower	D1: Fermentor D2: Boiler D3: Cooling tower	S1: Rec column S2: Beer column
Ligno: thermochemical	P1: Washing P2: Cooling tower	D1: Gasifier D2: Cooling tower	S1: Condensation
Ligno: thermobiochemical	P1: Washing P2: Boiler P3:Cooling tower	D1: Fermentor D2: Boiler/gasifier D3: Cooling tower	S1: Beer column S2: Condensation
Ligno: biochemical	P1: Washing P2: Boiler P3:Cooling tower	D1: Fermentor D2: Boiler D3: Cooling tower D4: Acid treatment	S1: Beer column

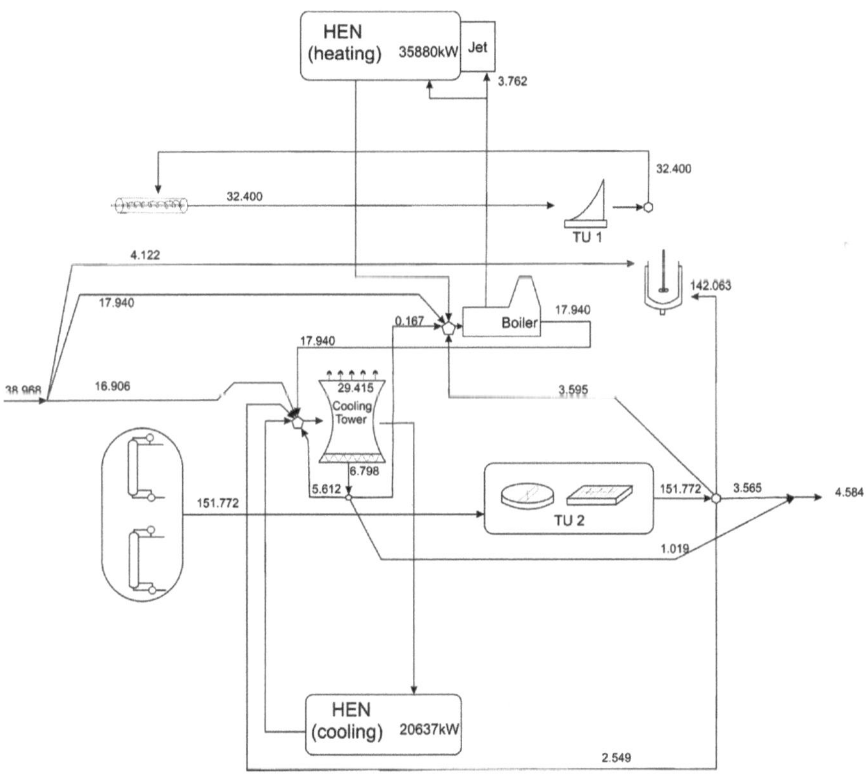

FIGURE 1.11

Optimal water network for corn based ethanol. With permission.

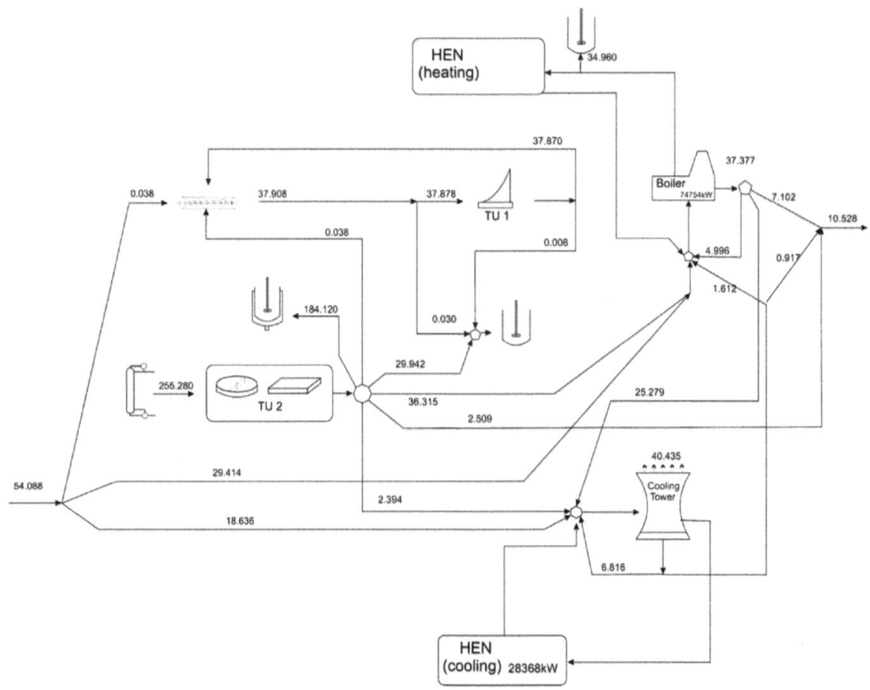

FIGURE 1.12

Optimal water network for second-generation bioethanol via biochemical processing. With permission.

is separated in the distillation columns and it becomes the largest source of water for the process once it is treated. It is not economically viable to use RO (Martín et al., 2011) to remove the salts from the streams exiting the cooling tower and the boiler and freshwater is used instead to dilute them before reuse.

In Figure 1.12, we present the optimal WN for second-generation bioethanol via biochemical processing where the freshwater is fed to the boiler and to the cooling tower, while the hydrolysis tank and the fermentor receive treated water from the one that is recovered at the distillation column.

Based on this procedure, we have computed the energy and water consumption for a number of biofuels. Tables 1.6(a) and 1.6(b) summarizes the results by Martín and Grossmann (2013a).

1.4.2 SIMULTANEOUS OPTIMIZATION AND WATER INTEGRATION

Recently Yang and Grossmann (2013) presented a procedure for the simultaneous flowsheet, heat and water optimization once the flowsheet structure is fixed where the detailed heat exchanger and WNs are synthesized and applied it to bioethanol

Table 1.6a Summary of Technoeconomic Details Using Switchgrass

	Ethanol (Hydrolysis)	Ethanol (Gasification and Catalysis)	Ethanol (Gasification and Fermentation)	FT-Diesel	H_2
Investment ($MM)	169	335	260	216	148
Capacity (MMgal/year)	60	60	60	60	60[a]
Biofuel yield (kg/kg_{wet})	0.28	0.20	0.33	0.24	0.11
Production cost ($/gal)	0.80	0.41	0.81	0.72	0.68[a]
Water consumption (gal/gal)	1.66	0.36	1.59	0.15	–
Energy consumption (MJ/gal)	−10.2	−9.5	27.2	−62.0	−3.84[a]

[a] *In kg instead of MMgal.*

Table 1.6b Summary of Technoeconomic Details for Biodiesel

	Biodiesel (Ethanol) (Algae)	Biodiesel (Methanol) (Cooking Oil)	Biodiesel (Methanol) (Algae)
Investment ($MM)	112	17	110
Capacity (MMgal/year)	68	72	69
Biofuel yield (kg/kg_{wet})	0.47	0.96	0.48
Production cost ($/gal)	0.54	0.66	0.42
Water consumption (gal/gal)	0.35	0.33	0.60
Energy consumption (MJ/gal)	1.93	1.94	1.94

production among other cases. Figures 1.13(a) and (b) show the resulting water and HENs yielding promising water and energy consumption values. The simultaneous approach improved the quality and the computational effort compared to the sequential targeting.

1.4.3 INDUSTRIAL RESULTS ON WATER OPTIMIZATION IN BIOFUEL PRODUCTION PROCESSES

There are not many industrial results on water consumption since only first-generation bioethanol plants are fully in operation. Water usage in bioethanol

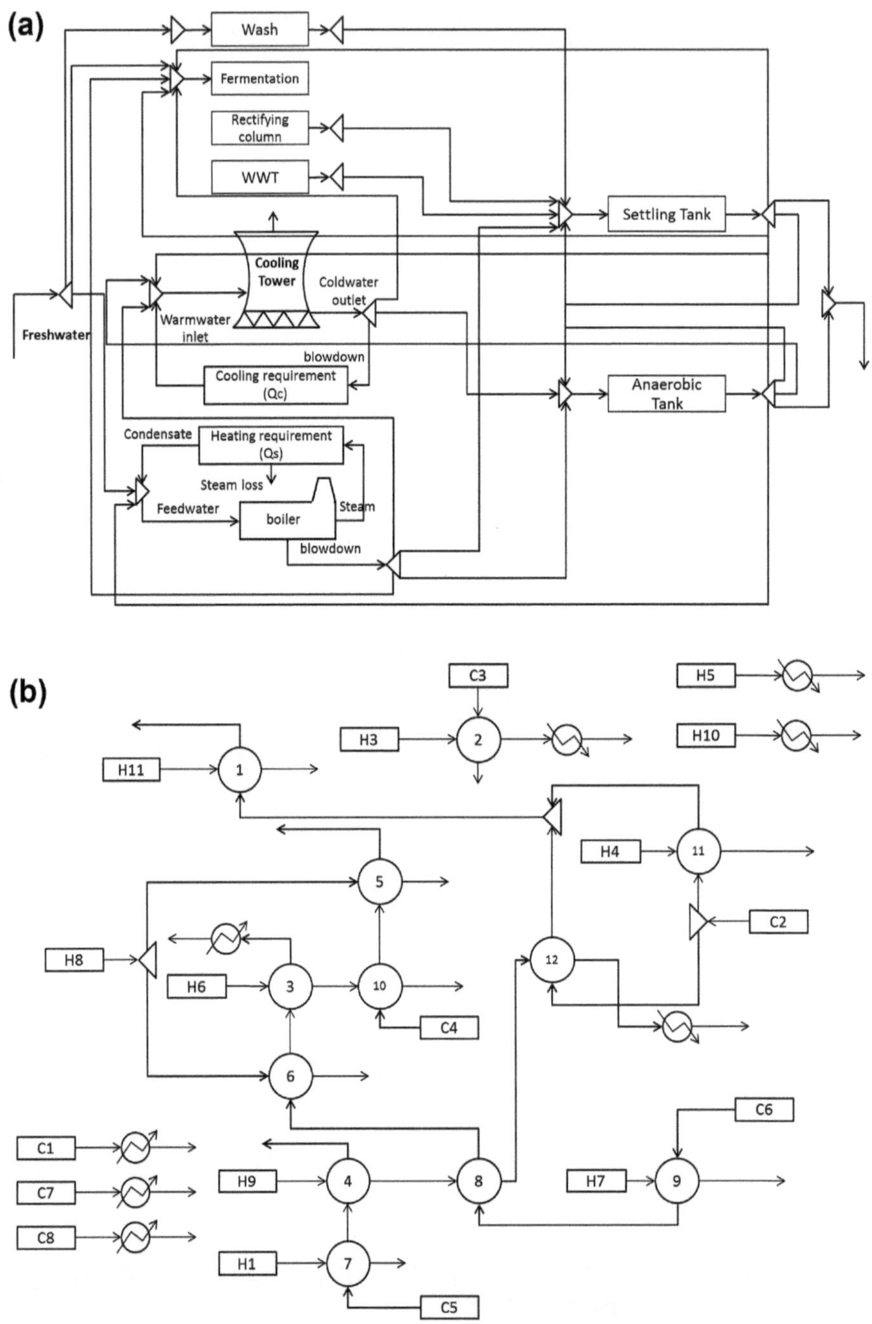

FIGURE 1.13

(a) Water network corn-based ethanol and (b) Water network corn-based ethanol.

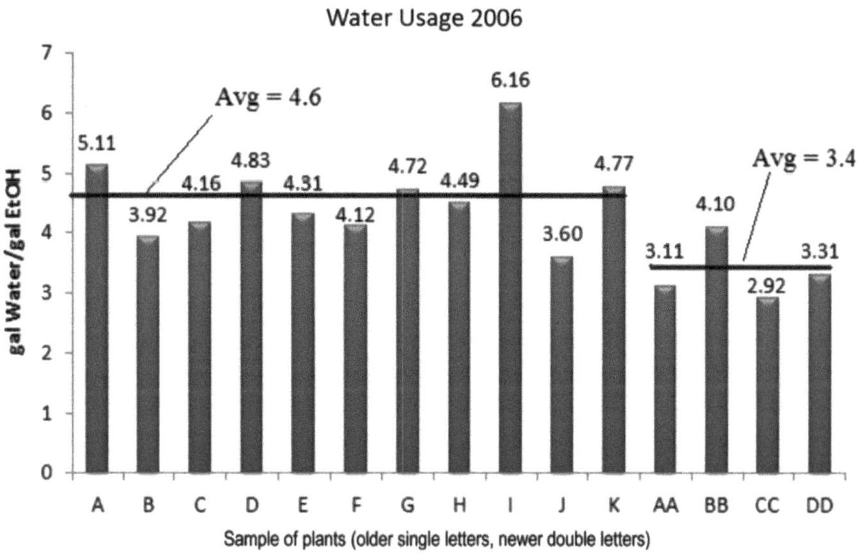

FIGURE 1.14

Water consumption in first-generation bioethanol plants.

Figure adapted from MTAP (2008).

plants is shown in Figure 1.14 as reported by MTAP (2008) on several plants. They classified the existing plants into two groups: old plants, before 2006 and new plants, from that point on. The variability of results from plant to plant and the improvement in the water efficiency obtained with the new designs can be seen in Figure 1.11. Further analysis is complex because the plants are not identified and the variability can also be due to their geographical location. While the analysis on water consumption for bioethanol is widely reported due to social and technical concerns, that is not the case for biodiesel production and Pate (2008) reported an interval range from 0.3 to 1 gal/gal, which actually can be reproduced in the simulations as presented in Table 1.6(b).

1.5 CONCLUSIONS

Water and energy are two major concerns in process design. While the production of energy and alternative fuels have been widely studied over the last couple of decades, water consumption issues associated with the production of biofuels has only recently been evaluated. There are two main sources of water consumption, the growing of the biomass and the process itself. The first source highly depends on the particular region. On the other hand, process water consumption can be minimized by reducing the cooling needs through energy integration and proper treatment and reuse. Mathematical optimization and heuristic methods have been developed over the last decades to address water consumption. Due to the limitations of heuristics, mathematical programming techniques have been used for the design of optimal processes and WNs with minimum water and energy consumption using a sequential approach based on superstructure optimization. However, simultaneous optimization of heat and water consumption is performed by a targeting method recently developed.

ACKNOWLEDGMENTS

The authors gratefully acknowledge funding from the NSF Grant CBET0966524 and from the Center for Advanced Process Decision-making at Carnegie Mellon University.

REFERENCES

Abanades, J.C., Alonso, M., Rodríguez, N., 2011. Biomass combustion with in situ CO_2 capture with CaO. I. Process description and economics. Ind. Eng. Chem. Res. 50, 6972–6981.

Aden, A., Foust, T., 2009. Technoeconomic analysis of the dilute sulfuric acid and enzymatic hydrolysis process for the conversion of corn stover to ethanol. Cellulose 16, 535–545.

Ahmetović, E., Martin, M., Grossmann, I.E., 2010. Optimization of energy and water consumption in corn–based ethanol plants. Ind. Eng. Chem. Res. 49 (17), 7972–7982.

Ahmetović, E., Grossmann, I.E., 2011. Global superstructure optimization for the design of integrated process water networks. AIChE J. 57, 434–457.

Alizadeh, H., Teymouri, F., Gilbert, T.I., Dale, B.E., 2005. Pretreatment of switchgrass by ammonia fiber explosion (AFEX). Appl. Biochem. Biotech. 124, 1133–1141.

Alva-Argaez, A., Vallianatos, A., Kokossis, A., 1999. A multi-contaminant transshipment model for mass exchange networks and wastewater minimization problems. Comput. Chem. Eng. 23, 1439–1453.

Bagajewicz, M.J., 2000. A review of recent design procedures for water networks in refineries and process plants. Comput. Chem. Eng. 24, 2093–2113.

Bagajewicz, M.J., Faria, D.C., 2009. On the appropriate architecture of the water/wastewater allocation problem in process plants. Comput. Aided Chem. Eng. 26, 1–20.

Baliban, R.C., Elia, J.A., Floudas, C.A., 2011. Optimization framework for the simultaneous process synthesis, heat and power integration of a thermochemical hybrid biomass, coal, and natural gas facility. Comput. Chem. Eng. 35 (9), 1647–1690.

Bogataj, M., Bagajewicz, M.J., 2008. Synthesis of non-isothermal heat integrated water networks in chemical processes. Comput. Chem. Eng. 32, 3130–3142.

Bridgewater, A.V., 2003. Renewable fuels and chemicals by thermal processing of biomass. Chem. Eng. J. 91, 87–102.

Brown, T.R., Zhang, Y., Hu, G., Brown, R.C., 2012. Techno-economic analysis of biobased chemicals production via integrated catalytic processing. Biofuels, Bioprod. Biorefin. 6 (1), 73–87.

Duran, M.A., Grossmann, I.E., 1986. Simultaneous optimization and heat integration of chemical processes. AIChE J. 32 (1), 123−138.

Dutta, A., Phillips, S.D., 2009. Thermochemical Ethanol via Direct Gasification and Mixed Alcohol Synthesis of Lignocellulosic Biomass. National Renewable Energy Laboratory, Golden, CO. No. NREL/TP-510−45913.

Dry, M.E., 2002. The Fischer−Tropsch process: 1950−2000. Catal. Today 71, 227−241.

El-Halwagi, M.M., 1992. Synthesis of reverse osmosis networks for waste reduction. AIChE J. 38, 1185−1198.

Foo, D.C.Y., 2009. State-of-the-art review of pinch analysis techniques for water network synthesis. Ind. Eng. Chem. Res. 48, 5125−5159.

FPS VERSION 21-10 GPSA Engineering Data Book, 2004.

Galan, B., Grossmann, I.E., 1998. Optimal design of distributed wastewater treatment networks. Ind. Eng. Chem. Res. 37, 4036−4048.

Grossmann, I.E., Martín, M., 2010. Energy and Water Optimization in Biofuel Plants. Chinese J. Chem. Eng. 18 (6), 914−922.

Hallale, N., 2002. A new graphical targeting method for water minimization. Adv. Environ. Res. 6, 377−390.

Huang, C.-H., Chang, C.-T., Ling, H.-C., Chang, C.-C., 1999. A mathematical programming model for water usage and treatment network design. Ind. Eng. Chem. Res. 38, 2666−2679.

Jacques, K., Lyons, T.P., Kelsall, D.R., 1999. The Alcohol Textbook, third ed. Nottingham University Press, United Kingdom.

Jezowski, J., 2010. Review of water network design methods with literature annotations. Ind. Eng. Chem. Res. 49, 4475−4516.

Ji, P., Feng, W., Chen, B., 2009. Production of ultrapure hydrogen from biomass gasification with air. Chem. Eng. Sci. 64, 582−592.

Karuppiah, R., Grossmann, I.E., 2006. Global optimization for the synthesis of integrated water systems in chemical processes. Comput. Chem. Eng. 30, 650−673.

Karuppiah, R., Grossmann, I.E., 2008. Global optimization of multiscenario mixed integer nonlinear programming models arising in the synthesis of integrated water networks under uncertainty. Comput. Chem. Eng. 32, 145−160.

Karuppiah, R., Peschel, A., Grossmann, I.E., Martín, M., Martinson, W., Zullo, L., 2008. Energy optimization of an ethanol plant. AIChE J. 54 (6), 1499−1525.

Karuppiah, R., Bury, S.J., Vazquez, A., Poppe, G., 2012. Optimal design of reverse osmosis-based water treatment systems. AIChE J. 58 (9), 2758−2769.

Kazi, F.K., Fortman, J.A., Anex, R.P., Hsu, D.D., Aden, A., Dutta, A., Kothandaraman, G., 2010. Technoeconomic comparison of process technologies for biochemical ethanol production from corn stover. Fuel 89, S20−S28.

Keshwani, D.R., Cheng, J.J., 2009. Switchgrass for bioethanol and other value-added applications: a review. Bioresour. Technol. 100, 1515−1523.

Khor, C.S., Foo, D.C.Y., El-Halwagi, M.M., Tan, R.R., Shah, N., 2011. A superstructure optimization approach for membrane separation-based water regeneration network synthesis with detailed nonlinear mechanistic reverse osmosis model. Ind. Eng. Chem. Res. 50, 13444−13456.

Ko, D., Siriwardane, R., Biegler, L.T., 2003. Optimization of a pressure-swing adsorption process using zeolite 13X for CO_2 sequestration. Ind. Eng. Chem. Res. 42 (2), 339−348.

Kuo, W.J., Smith, R., 1997. Effluent treatment system design. Chem. Eng. Sci. 52, 4273−4290.

Lee, S., 2007. Methanol synthesis from syngas. Handbook of Alternative Fuel Technologies. Taylor and Francis, Boca Raton, Florida.

Lu, Y.-Y., Hu, Y.-D., Xu, D.-M., Wu, L.-Y., 2006. Optimum design of reverse osmosis seawater desalination system considering membrane cleaning and replacing. J. Membr. Sci. 282, 7–13.

Martín, M., Ahmetovic, E., Grossmann, I.E., 2011. Optimization of water consumption in second generation bioethanol plants. Ind. Eng. Chem. Res. 50, 3705–3721.

Martín, M., Grossmann, I.E., 2011. Energy optimization of lignocellulosic bioethanol production via gasification. AIChE J. 57 (12), 3408–3428.

Martín, M., Grossmann, I.E., 2012. Simultaneous optimization and heat integration for biodiesel production from cooking oil and algae. Ind. Eng. Chem. Res. 51 (23), 7998–8014.

Martín, M., Grossmann, I.E., 2013a. On the systematic synthesis of sustainable biorefineries. Ind. Eng. Chem. Res. 52 (9), 3044–3064.

Martín, M., Grossmann, I.E., 2013b. Optimal use of hybrid feedstock, switchgrass and shale gas for the simultaneous production of hydrogen and liquid fuels. Energy 55, 378–391.

Martín, M., Grossmann, I.E., 2013c. Optimal engineered algae composition for the integrated simultaneous production of bioethanol and biodiesel. AIChE J. 59 (8), 2872–2883.

MTAP, 2008. Ethanol Benchmarking and Best Practices: The Production Process and Potential for Improvement. Minnesota Technical Assistance Program, University of Minnesota, Minneapolis, MN.

Pate, R., Apr 11, 2008. Biofuels and the Energy–Water Nexus. Presented at AAAS/SWARM, Albuquerque, NM.

Phillips, S., Aden, A., Jechura, J., Dayton, D., Eggeman, T., April 2007. Thermochemical Ethanol via Indirect Gasification and Mixed Alcohol Synthesis of Lignocellulosic Biomass. No. NREL/TP-510–41168. National Renewable Energy Laboratory, Golden, CO.

Rosegrant, M.W., Cai, X., Cline, S.A., 2002. Food Policy Report: Global Water Outlook to 2025 (Averting an Impending Crisis). International Food Policy Research Institute, Washington, DC.

Savelski, M.J., Bagajewicz, M.J., 2000. On the optimality conditions of water utilization systems in process plants with single contaminants. Chem. Eng. Sci. 55 (21), 5035–5048.

Savelski, M.J., Bagajewicz, M.J., 2003. On the necessary conditions of optimality of water utilization systems in process plants with multiple contaminants. Chem. Eng. Sci. 58 (23–24), 5349–5362.

Savulescu, L.E., Smith, R., 1998. Simultaneous energy and water minimization. In: AIChE Annual Meeting, Miami Beach, FL.

Severson, K., Martín, M., Grossmann, I.E., 2013. Process optimization biodiesel production using bioethanol. AIChE J. 59 (3), 834–844.

Shabaker, J.W., Huber, G.W., Dumesic, J.A., 2004. Aqueous-phase reforming of oxygenated hydrocarbons over Sn-modified Ni catalysts. J. Catal. 222, 180–191.

Sun, Y., Cheng, J., 2002. Hydrolysis of lignocellulosic materials for ethanol production: a review. Bioresour. Technol. 83, 1–11.

Takama, N., Kuriyama, T., Shiroko, K., Umeda, T., 1980. Optimal water allocation in a petroleum refinery. Comput. Chem. Eng. 4, 251–258.

Tan, Y.L., Foo, D.C., Tan, R.R., Ng, D.K.S., 2007. Approximate graphical targeting for water network with two contaminants. Chem. Eng. Trans. 12, 347–352.

Ullmer, C., Kunde, N., Lassahn, A., Gruhn, G., Schulz, K., 2005. WADO™: water design optimization methodology and software for the synthesis of process water systems. J. Cleaner Prod. 13, 485–494.

UNEP, 2008. Vital Water Graphics: An Overview of the State of the World's Fresh and Marine Waters, second ed. http://www.unep.org/dewa/vitalwater/article42.html.

Wang, Y.P., Smith, R., 1994. Wastewater minimization. Chem. Eng. Sci. 49 (7), 981–1006.

Yang, L., Grossmann, I.E., 2013. Water targeting models for simultaneous flowsheet optimization. Ind. Eng. Chem. Res. 52 (9), 3209–3224.

Yang, L., Salcedo, R., Grossmann, I.E., 2014. Water network optimization with wastewater regeneration models. Ind. Eng. Chem. Res. 53 (45), 17680–17695.

Yee, T.F., Grossmann, I.E., 1990. Simultaneous optimization models for heat integration. II. Heat exchanger network synthesis. Comput. Chem. Eng. 14 (10), 1165–1184.

CHAPTER

Energy Efficient Chiller Configuration—A Design Perspective

2

Kriti Kapoor, Thomas F. Edgar[1]

McKetta Department of Chemical Engineering, The University of Texas at Austin, Austin, TX, USA

[1]*Corresponding author: E-mail: tfedgar@austin.utexas.edu*

2.1 INTRODUCTION

Energy efficient manufacturing has emerged as an important part of the solution to the problem of rising energy demand. Energy efficient manufacturing not only saves energy, but also has the potential to reduce pollutant emissions, reduce the carbon footprint, improve yields, and hence make the overall process more profitable and sustainable. This study is focused on energy efficient cooling systems.

Large-scale cooling systems account for a significant portion of the electrical energy consumed by most industrial, residential, and university campuses. By some estimates, the cooling of buildings contributes up to 35% of the total electrical demand in the United States (Henze, 1995). Depending on a building's heating ventilation and air conditioning system, a building may require heating and cooling year round. In the summer, air may be cooled to lower than required room temperatures in order to remove humidity, and then reheated to bring it back up to the desired temperature. In the winter, thermal zones in the middle of large buildings require cooling because they are not exposed to ambient conditions, so the thermal needs are driven by the internal gains of the zone. Chillers are generally used to meet building cooling needs, and boilers are often used to provide heating. Chillers are also used as part of large chiller plants to satisfy the cooling demand of manufacturing sites in various industries, such as, semiconductors and pharmaceuticals.

The operation of a typical chiller plant has design flexibility to minimize its power consumption. One approach to improve energy efficiency of a complex process is to create an accurate system model, and then use optimization algorithms to determine more efficient operating strategies for the system (Kapoor et al., 2013; Powell et al., 2013; Chang, 2000). Energy efficiency can also be improved by optimally designing the plant configuration. Optimal plant design is a more important problem compared to optimal operation, especially for chiller plants. This is due to its impact on electrical energy costs over a long period of time and high retrofitting costs involved in a change in plant design.

Computer Aided Chemical Engineering, Volume 36. ISSN 1570-7946. http://dx.doi.org/10.1016/B978-0-444-63472-6.00002-1

Several factors affect the way chillers are arranged in the plant relative to one another. The chiller arrangement normally depends on the peak and average cooling load demand, chilled water flow rate, chilled water supply temperature, chilled water return temperature, and chiller design capacity. This paper discusses different ways in which chillers can be arranged in a plant followed by illustrating the effect of chiller plant design on its overall energy efficiency.

Accurate mathematical models play a vital role in estimating a chiller's energy efficiency as a function of several flow parameters that vary with the way chillers are arranged. We demonstrate development of chiller models and propose modifications to already existing models based on year round plant data obtained from the University of Texas at Austin and a semiconductor fab. These facilities have been referred to as Campus A and Campus B, respectively, in this paper. We evaluate the advantages and disadvantages of chiller configuration and quantify the benefits using these models.

2.2 CHILLER ARRANGEMENTS

Chillers in a multichiller plant can be configured in three possible ways—series, parallel, and hybrid. In a hybrid configuration, a set of chillers are arranged in series while operating in parallel with another set of chillers. While a majority of chiller plants have chillers operating in parallel, a number of plants do use hybrid or series arrangement. In the following section, series and parallel chiller arrangements are discussed in detail using a two-chiller system as an example.

2.2.1 CHILLERS IN SERIES

Figure 2.1 shows a schematic of a simple two-chiller system arranged in series. In such an arrangement, a common stream of chilled water passes through the chillers sequentially. Hence, flow rate of the chilled water is constant across all the chillers in series. But the temperature drop for the chilled water across each chiller can be different. The temperature of the chilled water stream is lowered from T_H to $T_{e,p}$ in two stages—(1) from T_H to $T_{e,1}$ by Chiller 1, and (2) from $T_{e,1}$ to $T_{e,p}$ by Chiller 2. Cooling load for each of the series chillers can be expressed as:

$$Q_{1,series} = F * \left(T_H - T_{e,1}\right) \tag{2.1a}$$

$$Q_{2,series} = F * \left(T_{e,1} - T_{e,p}\right) \tag{2.1b}$$

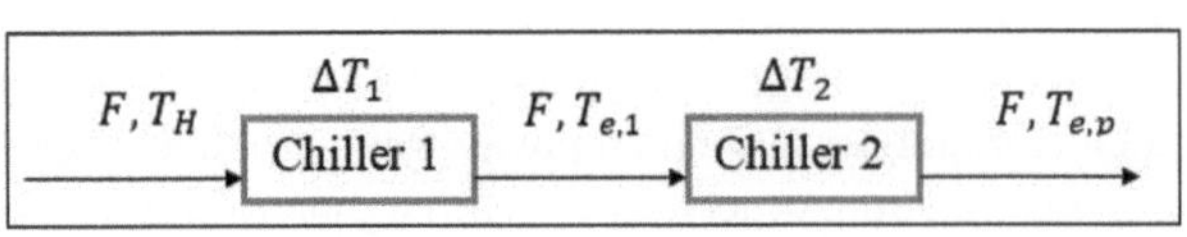

FIGURE 2.1

Schematic of a two-chiller system in series configuration.

By adding Eqns (2.1a) and (2.1b), the total cooling load of the chiller plant is:

$$Q_{Total,series} = F * \left(T_H - T_{e,p}\right) \tag{2.1c}$$

From Eqn (2.1c), it can be inferred that the total cooling load in the case of series chillers depends only on the properties of the chilled water stream (flow rate and temperature) entering and leaving the system. The cooling load is independent of the state of intermediate streams (chilled water at stages in between the series chillers).

2.2.2 CHILLERS IN PARALLEL

A schematic of a two-chiller system in a parallel configuration is depicted in Figure 2.2. For all the chillers in parallel arrangement, the input and output streams of chilled water have the same temperature (T_H and $T_{e,p}$, respectively) and hence the temperature drops across the chillers are identical. But the flow rate of chilled water may vary across chillers. In a two-chiller parallel arrangement, the input chilled water flow, F, is divided into two parallel streams with flow rate F_1 and $(F - F_1)$ respectively. Each stream passes through a separate chiller where it is cooled from T_H to $T_{e,p}$. The cooling load for each chiller in this system is given by the following equations:

$$Q_{1,parallel} = F_1 * \left(T_H - T_{e,p}\right) \tag{2.2a}$$

$$Q_{2,parallel} = (F - F_1) * \left(T_H - T_{e,p}\right) \tag{2.2b}$$

By adding Eqns (2.2a) and (2.2b), the total cooling load for this system is:

$$Q_{Total,parallel} = F * \left(T_H - T_{e,p}\right) \tag{2.2c}$$

Even though Eqns (2.1c) and (2.2c) are derived for a two-chiller system, they can be assumed true for any number of chillers as the resultant expressions are independent of the number of chillers. It is evident from comparing these two equations that the total plant cooling load is a function of the flow rate and temperatures of the input and output chilled water streams, irrespective of the chiller arrangement in the plant (series, parallel, or hybrid). Chiller arrangement in Campus B is of hybrid type, which consists of five parallel chilled water streams and each stream has two chillers connected in series (Figure 2.3).

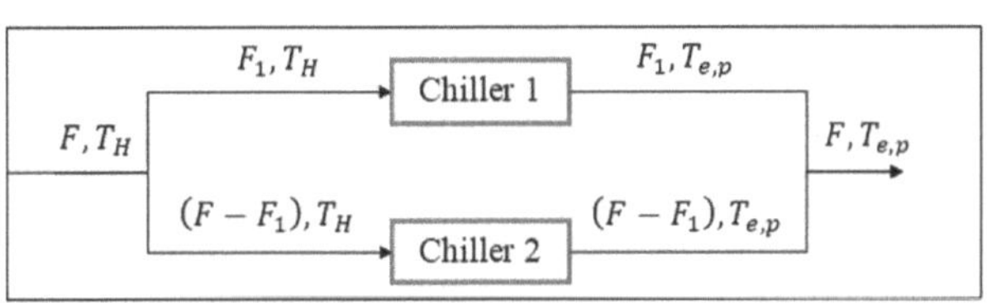

FIGURE 2.2

Schematic of a two-chiller system in parallel configuration.

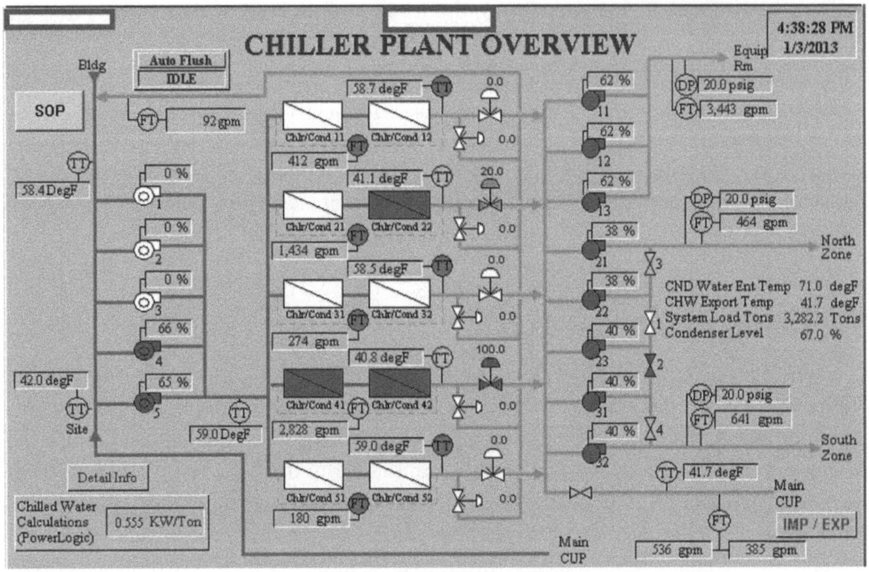

FIGURE 2.3

Screenshot of a multichiller plant at Campus B. Two white blocks have been put on the figure to hide the fab and chiller plant names so that anonymity of the data source is maintained.

In order to quantitatively compare different chiller arrangements, accurate models need to be developed that estimate individual chiller power consumption as a function of its flow parameters. The next section analyzes the plant data and discusses model development in detail.

2.3 MODELING ENERGY EFFICIENCY OF A CENTRIFUGAL CHILLER

The power consumption of a chiller is usually derived from its energy efficiency, or alternatively the coefficient of performance (COP). COP of a chiller is defined as the ratio of the heat removed (i.e., cooling load) to the power input to the compressor.

$$COP = \frac{Q}{P} \tag{2.3}$$

Steady state chiller models estimating a chiller's COP as a function of its operating conditions have been used extensively for a variety of chiller types and sizes (Gordon et al., 1995; Browne and Bansal, 1998; Le et al., 2004; Yik and Lam, 1998; Cui and Wang, 2005; Reddy et al., 2003; Reddy and Andersen, 2002; Hydeman and Gillespie, 2002; Hydeman et al., 2002; Swider, 2003; Monfet and Zmeureanu, 2012; Katipamula and Brambley, 2005; Gordon and Ng, 1994, 1995).

Lee et al. (2012) identified 11 types of centrifugal chiller models that have been used in the literature. Chiller models are increasingly being used to determine the best operating conditions for a chiller, as illustrated by Ng et al. (1997), who used a thermodynamic chiller model to determine the optimal chiller operating points. Chiller plant design can have an equal or more significant impact on the overall energy efficiency than optimizing operating conditions. Modified Gordon–Ng model for centrifugal chillers (Gordon et al., 1995) is used for this purpose. This semiempirical model for centrifugal chillers is based on an energy balance equation around the refrigerant's vapor compression cycle. It computes the power consumed by a chiller (P) as a function of its cooling load (Q), condenser water return temperature (referred to as condenser water temperature for simplicity) (T_c), and chilled water temperature setpoint (T_e) (see Eqns (2.4) and (2.5)). The parameters represented in bold font in Eqn (2.4) are the four model parameters ($\boldsymbol{M_c}$, $\boldsymbol{M_e}$, $\boldsymbol{q_c}$, and $\boldsymbol{q_e}$) that have different values for actual chillers. $\boldsymbol{M_c}$ and $\boldsymbol{M_e}$ are the condenser and evaporator heat exchanger coefficients (W/K) respectively, while $\boldsymbol{q_c}$ and $\boldsymbol{q_e}$ are the rates of internal energy losses (kW) at condenser side and evaporator side, respectively. All variables in the Gordon–Ng model are in SI units.

$$\frac{1}{COP} = -1 + \frac{T_c}{T_e} + \left(\frac{1}{Q}\right)\left(\frac{\boldsymbol{q_e}T_c}{T_e} - \boldsymbol{q_c}\right) + \left(\frac{1}{Q}\right)\left(\frac{\boldsymbol{q_e}}{\boldsymbol{M_c}T_e}\right)\left(\frac{\boldsymbol{q_e}T_c}{T_e} - \boldsymbol{q_c}\right)$$
$$+ \left(\frac{Q}{T_e}\right)\frac{T_c}{T_e}\left(\frac{1}{\boldsymbol{M_c}} + \frac{1}{\boldsymbol{M_e}}\right) + \frac{\frac{\boldsymbol{q_c}}{\boldsymbol{M_e}} + \frac{\boldsymbol{q_e}T_c}{T_e\boldsymbol{M_c}} + \left(\frac{T_c\boldsymbol{q_e}}{T_e} - \boldsymbol{q_c}\right)\left(\frac{1}{\boldsymbol{M_c}} + \frac{1}{\boldsymbol{M_e}}\right)}{T_e} \tag{2.4}$$

$$P = Q * (1/COP) \tag{2.5}$$

While the Gordon–Ng model provides a good representation for almost all real data sets obtained from a centrifugal chiller over its entire operating range, the Gordon–Ng model fitted to the data obtained from chillers at Campus A and Campus B, several limitations were observed, resulting in modifications to the original Gordon–Ng model.

2.3.1 MODIFICATIONS TO GORDON–NG MODEL FOR CENTRIFUGAL CHILLERS

The Gordon–Ng model assumes that its four model parameters ($\boldsymbol{M_c}$, $\boldsymbol{M_e}$, $\boldsymbol{q_c}$, and $\boldsymbol{q_e}$) are constant for a given chiller. The purpose of using a chiller model is to determine the power consumption of a chiller as a function of ambient weather conditions and its cooling load. Therefore, the model parameters are assumed to be constant with respect to chiller cooling load and ambient conditions.

Data from nine centrifugal chillers (Campus A) were independently fitted against the Gordon–Ng model. The model and data seemed to be in good agreement for each chiller for most of the operating range, except at the two extremes of the cooling load range (Figure 2.5). To explain this behavior, it was hypothesized that the rate of internal energy losses is a function of the chiller cooling load, and not a constant parameter. This dependence was assumed to be linear for simplicity. Therefore, the following equations are used to modify the Gordon–Ng model equations:

$$q_e = \boldsymbol{q_{e_m}} + \boldsymbol{a} * Q \tag{2.6}$$

$$q_c = \boldsymbol{q_{c_m}} + \boldsymbol{b} * Q \tag{2.7}$$

This model is referred to as the Modified Gordon–Ng model 1 (MGN1), is given by Eqns (2.4)–(2.7). All symbols represented in bold font are model parameters.

In most large-scale chiller plants, all chillers are operated in parallel mode. That is, the temperature difference ($\Delta T = T_H - T_e$), is the same across all chillers. Therefore, the cooling load across any chiller is directly proportional to the rate at which chilled water flows through its evaporator. Models for multichiller systems with chillers working in parallel do not consider cooling load (Q) and chilled water flow rate (F) as independent variables.

However, in some chiller plants (for instance at Campus B) some of the chillers are arranged in series (see Figure 2.3). For chillers in series, the chilled water flow rate remains the same and hence the cooling load is divided among them by proportionally reducing the total ΔT. For cooling systems with mixed (both series and parallel) chiller arrangements, the cooling load and chilled water flow rate should be treated as independent variables. Chiller cooling load, temperature difference across the chiller, and chilled water flow rate are all related to each other by the definition of cooling load, given by Eqn (2.8). In other words, such systems may have one additional degree of freedom in chiller operation.

$$Q = F * C_p * (T_H - T_e) \tag{2.8}$$

Data from nine chillers (Campus B) were studied and it was found that the ΔT values were nearly constant (range ~0.5 °F) for five of them. However, for the rest of the four chillers ΔT was observed to be within two distinct ranges of values (see Figure 2.4). These four chillers were operated independently at various times and operated in series with another chiller at other times. The data from each of these four chillers were divided into two parts based on its mode of operation. The original Gordon–Ng model was fitted separately against the two data sets for each of the four chillers using least squares minimization. Different values of fitting parameters were obtained for different data sets for each individual chiller (Tables 2.1 and 2.2). Two separate models described each chiller's behavior depending on whether its mode of operation was independent or in series with another chiller. It was concluded that a chiller performs differently in terms of energy efficiency when operated independently versus when operated in series with another chiller. Data also confirmed that for the same values of cooling load, condenser water temperature and chilled water temperature, a chiller always consumed more energy when operated in series than when operated independently.

An additional modification to the Gordon–Ng model was made to cover parallel versus series operations. The energy losses at condenser and evaporator sides, i.e., $\boldsymbol{q_c}$ and $\boldsymbol{q_e}$, are generated from internal dissipation, including fluid friction among

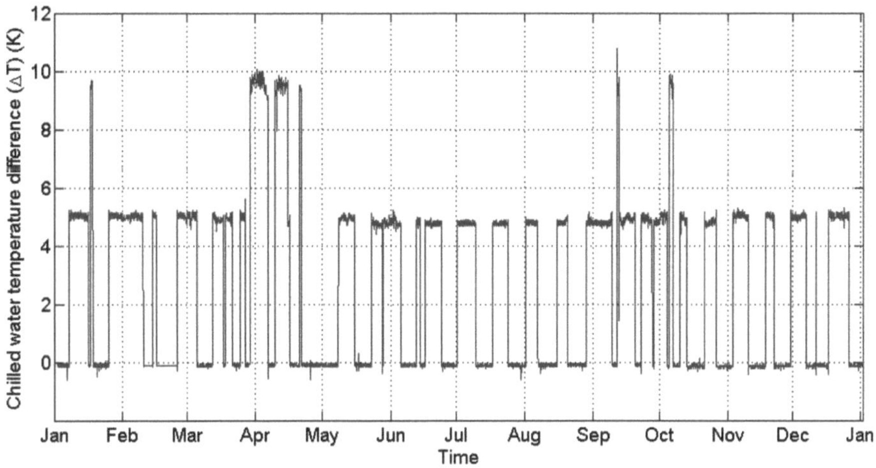

FIGURE 2.4

Two different nonzero ΔT values across one chiller at Campus B.

Table 2.1 MGN1 Model Fitting Parameters for Chiller 12

Fitting Parameter	Series Mode	Parallel Mode
M_c	277.6358	277.6363
M_e	13.93207	13.93432
q_{e_m}	494.6412	307.7059
q_{c_m}	7.53E−09	7.53E−09
a	1.369754	0.514203
b	2.308691	0.94418

Table 2.2 MGN1 Model Fitting Parameters for Chiller 32

Fitting Parameter	Series Mode	Parallel Mode
M_c	803.635	803.635
M_e	195.578	195.578
q_{e_m}	0.0007	0.0007
q_{c_m}	2.8E−07	2.28E−07
a	2.206	1.736
b	1.955	1.489

several other sources (Gordon et al., 1995). Therefore, it was proposed that the internal losses vary as a function of the chilled water flow rate in addition to the cooling load. This leads to addition of one more term to Eqns (2.6) and (2.7) as follows:

$$q_e = q_{e_m} + a * Q + d_1 * F \tag{2.9}$$

$$q_c = q_{c_m} + b * Q + d_2 * F \tag{2.10}$$

Here q_{e_m}, q_{c_m}, a, b, d_1, d_2 are fitting parameters.

The resulting chiller model for operation in a nonparallel arrangement, is referred to as the Modified Gordon–Ng model 2 (MGN2).

Two data sets for one chiller were fitted against the MGN2 model, i.e., Eqns (2.4), (2.5), (2.9) and (2.10), resulting in the values presented in Table 2.4.

2.3.2 VALIDATION OF MODIFIED GORDON–NG CHILLER MODELS

This section compares the chiller model with respect to quality of fit against real chiller plant data collected over 1 year from Campuses A and B. Each model was compared against the Gordon–Ng model (or with MGN1 model), which is the most commonly used first principles model for centrifugal chillers. The Campus A chiller plant consisted of nine separate electric centrifugal chillers with varying efficiencies and operating load ranges. They are numbered after their respective cooling stations—6.1, 6.2, and 6.3 from Station 6; 5.1, 5.2, and 5.3 from Station 5; and 3.1, 3.2, and 3.3 from Station 3. The data obtained from each chiller was fitted using the Gordon–Ng model, which demonstrated a good fit for the most part over the operating load range for all chillers. However, the predicted chiller power consumption deviated from the actual data for very low and very high cooling loads for all chillers.

Figure 2.5 plots the model predicted chiller power versus the actual chiller power values for Chiller 6.1. In case of a perfect fit, all the points in such a plot should lie on

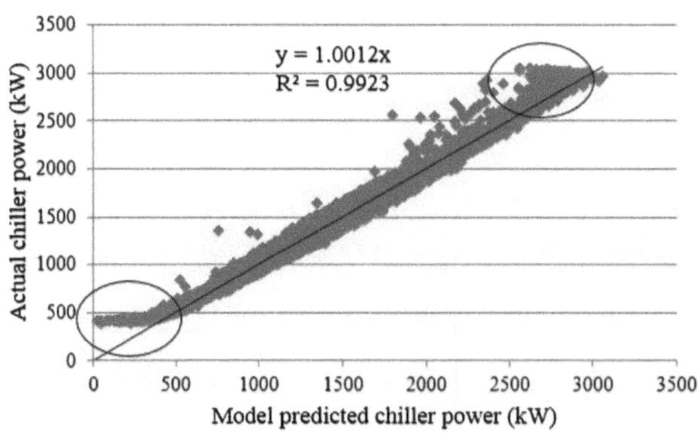

FIGURE 2.5

Gordon–Ng model predicted chiller power consumption versus data for Chiller 6.1.

the straight line y = x. However, in Figure 2.5 a large number of points, especially at the extremes, deviate from this straight line. It was observed that the model does not agree well with the actual data for power <500 kW and for power >2500 kW, which we term extreme load discrepancy.

The Gordon–Ng model uses rate of internal heat losses as model parameters, suggesting that they are independent of the chiller cooling load. But the comparisons of the model predictions with the data for very low and very high load values suggested that internal heat losses should be higher for high-cooling load and lower for low-cooling load (accounted for in MGN1 model).

Figure 2.6 illustrates the Modified Gordon–Ng model 1 fit for the same chiller 6.1. The comparison between Figure 2.5 and Figure 2.6 shows that the MGN1 model fits much better to the data from Chiller 6.1. The data points in Figure 2.6 were compared to a straight line (y = x). It is also evident from Figure 2.6 that the MGN1 model has a good fit over the entire range of cooling load, thus avoiding the extreme load discrepancy behavior.

Similar comparisons were done by fitting data from rest of the eight chillers at University of Texas, Austin. Every chiller, when fitted to the Gordon–Ng model, exhibited extreme load discrepancy behavior similar to Figure 2.5. However, when using the MGN1 model, the discrepancies disappeared. The model fits resulted in coefficient of determination (R^2) values ranging from 0.94 to 0.999 with one exception of $R^2 = 0.84$ for Chiller 3.1.

Year-long data collected from nine separate chillers in Campus B were used for comparing the two MGN models. These chillers are numbered as 11, 12, 21, 22, 31, 32, 42, 51, and 52. The chillers are arranged such that four pairs of chillers—11 and 12, 21 and 22, 31 and 32, 51 and 52—are operated in series with one another. Data collected from the chillers reveals that among the chillers in series pairs, the second chiller in each pair (Chiller 12, 22, 32, and 52) has been operated in two distinct modes—(1) when the first chiller among the pair (Chiller 11, 21, 31, and 41) is on and

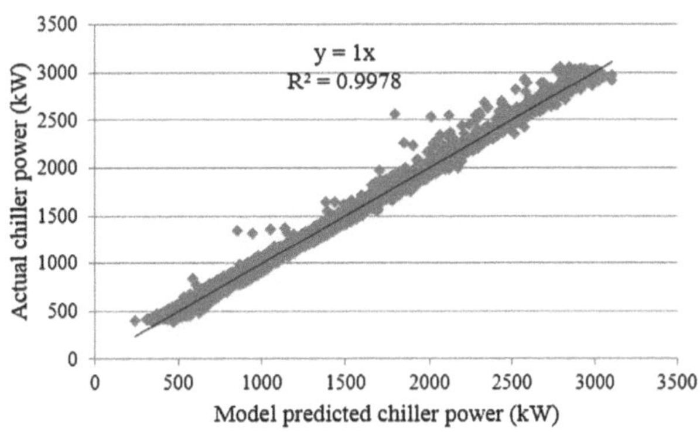

FIGURE 2.6

MGN1 model predicted chiller power consumption versus data.

(2) when the first chiller is off. But in both cases, the range of cooling load placed on the second chiller remained the same. In the first mode, both series chillers are on (say Chiller 11 and 12), the chilled water temperature difference is distributed between the two chillers. While in the second mode, Chiller 11 is off and Chiller 12 is operated in parallel with all other chillers, the chilled water temperature difference across Chiller 12 is greater because Chiller 11 is not contributing anything toward lowering the chilled water temperature. Hence, with the cooling load on Chiller 12 kept at a similar range in both modes, the chilled water flow rate is much lower in the second mode as compared to the first (see Eqn (2.8)). The two modes of operation, applicable for the second chiller in every series pair, are referred to as the series mode and the parallel mode, respectively, in this dissertation.

Data from both operating modes of Chiller 12 was fitted using the MGN1 model and their respective modeling parameters were obtained (Table 2.1). Similarly, Table 2.2 shows two sets of fitting parameters obtained for the two modes of Chiller 32.

It is evident from Tables 2.1 and 2.2 that two different sets of fitting parameters characterize the energy efficiency curve for the same chiller operating in two different modes. However, an interesting observation is that the parameters that define the heat transfer coefficients for the evaporator and condenser ($\boldsymbol{M_e}$ and $\boldsymbol{M_c}$) do not vary between series and parallel modes. This observation is consistent with the fact that heat transfer coefficient of a heat exchanger is a constant and does not vary with fluid flow rate or rate of heat transfer. On the other hand, the set of parameters which describe the rate of energy losses at evaporator and condenser have different values for different modes. The values of the fitting parameters suggest that internal losses are higher in series mode as compared to parallel mode. This can be partially explained by the fact that higher water flow rate in series mode leads to higher shear viscous dissipation, which is a contributor to internal energy losses.

The hypothesis suggesting that internal losses vary with chilled water flow rate, defined in Eqns (2.9) and (2.10), was validated by using data from Chiller 12 and Chiller 32. Data related to different modes of operation were combined to obtain one data set for each chiller. This data was fitted against both MGN1 and MGN2 models. Table 2.3 presents the sum of squared errors (SSE), mean absolute error (MAE), and mean percentage error (MPE) in both cases to compare the accuracy of model fits for Chiller 12.

From Table 2.3, it is clear that the MGN2 model estimates the chiller power more accurately than the MGN1 model, without having prior knowledge about the mode

Table 2.3 MGN1 Model versus MGN2 Model Fitting for Chiller 12

Error Metric	MGN1	MGN2
SSE	76694 kW^2	33653 kW^2
MAE	7.07 kW	4.41 kW
MPE	1.32%	0.83%

SSE, sum of squared errors; MAE, mean absolute error; MPE, mean percentage error.

Table 2.4 MGN2 Model Fitting Parameters

Fitting Parameter	Chiller 12	Chiller 32
M_c	277.7234	803.1358
M_e	14.02435	190.8316
q_{e_m}	304.4677	2.057007
q_{c_m}	7.53E-09	0
a	0.507089	1.677012
b	0.960825	1.472803
d_1	0.01816	0.153584
d_2	0	0.135811

of chiller operation. Table 2.4 presents the MGN2 model fitting parameters obtained after minimizing the SSE, for both Chillers 12 and 32. As expected, the heat transfer coefficients (M_e and M_c) obtained for MGN2 model for both Chiller 12 and Chiller 32 are similar to the ones from the MGN1 models. However, the parameters involved in the estimation of internal energy losses have different values from Tables 2.1 and 2.2. These observations support the hypothesis behind the MGN2 model.

The above comparison between chiller performances in different configurations is key to the analysis presented in the next section. Models developed for Campus B chillers are used to determine the effect of chiller arrangement on its energy efficiency.

2.4 COMPARISON OF ENERGY EFFICIENCY—SERIES VERSUS PARALLEL

The way of distributing cooling load in a chiller plant is strongly connected to its chiller arrangement. As discussed in Section 2.2, the chilled water flow rate is divided among chillers if they are arranged in parallel. On the other hand, the chilled water temperature drop ($\Delta T = T_H - T_e$) is divided among the chillers in a series configuration. In case of a hybrid arrangement, first the chilled water flow rate is divided among the number of total parallel streams and then each stream consisting of more than one chiller is treated like an independent series arrangement.

Even though the cooling capacity of a chiller is irrespective of its configuration (series/parallel or hybrid), its power consumption is not. In other words, for the same cooling load, chilled water outlet temperature and condenser water inlet temperature, the power consumed by a chiller may vary with its relative position in the overall plant arrangement. Hence, the total power consumed by a plant for a given total cooling load depends on two main factors—(1) cooling load distribution among its chillers, and (2) chiller arrangement in the plant. While the former is an operational decision which is to be taken several times in a day, the latter is a design decision which is taken either when a new plant is designed or when an existing plant undergoes retrofitting. Optimal chiller loading (Kapoor et al., 2013; Powell et al., 2013; Chang, 2000) aims at making the decision of cooling load distribution to maximize the overall plant energy efficiency for a fixed chiller arrangement. However, this section presents a study that compares energy efficiency of a chiller plant in both series and parallel arrangements.

2.4.1 DIFFERENCE IN CHILLED WATER FLOW RATE

The key difference between operating conditions of a chiller working in series or parallel configuration is its chilled water flow rate and ΔT for a particular cooling load. As described in the previous section, chilled water flow rate is higher when a chiller is working in series as compared to parallel (where the overall flow rate gets distributed among the chillers), assuming same cooling load in both cases. By making modifications to the Gordon–Ng model for centrifugal chillers (Eqns (2.9) and (2.10) of MGN2 model), it has been asserted that the chiller power consumption increases with chilled water flow rate. This implies that a chiller will consume more power in series mode than in parallel mode for the same amount of cooling load. The difference in chiller power consumption in series and parallel mode was quantified by analyzing MGN1 chiller models for Chillers 12 and 32 (Campus B). Separate MGN1 model parameters for series and parallel mode of these chillers were obtained from Tables 2.1 and 2.2.

Chiller power consumption was plotted against cooling load for Chiller 12 (Figure 2.7) and Chiller 32 (Figure 2.8) in both operating modes—series and

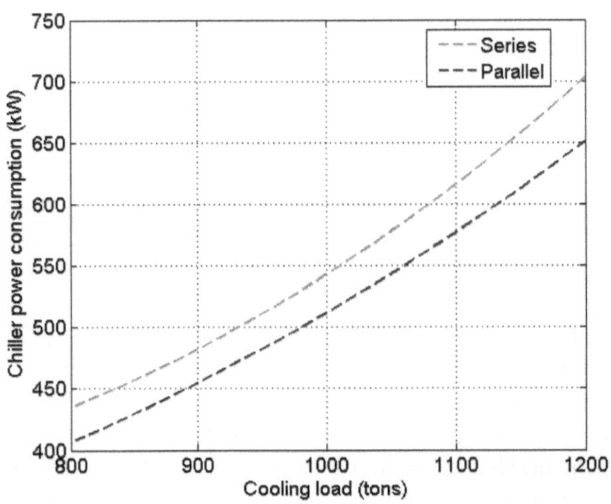

FIGURE 2.7

Chiller power variation with cooling load (series vs. parallel) for Chiller 12.

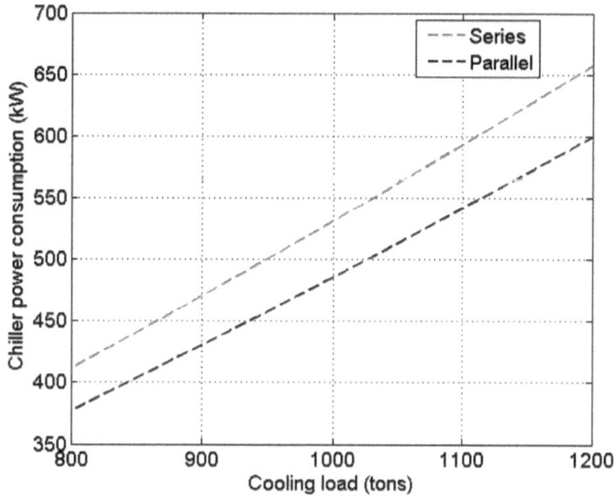

FIGURE 2.8

Chiller power variation with cooling load (series vs. parallel) for Chiller 32.

parallel). The values of T_e and T_c were kept constant for this analysis at 278 and 295 K, respectively. The calculations predict 9.13–9.62% of energy savings for Chiller 32 by switching from series to parallel mode of operation whereas Chiller 12 can save between 5.97% and 7.97% of its energy usage by switching the mode.

The above analysis using real year-long data from Campus B chiller plant illustrates the extent by which energy efficiency of a chiller can vary by simply changing its mode of operation from series (high F, low ΔT) to parallel (low F, high ΔT).

2.4.2 DIFFERENCE IN CHILLED WATER OUTLET TEMPERATURE

This section illustrates the effect of chiller arrangement on chilled water outlet temperature for each individual chiller and how the outlet temperature in turn affects the energy efficiency. Taking the two-chiller system in Figure 2.1 as an example, it is observed that the value of T_e differs among chillers in a series arrangement. While Chiller 2 (the last chiller) generates chilled water with $T_e = T_{e,p}$, Chiller 1 has $T_e = T_{e,1}$ where $T_{e,1} > T_{e,p}$. By extrapolation, the following relation is established for a series of m chillers where $T_{e,j}$ is the chilled water outlet temperature for the jth chiller.

$$T_H \geq T_{e,1} \geq T_{e,2} \geq \cdots \geq T_{e,j} \geq \cdots \geq T_{e,(m-1)} \geq T_{e,p} \tag{2.11}$$

On the other hand, all chillers in a parallel arrangement have a common value of $T_{e,j} = T_{e,p}$. Therefore, for each chiller j, the following relation holds.

$$T_{e,j}(series) \geq T_{e,j}(parallel) \tag{2.12}$$

The dependence of a chiller's COP on its T_e was studied using the Gordon–Ng model (Eqn (2.4)) and can be represented in the form of following equation.

$$\frac{1}{COP} = \gamma_0 + \gamma_1 * \frac{1}{T_e} + \gamma_2 * \frac{1}{T_e^2} \tag{2.13a}$$

$$\gamma_0 = -1 + \left(\frac{1}{Q}\right)(-q_c) \tag{2.13b}$$

$$\gamma_1 = \left(T_c - \frac{q_c}{M_c}\right)\left(1 + \frac{q_e}{Q}\right) \tag{2.13c}$$

$$\gamma_2 = \left(\frac{q_e}{Q} + 1\right)\frac{q_e T_c}{M_c} + T_c * (Q + q_e)\left(\frac{1}{M_c} + \frac{1}{M_e}\right) \tag{2.13d}$$

Because of positive values attained by heat transfer coefficients, rates of internal energy losses, cooling load and condenser water temperature, γ_2 is always positive (Eqn (2.13d)). Additionally, from the model fitting parameters of chillers at Campus A and Campus B it was observed that the order of magnitude of q_c/M_c is 10^0 whereas T_c is of the order of magnitude of 10^2. From Eqn (2.13c), this implies that the value of γ_1 is also positive. Therefore, chiller efficiency increases as T_e increases if all other factors are kept constant. To simulate this effect, Chiller 11 (Campus B) power consumption was plotted against its cooling load for $T_e = 277$ K (parallel mode) and $T_e = 282$ K (series mode) in Figure 2.9. It shows clearly that Chiller 11 is more energy efficient at a higher T_e setpoint. This analysis estimated that by changing the value of T_e from 282 K to 277 K, the COP drops and hence the power consumption rises by 11.96–12.26%. Combining this result with Eqn (2.12), this section

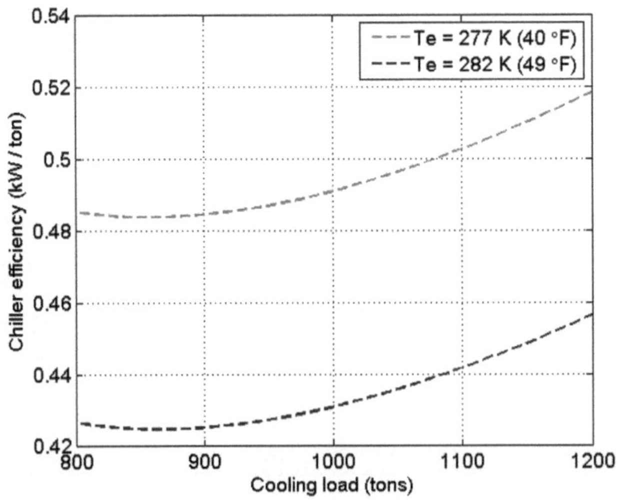

FIGURE 2.9

Chiller efficiency variation with cooling load for Chiller 11 for constant T_e.

concludes that a series arrangement adds to the plant energy efficiency by using a higher chilled water outlet temperature for most chillers.

2.5 CONCLUSIONS

This paper discusses different types of chiller arrangements that can be employed in a multichiller plant—series, parallel, and hybrid. These configurations are referred to as different modes of operation from the perspective of each individual chiller. Series and parallel chiller configurations are discussed in detail and their benefits with respect to overall plant energy efficiency are compared. The effects of two main differences between series and parallel arrangements (flow rate and temperature difference across individual chillers) on chiller power consumption are analyzed for this purpose.

For the systems studied here, the switch from parallel mode to series mode increases the chilled water flow rate by almost 100%, which adversely affects the energy efficiency. The analysis showed that the power consumption of a chiller can rise by 9.6% by switching its mode of operation from parallel to series, while keeping the cooling load, chilled water outlet temperature, and condenser water inlet temperature constant. It was shown by using the Gordon–Ng model that a higher T_e value increases the chiller efficiency. This analysis showed that efficiency of Chiller 11 can drop by 12.26% by switching its mode of operation from series (current) to parallel (hypothetical).

Series and parallel arrangements have their own pros and cons. In series configuration, the high-chilled water flow rate adversely affects the efficiency, but relatively the higher chilled water outlet temperature (T_e) reduces the power consumption. However, the magnitude of impact on energy efficiency depends on several factors such as the model parameters of each chiller, total cooling load requirement, the number of parallel streams in a parallel or hybrid chiller plant, and the number of chillers in each series arrangement.

NOMENCLATURE

T_H	Chilled water inlet temperature (K)
F	Chilled water mass flow rate (kg/sec)
T_e	Chilled water outlet temperature (K)
Q	Chiller cooling load (Tons or kW)
P	Chiller power consumption (kW)
COP	Coefficient of performance (Tons/kW)
T_c	Condenser water return temperature (K)
M_c	Heat transfer coefficient of condenser (WK^{-1})
M_e	Heat transfer coefficient of evaporator (WK^{-1})
q_c	Rate of internal condenser heat loss (kW)
q_e	Rate of internal evaporator heat loss (kW)

REFERENCES

Browne, M.W., Bansal, P.K., 1998. Steady-state model of centrifugal liquid chillers: Modèle pour des refroidisseurs de liquide centrifuges en régime permanent. Int. J. Refrig. 21, 343–358.

Chang, Y.C., 2000. A novel energy conservation method – optimal chiller loading. Electr. Pow. Syst. Res. 69 (2–3), 221–226.

Cui, J., Wang, S., 2005. A model-based online fault detection and diagnosis strategy for centrifugal chiller systems. Int. J. Therm. Sci. 44, 986–999.

Le, C.V., Bansal, P.K., Tedford, J.D., 2004. Three-zone system simulation model of a multiple-chiller plant. Appl. Therm. Eng. 24, 1995–2015.

Gordon, J.M., Ng, K.C., Chua, H.T., 1995. Centrifugal chillers: thermodynamic modeling and a diagnostic case study. Int. J. Refrig. 18, 253–257.

Gordon, J.M., Ng, K.C., 1994. Thermodynamic modeling of reciprocating chillers. J. Appl. Phys. 75, 2769–2774.

Gordon, J.M., Ng, K.C., 1995. A general thermodynamic model for absorption chillers: theory and experiment. Heat Recov. Syst. CHP 15, 73–83.

Henze, G.J., 1995. Evaluation of Optimal Control for Ice Storage Systems (Ph.D. dissertation), Department of Civil, Environmental and Architectural Engineering, University of Colorado, Boulder, CO.

Hydeman, M., Gillespie, K.L., 2002. Tools and techniques to calibrate electric chiller component models. ASHRAE Trans. 108, 733–741.

Hydeman, M., Sreedharan, P., Webb, N., Blanc, S., 2002. Development and testing of a reformulated regression-based electric chiller model. ASHRAE Trans. 108, 1118–1127.

Kapoor, K., Powell, K.M., Cole, W.J., Kim, J.S., Edgar, T.F., 2013. Improved large scale process cooling operation through energy optimization. Processes 1, 312–329.

Katipamula, S., Brambley, M.R., 2005. Methods for fault detection, diagnostic, and prognostics for building systems – a review, Part I. HVAC Res. 11, 3–25.

Lee, T.-S., Liao, K.-Y., Lu, W.-C., 2012. Evaluation of the suitability of empirically-based models for predicting energy performance of centrifugal water chillers with variable chilled water flow. Appl. Energy 93, 583–595.

Monfet, D., Zmeureanu, R., 2012. Ongoing commissioning of water-cooled electric chillers using benchmarking models. Appl. Energy 92, 99–108.

Ng, K.C., Chua, H.T., Ong, W., Lee, S.S., Gordon, J.M., 1997. Diagnostics and optimization of reciprocating chillers: theory and experiment. Appl. Therm. Eng. 17, 263–276.

Powell, K.M., Cole, W.J., Ekarika, U.F., Edgar, T.F., 2013. Optimal chiller loading in a district cooling system with thermal energy storage. Energy 50, 445–453.

Reddy, T.A., Niebur, D., Andersen, K.K., Pericolo, P.P., Cabrera, G., 2003. Evaluation of the suitability of different chiller performance models for on-line training applied to automated fault detection and diagnosis (RP-1139). HVAC Res. 9, 385–414.

Reddy, T.A., Andersen, K.K., 2002. An evaluation of classical steady-state off-line linear parameter estimation methods applied to chiller performance data. HVAC Res. 8, 101–124.

Swider, D.J., 2003. A comparison of empirically based steady-state models for vapor-compression liquid chillers. Appl. Therm. Eng. 23, 539–556.

Yik, F.H.W., Lam, V.K.C., 1998. Chiller model for plant design studies. Build. Serv. Eng. Res. Technol. 19, 233–241.

CHAPTER

Modeling Sustainable Chemical Processes for Biofuels

3

Eric H. Dunlop*[,1], A. Kimi Coaldrake*, Cory S. Silva[§], Warren D. Seider[§]
**Pan Pacific Technologies, Adelaide, Australia*
[§]Department of Chemical and Biomolecular Engineering, University of Pennsylvania, Philadelphia, PA, USA
[1]Corresponding author: E-mail: edunlop@panpacific.com.au

3.1 INTRODUCTION

The chemical engineering profession has made significant efforts in recent years to develop sustainable processes (Averill, 2011; El-Halwagi, 2012; Evans, 2007), often to improve energy security and reduce climate change (Deutch et al., 2007; Stoner, 2013), and to produce "green" products and processes (Parrish and Abraham, 2014). There is a clear trend in the chemical industry toward product and process design solutions that address growing energy demand in a safe and enduring manner.

In the design or redesign of conventional processes, it is often assumed that optimized processes are near-optimal and not likely to be improved significantly. By comparison, designs for sustainable bioprocesses are often in the early stages of development and are less well understood. This is particularly true for processes that generate biofuels from algae or even ethanol from cellulosic materials. Here, the chemical formula of the biological matter is available as crude stoichiometric formulas, e.g., for algae, $C_{0.8}H_{1.92}N_{0.63}P_{0.05}S_{0.6}$. Even small temperature changes often degrade or denature biomaterials without changing the apparent stoichiometry. Furthermore, biomaterial composition varies with growing season or the time in the growth cycle at which it is obtained. Even pilot-plant data are notoriously unstable and nonrepeatable, subject to the vagaries of weather. The concept of a "steady state," so fundamental to process engineering, is often not observed. Moreover, thermodynamic properties are usually unavailable; enthalpies of combustion are often the best available data, obtained under poorly defined conditions, while entropy data are almost nonexistent. Even when thermodynamic data are available, they are inconsistent due to different experimental conditions or reference states. Nevertheless, chemical engineers successfully design and model bioprocesses.

At a superficial level, especially early in process design (when detailed mechanisms often have not been formulated), like other chemical processes, biofuel

Computer Aided Chemical Engineering, Volume 36. ISSN 1570-7946. http://dx.doi.org/10.1016/B978-0-444-63472-6.00003-3

processes need to be modeled. However, difficulties often arise due to the limited cross-disciplinary interactions of biologists and chemical engineers working together. Here, lab or bench-scale processing steps often do not translate to industrial plants; that is, they are not scalable. Also, lab-scale researchers normally do not place emphasis on fully converged mass and energy balances. They do not use major process simulators, such as Aspen Plus, and do not prepare business plans to guide project management—estimating equipment sizes, purchase and installation costs, operating costs, and profitability analyses. For this purpose, simulation and related economics programs play an important role. Figure 3.1 shows AspenOne® as the simulation backbone, including RateFrac (e.g., to design CO_2 absorbers), with many related software packages. These are the Aspen Process Economic Analyzer (APEA – for equipment sizing, equipment cost estimation, and profitability analysis), Comsol Multiphysics (e.g., to model algae cell growth and mass transfer), Aspen Custom Modeler (e.g., to estimate algae pond area and evaporation rates), and EXCEL (to tabulate results).

Initially, this chapter shows how key missing parameters can be generated; e.g., the thermodynamic properties of algae cells having variable oil content using "parameter lumping." Its emphasis is on the use of the first principles of the underlying thermodynamics to build up a data bank of usable information for process design. When the analysis progresses to the stage in which a flowsheet emerges, several challenges arise making it difficult to use conventional, commercial,

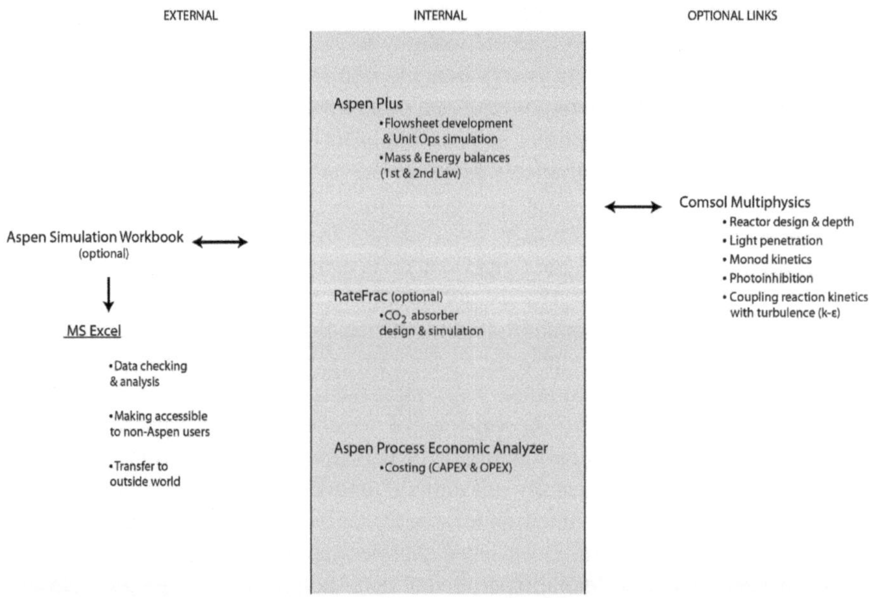

FIGURE 3.1

Integrated process systems software using AspenOne as the backbone for design, development, costing, and analysis of algal biofuels processes.

computer-aided modeling packages, such as AspenOne® and Comsol Multiphysics®. For example, it is customary to input the known data into Aspen Plus and proceed to an answer(s). Here, the opposite strategy is illustrated, letting Aspen Plus run "in reverse" to solve the mass and energy balances to locate feasible solutions.

Then, methods for estimating sustainability metrics, such as Life Cycle Analyses, for example, using GREENSCOPE (Gauging Reaction Effectiveness for the Environmental Sustainability of Chemistries with a multi-Objective Process Evaluator) and GREET (Greenhouse Gases, Regulated Emissions, and Energy Use in Transportation), are examined. These offer insights into energy and exergy consumption, and carbon and water management. This chapter shows how technoeconomic models that incorporate these sustainability measures provide greater insights to inform decision-making. It thus demonstrates how an integrated-modeling, systems approach delivers converged mass and energy balances and estimates sustainability indicators (SIs).

3.2 INTRODUCTION TO ALGAE PROCESSING

Microalgae are unicellular organisms that operate autotrophically or heterotrophically; that is, grow by either harnessing the radiant energy from sunlight or the chemical energy from organic molecules. To grow autotrophically (using sunlight), algae sequester carbon dioxide from the air and absorb water from their surroundings, producing organic molecules—like sugars and oils. These oils can be used for nutritional purposes or converted to transportation-grade fuels. The algae cells' ability to consume organic molecules (heterotrophically) makes them ideal for consuming waste products, like unused biomass or glycerol. In addition, algae are the fastest growing autotrophic organism, can grow in either salt or fresh water, and accumulate oils until they are more than 50 dry wt% of the cell under certain conditions (Silva et al., 2014).

The attractiveness of algae as a fuel source has been explored numerous times in the twentieth and twenty-first centuries, beginning in 1941 with Japan using algae for fuel production during World War II. The United States adopted the program in the 1950s (Burlew, 1953), but abandoned it due to the availability of inexpensive petroleum. Efforts to study algae were revived in the 1970s, as the Aquatic Species Program. However, again, algae were found to be too expensive to compete with traditional fuels. The most recent exploration of algae as a fuel resource was under the National Alliance for Advance Biofuels and Bioproducts (NAABB), from 2010 to 2013 (National Alliance for Advanced Biofuels and Bioproducts Final Report (Online)). The NAABB divided algae research into five key areas: algal biology, algae cell cultivation, algae slurry dewatering, oil extraction, and oil conversion. While the first area (algal biology) is confined to laboratory experimentation, the others give a clear picture of the processing steps that must be undertaken to generate transportation fuels from algae.

For the process presented herein (Dunlop et al., 2013), the *Nannochloropsis salina* algae species is selected because of its ability to grow in salty and polluted water. Figure 3.2 begins with a smokestack (carbon dioxide source), which can be from a power plant, a cement works, or other industrial operation. For this analysis, the source of the carbon dioxide is unimportant. The carbon dioxide is absorbed, not bubbled, into the following two reaction operations. The absorber is modeled as rate-based (using RateFrac in Aspen Plus), which is more accurate than equilibrium-staged, and allows the absorption rate constants to be adjusted to reflect mass transfer performance. The first reaction operation models just algae growth, and a separate reaction operation models lipid production. Note that it was helpful to have these last three operations totally independent, although all three operations could be carried out in a single equipment item. At this stage of process development, the type of reactor is not specified; i.e., raceways or glass tubes (e.g., photo-bioreactors). It was also deemed advisable to retain maximum flexibility in the modeling process by having recycle streams available to suit different design concepts. Water, biomass, and glycerol may or may not be recycled, according to different needs (O'Grady and Morgan, 2011). An additional possibility was created to bring in glycerol that has originated outside of the defined flow sheet, as the additional glycerol can be viewed as a supplemental carbon source. This may be helpful when there is an excess availability of glycerol from other processes. Also, bleed streams are provided to purge inert species; e.g., build-up of salt as evaporation occurs.

Accurate species diagrams are needed (Figure 3.3), not just for carbon dioxide and water, but also for phosphate. These include the ionic species, especially the bicarbonate ion, as the industry returns to the higher pH (9–10) used in food processing to control contamination.

The cultivation section is simulated using two reaction operations, one for algal cell generation and the other for lipid production. Both use information recycle loops (described in the section, *Running ASPEN in Reverse*) with blocks that account for solar energy input, and that estimate the conversion of the limiting reactant, the extents of reactions, and the heats of the reactions, and that account for energy losses due to water evaporation. Specifications for the model are measured local rates of evaporation, local solar inputs, pond depth, and photosynthetic efficiency (typically 4%).

The first reaction operation models just algae growth, and a separate reaction operation models lipid production. Note that it was helpful to have these last three operations totally independent, although all three operations could be carried out in a single equipment unit. At this stage of process development, the type of reactor is not specified; that is, raceways or glass tubes (e.g., photo-bioreactors). It is also deemed advisable to retain maximum flexibility in the modeling process by having recycle streams available to suit different design concepts. Water, biomass, and glycerol may or may not be recycled, according to different needs. An additional possibility is created to bring in glycerol that has originated outside of the defined flow sheet,

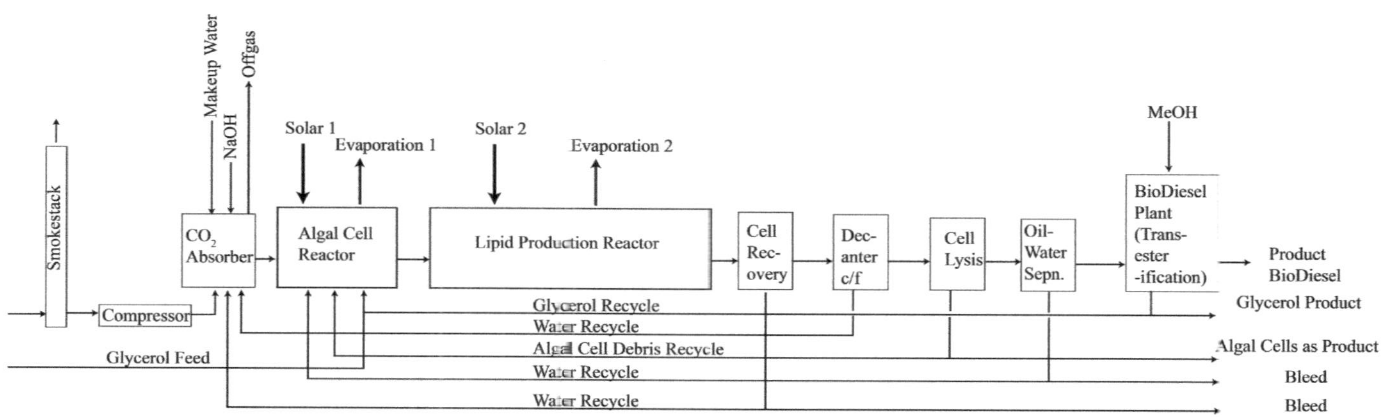

FIGURE 3.2

Generalized process block flow diagram for an algae-to-biodiesel process.

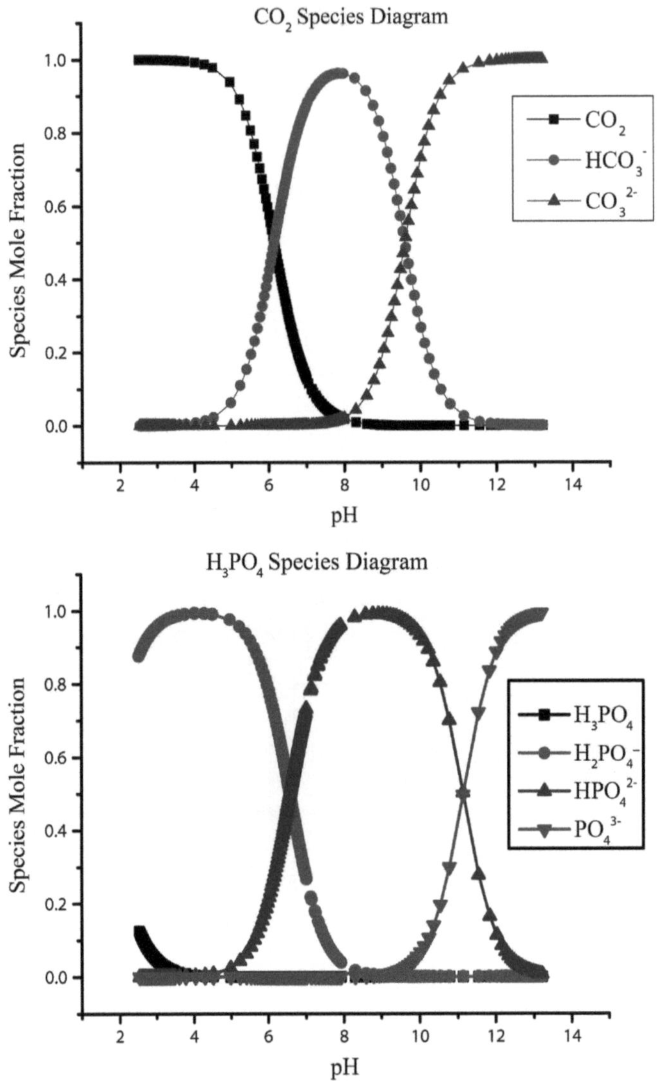

FIGURE 3.3

Species diagrams for the carbon dioxide and phosphate systems.

as the additional glycerol can be viewed as a supplemental carbon source. This may be helpful when there is an excess availability of glycerol from other processes. Also, bleed streams are provided to purge inert species; for example, build up of salt as evaporation occurs.

After emerging from the cultivation reaction operations, the algae slurry has a concentration of approximately 1 g (dry weight) per liter of water. Since oil-extraction methods often rely on solvents, a more concentrated feedstock (about 60–100 g of algae per L of water) is required. The most common methods of dewatering involve flocculation (Smith and Davis, 2012; Weissman and Goebel, 1987), which cause the algae to aggregate and thereby increase the efficiency of settling, clarification, filtration, and centrifugation operations (Flocculants Info (Online)). Flocculants can either be minerals (Weissman and Goebel, 1987), natural polymers (polysaccharides) (Origin Oil Single-Step Extraction (Online)), or synthetic polymers. Mineral flocculants leave residual metal ions in the biomass when the dosing rates are not optimized, which can cause problems with downstream processing. Likewise, synthetic polymer flocculants can leave residual carcinogens in the biomass, rendering it unusable (Flocculants Info (Online)). Synthetic flocculants also require a fixed pH and salinity to operate effectively. After algae have been flocculated (using one of the three options), they can be harvested using dissolved air flotation and/or centrifugation. Dissolved air flotation dissolves pressurized air in water, forming microbubbles. The bubbles stick to the solids, which cause them to float to the surface, where a skimmer can harvest them (Silva et al., 2014). The algae cells are sent to an oil-extraction step, while the water is recycled to the cultivation stage.

The oil must be extracted from the concentrated algae slurry. Thus far, this has proved to be the least reliable operation, with cost estimates spanning orders of magnitude (Silva et al., 2014). Ball mills are traditionally used to lyse the cells. However, ball mills require a dryer product than most other extraction methods (Brentner et al., 2011), and pulverizing the cells often leads to an oil feedstock contaminated with free fatty-acids, trace metals, proteins, and steroids (Silva et al., 2014). Other methods are in development which leach the oil from the cells (Czartoski et al., 2011; SRS Energy, Algae Fractionation (Online)) or rupture the cells using electric fields (Eckelberry et al., 2010; Origin Oil Single-Step Extraction (online)), ultrasonic waves (Lee et al., 2010), microwaves (Iqbal and Theegala, 2013; Terigar et al., 2011), or supercritical fluids (Soh and Zimmerman, 2011). However, these methods are still in the research and development phase. After the cells have been lysed, the oil is extracted, typically using hexane. The oil is then sent to the oil-conversion stage, while the residual biomass is recycled to the cultivation stage as heterotrophic feed for the algae cells.

The final stage involves oil upgrading to biofuels. The algae oils can either be hydrogenated to green diesel or green jet fuel, or transesterified (using methanol) to fatty-acid methyl-esters (FAME), also known as biodiesel. Herein, the second pathway is examined in more detail because it is more common. Transesterification occurs according to the three reversible reactions, shown in Figure 3.4—where TG is triglyceride, MEOH is methanol, DG is diglyceride, MG is monoglyceride, GLY is glycerol, and FAME is fatty-acid methyl-ester. These reactions are typically carried out at mild temperatures and pressures with either an acidic, basic, or enzymatic catalyst (Vyas et al., 2009). However, recent studies have examined the use of supercritical fluids (Duan and Savage, 2010; Levine et al., 2010; Soh and Zimmerman, 2011). In any case, biodiesel is the finished product, while the side product, glycerol, can either be sold or recycled to the cultivation ponds as algae feed.

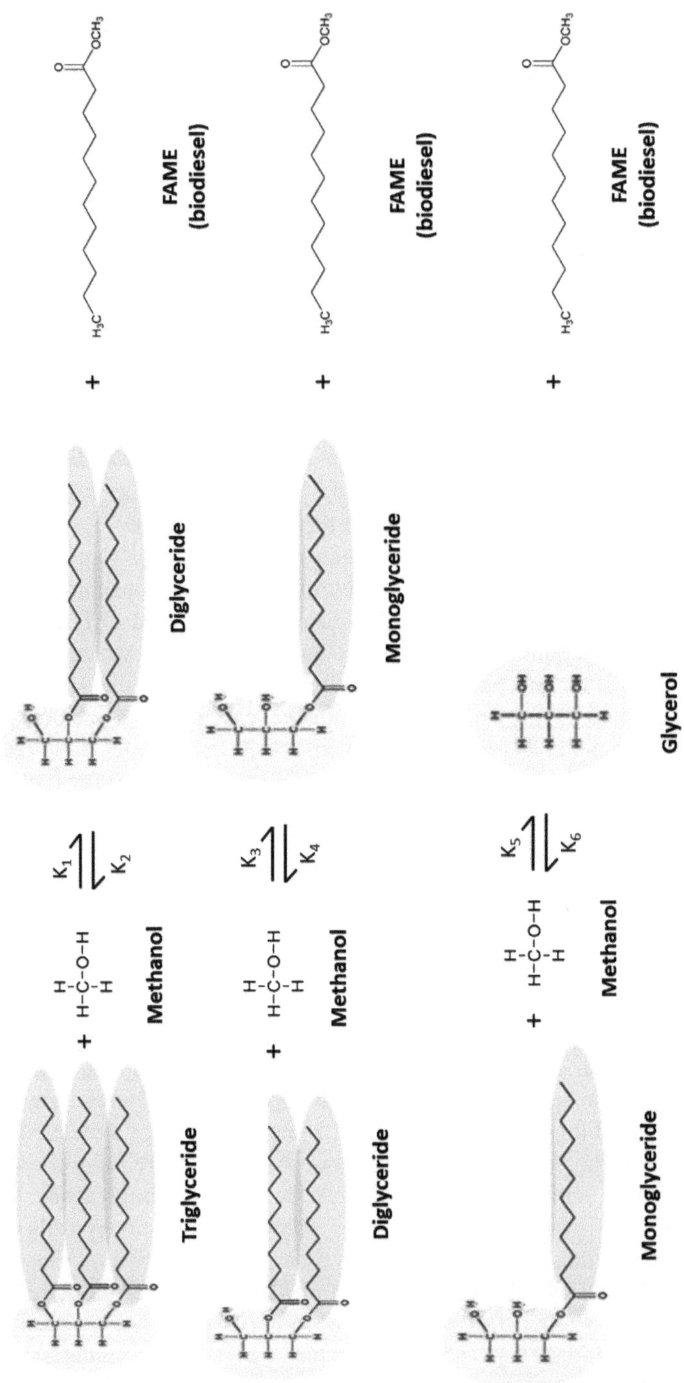

FIGURE 3.4

Transesterification reaction scheme.

3.3 PHYSICAL PROPERTIES

The formulation of a commercial-scale, systems-based simulation requires that the chemical species within the system be well-defined and have known thermophysical properties. The molecules in this system fall into three different categories, those that are well known, those for which limited data are available, and those that are undefined.

The first category contains basic chemical species, like methanol, carbon dioxide, urea, oxygen gas, nitrogen gas, and water. These chemical species are well-defined, and an abundance of thermophysical-property data (enthalpies and entropies of formation, heat capacities, etc.) are available.

Limited thermodynamic data are available for FAMEs and triglycerides. Triglycerides, also known as triacylglycerols (TAGs), are the primary species in which algae cells accumulate and store oil. FAMEs are the biodiesel molecules that result when TAGs react with methanol (Silva et al., 2014). However, algae cells produce far too many species of triglyceride to be incorporated into a systems-based model (Dunlop et al., 2013). Consequently, it is convenient to combine the most predominant nine triglycerides, using weighted averaging, into a species called TAG9 that represents triglycerides in this work and the accompanying calculations. Data for TAG9 and the accompanying FAME species are shown in Table 3.1.

The most difficult species to model are the algae cells and their derivatives. Algae cells are composed of a wide array of molecules, from long-chain polypeptides (proteins) and nucleic acids (DNA and RNA), to small nonpolar molecules (steroids). The true compositions of cells likely vary with time of day, sunlight intensity, humidity, cellular genetics, and cell density, among other factors. For some simulations, some of these factors are incorporated. However, for commercial-scale simulations, it is often infeasible to represent this level of complexity. Rather, algae cells are represented as a pseudo-component having an "equivalent formula," adopted to capture the average composition of the cells on an algae farm. Because of the diversity of cells, molecular formula and thermodynamic data are scarce. Thus, initially herein, data for baker's yeast (*Saccharomyces cerevisiae*) was used to represent the algae cells (Dunlop et al., 2013; Kemp, 1999)—having a pseudo-component formula, $C_{50}H_{50}O_{30}N_7$. As more data became available through the National Alliance for Advanced Biofuels and Bioproducts (NAABB) (National Alliance for Advanced Biofuels and Bioproducts Final Report (Online)), the formula for algae was modified to include both sulfur and phosphorus giving $C_{46.2859}H_{87.9432}O_{18.5143}N_{3.8417}S_{0.0787}P_{0.0926}$, with an enthalpy of formation of 13.0 MJ/kg and an entropy of formation of 6.04 kJ/kg-K, respectively, which is shown in Table 3.2 (Lammers, 2014).

In addition to standard algae cells, it is necessary to create species that represent cells accumulating oil. All species of cells naturally have some baseline quantity of oil, which is used for structural purposes. Algae cells, however, have the ability to

Table 3.1 Data for Triglycerides and TAG9

Methyl Esters (FAME)										
						MJ/kmol	MJ/kmol	MJ/kmol	MJ/kmol	
Type	Common Name	Aspen Name	CAS No.	Formula	MWt	ΔH Combustion	ΔH_f Formation	$T\Delta S_f$	ΔG_f Formation	Mole Fraction
C14:0	Methyl myristate	METHY-01	124-10-7	$C_{15}H_{30}O_2$	242.3975	−9,431	−759	149	−908	0.023
C16:0	Methyl palmitate	METHY-02	112-39-0	$C_{17}H_{34}O_2$	270.4507	−10,107	−1,421	149	−1,591	0.373
C16:1	Methyl palmitoleate	METHY-05	1120-25-8	$C_{17}H_{32}O_2$	268.4348	−10,548	−674	149	−823	0.387
C18:0	Methyl stearate	METHY-03	112-61-8	$C_{19}H_{38}O_2$	298.5038	−11,962	−945	149	−1,095	0.018
C18:ln9	Methyl oleate	METHY-04	2462-84-2	$C_{19}H_{36}O_2$	296.4879	−11,832	−789	149	−938	0.084
C18:2n6	Methyl linoleate	METHY-06	112-63-0	$C_{19}H_{34}O_2$	294.4721	−11,690	−605	149	−754	0.028
C20:4n6	Methyl arachidonate	METHY-07	2566-89-4	$C_{21}H_{34}O_2$	318.4935	−13,263	−1,003	149	−1,152	0.017
C20:5n3	Methyl eicosapen-tanoate	METHY-08	2734-47-6	$C_{21}H_{32}O_2$	316.4776	−12,659	−187	149	−336	0.061
C22:1	Methyl erucinate	METHY-09	1120-34-9	$C_{23}H_{44}O_2$	352.5943	−14,451	−887	149	−1,037	0.008
										1.0000

Triglycerids								
					MJ/kmol	MJ/kmol	MJ/kmol	MJ/kmol
Common Name	Aspen Name	CAS No.	Formula	MWt	ΔH Combustion	ΔH_f Formation	$T\Delta S_f$	ΔG_f Formation
Trimyristin	TRIMY-01	555-45-3	$C_{45}H_{86}O_6$	723.1925	−27,644	−2,355	371	−2,726
Tripalmitin	TRIPA-01	555-44-2	$C_{51}H_{98}O_6$	807.3521	−31,606	−2,469	414	−2,883
Tripalmitolein	TRIPA-02	118450-52-5	$C_{51}H_{92}O_6$	801.3044	−31,179	−2,038	411	−2,449
Tristearin	TRIST-01	555-43-1	$C_{57}H_{110}O_6$	891.5114	−35,807	−2,344	457	−2,801
Triolein	TRIOL-01	122-32-7	$C_{57}H_{104}O_6$	885.4637	−35,100	−2,194	454	−2,648
Trilinolein	TRILI-01	537-40-6	$C_{57}H_{98}O_6$	879.4163	−34,556	−1,880	451	−2,331
Triarachi-donate	TRIAR-02	23314-57-0	$C_{63}H_{98}O_6$	951.4805	−39,468	−2,759	500	−3,260
Trieicosapen-tanoate	TRAEI-01	99660-94-3	$C_{63}H_{92}O_6$	945.4328	−38,761	−1,891	485	−2,376
Trierucin	TRIEU-01	2752-99-0	$C_{69}H_{128}O_6$	1053.7829	−42,802	−2,643	540	−3,183

Table 3.2 Thermophysical Property Data for Algae Cells and Their Derivatives

	Formula	ΔH_f kJ/mol	ΔS_f kJ/mol-K	ΔG_f kJ/mol	MWt
Algae ("Nanno")	$C_{46.2859}H_{87.9432}O_{18.5143}N_{3.3417}S_{0.0787}P_{0.0926}$	−12,907	−6.04	−11,106	1000.00
OC5	$C_{47.1631}H_{89.3564}O_{17.9518}N_{3.6496}S_{0.0748}P_{0.0879}$	−12,384	−5.65	−10,699	1000.00
OC35	$C_{52.4265}H_{97.8357}O_{14.5763}N_{2.4971}S_{0.0511}P_{0.0602}$	−9,250	−3.32	−8,258	1000.00
OC70	$C_{58.5671}H_{107.7282}O_{10.6383}N_{1.1525}S_{0.0236}P_{0.0278}$	−5,592	−0.61	−5,411	1000.00
TAG9	$C_{52.7316}H_{96.0013}O_{6.0000}$	−2,030	1.72	−2,454	826.12
Debris	$C_{46.2859}H_{87.9432}O_{18.5143}N_{3.3417}S_{0.0787}P_{0.0926}$	−15,355	−0.13	−15,317	1000.00

overaccumulate triglycerides, making them excellent candidates for commercial oil generation. Three species were invented herein to represent oil cells; OC5, OC35, and OC70, which represent cells containing 5, 35, and 70 wt% TAG9, respectively. Their thermophysical properties were calculated as weighted averages of the thermophysical properties of algae cells and TAG9. Note that algae slurries with oil compositions between 5 and 70 wt% can be represented simply using mixtures of OC5 and OC70. This is shown in Figure 3.5.

Oil cells from which the oil has been extracted leave behind solid debris that can either be used as a coproduct or fed to the growing algae cells. The debris is assumed to have the same chemical formula as algae, but only 80% of the heat of combustion; a heat of formation for debris is calculated on this basis. These assumptions were necessary because no data were available. Thermophysical property data for algae cells and their derivatives are shown in Table 3.2.

With the mass, enthalpy, and entropy data for the pseudo-species complete, reactions can be postulated to generate the desired species. However, because the reactions involve pseudo-components, like algae, the stoichiometry is often difficult to calculate. Therefore, a matrix approach is used to determine the stoichiometric coefficients in the reactions, as shown in Figure 3.6. For this work, all the matrices are 6 × 6, as the species involve six atoms (C, H, O, N, S, and P). In Figure 3.6, *A* is the atom matrix, with the rows and columns representing the atoms and the chemical components. The weight percent of C, H, and O are shown in the desired product, TAG9, vector. Then, the stoichiometric coefficients, a, b,..., f, in the reaction are

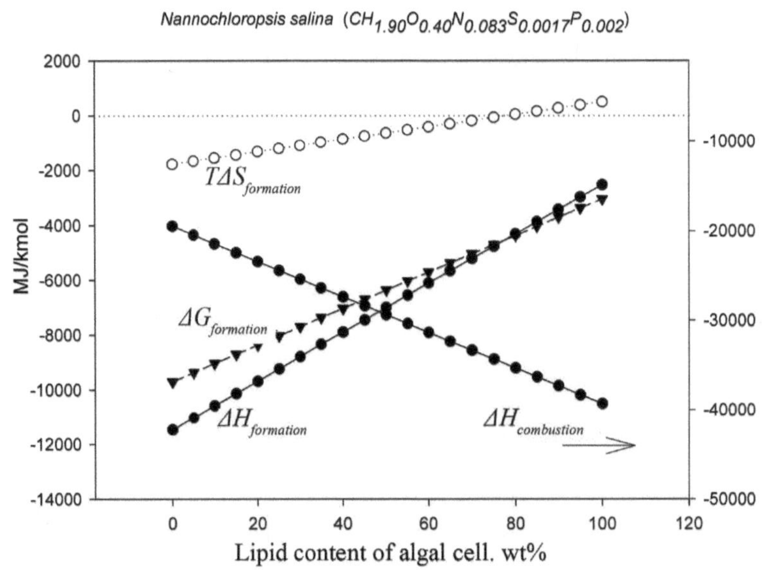

FIGURE 3.5

Thermodynamic properties of algae cells of varying lipid compositions.

Example: OC35 OIL CELLS to TAG9 and DEBRIS

a OC35 +b Debris +c Oxygen +d Water +e SO_4^{2-}+f HPO_4^{2-} → TAG9

INPUT MATRIX [**A**]
Raw materials

RESULT VECTOR [**B**]
Desired product

	OC35	Debris	O_2	H_2O	SO_4^{2-}	HPO_4^{2-}	TAG9
C	52.4265	46.2859	0	0	0	0	63.83
H	97.9892	87.9432	0	2	0	2	116.21
O	114.575	18.5143	2	1	4	4	7.263
N	2.4971	3.8417	0	0	0	0	0
S	0.0511	0.0787	0	0	1	0	0
P	0.0602	0.0926	0	0	0	1	0
	a	b	c	d	e	f	

INVERSE MATRIX [1/**A**]

SOLUTION VECTOR [**X**]
Stoichiometric coefficients

0.04476	0.0000	0.0000	-0.5393	0.0000	0.0000	2.8571	OC35	1
-0.0291	0.0000	0.0000	0.6108	0.0000	0.0000	-1.8571	Debris	-0.65
-1.8381	-0.2500	0.5000	25.5361	-2.0000	-1.5000	-142.74	O2	-49.96
-0.9137	0.5000	0.0000	-0.4130	0.0000	-1.0000	-0.2176	H20	-0.076
2.45E-06	-2.168E-19	5.421E-20	-0.0205	1	-3.2E-19	0.0002	SO4	5.5E-05
-4.68E-07	-1.084E-19	0	-0.0241	0	1	0.0001	H2PO4	1.0E-05
						-1.0000	TAG9	-0.350

Determinant = -343.3 (i.e. non-zero)

[**A**][**X**]=[**B**] or [**X**]=[**A**]$^{-1}$[**B**]

A positive coefficient in the solution vector denotes a reactant, while a negative coefficient denotes a product.

Thus the balanced equation for this example is :-

OC35 +5.5E-05 SO_4^{2-}+1.0E-5 HPO_4^{2-}→ 0.350 TAG9 + 0.65 Debris +49.96 Oxygen +0.076 Water

FIGURE 3.6

Stoichiometry generator for balancing fractional reactions.

in the X vector, which is computed by mass balance, $X = A^{-1}B$. The resulting reaction, in which negative stoichiometric coefficients denote products and positive ones denote reactants, is shown at the bottom of Figure 3.6. The full set of reactions generated using this method is shown in Table 3.3. While this is the opposite of the normal conventional reaction chemistry, it is the way that Excel chooses to define its matrix algebra. The addition of the determinant is useful if it is nonzero, since it will confirm the independence of the input equations.

It would be infeasible to include the plurality of reactions that occur within algae cells. Therefore, the reactions in Table 3.3 are "lumped," such that each represents a specific phenomenon. For example, reaction 1 represents algae cell growth using carbon dioxide, water, and sunlight. Reactions 2 and 3 represent algae cell growth using organic carbon (glycerol in reaction 2, debris in reaction 3). Reactions 4 and 5 (same as 12 and 13) represent cell death and cannibalism. Reactions 6–11 represent oil generation using sunlight (6, 7, 10) or glycerol (8, 9, 11). Finally, reactions 14–16 represent the bursting of oil cells to liberate TAG9.

Table 3.3 Reactions

Reaction Number		ΔH_R kJ/mol	ΔG_R kJ/mol
Reaction Operation 1			
1	$46.29\ HCO_3^- + 3.84\ NO_3^- + 0.079\ SO_4^{2-} + 0.093\ HPO_4^{2-} + 46.02\ H_2O \rightarrow 1\ Algae + 64.05\ O_2 + 50.46\ OH-$	18,930	14,752
2	$15.4286\ Glycerol + 3.8417\ NO_3^- + 0.0926\ SO_4^{2-} + 0.0926\ HPO_4^{2-} \rightarrow 1\ Algae + 10.0502\ O_2 + 15.6971\ H_2O + 4.1842\ OH^-$	−6,517	−7,501
3	Debris → Algae	3,600	5,914
4	$0.0216\ Algae + 0.0074\ NO_3^- + 1.2708\ O_2 \rightarrow 1\ CO_2 + 0.0017\ SO_4^{2-} + 0.002\ HPO_4^- + 0.949\ H_2O + 0.0452\ N_2$	−385	−331
5	$0.1296\ Algae + 0.0444\ NO_3^- + 1.6248\ O_2 + 0.306\ H_2O \rightarrow 1\ SOLC + 0.0102\ SO_4^{2-} + 0.012\ HPO_4^{2-} + 0.2712\ N_2$	467	630
Reaction Operation 2			
6	$0.65\ Algae + 22.3292\ HCO_3^- + 20.4131\ H_2O \rightarrow 1\ OC35 + 31.2653\ O_2 + 22.3292\ OH^- + 22.3292\ OH^-$	13,772	9,891
7	$0.3\ Algae + 44.6585\ HCO_3^- + 40.8261\ H_2O \rightarrow OC70 + 62.5306\ O_2 + 44.6585\ OH^-$	27,545	21,416
8	$0.65\ Algae + 7.4431\ Gly + 0.9942\ H_2O \rightarrow 1\ OC35 + 9.3593\ H_20 + 5.2145\ O_2$	1,496	19
9	$0.3\ Algae + 14.88616\ Gly \rightarrow 1\ OC70 + 18.7185\ H_2O + 10.4291\ O_2$	2,992	38
10	$0.95\ Algae + 2.91615\ H_2O + 3.18989\ HCO_3^- \rightarrow 1\ OC5 + 4.4665\ O_2 + 3.18989\ OH^-$	1,967	1,530
11	$0.95\ Algae + 1.063297\ Gly \rightarrow 1\ OC5 + 0.7449\ O_2 + 1.3370H_2O$	214	3
12	$0.0217\ Algae + 0.0074\ NO_3^- + 1.2708\ O_2 \rightarrow 1\ CO_{2-} + 0.0017\ SO_4^{2-} + 0.002\ HPO_4^{2-} + 0.949\ H_2O + 0.0452\ N_2$	385	−331
13	$0.1296\ Algae + 0.0444\ NO_3^- + 1.6248\ O_2 + 0.306\ H_2O \rightarrow 1\ SOLC + 0.0102\ SO_4^{2-} + 0.012\ HPO_4^{2-} + 0.2712\ N_2$	467	630
Lysis Operation			
14	$OC35 + 0.1880\ H_2O \rightarrow 0.7143\ Debris + 0.2068\ O_2 + 0.3910\ TAG9$	−5,783	−9,407
15	$1.4\ OC70 + 0.4\ H_2O \rightarrow 0.4\ Debris + 0.3\ O_2 + TAG9$	−1,532	−2,425
16	$OC5 + 0.0313\ H_2O \rightarrow 1.2857\ Debris + 0.3133\ O_2 + 0.0651\ TAG9$	−56,788	−93,143

**Reaction 12 is a duplicate of Reaction 4. **Reaction 13 is a duplicate of Reaction 5.*

3.4 RUNNING ASPEN IN REVERSE

To solve large numbers of mass and energy balances, process simulators are typically used, with AspenOne's Aspen Plus being very widely used. In traditional chemical processes, extensive data are normally available long before process simulation or design is contemplated. However, in biochemical processes such as biofuel production, this is frequently not so. In many of these simulations, highly uncertain bioreactor sizes, carbon conversion of CO_2 to algal cells, carbon conversion of CO_2 to triglyceride oils, and rate constants, were assumed. Even the chemical formulas for algae, and its products were unreliable. In other words, data provided for process simulations were often far less accurate when simulating biological processes. Through the use of information flows that is normally used to simulate traditional chemical plants, the process flow sheet was described and equipment items were simulated sequentially, yielding internal and product stream states and specific equipment parameters (e.g., heat duties).

These traditional information flows, however, proved difficult when modeling emerging bioprocesses (Bhattacharya et al., 1986; Evans, 1988), which led us to introduce a preferable proactive approach (Dunlop et al., 2013), one in which the output streams are specified, with Aspen calculating the unknowns; that is, having Aspen operate "in reverse." As will be shown next, when designing a complete algae-to-biodiesel process, such as the one presented in Figure 3.2, given specifications for the biodiesel product, Aspen is used to compute the pond area, productivities, oil content of algal cells, and many of the operating conditions within the process such as total dissolved solids, pH, salinity, and ionic strength of key streams.

Examining Aspen Plus in the equation-oriented mode shows that the simulator is solving approximately 20,000 to 50,000 simultaneous equations. By restricting the degrees of freedom, the engineer is responsible for a well-posed formulation, which imposes carefully designed mass and energy balances that have only one possible solution; i.e., yielding key unknown design variables such as the oil contents, cell growth rate, pond area, and evaporation rate. One of the most prominent advantages of this approach is the flexibility of its design. For example, streams can be introduced as needed and recycled if necessary or desired. A comparison of these two approaches applied to modeling an algae growth reactor is shown in Figure 3.7. Under "normal" Aspen operation, it is necessary to specify the pond area, oil production rate, harvesting cell density, reaction kinetics, but, while these are available for existing processes, in the early design stages, these are difficult to specify. To the contrary, in "reverse" Aspen operation, given the feed and product stream specifications, these are computed by Aspen. In this case, for unreasonable specifications, Aspen has difficulty solving thousands of equations. For slow or unsuccessful convergence, improved specifications can be formulated.

As mentioned above, the cultivation section is simulated using two reaction operations, one for algal cell generation and the other for lipid production. Both use information recycle loops, as illustrated in Figure 3.8, with blocks that account for solar energy input, that estimate the conversion of the limiting reactant, the

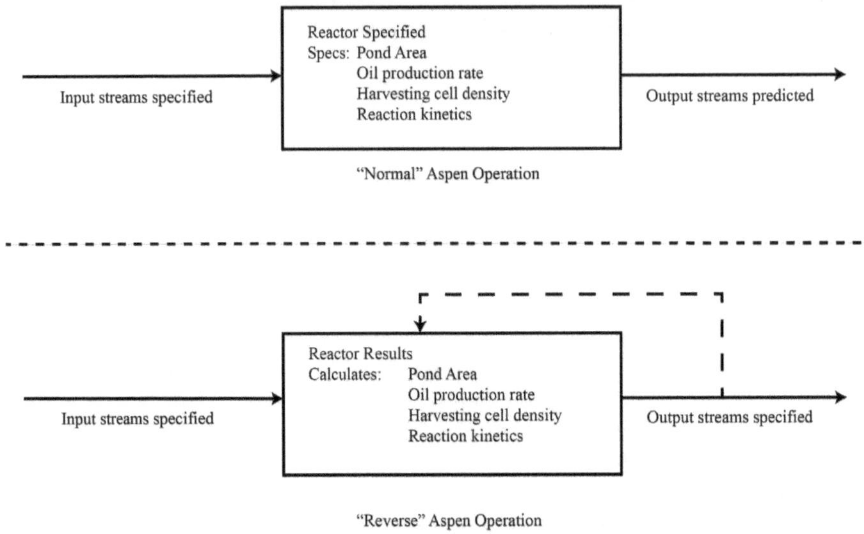

FIGURE 3.7

Comparison of reactive and proactive simulation approaches.

extents of reactions, and the heats of the reactions, and that account for energy losses due to the evaporation of water.

Specifications for the model are measured local rates of evaporation, local solar inputs, pond depth, and photosynthetic efficiency (typically 4 percent). Incident light, in kW, is one of the most fundamental variables in this process, and setting up the simulation flow sheet to run in light limitation is critical.

Returning to Figure 3.2, the block diagram shows the key information flows in the heuristic model for the entire process to grow and convert algae-to-biodiesel. As mentioned above, for the first reaction operation (algal cell generation), the information recycle loop in Figure 3.8 is used. Here, the material inputs (HCO_3^-, glycerol, cell debris, water, urea, sulfate, and phosphate) are mixed with recycle stream, S8. The combined stream, S1, is sent to a single-stream heat exchanger (implemented using the HEATER block), E-100, where 100% of the solar energy flux (kW/m^2) multiplied by a guess for the cultivation area, A_1^*, is the heat duty added to S1. The effluent, S2, is sent to an RSTOIC block, R-100, that models reactions 1–5 in Table 3.3, and algae biomass is produced until the limiting reagent (urea in this case) is entirely consumed. Meanwhile, the extents of the other reactions are estimated and stored in the Calculator block.

In E-101, the extents of reaction are multiplied by the heats of reaction to determine the amount of energy consumed in biomass production. Also, the area for the next iteration is determined, using Eqn (3.1):

$$A_1 = \frac{\sum_{i=1}^{5} \xi_i \Delta H_{R_i}}{\Phi_s \phi} \tag{3.1}$$

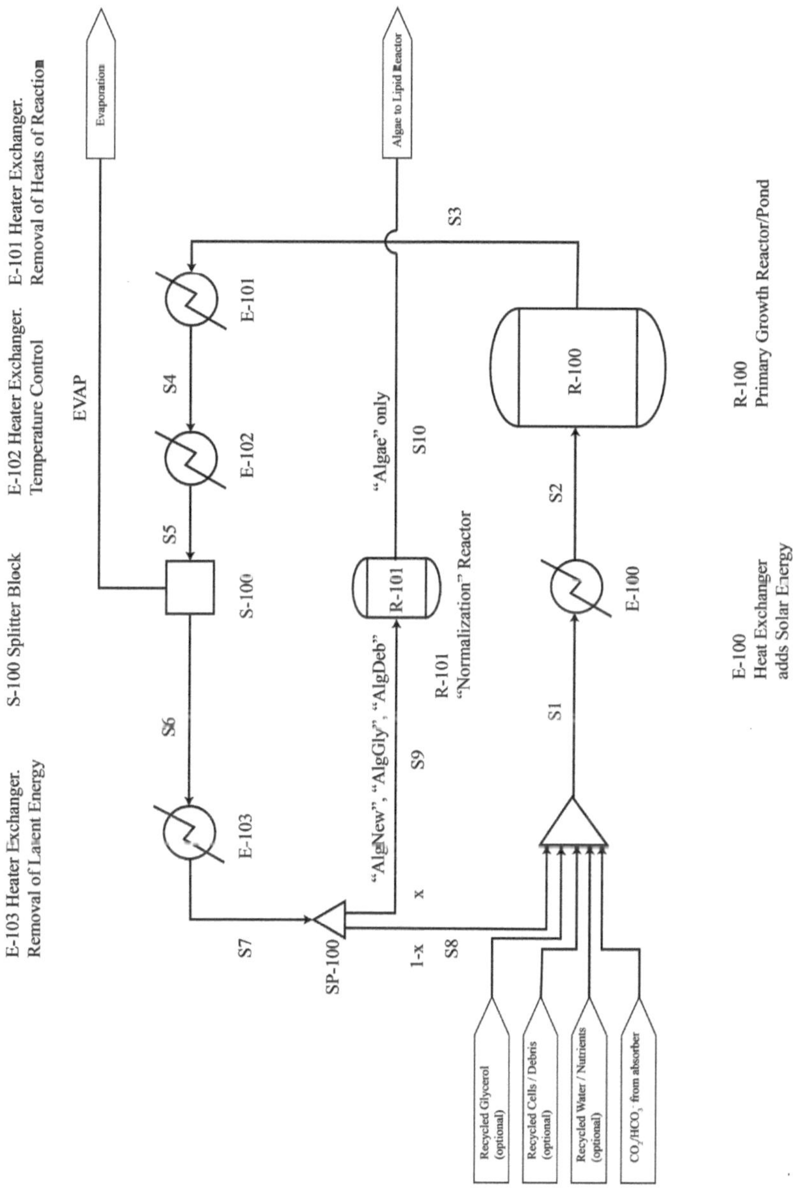

FIGURE 3.8

Example flow diagram for algae cell generation.

where A_1 is the area for biomass generation, ξ_i is the extent of reaction i, $\Delta H_{R,i}$ is heat of reaction i, Φ_s is solar energy flux, and φ is the photosynthetic efficiency.

The effluent, S4, with the heats of reaction removed, is sent to heat exchanger block, E–102, which cools/heats it to a specified temperature—that is maintained using utilities. A Calculator block multiplies the specified evaporation flux (Kg/s-m^2) by the pond area to give the isothermal evaporation rate. Then, Separator block, S-100, removes water in the EVAP stream. A Calculator block computes the heat lost to evaporation, and a heat exchanger block, E-103, adjusts the enthalpy of S6. Finally, the splitter block, SP-100, sends 100x percent of S7 to R-101, a normalization operation to form the "algae" species which is sent to the second reaction operation. The remaining $100(1-x)$ percent is recycled to the beginning of the "reaction operation 1." For the results presented herein, $x = 0.25$. Note that SP-100 simulates the action of a raceway in which the bulk of the algae slurry is recycled. Iterations about this information recycle loop are repeated until convergence is achieved; that is, until the relative change of the pond area $\left(\frac{A_1 - A_1^*}{A_1^*}\right)$ is negligible.

The lipid production reaction operation is similar to Figure 3.8. While not shown, its units are comparable, E-200, R-200, E-201, E-202, S-200, E-203, SP-200, and R-201. Note that the lipid production reaction operation, modeled with the RSTOIC block in R-200, requires a cultivation area, shown in Eqn (3.2). A condensed picture of the cell and oil generation reactors is shown as Figure 3.9.

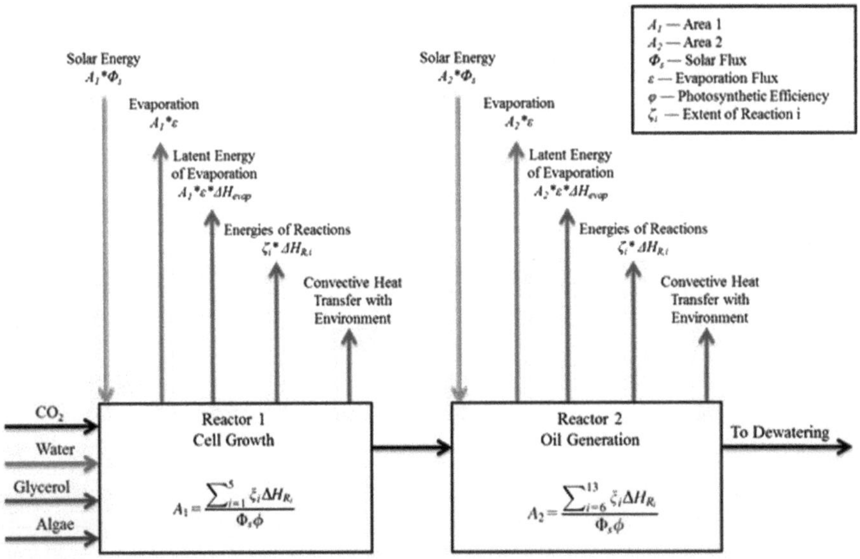

FIGURE 3.9

Cell growth and oil generation reactors.

Table 3.4 Summary of Simulation Results

Selected Simulation Results		
Full Recycle, No Lipid Loss		
Process operation	7,884	hour/year (90%)
CO_2 removal	1,000,000	tonne/year CO_2
Insolation (Corpus Christi, TX)	4.2	kWh/m^2/day
Evaporation (Corpus Christi, TX)	1.27	m^3/hr/hc
Carbon conversion to biodiesel (9 methyl esters)	95	wt%
Biodiesel produced (as 9 methyl esters)	47.4	tonne/hr
	2,630,450	Barrel oil equivalent/yr
Total area	14,810	Hc
Reaction operation 1 area for algae generation	365	Hc
Reaction operation 2 area for oil generation	14,445	Hc
Oil productivity (based on total area)	7.7	g/m^2/day
Effective oil content	58%	wt%
Genetic stability requirements Number of generations/year (binary fission)	43.6	

$$A_2 = \frac{\sum_{i=6}^{13} \xi_i \Delta H_{R_i}}{\Phi_S \phi} \tag{3.2}$$

After emerging from the reaction operations in Figure 3.9, the oil-containing algae cells are concentrated to a specified density, and the remaining water is recycled back to the cultivation ponds. This is modeled using a combination of heuristics and the standard solid-separation operations in Aspen Plus. Electricity from the grid is used to drive the separation. Another RSTOIC reactor block carries out reactions 14–16 in Table 3.3. These reactions represent the lysing of the cells, and the separation of the oil–cell into triglycerides and debris. Again, this operation consumes electricity. The debris is recycled back to reaction operation 1 while the triglycerides are converted to biodiesel and glycerol in a final RSTOIC reactor block. The glycerol is recycled to reaction operations 1 and 2, whereas, the biodiesel is the finished product. The results that emerge from the complete simulation are shown in Table 3.4. Note the oil production rate that results is 7.7 g/m^2/day with an equivalent of 58 wt% oil in the cell.

3.5 SUSTAINABILITY MODELS

With the mass and energy balances converged, it becomes possible and advantageous to examine SIs and life cycle analyses. A life cycle analysis (LCA) for this system must consider the needs for nutrients (phosphorous, sulfur, and nitrogen), as well as the energy requirements and the handling of by-products (glycerine and debris).

First, it is important to consider the definition of a sustainable product. A sustainable product is one that meets the current environmental, social, and economic requirements of today, while not negatively impacting the capabilities of future generations to meet their needs. Essentially, the object is to produce high-quality products, using small amounts of raw material and energy and generating small amounts of waste. More in-depth definitions of sustainable products exist in the literature (Anastas and Zimmerman, 2003); however, this chapter focuses primarily on quantitative measures of sustainability that apply to the algae-to-biodiesel venture described above, and similar biofuels processes.

Authors at the U.S. Environmental Protection Agency have created GREENSCOPE , a methodology that evaluates processes with 140 different SIs at five levels (Allen and Shonnard, 2002; Ruiz-Mercado et al., 2012a,b,c; Seider et al., 2015). Level 1 is comprised of facility compliance/conformance indicators dealing with environment, health, and safety. Level 2 indicators measure material and energy use, as well as rates of customer complaints and returns. Level 3 indicators measure facility effects, while level 4 indicators involve supply chains and product life cycles. Level 5 indicators measure system effects; e.g., ecological cumulative exergy consumption. For a process of this nature (algae growth and conversion of triglycerides to biofuels), the indicators related to the 4 E's (environmental, efficiency, economic, and energy) tend to be the most meaningful.

To calculate a SI, two reference values are required: the best-target and worst-case state. Then, the percent score is calculated using Eqn (3.3). Once several individual indicators, SI_i, have been analyzed, an overall sustainability index, OSI, is calculated using Eqn (3.4):

$$\text{SI} = \frac{\text{Actual} - \text{Worst}}{\text{Best} - \text{Worst}} \times 100\% \tag{3.3}$$

$$\text{OSI} = \sum_{i=1}^{NI} w_i \times \text{SI}_i \tag{3.4}$$

where the Best score is at 100% sustainability and the Worst score is at 0% sustainability, w_i are weighting factors, NI is the number of SIs, and OSI is the overall sustainability index. Most indicators take the form shown in Eqn (3.5):

$$\text{Actual} = \frac{X}{Y} \tag{3.5}$$

where X and Y are specific to each indicator. For example, the Carbon Efficiency indicator, shown in Eqn (3.6), is of particular importance to the algae-to-biodiesel processes:

$$\text{Carbon Efficiency} = \frac{m_{\text{C,prod}}}{m_{\text{C,reactants}}} \tag{3.6}$$

where Carbon Efficiency is the efficiency of carbon atoms utilization, $m_{C,prod}$ is the mass of carbon atoms in the products, and $m_{C,reactants}$ is the mass of carbon atoms in the reactants. Note that, herein, the flow of carbon in algae-to-biodiesel production is described in the "Carbon and Water Management" section. The Solvent Recovery Energy indicator is another example:

$$\mathrm{SRE} = \frac{Q_{\mathrm{solv\ recov}}}{m_{\mathrm{prod}}} \tag{3.7}$$

where $Q_{\mathrm{solv\ recov}}$ is the energy required to recover the solvent—also, the *Specific Energy Intensity indicator*:

$$\mathrm{SEI} = \frac{Q_{\mathrm{net\ used\ prim\ fuel\ equiv}}}{m_{\mathrm{prod}}} \tag{3.8}$$

where $Q_{\mathrm{net\ used\ prim\ fuel\ equiv}}$ is the net energy consumed as a primary fuel equivalent.

An alternative to GREENSCOPE is the GREET package, which was developed for the U.S. Department of Energy's Office of Energy Efficiency and Renewable Energy (EERE) by the Argonne National Laboratory (GREET (Online)). The GREET model is particularly useful for algae-to-biodiesel processes because it was developed to analyze energy and emissions impacts for advanced vehicles and new transportation fuels—evaluating the fuel cycle "from well to wheel" (GREET (Online)). GREET has the ability to calculate energy consumption of both renewable and nonrenewable sources, emissions of greenhouse gases (carbon dioxide, methane, and nitrous oxide), and the emissions of six criteria pollutants (volatile organic compounds (VOCs), carbon monoxide (CO), nitrogen oxide (NOx), particulate matter with size smaller than 10 μm (PM10), particulate matter with size smaller than 2.5 μm (PM2.5), and sulfur oxides (SOx)). (GREET (Online)).

3.6 CARBON AND WATER MANAGEMENT

As mentioned above, there are five crucial nutrients that must be added to the algae inoculant for the cells to grow to a high density. These are carbon dioxide, water, urea (or an equivalent nitrogen source), phosphate, and sulfate (or an equivalent sulfur source). The first two are required in the greatest quantity and are likely to have the most crucial sustainability impact.

Carbon flow and water flow diagrams were generated from the mass and energy balances described earlier; they are shown in Figures 3.10 and 3.11. It is apparent that almost all of the carbon that enters the system (as carbon dioxide or methanol) is transferred to the final biodiesel product; whereas, essentially all of the water loss is either to evaporation or purges. Note that the scale of the water flow diagram (in tonne/h) is 1000 times the scale of the carbon flow diagram (in kg/h). Because of the low density of the algae culture (typically 0.5–1.0 g/L) during cultivation, massive amounts of water must be supplied to and circulated through the ponds. The

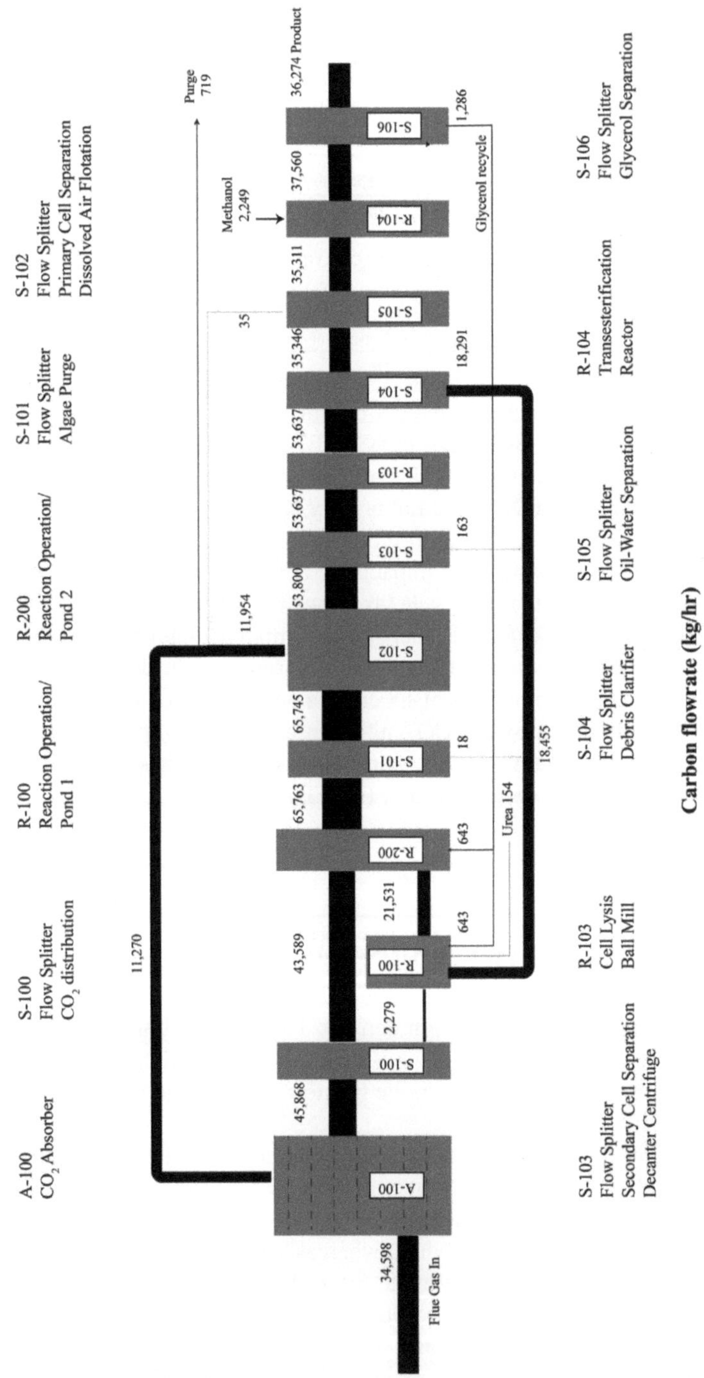

FIGURE 3.10

Carbon flows through the process.

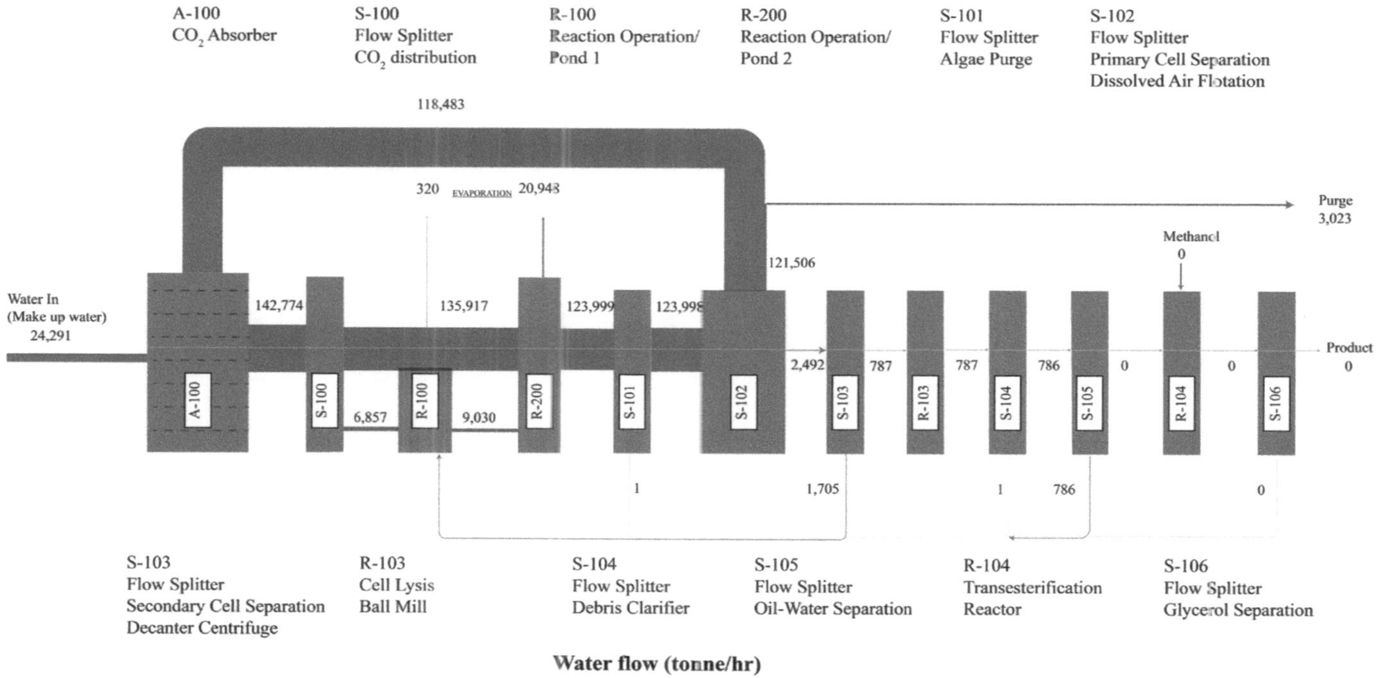

FIGURE 3.11

Water flows through the system.

quantities of water required, as well as the cultivation pond area, have been the most substantial barrier to commercial-scale production thus far.

Another key issue is evaporation, which removes water from the ponds, having an undesirable effect; however, without evaporation, heat would accumulate in the ponds, raising them to temperatures harmful to the algae. Note that rates of evaporation are recorded by the farming community, and are readily available. Evaporation fluxes typically vary between 1–5 m^3/m^2/year in regions where algae are likely to be grown. However, for evaporation to completely cancel out thermal buildup, an evaporation flux of 3–4 m^3/m^2/year is required. Otherwise, temperatures can reach high levels.

3.7 ENERGY MANAGEMENT AND LOST WORK

There are four energy flows for each reactor (Dunlop et al., 2013), each of which is represented by one of the four heat exchangers in Figure 3.8. They are the incoming solar energy (E100/E200), the heats of reaction (E101/E201), the temperature control using utilities (E102/E202), and the energy lost to the latent heat of evaporation of water (E103/E203). It is helpful to examine the effects that each of these phenomena has, by turning them on and off and observing the results. For example, switching off evaporation instantly halves the cost of production without adjusting any other parameters, due to the reduction of water pumped through the system. This is probably the greatest incentive for moving from open raceways to closed photobioreactors. Clearly, the increased cost of closed photobioreactors needs to be considered.

It is also instructive to examine the exergy flows through the process. Exergy (B) is a thermodynamic property that expresses the maximum (reversible) work necessary to produce a material in its specified state from components common in the natural environment (i.e., carbon dioxide and water), heat being exchanged only with the environment (Szargut, 2005). Stated differently, exergy is a measure of the quality of energy, obtained by analyzing the effects of entropy. Exergy is well-defined concept that has been used by engineers for many decades (Keenan, 1951; Seider et al., 1999; Sussman, 1980).

The exergy (also called thermodynamic availability) is defined as $B = H - T_0S$, where T_0 is a reference temperature (298.15 K herein). H and S are the enthalpy and entropy, respectively, and they are estimated by Aspen Plus for each stream. A diagram of an example of the exergy flows through a system is shown as Figure 3.12. Clearly, the largest lost work is in evaporation, and consequently, a very small proportion of sunlight is carried forward in a thermodynamically useful form in the biodiesel.

Note that in this example the sum of inlet stream exergies is 27,614,300 kW and the sum of outlet stream exergies is 27,558,600 kW, yielding a net decrease of 55,700 kW. The solar work done on the system in Reaction Operation

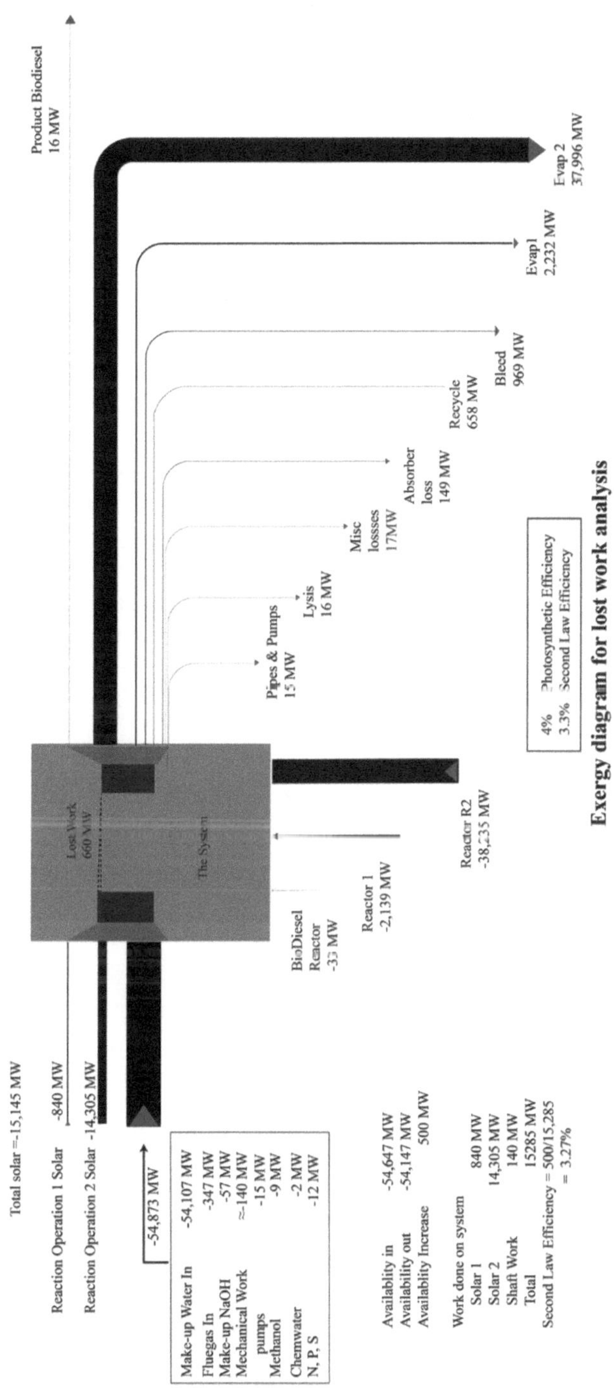

FIGURE 3.12
Exergy flow diagram.

1 (187,600 kW) and in Reaction Operation 2 (1,204,000 kW), minus the shaft work needed for compressors and pumps (3100 kW) yields a net increase of 1,395,000 kW. Thus 1,339,000 kW is the total lost work, yielding an exergy efficiency of just under 4%. Consequently, the process efficiency is essentially equal to the photosynthetic efficiency.

3.8 TECHNOECONOMIC MODELS

Technoeconomic modeling has gained greater attention in recent years, and a range of models are available for the evaluation of commercially viable bioprocess systems. The major processing steps from algae-to-biofuels have been examined, either partially or totally, in several studies. Richardson et al. compared the use of open raceways with photobioreactors for algae cultivation (Richardson and Johnson, 2014; Richardson et al., 2012, 2014). Jones et al. and Zhu et al. examined the process of hydrothermal liquefaction to convert algae cells to a petroleum-like blend (Jones et al., 2014; Zhu et al., 2014). Silva et al. (2014), Davis et al. (2012), Martin and Grossmann (2012) and Lundquist et al. (2010) assembled computational models of complete algae-to-biofuels ventures using different assumptions, methods, and processing cost. Beyond this, Bryant and Gebrelassie examined bioproducts and biorefineries (Bryant et al., 2012; Gebreslassie et al., 2013). These technoeconomic studies had differing assumptions and foci, leading to wide variability in costing and level of analysis.

Monte-Carlo simulations often help to identify the drivers of the economic outcomes and risks in the development of commercial-scale operations (Richardson and Johnson, 2014; Richardson et al., 2014). In such simulations, as many as 20 variables are evaluated exhaustively on the order of 1-million times to yield probability density functions. These identify key variables, and their magnitudes (often displayed in Tornado graphs), which are used to prioritize Work Breakdown Structures for improving ongoing technical and economic outcomes.

3.9 OVERALL SUSTAINABILITY INDEX

Finally, to complete this discussion of SIs, consider the study using GREENSCOPE to evaluate the sustainability of a process to convert vegetable oils to biodiesel using a caustic (NaOH) catalyst (Ruiz-Mercado et al., 2012a,b; 2013). To illustrate its approach, GREENSCOPE begins with the CHEMCAD simulation results for a process to transesterify the oil (assumed to be pure oleic acid) to methyl oleate, a FAME, which is typical of biodiesel. Space is not available herein to define the 66 environmental indicators computed by GREENSCOPE, but Figure 3.13 shows the distribution of indicator results. Clearly, most of the indicators approach unity, showing good sustainability. But, the $N_{\text{haz mat}}$ indicator (1) dips to 40%, which shows

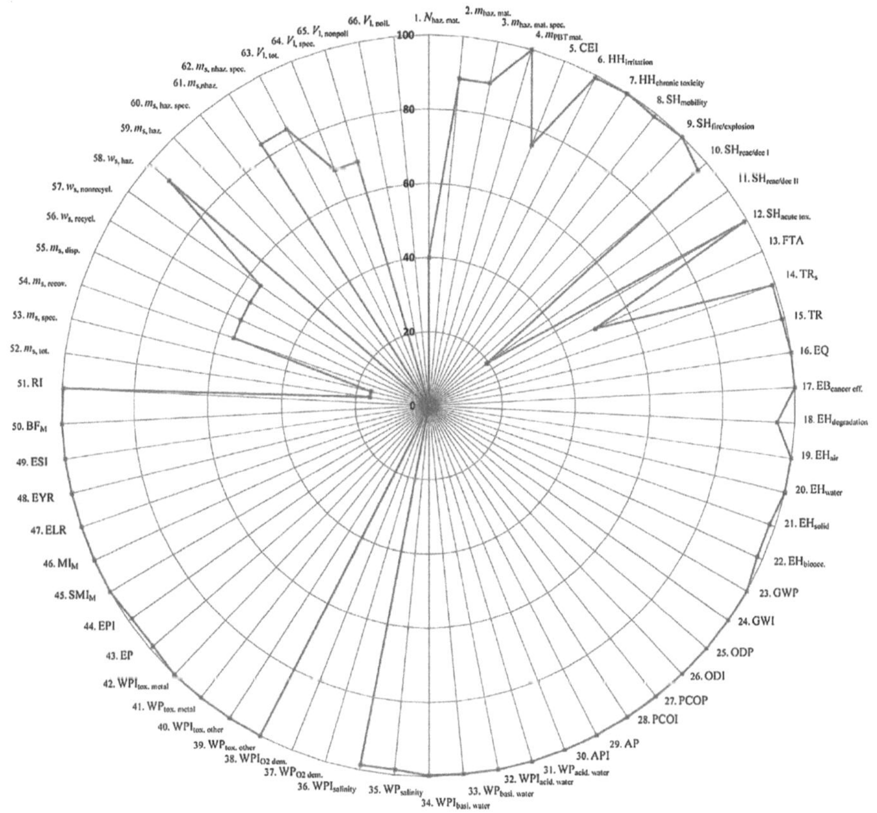

FIGURE 3.13

GREENSCOPE environmental indicator results for alkali-catalyzed manufacture of biodiesel (Ruiz-Mercado et al., 2012a,b; 2013).

Reproduced with permission.

a high ratio of hazardous inputs to the total number of process inputs; $m_{\mathrm{s\ spec}}$ (53) dips to ~15%, which shows a high ratio of the mass of solid waste to the mass of biodiesel product; *FTA* (13) dips to ~50%, which shows a high probability of system failures, given estimates of the component reliabilities; among a few other significant dips. These draw attention to design aspects that should be reconsidered to improve the environmental sustainability of the proposed biodiesel process. Similar charts are presented to display the efficiency, economic, and energy indicators.

A related approach to assessing the sustainability of potential product and/or process designs is to implement an LCA. For chemical products, the analysis begins with the selection of raw materials and their harvesting techniques. Then, it moves to manufacturing processes that produce the product, which is used by the customer, and then disposed of or recycled. Throughout the life cycle, energy is consumed, and wastes and emissions are generated. For each product and/or process design, various SIs for raw materials utilization, energy consumption, and the like, are evaluated to assess its sustainability throughout its life cycle. For a good introduction to LCA, the reader is referred to Chapter 13, "Life-Cycle Concepts, Product Stewardship and Green Engineering," in *Green Engineering* (Allen and Shonnard, 2002). Also, see papers that focus on LCA for the design of biofuels produced by algae (Brentner et al., 2011; Frank et al., 2011; Handler et al., 2012; Stephenson et al., 2010).

3.10 CONCLUSIONS

The imperative to meet increasingly stringent government and industry standards will continue to drive efforts to implement sustainable practice. However, the development of sustainable process models for biofuels continues to present great challenges to the chemical engineering community. This chapter has demonstrated how an integrated modeling-systems approach, using software packages like AspenOne and sustainability metrics, can play a significant role in the design of biofuels processes. To achieve this, effective methods for estimating key missing parameters (e.g., the thermodynamic properties and chemical composition of biological compounds) are presented. This integrated approach is thus capable of delivering converged mass and energy balances that inform technoeconomic modeling and sustainability metrics for commercial-scale biofuel production.

REFERENCES

Allen, D.T., Shonnard, D.R., 2002. Green Engineering: Environmentally Conscious Design of Chemical Processes. Prentice-Hall, Englewood Cliffs, New Jersey.

Anastas, P.T., Zimmerman, J.B., 2003. Through the 12 principles of green engineering. Environ. Sci. Tech. 37, 94A−101A.

Averill, D., 2011. Lean Sustainability: Creating Safe, Enduring, and Profitable Operation. CRC Press, Boca Raton, Florida.

Bhattacharya, A., Motard, R.L., Dunlop, E.H., 1986. BIOASPEN: System for Technology Development. NASA Jet Propulsion Laboratory, Pasadena. NASA CR-181086.

Brentner, L.B., Eckelman, M.J., Zimmerman, J.B., 2011. Combinatorial life cycle assessment to inform process design of industrial production of algae biodiesel. Environ. Sci. Tech. 45, 7060−7067.

Bryant, H.L., Gogichaishvili, I., Anderson, D., Richardson, J.W., Sawyer, J., Wickersham, T., Drewery, M.L., 2012. The value of post-extracted algae residue. Algal Res. 1, 185−193.

Burlew, J.S., 1953. Algal Culture: From Laboratory to Pilot Plant, Carnegie Institution of Washington Publication 600. D. C., Kirby Lithographic Co, Washington.

Czartoski, T.J., Perkins, R., Villaneuva, J.L., Richards, G.. (2011). Algae Biomass Fractionation. (US patent application).

Davis, R., Fishman, D., Frank, E.D., Wigmosta, M.S., 2012. Renewable Diesel from Algal Lipids: An Integrated Baseline for Cost, Emissions, and Resource Potential from a

Harmonized Model. Argonne National Laboratory; Golden, CO: National Renewable Energy Laboratory, Argonne, I. Pacific Northwest National Laboratory, Richmond, WA. ANL/ESD/12–4; NREL/TP-5100-55431; PNNL-21437.

Deutch, J., Lauvergeon, A., Prawiraatmadja, W., 2007. Energy Security and Climate Change. The Trilateral Commission, Washington, Paris, Tokyo.

Duan, P., Savage, P.E., 2010. Hydrothermal liquefaction of a Microalga with Heterogeneous catalysts. Ind. Eng. Chem. Res. 50, 52–61.

Dunlop, E.H., Coaldrake, A.K., Silva, C.S., Seider, W.D., 2013. An energy-limited model of algal biofuel production: towards the next generation of advanced biofuels. AIChE J. 59, 4641–4654.

Eckelberry, N.D., Green, M.P., Fraser, S.A., 2010. Systems, Apparatus and Methods for Obtaining Intracellular Products and Cellular Mass and Debris from Algae and Derivative Products and Process of Use Thereof.

El-Halwagi, M.M., 2012. Sustainable Design through Process Integration. Elsevier, Waltham, Massachusetts.

Evans, L.B., 1988. Bioprocess simulation: a new tool for process development. Nat. Biotechnol. 6, 200–203.

Evans, R.L., 2007. Fueling Our Future: An Introduction to Sustainable Energy. Cambridge University Press, Cambridge.

Flocculants Info. Available from: http://www.flocculants.info/ (accessed 10.12.12).

Frank, E.D., Han, J., Palou-Rivera, I., Elgowainy, A., Wang, M.Q., 2011. User Manual for Algae Life-cycle Analysis with GREET: Version 0.0. Energy Systems Division. Argonne National Laboratory.

GREET. Available from: https://greet.es.anl.gov/ (accessed 22.07.14).

Gebreslassie, B.H., Waymire, R., You, F., 2013. Sustainable design and synthesis of algae-based biorefinery for simultaneous hydrocarbon biofuel production and carbon sequestration. AIChE J. 59, 1599–1621.

Handler, R.M., Canter, C.E., Kalnes, T.N., Lupton, F.S., Kholiqov, O., Shonnard, D.R., Blowers, P., 2012. Evaluation of environmental impacts from microalgae cultivation in open-air raceway ponds: analysis of the prior literature and investigation of wide variance in predicted impacts. Algal Res. 1, 83–92.

Iqbal, J., Theegala, C., 2013. Microwave assisted lipid extraction from microalgae using biodiesel as a co-solvent. Algal Res. 2, 34–42.

Jones, S., Zhu, Y., Anderson, D., Hallen, R., Elliott, D., Schmidt, A., Albrecht, K., Hart, T., Butcher, M., Drennan, C., Snowden-Swan, L., Davis, R., Kinchin, C., 2014. Process Design and Economics for the Conversion of Algal Biomass to Hydrocarbons: Whole Algae Hydrothermal Liquefaction and Upgrading. Pacific Northwest National Laboratory. PNNL-23227.

Keenan, J.H., 1951. Availability and Irreversibility in thermodynamics. Brit. J. Appl. Phys. 2, 183–192.

Kemp, R.B., 1999. Macromolecules to Man. Elsevier, Amsterdam.

Lammers, P., 2014. Private Communication with Peter Lammers of NMSU.

Lee, J.Y., Yoo, C., Jun, S.Y., Ahn, C.Y., Oh, H.M., 2010. Comparison of several methods for effective lipid extraction from microalgae. Bioresour. Technol. 101, S75–S77.

Levine, R.B., Pinnarat, T., Savage, P.E., 2010. Biodiesel production from wet algal biomass through in situ lipid hydrolysis and supercritical transesterification. Energy Fuels 24, 5235–5243.

Lundquist, T.J., Woertz, I.C., Quinn, N.W.T., Benemann, J.R., 2010. "A Realistic Technology and Engineering Assessment of Algae Biofuel Production" Energy Biosciences Institute.

Martin, M., Grossmann, I.E., 2012. Simultaneous optimization and heat integration for biodiesel production from cooking oil and algae. Ind. Eng. Chem. Res. 51, 7998–8014.

National Alliance for Advanced Biofuels and Bioproducts Final Report. Available from: http://energy.gov/eere/bioenergy/downloads/national-alliance-advanced-biofuels-and-bioproducts-synopsis-naabb-final (accessed 10.10.14).

Origin Oil Single-Step Extraction. Available from: http://www.originoil.com/technology/single-step-extraction.html (accessed 10.12.12).

O'Grady, J.P., Morgan, J.A., 2011. Heterotrophic growth and lipid production of chlorella protothecoides on glycerol. Bioprocess Biosyst. Eng. 34, 121–125.

Parrish, A.E., Abraham, M.A., 2014. Green Chemistry and Engineering: A Pathway to Sustainability. John Wiley& Sons, Hoboken, New Jersey.

Richardson, J.W., Johnson, M.D., 2014. Economic viability of a reverse engineered algae farm (REAF). Algal Res. 3, 66–70.

Richardson, J.W., Johnson, M.D., Outlaw, J.L., 2012. Economic comparison of open pond raceways to photo bio-reactors for profitable production of algae for transportation fuels in the southwest. Algal Res. 1, 93–100.

Richardson, J.W., Johnson, M.D., Zhang, X., Zemke, P., Chen, W., Hu, Q., 2014. A financial assessment of two alternative cultivations systems and their contributions to algae biofuel economic viability. Algal Res. 4, 96–104.

Ruiz-Mercado, G.J., Smith, R.L., Gonzalez, M.A., 2012a. Sustainability indicators for chemical processes: I. Taxonomy. Ind. Eng. Chem. Res. 51, 2309–2328.

Ruiz-Mercado, G.J., Smith, R.L., Gonzalez, M.A., 2012b. Sustainability indicators for chemical processes: II. Data needs. Ind. Eng. Chem. Res. 51, 2329–2353.

Ruiz-Mercado, G.J., Gonzalez, M.A., Smith, R.L., 2013. Sustainability indicators for chemical processes: III. Biodiesel case study. Ind. Eng. Chem. Res. 52, 6747–6760.

SRS Energy, Algae Fractionation. Available from: http://www.solutionrecovery.com/algae/ (accessed 10.12.12).

Seider, W.D., Lewin, D.R., Seader, J.D., Widagdo, S., Gani, R., Ng, K.M., 2015. Essentials of Chemical Engineering Design, fourth ed. Wiley.

Seider, W.D., Seader, J.D., Lewin, D.R., 1999. Process Design Principles: Synthesis, Analysis, and Evaluation. Wiley, Hoboken, NJ.

Silva, C.S., Soliman, E., Cameron, G., Fabiano, L.A., Seider, W.D., Dunlop, E.H., Coaldrake, A.K., 2014. Commercial-scale biodiesel production from algae. Ind. Eng. Chem. Res. 53, 5311–5324.

Smith, B.T., Davis, R.H., 2012. Sedimentation of Algae Flocculated Using Naturally-Available Magnesium-Based Flocculants, vol. 1, pp. 32–39.

Soh, L., Zimmerman, J., 2011. Biodiesel production: the potential of algal lipids extracted with supercritical carbon dioxide. Green Chem. 13, 1422–1429.

Stephenson, A.L., Kazamia, E., Dennis, J.S., Howe, C.J., Scott, S.A., Smith, A.G., 2010. Life-cycle assessment of potential algal biodiesel production in the United Kingdom: a comparison of raceways and air-lift tubular bioreactors. Energy Fuels 24, 4062–4077.

Stoner, T.H., 2013. Small Change, Big Gains. Greenleaf Book Group Press, Austin, Texas.

Sussman, M.V., 1980. Availability (Exergy) Analysis. Mulliken House, Lexington, MA.

Szargut, J., 2005. Exergy Method, Technical and Ecological Applications. WIT Press, Billerica, MA.

Terigar, B.G., Balasubramanian, C.M., Lima, M., Boldor, D., 2011. Soybean and rice bran oil extraction in a continuous microwave system: from laboratory to pilot-scale. J. Food Eng. 104, 208–217.

Vyas, A.P., Verma, J.L., Subrahmanyam, N.A., 2009. Review on FAME production processes. Fuel 89, 1–9.

Weissman, J.C., Goebel, R.P., 1987. Design and analysis of microalgal open pond systems for the purpose of producing fuels. In: Institute, S.E.R. (Ed.). Available from: http://www.nrel.gov/docs/legosti/old/2840.pdf (accessed 26.08.13).

Zhu, Y., Biddy, M.J., Jones, S.B., Elliott, D.C., Schmidt, A., 2014. Techno-economic analysis of liquid fuel production from woody biomass via hydrothermal liquefaction (HTL) and upgrading. J. Appl. Energy 129, 384–394.

CHAPTER

4

Sustainable Design and Operation of a Reactive Distillation System Used for the Production of Cosmetic Ingredients

Edwin Zondervan[§,1], Aaron D. Bojarski*, Antonio Espuña*, Luis Puigjaner*

Department of Chemical Engineering, Universitat Politècnica de Catalunya, Barcelona, Spain

[§]Department of Chemistry and Chemical Engineering, Eindhoven University of Technology, Eindhoven, The Netherlands

[1]Corresponding author: E-mail: e.zondervan@tue.nl

4.1 INTRODUCTION

One of the major tasks of the chemical engineer has been how to find the optimal design and/or operation of a chemical process. In the process of optimization, often economic criteria were used as optimization objectives. Nowadays economic feasibility is no longer the only objective that needs to be satisfied. The design and/or operation also have to fulfill environmental and societal needs, which require a holistic view (Puigjaner and Heyen, 2009). In recent history many tools have become available that offer such methodology, e.g., life cycle assessment (LCA), industrial ecology, pollution prevention, and "green chemistry and engineering principles" (Dimian and Bildea, 2008; Anastas and Zimmerman, 2003). It is evident that the sustainability considerations should be evaluated at the design stage of a process (Yang and Shi, 2000; Rebitzer et al., 2004).

By integrating the different sustainability concerns the system boundaries are extended and this gives rise to multiobjectivity and additional constraints in the formulation of the design problem (Azapagic et al., 2006). The design process should not be limited to the process itself but also has to take into account the environment, societal factors, and other processes that are linked to the process upstream as well as downstream.

Especially the environmental and societal impacts are in the current methodologies still considered as an "after-thought" where the priorities are on the technical and economic components of the design. Such an approach inevitably leads to suboptimal performance of the plant as the design choices are limited after construction

Computer Aided Chemical Engineering, Volume 36. ISSN 1570-7946. http://dx.doi.org/10.1016/B978-0-444-63472-6.00004-5

and may not allow more sustainable process alternatives. This integration issue has been identified and discussed by several authors (Bakshi and Fiksel, 2003; Chen and Shonnard, 2004; Diwekar, 2005; Clift, 2006).

Closely related to sustainable design is process intensification (PI), which is aiming to develop or improve processes of higher flexibility, reduced environmental impact, improved safety, and higher energy efficiency (Stankiewicz and Moulijn, 2000). A well-known example of PI is the concept of reactive distillation (RD), where separation and reaction are combined into one compact unit operation. As reactants are transformed into products, the product is immediately purified by means of distillation. This fast product removal is advantageous for the conversion, as the chemical equilibrium is shifted to the right side of the reaction. An illustrating example is the commercialization of the Eastman Chemical Methyl Acetate Process that was traditionally operated in a reactor and a separation train of nine distillation units. With the aid of RD, the process could be intensified; where a single RD unit could generate the same product quality and capacity (Krishna, 2002).

RD has become a mature technology and in these days new applications are found in the specialty chemicals. RD is proven useful for biofuel production (Mahfud et al., 2007), but also for fatty-acid production (de Jong et al., 2009) and polyester production (Shah et al., 2009). But, RD processes are difficult to design and control, as reaction and separation have complicated interactions.

To attack the issues with the integration of economic and environmental metrics, we are going to unite concepts from sustainability-related methodologies (LCA) and PI. The application is done for the evaluation of a RD process used to produce fatty acid esters (ingredients for cosmetics) by an esterification reaction between isopropanol and myristic acid.

4.2 PROPOSED METHODOLOGY

In our method we will use a unique 3-points procedure that is based on life cycle concepts which are embedded in the LCA methodology. The first step in the procedure is building a plausible model for generating the data needed for calculation of the metrics (Bojarski et al., 2008). In this case the model is constructed using commercial process simulation tools.

4.2.1 GOAL DEFINITION AND MODEL BUILDING

From the problem statement, first the goals have to be defined and the model has to be built. A key point in this study is how different input variables affect the model outputs. The model outputs are typically economic, environmental, and operational in its nature.

A common way to evaluate economic metrics is by the total annualized costs (TAC), which can be used to analyze the economic performance of a system, according to (Biegler et al., 1997; Peters and Timmerhaus, 1991):

$$TAC = S - C - \frac{I}{n} \tag{4.1}$$

where S are the product sales in (€/year), C the operational costs in (€/year), I the investment costs (€), and n the years considered for the project lifetime. TAC considers operational costs that are normally associated to the use of utilities such as water and steam. It also includes the consumption or raw materials. It is noted that the more positive TAC is, the more profitable the design becomes.

For the quantification of the environmental impact (emissions or natural raw material consumptions) several possibilities exist, depending on which position along the environmental damage chain (environmental mechanism) is chosen. Normally the environmental metrics are related to impacts as endpoints or midpoint metrics which can be obtained from characterization factors (CFs) (de Haes et al., 1999).

Two good methods to determine the CFs are in place (Finnveden et al., 2009), the problem impact assessment methods (e.g., the impact method by Center of Environmental Science at Leiden University), which restrict quantitative modeling to early stages in the cause—effect chain (midpoints) and damage-oriented methods (e.g., eco-indicator 99), which model the cause—effect chain up to the damage itself (endpoints). In this work we will use a hybrid method proposed in Humbert et al. (2005) that evaluates the midpoints as well as the endpoints. Fifteen midpoint categories are linked to four damage categories (Humbert et al., 2005). For each environmental impact two CFs are proposed which will simplify the model implementation. We will further aggregate the categories into a single overall metric and endpoint metrics, since these metrics are easier to comprehend as compared to the individual midpoint values.

4.2.2 SENSITIVITY ANALYSIS

Sensitivity analysis (SA) can be used to study how a change in the inputs of a model influences the outputs, or more formal: SA is the study of how the variation in the output of a model can be apportioned, qualitatively or quantitatively, to different sources of variation and of how the given model depends upon the information fed into it. In principle three SA methods exist: (1) screening (Campolongo et al., 2000), (2) local sensitivity analysis (Turanyi and Rabitz, 2000), and (3) global sensitivity analysis (Saltelli et al., 2008). In this work we will use SA to study how engineering decision variables affect the overall optimization objective. Commonly, an objective function (OF) has the following shape:

$$OF_k = \sum_{i=1}^{nTerms_k} OFterm_{ik} \tag{4.2}$$

where each $OFterm_{ik}$ is a given function of the given engineering decision variables x_j. SA can be used to determine how $OFterm_{ik}$ influences the overall OF_k and how each decision variable influences $OFterm_{ik}$. The outcome of the SA enables us to tune our model and set priorities.

In addition to SA we will use rank and proximity criteria (Fisher et al., 1985) to get a better grip on how far from the optimal conditions a given design is and how each input variable affects a given objective function. The rank order parameter is given as:

$$r_{jk} = \sum_{i=1}^{nTerms_k} \left| \frac{\partial OFterm_{ik}}{\partial x_j} \right| \Delta x_j^{MAX_k} \tag{4.3}$$

which indicates whether large positive incremental changes are being traded off by larger negative changes. The proximity parameter is given as:

$$p_{jk} = \frac{\left| \sum_{i=1}^{nTerms_k} \frac{\partial OFterm_{ik}}{\partial x_j} \Delta x_j^{MAX_k} \right|}{\sum_{i=1}^{nTerms_k} \left| \frac{\partial OFterm_{ik}}{\partial x_j} \right| \Delta x_j^{MAX_k}} \tag{4.4}$$

The proximity parameter tells us how far a design is away from its optimum, as it goes to zero it will be very close to optimum, and as it moves toward unity it will be away from the optimum.

4.2.3 INTERPRETATION

Once the design objectives have been set, a proper model has been formulated and sensitivity analysis is done, data can be collected, and interpretation of the data has to be done. Different multicriteria decision analysis (MCDA) techniques could be used to rank the obtained design alternatives. However in order to prune the number of alternatives for analysis the decision-maker should only focus on the Pareto efficient ones.

Pareto analysis identifies the set of nondominated alternatives if the problem has multiple objectives. Such alternatives are known as the Pareto set of noninferior alternatives, or Pareto front (PF) (Stefanis et al., 1997). A dominated alternative is one that is inferior to another feasible alternative in the set with respect to all attributes under consideration. This means that for each dominated alternative there is at least one win–win alternative that can be attained without sacrificing achievement in any of the design objectives. The set of alternatives that remains after all the dominated alternatives have been removed is called the set of nondominated alternatives.

Within the PF, the single objective best solutions or extreme points (OF_k^*) can be used to generate a solution of compromise. This solution can be obtained by the assumption that all objectives share the same importance. In such case, the compromise solution ($OF_{compromise}$) can be thought of as the one that redistributes the percentage by which all objectives differs from their single objective optimum values. This condition is mathematically expressed as:

$$OF_{compromise} = \sum_{k=1}^{nOFs} \alpha_k \frac{OF_k - OF_k^*}{OF_k^*} \tag{4.5}$$

This compromise solution will yield a single design that takes economic and environmental metrics into account with the same importance if all α_k are the same. Similarly, if the utopian point is defined as the one formed by the best solutions, then if these distances are normalized the single design that minimizes the distance to this utopian point is the same as the one obtained in Eqn (4.5), provided all α_k are the same. The decision-maker could use any different set of α_k, and each combination will provide with a different compromise solution. This rough selection procedure can be changed if other more complex MCDA method is applied.

Figure 4.1 illustrates the whole procedure; step 1 sets the overall modeling hypothesis and the metrics that will measure the goodness of a design. Step 2, which encompasses the sensitivity analysis, uses the metrics defined in Eqns (4.3) and (4.4) to order in terms of importance the input variables and how they affect the design metrics. Last step considers the generation of Pareto points considering the most influential variables and other considerations, from which a compromise solution will be selected.

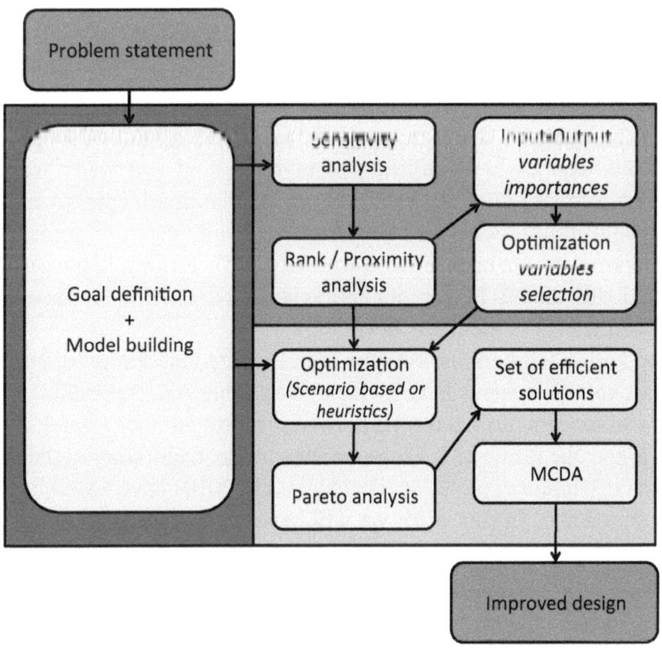

FIGURE 4.1

Proposed strategy for sustainable design.

4.3 RESULTS AND DISCUSSION

To test the methodology proposed above we have selected a case for the RD of isopropyl myristate, a fatty acid ester which is used in cosmetics as oil component. Nowadays fatty acid esters are produced in batch reactors using strong acids like sulfuric acid. Moreover, their production processes encompass costly separations, large energy requirements, and the production of polluting by-products, which in this case are mixture of water and unreacted alcohol. Because of equilibrium limitations high conversions can only be obtained by using a large excess of reactants (Dimian et al., 2004).

4.3.1 GOAL DEFINITION AND MODEL BUILDING

The model is developed in Aspen Plus and contains the thermodynamic and unit operations models. The economic and environmental metrics were coded in Matlab, which can be connected to Aspen Plus using the Windows-COM interface. Reaction data was retrieved from de Jong et al. (2009). The reaction is first-order in each of the reactive components. In both reactions, the Aspen Plus preexponential constant was calculated using a given molar concentration of catalyst. The catalyst used for ester synthesis is *para*-toluene sulfonic acid (pTSA). For the reaction of the fatty acid and isopropanol the $Ea_d = 58.9 \cdot 10^3$ (kJ/kmol) and $k_d = 3.33 \cdot 10^5$ (molarity). And the kinetics for the reverse reaction are given as $Ea_r = 45.9 \cdot 10^3$ (kJ/kmol) and $k_r = 2.18 \cdot 10^3$ (molarity).

All thermodynamic and transport data for the process was obtained from the Aspen Properties database. pTSA was assumed to be nonvolatile. To describe phase separation phenomena of the process V–L equilibrium data of isopropanol and water was used. The L–L equilibrium of myristic acid and water was regressed from the literature (Maeda et al., 1997). If we assume that isopropyl myristate has similar L–L equilibrium behavior as myristic acid with water, all remaining binary interaction parameters can be estimated with UNIFAC. The liquid phase equilibrium was calculated with the NRTL activity coefficient model, while the vapor phase data can be obtained from the ideal gas law.

From the boiling temperatures can be seen that the reaction products are not the most and least volatile species in the system (See Table A.1, Appendix). In this case there exists a homogeneous azeotrope between isopropanol and water. The appearance of this azeotrope is one of the reasons for adding an entrainer to the RD system. Several authors (Dimian et al., 2004; de Jong et al., 2010) have tested the feasibility of entrainer-based RD. In this work we will have a different approach to deal with the azeotrope, where we operate the column at higher pressure as a normal distillation column.

In Figure 4.2 the process flow sheet is given where the main model block is the reactive distillation column (RDCOL). The RD column is modeled using RADFRAC. It calculates QCONCD and QREB, which model the condenser and reboiler heat duties, which are fed to two HEATER models to represent the column

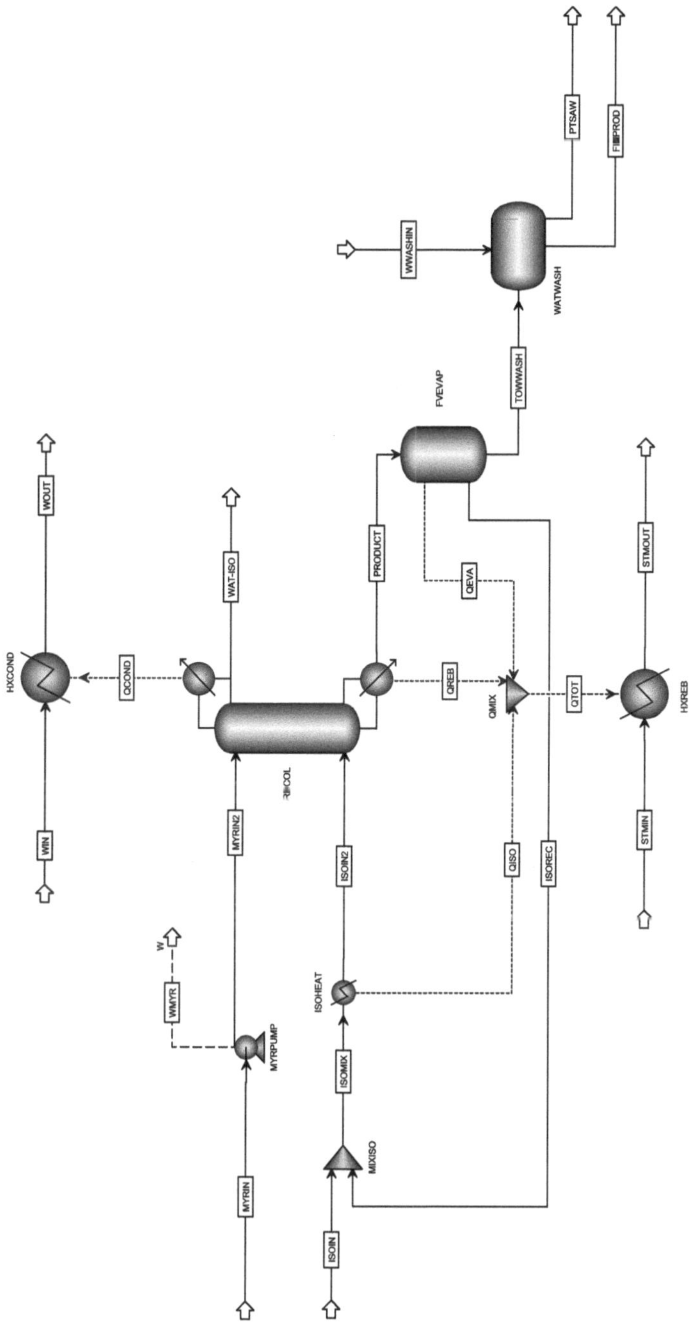

FIGURE 4.2
Reactive distillation flow sheet.

condenser and reboiler. As can be seen from Figure 4.2, streams MYRIN2 and ISOIN2 represent the inlet flows of myristic acid and isopropanol while the WAT-ISO and PRODUCT streams represent the outlet flows for the produced water and ester. The PRODUCT stream is connected to a falling film evaporator which is modeled as a two phase flash model (FLASH2), where feed pressure is decreased to 0.5 bar.

A certain amount of heat is added to allow further separation of isopropanol from the ester stream. The ester stream (TOWWASH) connected to a water contactor which is modeled as a liquid–liquid DECANTER block, in which a certain amount of water is added by means of a WWASHIN stream, to remove the catalyst that is still present in the ester. The product stream is cleaned from the catalyst yield the final product (FINPROD), while the water used for washing is removed as waste (PTSAW).

In WWASHD a design specification is given to set the amount of water that is used for washing out the catalyst, in such way a recovery of pTSA from the product stream of 99% percent is enforced.

The ISOREC stream which mainly contains isopropanol at the bottom of the RD can be recycled into the column. To set the ratio of myristic acid and isopropanol a calculator block (FEEDRAT) is used which takes the available isopropanol flow in the recycle streams to set the fresh isopropanol flow (ISOIN) into account.

Reaction is supposed to take place in the liquid phase only. Within the column no L–L equilibrium behavior occurs and the isopropanol feed stream to the RD is assumed to be a vapor stream of isopropanol and water. The isopropanol–water stream is vaporized to meet the RD bottoms temperature and pressure. A HEATER model (ISOHEAT) is used to calculate the duty requirements. Myristic acid is fed at the top as a liquid under condenser pressure and temperature. A pump model (MYRPUMP) is used to calculate the pumping requirements. The column consists of a single vessel with the same diameter for its whole length. Also the tray liquid holdup is assumed to be the same on all trees. Aspen Plus provides a stage design utility which is used for tray sizing.

With DIAMSET the tray volume can be calculated and updated for each calculation until the proposed value is within a tolerance value. Lastly, to calculate the catalyst molarity the REACTION block is added to the flow sheet.

4.3.2 SENSITIVITY ANALYSIS

The first step in the SA is to determine which factors in the operational and investment costs are significant, i.e., the economic impact. The operational costs are calculated from the basic pricing of the reactants and of the utilities. The investment amounts are associated to the column vessel and heat exchanger capacities and are calculated using Biegler's method (Biegler et al., 1997) (see Table A.2, Appendix).

The stages in the column are considered to be sieve trays, and its number is N_{st}. The catalyst costs are also included in the calculations under the assumption that

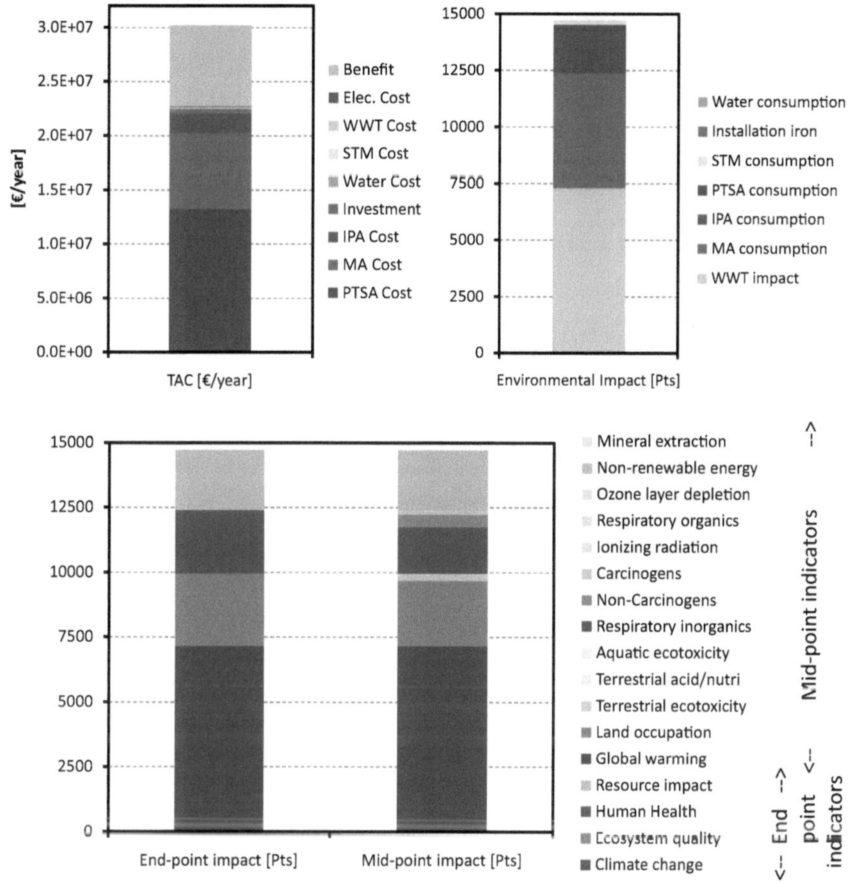

FIGURE 4.3

Distribution of operational costs and environmental impacts for the base case. STM, steam; PTSA, para toluene sulphonic acid; IPA, isopropyl alcohol; MA, myristic acid; WWT, waste water treatment.

no recovery of the catalyst is possible and that it can be washed from the product stream at a 99% removal. Lastly, the *Fp* factor is calculated from the procedure by Doherty and Malone (2001) to account for pressure effects on investment costs, see Eqn (A.1) in the Appendix. Figure 4.3 shows how annualized costs and environmental impacts for the base case are divided. The base case was compared to four other cases, where the number of stages, the condenser pressure (P_{cond}), the ratio of reactants in the feed and the conversion were varied as listed in Table 4.1. Each column shows the different scenarios where a single variable is modified. These values were selected to explore the shape and effect of the input variables changes on the model's output, aiming at answering which and how much the model's output depend. The base case value represents a mid value in the possible

Table 4.1 Base Case and Single Variable Changes

	N_{st}	P_{cond}	Ratio	Conversion
Case 1	40	4000	1.20	0.927
Case 2	45	4500	1.35	0.990
Base case	50	5000	1.50	0.995
Case 3	55	5500	1.65	0.999
Case 4	60	6000	1.80	–

optimization variable's interval, where the column roughly produces 900 kg/h of isopropyl myristate.

The environmental impacts of products are normally not incorporated in the life cycle analysis as there is a large amount of possible products that can be formed from isopropyl myristate, which cannot be considered; for this reason the impacts that are considered can be categorized as follows: (1) use of raw materials, (2) utilities consumption, and (3) infrastructure. The appropriate environmental information was taken from the ecoinvent database (Ecoinvent V. 2.0, 2008). Fugitive emissions of isopropanol, myristic acid, and isopropyl myristate are neglected, due to the lack of appropriate characterization factors in the selected impact characterization model.

The largest contribution to the annualized costs is the consumption of the catalysts, while the wastewater treatment is the largest contributor to the environmental impact, see top left and right graphs in Figure 4.3. The results show that the catalyst requirement is the major contributor to the costs, as it is assumed that it cannot be easily recovered. Bottom Figure 4.3, shows the distribution of environmental impact along the different mid- or endpoint impact categories, which are available in the IMPACT 2002+ methodology. It can be appreciated that most environmental impact arises from global warming and consequently climate change, while the remaining endpoint impact categories share almost equal importance.

The analysis of input—output variables is incorporated by varying the number of stages in the column, the operational pressure of the condenser, the ratio of raw materials inlet, and the conversion. In Figure 4.4 the outcomes are displayed.

Figure 4.4 shows that conversion influences the TAC, where the process is only economically attractive for conversions of higher than 99.5%. It also shows clearly that at higher conversion the sales of isopropyl myristate increases, but that this increment cannot be compensated by the large amount of catalyst that is required. Regarding the operating pressure and number of stages it shows that higher number of stages and higher operating pressure render more profitable results.

From Figure 4.4 can also be seen that at lower conversion ranges the wastewater treatment impacts are higher and steam-related impacts are lower. In the case of a high conversion requirement, the wastewater treatment impact decreases while the steam-related impact increases. The difference gradually disappears as

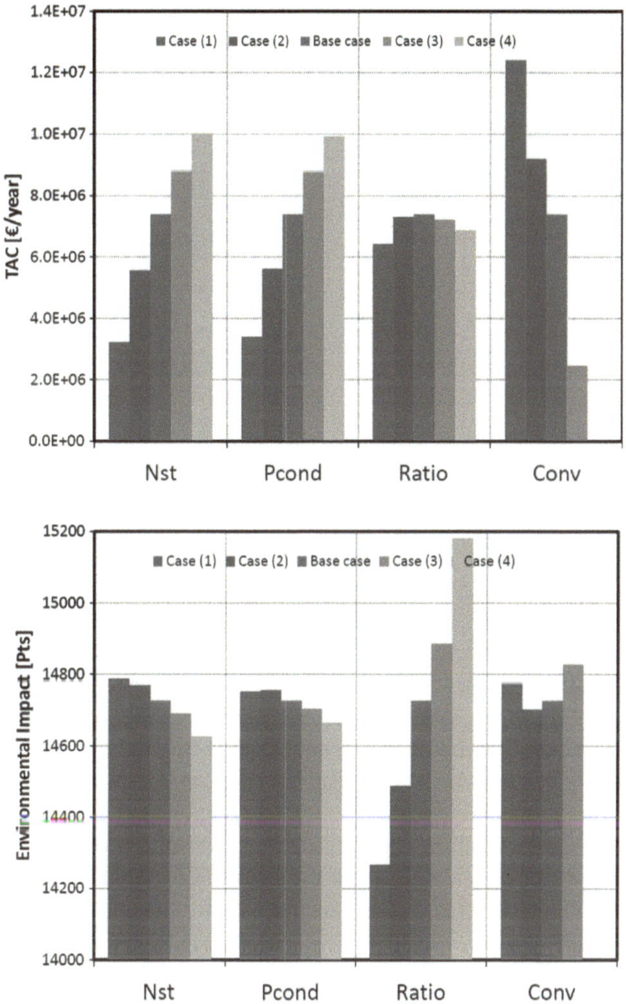

FIGURE 4.4

Top: Annualized cost distribution for a varying number of stages, condenser pressures, ratios, and conversions. Bottom: The environmental impact distribution for a varying number of stages, condenser pressures, ratios, and conversions.

operational pressure increases. The rank and proximity values for the economic and environmental impacts are calculated and shown in Table 4.2. The values of $r_{TAC,j}$ calculated from Eqn (4.3), in relation to economic metric (*TAC*) show that the rank of variables in order of importance is: Conversion > N_{st} ≈ P_{cond} > reactant ratio. In the case of the $r_{EI,j}$ in relation to environmental impact (*EI*) shows that the rank of variables in order of importance is: Conversion > reactant ratio > P_{cond} ≈ N_{st}. Note that these values have been normalized to the base case

Table 4.2 Rank Order and Proximity Parameter Values for the Input–Output Variables Relationships Studied

	Rank Value ($\bar{r}_{jk}$)		Proximity Value (p_{kj})	
Input Variable	*TAC*	*EI*	*TAC*	*EI*
P_{cond}	1.108	0.028	0.968	0.162
N_{st}	1.169	0.010	0.956	0.319
Reactant ratio	0.509	0.070	0.155	0.887
Conversion	11.397	0.125	0.569	0.163

value (see Eqn (A.2) in the Appendix). Interestingly for both OFs, an order of magnitude in terms of the rank parameter is observed, showing that very small changes in the value of the conversion will impact much more than changes in the number of stages or column pressure.

The p_{kj} values, calculated from Eqn (4.4), show that none of the variables is very close to the optimal values (for which p_{kj} takes values of unity) conditions; being the closer ones P_{cond} and N_{st} in the case of *TAC*, while the reactant ratio for the case of *EI*.

To continue with the analysis the most influential variable Conversion has been fixed to 0.995 while the ratio has been fixed to 1.5 which showed a close proximity to the optimal value in terms of environmental impact, consequently the number of stages and the operating condenser pressure are left free to be changed.

As can be seen from Figure 4.5, the optimal economic impact (expressed as *TAC*) and environmental impact do not coincide. These solutions can be seen as optimal if the optimization method applied is the full tree exploration (brute force method), given that are the maximum (*TAC*) and minimum (*EI*) obtained values. Note that each one of the modeled scenarios included the optimization within the simulation of the ratio of inlet raw materials and the conversion fixed to base case value. The *TAC* optimum resulted in a column design of 90 stages and a condenser pressure of 7000 mmHg, while the environmental impact optimal design was found for a column of 30 stages and a condenser pressure of 2000 mmHg. In the case of Investment it is seen that its minimum value lies at the shortest column ($N_{st} = 30$), showing that the investment cost is heavily influenced by the column size and not due to the reboiler and condenser's area requirements. Please note that the highest investment is found at higher pressures and with the tallest column is around 10^6€ which is small compared to the operating benefit and *TAC* which at the same point show values around 10^7€/year. The former values clearly demonstrate that, in the case of *TAC*, the same relationships among operating costs and investment, shown in Figure 4.3, holds. Regarding the endpoint metrics the same relationship as in the base case holds, being the largest contribution to overall environmental impact the climate change, followed by ecosystem quality human health impact and resource use. Most interestingly the optimal scenarios only coincide for the case of climate

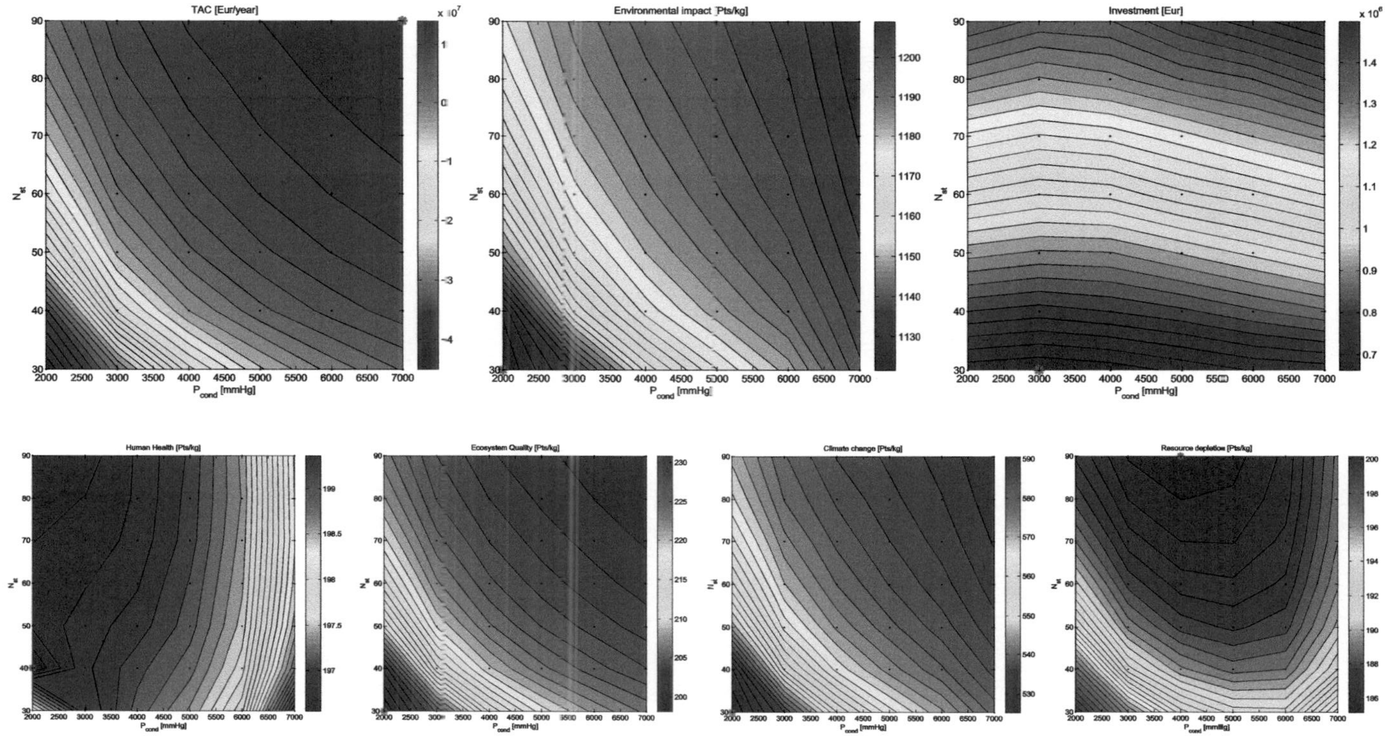

FIGURE 4.5

Relationships for different economic and fixed conversion and the ratio of inlet materials for different number of stages and condenser pressures. The black dots show the simulated results and the red star denotes the single objective best scenario settings.

change and ecosystem quality, these optima are the same point that the one found for the case of overall environmental impact. In the case of Human Health the scenario with lowest value was found for $N_{st} = 30$ and condenser pressure of 2000 mmHg, while in the case of resource use the optimal value lies at $N_{st} = 30$ and condenser pressure of 4000 mmHg. The figures were generated using the *contour* Matlab function which uses the simulation results as inputs, and provides with the appropriate isolines.

4.3.3 INTERPRETATION

To further balance the interpretation of the economic and environmental impact Pareto-efficiency plots are generated. If we consider TAC, and the other four endpoint environmental indicators, then 10 binary comparisons are possible, which are shown in Figures 4.6 and 4.7.

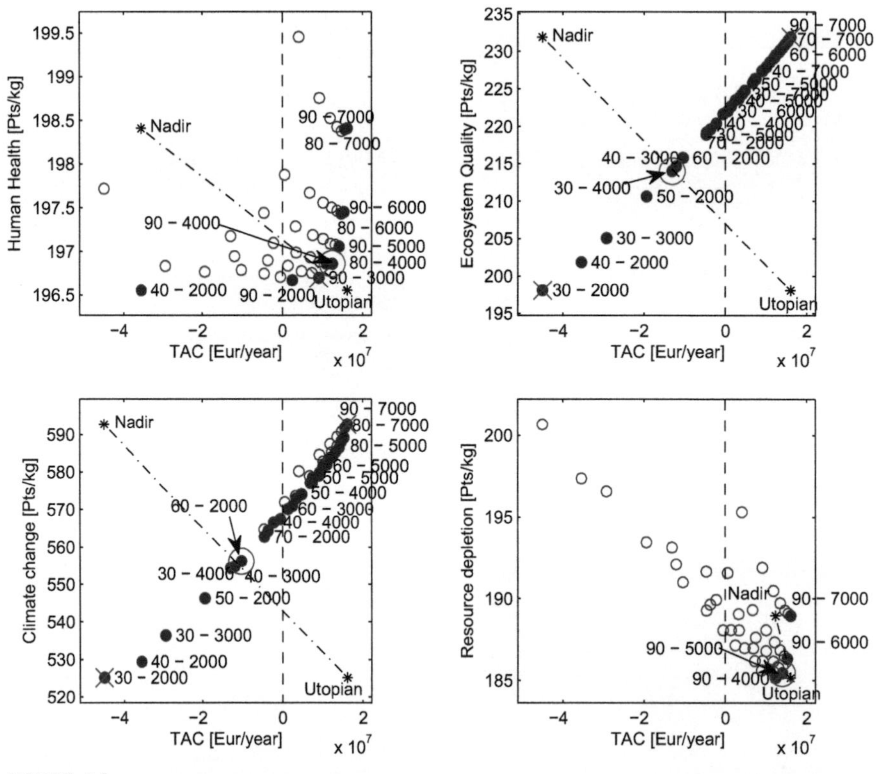

FIGURE 4.6

Pareto plot of different obtained solutions in terms of economic and different endpoint environmental impacts. Labels indicate number of stages and condenser operating pressure. Wide circles emphasize solutions which are closer to utopian point, while large crosses show solutions farthest from nadir point. Not all scenario labels have been plotted.

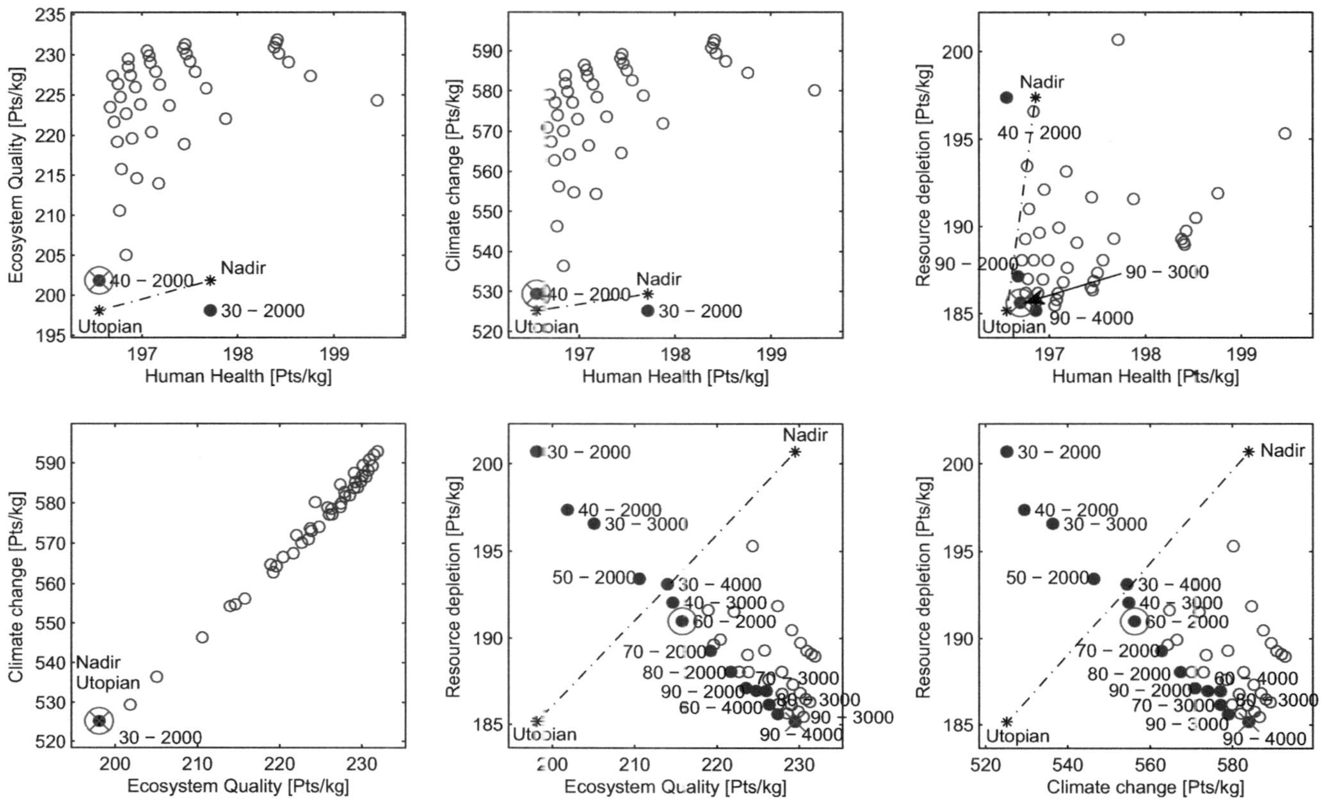

FIGURE 4.7

Pareto plot of different obtained solutions in terms of different endpoint environmental impacts. Labels indicate number of stages and condenser operating pressure. Wide circles emphasize solutions which are closer to utopian point, while crosses show solutions farthest from nadir point.

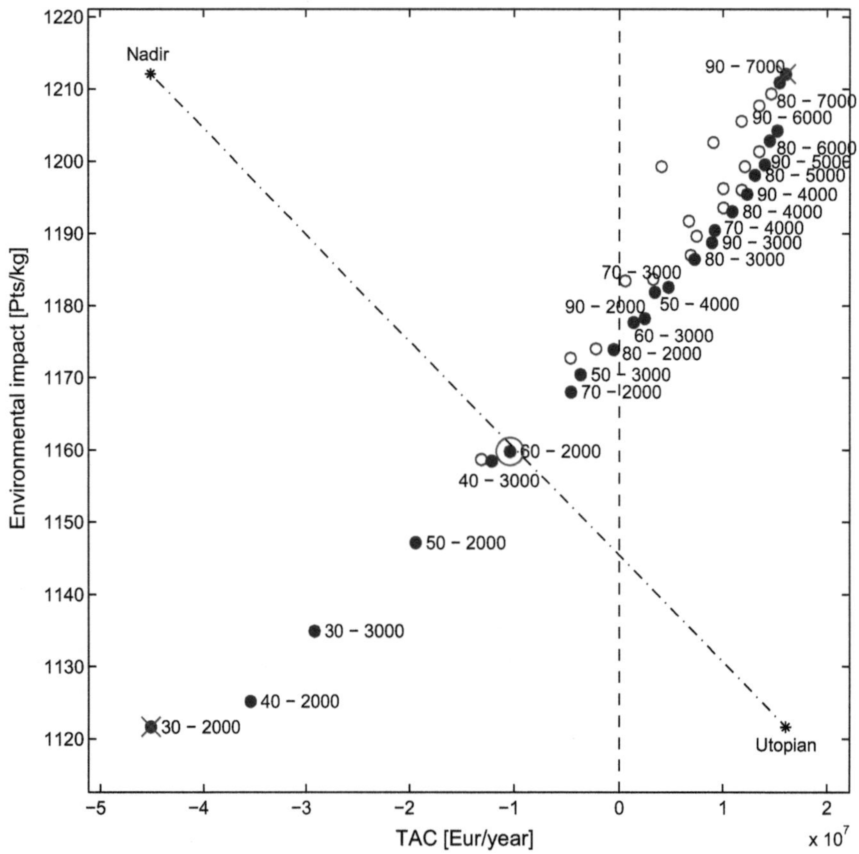

FIGURE 4.8

Pareto plot of different obtained solutions in terms of economic metric and overall environmental impact. Labels indicate number of stages and condenser operating pressure. Wide circles emphasize solutions which are closer to utopian point, while crosses show solutions farthest from nadir point.

From Figures 4.6–4.8 follows that the set of efficient solutions is different depending on the metrics considered. In the case of endpoint metrics compared to TAC, the profiles of climate change and ecosystem quality versus TAC are very similar spanning the PF along numerous solutions, in both cases the closest solution to the utopian point is (30–4000). The other PF for human health and resource consumption versus TAC contains less solutions and both give as closest to utopian point solution a column which has 90 stages but working at 4000 and 5000 mmHg, respectively. In the case of the comparison of endpoint metrics, also other PFs are obtained. The PF comparing climate change and ecosystem quality shows only one solution which coincides with the best solution for both indicators previously shown in Figure 4.5. The PFs of these indicators compared to the other metrics also show the same tendencies. The PF considering human health and resource depletion shows most solutions containing 90 or 40 stages high columns working at pressures below 4000 mmHg. To end the Pareto efficiency analysis the comparison of the overall environmental impact against TAC is performed, the PF found resembles the one obtained for climate change and ecosystem quality, this is due to the large weight that the IMPACT2002+ methodology considers for such endpoint categories. For this case the closest to the utopian point is the design which considers a column with 60 stages and works at very low pressure 2000 mmHg. However it is interesting to note that this solution selected using a very simple MCDA technique leads to a negative TAC, and consequently in order to make such scenario profitable it will require to increase the product sales price or to diminish the operating costs.

The different PFs obtained show how the process has a different behavior depending on which metric is being used, and that there is a need for such analysis in order to consider possible tradeoffs. Each one of the compromise solutions obtained for each binary metric comparison shows a different decision-maker point of view where only two criteria are considered to have the same importance, and hence the same weight as discussed in Eqn (4.5). Furthermore Figure 4.8 clearly shows the dangers of using such simplistic MCDA technique for selecting possible solutions, and shows that more information into the MCDA has to be considered, for example if negative TACs scenarios should be considered or not.

4.4 CONCLUSIONS

The current methodology has been implemented as scenario-based optimization studying the combination of integer and continuous process variables, which were shown to be the most influential on the objective functions; however, the use of derivative-free optimization methods can be readily used together with process simulation methods and is part of future work that could be undertaken.

APPENDIX

The thermodynamic behavior of a system is within the expected for the kind of estimations that the universal functional activity coefficient provides, as can be seen in Table A.1 and Figure A.1.

In order to tackle with pressure effects in the estimation of vessel costs, it is customarily to affect the investment associated to a given pressure by a correction factor that accounts for investment increases due to pressure increases. In

Table A.1 Main Thermodynamic Data, $p = 1$ atm, the Aspen Split Results for the Four Components of the System Under Study, Node Composition is Given in Molar Fraction. Bold Rows Indicate Pure Components

Temp. [C]	Classification	Type	Number of Components	Myristic Acid	Isopropanol	Isopropyl Myristate	Water
80.37	Unstable node	Homogeneous	2	0.000	0.666	0.000	0.334
82.35	**Saddle**	**Homogeneous**	**1**	**0.000**	**1.000**	**0.000**	**0.000**
99.58	Saddle	Homogeneous	2	0.000	0.000	0.005	0.995
100.00	Saddle	Homogeneous	2	0.002	0.000	0.000	0.998
100.02	**Stable node**	**Homogeneous**	**1**	**0.000**	**0.000**	**0.000**	**1.000**
315.30	Saddle	Homogeneous	2	0.041	0.000	0.959	0.000
315.32	**Stable node**	**Homogeneous**	**1**	**0.000**	**0.000**	**1.000**	**0.000**
325.83	**Stable node**	**Homogeneous**	**1**	**1.000**	**0.000**	**0.000**	**0.000**

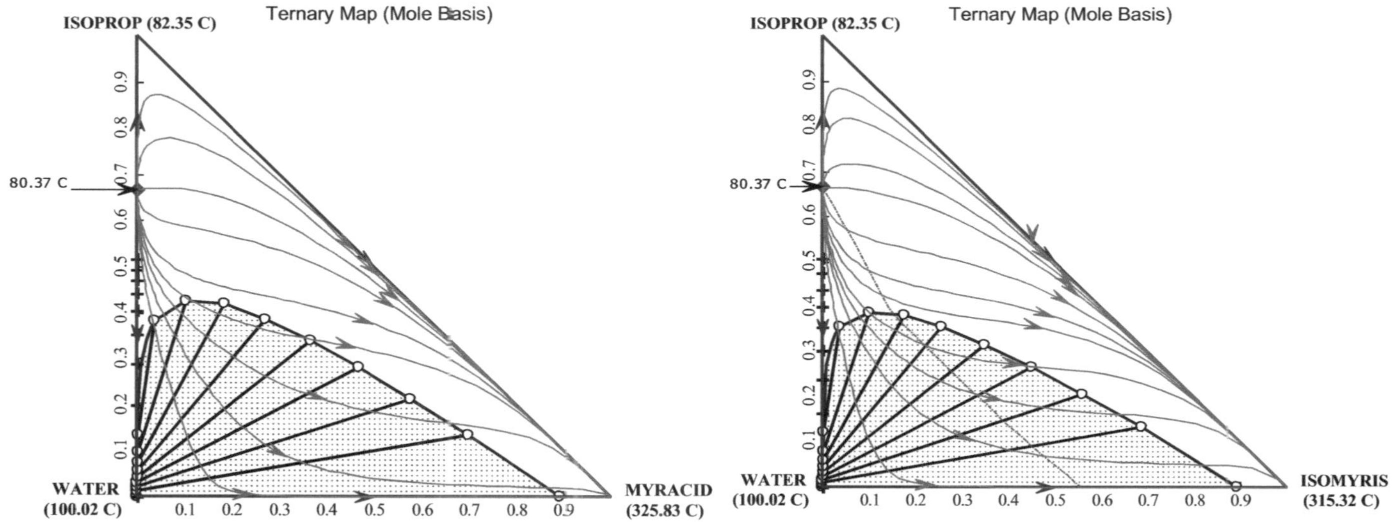

FIGURE A.1

Residue curve maps for two possible ternary mixtures encountered in the reactive distillation column. Brown lines (light gray in print versions) correspond to residue curves, while blue (black in print versions) correspond to the possible partial miscibility boundary and connection lines for Myristic acid or Isopropyl myristate and water.

Table A.2 Different Raw Material, Product, and Utilities Costs

Compound	€/ton	(€/mol)·10^{-3}	Utilities	Value
Myristic acid	1076	246	Steam (€/ton)	6.32
Ester	4145	1121	Industrial water (€/m^3)	1.10
Isopropanol	737	44	Electricity (€/kWh)	0.08
PTSA	24,133	4591	Waste water treatment (€/m^3)	0.54

the case adopted, this factor increases if the column's pressure is higher than 4.5 bar.

$$F_p = \begin{cases} 1 & p < 4.5 \text{ bar} \\ 1 + t\left(1 + e^{-t/2}\right) & p \geq 4.5 \text{ bar} \end{cases} \tag{A.1}$$

where t is defined as $t = k_p\left(\frac{P-4.5}{4.5}\right)$ k_p is 0.13 and P has to be measured in bars. Further details can be found in Chapter 6 of Doherty and Malone (2001). In the case of r_{jk} values of Eqn (4.3), the normalization is done as in Eqn (A.2) where the normalized values are calculated ($\tilde{r}_{jk}$), dividing r_{jk} by the value of the objective function at the base conditions ($OF_k^{BaseCase}$).

$$\tilde{r}_{jk} = \frac{r_{jk}}{OF_k^{BaseCase}} = \frac{\sum_{i=1}^{nTerms_k} \left|\frac{\partial OFterm_{ik}}{\partial x_j}\right| \Delta x_j^{MAX_k}}{OF_k^{BaseCase}} \tag{A.2}$$

LIST OF ACRONYMS USED

CF	characterization factors
EI	environmental impact
LCA	life cycle assessment
L–L	liquid–liquid
MCDA	multicriteria decision analysis
NRTL	nonrandom two liquid
OF	objective function
PF	Pareto front
PI	process intensification
pTSA	para-toluene sulphonic acid
RD	reactive distillation
SA	sensitivity analysis
TAC	total annualized costs
UNIFAC	UNIversal Functional Activity Coefficient
V–L	vapor–liquid

LIST OF SYMBOLS USED

C	operational costs (€/year)
$Ea_{d,r}$	reaction activation energy (d: direct, r: reverse) (kJ/kmol)
F_p	investment cost pressure factor (Dimensionless)
I	investment costs (€)
$k_{d,r}$	reaction arrhenius pre-exponential (d: direct, r: reverse) (Molarity)
n	years considered for the project lifetime (year)
$nOFs$	number of objective functions
N_{st}	number of distillation column stages
$nTerms_k$	number of terms that the *k*-th OF has
$OF_{compromise}$	compromise solution
$OF_k^{BaseCase}$	*k*-th objective function value for the base case
OF_k	*k*-th objective function
OF_k^*	best *k*-th OF value
$OFterm_{ik}$	*i*-th term of *k*-th objective function
P_{cond}	columns condenser pressure (mmHg)
p_{jk}	proximity parameter, for variable *j*-th and *k*-th OF (Dimensionless)
r_{jk}	rank order parameter, for variable *j*-th and *k*-th OF
$\tilde{r}_{jk}$	normalized rank order parameter, for variable *j*-th and *k*-th OF (Dimensionless)
S	Sales (€/year)
x_j	*j*-th decision variable
α_k	compromise weight for the *k*-th objective function (Dimensionless)

ACKNOWLEDGMENTS

The Agencia de Gestiò d'Ajuts Universitaris I de Recerca (AGUAR) from generalitat de Catalunya and Fons Social Europeu (EU), through FI and BE-DGR grants. And the financial support received from projects TOLERANT (DPI2006-05673) and EHMAN (DPI2009-09386) co-financed by the Spanish government (MICINN) and the European Commission (FEDER) are fully acknowledged.

REFERENCES

Anastas, P.T., Zimmerman, J.B., 2003. Design through the 12 principles of green engineering. Environ. Sci. Technol. 1, 95A–101A.

Azapagic, A., Millington, A., Collet, A., 2006. A methodology for integrating sustainability considerations into process design. Chem. Eng. Res. Des. 84 (A6), 439–452.

Bakshi, B.R., Fiksel, J., 2003. The quest for sustainability: challenges for process systems engineering. AICHE J. 49 (6), 1350–1358.

Biegler, L.T., Grossmann, I.E., Westerberg, A.W., 1997. Systematic Methods of Chemical Process Design (Chapter 5). Prentice Hall.

Bojarski, A.D., Guillén-Gosálbez, G., Jiménez, L., Espuña, A., Puigjaner, L., 2008. Life cycle assessment coupled with process simulation under uncertainty for reduced environmental impact: application to phosphoric acid production. Ind. Eng. Chem. Res. 47, 8286–8300.

Campolongo, F., Kleijnen, J., Andres, T., 2000. Screening methods (Chapter 4). In: Sensitivity Analysis. Wiley and Sons, pp. 65–80.

Chen, H., Shonnard, D.R., 2004. Systematic framework for environmentally conscious chemical process design: early and detailed design stages. Ind. Eng. Chem. Res. 43, 535–552.

Clift, R., 2006. Sustainable development and its implications for chemical engineering. Chem. Eng. Sci. 61, 4179–4187.

de Jong, M., Feijt, R., Zondervan, E., Nijhuis, T., de Haan, A., 2009. Reaction kinetics of the esterification of myristic acid with isopropanol and n-propanol using p-toluene sulphonic acid as catalyst. Appl. Catal. A 365 (1), 141–147.

de Jong, M.C., Zondervan, E., Dimian, A.C., de Haan, A.B., 2010. Entrainer selection for the synthesis of fatty acid esters by entrainer based reactive distillation. Chem. Eng. Res. Des. 88 (1), 34–44.

de-Haes, H.A.U., Jolliet, O., Finnveden, G., Hauschild, M.Z., Krewitt, W., Muller-Wenk, R., 1999. Best available practice regarding impact categories and category indicators in life cycle impact assessment, background document for the second working group on life cycle impact assessment of SETAC-Europe (WIA-2). Int. J. Life Cycle Assess. 74, 66–74.

Dimian, A.C., Bildea, C.S. (Eds.), 2008. Chemical Process Design Computer-aided Case Studies. Wiley-vch Verlag GmbH & Co. KGaA, Weinheim.

Dimian, A.C., Omota, F., Bliek, A., 2004. Entrainer-enhanced reactive distillation. Chem. Eng. Process 43, 411–420.

Diwekar, U., 2005. Green process design, industrial ecology and sustainability: a systems perspective. Resour. Conserv. Recycl. 44, 215–235.

Doherty, M.F., Malone, M.F., 2001. Conceptual Design of Distillation Systems. McGraw-Hill, USA.

EcoinventV2.0, 2008. The Ecoinvent Database V2.0. Swiss Centre for Life Cycle Inventories Inc, Switzerland.

Finnveden, G., Hauschild, M.Z., Ekvall, T., Guinee, J.B., Heijungs, R., Hellweg, S., Koehler, A., Pennington, D.W., Suh, S., 2009. Recent developments in life cycle assessment. J. Environ. Manage. 91, 11–21.

Fisher, W.R., Doherty, M.F., Douglas, J.M., 1985. Evaluating significant economic trade-offs for process design and steady-state control optimization problems. AICHE J. 31 (9), 1538–1547.

Humbert, S., Margni, M., Jolliet, O., 2005. Impact 2002+: User Guide Draft for Version 2.1. Technical Report, Industrial Ecology & Life Cycle Systems Group, GECOS. Swiss Federal Institute of Technology, Lausanne (EPFL), Lausanne, Switzerland.

Krishna, R., 2002. Reactive separations: more ways to skin a cat. Chem. Eng. Sci. 57 (9), 1492–1504.

Maeda, K., Yamada, S., Hirota, S., 1997. Binodal curve of two liquid phases and solid-liquid equilibrium for water + fatty acid + ethanol systems and water + fatty acid + acetone systems. Fluid Phase Equilib. 130, 281–294.

Mahfud, F., Melin-Cabrera, I., Manurung, R., Heeres, H., 2007. Biomass to fuels: upgrading of flash pyrolysis oil by reactive distillation using a high boiling alcohol and acid catalysts. Process Saf. Environ. Prot. 85 (5), 466–472.

Peters, M.S., Timmerhaus, K.D., 1991. Plant Design and Economics for Chemical Engineers (Chapter 10), fourth ed. McGraw-Hill Book Co, Singapore.

Puigjaner, L., Heyen, G. (Eds.), 2009. Computer Aided Process and Product Engineering. Chapter: Modeling in the Process Life Cycle. Section 4.2. Wiley-vch Verlag GmbH & Co. KGaA, Weinheim, pp. 667–693.

Rebitzer, G., Ekvall, T., Frischknecht, R., Hunkeler, D., Norris, G., Rydberg, T., Schmidt, W.P., Suh, S., Weidema, B.P., Pennington, D.W., 2004. Life cycle assessment part 1: framework, goal and scope definition, inventory analysis and applications. Environ. Int. 30, 701–720.

Saltelli, A., Ratto, M., Andres, T., Campolongo, F., Cariboni, J., Gatelli, D., Saisana, M., Tarantola, S., 2008. Global Sensitivity Analysis, the Primer, first ed. Wiley.

Shah, M., Zondervan, E., de Haan, A., 2009. Modelling and simulation of an unsaturated polyester process. In: Proceedings of the International Conference on Chemical Engineering and Biotechnology, Malaysia.

Stankiewicz, A.I., Moulijn, J.A., 2000. Process intensification: transforming chemical engineering. Chem. Eng. Prog. 96 (1), 22–33.

Stefanis, S.K., Livingston, A.G., Pistikopoulos, E.N., 1997. Environmental impact considerations in the optimal design and scheduling of batch processes. Comput. Chem. Eng. 21 (10), 1073–1094.

Turanyi, T., Rabitz, H., 2000. Local methods (Chapter 5). In: Sensitivity Analysis. Wiley and Sons, pp. 81–99.

Yang, Y., Shi, L., 2000. Integrating environmental impact minimization into conceptual chemical process design—a process systems engineering review. Comput. Chem. Eng. 24, 1409–1419.

CHAPTER

5

A Graphical Method for Hydrogen Network Integration with Purification Reuse

Xiao Feng*[,1], Qiao Zhang[§], Guilian Liu[§]

**Institute of New Energy, China University of Petroleum, Beijing, China*

[§]Department of Chemical Engineering, School of Chemical Engineering and Technology, Xi'an Jiaotong University, Xi'an, China

[1]Corresponding author: E-mail: xfeng@cup.edu.cn

5.1 BACKGROUND

Hydrogen, an important secondary resource, is widely used in chemical industries because of its benign physical and chemical properties. Particularly for refineries, hydrogen is widely used in hydroprocessing operations (Zagoria and Huycke, 2003), such as hydrotreating, hydrocracking, hydrodesulfurization, hydrofining, and hydrodeoxygenation. In recent years, global economy increase leads to increasing consumption of crude oil, whose heavier composition and higher sulfur content require higher hydrogen consumption per unit oil production. Besides, stricter environmental regulations on clean fuels also impel refineries to increase their hydrotreating depth to get short-chain hydrocarbons. As a result, hydrogen demand has been increasing sharply during the past two decades.

As secondary resource, hydrogen needs special feed and production pathway, such as steam methane reforming, autothermal reforming, partial oxidation, crude bio-oil steam reforming, coal gasification, and biomass gasification. Hydrogen production consumes resource and energy, and also emits greenhouse gas as well as other waste to the environment. Therefore, hydrogen conservation can both save nonrenewable resource and reduce emissions, improving the overall sustainability.

Hydrogen network integration considers all hydrogen-producing and hydrogen-consuming units as a whole, so as to reduce the fresh hydrogen consumption by maximizing the reuse of internal sources. It is an effective way to conserve hydrogen for refineries, and thus has been a hot research focus.

As a pinch-based graphical method, value composite curve (Towler et al., 1996) is first introduced for hydrogen networks. Later, the source−sink mapping diagram (El-Halwagi and Spriggs, 1998) is developed for mass integration. After these early works, several methods based on the shifting of composite curves, such as the

Computer Aided Chemical Engineering, Volume 36. ISSN 1570-7946. http://dx.doi.org/10.1016/B978-0-444-63472-6.00005-7

hydrogen surplus diagram (Alves and Towler, 2002), material recovery pinch diagram (El-Halwagi et al., 2003; Almutlaq et al., 2005; Zhao et al., 2006), as well as some other methods (Bandyopadhyay, 2006; Wan Alwi et al., 2009) were successively developed to determine the minimum fresh hydrogen consumption and waste discharge.

Although the concepts and targeting processes of the aforementioned methods are clear and simple, purification reuse was not considered. As an effective way to conserve fresh hydrogen, purification processes have been widely used in refineries to recover low-concentration hydrogen sources (Bollinger et al., 1984). As part of hydrogen networks, purification reuse should be integrated together with hydrogen sources and sinks. By constructing limiting composite curve, a unified pinch-based conceptual method (Agrawal and Shenoy, 2006) is developed for targeting water and hydrogen networks. In such diagram, water regeneration and hydrogen purification can be taken into consideration to get further reduction of fresh resources. Afterward, the integration of hydrogen networks with purification reuse on optimizing the purification feed flow rate, as well as purification feed and product purities (Liu et al., 2012, 2013) are studied to reveal the relationship between fresh hydrogen consumption and purification performance. By taking different types of purifiers into consideration, a pinch sliding graphical solution (Lou et al., 2013) is developed to identify the target of both hydrogen and water networks with purification/regeneration reuse. The triangle-rule-based purification reuse hydrogen networks (Yang et al., 2014) is enlarged by considering the feed location and flow rate, so as to give the optimal and economical purification performance.

This chapter addresses hydrogen network integration with purification reuse through visual solution: pinch-based graphical approach. By constructing source and sink composite curves and purification polygon in hydrogen load versus flow rate diagram, a shifting procedure is proposed to maximize purification reuses, achieving the minimization of fresh hydrogen.

5.2 HYDROGEN NETWORK INTEGRATION

In hydrogen network, streams containing hydrogen are termed as hydrogen sources, while units needing hydrogen supply are hydrogen sinks. Hydrogen network consists of all sources, sinks, and connections among them. As an important part of hydrogen network, a purification unit is placed to upgrade the hydrogen concentration of certain sources to reuse them. In order to reduce fresh hydrogen consumption, the values of purification streams, including feed, purified product, as well as tail gas, should be part of hydrogen network integration.

Figure 5.1 is a typical refinery hydroprocessing flow sheet. The liquid oil is fed to the reactor together with the mixture hydrogen source, encompassing recycle and fresh hydrogen. After hydrogenation, the product oil is obtained at the bottom of the separator and the unreacted hydrogen is collected on its top. Afterward, the

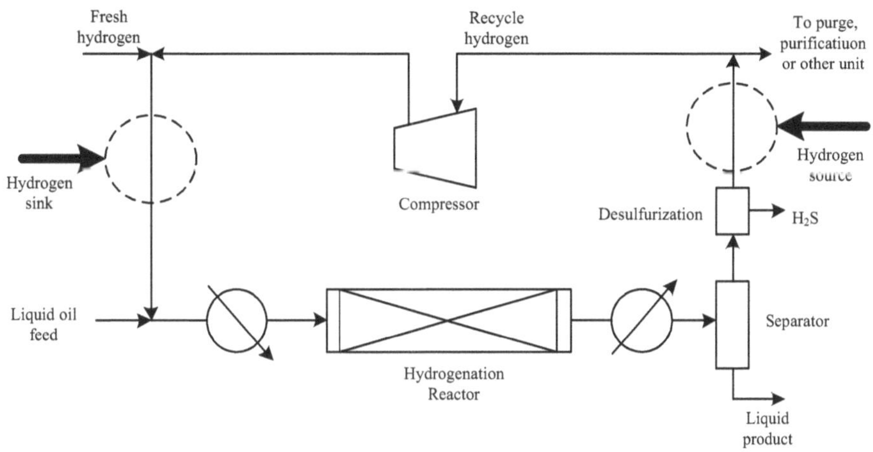

FIGURE 5.1

A typical refinery hydroprocessing flow sheet.

unreacted hydrogen source is partly sent to other units, purification systems, or fuel gas systems, and the majority is recycled to the reactor again by compressing. The modern scale refinery often contains many such hydroprocessing units, so generally a number of hydrogen sources and sinks form the corresponding hydrogen network.

5.3 GRAPHICAL METHOD

The graphical method in this chapter is performed in the hydrogen load versus flow rate diagram (Zhao et al., 2006). A brief overview of this visual method is described below.

Before integration, the flow rate and concentration of each hydrogen source or sink in a given hydrogen network are known, and the minimum flow rate of the fresh hydrogen is the target to be determined. In the hydrogen load versus flow rate diagram, each source or sink is presented by a segment of straight line, as shown in Figure 5.2(a). The horizontal projection between the two ends represents the flow rate, while that of the vertical coordinate represents its hydrogen flow rate, or named hydrogen load, of the corresponding stream. Thus, the slope of such line describes the hydrogen concentration of the given stream.

The advantage of such graphical presentation is that the line can move randomly in both horizontal and vertical directions without changing its flow rate, hydrogen load, and hydrogen concentration on the given stream. For example, if the line shown in Figure 5.2(a) moves to the new position shown in Figure 5.2(b), it is still the same stream because the projections on the two coordinates and the slope are all intact. In a word, such a segment of a line with a certain slope quantifies a unique source or sink wherever it is in the hydrogen load versus flow rate diagram.

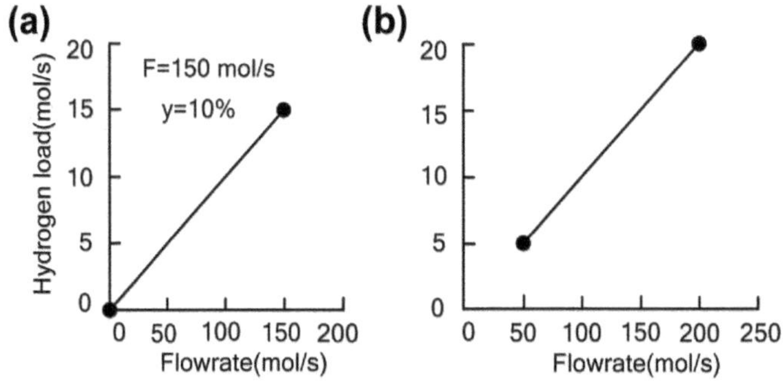

FIGURE 5.2

Graphical presentation of a hydrogen source or sink. (a) The initial location with certain flow rate and hydrogen load; (b) The shifted location with the same flow rate and hydrogen load.

Based on the graphical presentation for each source/sink stream introduced in hydrogen load versus flow rate diagram, the source and sink composite curves can be constructed with all these single source/sink lines connected in the order of increasing hydrogen concentration. These composite curves are broken lines: each segment represents a source or sink with slope as its hydrogen concentration, as shown in Figure 5.3. Similar to a single source or sink, the composite curve can also move randomly in both directions without causing any changes to the properties of sources or sinks because the projections of each source or sink on the two coordinates are the same as the original. Note that fresh hydrogen is not necessary always the highest hydrogen concentration source, i.e., in Figure 5.3, SR2 (short for source 2) is the fresh hydrogen.

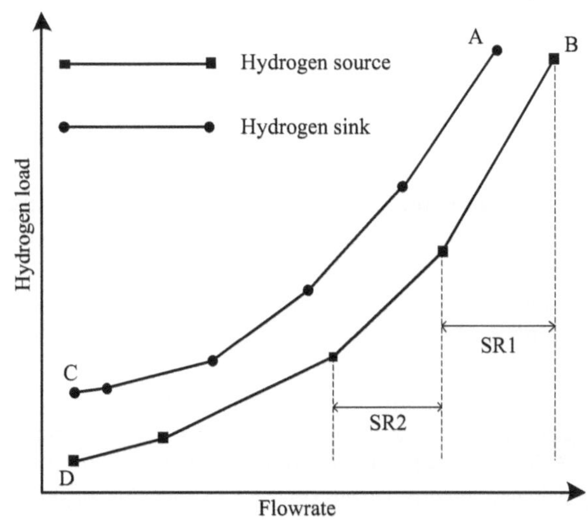

FIGURE 5.3

Composite curves for hydrogen sources and sinks.

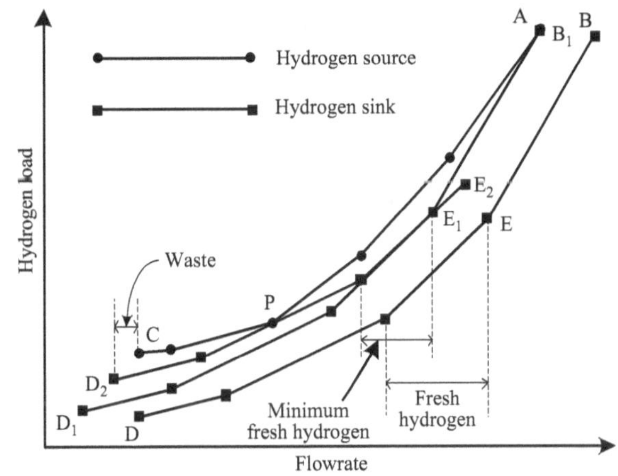

FIGURE 5.4

SR2 is fresh hydrogen.

The minimum fresh hydrogen consumption target can be identified by shifting the source composite curve. First, the source composite curve BD is moved to B_1D_1, with the top end of the source composite curve B_1 coinciding with that of the sink composite curve. After B_1 and A coincide, the source streams whose hydrogen concentrations are higher than that of fresh hydrogen, B_1E_1, are kept stationary, and then the lower section of the source composite curve E_1D_1 is moved along the fresh hydrogen to decrease the fresh hydrogen flow rate, until the location E_2D_2, where another intersection point P appears, as shown in Figure 5.4, which is the pinch.

Note that, a feasible hydrogen network requires the sink composite curve to lie above the source composite curve.

5.4 GRAPHICAL METHOD WITH FIXED PURIFICATION PARAMETERS (ZHANG ET AL., 2011)

Purification is a unit placed in refinery hydrogen networks to purify source streams, generating a high-purity source and a low-purity tail gas. It is a very important process in refineries, as it can upgrade the purity of sources so as to reuse it, improving the efficiency of hydrogen utilization. To minimize hydrogen consumption, it is necessary to optimize the hydrogen network with purification as a whole. Similar to a heat exchange network, the part below the pinch point of a hydrogen network can be taken as a net hydrogen source with hydrogen surplus, while that above the pinch point can be taken as a net sink, which needs to consume fresh hydrogen. Therefore, to reduce the fresh hydrogen consumption as well as waste the surplus stream below the pinch point can be purified to supply the corresponding purified product to the sink above the pinch point.

5.4.1 MATERIAL BALANCE AND CHARACTERISTICS OF PURIFICATION PROCESS

A purification process generates two outputs from one feed (F_{in}, y_{in}): the one with higher hydrogen concentration is the purified product and the other with lower hydrogen concentration is named tail gas, as shown in Figure 5.5.

The material balances of such a purification process are as follows.

$$F_{in} = F_{pur} + F_r \tag{5.1}$$

$$F_{in}y_{in} = F_{pur}y_{pur} + F_r y_r \tag{5.2}$$

$$R = \frac{F_{pur}y_{pur}}{F_{in}y_{in}} \tag{5.3}$$

where F and y denote the flow rate and hydrogen concentration, respectively; subscripts *in*, *pur*, and r represent the feed, purified product; and tail gas, respectively; and R represents the hydrogen recovery ratio of the purification process. These three equations are the design equations for a purifier.

There are three concentration variables, three flow rate variables, and a recovery ratio R in design equations. Generally, for certain feed concentration y_{in}, purifiers have technical specifications about the purified concentration, y_{pur}, and recovery ratio, R. Therefore, the designing equations can be rearranged as:

$$R = \frac{y_{in} - y_r}{y_{in}\left(1 - \frac{y_r}{y_{pur}}\right)} \tag{5.4}$$

$$F_{pur} = \frac{F_{in}y_{in}R}{y_{pur}} \tag{5.5}$$

$$F_r = F_{in} - F_{pur} \tag{5.6}$$

From Eqn (5.4), it can be seen that R and the three concentration variables are independent of the flow rate variables. F_{in} and y_{in} are specifications of a purifier. Once F_{in}, y_{in}, y_{pur}, and R are given, the tail gas purity y_r can be determined, and other flow rates can then be calculated according to Eqns (5.5) and (5.6).

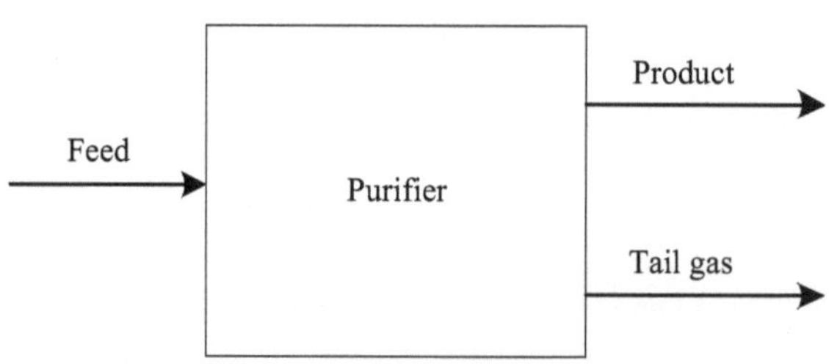

FIGURE 5.5

A purification process.

5.4.2 TRIANGLE RULE

The material balance introduced above can be expressed in the hydrogen load versus flow rate diagram. The feed can be represented by the straight line MN, as shown in Figure 5.6. Similar to each single line in hydrogen load versus flow rate diagram, the horizontal distance between the two ends of the feed represents its flow rate, while the vertical distance represents its hydrogen load.

For a purification process with given feed (F_{in}, y_{in}), as long as two of the three variables (y_{pur}, R, and y_r) are specified, the material balance triangle can be constructed, and the other two variables can be determined. For example, with known y_{in}, if y_{pur} and R are given, then y_r can be calculated by Eqn (5.4). The other two flow rate variables, F_{pur} and F_r, can be identified according to Eqns (5.5) and (5.6). First, according to F_{in} and y_{in}, feed line MN can be constructed. Next, from point M a straight line MY with the purified hydrogen concentration y_{pur} as slope can be drawn. Similarly, another straight line NX with the tail gas concentration y_r as slope can also be constructed. These two lines will interact at point V with MV and NV representing purified product and tail gas, respectively. The three lines represent the material balance of the purification process. According to the distance between the two ends of lines MV and NV in the horizontal direction, the flow rates of these two products can be identified.

From the above example, it can be seen that, for a purification process, the material balance between its feed and products can be well represented by a triangle plotted in the hydrogen load versus flow rate diagram. Based on this observation, the triangle rule for purification process is proposed: for purification process with one feed and two output products, its mass balance can be represented by a triangle constructed in the hydrogen load versus flow rate diagram.

In the above exposition, the purification process has only one feed stream. Practically, the feed can be a mixture of some feed streams mixed before entering a

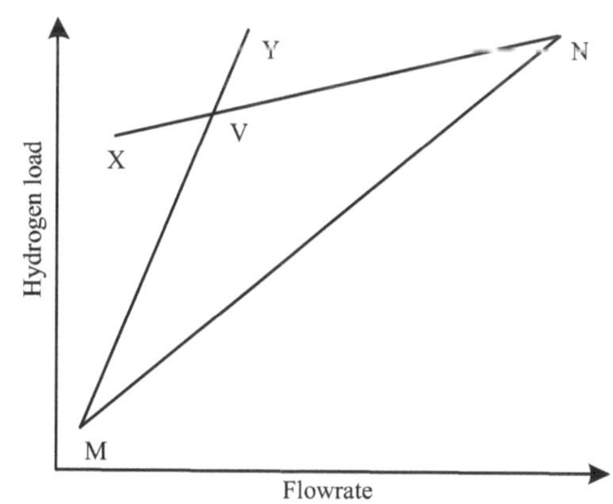

FIGURE 5.6

Triangle rule for purification process.

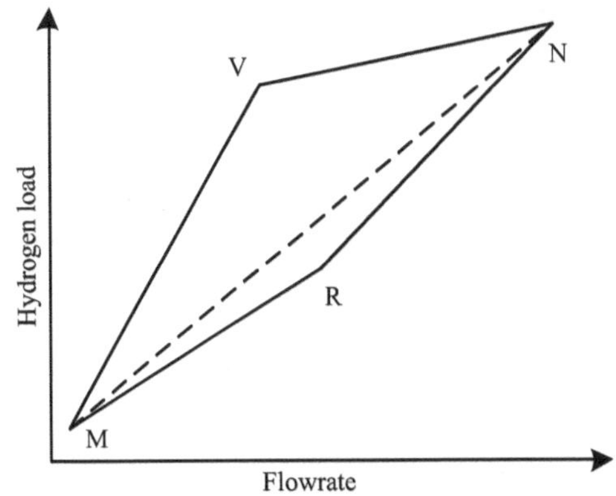

FIGURE 5.7

Polygon rule with mixture source.

purifier, and the material balance can also be solved according to the triangle rule. As shown in Figure 5.7, MR and RN are two different feed streams, triangle MRN satisfies the material balance in mixing, and MN is the equivalent feed after mixing. Through the purification process, MN can be further separated into two streams, the purified product MV and the tail gas NV. On the whole, the actual purification could be described by quadrangle MRNV. Thus, the triangle rule could be generalized to polygon rule: for a purification process, the material balance between its feeds and products can be well represented by a polygon plotted in the hydrogen load versus flow rate diagram. The triangle rule is a special case of the polygon rule. This generalized polygon rule could also be used to analyze the regeneration process of water networks.

5.4.3 GRAPHICAL METHOD

For a hydrogen network without purification reuse, its composite curves and pinch are shown in Figure 5.8. The top end of the sink composite curve A overlaps with that of the source composite curve B. AH represents the minimum fresh hydrogen consumption; point P is the pinch point; and C and D are the lower ends of sink and source composite curves, respectively. There is a net source below P, while there is a net sink above P. Purification could be done to purify the source streams below the pinch and provide purified product with high hydrogen concentration to the sink streams above the pinch, saving the fresh hydrogen.

To systematically analyze hydrogen networks with purification reuse, the purification triangle and the composite curves should be coupled together to identify the target. Note that purifiers have technical specifications about the purified concentration y_{pur} and recovery ratio R. Therefore, if y_{pur} and R are specified first, then y_r can be determined by Eqn (5.4). With y_{pur} and y_r, two straight lines with concentrations of the purified product and the tail gas can be constructed to form an initial purification polygon, starting the graphical targeting process.

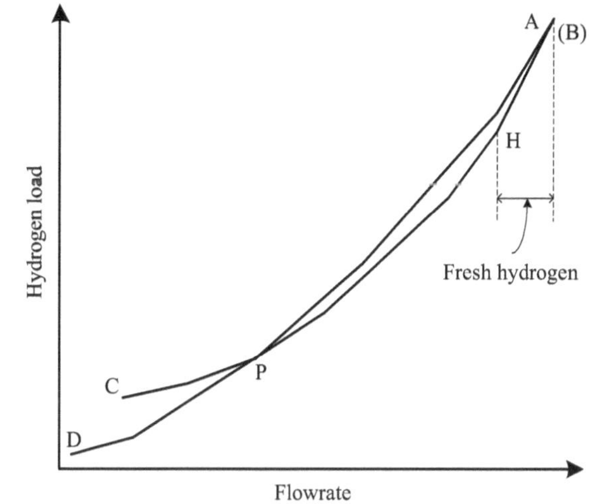

FIGURE 5.8

Initial pinched hydrogen network without purification reuse.

Specifically, in Figure 5.9, once y_{pur} and y_r are specified, straight line DR with y_{pur} as its slope can be constructed to represent the purified product; a similar case is applied to straight line O_1L_1 with y_r as its slope. Next, slide O_1L_1 upward until it intersects the sink composite curve at point P_1, making the line intersect the product line DR and the source composite curve at points E and F, respectively. An initial purification polygon DEFG is formed.

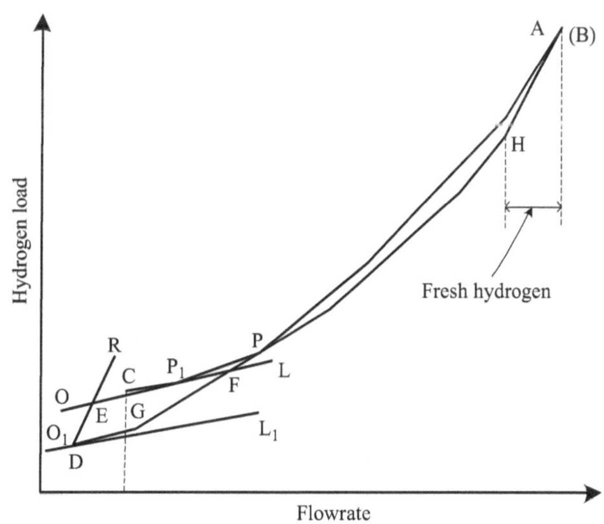

FIGURE 5.9

Initialization of purification polygon.

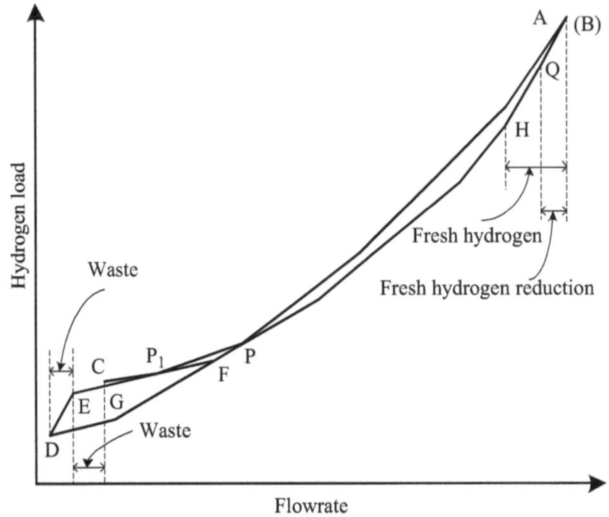

FIGURE 5.10

Pinch identification with purification reuse (purified product purity equals fresh hydrogen purity).

If the hydrogen concentration of the purified product equals that of the fresh hydrogen, then DEFG is the purification polygon. The mixture of DG and GF is the feed; DE and EF represent the purified product and tail gas, respectively. Since the purified concentration equals that of fresh hydrogen, the purified product line (PPL) DE can be shifted upward, intersecting the sink composite curve at point A and the source composite curve at point Q, as shown in Figure 5.10. Then, $AQHPFP_1E$ is the source composite curve after purification and P and P_1 are the pinch points of the system with purification reuse. The fresh hydrogen consumption is reduced from AH to HQ, with EG representing the waste gas of the system with purification reuse.

If the purified concentration is higher than that of the fresh hydrogen, the targeting process will be different. In such case, an assistant purified product line (APPL) (AI) with concentration of the purified product is drawn from the upper end of the fresh hydrogen, point A, as shown in Figure 5.11. This line is parallel to the PPL DR. Next, the source composite curve together with the PPL BDR can be shifted to increase the participating flow rate of AI in source composite curve. When BDR is located at $B_1D_1R_1$, intersecting the tail gas line at E_1 and line AI at point I_1, and the length of D_1E_1 equals that of AI_1, the target is identified. The source composite curve intersects the tail gas line at point F_1, $E_1F_1I_1A$ is the source composite curve after purification reuse, and P_1 is the pinch point of the system with purification reuse. On the source composite curve, AI_1 represents the purified product, and E_1F_1 represents the tail gas of the purification. $D_1G_1F_1$ represents the feed of the purification, and $D_1E_1F_1G_1$ is the purification polygon. With

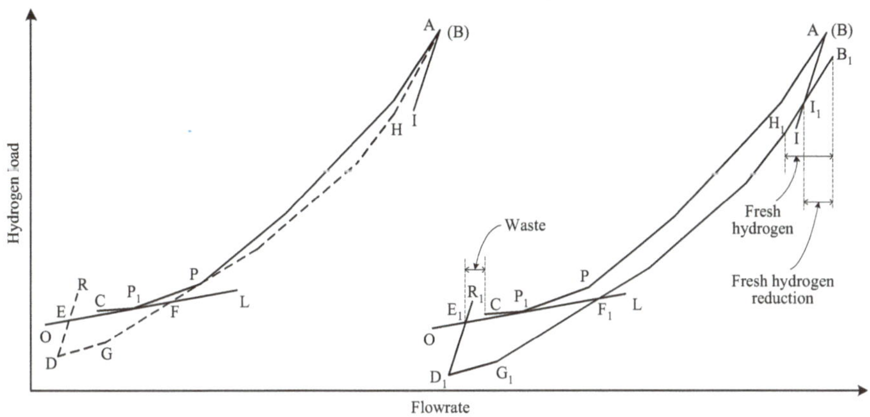

FIGURE 5.11

Pinch identification with purification reuse (purified product purity is higher than fresh hydrogen purity).

purification reuse, the fresh hydrogen reduction is represented by B_1I_1. Finally, the waste can be determined, as shown in Figure 5.11. After purification reuse, the purified product D_1E_1 is reused in terms of AI_1 above the pinch, saving fresh hydrogen B_1I_1.

When the purified concentration is lower than that of the fresh hydrogen, the targeting procedure is similar to that of the situation in Figure 5.11. The difference is that the APPL is not drawn from the upper end of the fresh hydrogen, but should be constructed from the lower end of the fresh hydrogen line, H.

Exactly, for the network shown in Figure 5.9, the APPL HI can be drawn from the lower end of fresh hydrogen, as shown in Figure 5.12(a). RDGHI is shifted to $R_1D_1G_1H_1I_1$, where it intersects the tail gas line at E_1 and line BH at point K_1, H_1K_1 equals D_1E_1, and the target is identified. $E_1F_1H_1K_1A$ is the source composite curve and P_1 is the pinch point of the system with purification reuse. $D_1E_1F_1G_1$ is the purification polygon. The fresh hydrogen consumption is represented by BK_1, and compared to the network without purification reuse, the fresh hydrogen consumption is reduced, as represented by HK_1. The waste can also be determined, as shown in Figure 5.12(a).

According to the hydrogen load versus flow rate diagram (Zhao et al., 2006), a hydrogen network is feasible only when the sink composite curve lies above the source composite curve. This is also applicable to the hydrogen network with purification reuse. However, when the purified concentration is lower than that of fresh hydrogen, it is possible that in Figure 5.12(a), the two composite curves intersect at point P_2 before H_1K_1 is equal to D_1E_1, as shown in Figure 5.12(b). Since H_1K_1 is not equal to D_1E_1, the network is infeasible under this situation. However, $R_1D_1G_1H_1I_1$ cannot be shifted anymore. Otherwise, part of the source composite curve will cross the sink composite curve, causing an infeasible network.

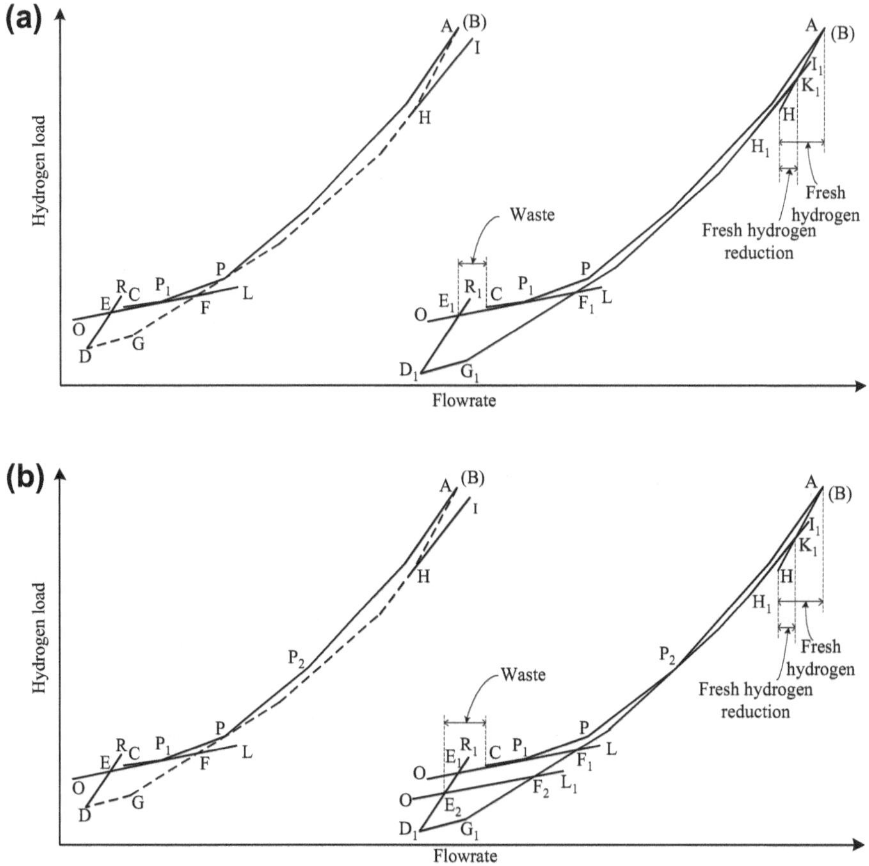

FIGURE 5.12

Pinch identification with purification reuse (purified product purity is lower than fresh hydrogen purity): (a) Case 1; (b) Case 2.

To target the minimum fresh hydrogen in this situation, the tail gas line OL can be slid downward to O_2L_2, making it intersect the product line and the source composite line at E_2 and F_2, respectively. If the length of D_1E_2 equals that of H_1K_1, shifting is stopped. $E_2F_2H_1K_1A$ is the source composite curve of the hydrogen network with purification reuse and P_2 is the corresponding pinch point. $D_1E_2F_2G_1$ is the purification polygon and HK_1 is still the fresh hydrogen reduction. The waste can also be identified, as shown in Figure 5.12(b). Since in such case the sink composite curve lies above the source composite curve, the network is feasible.

The feature of the cases stated above is that the purified product concentration is in neighboring of the fresh hydrogen. To generalize the method, consider the case

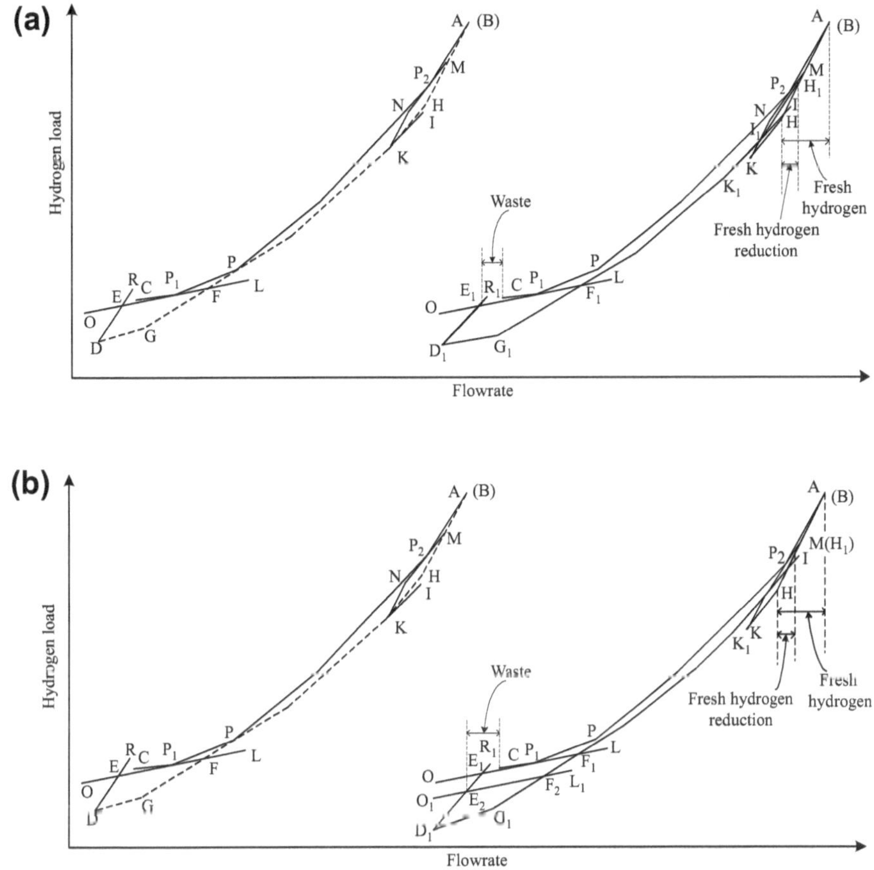

FIGURE 5.13

Pinch identification with purification reuse (purified product purity is not neighboring to fresh hydrogen purity) (a) Case 1; (b) Case 2.

that the concentrations of the purified product and fresh hydrogen are not neighboring to each other, as shown in Figure 5.13(a). AH and KI represent fresh hydrogen and APPL, respectively, and there is hydrogen source KH between them. In this case, the source composite curve can be divided into three parts: fresh hydrogen AH, source KH, and the lower section of the composite curve including the APPL (KI), as shown in Figure 5.13(a). The upper section AH should be kept intersecting the sink composite curve at their upper ends, and KH and the lower section of source composite curve can move respectively. However, to identify the fresh hydrogen target on condition that with the purified product equals to the source represented by the APPL, it is difficult to move these two sections simultaneously. To solve this problem, the feasible range of KH can be determined first, and then the lower section can be moved. The targeting procedure is as follows.

First, KH is slid up along the fresh hydrogen line until it intersects the sink composite curve at point P_2, so the track of thé moving process is a parallelogram KHMN, which is the feasible range of KH. Then, move the lower section along KN until the location at which K_1I_1 equals D_1E_1. At last, draw a line parallel to KH from I_1, and it will intersect MH at point H_1. So I_1H_1 is the location of KH after moving, $AH_1I_1K_1F_1E_1$ is the source composite curve after purification, and P_1 is the new pinch point. In this case, $D_1E_1F_1G_1$ is the purification polygon, D_1E_1 is the purified product, and HH_1 is the fresh hydrogen reduction. The waste can also be identified, as shown in Figure 5.13(a).

The same to the difference between Figure 5.12(a) and (b), Figure 5.13(b) is also possible compared to Figure 5.14(a). The moving of the lower section should not be stopped until I_1 arrives at point N. At this location, N and I_1 overlap, and M and H_1 overlap. K_1I_1 is its maximum, but it may still be less than D_1E_1. To make D_1E_1 smaller, the tail gas line OL should be slide down to O_1L_1, which intersect D_1R_1 at point E_2 and D_1E_2 equals K_1I_1. In this case, P_2 is the new pinch point after purification reuse, $D_1E_2F_2G_1$ is the purification polygon, and MH is the fresh hydrogen reduction.

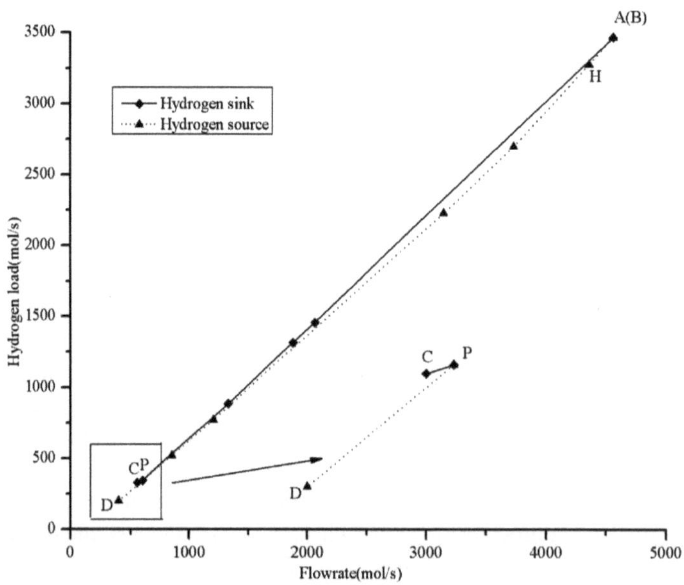

FIGURE 5.14

The pinch diagram of the hydrogen network without purification reuse.

5.5 CASE STUDY FOR GRAPHICAL METHOD WITH FIXED PURIFICATION PARAMETERS (ZHANG ET AL., 2011)

In this section, we use an example to illustrate the proposed graphical method, and the data is given by Table 5.1. In this case, there are six sources, including fresh hydrogen, and five sinks. The hydrogen concentration in mole percent and the flow rate and hydrogen load in moles per second are listed from the second to the fourth columns, respectively. First, all hydrogen sources and sinks have been ranked in the order of increasing hydrogen purity. The maximum available fresh hydrogen flow rate is 400 mol/s, and its purity is 95%.

For the case given in Table 5.1, the minimum fresh hydrogen consumption without purification reuse can be determined by the hydrogen load versus flow rate diagram shown in Figure 5.14. P is the pinch point. From this figure, the minimum consumption of fresh hydrogen, represented by AH, is 200 mol/s. The waste is represented by the tail gas line located between the horizontal difference of points C and D, whose flow rate is 155 mol/s and purity is 70%.

When purification reuse is applied, the integration of the hydrogen network will be affected. This will be analyzed by the proposed graphical method with four different purified purities, as shown in Table 5.2.

When the purified product purity is the same as the purity of fresh hydrogen, 95%, and the tail gas purity is 32%, the pinch diagram with purification reuse is shown in Figure 5.15. The purification triangle is DEP, the purification feed

Table 5.1 Sources and Sinks Data of a Hydrogen Network as Case One

Streams	Hydrogen Concentration (mol%)	Flow Rate (mol/s)	Hydrogen Load (mol/s)
Hydrogen Sources			
Fresh hydrogen	95	400	380
SR1	93	623.8	580.13
SR2	80	587.1	469.68
SR3	75	1940.5	1455.38
SR4	73	346.5	252.95
SR5	70	457.4	320.18
Hydrogen Sinks			
SK1	81	2495.0	2011.22
SK2	79	180.2	142.09
SK3	78	554.4	430.05
SK4	75	720.7	541.53
SK5	30	50	15.00

Table 5.2 Different Purified Product Options and Fresh Hydrogen Reductions

Feed Concentration (%)	Purified Product Purity (%)	Tail Gas Purity (%)	Hydrogen Recovery (%)	Minimum Fresh Hydrogen (mol/s)	Fresh Hydrogen Reduction (mol/s)
None	None	None	0	200	0
70	95	32	81.9	76.3	123.7
70	99	37	74.8	84	116
70	92	24	88.9	71.4	128.6

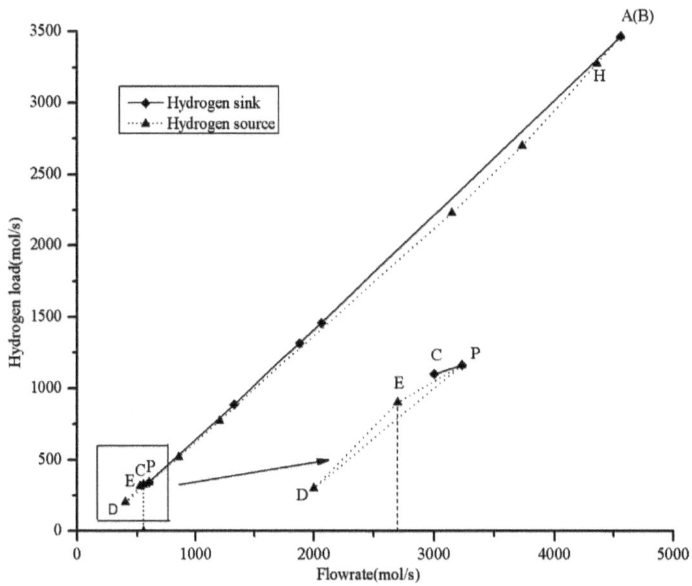

FIGURE 5.15

The pinch diagram of the hydrogen network with purification reuse ($y_{pur} = 0.95$, $y_r = 0.32$).

(represented by PD) is 205 mol/s with a purity of 70%, and the recovery ratio R is 81.9%. The minimum fresh hydrogen flow rate is 76.3 mol/s and its reduction is 123.7 mol/s (represented by DE). The waste is decreased from 155 mol/s with a purity 70% to 31.3 mol/s (the difference of E and C in the horizontal direction) with a purity 32%.

When the purified concentration is taken as 99%, which is higher than that of fresh hydrogen, and the tail gas concentration is 37%, the corresponding composite curves and the purification polygon is shown in Figure 5.16. The input of the purification, represented by PD in Figure 5.16, is still 205 mol/s, but R is decreased

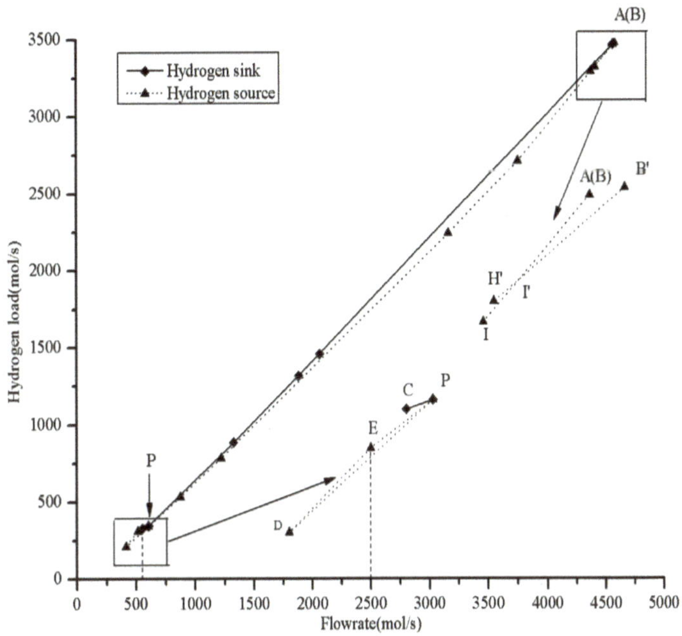

FIGURE 5.16

The pinch diagram of the hydrogen network with purification reuse ($y_{pur} = 0.99$, $y_r = 0.37$).

to 74.8%. The reason is that for purifiers, the higher the purified concentration is, the smaller the recovery ratio R will be. In this case, the flow rate of the purified stream is 100 mol/s, the minimum fresh hydrogen consumption is 84 mol/s, and its reduction, represented by B′I′, is 116 mol/s. With the purification reuse, the flow rate of the waste (the difference of E and C in horizontal direction in Figure 5.16), changes from 155 mol/s with a hydrogen purity of 70% to 39 mol/s with a hydrogen concentration of 37%.

Figure 5.17 shows the situation when the purified concentration is taken as 92% and the tail gas concentration is 24%. The result shows that the input of the purification, represented by DF, is 216.1 mol/s with a purity of 70%. Point C is the pinch point, lower than the original pinch point P. R is 88.9%, higher than the above cases. The reason is that lower purified concentration causes higher R for purifiers. From Figure 5.17, the flow rate of the purified product is 146.2 mol/s, and the minimum fresh hydrogen consumption is 71.4 mol/s, whose reduction, represented by HH′, is 128.6 mol/s. The flow rate and hydrogen concentration of the waste, represented by EC in Figure 5.17, changes to 26.4 mol/s with a purity 24%.

From this case study, it can be seen that, with purification reuse, both the fresh hydrogen consumption and waste discharge can be reduced. And the waste discharge is decreased in both flow rate and purity.

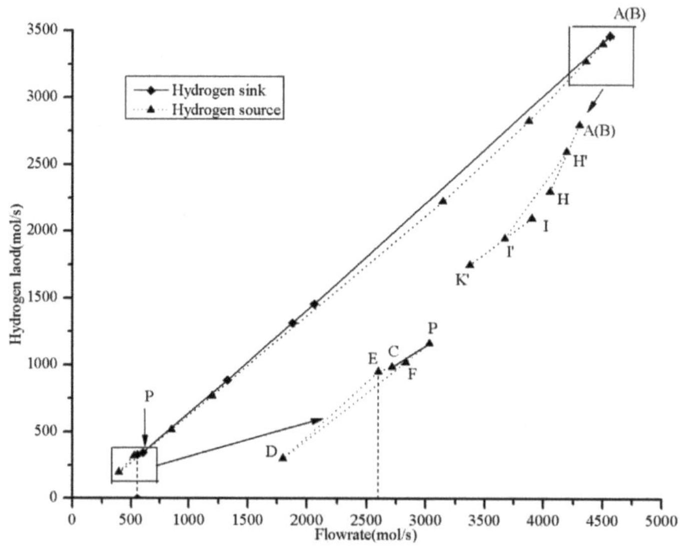

FIGURE 5.17

The pinch diagram of the hydrogen network with purification reuse ($y_{pur} = 0.92$, $y_r = 0.24$).

Table 5.2 summarizes the results of all cases of the hydrogen network in this case study, including purification reuse or not. After purification reuse, the fresh hydrogen consumption can be reduced, and furthermore, different purified product purities lead to different hydrogen reduction.

In Table 5.2, it seems that lower purified product purity can save more fresh hydrogen. For any purifier, lower product purity means higher hydrogen recovery, indicating more hydrogen load in the purified product. Thus, more hydrogen can be reused as high-quality product, and then the fresh hydrogen can be saved more. However, product purity cannot be of any value because of technical specification of purifiers, as well as other network constraints and considerations. Consequently, the purified product purity lies between a minimum value and fresh hydrogen purity $[y_{min}, y_U]$. In summary, the product purity should be optimized within its feasible region to save fresh hydrogen as much as possible.

5.6 GRAPHICAL METHOD FOR THE OPTIMIZATION OF PURIFICATION REUSE (ZHANG ET AL., 2014)

From the conclusion in Table 5.2, to maximize the fresh hydrogen reduction, the purity of purified product should be optimized within its feasible region. In this section, a graphical method optimizing the purification performance is presented to get the maximum fresh hydrogen conservation.

5.6.1 MAXIMUM HYDROGEN SURPLUS

Fundamentally, a feasible hydrogen network should satisfy the following material constraints:

$$\sum_{i=1}^{n_{SR}} F_{SR,i} \geq \sum_{j=1}^{n_{SK}} F_{SK,j} \tag{5.7}$$

$$\sum_{i=1}^{n_{SR}} F_{SR,i} y_{SR,i} \geq \sum_{j=1}^{n_{SK}} F_{SK,j} y_{SK,j} \tag{5.8}$$

where $F_{SR,i}$ (including fresh hydrogen) and $F_{SK,j}$ denote the flow rates of source i and sink j, respectively; $y_{SR,i}$ and $y_{SK,j}$ are the purity of source i (including fresh hydrogen) and sink j, respectively; and n_{SR} and n_{SK} represent the numbers of sources and sinks, respectively.

The material constraints given by Eqns (5.7) and (5.8) mean that both flow rate and hydrogen load supplied from all sources should not be less than the demand of all sinks. In a hydrogen network without purification reuse, there is a net hydrogen sink above the pinch because of fresh hydrogen consumption while below the pinch there is a net source, which usually has excess hydrogen. When purifiers are applied, the excess hydrogen below the pinch can be upgraded and then reused above the pinch point. Hence, both the excess hydrogen below the pinch and the demand of fresh hydrogen above the pinch will be decreased. Note that the upper limit of the source reuse is reached when the excess hydrogen below the pinch equals zero after purification reuse. Consequently, to minimize the fresh hydrogen consumption via purification reuse, it is crucial to identify the maximum excess hydrogen below the pinch.

For each sink, the ideal match status is that the hydrogen flow rate and purity supplied by sources are identical to its demands, and such status is named as "sharp match." According to material balance, the excess hydrogen reaches the maximum at sharp match status, and it is defined as the maximum hydrogen surplus (MHS), which can be deduced from Eqns (5.9) and (5.10).

$$F_S = \sum_{i=1}^{n_{SR-1}} F_{SR,i} + F_U - \sum_{j=1}^{n_{SK}} F_{SK,j} \tag{5.9}$$

$$F_S y_S = \sum_{i=1}^{n_{SR-1}} F_{SR,i} y_{SR,i} + F_U y_U - \sum_{j=1}^{n_{SK}} F_{SK,j} y_{SK,j} \tag{5.10}$$

where F_U is the flow rate of fresh hydrogen and F_S and y_S are the flow rate and the purity of MHS, respectively.

The material balance described by Eqns (5.9) and (5.10) can also be presented in the hydrogen load versus flow rate diagram, as shown in Figure 5.18(a) and (b). In Figure 5.18(a), AEPD is the source composite curve, APB is the sink composite curve, and AE is the minimum fresh hydrogen without purification. P is the pinch point and CD corresponds to the waste. In Eqns (5.9) and (5.10), $\sum_{j=1}^{n_{SK}} F_{SK,j}$ and

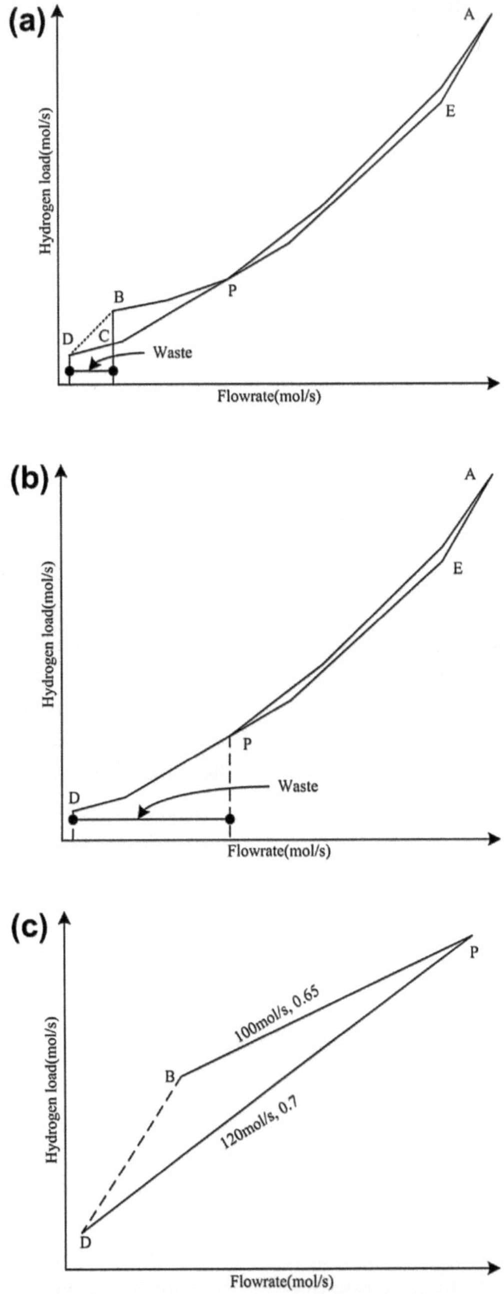

FIGURE 5.18

Maximum hydrogen surplus of a hydrogen network (a) MHS is pseudo source(s); (b) MHS is real source(s); (c) an example for determination of y_S.

$\sum_{j=1}^{n_{SK}} F_{SK,j}y_{SK,j}$ represent the horizontal and vertical distance between A and B, while $\sum_{i=1}^{n_{SR}-1} F_{SR,i}y_{SR,i} + F_U y_U$ and $\sum_{i=1}^{n_{SR}-1} F_{SR,i} + F_U$ are the horizontal and vertical distances between A and D, respectively. Thus, the connection of the two end points of the sink and source composite curves (B and D) represents the MHS, because the horizontal and vertical differences between these two points are F_S and $F_S y_S$ (hydrogen load of the MHS), respectively. Similarly, in Figure 5.18(b), AEPD is the source composite curve, AP is the sink composite curve, AE is the minimum fresh hydrogen consumption of the hydrogen network without purification, P is the pinch point, and PD represents the waste. In this network, PD can be used to represent the MHS. Thus, the MHS can be taken as the "source" connecting the lower ends of the two composite curves.

Comparing the wastes with the corresponding MHS in Figure 5.18, it can be seen that the waste is always identical to MHS in flow rate but not always the same in purity. Furthermore, the hydrogen purity (y_S) of MHS is not lower than that of the waste.

The MHS can be a real hydrogen source or source mixture, and also possibly a pseudostream. As shown in Figure 5.18(b), PD is the MHS of this network. The average hydrogen purity of this section is y_S, which lies between [0, y_p]. However, as shown in Figure 5.18(a), BD is the MHS, and its purity (y_S) may be larger than y_p or even larger than y_U. The reason is that it may be impossible to achieve the sharp matches between the sink and source streams below the pinch point even after purification reuse. The definition of MHS excludes this case and thus makes the hydrogen load surplus become greater than its achievable level. For example, in Figure 5.18(c), P is the pinch and below it, there is only one sink (100 mol/s, hydrogen purity is 65%) represented by PB and only one source represented by PD (120 mol/s, hydrogen purity is 70%). Without any other processing, i.e., purification, the demand of sink PB can only be satisfied by part of source PD (100 mol/s), and the flow rate and purity of the waste are 20 mol/s and 70%, respectively. Hence, it is impossible to achieve the sharp match between the source and the sink. However, according to the definition of MHS, F_S is 20 mol/s, hydrogen load surplus is 19 mol/s, so y_S is 95%, which is larger than 70%.

In summary, MHS cannot be described by waste, and it corresponds to the maximum possible hydrogen surplus that could be reused. Therefore, it is more reasonable to use MHS rather than the waste.

5.6.2 MASS TRANSFER TRIANGLE

The material constraints defined by Eqns (5.9) and (5.10) are the feasible premises of hydrogen networks no matter whether purification reuse is included. Purifiers can change the distribution of hydrogen sources and thus increase their reuse to reduce the fresh hydrogen consumption and waste discharge. Therefore, F_U, F_S, and y_S will be different values with purification reuse. Assuming F_{Ub}, F_{Sb}, y_{Sb} and F_{Ua}, F_{Sa}, y_{Sa} are the flow rate of fresh hydrogen, the flow rate of maximum hydrogen surplus and the hydrogen purity before and after purification reuse, respectively. It is obvious that F_U, F_S, and the hydrogen load of MHS will decrease after purification, i.e., $F_{Ub} > F_{Ua}$, $F_{Sb} > F_{Sa}$ and $F_{Sb}y_{Sb} > F_{Sa}y_{Sa}$.

As a result, Eqns (5.9) and (5.10) can be rearranged as follows.

$$\sum_{i=1}^{n_{SRb-1}} F_{SRb,i} + F_{Ub} - \sum_{j=1}^{n_{SK}} F_{SK,j} = F_{Sb} \tag{5.11}$$

$$\sum_{i=1}^{n_{SRb-1}} F_{SRb,i}y_{SRb,i} + F_{Ub}y_U - \sum_{j=1}^{n_{SK}} F_{SK,j}y_{SK,j} = F_{Sb}y_{Sb} \tag{5.12}$$

$$\sum_{i=1}^{n_{SRa-1}} F_{SRa,i} + F_{Ua} - \sum_{j=1}^{n_{SK}} F_{SK,j} = F_{Sa} \tag{5.13}$$

$$\sum_{i=1}^{n_{SRa-1}} F_{SRa,i}y_{SRa,i} + F_{Ua}y_U - \sum_{j=1}^{n_{SK}} F_{SK,j}y_{SK,j} = F_{Sa}y_{Sa} \tag{5.14}$$

Although purification can change the distribution of sources except fresh hydrogen, the flow rate and hydrogen load of these hydrogen sources are the same before and after purification reuse.

$$\sum_{i=1}^{n_{SRb-1}} F_{SRb,i} = \sum_{i=1}^{n_{SRa-1}} F_{SRa,i} \tag{5.15}$$

$$\sum_{i=1}^{n_{SRb-1}} F_{SRb,i}y_{SRb,i} = \sum_{i=1}^{n_{SRa-1}} F_{SRa,i}y_{SRa,i} \tag{5.16}$$

With Eqns (5.15) and (5.16), the difference of Eqns (5.11) and (5.13) as well as that of Eqns (5.12) and (5.14) show the following results:

$$F_{Ub} - F_{Ua} = F_{Sb} - F_{Sa} \tag{5.17}$$

$$(F_{Ub} - F_{Ua})y_U = F_{Sb}y_{Sb} - F_{Sa}y_{Sa} \tag{5.18}$$

The left side of Eqn (5.17) represents the flow rate reduction of fresh hydrogen consumption, while the right side is the flow rate reduction of MHS. Meanwhile, the left side of Eqn (5.18) is the hydrogen load reduction of fresh hydrogen, while the right side is the hydrogen load reduction of MHS. Therefore, the decrease of MHS equals the fresh hydrogen reduction. Purification can thus be viewed as a "transformation" process, converting hydrogen from MHS to fresh hydrogen. This process can also be visualized in the hydrogen load versus flow rate diagram, as shown in Figure 5.19.

In Figure 5.19(a), LT is the MHS of a hydrogen network without purification. To illustrate the transformation process, a line from L with fresh hydrogen purity as slope and a horizontal line from T are drawn, and the two lines intersect at point P, forming a mass transfer triangle (MTT). In this triangle, the horizontal line PT represents the limiting case, in which the hydrogen purity of the tail gas is zero.

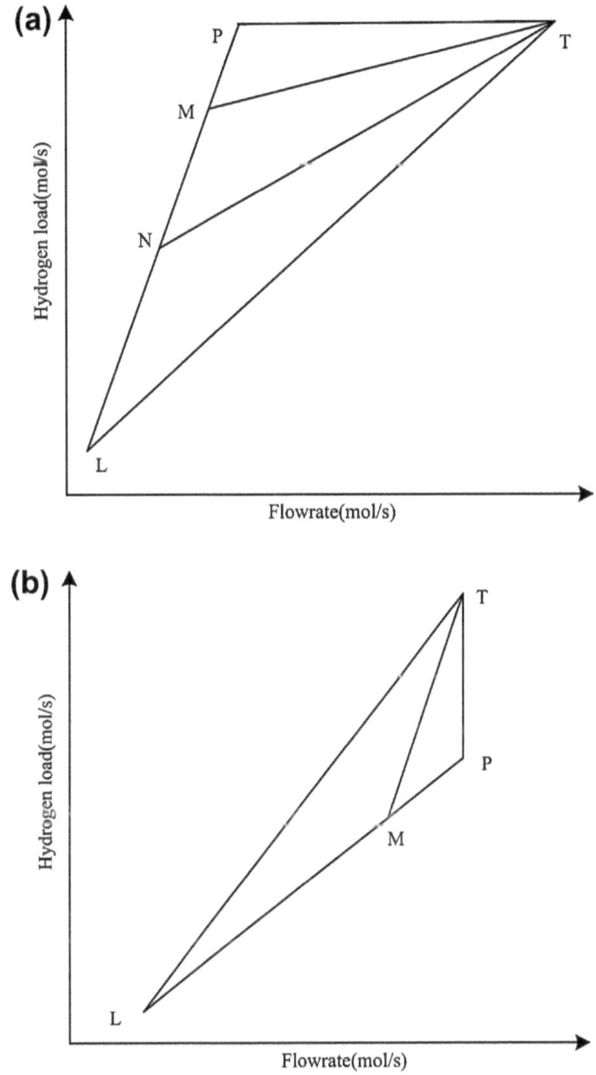

FIGURE 5.19

Mass transfer triangle (a) $y_S < y_U$; (b) $y_S > y_U$.

Correspondingly, the line LP, whose slope equals fresh hydrogen purity, is the maximum fresh hydrogen reduction with hydrogen load in the MHS recovered completely by purification reuse. Totally, LT and PT are the two limiting cases of MHS with 0% and 100% reuse of hydrogen load in the MHS. As a result, the practical reuse lies between them. The real reduction of fresh hydrogen is generally less than LP, such as LM or LN in Figure 5.19(a), and the corresponding MHS after reuse is represented by TM or TN. Since limiting cases can conveniently be described in the triangle, such graphical "transformation" process is termed as "mass transfer triangle." In the MTT, the reduction of the MHS (LM or LN) is identical to that of the fresh hydrogen. This is in accordance with the mass balance described by Eqns (5.15) and (5.16).

In the case shown in Figure 5.19(a), the hydrogen purity of the MHS before purification (y_S) is less than fresh hydrogen purity, i.e., $y_S < y_U$. However, y_S may be larger than y_U. In this case, similar to Figure 5.19(a), the appropriate MTT is LPT in Figure 5.19(b). LT is the MHS of a hydrogen network without purification and LP is still a line with purity of fresh hydrogen, but its maximum flow rate is F_S (as F_S is the flow rate of MHS). This indicates 100% recovery of the flow rate in MHS, and LP is the limiting case for fresh hydrogen reduction.

Note that the MHS is the lower end connection of the two composite curves. After the MHS is changed from LT to MT or NT, M or N is the lower end of the source composite curve after purification reuse. For different purification cases, point M and N will change along LP. In other words, LP is the moving track of the lower end of the source composite curve when performing purification reuse.

Note that the MTT is not the real purification process. However, the MTT describes the mass transfer process from material balance consideration. Whatever the purification performance is, the fresh hydrogen reduction should be within the MTT and the attainable reduction can be determined by the MTT in the targeting process. In other words, the MTT is a guide for analyzing practical purification reuse and can thus be used to optimize the practical purification process, as will be discussed.

5.6.3 THE GRAPHICAL APPROACH

To optimize purification reuse, both the feasible operating ranges of the purifier and the feasibility of the network should be satisfied. The feasible operating ranges of purifiers mean that the separation performance can be realized by a selected purifier. As for the feasibility of the network, it is indicated by the location of the composite curves, i.e., the sink composite curve should lie above the source composite curve in the hydrogen load versus flow rate diagram.

The optimal purification performance should result in the maximum fresh hydrogen reduction. However, the maximum fresh hydrogen reduction target and corresponding optimal purification performance cannot be simultaneously identified simply by shifting the two composite curves. Their determination should be performed separately by a trial and error method.

The strategy of this work is to set a fresh hydrogen reduction first, and then search for corresponding purification performance within feasible operating ranges of the selected purifier and feasibility of the hydrogen network. For a given initial fresh hydrogen reduction, if the feasible purification performance can be found, this initial reduction is feasible but normally not the maximum value. Thus, a larger fresh hydrogen reduction should be selected to perform the same process. For every larger one, if the feasible purification performance can be found, then the same targeting procedure is repeated. This trial and error procedure continues until the feasible purification performance cannot be found for the selected purifier within feasible ranges of selected purifier and feasibility of the network. Finally, the largest reduction with feasible purification performance and network is the solution.

Conversely, it is also possible that the feasible purification performance cannot be found for the initial fresh hydrogen reduction. In this case, the initial reduction is infeasible and should be reduced. Hence, a smaller one, rather than a larger one, should be used to repeat the targeting process until the feasible purification performance is found.

Furthermore, for feasible purification performance, both the flow rate and purity of relative streams should be optimized, so as to serve the target of maximizing fresh hydrogen reduction. The steps are described in detail below. To illustrate the whole method clearly, the hydrogen system shown in Figure 5.24 is taken as an illustrative example.

5.6.3.1 Setting the initial fresh hydrogen reduction

According to the concept of MTT, the initial fresh hydrogen reduction can be chosen as a half of the maximum fresh hydrogen reduction. For the case shown in Figure 5.20, APC is the sink composite curve, ABHD is the source composite curve, AB is the fresh hydrogen consumption without purification, and P is the pinch. First,

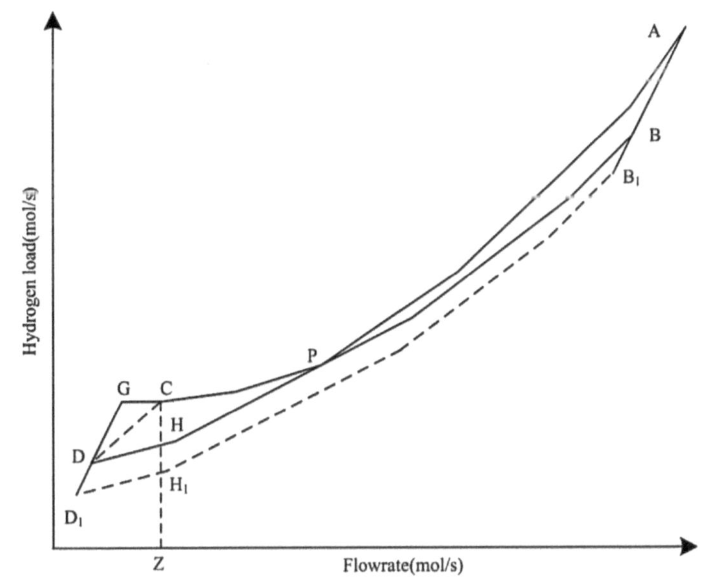

FIGURE 5.20

Determination of MHS and MTT and initial flow rate set.

the MHS is obtained by connecting the two lower ends C and D. Line DG with slope equaling the hydrogen purity of fresh hydrogen is drawn from point D and CDG is the MTT. Here DG is the maximum fresh hydrogen reduction, and the initial fresh hydrogen reduction is taken as (1/2)·DG.

5.6.3.2 Identifying the initial purified product flow rate

For resource conservation networks, there is a common conclusion that the concentration of the purified stream is generally lower than the concentration of the fresh resource. Thus, the purity of the purified product should be lower than the purity of fresh hydrogen. Hence, the flow rate of the purified product must be larger than the fresh hydrogen flow rate that could be saved. Otherwise, the hydrogen network may be infeasible. For example, the two composite curves shown in Figure 5.20 overlap each other at the pinch point P. So for sources and sinks, the total hydrogen load within top point A and pinch P of the sinks and sources are the same. This means that sources can just satisfy sinks within these two points without any excess hydrogen. After the purification reuse is applied, the purified product lies above the pinch, and the tail gas of purification lies below the pinch. Since the purity of the purified product is lower than that of the fresh hydrogen, the hydrogen load supplied by the sources within A and P will not be sufficient to satisfy the sinks when the flow rate of the purified product is the same or even lower than that of the saved fresh hydrogen. Consequently, to keep the network feasible and avoid increasing the fresh hydrogen consumption, the flow rate of purified product should be larger than the fresh hydrogen flow rate reduction when the purified product purity is less than the purity of fresh hydrogen.

Based on the above observation, the section of the source composite curve between B and D should be shifted downward to B_1 and D_1, with the extension selected as a half of the maximum reduction DG, i.e., $D_1D = BB_1 = (1/2)\cdot DG$, as shown in Figure 5.20. After such an extension, the flow rate of the product is the sum of the flow rate of fresh hydrogen reduction and the flow rate indicated by DD_1. As a result, the initial product flow rate is two times the flow rate of fresh hydrogen reduction, and can ensure a safe and conservative product flow rate. The reason for such extension will be explained in the next section.

5.6.3.3 Flow rate and purity optimization

Since the product purity is generally not the same as the purity of fresh hydrogen but has lower value, the product purity as well as flow rates of the feed and product should be optimized. The aim is to minimize the fresh hydrogen consumption of the hydrogen network. The procedure is as follows.

Based on the initial flow rate set above, first the purified product purity is specified as the fresh hydrogen purity y_U, as shown in Figure 5.21, and the fresh hydrogen reduction is chosen as (1/2)·DG, which is TD in Figure 5.21. Meanwhile, as stated above, the initial flow rate of the purified product TD1, the sum of DD_1 and DT, is the PPL. Thus, identify a part UB_1 equaling TD_1 on AB_1, then UB_1 is the APPL, as shown in Figure 5.21. Next, connect points T and an intersection point on the sink

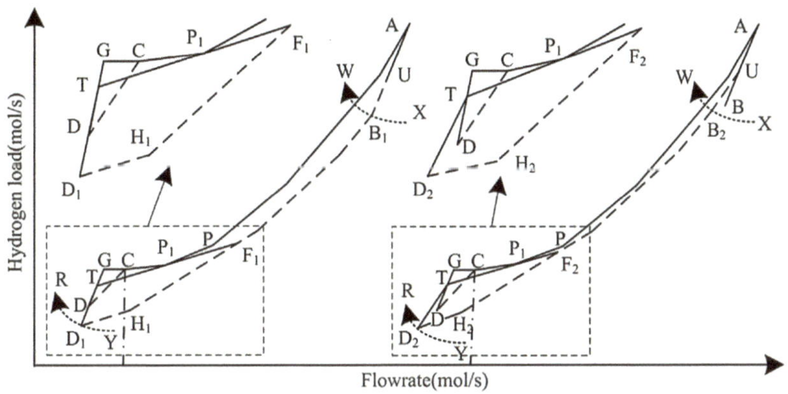

FIGURE 5.21

The feasible case after flow rate and purity optimization ($y_S < y_U$).

composite curve, ensure that the connection line only touches the sink composite curve at this point (P_1), and then prolong the connection to intersect source composite curve at point F_1. This leads to a polygon $TD_1H_1F_1$, which touches the sink composite curve at point P_1. In such case, if the purification performance determined by polygon $TD_1H_1F_1$ is feasible for the selected purifier, then the initial reduction is feasible. Otherwise, if the purification performance is infeasible, then the initial reduction is infeasible. This attempt to determine the feasible purification performance for the initial fresh hydrogen reduction is termed the first step. If a feasible purification performance for the initial reduction exists, then the next reduction for evaluation should be a larger one, otherwise it should be a smaller one.

For each larger one, this method still introduces half of the unreached reduction: the next reduction is the sum of the previous one and half of the distance between TG. Consequently, in the second step, the purified product flow rate is increased by $(1/2)\cdot(\text{DG} - (1/2)\cdot\text{DG}) = (1/4)\cdot\text{DG}$, and in the third step, it is increased by $(1/8)\cdot\text{DG}$, etc. Thus, the fresh hydrogen reduction in the nth step is increased by $(1/2)^n\cdot\text{DG}$ and thus,

$$FHR^n = FHR^{n-1} + (1/2)^n\cdot \text{DG} \tag{5.19}$$

where FHR^n represents the fresh hydrogen reduction in the nth step. From Eqn (5.19), it can be seen that the reduction gradually approaches its maximum value. During the process of increasing fresh hydrogen reduction, since the purity of the product is that of the fresh hydrogen, the flow rate of the purified product will increase, while the recovery will increase gradually according to Eqn (5.19). Hence, if at the nth step, the recovery of the purification performance determined by the formed polygon is not feasible for the selected purifier, then the flow rate should be the current value, while the purified product purity cannot be y_U but will take on a lower value so as to decrease the recovery. In such case, the purified product purity should be decreased and the purity optimization should then restart.

Conversely, for the case of selecting a smaller reduction, the reduction in the *n*th step is decreased by $(1/2)^n \cdot \mathrm{DG}$ and thus the fresh hydrogen reduction in such case is,

$$FHR^n = FHR^{n-1} + (1/2)^n \cdot \mathrm{DG} \tag{5.20}$$

For the case of decreasing fresh hydrogen reduction, the flow rate of the purified product will decrease, while the recovery will decrease gradually according to Eqn (5.20). Hence, if at the *n*th step, the recovery of the purification performance determined by the formed polygon is feasible for the selected purifier, then the reduction flow rate is feasible. Consequently, the purity optimization can be performed next.

In the above step, the purified product flow rate is optimized. To identify the maximum hydrogen utility reduction, the purified product purity optimization can be performed afterward. The optimization strategy is illustrated with Figure 5.21 as example and is given below.

To optimize the purified product purity, the APPL and PPL should be rotated to change their slopes. In Figure 5.21, D_1T is the PPL and UM_1 is the APPL. Since APPL and PPL are always identical to each other, when the product purity decreases, the APPL and PPL should be changed simultaneously. This can be done by introducing an assistant line, as shown in Figure 5.21. First, draw two arcs RY and WX with T and U as circle centers and the length of D_1T as radius. Next, keep circle centers T and U stationary and rotate D_1T and UM_1 along RY and WX simultaneously to decrease the purified product purity. During the rotating process, wherever D_1T and UM_1 are, the source composite curve between D_1T and UM_1 is shifted along horizontal and vertical directions accordingly to maintain their original intersection points, but without any rotation.

After rotating PPL and APPL, the purified product purity is decreased and thus the recovery is decreased, converting the infeasible purification performance to a feasible one. If the result of the *n*th step is still feasible after rotation, meaning that the purity optimization can fulfill the current reduction, then the fresh hydrogen reduction should still be increased by Eqn (5.19), and the (n+1)th step should be performed. Only when the following situation is reached can the optimization stop: the *k*th step has feasible purification performance but at the (*k*+1)th step, the feasible purification performance and the feasibility of the network cannot be satisfied simultaneously. For example, before feasible purification performance appears, the two composite curves touch each other at another point, i.e., P_2, as shown in Figure 5.22. Due to the feasibility of the hydrogen network stated above, the APPL and PPL cannot be rotated anymore, and then the purity cannot be lowered further. Finally, the result of the (*k*+1)th step is infeasible and the previous one, the *k*th step is the final solution, as shown in Figure 5.21.

From the targeting process, the flow rate and purity of the purified product are both optimized, and the final fresh hydrogen target is the attainable minimum. After considering purification reuse, the purification performance is described by the polygon $TF_2H_2D_2$, in which the feed is $F_2H_2D_2$, the purified product is TD_2, and the tail gas is TF_2. The source composite curve is modified from ABHD in Figure 5.20 to $AUB_2F_2H_2D_2$ in Figure 5.21. Since the lower end of the source composite curve

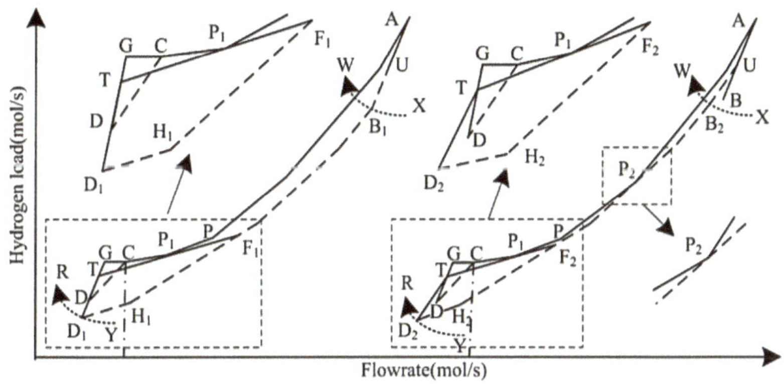

FIGURE 5.22

Infeasible case ($y_S < y_U$).

is nearest to its top position, the fresh hydrogen reduction is the attainable maximum and the MHS is also decreased to the minimum value. The final reduction DT shows the guiding function of MTT. In addition, only the case with y_S lower than y_U has been discussed above. When y_S is higher than y_U, the targeting process is the same as that shown in Figures 5.21 and 5.22.

The results reported here are significantly influenced by purification performance. The higher the maximum recovery is, the larger fresh hydrogen reduction will be, and the smaller the MHS and the waste will be. When the recovery for a given feed and product purity is at the technically maximum level, then the fresh hydrogen consumption is the attainable minimum. In other words, the available fresh hydrogen reduction depends on the selected purifier.

In summary, the targeting procedure for the proposed trial and error method is shown in Figure 5.23. In the targeting process, after the prior two steps of preparation, step 3 is involves flow rate optimization which can identify a suitable fresh hydrogen reduction, and the purity optimization is followed to fulfill the reduction. Every time when the reduction is feasible after purity optimization, the flow rate optimization continues to increase the reduction until it reaches the upper limit under feasibility of operating ranges and the network. Thus, the optimization process is performed in the order flow rate—purity—flow rate—purity, and so on, repeatedly, until maximum reduction is found. From Figure 5.23, it can be seen that the targeting procedure is performed in a sequential way, so it is easy to operate.

The targeting process is easy to implement by software, such as Matlab and Origin Pro 8.0. First, use the given data to draw source and sink composite curves, and then identify the pinch by changing the coordinate data of the source composite curve. Next, the APPL, the PPL, and purification polygon are constructed by specifying coordinates data for lines at certain location. As for the targeting process, the trial and error can also be performed by changing coordinates data of relative lines, until the fresh hydrogen reduction reaches the maximum.

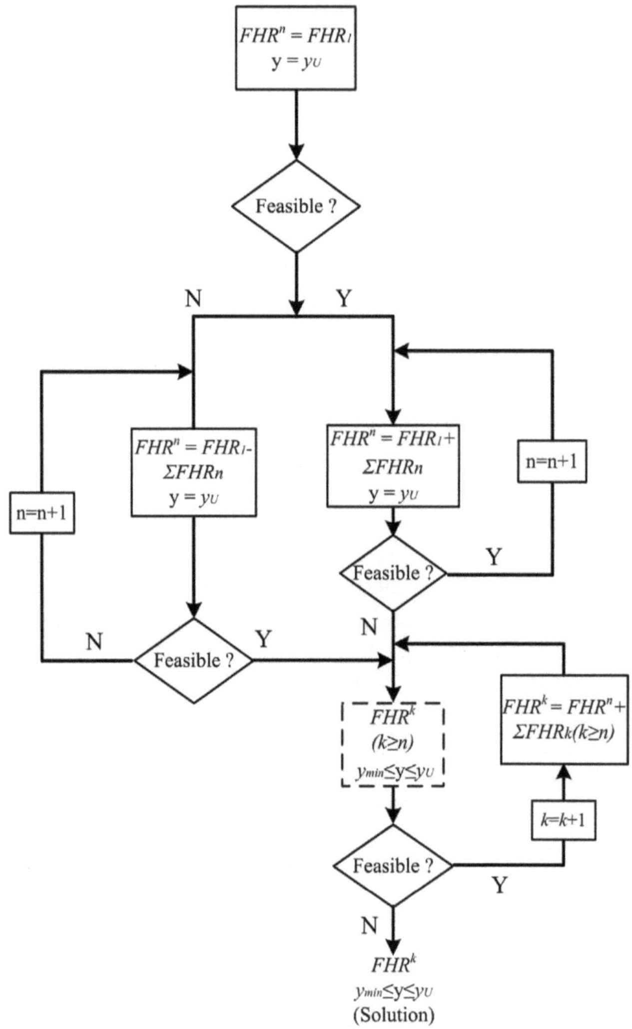

FIGURE 5.23

Graphical targeting procedure.

5.7 CASE STUDY FOR GRAPHICAL METHOD WITH OPTIMIZED PURIFICATION PERFORMANCE (ZHANG ET AL., 2014)

The stream data for all sources and sinks of this case study are listed in Table 5.3. There are six sources and six sinks. The fresh hydrogen has a purity of 99%. Without purification, the pinch point is the intersection point of SK4 and SK5 on the sink

Table 5.3 Sources and Sinks Data of a Hydrogen Network as Case Two

Streams	Hydrogen Concentration (mol%)	Flow Rate (mol/s)	Hydrogen Load (mol/s)
Hydrogen Sources			
Fresh hydrogen	99	180	178.2
SC1	93	546.13	507.9
SC2	80	744.45	595.56
SC3	75	1940.5	1455.38
SC4	73	346.5	252.95
SC5	70	473.47	331.43
Hydrogen Sinks			
SK1	81	2495.0	2011.22
SK2	79	180.2	142.09
SK3	78	554.4	430.05
SK4	75	720.7	541.53
SK5	30	40	12.00
SK6	20	20	4.00

composite curve while it lies on the last source (SC6) of the source composite curve. The minimum fresh hydrogen consumption without purification is 180 mol/s and the waste flow rate and purity are 220.7 mol/s and 70%, respectively. The corresponding hydrogen network is shown in Figure 5.24.

When there is no purification in the hydrogen network, the MHS is given by DC and its flow rate and purity are 220.7 mol/s and 81.8%. The maximum fresh hydrogen reduction is 182.4 mol/s, as shown by DG in Figure 5.25. In this case, the MHS is not the same as waste, and the triangle DCG is the MTT, as shown in Figure 5.25. Similarly, the purity of MHS in this case is also lower than that of fresh hydrogen (99%), and thus the targeting process is essentially the same as that in Figure 5.21.

Here the same pressure swing adsorption (PSA) unit is still selected as the purifier. According to the targeting procedure shown by Figure 5.23, the initial fresh hydrogen reduction is a half of the maximum reduction (182.4 mol/s) identified above: 91.2 mol/s, and its purity is the fresh hydrogen purity: 99%, which is the same as fresh hydrogen purity. At first, the feasible purification performance could be found, and the fresh hydrogen reduction could be increased by Eqn (5.19) in the next step. After flow rate optimization, the purified product purity (99%), which is higher than the minimum value (95%), can be optimized to get further fresh hydrogen reduction. After flow rate and purity optimization, the maximum fresh hydrogen reduction is 168.8 mol/s and the purified product purity is 95% after seven iterations. The resulting separation performance for the selected PSA purifier is

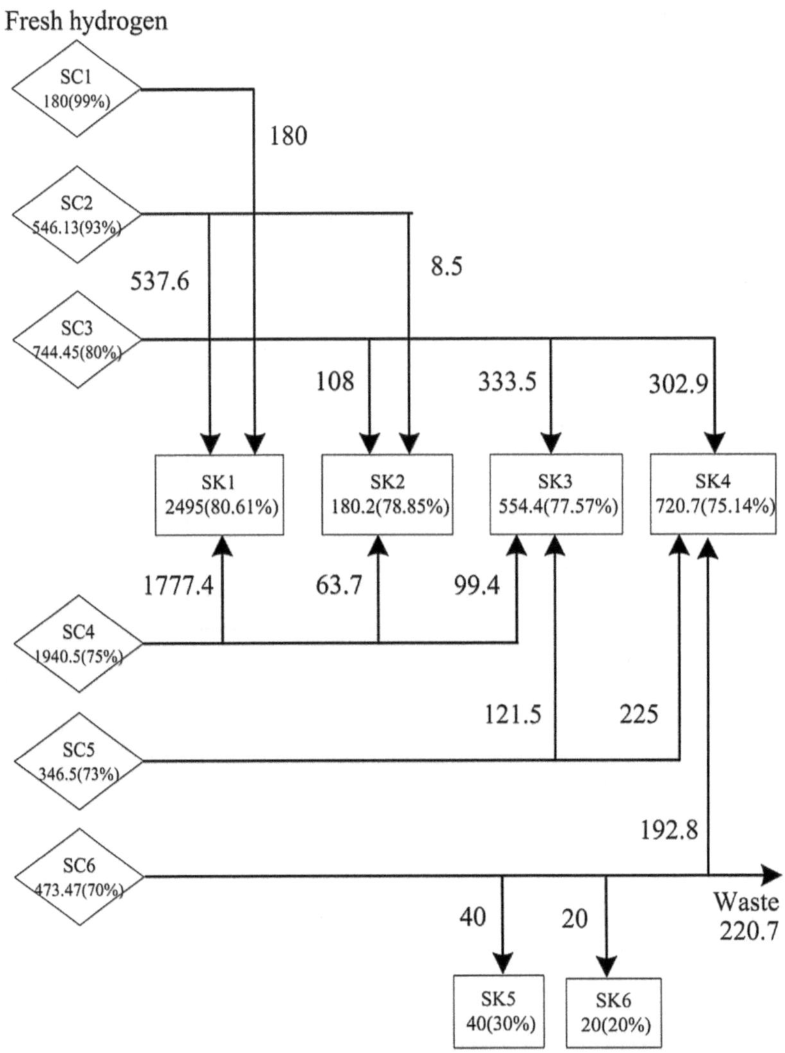

FIGURE 5.24

The current hydrogen network of Case 2 (the numbers in bold represent the flow rate (mol/s)).

represented by the purification triangle D′TF in Figure 5.25. In D′TF, the flow rate and purity of the feed stream (D′F) are 401.9 mol/s and 70%, those of the purified product (D′T) are 260 mol/s and 95%, and those of the tail gas (TF) are 141.9 mol/s and 24.2%, respectively. The purification parameters are the limiting values.

As shown in Figure 5.25, after purification reuse the minimum fresh hydrogen consumption (AU) is 11.2 mol/s, so the corresponding reduction (UM) of fresh

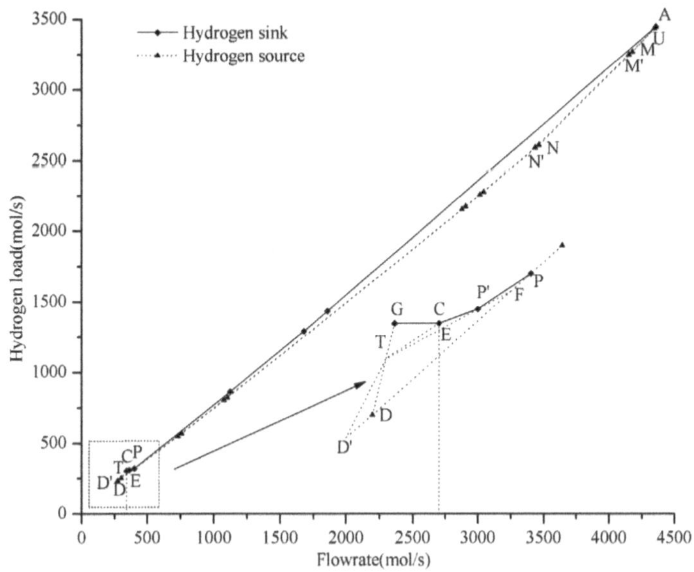

FIGURE 5.25

Optimization results for the case study.

hydrogen is 168.8 mol/s, which equates to 93.8% of the current consumption (180 mol/s); the flow rate and purity of the waste (ET) are 51.9 mol/s and 24.2%, respectively, and its reduction in flow rate is also 168.8 mol/s. It can be seen that, with the purification applied, the MHS is decreased from DC (220.7 mol/s, 81.8%) to CT (51.9 mol/s, 25.9%). The improved network is shown in Figure 5.26 with the difference from Figure 5.24 illustrated by dotted lines.

If other methods are used to analyze the case study, a different minimum fresh hydrogen consumption will be obtained. Since the method of Agrawal and Shenoy (2006) can only be used for systems with specified purified product purity and recovery, these two parameters are assigned, respectively, the values of 0.95% and 87.2%, which are used in this study. The result shows that the minimum fresh hydrogen consumption is 23.3 mol/s and the corresponding reduction is 156.7 mol/s. Similarly, the graphical method with fixed purification parameters can be used for systems with specified purified product purity and tail gas purity. Assigning 0.95 to the purified product purity and 0.32 to the tail gas purity, the result obtained shows that the minimum fresh hydrogen consumption is 5.4 mol/s, and the corresponding reduction is 162.6 mol/s.

Table 5.4 summarizes the optimization results obtained from different methods for case 2. The fresh hydrogen cost reduction is calculated according to the price of fresh hydrogen in the literature of Alves and Towler (2002). Note that the result obtained from the method of Alves and Towler (2002), which does not consider purification reuse, is also included in the table for comparison. It is clear that the

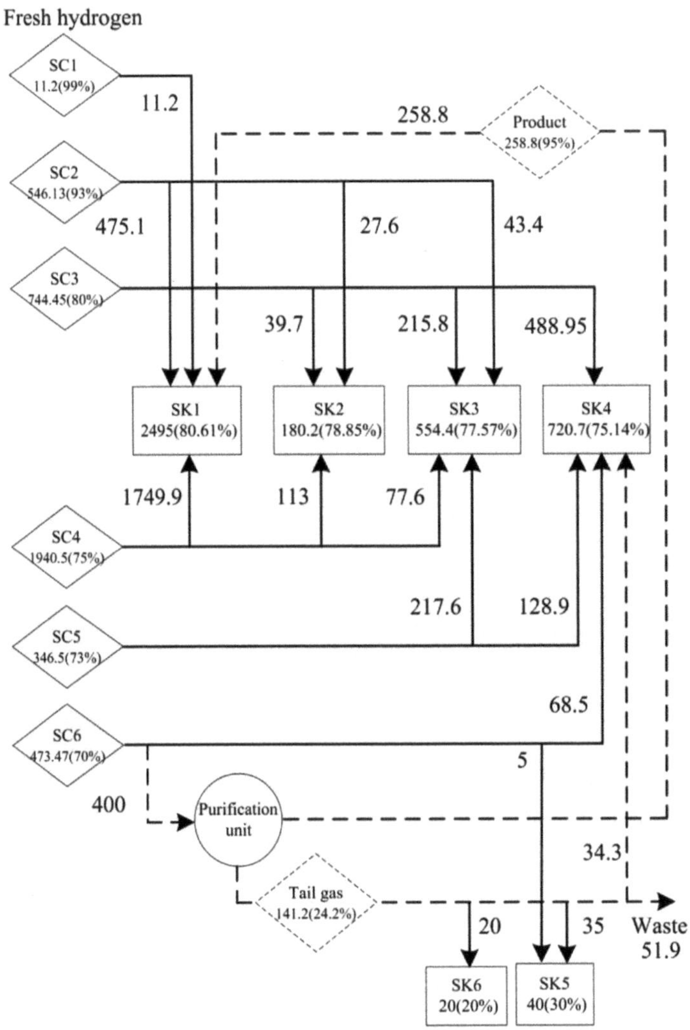

FIGURE 5.26

The improved hydrogen network of Case 2 (the numbers in bold represent the flow rate (mol/s)).

minimum fresh hydrogen identified by the methods of Agrawal and Shenoy (2006) and the graphical method with fixed purification parameters is more than the minimum fresh hydrogen identified by the method proposed here. The reason for these differences is that the optimization of purity and flow rate of all streams in systems with purification reuse is neglected in the former two methods. In summary, the proposed method of the present study offers significant improvements over the two previous methods.

Table 5.4 Comparison of Results for Case Study 2 Obtained From Different Methods

Method	Alves and Towler (2002)	Agrawal and Shenoy (2006)	Fixed Purification Performance	Optimized Purification Performance
Purification Reuse	No	PSA	PSA	PSA
Minimum fresh hydrogen (mol/s)	180	23.3	17.4	12
Reduction (mol/s)	0	156.7	162.6	168
Cost reduction (MM$/year)	0	9.1	9.45	9.76

5.8 CONCLUSIONS

This chapter aims to conserve fresh resource of refinery hydrogen networks through process integration. Triangle rule is employed to illustrate the mass balance of purification process. By incorporating triangle rule and hydrogen load versus flow rate diagram, the minimum fresh hydrogen consumption can be determined through shifting composite curve and purification polygon. For fixed purified product purity, the minimum fresh hydrogen demand and waste discharge with purification reuse can be determined when the flow rates of purified product and assistant purified product equal each other. To optimize both concentration and flow rate of the purified product, MHS and MTT are introduced. Together with shifting the composite curve, purified product, and assistant purified product in certain order, the minimum fresh hydrogen demand and waste discharge can be determined. The results show that the attainable minimum fresh hydrogen consumption after purification reuse depends on separation ability of the selected purifier.

REFERENCES

Agrawal, V., Shenoy, U.V., 2006. Unified conceptual approach to targeting and design of water and hydrogen networks. AIChE J. 52, 1071–1082.

Almutlaq, A.M., Kazantzi, V., El-Halwagi, M.M., 2005. An algebraic approach to targeting waste discharge and impure fresh usage via material recycle/reuse networks. Clean Technol. Environ. Policy 7, 294–305.

Alves, J.J., Towler, G.P., 2002. Analysis of refinery hydrogen distribution systems. Ind. Eng. Chem. Res. 41, 5759–5769.

Bandyopadhyay, S., 2006. Source composite curve for waste reduction. Chem. Eng. J. 125, 99–110.

Bollinger, W., Long, S., Metzger, T., 1984. Optimizing hydrocracker hydrogen. Chem. Eng. Prog. 80, 51–57.

El-Halwagi, M., Gabriel, F., Harell, D., 2003. Rigorous graphical targeting for resource conservation via material recycle/reuse networks. Ind. Eng. Chem. Res. 42, 4319–4328.

El-Halwagi, M.M., Spriggs, H.D., 1998. Solve design puzzles with mass integration. Chem. Eng. Prog. 94, 25–44.

Liu, G., Li, H., Feng, X., Deng, C., 2013. Novel method for targeting the optimal purification feed flow rate of hydrogen network with purification reuse/recycle. AIChE J. 59, 1964–1980.

Liu, G., Li, H., Feng, X., Deng, C., Chu, K.H., 2012. A conceptual method for targeting the maximum purification feed flow rate of hydrogen network. Chem. Eng. Sci. 88, 33–47.

Lou, J., Liao, Z., Jiang, B., Wang, J., Yang, Y., 2013. Pinch sliding approach for targeting hydrogen and water networks with different types of purifier. Ind. Eng. Chem. Res. 52, 8538–8549.

Towler, G.P., Mann, R., Serriere, A.J., Gabaude, C.M., 1996. Refinery hydrogen management: cost analysis of chemically-integrated facilities. Ind. Eng. Chem. Res. 35, 2378–2388.

Wan Alwi, S.R., Aripin, A., Manan, Z.A., 2009. A generic graphical approach for simultaneous targeting and design of a gas network. Resour. Conser. Recy. 53, 588–591.

Yang, M., Feng, X., Chu, K., Liu, G., 2014. Graphical method for integrating purification processes in hydrogen systems with constraints of flow rate and concentration. Ind. Eng. Chem. Res. 53, 3246–3256.

Zagoria, A., Huycke, R., 2003. Refinery hydrogen management: the big picture: clean fuels. Hydrocarb. Process. 82, 41–46.

Zhang, Q., Feng, X., Liu, G., Chu, K.H., 2011. A novel graphical method for the integration of hydrogen distribution systems with purification reuse. Chem. Eng. Sci. 66, 797–809.

Zhang, Q., Liu, G., Feng, X., Chu, K.H., Deng, C., 2014. Hydrogen networks synthesis considering separation performance of purifiers. Int. J. Hydrogen Energ. 39, 8357–8373.

Zhao, Z., Liu, G., Feng, X., 2006. New graphical method for the integration of hydrogen distribution systems. Ind. Eng. Chem. Res. 45, 6512–6517.

CHAPTER

6 A New Biochemical Fermentation Process Design Method Considering Bifurcations and Oscillations

Wang Hangzhou, Chen Bingzhen, Qiu Tong, Zhao Jinsong[1]

Department of Chemical Engineering, Tsinghua University, Beijing, China

[1] *Corresponding author: E-mail: jinsongzhao@tsinghua.edu.cn*

6.1 INTRODUCTION

Bioethanol is an important renewable and sustainable alternative fuel source (Abashar and Elnashaie, 2011). It can be directly used as fuel ethanol or blended with gasoline. Agricultural feed reactants containing sugars, starches, and cellulosic materials are suitable for bioethanol-producing fermentation processes. The yeast *Saccharomyces cerevisiae* (Sridhar, 2011; Paz Astudillo and Cardona Alzate, 2011; Shen et al., 2009; Bai et al., 2009b) and the bacterium *Zymomonas mobilis* (Paz Astudillo and Cardona Alzate, 2011; Shen et al., 2009; Bai et al., 2008, 2009b) are currently the most important microorganisms for industrial production of ethanol. *Z. mobilis* produces less biomass than *S. cerevisiae*, and uses more carbon for ethanol fermentation. The ethanol:glucose yield of *Z. mobilis* has been reported to reach 97% of the theoretical yield (Sprenger, 1996), whereas the maximum yield of *S. cerevisiae* is only 90–93% (Bai et al., 2008). Moreover, the *higher* glucose metabolic flux in *Z. mobilis* guarantees higher ethanol productivity (Sprenger, 1996). Thus, *Z. mobilis* has been promoted as a more promising microorganism than yeast for the industrial production of ethanol (Li, 2009).

Continuous fermentation is more productive (i.e., requires less labor, fewer maintenance costs, and less capital investment in production facilities) than batch operation. As such, it is exploited for large-scale industrial production of fuel ethanol (Wang et al., 2013b). However, continuous ethanol fermentation by *Z. mobilis* is prone to dynamic instability, observed as sustainable or damped periodic oscillations in the concentrations of product, biomass, and substrate. Oscillations are mainly caused by product inhibition of enzymatic reactions, which may be enhanced by low growth rates and high ethanol concentrations (Ghommidh et al., 1989).

Computer Aided Chemical Engineering, Volume 36. ISSN 1570-7946. http://dx.doi.org/10.1016/B978-0-444-63472-6.00006-9

The oscillations are characterized by long period and large amplitude. Under certain operational conditions of continuous ethanol fermentation by *Z. mobilis*, the oscillations in the glucose, ethanol, and biomass concentrations are sustained for 50 h, and their amplitudes reach 20–100 kg/m^3 for glucose and 50–90 kg/m^3 for ethanol (McLellan et al., 1999).

These oscillations negatively affect ethanol fermentation performance, increasing the residual sugar and decreasing the ethanol yield (Bai et al., 2009a). Oscillations occurring in one industrial process are propagated to downstream processes. This is particularly serious in ethanol distillation, which requires a relatively constant ethanol concentration in the fermentation broth (Bai et al., 2009b). The fluctuating ethanol concentration at the outlet significantly affects the subsequent ethanol distillation step. Moreover, oscillatory conditions can discharge sugars along with the effluent. Because the ethanol yield calculation is based on the sugars fed into the fermentation system, the unfermented discharged sugars must be deducted from the calculation (Bai et al., 2004). The increased average residual sugar correspondingly decreases the ethanol yield (Bai et al., 2008), so these oscillations need to be attenuated.

To this end, we must elucidate the mechanisms leading to oscillatory behavior. The causes of oscillatory behaviors in fermentation processes have been extensively investigated, both experimentally and numerically.

According to Lee et al. (1979, 1980), *Z. mobilis* is highly suitable for ethanol production because it has high specific growth rate and ethanol production rate, combined with high ethanol tolerance. After studying a continuous *Z. mobilis* culture in glucose media, they concluded that oscillatory behavior results from inhibition of growth by the accumulated alcohol. Jöbses et al. (1986) also assumed that oscillations were mainly caused by product inhibition of enzymatic reactions. They reported that inhibitory effects might be enhanced at high temperatures, and that oscillations are difficult to avoid at high product concentrations. Ghommidh et al. (1989), who introduced structuration of the cell population into a mathematical model, reported that a continuous culture could range from completely stable to completely unstable, with oscillatory steady states occurring at low dilutions. Bruce et al. (1991) improved the performance of continuously cultured *Z. mobilis* during the conversion of concentrated substrates to ethanol by extractive fermentation. This fermentation also eliminates the oscillatory behavior that routinely arises in conventional *Z. mobilis* fermentations. Jarzebski (1992) developed a tricompartment cell population model that predicts how product inhibition decreases with decreasing substrate concentration. Depending on substrate concentration, this model portrays all the dynamic situations observed in continuous ethanol fermentations, from sustained oscillations to completely stable cultures. Hobley and Pamment (1994) conjectured that the oscillatory mechanism is driven by the ethanol concentration history. Li et al. (1995) observed that although the ethanol concentration history did not alter the fermentation capacity of *Z. mobilis*, increasing ethanol concentrations exerted a strong inhibitory effect. The mathematical models of Daugulis et al. (1997) and McLellan et al. (1999) yielded numerical results that perfectly matched their experimental results. According to Garhyan et al. (Garhyan et al.,

2003; Garhyan and Elnashaie, 2004a,b), the ethanol yield and production rate is favored by minimizing ethanol inhibition. They explored the conditions for increasing the ethanol productivity and yield by fermentation using mathematical modeling and nonlinear analysis. They increased the ethanol productivity by positioning membranes in the fermentor, which continuously removed the ethanol and also stabilized the system by removing oscillations. Experimental and theoretical studies have also highlighted the importance of variables such as dilution rate and feed substrate concentration on the static and dynamic behaviors in continuous stirred tank fermentors. Bai et al. (2004, 2009b) showed that parameter variations could generate oscillations in continuous ethanol fermentations in media with very high gravity. In their experiments, oscillations were attenuated and fermentation performance was improved only by Intalox ceramic saddles and wood chips. Diehl and Trierweiler (2009) analyzed the nonlinear dynamic behavior during continuous ethanol fermentation by *Z. mobilis*, and presented a multivariable control strategy in a high-yield solution branch inside the steady-state multiplicity region, which maintains process stability. Li (2009) developed three models for investigating the inhibitory effects leading to oscillatory behavior in *Z. mobilis* alcohol fermentation. They reported that periodic behaviors are not sourced from ethanol concentration alone, but develop under certain circumstances. According to Sridhar (2011), oscillatory behavior emerges in the presence of Hopf bifurcations. Sridhar proposed simple strategies by which the Hopf bifurcation points (and hence oscillations) are eliminated by very minor changes in the input conditions. In our previous work (Wang et al., 2012c), we analyzed and calculated the distribution of the Hopf points in a *Z. mobilis* continuous fermentation process, and thereby enhanced and maintained the stability of the fermentation process.

As highlighted above, continuous ethanol fermentation processes involving *Z. mobilis* show complex nonlinear dynamics (Mahecha-Botero et al., 2006), and are especially prone to unfavorable oscillatory phenomena. The operating parameters, such as the dilution rate and feed substrate concentration, influence both the static and dynamic behavior of the system. Previous experimental and numerical researches have shown that oscillatory behavior is avoided by optimizing the operating points and regions. Moreover, optimal operating conditions greatly reduce the influence of the Hopf points and oscillatory phenomena, directly improving the stability and productivity of the fermentation process.

Nonetheless, oscillatory phenomena are unavoidable under certain situations, especially when the Hopf points lie very close to the maximum product concentration operation point. Supercritical Hopf bifurcation will likely generate sustainable oscillation. Consequently, these oscillations must be attenuated to improve the ethanol fermentation performance. To this end, we must elucidate the relationships between the operational conditions and the characteristics of the dynamic behavior of oscillatory phenomena. However, these relationships have not been investigated in the published literature. Therefore, here we examine the oscillatory phenomena and formulate the relationships between the amplitude/period and operational conditions. This quantitative relationship model will assist the optimization of fermentation processes.

There are two main shortcomings of experimental research: (1) it cannot locate all points corresponding to certain dynamic behaviors because fermentation is a slow process and (2) the number of experiments that can be performed is limited, whereby some dynamic behaviors may be ignored or overlooked. Both shortcomings can be overcome by numerical analysis, which enables complete and thorough exploration of dynamic systems. In our present research, we numerically simulated the oscillations during a continuous fermentation process, and quantified the relationship between the amplitude/period and operating conditions.

In this article, we introduce our numerical model of a continuous *Z. mobilis* fermentation process in Section 6.2. Section 6.3 describes the Hopf points in steady-state solutions as well as the limit cycles, and Section 6.4 simulates the oscillatory phenomena under different operating conditions. From these simulations, we acquire a series of data and formulate a quantitative relationship between the amplitude/period and the operating conditions in Section 6.5. In Section 6.6, this formulation is used as a constraint in an optimization problem. Discussion and conclusions are presented in Sections 6.7 and 6.8, respectively.

6.2 DYNAMIC MODEL OF FERMENTATION PROCESS

Dynamic mathematical models of *Z. mobilis* fermentation processes have been developed over the years (Jarzebski, 1992; Ghommidh et al., 1989; Daugulis et al., 1997; Jöbses et al., 1986; McLellan et al., 1999; Sridhar, 2011; Mahecha-Botero et al., 2006). The calculations in the present article are based on the model of Mahecha-Botero et al. (2006), who incorporated a product-separating membrane to mimic real-time approaches.

The continuous flow fermentor is schematized in Figure 6.1, and the numerical simulation data are listed in Table 6.1. The model adopts the kinetics of *Z. mobilis* bioethanol production developed by Jöbses et al. (Abashar and Elnashaie, 2011).

Ethanol production by *Z. mobilis* is subject to end product inhibition. To prevent ethanol inhibition and increase the productivity and efficiency of the fermentation process, we propose a coupled membrane separation system that removes the accumulated ethanol.

6.2.1 MODEL ASSUMPTIONS

Our model is based on the following assumptions (Abashar and Elnashaie, 2010):

1. The fermentor is a well-mixed system and is not controlled by diffusion.
2. The feed contains only nutrient (substrate).
3. The specific growth rate is defined by a Monod-type equation.
4. The maximum specific growth rate is proportional to the concentration of the main component.

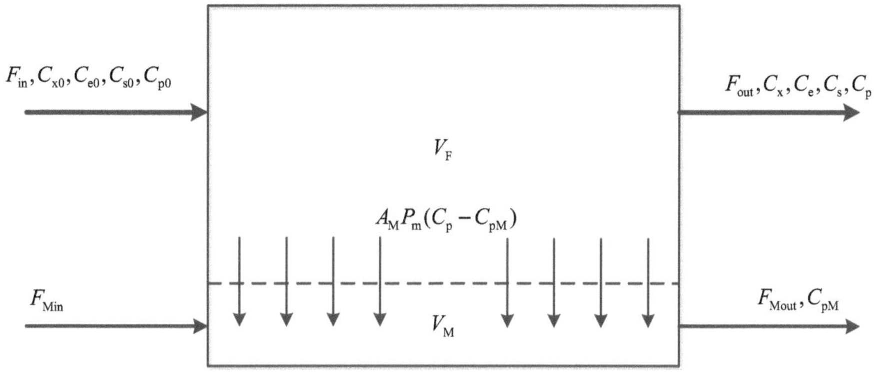

FIGURE 6.1

The fermentation process (Wang et al., 2012c).

Table 6.1 Parameters in the Fermentation Process Model (Mahecha-Botero et al., 2006)

Parameter	Value	Unit	Parameter	Value	Unit
C_{s0}	140	kg/m^3	Y_{sx}	0.0244	kg/kg
D_{Min}	0.5	h^{-1}	Y_{px}	0.0526	kg/kg
A_M	0.24	m^2	C_{p0}	0	kg/m^3
D_{in}	0.04	h^{-1}	C_{e0}	0	kg/m^3
K_s	0.5	kg/m^3	C_{x0}	0	kg/m^3
k_1	16	h^{-1}	C_{pM0}	0	kg/m^3
k_2	0.497	m^3/kg h	ρ	789	kg/m^3
k_3	0.00383	m^6/kg^2 h	V_M	0.0003	m^3
m_s	2.16	kg/kg h	V_F	0.003	m^3
m_p	1.1	kg/kg h	P_m	0.1283	m/h

5. Besides the empirical constants (e.g., k_1, k_2, k_3) and the Monod constant (K_s), the maintenance factors based on substrate requirement (m_s) and product formation (m_p), and the yield factors of biomass on substrate (Y_{sx}) and product (Y_{px}), are constants of the model.

6.2.2 MODEL DEVELOPMENT AND DISCUSSION

One of the most accepted models of fermentation processes is the maintenance model, in which the substrate consumption is given by

$$r_s = \left(\frac{\mu}{Y_{sx}} + m_s\right) C_x,$$

where μ is the specific growth rate and C_x is the concentration of the organism. The first and second terms represent the growth rate and maintenance, respectively. Similarly, the rate of product generation is given by

$$r_p = \left(\frac{\mu}{Y_{px}} + m_p\right) C_x.$$

The growth rate of biomass r_x is defined in the classic sense and is expressed as

$$r_x = \mu C_x.$$

In subsequent formulas, the subscripts s and p refer to the substrate and product, respectively. The formation rate of the main compound r_e is formulated as

$$r_e = f(C_s) f(C_p) C_e.$$

The formation rate of the main component is a function of the substrate concentration and product concentration (Jöbses et al., 1986).

$f(C_s)$ is given by the Monod equation:

$$f(C_s) = \frac{C_s}{K_s + C_s}.$$

The function $f(C_p)$ is a polynomial, given by

$$f(C_p) = k_1 - k_2 C_p + k_3 C_p^2.$$

Substituting these expressions into r_e, we obtain:

$$r_e = \frac{k_1 - k_2 C_p + k_3 C_p^2}{K_s + C_s} C_s C_e.$$

The base set of parameters in the current investigation is listed in Table 6.1 (Mahecha-Botero et al., 2006).

The concentration dynamics of the substrate (s), microorganism (x), and main compound (e) in the continuous fermentation process are given by the following set of ordinary differential equations:

$$\frac{dC_x}{dt} = \mu C_x + D_{in} C_{x0} - D_{out} C_x$$

$$\frac{dC_s}{dt} = -\left(\frac{\mu}{Y_{sx}} + m_s\right) C_x + D_{in} C_{s0} - D_{out} C_s$$

$$\frac{dC_e}{dt} = \frac{k_1 - k_2 C_p + k_3 C_p^2}{K_s + C_s} C_s C_e + D_{in} C_{e0} - D_{out} C_e$$

The product (ethanol) on the fermentor side (p) and on the membrane permeate side (pM), and the mass balance of the product (ethanol) over the fermentor side is given by

$$\frac{dC_p}{dt} = \left(\frac{\mu}{Y_{px}} + m_p\right) C_x + D_{in} C_{p0} - D_{out} C_p - \frac{A_M P_m}{V_F}\left(C_p - C_{pM}\right).$$

The mass balances of the product (ethanol) over the membrane are given by

$$\frac{dC_{pM}}{dt} = \frac{A_M P_m}{V_F}(C_p - C_{pM}) + D_{Min} C_{pM0} - D_{Mout} C_{pM},$$

where A_M is the permeation area and P_m is the ethanol permeability of the membrane.

The dilution rate of the exit fermentor is given by

$$D_{out} = D_{in} - \frac{A_M P_m}{V_F \rho}(C_p - C_{pM})$$

and the corresponding dilution rate across the exit membrane becomes

$$D_{Mout} = D_{Min} + \frac{A_M P_m}{V_M \rho}(C_p - C_{pM}),$$

where the subscripts D_{in}, D_{out}, D_{Min}, and D_{Mout} represent the respective dilution rates, and Y_{SX} and Y_{PX} are the yield factors of biomass on substrate and product, respectively. k_1, k_2, and k_3 are empirical constants, V is the volume, and m_S and m_P are maintenance factors based on substrate requirements and product formation, respectively. The membrane and fermentation sides are, respectively, distinguished by subscripts M and F.

Now let us explore the dynamic behavior and singularity operating points of this model.

6.3 IDENTIFICATION OF HOPF SINGULARITIES

As mentioned above, the *Z. mobilis* fermentation process exhibits rich dynamic behavior by virtue of the Hopf singularities. A Hopf bifurcation is a local bifurcation in which a fixed point of the dynamical system loses stability as a pair of complex conjugate eigenvalues of the linearization around the fixed point crosses the imaginary axis of the complex plane. A Hopf bifurcation should exhibit a small-amplitude limit cycle branching from the fixed point. The methods for identifying Hopf singularities and calculating the limit cycles (Wang et al., 2012c) are presented in the Appendix.

Locating the singular points and calculating the limit cycles are complex procedures. Fortunately, the numerical calculation is aided by numerous software packages, such as Auto (Doedel, 1981; Doedel and Kernevez, 1986; Doedel et al., 1997), content (Kuznetsov and Levitin, 1997), matcont (Dhooge et al., 2003, 2004), DsTool (Back et al., 1992), PyDSTool (Clewley et al., 2007), XPPAUT (Ermentrout, 2002), and by our previous work (Wang et al., 2010b; Yuan et al., 2009; Wang et al., 2009, 2010a, 2011, 2012a,b,c; Wang et al., 2013a; Wang et al., 2014b).

The strategy for locating Hopf points in a dynamic system is summarized in Figure 6.2 (Wang et al., 2012c). Beginning with a fermentation process model, the steady-state solution of the dynamic system was calculated, and examined for

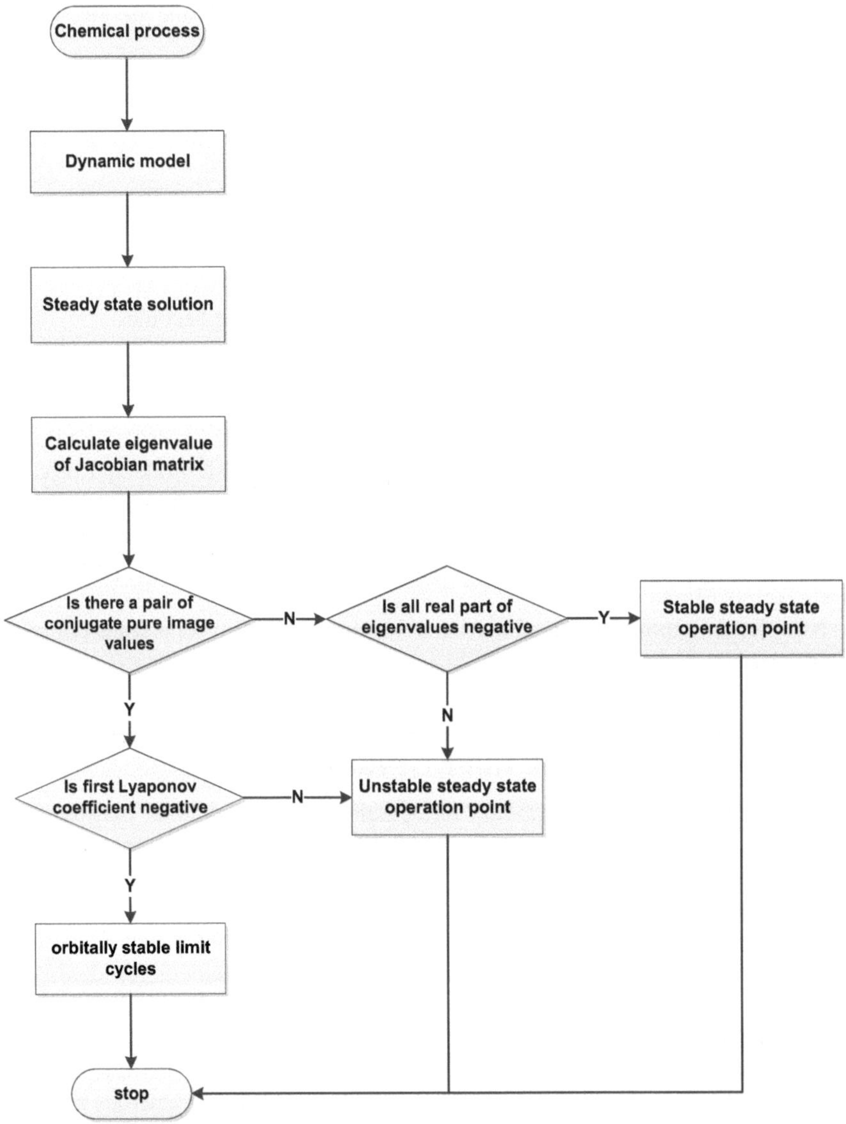

FIGURE 6.2

Flowchart of numerical Hopf analysis (Wang et al., 2012c).

the presence of Hopf points. If there exists a Hopf point and the first Lyapunov coefficient is negative, orbital stable limit cycles and corresponding oscillatory behavior will certainly exist in the dynamic system. By varying the parameters of this ergodic process, we identified all Hopf singularities, providing a valuable reference for optimizing the operating conditions.

6.4 HOPF SINGULARITIES IN THE *Z. MOBILIS* FERMENTATION PROCESS

In this section, we calculate the Hopf singularities on the steady-state solution curve. The Hopf curve was plotted on a plane constructed from two variable continuous operating conditions. By altering the parameters, we obtain additional Hopf curves. Here, we demonstrate that there are many Hopf singular points in the system. Importantly, we must calculate the Hopf points around all selected operating points.

6.4.1 HOPF SINGULAR POINTS

Figure 6.3 shows the steady-state points plotted on a plane constructed from product concentration and dilution rate. The Hopf singularity is indicated in red.

The Hopf singular point occurs at $C_{s0} = 150.3$ kg/m^3 and $D_{in} = 0.052$ h^{-1}, consistent with reported values (Sridhar, 2011; Garhyan and Elnashaie, 2004a). At this point, the product concentration C_p is 58.050 kg/m^3, approaching the point of maximum product yield ($D_{in} = 0.046$ h^{-1}, $C_p = 58.103$ kg/m^3). This finding highlights the importance of Hopf point investigation in the present dynamic system, in which the Hopf points approach the sought maximum conversion point. In Figure 6.3, the first Lyapunov coefficient l_1 is −0.00,153. The negative sign indicates a supercritical Hopf bifurcation and an orbital stable limit cycle.

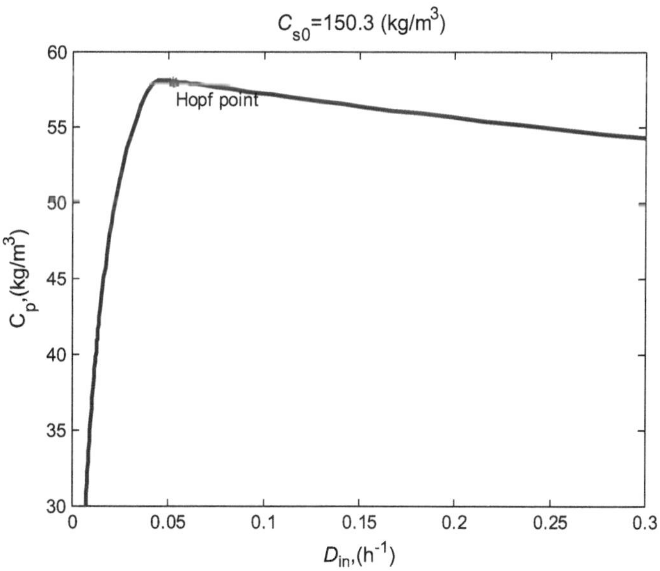

FIGURE 6.3

Product concentration as a function of dilution rate.

The abscissa in Figure 6.3 denotes the feedstock dilution rate, which ranges from zero to a reasonable maximum. Because it specifies the feedstock volume diffusing into the reactor volume, the dilution rate is relatively small in practice. Therefore, our analysis covers a limited range of dilution rates.

6.4.2 A CURVE OF HOPF POINTS

Bifurcation theory states that oscillations in a dynamic system begin and end at certain critical points in the system. As mentioned above, the *Z. mobilis* fermentation process demonstrates static and dynamic bifurcation behaviors over a wide range of operating parameters. Therefore, we must elucidate the distribution of Hopf points in this system. Figure 6.3 reveals a Hopf singular point at a specified initial substrate concentration. By changing the initial substrate concentration, we can shift the location of the Hopf point. That is, we can construct curves of substrate concentration versus dilution rate to identify different Hopf points. This numerical method reveals all Hopf singularity points, which can be assembled as a Hopf point curve in the operating parameter panel. The result is shown in Figure 6.4.

Each point in Figure 6.4 is an operating point at the Hopf singularity, capable of generating oscillatory phenomena. The Hopf points respond to variations in the feedstock dilution rate and initial substrate concentration. At low feedstock dilution rates or initial substrate concentrations, Hopf points are generated with higher probability. Notably, Hopf points appear over narrow alterations of one parameter, provided that the other is varied more aggressively. This indicates that no operating

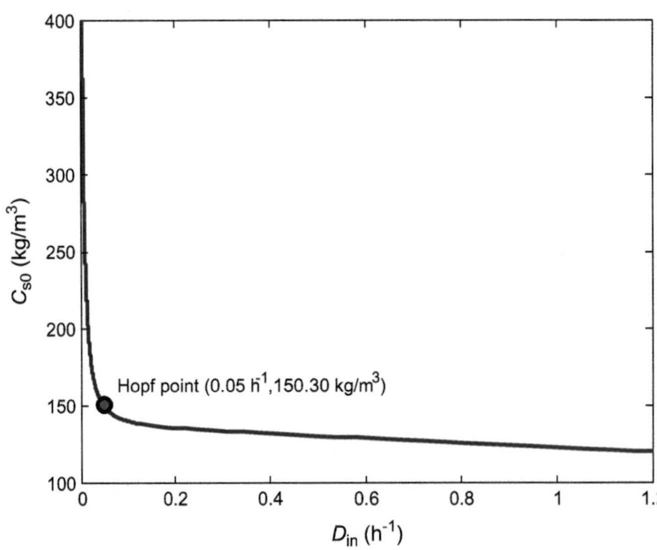

FIGURE 6.4

Hopf point curve plotted in operating condition panel.

region can optimize every parameter. The red Hopf point in Figure 6.4 is that highlighted in Figure 6.3.

The Hopf point is highly sensitive to the operating conditions, enabling the discovery of new Hopf singular point curves.

6.4.3 MULTIPLE HOPF CURVES

In the *Z. mobilis* fermentation process, the Hopf point curve is detectable in plots of initial substrate rate versus dilution rate, given some dilution rate of the membrane-side feedstock. Therefore, by changing the dilution rate of the membrane-side feedstock, we should identify other Hopf point curves in the operating conditions panel.

Figure 6.3 illustrates the Hopf point in a steady-state curve constructed from a single varying parameter. Figure 6.4 plots the Hopf singularity points in an operating conditions panel. However, operating parameters are generally interdependent and change simultaneously. Thus, we examined another important parameter, the membrane-side feedstock dilution rate. Specifically, we constructed Hopf curves for varying D_{Min} (under 0.1, 0.5, and 0.9 h^{-1}). The resulting curves are plotted in Figure 6.5.

In Figure 6.5, because we vary an additional parameter (namely, the dilution rate of the membrane-side feedstock), we generate more Hopf curves. As expected, the Hopf curves alter with dilution rate at the membrane side.

From the above analyses, we conclude that among the operating parameters, the initial substrate concentration, and the dilution rates on both feedstock and

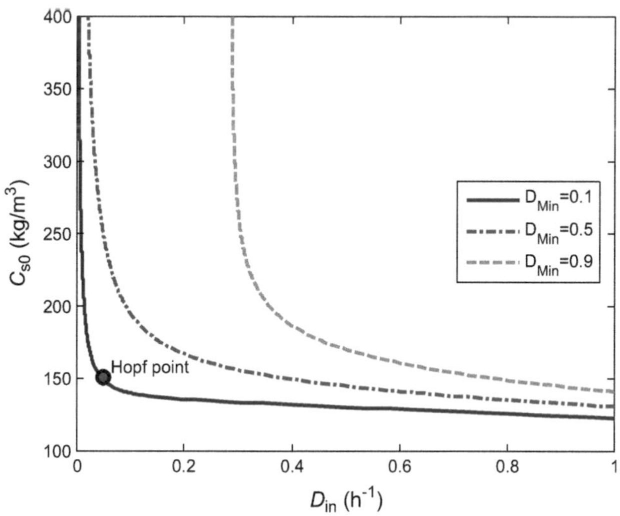

FIGURE 6.5

Hopf curves in the parameter panel.

membrane sides, most sensitively affect product yield and quality. These parameters also significantly influence the location of the Hopf points. Hopf points likely appear when either the feedstock dilution rate or the initial substrate concentration is low; thus there is no universally safe operating region for a single parameter. This section focused on locating the Hopf singularity points in the *Z. mobilis* fermentation process. Such points were identified within a large operational space, indicating that their dynamic behavior requires a careful investigation before deciding the operating conditions of the process. In some cases, the oscillatory phenomena can be attenuated by a regulatory method. Furthermore, by relating the characteristics of the oscillatory phenomena to the operating conditions, we can begin to optimize the process operating conditions.

6.5 REGULATION OF OSCILLATORY PHENOMENA

As demonstrated above, there are many Hopf singularities in the operating conditions panel. As each of these Hopf points may generate an oscillatory phenomenon, the characteristics of the oscillatory phenomenon must be properly related to the operating conditions. The operating relationship model could then predict the operating conditions that attenuate the unfavorable oscillatory phenomenon. Moreover, by quantifying how oscillatory phenomena such as amplitude and period depend on the operating conditions, we could formulate an optimization of *Z. mobilis* fermentation. Toward this goal, we now investigate the relationship between the oscillation characteristics and the operating conditions.

Figure 6.6 summarizes the basic procedures of regulating the oscillatory phenomena in a dynamic system (Wang et al., 2014a). For different parameters in the fermentation process model, this method calculates the steady-state solution of the dynamic system and identifies any Hopf points. If Hopf singularities exist, their locations are determined. The relationships between the oscillation parameters and operating condition are then determined by simulating the oscillatory phenomena near the Hopf singularity. The formulated model guides the practical implementation of the fermentation process. If the simulated result deviates from the experimental result, additional experimental data are gathered, the dynamic model of the chemical process is updated, and the coefficients of the dynamic model are regressed against the obtained data. These steps are iterated until the calculated and experimental results are acceptably converged.

According to bifurcation theory, oscillations in a dynamic system initiate from certain critical points in the system. From this singular point, sustained oscillations develop for various initial values of the parameters.

As mentioned earlier, the *Z. mobilis* continuous fermentation process is characterized by many singular points and limit cycles, and is thus highly susceptible to oscillatory behavior. The values of this Hopf point introduced in Figure 6.3 are listed in Table 6.2.

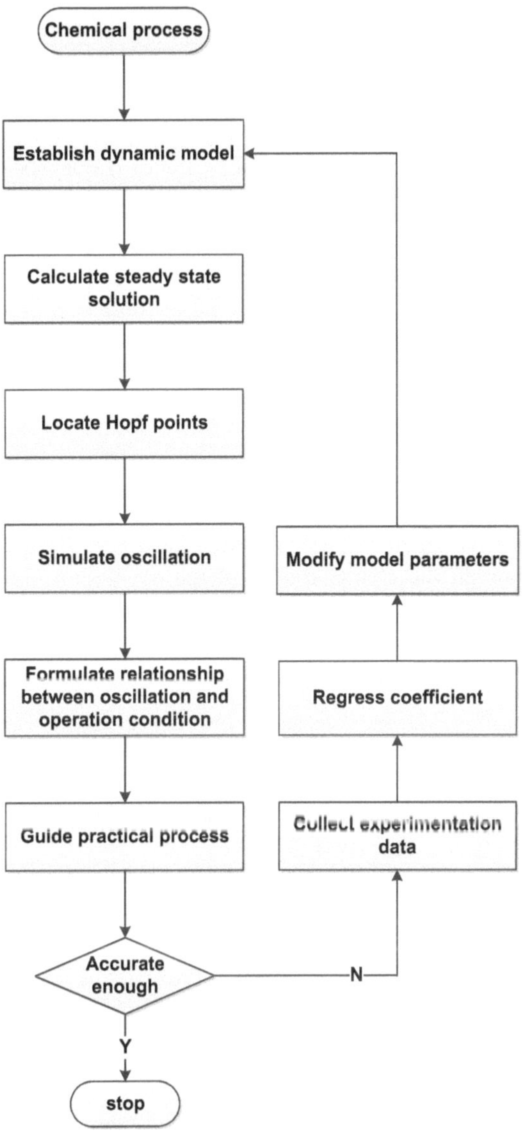

FIGURE 6.6

Flowchart of regulation method of oscillatory phenomena (Wang et al., 2014a).

Bifurcation theory suggests that limit cycles, which represent oscillatory phenomena, are intrinsically concomitant with Hopf singularity points. More specifically, limit cycles indicate sustained oscillations, as shown in Figure 6.7. The red star in Figure 6.7 is the Hopf singularity point, the blue line traces the steady-state

Table 6.2 Values of Hopf Point

Parameter	Value	Unit
D_{in}	0.05	h^{-1}
C_e	0.06	kg/m^3
C_p	58.05	kg/m^3
C_{pM}	57.98	kg/m^3
C_{s0}	150.30	kg/m^3
C_s	6.10	kg/m^3
C_x	1.77	kg/m^3

operating conditions, and the blue circles are the limit cycles established under different initial operating conditions.

Oscillatory phenomena manifest as periodic variations in the product, biomass, and substrate concentrations. The characteristics of the oscillation depend on the operating condition parameters. Therefore, we must scrutinize the effect of the operating conditions on the characteristics of the oscillation phenomena. To this end, we generated a suitable data set from numerically simulated oscillations. After analyzing these data, we modeled the relationship between the oscillatory amplitude/period and the operation parameter values.

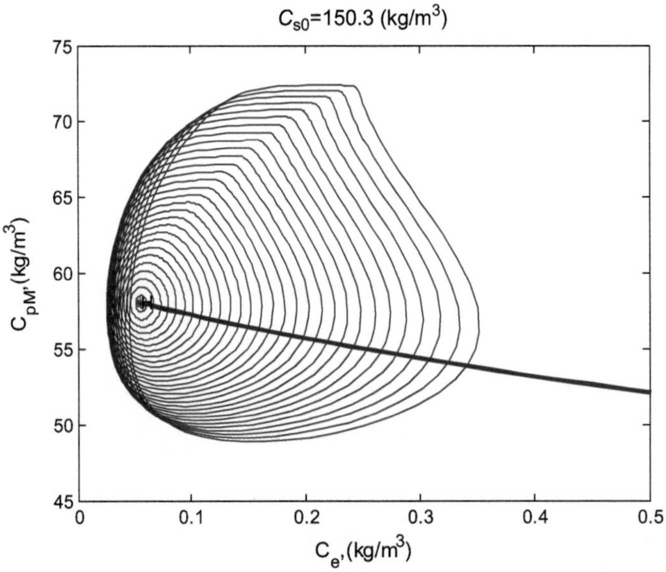

FIGURE 6.7

Limit cycles around Hopf point.

To gather the data, we gradually increased the dilution rate, ran the simulations, and recorded the oscillatory characteristics.

Figure 6.8 shows the oscillatory dynamic behaviors for $D_{in} = 0.1\ h^{-1}$ and $0.2\ h^{-1}$. The Hopf singularity induced unfavorable oscillations at $D_{in} = 0.1\ h^{-1}$; the same situation also can be found at $D_{in} = 0.2\ h^{-1}$.

From Figure 6.8(a), we observe that when the initial point in the mathematical model of the continuous *Z. mobilis* fermentation process is the Hopf point, the system dynamics behave as expected. Oscillations are established in the system, and the product concentration periodically varies at constant amplitude. Similar behaviors are established in the biomass and substrate concentrations. From the dynamics of Figure 6.8(a), we can extract the oscillation characteristics. We find that the product concentration fluctuates with an amplitude $A_{mp,C_{pM}}$ of $12.19\ kg/m^3$ and a period T of 15.12 h.

Because unfavorable oscillations were generated near the Hopf singular point, we took the parameters at the Hopf point as the initial values in the dynamic simulations of our ordinary differential equation (ODE) model. We set a new operating condition by changing the dilution rate from $0.05\ h^{-1}$ at the Hopf point to $0.1\ h^{-1}$, and studied the oscillatory phenomena under this new condition. The oscillations remained but their characteristics were altered.

The dilution rate was then raised to $0.2\ h^{-1}$, and the dynamic simulation was repeated. The dynamic behavior is displayed in Figure 6.8(b).

After a short adjustment time, the new system developed sustained oscillations in the product concentration. Under the new condition, $A_{mp,C_{pM}}$ was $17.42\ kg/m^3$ and the period T was 9.46 h.

Next, maintaining the other initial parameters at their Hopf point values, we gradually increased the initial dilution rate, and mathematically analyzed the continuous fermentation process by *Z. mobilis*, thereby obtaining the amplitude and period of each new oscillation. The data are presented in Table 6.3. Here, the dilution rate was varied from $0.1\ h^{-1}$ to $0.9\ h^{-1}$ in $0.1\text{-}h^{-1}$ increments.

Table 6.3 reveals changes in the amplitude and period as the dilution rate is varied. From these data, we can establish the relationship between amplitude/period and the operating conditions. To increase the accuracy of our model, we enriched our dynamic behavior data set by performing additional simulations. After analyzing the enriched data set, we formulated the relationships between the dilution rates and the oscillatory characteristics (i.e., amplitude and period) as follows:

The dilution rate D_{in} is a cubic polynomial function of the oscillation amplitude of the product concentration, $A_{mp,C_{pM}}$:

$$D_{in} = 0.000037A^3_{mp,C_{pM}} - 0.00012A^2_{mp,C_{pM}} - 0.00028A_{mp,C_{pM}} + 0.054.$$

In this polynomial fitting, $R^2 = 0.9996$.

The oscillation period is related to the dilution rate by:

$$T = 3.03\ D_{in}^{-0.7},$$

with $R^2 = 0.9998$.

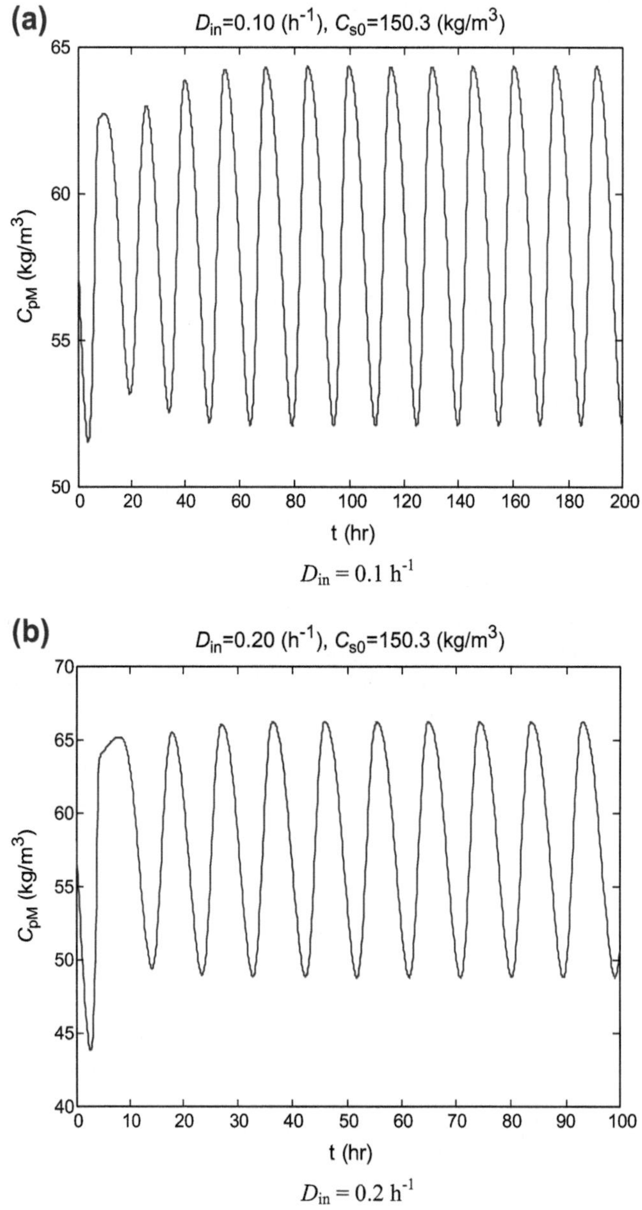

FIGURE 6.8

Numerical run illustrating limit cycle behavior: $D_{in} = 0.1\ h^{-1}$ (top panel) and $0.2\ h^{-1}$ (bottom panel), $C_{s0} = 150.3\ kg/m^3$, and other conditions are listed in Table 6.1.

Table 6.3 Response of Oscillatory Amplitude and Period to Initial Dilution Rate

D_{in} (h^{-1})	$A_{mp,C_{pM}}$ (kg/m^3)	T (h)
0.10	12.19	15.12
0.20	17.42	9.46
0.30	20.14	7.09
0.40	22.20	5.76
0.50	23.97	4.91
0.60	25.55	4.32
0.70	26.99	3.88
0.80	28.32	3.53
0.90	29.57	3.26

The formulated relationships between the amplitude and dilution rate, and between the period and the dilution rate, are plotted in Figures 6.9 and 6.10, respectively.

The above investigation revealed how the product concentration behaves under oscillation-inducing conditions. The important results are summarized below.

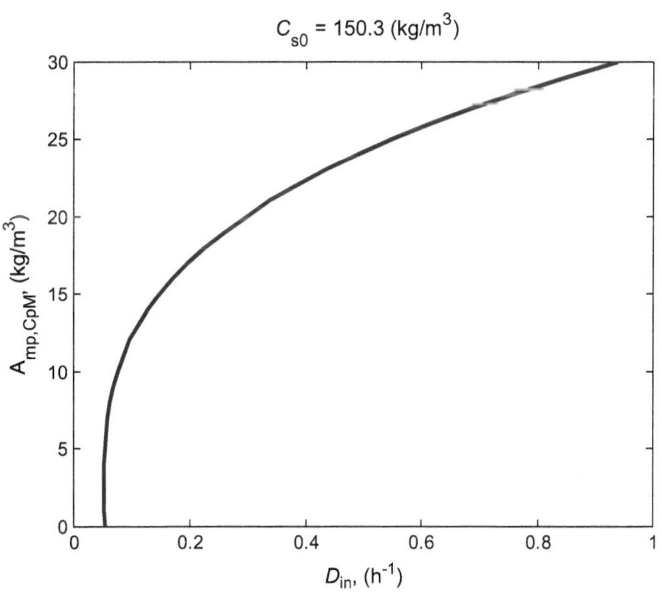

FIGURE 6.9

Relationship between amplitude and dilution rate.

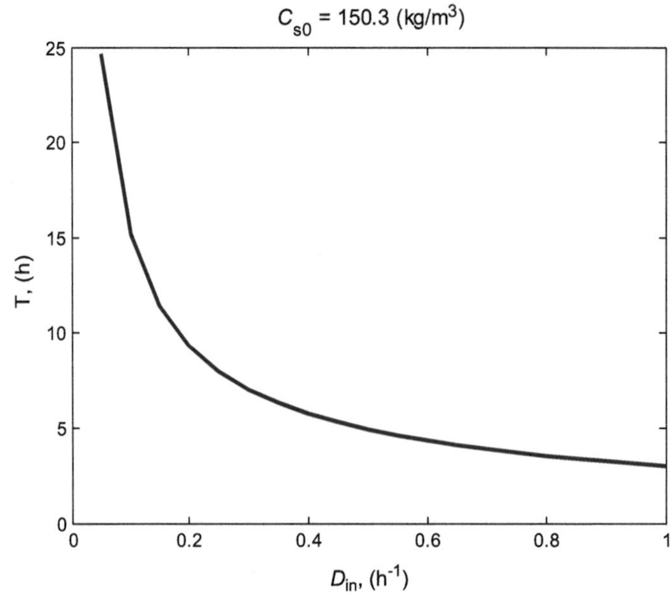

FIGURE 6.10

Relationship between period and dilution rate.

At a given Hopf singularity, the amplitude of the product concentration responds to the varying dilution rate D_{in}. The result is displayed in Figure 6.11. Clearly, increasing the dilution gradually enhances the amplitude of the product concentration.

Figure 6.11 reveals the maximum and minimum concentrations of the product at various dilution rates. Note that (1) the amplitude increases with increasing dilution rate; (2) the maximum product concentrations are upper-bounded, implying a limitation in this fermentation process; (3) the minimum concentrations decrease with increasing dilution rate; and (4) the average product concentration decreases with increasing dilution rate. We conclude that severe oscillatory phenomena should be avoided (the maximum amplitude under oscillatory conditions should not exceed 50% of the product concentration), and that oscillations reduce the quantity of total product (because they lower the average product concentration).

In this section, we studied and formulated the relationships between the oscillatory characteristics (namely, the amplitude and period) and the operating conditions. Using this model, we can regulate the oscillation behavior. We also confirmed the negative influence of oscillations on the average product concentration. When oscillations cannot be avoided, they can be attenuated to increase the product concentration as far as possible.

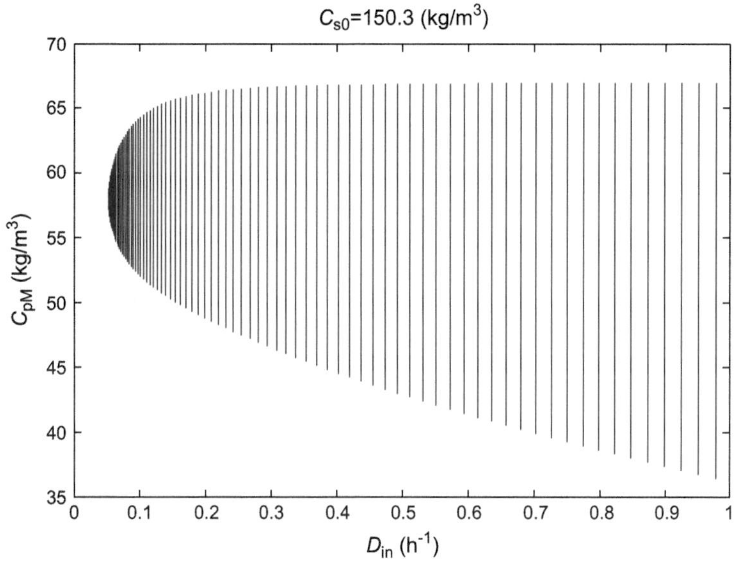

FIGURE 6.11

Amplitudes of the product concentration at different dilution rates.

6.6 OPERATING CONDITION OPTIMIZATION

This section, which introduces the optimization process, is the crux of the present paper. In the previous sections, we located all Hopf singular points and formally related the oscillatory phenomena to the operating conditions. We now use these results to formulate the optimization problem. The optimization reveals the best operating conditions under the constraint of oscillatory phenomena. Specifically, we seek to attenuate the oscillation amplitude of the product concentration and thereby reduce the negative effects of oscillation.

6.6.1 SCHEMATIC OF THE PROCESS DESIGN

The process design for optimizing the fermentation operating conditions is schematized in Figure 6.12. First, we formulate dynamic mathematical models based on the biochemical processes. We then define the optimization problem, select the operating conditions, obtain the optimal result, and determine the presence of nearby Hopf singularities. If there are no close Hopf singularities, we retrieve the optimized result. Otherwise, the Hopf singularity points are identified and the relationship between the oscillatory amplitude/period and the operating conditions is obtained by simulation. We also check whether the original optimal operating point will be influenced by unavoidable oscillatory phenomena. If the optimal design is little affected by the oscillations, it is accepted. On the other hand, if the optimal design

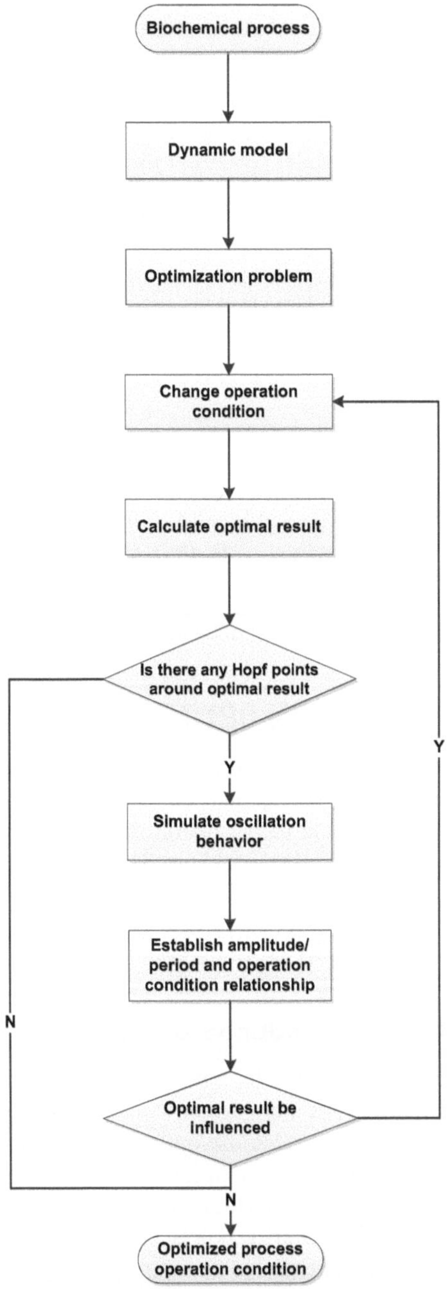

FIGURE 6.12

Schematic of optimization design.

is appreciably degraded by the oscillations, the operating condition is changed, and the process design is repeated until a satisfactory operating point is reached.

6.6.2 OPTIMIZATION OF THE *Z. MOBILIS* FERMENTATION PROCESS

Based on the above-illustrated approach, we established the optimization model as follows. The optimization problem seeks to maximize the product concentration. The constraints contain the mass balances, the initial substrate concentration, and the relationships between the oscillatory characteristics and operating conditions.

$$\max C_{\mathrm{pM}}$$

$$\text{s.t.}$$

$$D_{\mathrm{out}} = D_{\mathrm{in}} - \frac{A_{\mathrm{M}}P_{\mathrm{m}}}{V_{\mathrm{F}}\rho}(C_{\mathrm{p}} - C_{\mathrm{pM}})$$

$$D_{\mathrm{Mout}} = D_{\mathrm{Min}} + \frac{A_{\mathrm{M}}P_{\mathrm{m}}}{V_{\mathrm{M}}\rho}(C_{\mathrm{p}} - C_{\mathrm{pM}})$$

$$\mu C_{\mathrm{x}} + D_{\mathrm{in}}C_{\mathrm{x0}} - D_{\mathrm{out}}C_{\mathrm{x}} = 0$$

$$-\left(\frac{\mu}{Y_{\mathrm{sx}}} + m_{\mathrm{s}}\right)C_{\mathrm{x}} + D_{\mathrm{in}}C_{\mathrm{s0}} - D_{\mathrm{out}}C_{\mathrm{s}} = 0$$

$$\frac{k_1 - k_2C_{\mathrm{p}} + k_3C_{\mathrm{p}}^2}{K_{\mathrm{s}} + C_{\mathrm{s}}}C_{\mathrm{s}}C_{\mathrm{e}} + D_{\mathrm{in}}C_{\mathrm{e0}} - D_{\mathrm{out}}C_{\mathrm{e}} = 0$$

$$\left(\frac{\mu}{Y_{\mathrm{px}}} + m_{\mathrm{p}}\right)C_{\mathrm{x}} + D_{\mathrm{in}}C_{\mathrm{p0}} - D_{\mathrm{out}}C_{\mathrm{p}} - \frac{A_{\mathrm{M}}P_{\mathrm{m}}}{V_{\mathrm{F}}}(C_{\mathrm{p}} - C_{\mathrm{pM}}) = 0$$

$$\frac{A_{\mathrm{M}}P_{\mathrm{m}}}{V_{\mathrm{F}}}(C_{\mathrm{p}} - C_{\mathrm{pM}}) + D_{\mathrm{Min}}C_{\mathrm{pM0}} - D_{\mathrm{Mout}}C_{\mathrm{pM}} = 0$$

$$C_{\mathrm{s0}} = 150.3$$

$$D_{\mathrm{in}} = 0.000037A_{\mathrm{mp},C_{\mathrm{pM}}}^3 - 0.00012A_{\mathrm{mp},C_{\mathrm{pM}}}^2 - 0.00028A_{\mathrm{mp},C_{\mathrm{pM}}} + 0.054, \quad \text{when } D_{\mathrm{in}} \geq 0.054, \text{ else } A_{\mathrm{mp},C_{\mathrm{pM}}} = 0$$

$$T = 3.03D_{\mathrm{in}}^{-0.7}, \quad \text{when } D_{\mathrm{in}} \geq 0.054, \text{ else } T \to \infty$$

$$D_{\mathrm{in}} \geq 0$$

$$A_{\mathrm{mp},C_{\mathrm{pM}}} \geq 0$$

$$A_{\mathrm{mp},C_{\mathrm{pM}}} \leq 0.1C_{\mathrm{pM}}$$

$$T \geq 24$$

The optimal result in the above model is $D_{\mathrm{in}} = 0.046\ \mathrm{h}^{-1}$ and $C_{\mathrm{pM}} = 58.033\ \mathrm{kg/m^3}$. Further testing revealed that under these conditions, the system is negligibly affected by oscillatory phenomena.

The proposed approach selects the operating points subject to oscillatory influence. Despite the presence of Hopf points near the operating point that maximizes the product concentration, the resulting oscillations do not compromise the optimal result. This finding highlights the advantages of our approach. Using this approach, we could also test the influence of potential oscillatory phenomena at the design stage, thereby maximizing both the yield and stability of the product concentration.

6.7 DISCUSSION

As illustrated above, the proposed optimization approach seeks and selects the best operating conditions. Two aspects of the approach require further discussion.

6.7.1 CONDITIONS OF OSCILLATORY PHENOMENA

The oscillation phenomenon has been widely reported in continuous fermentation processes by *Z. mobilis*. Presumably, sustained oscillations naturally arise from product inhibition. In previous research, oscillatory phenomena have been ascribed to Hopf singularities and their bifurcations. By considering the effects of these oscillations on product yield, we have improved the biochemical process design over previous works. We also simulated the oscillatory dynamics of the fermentation system. Here we emphasize that the Hopf singularity is a trigger point that automatically generates oscillations for any combination of variables lying on that point. The detailed characteristics of the oscillation are determined for the initial values of the process. In this chapter, we related the amplitude/period of the oscillation to the operating conditions; specifically, we altered the dilution rate of the feedstock and observed the changes in the amplitude and period, maintaining the other variables at their Hopf singularity values. We found that oscillatory phenomena respond to slight changes in the feedstock dilution rate. It should be noted that we varied the feedstock dilution rate because this parameter is easily manipulated.

We conclude that whereas oscillation is triggered by a Hopf singularity, the dynamic behavior (amplitude and period of the oscillations) is determined by small deviations of the initial values from the Hopf singularity. This indicates that during the design process, the operating point must avoid not only the Hopf singularity (which triggers oscillations) but also regions close to the Hopf singularity, conditional on the effects of those oscillations (i.e, oscillations exerting small effects are permitted).

6.7.2 DECREASED NEGATIVE INFLUENCE OF THE OSCILLATORY PHENOMENA

Since we have calculated the distribution of the Hopf singular points, oscillatory phenomena are most simply avoided by maintaining the operating conditions far from these points. Far from the Hopf singularities, the probability of oscillatory phenomena is low. However, this simple approach is disfavored for three reasons. First, the product concentration may be maximized close to some Hopf singularities. Maintaining the operating conditions distant from these highly productive points would greatly reduce the productivity of the fermentation. Second, even if the operating point is distant from the Hopf singularities, it may deviate little from a

Hopf singularity, and therefore develop inevitable oscillations. Third, because our approach accepts benign oscillatory phenomena, it maximizes the productivity while limiting the effects of the oscillation to an acceptable range.

Considering the above concerns, we designed an improved biochemical process that first optimizes the operating conditions, and then tests the likelihood and effect of the oscillations. Unavoidable oscillations need to be attenuated to an acceptable level.

6.8 CONCLUSION

We propose a novel biochemical fermentation design method that permits constrained bifurcations and oscillations. The method is applied to typical fermentation by the ethanol producer *Z. mobilis*. Experimentally and in simulations, continuous *Z. mobilis* fermentation undergoes periodic fluctuations in biomass, product, and substrate concentrations under certain fermentation conditions. These oscillations have previously been attributed to Hopf singularities in the fermentation process. Because they reduce the fermentation yield, system stability, and controllability (and thereby the product quality), they must be suppressed by a suitable approach. The proposed design method first locates all Hopf singularities in this dynamic process, while varying the operating parameter values. It then simulates the oscillatory behavior to formulate the relationship between the amplitude/period of the oscillating product concentration and the operating conditions. Finally, it incorporates the oscillatory effects into an optimization model, and calculates the optimal result. The new method accepts solutions close to Hopf points, provided that the oscillatory behavior remains within an acceptable level. This approach differs from previously published methods, which maintain the operating point far from the Hopf curves. The new method offers a deep analysis of the dynamic behavior, and may improve the productivity, product quality, and stability of dynamic fermentation processes. The method is also extendible to other fermentation processes developing sustained oscillatory phenomena.

APPENDIX

A.1 INTRODUCTION TO HOPF POINTS AND NUMERICAL ANALYSIS METHODS

A generalized dynamic system of chemical processes is modeled by the following equation:

$$\begin{cases} \dfrac{d\mathbf{x}}{dt} = \mathbf{F}(\mathbf{x}, \alpha) \\ \mathbf{x}(\mathbf{0}) = \mathbf{x_0} \end{cases}, \tag{6.1}$$

where $\mathbf{x}$ denotes the state variables (such as temperature, pressure, and component concentrations) ($\mathbf{x} \in \mathrm{R}^n$), and α represents the operating condition values or design variables, for instance, the flow rate, initial feedstock concentration, and reactor volume. Equation (6.1) describes the dynamic behavior of the state variables under specified operating conditions, by which we can better understand a system's behaviors and optimization. By solving Eqn (6.2), we obtain the steady-state solution that describes the state variables under steady operating conditions.

$$\mathbf{F}(\mathbf{x}, \alpha) = 0 \tag{6.2}$$

The steady-state solution inevitably contains singular points that are worth investigating to smooth the fermentation operation.

A.2 DETECTION AND LOCATION OF HOPF POINTS

A Hopf point is characterized by a special pair of conjugate pure image eigenvalues associated with the Jacobian matrix at the steady-state point. Theoretically, a Hopf point can be detected by calculating the Jacobian eigenvalues. More commonly, Hopf points and their corresponding unique stable steady states are identified by a detection function. Consider the following test function:

$$\psi_{\mathrm{H}}(\mathbf{x}, \alpha) = \prod_{\mathrm{i}>\mathrm{j}} \left(\lambda_{\mathrm{i}}(\mathbf{x}, \alpha) + \lambda_{\mathrm{j}}(\mathbf{x}, \alpha)\right), \tag{6.3}$$

where λ_{i} are eigenvalues of the Jacobian matrix at a specified steady-state point. According to the theory, this function value vanishes at a Hopf bifurcation point, which is associated with a complex conjugate pair of pure image eigenvalues $\lambda_{1,2} = \pm i\omega_0$. Note that the paired real eigenvalues $\lambda_1 = \kappa, \lambda_2 = -\kappa$ are a special case of $\psi_{\mathrm{H}} = 0$, which corresponds to a neutral saddle situation rather than the Hopf point of our present concerns. These neutral saddle points must be carefully distinguished from Hopf points at the zero point of the detection function.

Hopf point detection is based on the bialternate product $\odot$ of the matrix:

$$\psi_{\mathrm{H}}(\mathbf{x}, \alpha) = \det\left(2\mathbf{F}_{\mathbf{x}}(\mathbf{x}, \alpha) \odot \mathbf{I}_{\mathrm{n}}\right) \tag{6.4}$$

where $\odot$ is formulated as:

$$(2\mathbf{A} \odot \mathbf{I}_{\mathrm{n}})_{(\mathrm{p,q}),(\mathrm{r,s})} = \begin{vmatrix} a_{\mathrm{pr}} & a_{\mathrm{ps}} \\ \delta_{\mathrm{qr}} & \delta_{\mathrm{qs}} \end{vmatrix} + \begin{vmatrix} \delta_{\mathrm{pr}} & \delta_{\mathrm{ps}} \\ a_{\mathrm{qr}} & a_{\mathrm{qs}} \end{vmatrix}. \tag{6.5}$$

In Eqn (6.5), $\delta_{\mathrm{ij}} = \begin{cases} 1, \mathrm{i} = \mathrm{j} \\ 0, \mathrm{i} \neq \mathrm{j} \end{cases}$. The matrix rows are labeled as (p, q) $(p = 2, 3,\ldots, n; q = 1, 2,\ldots, p-1)$, and the columns are labeled as (r, s) $(r = 2, 3,\ldots, n; s = 1, 2,\ldots, r-1)$. The determinants are evaluated as:

$$(2\mathrm{A} \odot \mathrm{I}_n)_{(\mathrm{p,q}),(\mathrm{r,s})} = \begin{cases} -a_{\mathrm{ps}}, & r = q \\ a_{\mathrm{pr}}, & r \neq q, s = q \\ a_{\mathrm{pr}} + a_{\mathrm{ps}}, & r = p, s = q \\ a_{\mathrm{qs}}, & r = p, s = q \\ -a_{\mathrm{qr}}, & p = s \\ 0, & \text{otherwise} \end{cases}. \tag{6.6}$$

A.3 LOCATION OF LIMIT CYCLES

In bifurcation theory, limit cycles are intrinsically associated with Hopf singularities. The shooting method, a widely accepted algorithm for calculating limit cycles, first converts the proposed question to the following periodic boundary-value problem for cycle continuation:

$$\begin{cases} \dfrac{d\mathbf{x}}{dt} = \dfrac{T}{2\pi} F(\mathbf{x}, \alpha) \\ \mathbf{x}(0) = \eta \end{cases}, \tag{6.7}$$

The parameter α is preset. To obtain a periodic solution to the problem, the algorithm seeks the initial value η and period T satisfying Eqn (6.7).

NOMENCLATURE

A_M	Permeation area (m^2)
$A_{mp,C_{pM}}$	Amplitude of oscillating product concentration (kg/m^3)
C_{e0}	Concentration of main component in fermentor feed (kg/m^3)
C_{p0}	Concentration of product in fermentor feed (kg/m^3)
C_{pM0}	Concentration of product in membrane-permeate-side feed (kg/m^3)
C_{s0}	Concentration of product in fermentor feed (kg/m^3)
C_{x0}	Concentration of biomass in fermentor feed (kg/m^3)
C_e	Concentration of main component in fermentor (kg/m^3)
C_p	Concentration of product in fermentor (kg/m^3)
C_{pM}	Concentration of product at membrane permeate side (kg/m^3)
C_s	Concentration of substrate in fermentor (kg/m^3)
C_x	Concentration of biomass in fermentor (kg/m^3)
D_{in}	Inlet dilution rate of fermentor (h^{-1})
D_{out}	Outlet dilution rate of fermentor (h^{-1})
D_{Min}	Inlet dilution rate at membrane permeate side (h^{-1})
D_{Mout}	Outlet dilution rate at membrane permeate side (h^{-1})
K_S	Monod constant (kg/m^3)
P_m	Ethanol permeability of the membrane (m/h)
T	Period of oscillation (h)
V_M	Volume of membrane permeate side (m^3)
V_F	Volume of fermentor (m^3)
Y_{px}	Yield factors of biomass on product (kg/kg)
Y_{sx}	Yield factors of biomass on substrate (kg/kg)
k_1	Empirical constant (h^{-1})
k_2	Empirical constant (m^3/kg h)
k_3	Empirical constant (m^6/kg^2 h)
m_p	Maintenance factors based on product formation (kg/kg h)
m_s	Maintenance factors based on substrate requirements (kg/kg h)
r_e	Production rate of the main compound (kg/m^3 h)
r_p	Formation rate of product (kg/m^3 h)
r_s	Substrate consumption rate (kg/m^3 h)

r_x	Biomass growth rate (kg/m^3 h)
ρ	Density of mixture (kg/m^3)
μ	Specific growth rate of fermenting organism (kg/m^3 h)

ACKNOWLEDGMENTS

The authors gratefully acknowledge financial support from the NSFC (No. 21306100) and the National Basic Research Program (No. 2012CB720500).

REFERENCES

Abashar, M., Elnashaie, S., 2010. Dynamic and chaotic behavior of periodically forced fermentors for bioethanol production. Chem. Eng. Sci. 65, 4894–4905.

Abashar, M.E.E., Elnashaie, S.S.E.H., 2011. Multistablity, bistability and bubbles phenomena in a periodically forced ethanol fermentor. Chem. Eng. Sci. 66, 6146–6158.

Back, A., Guckenheimer, J., Myers, M., Wicklin, F., Worfolk, P., 1992. DsTool: computer assisted exploration of dynamical systems. Notices Amer. Math. Soc 39, 303–309.

Bai, F., Ge, X., Anderson, W., Moo-Young, M., 2009a. Parameter oscillation attenuation and mechanism exploration for continuous VHG ethanol fermentation. Biotechnol. Bioeng. 102, 113–121.

Bai, F.W., Anderson, W.A., Moo-Young, M., 2008. Ethanol fermentation technologies from sugar and starch feedstocks. Biotechnol. Adv. 26, 89–105.

Bai, F.W., Chen, L.J., Anderson, W.A., Moo-Young, M., 2004. Parameter oscillations in a very high gravity medium continuous ethanol fermentation and their attenuation on a multistage packed column bioreactor system. Biotechnol. Bioeng. 88, 558–566.

Bai, F.W., Ge, X.M., Anderson, W.A., Moo-Young, M., 2009b. Parameter oscillation attenuation and mechanism exploration for continuous VHG ethanol fermentation. Biotechnol. Bioeng. 102, 113–121.

Bruce, L., Axford, D., Ciszek, B., Daugulis, A., 1991. Extractive fermentation by *Zymomonas mobilis* and the control of oscillatory behavior. Biotechnol. Lett. 13, 291–296.

Clewley, R., Sherwood, W., Lamar, M., Guckenheimer, J., 2007. PyDSTool, a Software Environment for Dynamical Systems Modeling. http://pydstool.sourceforge.net.

Daugulis, A.J., Mclellan, P.J., Li, J.H., 1997. Experimental investigation and modeling of oscillatory behavior in the continuous culture of *Zymomonas mobilis*. Biotechnol. Bioeng. 56, 99–105.

Dhooge, A., Govaerts, W., Kuznetsov, Y.A., 2003. MATCONT: a MATLAB package for numerical bifurcation analysis of ODEs. ACM Trans. Math. Software (TOMS) 29, 141–164.

Dhooge, A., Govaerts, W., Kuznetsov, Y.A., 2004. MATCONT: a Matlab package for numerical bifurcation analysis of ODEs. ACM SIGSAM Bull. 38, 21–22.

Diehl, F.C., Trierweiler, J.O., 2009. Control strategy for a *Zymomonas mobilis* bioreactor used in ethanol production. Comput. Aided Chem. Eng. 27, 1605–1610.

Doedel, E., Champneys, A., Fairgrieve, T., Kuznetsov, Y., Sandstede, B., Wang, X., 1997. AUTO 97: Continuation and Bifurcation Software for Ordinary Differential Equations,

User's Manual. Center for Research on Parallel Computing, California Institute of Technology, Pasadena.

Doedel, E., Kernevez, J.P., 1986. AUTO: Software for Continuation and Bifurcation Problems in Ordinary Differential Equations: Including the AUTO 86 User Manual. Applied Mathematics, California Institute of Technology.

Doedel, E.J., 1981. AUTO: a program for the automatic bifurcation analysis of autonomous systems. Congr. Numer. 30, 265–284.

Ermentrout, B., 2002. Simulating, Analyzing, and Animating Dynamical Systems: A Guide to XPPAUT for Researchers and Students. Society for Industrial Mathematics.

Garhyan, P., Elnashaie, S., 2004a. Static/dynamic bifurcation and chaotic behavior of an ethanol fermentor. Ind. Eng. Chem. Res. 43, 1260–1273.

Garhyan, P., Elnashaie, S.S.E.H., 2004b. Experimental investigation and confirmation of static/dynamic bifurcation behavior in a continuous ethanol fermentor. Practical relevance of bifurcation and the contribution of Harmon Ray. Ind. Eng. Chem. Res. 44, 2525–2531.

Garhyan, P., Elnashaie, S.S.E.H., AL-Haddad, S.M., Ibrahim, G., Elshishini, S.S., 2003. Exploration and exploitation of bifurcation/chaotic behavior of a continuous fermentor for the production of ethanol. Chem. Eng. Sci. 58, 1479–1496.

Ghommidh, C., Vaija, J., Bolarinwa, S., Navarro, J.M., 1989. Oscillatory behaviour of *Zymomonas* in continuous cultures: a simple stochastic model. Biotechnol. Lett. 11, 659–664.

Hobley, T., Pamment, N., 1994. Differences in response of *Zymomonas mobilis* and *Saccharomyces cerevisiae* to change in extracellular ethanol concentration. Biotechnol. Bioeng. 43, 155–158.

Jöbses, I., Egberts, G., Luyben, K., Roels, J., 1986. Fermentation kinetics of *Zymomonas mobilis* at high ethanol concentrations: oscillations in continuous cultures. Biotechnol. Bioeng. 28, 868–877.

Jarzebski, A.B., 1992. Modeling of oscillatory behavior in continuous ethanol fermentation. Biotechnol. Lett. 14, 137–142.

Kuznetsov, Y., Levitin, V., 1997. CONTENT: Dynamical System Software. http://www.computeralgebra.nl/systemsoverview/special/diffeqns/content/gcbody.html.

Lee, K., Skotnicki, M., Tribe, D., Rogers, P., 1980. Kinetic studies on a highly productive strain of *Zymomonas mobilis*. Biotechnol. Lett. 2, 339–344.

Lee, K.J., Tribe, D., Rogers, P., 1979. Ethanol production by *Zymomonas mobilis* in continuous culture at high glucose concentrations. Biotechnol. Lett. 1, 421–426.

Li, C.C., 2009. Mathematical models of ethanol inhibition effects during alcohol fermentation. Nonlinear Anal.: Theory, Methods Appl. 71, 1608–1619.

Li, J., Mclellan, P.J., Daugulis, A.J., 1995. Inhibition effects of ethanol concentration history and ethanol concentration change rate on *Zymomonas mobilis*. Biotechnol. Lett. 17, 321–326.

Mahecha-Botero, A., Garhyan, P., Elnashaie, S., 2006. Non-linear characteristics of a membrane fermentor for ethanol production and their implications. Nonlinear Anal. Real World Appl. 7, 432–457.

Mclellan, P.J., Daugulis, A.J., Li, J.H., 1999. The incidence of oscillatory behavior in the continuous fermentation of *Zymomonas mobilis*. Biotechnol. Prog. 15, 667–680.

Paz Astudillo, I.C., Cardona Alzate, C.A., 2011. Importance of stability study of continuous systems for ethanol production. J. Biotechnol. 151, 43–55.

Shen, Y., Zhao, X.Q., Ge, X.M., Bai, F.W., 2009. Metabolic flux and cell cycle analysis indicating new mechanism underlying process oscillation in continuous ethanol fermentation with *Saccharomyces cerevisiae* under VHG conditions. Biotechnol. Adv. 27, 1118–1123.

Sprenger, G.A., 1996. Carbohydrate metabolism in *Zymomonas mobilis*: a catabolic highway with some scenic routes. FEMS Microbiol. Lett. 145, 301–307.

Sridhar, L.N., 2011. Elimination of oscillations in fermentation processes. AIChE J. 57, 2397–2405.

Wang, H.Z., Chen, B.Z., He, X.R., Qiu, T., Zhao, J.S., 2009. Singularity theory based stability analysis of reacting systems. Comput. Aided Chem. Eng. 27, 645–650.

Wang, H.Z., Chen, B.Z., He, X.R., Zhao, J.S., Qiu, T., 2010a. Modeling, simulation and analysis of the liquid-phase catalytic oxidation of toluene. Chem. Eng. J. 158, 220–224.

Wang, H.Z., Chen, B.Z., He, X.R., Zhao, J.S., Qiu, T., 2010b. Numerical analysis tool for obtaining steady-state solutions and analyzing their stability characteristics for nonlinear dynamic systems. J. Chem. Eng. Jpn. 43, 394–400.

Wang, H.Z., Chen, B.Z., Qiu, T., He, X.R., Zhao, J.S., 2012a. An approach considering both operation stability and system's Hopf bifurcations to chemical process design. In: 2012 AIChE Annual Meeting. Pittsburg, PA.

Wang, H.Z., Chen, B.Z., Qiu, T., He, X.R., Zhao, J.S., 2012b. An integrated quantitative index of stable steady state points in chemical process design. In: 11th International Symposium on Process Systems Engineering-PSE2012. Singapore.

Wang, H.Z., Yuan, Z.H., Chen, B.Z., He, X.R., Zhao, J.S., Qiu, T., 2011. Analysis of the stability and controllability of chemical processes. Comput. Chem. Eng. 35, 1101–1109.

Wang, H.Z., Zhang, N., Qiu, T., Zhao, J.S., Chen, B.Z., 2014a. Method for regulating oscillatory dynamic behavior in a *Zymomonas mobilis* continuous fermentation process. Ind. Eng. Chem. Res. 53, 12399–12410.

Wang, H.Z., Zhang, N., Qiu, T., Zhao, J.S., He, X.R., Chen, B.Z., 2012c. Analysis of Hopf points for a *Zymomonas mobilis* continuous fermentation process producing ethanol. Ind. Eng. Chem. Res. 52, 1645–1655.

Wang, H.Z., Zhang, N., Qiu, T., Zhao, J.S., He, X.R., Chen, B.Z., 2013a. A process design framework for considering the stability of steady state operating points and Hopf singularity points in chemical processes. Chem. Eng. Sci. 99, 252–264.

Wang, H.Z., Zhang, N., Qiu, T., Zhao, J.S., He, X.R., Chen, B.Z., 2014b. Optimization of a continuous fermentation process producing 1,3-propane diol with Hopf singularity and unstable operating points as constraints. Chem. Eng. Sci. 116, 668–681.

Wang, L., Zhao, X.Q., Xue, C., Bai, F.W., 2013b. Impact of osmotic stress and ethanol inhibition in yeast cells on process oscillation associated with continuous very-high-gravity ethanol fermentation. Biotechnol. Biofuels 6, 133.

Yuan, Z.H., Wang, H.Z., Chen, B.Z., Zhao, J.S., 2009. Operating zone segregation of chemical reaction systems based on stability and non-minimum phase behavior analysis. Chem. Eng. J. 155, 304–311.

PART

Carbon capture, utilization and sequestration

II

CHAPTER

7 Sustainable Process Design: Sustainable Process Networks for Carbon Dioxide Conversion

Rebecca Frauzem*, Pichayapan Kongpanna§, Kosan Roh¶, Jay H. Lee¶, Varong Pavarajarn||, Suttichai Assabumrungrat§, Rafiqul Gani*,[1]

**Department of Chemical and Biochemical Engineering, Technical University of Denmark, Kongens Lyngby, Denmark*

§Center of Excellence in Catalysis and Catalytic Reaction Engineering, Department of Chemical Engineering, Faculty of Engineering, Chulalongkorn University, Bangkok, Thailand

¶Department of Chemical and Biomolecular Engineering, Korea Advanced Institute of Science and Technology, Yuseong-gu, Daejeon, Republic of Korea

||Center of Excellence in Particle Technology, Department of Chemical Engineering, Faculty of Engineering, Chulalongkorn University, Bangkok, Thailand

[1]Corresponding author: E-mail: rag@kt.dtu.dk

7.1 INTRODUCTION

With an increase in population and industry throughout the world, comes an increasing need for energy, electricity, fuel, and heat requirements (EIA, 2013). The amount of energy used around the world is expected to increase by 56% before 2040 (EIA, 2013). A complex network linking the initial energy carrier with the consumed product describes the energy consumption. At various stages of the energy network, there are processes which result in emissions; for example, burning fuel results in the emission of flue gas (US Department of State, 2010). Consequently, the amount of emissions continues to increase with the increased energy demand.

Global warming is a primary environmental concern that is attributed to emissions (Dincer, 2010). Of these emissions, the compounds that are of concern are greenhouse gases (Dincer, 2010). Greenhouse gases are a selection of gases that trap heat that results in global warming; they include water vapor, carbon dioxide (CO_2), methane (CH_4), nitrous oxide (N_2O), ozone (O_3), and various types of halogenated substances containing fluorine, chlorine, and bromine (such as chlorofluorocarbons (CFCs) and hydrochlorofluorocarbons (HCFCs)) (Dincer, 2010). Of these compounds, carbon dioxide represents more than 85% of

Computer Aided Chemical Engineering, Volume 36. ISSN 1570-7946. http://dx.doi.org/10.1016/B978-0-444-63472-6.00007-0

greenhouse gas emissions (US Department of State, 2010). Therefore, the reduction of carbon dioxide emissions is a primary target of environmental policy and regulation (IPCC, 2007).

There are two long-term methods of CO_2 reduction that are under investigation and slowly being implemented: carbon capture and storage (CCS) and conversion (IPCC, 2007). As a result of the longer investigation time and the support of governments, CCS is the more prevalent of the two (IPCC, 2007). Recently, however, areas for improvement have arisen in the capture technology and the storage methods (Olah et al., 2009). There are a variety of capture technologies that exist: industrial, postcombustion, precombustion, and oxy-fuel (Wilcox, 2012). However, in order to achieve the desired separation, an input of energy, raw materials, or solvents is required. These inputs have a resulting environmental impact that should be avoided as much as possible (production of energy releases CO_2). In addition to these issues, the storage methods are also controversial. Storage is done in geological formations or in vessels for further use. But, there is concern that the technology is immature and still needs to be further developed (IPCC, 2005).

In contrast, conversion utilizes the emitted CO_2 to produce other valuable products (Aresta, 2010). This is achieved through chemical reactions that are often performed catalytically. The selection of processes is limited by the need for processes with high conversion and yield. This is hindered by the stability of CO_2 and the difficulty in selecting reactions that are favorable (Aresta, 2010). However, the processes that fulfill these requirements are desirable as they are able to not only produce desired products, but are also able to utilize some of the CO_2 emitted into the atmosphere (Aresta et al., 2013). The amount of emissions can thereby be reduced if the amount utilized through the conversion process is greater than the emissions from consumed energy and utilities. Therefore, work focusing on conversion processes shows great promise; the transformation of carbon dioxide into other compounds could help reduce the emissions and mitigate the storage concerns.

In order to describe conversion processes, the feed source (emissions) need to be detailed. Emissions are primarily the result of energy production and manufacturing processes; these emission sources represent 80% of the emissions globally (IPCC, 2007). The remaining sources are the result of natural phenomena such as decay and are therefore not considered as carbon dioxide sources for reduction methods. In addition to the amounts being emitted, the quality of the emissions needs to be described (Last and Schmick, 2011). Though high in quantity, flue gas emissions from energy production contain low concentrations of CO_2. In order for these emissions to be utilized, they must be purified using carbon capture technology (Wilcox, 2012). Alternatively, high-purity emissions can also be used directly (Zakkour and Cook, 2010). The amounts and processes that produce these emissions are listed in Table 7.1.

As Table 7.1 reveals, there is a large supply of high-purity emissions that can be used (approximately 300 million tons of CO_2 annually ($MtCO_2$/year)). However, in order to address the global issues, a combination of high-purity and low-purity emissions must be eliminated as high-purity emissions only represent 10% of all

Table 7.1 High-Purity Emission Sources (Frauzem, 2014)

CO_2 Source	Nonutilized Emitted Amount ($MtCO_2$/Year)	Process	Emission Information
Natural gas production	150	Natural gas sweetening	Acid gas purge vent gas
Ammonia production	119	Ammonia production with on-site hydrogen production	Purge gas from CO_2 scrubbing Vent gas
Organic chemical production process	6.3	Ethylene oxide production	Vent gas
Synthetic fuel production	27.6	Fischer–Tropsch	Vent gas from CO_2 scrubbing
		Methanol/DME	Vent gas

emissions (3500 $MtCO_2$/year) (IPCC, 2007). Therefore, additional work is being implemented in feasible carbon capture methods; these are limited to those that emit less CO_2 through energy consumption and processing than would be emitted without capture technology. Currently, amine stripping is one of the more prominent methods (Wilcox, 2012); this uses an amine, primarily monoethylamine (MEA), as a solvent to remove the carbon dioxide from the emission streams. This technology must be made more sustainable (emits less carbon dioxide in the process, both emission streams and energy consumption, than is captured through the process). Primarily, the reduction of the purity requirements in the separations is being investigated to help achieve more sustainable capture technology (Knudsen et al., 2007). With sustainable capture technology, it becomes possible to target a much higher amount of emissions for use in conversion processes.

Environmental concerns are the motivation for discovering and implementing long-term carbon dioxide reduction methods. With a proper understanding of the emissions and the possible reactions, conversion is a promising aspect of this solution. This chapter focuses on the formulation of a network of various conversion pathways to help reduce CO_2 and create valuable products. In the subsequent sections, the method implemented for superstructure generation and analysis, the network formulated, and case studies for methanol and DMC production are described.

7.2 IDENTIFYING MORE SUSTAINABLE DESIGNS

Knowing the background and motivation for the use of conversion processes to help combat the CO_2 problem, it is possible to follow a methodology to create a superstructure that networks various conversion processes. The objective of this chapter is the sustainable CO_2 conversion process design for the production of valuable products. Sustainability is determined by finding a nontrade-off solution compared with a base case design. This is achieved by a reduction in energy consumption, increased profit and reduced environmental impact relative to this original base case design. In addition, there is a second aim of overall carbon dioxide reduction within the process, as defined by the subsequent equation.

$$\text{NET } CO_2 \text{ flow} = \sum_{n}^{i} CO_2 \text{ generated} - \sum_{n}^{i} CO_2 \text{ utilized} \tag{7.1}$$

Here the amount input is that from the feed and the amount generated is the amount emitted through purge, for heating requirements and for electrical supply in the process. This is determined and compared by determining the net carbon dioxide relative to the amount of product produced ($kg_{CO_2,net}/kg_{Product}$) and should be negative (more input than output resulting in reduction).

The methodology followed consists of three stages: first, superstructure-based identification of feasible processing networks and optimal networks; second, base case design of a CO_2 conversion process; third, identification of more sustainable designs. The subsequent sections will detail the method followed for the formation of the network of sustainable CO_2 conversion paths, their selection and design as a base case, and their improvement as more sustainable designs.

7.2.1 STAGE 1—SUPERSTRUCTURE-BASED OPTIMAL PROCESSING NETWORK

The first aspect is the description of the generic form of the superstructure and its generation. Figure 7.1 shows the generic form of the superstructure interface. The intervals shown are the steps linking raw materials and products. This linkage is formed through a variety of processing paths shown as intervals.

This superstructure needs to be described mathematically to provide a resulting problem formulation that can be solved. The framework employs a systematic approach for solving both the economic evaluation and the engineering aspects of problems simultaneously so that a large number of alternatives can be compared at their optimal points by using mixed integer nonlinear programming (MINLP) (Quaglia et al., 2012; Zondervan et al., 2011). This mathematical description becomes nonlinear as a result of the complexity of the process models involved; for example, processes with recycle loops create a nonlinear set of equations. The result from this part is the optimal process technology together with the optimal material flows through the network, from which the corresponding performance and sustainability matrices can be calculated.

The process superstructure consists of processing steps and process intervals. A process interval presented in Figure 7.2 is defined as a process section, which can perform a certain processing task, e.g., mixing, reaction, separation, and transportation.

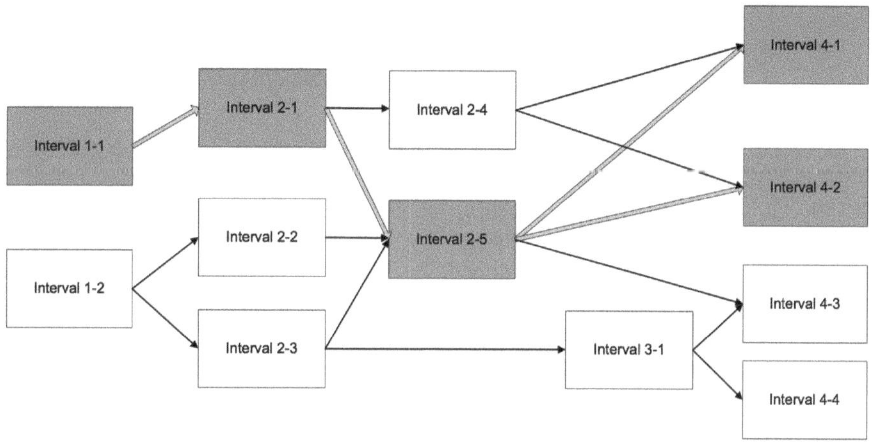

FIGURE 7.1

The generic form of superstructure and GAMS user interface for superstructure generation. In the blue (gray in print versions) boxes, the optimal network is indicated (Quaglia et al., 2012).

The important data for each of the process intervals are collected and organized in a predefined knowledge structure. At this step, collection of available data (in database as well as in literature) is prepared. Information on raw materials, main products, side products, reactions, catalysts, processing steps, reaction conversion,

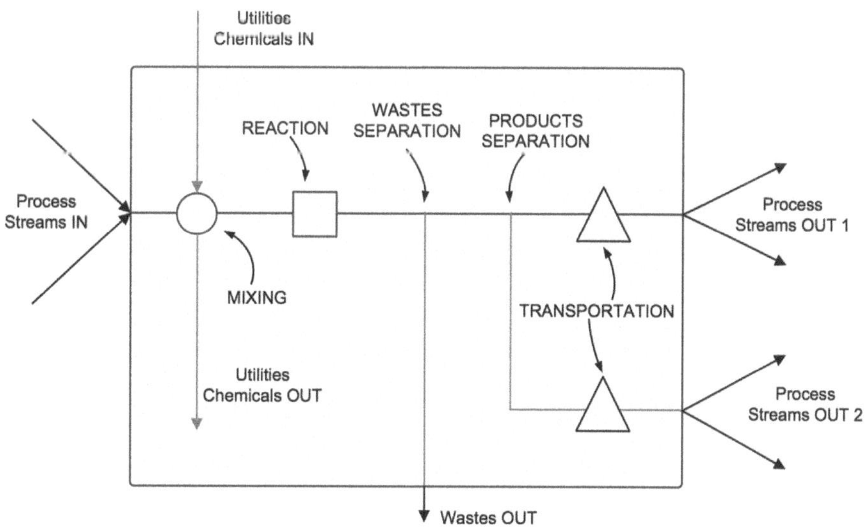

FIGURE 7.2

Process interval of superstructure (Quaglia et al., 2012).

energy requirements, and operating conditions is collected. The chemical and physical properties of substances related to the process, such as solubility, vapor pressure, azeotropes, heat of vaporization, boiling point, critical properties, and molar volume, are also important for generation of the process superstructure.

Then, the entire superstructure has to be described mathematically and formulated as a mixed integer problem. This is described by the set of equations shown in Table 7.2.

The combination of all this creates a mixed integer linear programming or MINLP model (the superstructure), which can be solved to find the feasible network of paths, including the optimal path or paths (Quaglia et al., 2012; Zondervan et al., 2011).

7.2.2 STAGE 2—BASE CASE DESIGN

The network of optimal or most promising processing paths found from the superstructure (Stage 1) contains various process steps linking the raw materials and products. Each of the processing steps contains a combination of unit operations. The next step is the detailed description of processing steps within the optimal path or most promising paths; this is done through creation of a flowsheet and then the subsequent analysis. From the results of the simulation, information about the process operation is obtained; the streams flow and composition are fully described and the operating conditions of the processing units are complete. This data is then implemented for analysis. This analysis is performed to determine the sustainability of the process via economic, environmental, and life cycle assessment (LCA) factors. With these factors, it is possible to set targets for improvements.

7.2.3 STAGE 3—IDENTIFICATION OF MORE SUSTAINABLE DESIGNS

Each of the targeted processing steps have been described in detail and areas for improvement are formed; this third step then takes these areas and generates improvements to the base case design. Tailor-made designs are obtained in a similar way as designing molecular chemical products with desired properties (Gani, 2004). The end result is a more sustainable design for the CO_2 conversion path.

7.3 METHOD AND TOOLS

The stages that are followed in the superstructure generation each require the additional support of methods and tools. A description of the workflow (described in Sections 7.2.1–7.2.3) and the dataflow are shown in Figure 7.3. The workflow follows the vertical stages, while the data flows horizontally as input and output of each stage.

Table 7.2 Mathematical Formulation of the Superstructure Network (Quaglia et al., 2012)

Objective function	$Max\ profit = \sum_{i,kk}(P3_{i,kk} \cdot F^{out}_{i,kk}) - \sum_{i,kk}(P2_{i,kk} \cdot R_{i,kk}) - \sum_{i,kk}(P1_{i,kk} \cdot F^{out}_{i,kk}) - W^{Price} \cdot \sum_{i,kk}(F^{R}_{i,kk} \cdot SW_{i,kk}) - \frac{CAPEX}{\#year}$	(7.2)
Raw material assignment	$F^{out}_{i,kk} = \phi_{i,kk}$	(7.3)
Utility consumption	$F^{M}_{i,kk} = \sum_{k}(F_{i,k,kk}) + \alpha_{i,kk} \cdot \mu_{i,kk} \cdot \sum_{i,k}(F_{i,k,kk})$	(7.4)
Reaction	$F^{R}_{i,kk} = F^{M}_{i,kk} + \sum_{rr,react}(\gamma_{i,kk,rr} \cdot \theta_{react,kk,rr} \cdot F^{M}_{react,kk})$	(7.5)
Waste separation	$F^{out}_{i,kk} = F^{R}_{i,kk} \cdot (1 - SW_{i,kk})$	(7.6)
	$F^{WASTE}_{i,kk} = F^{R}_{i,kk} - F^{out}_{i,kk}$	(7.7)
Product–product separation	$F^{out1}_{i,kk} = F^{out}_{i,kk} \cdot \sigma_{i,kk};\ F^{out2}_{i,kk} = F^{out}_{i,kk} \cdot (1 - \sigma_{i,kk})$	(7.8)
Superstructure flow model	$F^{1}_{i,k,kk} \le F^{out1}_{i,kk} \cdot \xi^{P}_{k,kk};\ F^{2}_{i,k,kk} \le F^{out2}_{i,kk} \cdot \xi^{S}_{k,kk}$	(7.9)
	$\sum_{k} F^{1}_{i,k,kk} = F^{out1}_{i,kk};\ \sum_{k} F^{2}_{i,k,kk} = F^{out2}_{i,kk}$	(7.10)
Superstructure logic model	$F^{R}_{i,kk} \le M \cdot y_{kk}$	(7.11)
	$\sum_{j}(y_k \cdot v_{st,k}) \le 1$	(7.12)
Throughput limitation	$\sum_{j} F^{R}_{i,kk} \le F^{MAX}_{KK}$	(7.13)
Piece-wise linear capital cost model	$CAPEX = \sum_{kk}\left[\sum_{j}(a_{j,kk} \cdot w_{j,kk} + \beta_{j,kk} \cdot Q_{j,kk})\right]$	(7.14)
	$F^{Thr}_{kk} = \sum_{j} Q_{j,kk};\ \sum_{j} w_{j,kk} = 1;\ Q^{0}_{j,kk} \cdot w_{j,kk} \le Q_{j,kk} \le Q^{0}_{j+1,kk} \cdot w_{j,kk}$	(7.15)

7.3.1 STAGE 1

In order to describe the superstructure and to solve the superstructure-based optimization, tools are required. First, a thorough collection of data and information from literature and databases is required. This is necessary to determine the raw materials, products, and the processing steps that link them. Then, this must be further defined mathematically using the equations described in Table 7.2. With this, the problem is

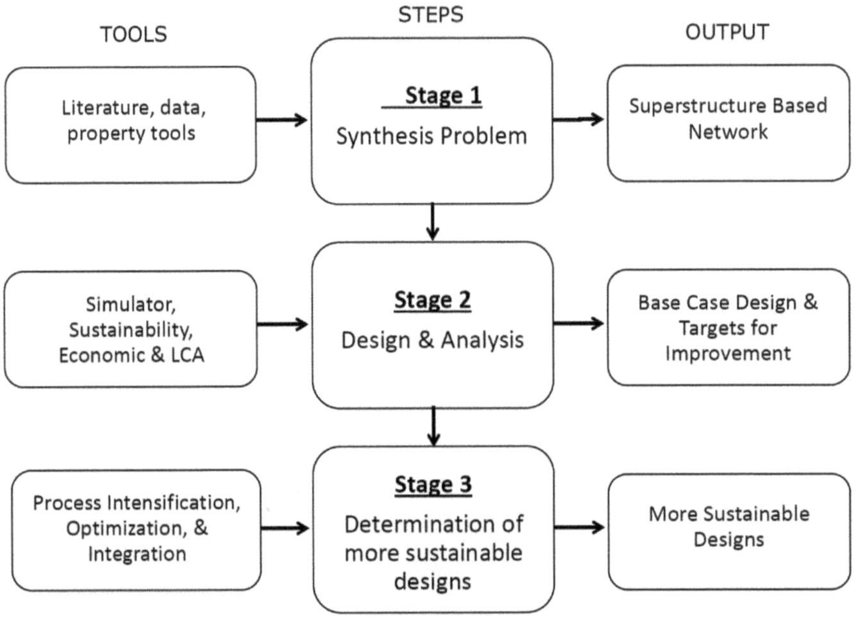

FIGURE 7.3

Description of workflow (vertical flow) and dataflow (horizontal flow) for formation of a network of sustainable carbon dioxide conversion processes.

defined and must be solved. In order to solve the problem formulated in this stage, GAMS is implemented (Quaglia et al., 2012). The result is a network of processes from the superstructure.

7.3.2 STAGE 2

The base case simulations for detailed description and analysis of the processing steps require the implementation of various tools to obtain the flowsheets and analysis factors. In order to achieve this, implementation of process simulators and analysis software is necessary. This includes simulators such as Aspen Plus® and PROII, and software tools developed at CAPEC-PROCESS such as ICAS, ECON, LCSoft, and SustainPro. With these tools, it is possible to describe the individual process steps in detail and determine the hot spots and points for improvement in the process.

7.3.3 STAGE 3

Within the final stage, the more sustainable design alternatives are found and evaluated. In order to achieve this, further tools and methods are required (tools from

ICAS). The goal of this final stage is the determination of more sustainable design alternatives to the base case for the processing steps. Therefore, tools and methods are necessary that aid in finding the alternatives that match targets determined upon analyzing the results from the base case simulations and software tools. Methods of process intensification, integration, and optimization are utilized. Process integration and optimization are common chemical engineering tools; heat and work are analyzed and integrated to minimize losses and operating conditions are manipulated to determine optimal operation according to various parameters. Process intensification uses target properties to determine more sustainable alternatives systematically (Babi et al., 2014). The flowsheet is defined by the tasks performed and then manipulated accordingly. Comparing the performance criteria then yields intensified alternatives to the base case. Each processing step can be analyzed for sustainability resulting in a network of more sustainable processes.

7.4 MORE SUSTAINABLE CO_2 CONVERSION PROCESS DESIGNS

The workflow is applied using the methods and tools described above (see Sections 7.3.1—7.3.3) for more sustainable carbon dioxide conversion paths.

7.4.1 CO_2 PROCESSING PATHS (STAGE 1)

Data and information is collected on the carbon dioxide feed source, feasible processing steps and possible products. The carbon dioxide feed can either be high-purity emissions directly or purified low-purity emissions. The low-purity emissions are purified using MEA stripping; the purity of the product is reduced to ensure the sustainability of the capture process. Then, a representative composition is described (Table 7.3).

From literature and available data, the proposed processes include the conversion to several chemicals such as polyurethane, salicylic acid, formic acid, methanol, dimethyl ether (DME), glycerol carbonate, styrene carbonate, urea, ethylene carbonate, propylene carbonate, and dimethyl carbonate (DMC) (Aresta et al., 2013). Combining all the information collected, the superstructure-based network for sustainable carbon dioxide conversion is described by Figure 7.4.

Table 7.3 Representative High-Purity CO_2 Emission Composition (Frauzem, 2014)

Mol%	CO_2	H_2	CH_4	H_2O	C_2H_6	Other
CO_2-rich	0.95	0	0.015	0.024	0.01	0.001

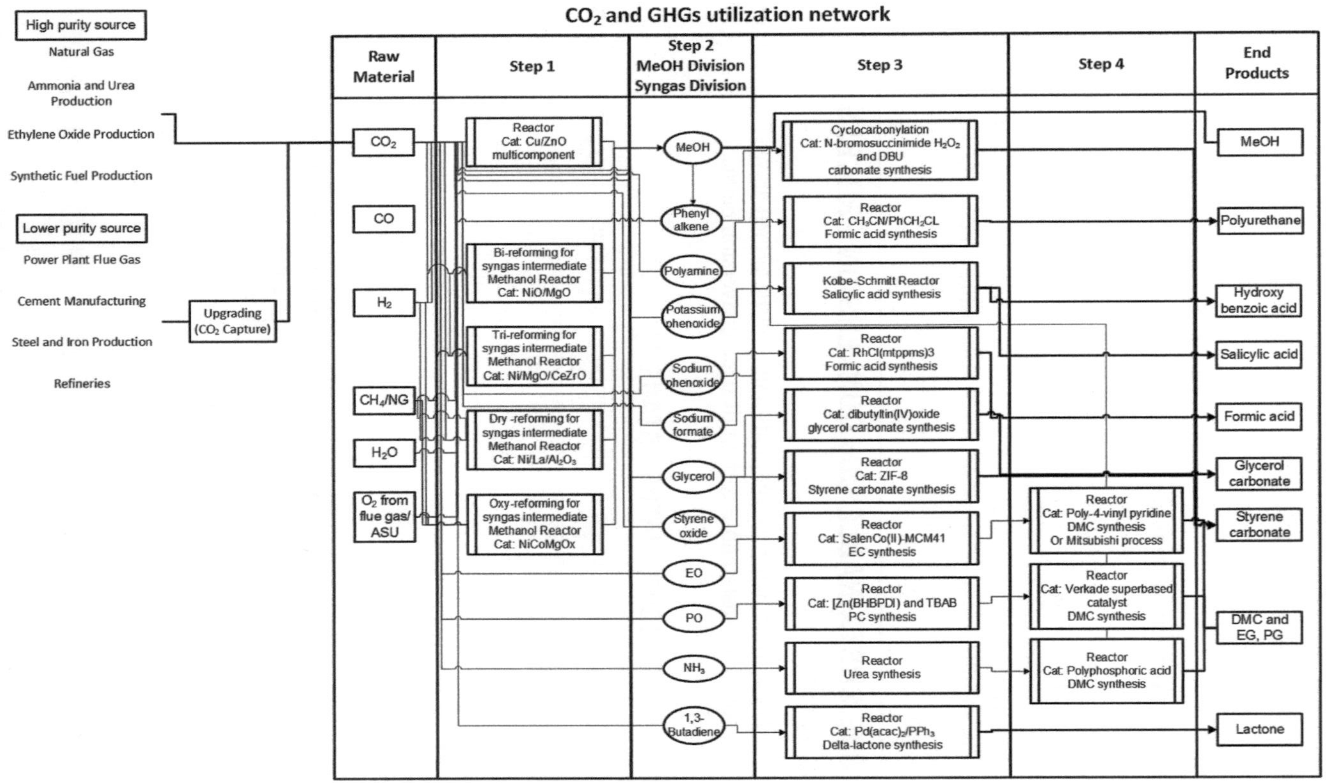

FIGURE 7.4

CO_2 conversion superstructure (Frauzem et al., 2015).

These steps can either take CO_2 to produce a desired product or to produce intermediates. In addition, various products can be used both as raw materials for other process steps and final product. There are two products (dimethyl carbonate (DMC) and methanol) that are highlighted in the following stage; for each there are numerous processing paths.

7.4.2 DETAILED FLOWSHEETS (STAGE 2)

DMC production is analyzed with a base case using the urea route. The bireforming route for methanol synthesis via synthesis gas (syngas) is also detailed. Finally, the production of methanol via direct hydrogenation of carbon dioxide is studied.

7.4.2.1 *Dimethyl carbonate production*

The DMC production pathways are all in the secondary processing steps of the superstructure. These processing steps involve previously produced inputs in addition to CO_2. Using thermodynamic analysis, DMC production via synthesis from urea is defined as the base case (Kongpanna et al., 2014). The detailed flowsheet of this process is shown in Figure 7.5.

With the flowsheet and the detailed simulations, the details of the conversion process to DMC are known.

7.4.2.2 *Methanol synthesis via synthesis gas*

One of the promising thermochemical CO_2 conversion reactions is methane reforming. This is a conventional processing way to prepare the syngas (mixture of hydrogen and carbon monoxide mainly, and small portion of carbon dioxide) at high temperature and pressure for producing a variety of chemicals such as methanol, gasoline, or diesel. There are four different kinds of CO_2-joining reforming pathways to produce the syngas: bireforming (or combined/mixed steam and dry reforming of methane), trireforming, dry reforming, and dry reforming with partial oxidation of methane. For this study, the base case is an industrial bireforming process (Holm-Larsen, 2001). This is simulated in detail using simulation software resulting in the flowsheet in Figure 7.6.

7.4.2.3 *Methanol production via direct hydrogenation*

Finally, the methanol production via direct carbon dioxide hydrogenation is analyzed. The current standard of production of methanol is via synthesis gas. However, often this process requires large amounts of energy resulting in high amounts of CO_2 emissions. Therefore, a direct method, which does not result in these emissions, is desirable. Direct hydrogenation is a conversion process that has been suggested as it uses carbon dioxide as a feed in a single reactor. In this process, there are three units: reactor, flash column, and distillation column. The reaction is performed in a packed bed reactor containing pellets of a copper/zinc oxide (Cu/ZnO) multicomponent catalyst (Saito et al., 1996). Using

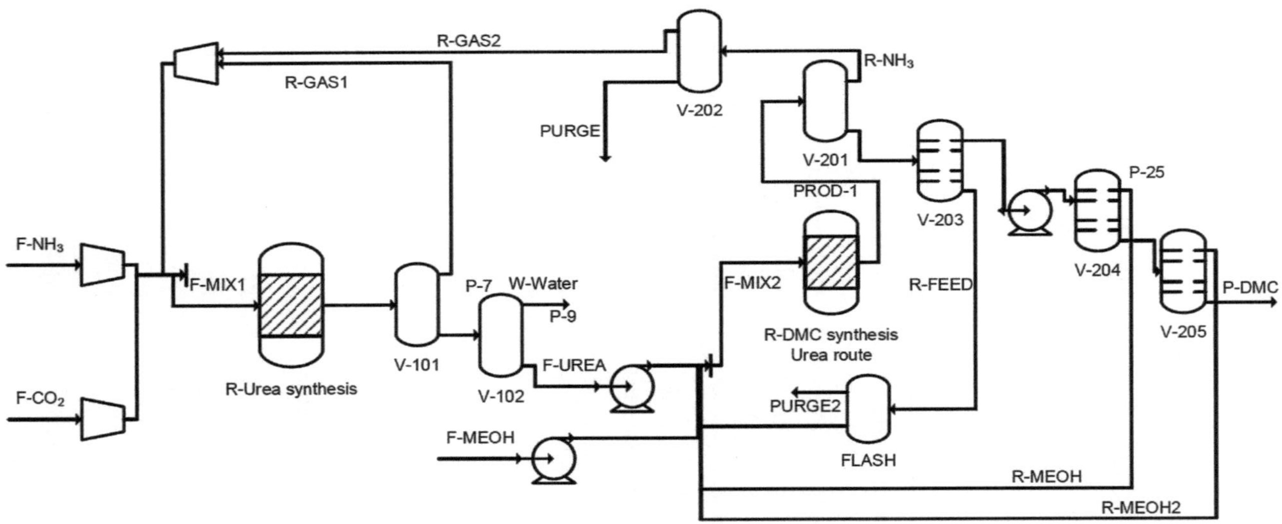

FIGURE 7.5

The DMC production process via urea route (base case design) (Kongpanna et al., 2014).

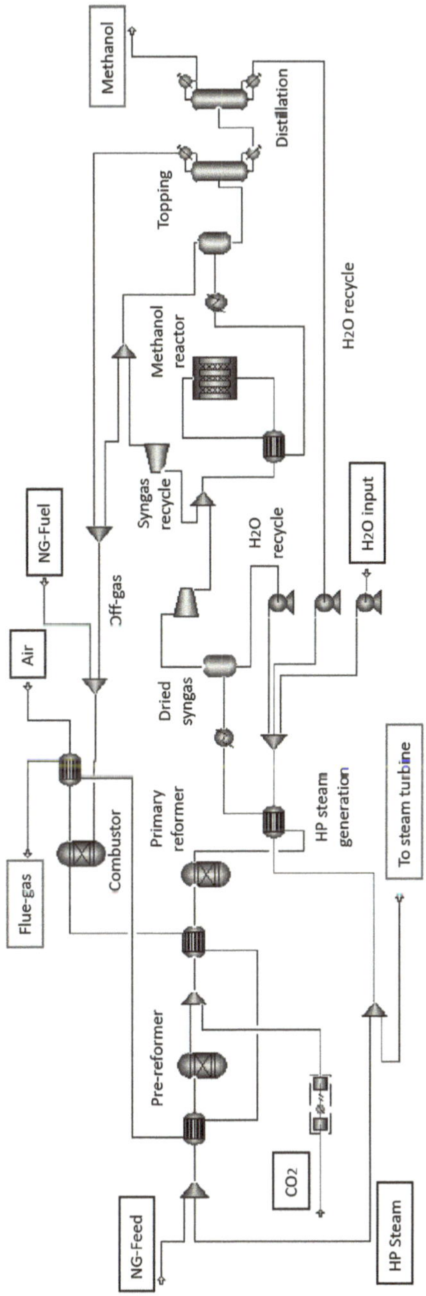

FIGURE 7.6

Flowsheet for the bireforming base case for methanol synthesis (Holm-Larsen, 2001).

thermodynamic and kinetic information collected from literature and data, the base case process is simulated resulting in Figure 7.7 (Kubota et al., 2001; Ushikoshi et al., 2000).

7.4.3 MORE SUSTAINABLE SOLUTIONS (STAGE 3)

Each of the base cases is examined thoroughly using economic factors, environmental parameters, and LCA factors. These reveal the hot spots of the processes and the locations for improvements. Using intensification methods, improvements are determined and compared for each base case.

7.4.3.1 Dimethyl carbonate production

From the base case simulation, sustainability factors show that the net CO_2 is low because this process is energy intensive in downstream separation. So, the process intensification by phenomena-based synthesis has been used in this work for generating other process candidates, which meet the target for improvement (Babi et al., 2014). From this, the base case is manipulated in three ways: incorporation of a pervaporation membrane, use of a membrane reactor, and the use of reactive distillation (Kongpanna et al., 2014). Using targets for the costs and environmental factors, it is possible to find a nontrade-off solution. The following radar chart (Figure 7.8) shows these factors for these improved designs compared with the urea base case design.

Additionally, the net carbon dioxide flow in the process is calculated for the various designs (Figure 7.9).

The radar chart and the net CO_2 calculations reveal that DMC production can be intensified with reactive distillation resulting in a more sustainable process. In addition, this reveals that conversion to DMC can reduce carbon dioxide emissions while producing a desirable product.

7.4.3.2 Methanol synthesis via synthesis gas

The second case, methanol synthesis via syngas, shows targets for improvement because it requires large amounts of energy to heat the reformer. This conventionally releases high amounts of carbon dioxide. An optimization is performed by manipulating the operating conditions and then comparing the net CO_2 with the base case. The result is a nontrade-off solution compared with conventional methanol production facilities as shown in the radar chart (Figure 7.10).

Then, the net CO_2 of the process is calculated. Relative to the current methanol production technology and the base case, the sustainability of the process is improved (Figure 7.11).

As these values show, the optimization of the bireforming-based methanol plant will result in a reduction compared with the conventional case. However, the process is still releasing CO_2 to the atmosphere (positive net CO_2). These results obtain

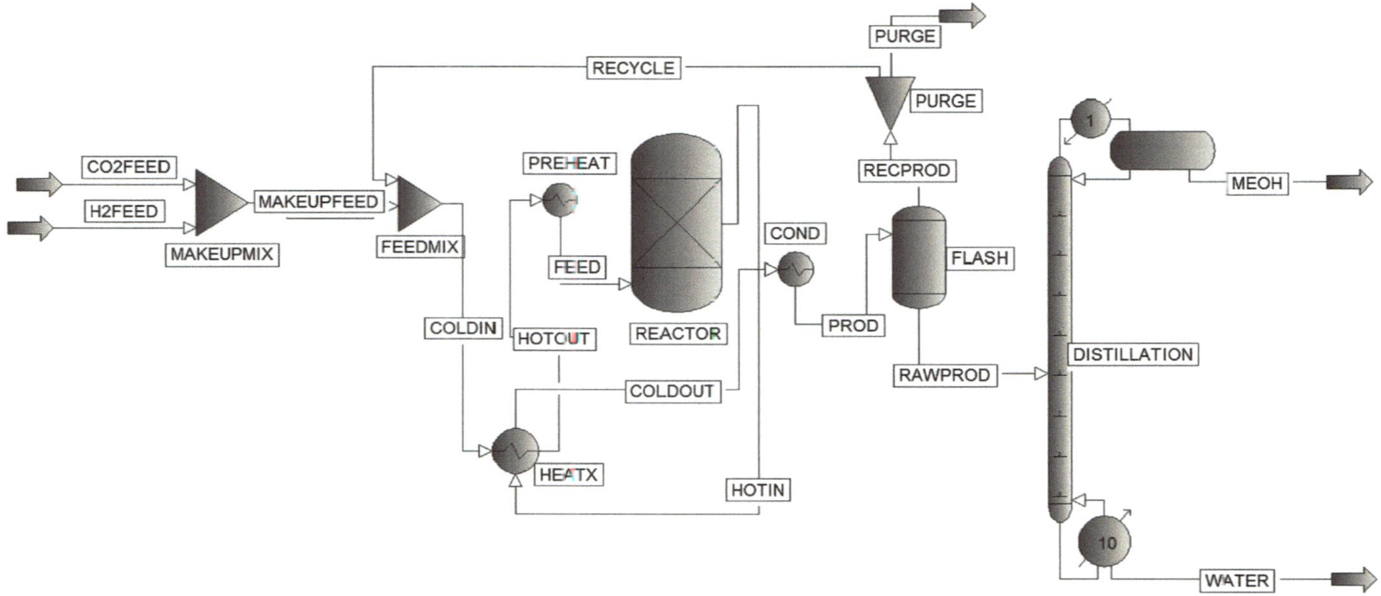

FIGURE 7.7

Flowsheet of the base case simulation for methanol production via carbon dioxide hydrogenation (Frauzem, 2014).

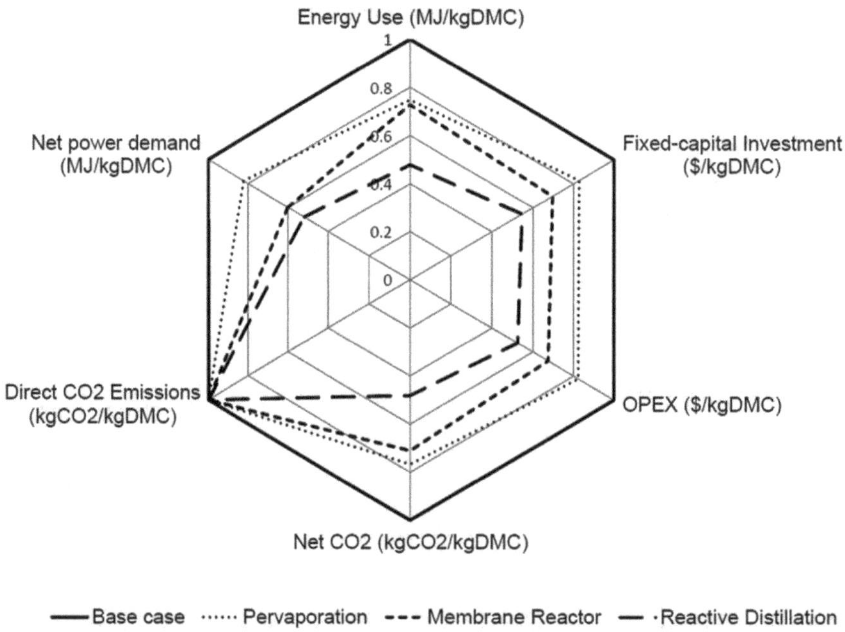

FIGURE 7.8

Radar chart comparing the various targeted values for the improved designs (Kongpanna et al., 2014).

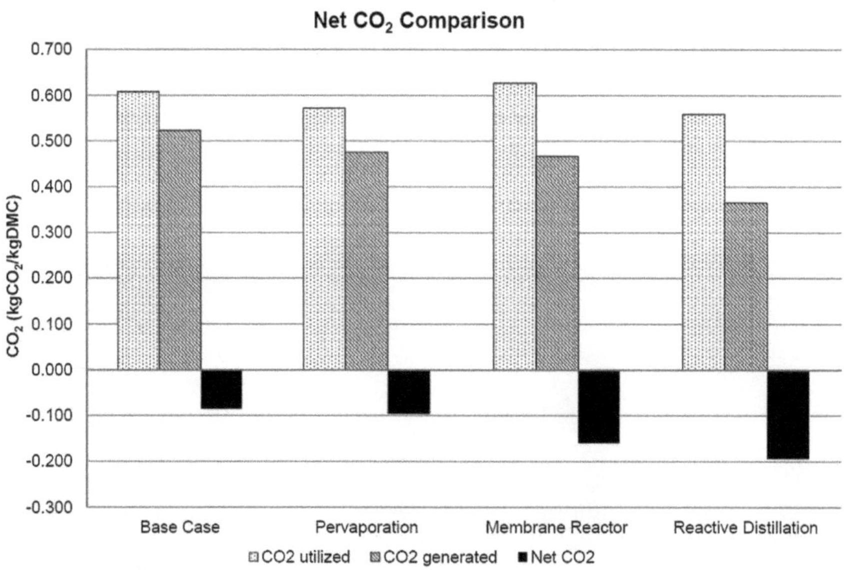

FIGURE 7.9

Net CO_2 for the base case compared with process intensification for a sustainable design of DMC production (Kongpanna et al., 2014).

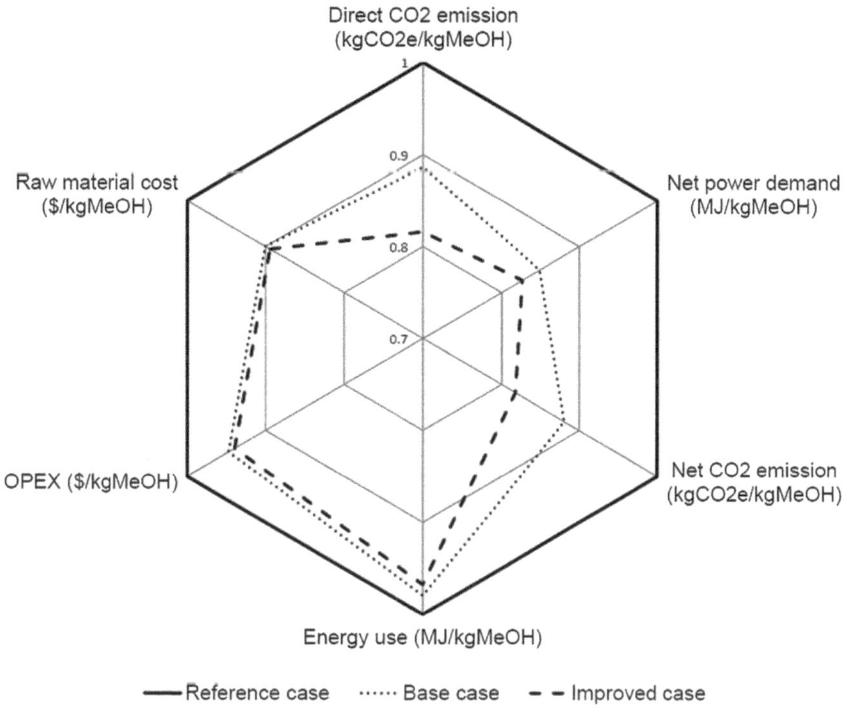

FIGURE 7.10

Radar chart comparing targeted factors for the conventional methanol plant with the base case and the optimization.

lower net CO_2 emission level as well as lower operating cost, but are still not a negative net CO_2 process. Therefore, the improvements are sustainable compared with the current production methods, but should be further optimized and intensified to yield a negative net CO_2 process.

7.4.3.3 Methanol production via direct hydrogenation

The third case study shows a hot spot in the energy consumption. Carbon dioxide hydrogenation to form methanol has low conversions to ensure that the selectivity is high. Therefore, there is a large recycle stream that results. Using heat integration it is possible to improve this hot spot to a net CO_2 of -1.125 kg CO_2/kg MeOH. In addition, the influence of the operating conditions and feed conditions is observed. These show the benefits of increasing the pressure, decreasing the temperature, and having pure feeds (Figures 7.12 and 7.13).

The economic factors for analysis show improvements compared with the base case with a reduction in the capital and operating costs. As this is new technology,

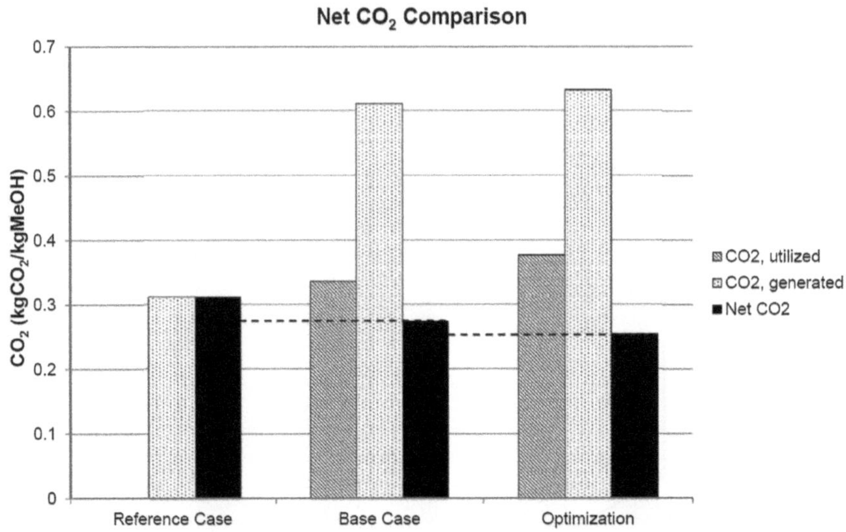

FIGURE 7.11

Comparison of sustainability of base case and improvements for methanol synthesis.

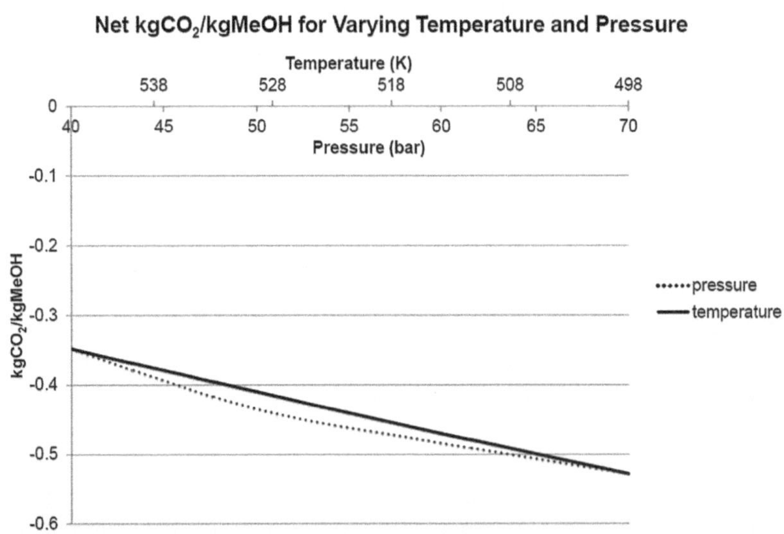

FIGURE 7.12

Indicators of the sustainability for various operating conditions (Frauzem, 2014).

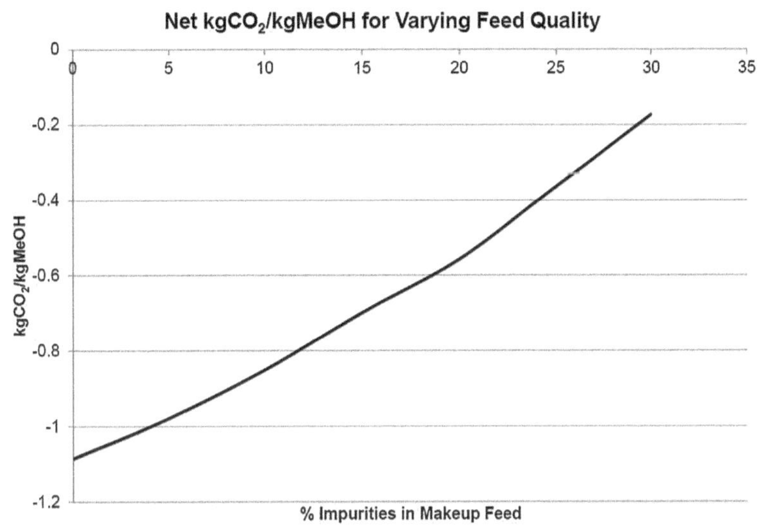

FIGURE 7.13

Indicators of the sustainability for various feed purities (Frauzem, 2014).

there is no reference point other than the base case design. Using heat integration, the base case is improved to make a more sustainable conversion process for the production of methanol.

7.5 CONCLUSIONS

In order to combat global warming and the influence of greenhouse gases, long-term solutions are required. Since carbon dioxide is the largest contributor to these phenomena, methods for reducing the emissions are necessary. This network of more sustainable conversion processes aims at utilizing carbon dioxide to create valuable products. The target is a nontrade-off solution that also shows an overall reduction of the emissions; the economic costs are reduced, environmental factors are improved, and the amount of carbon dioxide fed into the process is greater than the amount emitted or generated to ensure sustainability. Utilizing the method described, it is possible to generate a network based on superstructure optimization. This network is further analyzed and optimized using process integration, optimization, and intensification. The three paths, DMC production, methanol synthesis from syngas, and methanol production via direct hydrogenation, are used as case studies for the detailed flowsheet creation, analysis, and improvement. Following the steps described, it is possible to create a network of more sustainable carbon dioxide conversion processes that yield a portfolio of valuable products.

REFERENCES

Aresta, M., 2010. Carbon Dioxide as Chemical Feedstock, Carbon Dioxide as Chemical Feedstock. http://dx.doi.org/10.1002/9783527629916.

Aresta, M., Dibenedetto, A., Angelini, A., 2013. The changing paradigm in CO_2 utilization. J. CO_2 Util. 3–4, 65–73. http://dx.doi.org/10.1016/j.jcou.2013.08.001.

Babi, D.K., Woodley, J.M., Gani, R., 2014. Achieving more sustainable designs through a process synthesis-intensification framework. Comput. Aided Chem. Eng. 33, 31–36. http://dx.doi.org/10.1016/B978-0-444-63456-6.50006-5.

Dincer, I., 2010. Global Warming, Green Energy and Technology. Springer, US, Boston, MA. http://dx.doi.org/10.1007/978-1-4419-1017-2.

EIA, 2013. International Energy Outlook 2013. US Energy Information Administration (EIA), Washington, DC.

Frauzem, R., 2014. Techno-economic Evaluation of a CO_2-conversion Process.

Frauzem, R., Kongpanna, P., Roh, K., Lee, J.H., Pavarajarn, V., Assbumrungrat, S., Gani, R., 2015. Formulation of a Network and the Study of Reaction Paths for the Sustainable Reduction of CO_2 Emissions (In progress).

Gani, R., 2004. Chemical product design: challenges and opportunities. Comput. Chem. Eng. 28, 2441–2457. http://dx.doi.org/10.1016/j.compchemeng.2004.08.010.

Holm-Larsen, H., 2001. CO_2 reforming for large scale methanol plants – an actual case. Stud. Surf. Sci. Catal. 441–446.

IPCC, 2005. Carbon Dioxide Capture and Storage, Intergovernmental Panel on Climate Change (IPCC). Cambridge University Press, New York.

IPCC, 2007. Climate Change 2007: Mitigation of Climate Change, Intergovernmental Panel on Climate Change (IPCC). Cambridge University Press, New York.

Knudsen, J.N., Jensen, J.N., Vilhelmsen, P., 2007. First year operation experience with a 1 t/h CO_2 absorption pilot plant at Esbjerg coal-fired power plant. In: European Congress of Chemical Engineering (ECCE-6). Copenhagen.

Kongpanna, P., Pavarajarn, V., Gani, R., Assabumrungrat, S., 2014. Techno-economic evaluation of different CO_2-based processes for dimethyl carbonate production. Chem. Eng. Res. Des. 93, 496–510. http://dx.doi.org/10.1016/j.cherd.2014.07.013.

Kubota, T., Hayakawa, I., Mabuse, H., Mori, K., Ushikoshi, K., Watanabe, T., Saito, M., 2001. Kinetic Study of Methanol Synthesis from Carbon Dioxide and Hydrogen, pp. 121–126.

Last, G., Schmick, M., 2011. Identification and Selection of Major Carbon Dioxide Stream Compositions. Pacific Northwest National Laboratory, Oak Ridge.

Olah, G.A., Goeppert, A., Prakash, G.K.S., 2009. Beyond Oil and Gas: The Methanol Economy. Wiley-VCH, Weinheim.

Quaglia, A., Sarup, B., Sin, G., Gani, R., 2012. Integrated business and engineering framework for synthesis and design of enterprise-wide processing networks. Comput. Chem. Eng. 38, 213–223. http://dx.doi.org/10.1016/j.compchemeng.2011.12.011.

Saito, M., Fujitani, T., Takeuchi, M., Watanabe, T., 1996. Development of copper/zinc oxide-based multicomponent catalysts for methanol synthesis from carbon dioxide and hydrogen. Appl. Catal. A Gen. 138, 311–318. http://dx.doi.org/10.1016/0926-860X(95)00305-3.

US Department of State, 2010. U.S. Climate Action Report. US Department of State (US DOS). US DOS, Washington, DC.

Ushikoshi, K., Mori, K., Kubota, T., Watanabe, T., Saito, M., 2000. Methanol synthesis from CO_2 and H_2 in a bench-scale test plant. Appl. Organomet. Chem. 14, 819–825. http://dx.doi.org/10.1002/1099-0739(200012)14, 12<819::AID-AOC90>3.0.CO;2-#.

Wilcox, J., 2012. Carbon Capture. Springer, New York. http://dx.doi.org/10.1007/978-1-4614-2215-0.

Zakkour, P., Cook, G., 2010. CCS Roadmap for Industry: High-purity CO_2 Sources. Carbon Counts Company (UK) Ltd.

Zondervan, E., Nawaz, M., de Haan, A.B., Woodley, J.M., Gani, R., 2011. Optimal design of a multi-product biorefinery system. Comput. Chem. Eng. 35, 1752–1766. http://dx.doi.org/10.1016/j.compchemeng.2011.01.042.

CHAPTER

Pressure Swing Adsorption Optimization Strategies for CO_2 Capture

8

Yajun Wang*, Alexander W. Dowling*, Carlos Krieft[§], Andrea Walther[§], Lorenz T. Biegler*[,1]

**Department of Chemical Engineering, Carnegie Mellon University, Pittsburgh, PA, USA*

[§]Institute for Mathematics, Universität Paderborn, Paderborn, Germany

[1]Corresponding author: E-mail: lb01@andrew.cmu.edu

8.1 INTRODUCTION

Global energy-related CO_2 emissions increased by 1.4% in 2012 and have continued to grow (IEA, 2012). Unless CO_2 and other green house gas emissions are reduced in the chemical and power industries, the increase in average global temperature is likely to exceed 2 °C (IEA, 2012). One possible avenue to reduce CO_2 emissions and improve sustainability is the wide-scale implementation of carbon capture and sequestration technologies, where CO_2 is removed from each process, purified and stored underground in geological formations. Each industrial process will require a customized carbon capture system, tuned especially to minimize the additional costs and energy penalties. Different separation technologies will be preferred for different carbon capture applications.

Pressure swing adsorption (PSA) is a promising technology for certain CO_2 capture applications, such as integrated gasification combine cycle (IGCC) power plants. In this particular application, CO_2 must be economically removed from shifted syngas in the power plant, with both high CO_2 and high H_2 recoveries. Over the past few decades, PSA has emerged as a preferred technology for H_2 separation in the oil and gas industries as these systems operate near ambient temperatures, are cost-effective, do not require corrosive solvents or reagents and are moderately simple to operate. Several researchers have considered the use of PSA systems for gas separation in both IGCC and conventional coal power plants (Agarwal et al., 2010; Chou and Chen, 2004; Dowling et al., 2012; Gomes and Yee, 2002; Ko et al., 2003, 2005; Liu et al., 2011; Takamura et al., 2001; Tlili et al., 2009).

PSA utilizes the difference in pressure-dependent adsorption equilibrium and kinetics (and/or sterics) to achieve gas separation. During the pressurization steps of a PSA cycle, the gas mixture is fed into a packed bed at high to moderate pressures, and the *heavy product* gas preferentially adsorbs, resulting in a *light product* enriched stream. For the IGCC application, H_2 and CO_2 are the light and heavy products, respectively. During depressurization steps, the pressure of the bed is reduced

Computer Aided Chemical Engineering, Volume 36. ISSN 1570-7946. http://dx.doi.org/10.1016/B978-0-444-63472-6.00008-2

and gases desorb, producing a *heavy product* enriched stream. Most PSA cycles also include pressure equalization and reflux steps to reduce energy usage and improve recovery, respectively. For example, one of the earliest PSA cycles, the Skarstrom cycle, consists of four steps: pressurization, adsorption at high pressure, blowdown/desorption and purge. For a comprehensive overview of PSA systems, see Ruthven et al. (1994).

There are five main challenges that must be carefully addressed when designing PSA systems: (1) integrating the highly nonlinear partial differential and algebraic equation (PDAE) bed models, (2) calculating cyclic steady state (CSS) operating points, (3) scheduling the PSA cycle, (4) selecting the appropriate adsorption material(s), and (5) implementing optimization strategies. These issues are especially challenging, as the nonlinear dynamics make simulations and derivative calculations computationally expensive.

Some of the earliest PSA optimization work utilized short-cut models to avoid the computationally expensive PDAE bed equations. Smith and Westerberg (1990, 1991) simplified the mass and energy balances and applied mixed integer linear programming to minimize both capital and operating cost while scheduling PSA steps. Similarly, many modelers neglect pressure drops and/or heating/cooling effects for lab-scale PSA columns (Chou and Chen, 2004; Takamura et al., 2001), or they assume simplified pressure profiles (e.g., linear) for certain steps (Ayoub et al., 2011). Other modelers assume simple kinetic and/or equilibrium models that allow for analytical solutions to be obtained using wave theory (Subramanian and Ritter, 1997). Unfortunately, these simplifications limit their use for many industrial systems.

One of the most important aspects of PSA simulation is the calculation of the CSS, for which PSA cycles are designed. Most PSA simulators use direct substitution (i.e., Picard iterations, fixed-point methods) to calculate CSS operating points. This is undesirable from an optimization perspective, as it may require hundreds of iterations to reach CSS. Other researchers have formulated CSS calculations as a dynamic optimization problem (Ayoub et al., 2011) or used (quasi)-Newton methods to accelerate convergence to CSS (Smith and Westerberg, 1992). Jiang et al. (2003, 2004) simultaneously converged the CSS conditions while optimizing a performance metric, using a reduced space sequential quadratic programming (SQP) method. This approach is analogous to simultaneously converging and optimizing a flow sheet using equation-based simulation/optimization tools (e.g., gPROMs, ASPEN/EO). They demonstrated significant improvements over "black box" direct substitution methods.

Determining the operating conditions (pressures, valve positions, etc.) for a PSA cycle is only half of the design problem. The designer must also select the type and sequence of PSA steps. Nikolic et al. (2009) proposed a systematic approach to determine an optimal sequence and optimal number of beds based on a state-task network method. However, for larger, more complex cycles, their formulation is computationally intractable, as it requires the solution of mixed integer nonlinear optimization problems that include highly nonlinear PSA bed models. Many optimization practitioners bypass this issue by simply assuming a PSA cycle configuration and only optimizing the operating conditions (Haghpanah et al., 2013; Hasan et al., 2012).

This approach is restrictive, as the selected cycle may not contain the best energy-saving steps (product purge/reflux, pressure equalization, etc.). Instead, Agarwal et al. (2010) proposed a PSA superstructure that allows for simultaneous optimization of cycle operating conditions and step selection/sequencing using only continuous variables. A modified version of this superstructure was used by Dowling et al. (2012) and in this study. In both studies, it was determined that the scarcely used heavy product reflux step is promising for CO_2/H_2 separations. This type of step is generally not used in PSA cycles that favor light product recovery. Additional results on this design are presented in Section 8.2.

Selecting the best solid sorbents for mixture separation is essential for PSA processes. Many practitioners either assume specific adsorbents or solve the adsorbent selection problem using simplified models. The exception to this trend is by Floudas and coworkers (First et al., 2014; Hasan et al., 2012, 2013), who recently proposed an advanced methodology for PSA adsorbent screening and optimization for CO_2 capture. By solving a sequence of optimization problems, geometric data for zeolites are used to predict adsorbent properties required as inputs for PSA simulators.

The final challenge for PSA design is to construct an optimization methodology that accommodates the previously mentioned challenges, namely, efficient optimization of the detailed PDAE bed models. Nilchan and Pantelides (1998) discretized the system in both temporal and spatial domains and applied a Newton-type method to solve the nonlinear optimization problem, yielding both single-bed rapid PSA and two-bed six-step PSA designs. Extremely fine resolution is required for simulating the rapid bed profile changes. However, the large number of discretized bed equations challenges the nonlinear solver. Ko et al. (2003, 2005) discretized the spatial domain by method of lines (MOL) and applied a penalty term in the objective function to guarantee CSS. They implemented this method to N_2/CO_2 separation cases, using reduced space SQP method in gPROMS. Jiang et al. (Biegler et al., 2005; Jiang et al., 2003, 2004, 2005) used a simultaneous tailored optimization approach for both unibed and multibed design. They also used Newton-based methods to quickly converge to CSS. To further accelerate the optimization, sensitivity equations were computed in parallel. Khajuria and Pistikopoulos (2013) provided a real-time optimization strategy to simultaneously design, operate, and control PSA processes. CSS was considered over time horizon. Path constraints were implemented to avoid nonrealistic values of intermediate variables. The dynamic feasibility test was performed to account for uncertain model parameters. However, the studies mentioned above all require step specification a priori; cycle scheduling is not considered in the optimization formulation. In contrast, Dowling et al. (2012) modified and applied Agarwal's superstructure-based approach (Agarwal et al., 2010) to optimize PSA processes for CO_2 capture. This approach simultaneously determines the optimal step types and other decision variables. They reported poor performance of the derivative-free solver bound optimization by quadratic approximation (BOBYQA) for the PSA problem. Performance of the optimization strategy improved significantly using first derivative information and the interior point optimizer (IPOPT) solver. Nevertheless, they observed heavy computational loads and difficulties in convergence to the optimum of the PSA process.

To decrease computational load, Hasan et al. (2012) used kriging models in the materials screening and PSA system design framework. Optimal operating conditions of a four-step PSA cycle are determined with simultaneous selection of the best sorbent. Similarly, Khalighi et al. (2014) coupled nonisothermal isobaric micropore diffusion models with a PSA simulator and constructed surrogate models. Although their use of detailed PSA models is promising, the use of genetic algorithms provides no convergence guarantee, even to a local optimum. Agarwal and Biegler (2013) also used surrogate models based on proper orthogonal decomposition for optimization of PSA systems. They developed a trust-region framework to provide guaranteed convergence within a neighborhood around the optimum of the original detailed model.

In this study, we address these challenges by designing a PSA system for carbon capture in an IGCC power plant. Emphasis is placed on using advanced modeling and dynamic optimization methods to simultaneously schedule the PSA cycle and optimize operating conditions (pressures, flow rates, etc.) to minimize separation energy. More specifically, a periodic boundary condition formulation is used that combines optimization and convergence to CSS. The superstructure methodology proposed by Agarwal et al. (2010) leads to optimal step selections. This study extends the work of Dowling et al. (2012) with four contributions. First, an alternative flux limit scheme (weighted essentially nonoscillatory, WENO) is compared with previously used van Leer flux limiter, ensuring accurate and computationally efficient integration of the PDAE bed model. Second, sparse linear algebra routines are used to speed up integration of the bed model and direct sensitivity equations. Third, convergence issues reported in Dowling et al. (2012) are investigated, highlighting the importance of proper scaling, the impact of low curvature in the Lagrange function and inaccurate derivative information on optimizer termination. Finally, an optimal design for high CO_2 purity (98.49%) and H_2 recovery (99.00%) is presented.

This paper is organized as follows. In Section 8.2, the PSA partial differential algebraic equation bed model is summarized, along with PSA superstructure and optimization formulation with constraints to enforce CSS. In Section 8.3, the efficacy and efficiency of two flux limiter schemes are compared. In Section 8.4, the impacts of sparse linear algebra routines and problem size on computation time are discussed. Next, a case study for the CO_2/H_2 separation with 92% CO_2 purity and 90% CO_2 recovery is presented to elucidate the difficulties in optimizer convergence. In Section 8.6, all these methods are applied in a case study to design PSA systems with high H_2 recovery and CO_2 purity requirements for advanced coal power plants. Conclusions and future research directions are presented in Section 8.7.

8.2 PSA SUPERSTRUCTURE AND MATHEMATICAL MODEL

This work uses the PSA bed model presented by Jiang et al. (2003, 2004), the superstructure developed by Agarwal et al. (2010) and the extensions by Dowling et al. (2012). For completeness, the superstructure and mathematical models are summarized below.

8.2.1 PSA SUPERSTRUCTURE

The PSA superstructure allows for simultaneous optimization of bed-operating conditions and selection of steps (pressurization, blowdown, etc.). Shown in Figure 8.1, the superstructure includes two coupled packed beds. In the co-current bed (CoB, left), feed enters from the bottom and adsorption occurs. It produces light product (H_2) and optional recycle. In contrast, gas enters the top of the counter-current bed (CnB, right) where desorption occurs, and it produces heavy product (CO_2) and optional recycle out of the bottom. Using this configuration, only half of a PSA cycle needs to be simulated/optimized, as the cycle is symmetric; the beds "switch roles" for the second half of the cycle. This allows for interactions between the beds to be taken into account without simulating intermediate storage tanks.

Using the orientation for the cycle in Figure 8.1 for CO_2/H_2 separation, shifted syngas is fed into the co-current bed (CoB), increasing the bed pressure. As this occurs, CO_2 preferentially adsorbs onto the solid sorbent, "filling" the bed from the bottom up. This can be seen in bed profiles as a wave of H_2-enriched gas moving through the column. Once the desired operating pressure is obtained and before the wave breaks through the top of the bed, the top valve opens and H_2 product is produced. Simultaneously, in the counter-current bed (CnB), the pressure is slowly reduced causing gas to desorb, which forms the CO_2-enriched product. More complex cycles involve interactions between the beds via recycles and purges.

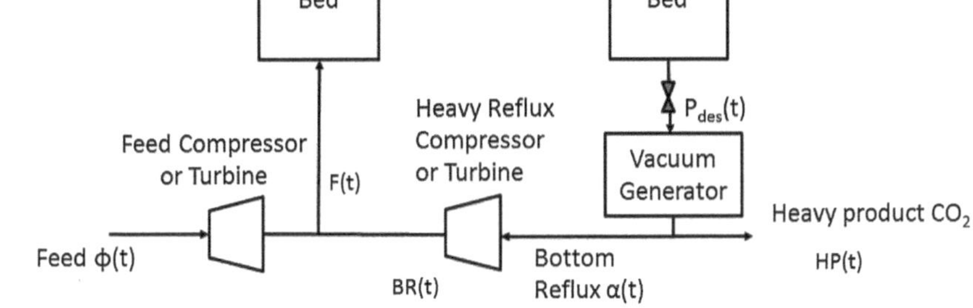

FIGURE 8.1

PSA superstructure.

The type of PSA step (pressurization, blowdown, light product recycle, etc.) and the cycle operating conditions are described using five continuous control variables in the superstructure: $\alpha(t)$, $\beta(t)$, $\phi(t)$, $P_{ads}(t)$, and $P_{des}(t)$, representing bottom reflux fraction, top reflux fraction, feed fraction, pressure at the top of CoB and pressure at the bottom of CnB, respectively. A PSA cycle consists of a sequence of steps. The control variables remain constant over each individual slot. There are two additional design variables, slot duration t_s and height of beds L_{bed}, which do not vary with time. When applying the superstructure, the designer chooses the number of PSA time slots a priori. This parameter can be considered in the optimization problem by resolving with various values and comparing the results.

8.2.2 MATHEMATICAL MODEL

The mathematical model of PSA process involves coupled PDAEs, which describe the dynamic behavior of the system. The model is based on several underlying physical phenomena, including mass, energy, and momentum balances, mass transfer, ideal gas thermodynamics, valve logic, and auxiliary equipment power consumption calculations. Additionally, sharp fronts in this process make the problem stiff, requiring an efficient numerical method to accurately solve these PDAEs.

8.2.2.1 Mass, energy, and momentum balance equations

8.2.2.1.1 Mass balance

The transient gas component mass balance includes adsorption kinetics and equilibrium, as well as fluid dynamics. The three terms in Eqn (8.1) are accumulation in gas phase, accumulation in solid phase, and gas convection. Consistent with other studies (Agarwal, 2010; Dowling et al., 2012; Jiang et al., 2004), radial variation and axial mixing are neglected, especially for models of large-scale industrial units (Ruthven et al., 1994).

$$\varepsilon_b \frac{\partial C_i}{\partial t} + (1-\varepsilon_b)\rho_s \frac{\partial q_i}{\partial t} + \frac{\partial (vC_i)}{\partial x} = 0 \quad i \in \{gas\ components\} \tag{8.1}$$

8.2.2.1.2 Energy balance

Because the adsorption (desorption) process is exothermic (endothermic), there are spatial and temporal variations in temperature. These variations are important to calculate with an energy balance because temperature impacts adsorption equilibrium and kinetics. For simplicity, the solid and gas phases are assumed to be at the same temperature.

$$\left(\varepsilon_t \sum_i C_i \left(C_{pg}^i - R\right) + \rho_s C_{ps}\right) \frac{\partial T}{\partial t} - \rho_s \sum_i \Delta H_i^{ads} \frac{\partial q_i}{\partial t} + \frac{\partial (vh)}{\partial x} + U_A (T - T_w) = 0 \tag{8.2}$$

where $C_{pg}^i = a_c^i + b_c^i T + c_c^i T^2 + d_c^i T^3 \quad h = \sum_i (C_i \int C_{pg}^i dT)$.

8.2.2.1.3 Momentum balance

Pressure drop is important to consider during PSA design, as it affects the compression (and energy) requirements of the system. Consistent with many other PSA studies (Alpay et al., 1993; Lee et al., 1999; Lu et al., 1993; Mohamadinejad et al., 2000), the Ergun equation is used to calculate the pressure drop.

$$-\frac{\partial P}{\partial x} = \frac{150\mu(1-\varepsilon_b)^2}{d_p^2\varepsilon_b^3}v + \frac{1.75}{d_p}\left(\frac{1-\varepsilon_b}{\varepsilon_b^3}\right)\left(\sum_i M_w^i C_i\right)v|v| \tag{8.3}$$

8.2.2.2 Adsorption phenomena

By considering both nonlinear adsorption isotherms and mass transfer, it is possible to design both kinetic and equilibrium dominated PSA systems with the same general mathematical model.

8.2.2.2.1 Adsorption isotherm

An extensive review of adsorption isotherms is presented by Ruthven et al. (1994). Here we apply the widely used dual-site Langmuir isotherm to represent the dependence of adsorption equilibrium on temperature, pressure, and the presence of competing molecules.

$$q_i^\star = \frac{q_{1i}^s b_{1i} C_i RT}{1+\sum_j b_{1j} C_j RT} + \frac{q_{2i}^s b_{2i} C_i RT}{1+\sum_j b_{2j} C_j RT} \tag{8.4}$$

where $q_{mi}^s = k_{mi}^1 + k_{mi}^2 T \quad b_{mi} = k_{mi}^3 exp\left(\frac{k_{mi}^4}{T}\right) \quad m = 1, 2.$

8.2.2.2.2 Adsorption kinetics

In the linear driving force equation, the mass transfer rate of adsorption/desorption is proportional to the difference between actual and equilibrium sorbent loadings (driving force). This approximation is widely used in PSA simulators and has been validated by Ruthven et al. (1994).

$$\frac{\partial q_i}{\partial t} = k_i(q_i^\star - q_i) \tag{8.5}$$

8.2.2.3 Ideal gas thermodynamics

This study assumes that all gases follow the ideal gas law, due to the moderate temperature and pressure ranges considered.

$$P = RT\sum_i C_i \tag{8.6}$$

For nonideal systems at higher temperature and pressure, a more complex equation of state can be considered.

8.2.2.4 Valve logic

The two valves in the superstructure (Figure 8.1) are used to maintain one-directional flow in each PSA column. Flow reversals are not considered in these bed models, as additional modeling effort would be required to ensure consistent derivatives for optimization (Jiang et al., 2003). In the CoB, the flow direction must remain from bottom to the top, and $P_{ads} \leq P_{CoB}(x,t)$. If P_{ads} is greater than P_{CoB} at the top of the bed, then (8.3) determines a negative velocity. To avoid this the actual velocity is forced to zero, using the *max* operator below. A zero velocity corresponds with the control valve at the top of the CoB being closed. Similarly, the *min* operator is applied for the CnB due to the negative velocity convention (Dowling et al., 2012). As a consequence, the valves are indirectly controlled via P_{ads} and P_{des}. Note that when P_{ads} is specified as much larger than the pressure in the bed, the valve above the CoB will remain closed and a pressurization step is realized.

$$v(t,x) = \begin{cases} max(0, v_k(t,x)) & if\ k = \text{CoB} \\ min(0, v_k(t,x)) & if\ k = \text{CnB} \end{cases} \tag{8.7}$$

Absolute value, min and max operators are nondifferentiable when velocity is zero. In order to facilitate optimization with derivative-based methods, these operators are smoothed using Eqn (8.8) (Balakrishna and Biegler, 1992).

$$\begin{aligned} & max(0,\xi) \approx 0.5\xi + 0.5\sqrt{\xi^2+\varepsilon^2} \\ & \left|\xi\right| \approx \sqrt{\xi^2+\varepsilon^2} \\ & min(0,\xi) = -max(0,-\xi) \end{aligned} \tag{8.8}$$

8.2.2.5 Auxiliary equations

Auxiliary equations are used to calculate purity and recovery of light and heavy products, feed flux and the power consumption/production of compressors, and expanders in the superstructure. These quantities are incorporated into the PSA design optimization problem as the objective function and constraints.

$$LP\ purity_{\mathrm{H_2}} = \frac{\int(1-\beta(t))v_a(t)C_{a,\mathrm{H_2}}(t)dt}{\int(1-\beta(t))v_a(t)\sum_i C_{a,i}(t)dt} \tag{8.9}$$

$$HP\ purity_{\mathrm{CO_2}} = \frac{\int(1-\alpha(t))(-v_d(t))C_{d,\mathrm{CO_2}}(t)dt}{\int(1-\alpha(t))(-v_d(t))\sum_i C_{d,i}(t)dt} \tag{8.10}$$

$$LP\ recovery_{\mathrm{H_2}} = \frac{\int(1-\beta(t))v_a(t)C_{a,\mathrm{H_2}}(t)dt}{Q_{feed,\mathrm{H_2}}} \tag{8.11}$$

$$HP\ recovery_{\mathrm{CO_2}} = \frac{\int(1-\alpha(t))(-v_d(t))C_{d,\mathrm{CO_2}}(t)dt}{Q_{feed,\mathrm{CO_2}}} \tag{8.12}$$

$$Q_{feed,i} = \int \phi(t) \upsilon_{feed} C_{feed,i} dt \tag{8.13}$$

$$E^k = \int \frac{\upsilon^k}{\eta^k} \left(\sum_i C_i^k \right) N_{stages}^k R T^k \left[\left(\frac{P_{out}^k}{P_{in}^k} \right)^{\frac{\gamma-1}{\gamma}} - 1 \right] dt \tag{8.14}$$

$$E = \frac{\sum_k E^k}{\int (1-\alpha(t))(-\upsilon_d(t)) C_{d,CO_2}(t) dt} \tag{8.15}$$

where $i \in \{Components\}$ $k \in \{Compressors/Expanders\}$.

8.2.3 SPATIAL DISCRETIZATION FOR NUMERICAL INTEGRATION

The above PSA bed model can only be solved analytically with several additional assumptions, including instantaneous equilibrium between the fluid and gas, negligible pressure drop, and linear isotherms (Cruz et al., 2005). These assumptions typically oversimplify the system model for industrial PSA applications, and the PDAE PSA model must be evaluated by numerical integration instead.

One popular approach for numerical integration of the bed model is to apply the method of lines and discretize in the spatial domain, producing a moderately sized differential algebraic equation system that can be numerically integrated with standard tools. Regarding the discretization step, there are several options including finite element, finite volume, and orthogonal collocation methods. For PSA systems, finite volume methods are advantageous due to their accuracy, stability, and computational efficiency. Because they carry the physical notion of flux, the values of the conserved variables are averaged in a volume and the conservation principle is always ensured (LeVeque, 2002; Cruz et al., 2005).

Figure 8.2 illustrates a finite-volume method for the one-dimensional case. u is the conserved variable. u_j represents the value of u at jth cell center and $u_{j\pm\frac{1}{2}}$

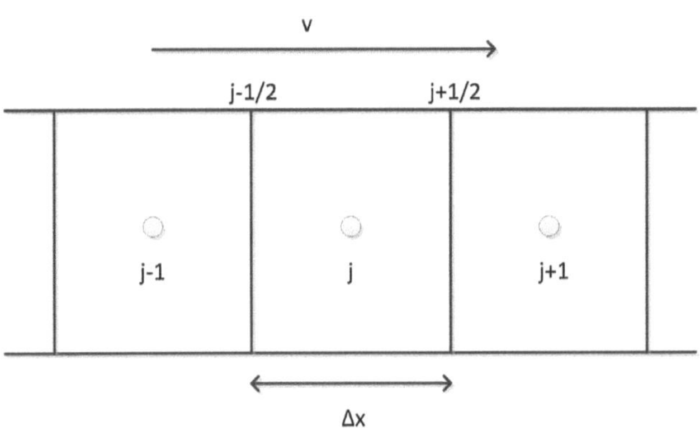

FIGURE 8.2

One-dimensional finite volume method.

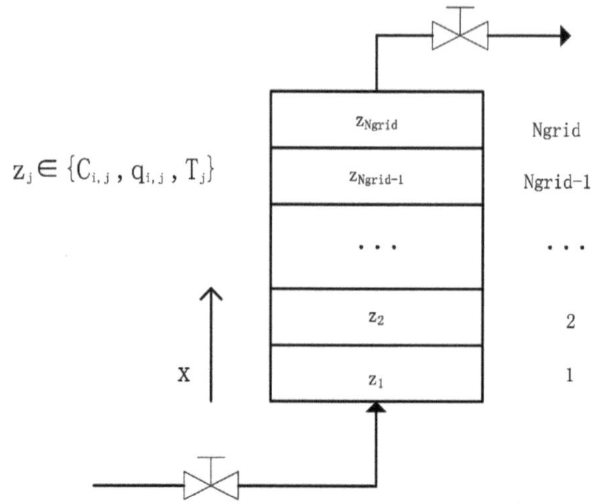

FIGURE 8.3

Discretize the bed by finite volume method.

represents the value of u at the wall. In this method, the conserved quantities are cell-average values, defined by

$$u_j(t) = \frac{1}{\Delta x} \int_{x_{j-\frac{1}{2}}}^{x_{j+\frac{1}{2}}} u(x,t)dx \tag{8.16}$$

where terms with respect to flux $f(u)$ are represented by values at the cell wall.

$$\frac{\partial f}{\partial x}(u) = \frac{1}{\Delta x_i}\left(f\left(u_{j+\frac{1}{2}}\right) - f\left(u_{j-\frac{1}{2}}\right)\right) \tag{8.17}$$

The PSA bed is divided into N_{grid} equal volume elements, as shown in Figure 8.3, and the partial differential equations (PDEs) are converted to ordinary differential equations. As a result, the state variables (C, q, and T) are now indexed in the spatial dimension. Moreover, selection of a good flux limiter is essential to maintaining accuracy during this discretization step, as addressed in Section 8.3.

8.2.4 CSS OPERATION

Different from many other chemical separation equipments, PSA systems are dynamic and typically operate at CSS. This is expressed using Eqn (8.18), in which the initial conditions of the CoB match the final conditions of CnB and vice versa. Note that only half of the PSA cycle is simulated because of symmetry in the superstructure.

$$z_{CoB}(x,0) = z_{CnB}(x,t_f), \quad z_{CnB}(x,0) = z_{CoB}(x,t_f) \tag{8.18}$$

8.2.5 PSA OPTIMIZATION AND NUMERICAL INTEGRATION

In this work, the objective of the PSA design problem is to minimize energy consumption per tonne of captured CO_2. This problem is subject to a number of constraints, including the PDAE bed model, CSS operations, CO_2 purity and recovery constraints, and bounds on the control/decision variables. These variables include $\alpha(t)$, $\beta(t)$, $\phi(t)$, $P_{ads}(t)$, $P_{des}(t)$ for each slot, along with slot duration t_s and bed dimension L_{bed}, which are time invariant. The reflux and feed fractions are bounded between 0 and 1, while other bounds are set by practical experience. This optimization formulation is shown in Eqn (8.19).

$$\begin{aligned}
&min && \Phi(z(x,t_f),p) \\
&s.t. && F\left(\frac{\partial z}{\partial t},\frac{\partial z}{\partial x},z(x,t),z(x,0),p\right)=0 \\
& && g(z(x,t),p)\geq 0 \\
& && 0\leq(\alpha(t),\beta(t),\phi(t))\leq 1 \\
& && b_L\leq(P_{ads}(t),P_{des}(t),t_s,L_{bed})\leq b_U \\
& && p=[\alpha(t),\beta(t),P_{ads}(t),P_{des}(t),t_s,L_{bed}]
\end{aligned} \tag{8.19}$$

Dowling et al. (2012) explored three approaches to formulate the PSA optimization problem: (1) a periodic boundary condition (PBC) formulation with direct sensitivity equations, (2) a fixed horizon (FH) formulation with adjoint sensitivity equations, and (3) a derivative-free optimization (DFO) approach using direct substitution to converge to CSS (within a tolerance). The first two approaches apply optimization algorithms that require first-order derivative information, whereas the DFO approach treats the problem as a black box; they found the DFO approach terminated at least 30+% higher objective function values than with the former. For binary separations with moderate number of finite volume cells, the periodic boundary condition formulation requires significantly less time to integrate the sensitivity equations. Furthermore, adjoint sensitivities obtained from discretized state equations may be inconsistent (Walther, 2007). For these reasons, we use the periodic boundary condition formulation in this study.

Gradient information, through $S(t)=\left(\frac{dz(t)}{dp}\right)^T$, is obtained by integrating the following direct sensitivity equations,

$$\frac{dS}{dt}=\frac{\partial f^T}{\partial z}S(t)+\frac{\partial f^T}{\partial p} \quad S(0)=\frac{\partial z_0^T}{\partial p} \tag{8.20}$$

$$\nabla_p\psi=S(t_f)^T\frac{\partial\psi}{\partial z}+\frac{\partial\psi}{\partial p} \tag{8.21}$$

where $p=\left[z_0^{(1)},\ldots,z_0^{(N_{states})},\alpha,\beta,\phi,P_{ads},P_{des},t_s,L_{bed}\right]^T$ (optimization variables) and $\psi=[\Phi,\ g_{pur},\ g_{rec},\ h_{PBC}]^T$ (objective function and constraints), in conjunction with the PSA bed model (f). This is done using a backward difference formula (BDF) integration method available in CVODES-2.7.0 (Hindmarsh and Serban,

2002), which natively supports these sensitivity equations. Derivatives $\left(\frac{\partial f^T}{\partial z}, \frac{\partial f}{\partial p}, \frac{\partial \psi}{\partial z}, \frac{\partial \psi}{\partial p}\right)$ are calculated using the automatic differentiation toolset ADOL-C-2.4.1 (Griewank et al., 1996), which is also used to provide exact (sparse) Jacobians of the bed model for the BDF method. Additional background on dynamic optimization using sensitivity equations can be found in the work of Biegler (2010) and Walther (2007).

8.3 COMPARISON OF FLUX LIMITERS

Steep adsorption fronts occur in PSA beds, especially with high adsorption pressures, high selectivities, and small dispersive effects (Cruz et al., 2005). Lower order methods used for spatial discretization of the PDE bed model (such as a central difference formula) lack accuracy and often flatten steep moving fronts. High-order methods offer better accuracy in smooth regions, but introduce oscillations near steep fronts. These oscillations lead to inaccurate profiles and infeasibility in the cyclic PSA simulation. The idea of a flux limiter is to combine both low- and high-order methods, leveraging the advantages of each. Ideally, this will provide at least second-order accuracy in smooth regions with well resolved, nonoscillatory behavior near discontinuous fronts. Flux limiters use a modified high-order method, in which additional calculations are performed near each discontinuity. This is commonly done by adding an "artificial velocity" to the higher-order method. Near a discontinuity, this additional term is large and it vanishes quickly in smooth regions (LeVeque, 2002).

8.3.1 VAN LEER FLUX LIMITER

In previous work (Dowling et al., 2012; Jiang et al., 2003), a modified version of the van Leer flux limiter was applied to a PSA optimization problem. With the van Leer flux limiter, $\Phi(r)$ is used to toggle on/off high-order terms, as shown below:

$$u_{j+\frac{1}{2}} = u_j + \frac{u_{j+1} - u_j}{2}\Phi(r') \quad if \;\; v_j \geq 0 \tag{8.22}$$

$$u_{j+\frac{1}{2}} = u_{j+1} + \frac{u_j - u_{j+1}}{2}\Phi(r') \quad if \;\; v_j < 0 \tag{8.23}$$

where u is the quantity to be conserved (such as mass flux). When $\Phi = 0$, the flux limiter simplifies to a first-order method. Φ is calculated as follows:

$$\Phi(r) = \frac{r + |r|}{1 + r} \tag{8.24}$$

$$r = \frac{u_j - u_{j-1}}{u_{j+1} - u_j} \quad if \;\; v_{j+\frac{1}{2}} \geq 0 \tag{8.25}$$

$$r = \frac{u_{j+1} - u_{j+2}}{u_j - u_{j+1}} \quad if \;\; v_{j+\frac{1}{2}} < 0 \tag{8.26}$$

At $|r| = 0$ the first derivative of flux limiter (Eqn (8.24)) is discontinuous, which leads to difficulties in integrating sensitivity equations to obtain derivative information for optimization. Thus, Jiang et al. (2003) proposed a smoothed version of the flux limiter, which uses a smoothed max operator.

$$r' = 0.5(r^2 + \varepsilon^2)^{\frac{1}{2}} + 0.5r \approx max(0, r), \quad \text{with } \Phi(r') = \frac{2r'}{1 + r'} \tag{8.27}$$

8.3.2 WENO FLUX LIMITER

In order to avoid the need for smoothing, the WENO method is also considered. The idea of WENO is to reconstruct the wall value $u_{j+\frac{1}{2}}$ by adaptive stencils to achieve high-order accuracy in smooth region and nonoscillatory property near discontinuous points. Haghpanah et al. (2013) developed Eqn (8.28) by considering two stencils and two cells in each stencil. Flux values at walls are calculated using

$$u_{j+\frac{1}{2}} = \frac{\alpha_{0,j}}{\alpha_{0,j} + \alpha_{1,j}}\left[\frac{1}{2}\left(u_j + u_{j+1}\right)\right] + \frac{\alpha_{1,j}}{\alpha_{0,j} + \alpha_{1,j}}\left[\frac{3}{2}u_j - \frac{1}{2}u_{j-1}\right] \tag{8.28}$$

where $\alpha_{0,j} = \frac{2/3}{(u_{j+1} - u_j + \delta)^4}$ $\alpha_{1,j} = \frac{1/3}{(u_j - u_{j-1} + \delta)^4}$.

Unlike the original van Leer flux limiter, the WENO method is differentiable and directly suitable for optimization.

8.3.3 COMPUTATIONAL COMPARISON

We compare the computational times and accuracy of these two flux limiters within the PSA bed model. The results for both a two component (H_2, CO_2) and five component (H_2, CO_2, N_2, CH_4, and CO) PSA separation system are shown in Table 8.1. Computations were performed using an Intel i5-2520M CPU (2.50 GHz) with 4.00 GB of RAM running Ubuntu Linux.

As seen in Table 8.1, WENO requires approximately 13% less computational time for two-component separation and 23% less time for five-component case

Table 8.1 Simulation Time to Obtain Cyclic Steady State

Flux Limiter	No Flux Limiter		van Leer		WENO	
Type	Time(s)	Iter	Time(s)	Iter	Time(s)	Iter
2 comps 11 grids	205.9	236	264.2	232	227.0	234
2 comps 21 grids	1064.4	202	1772.7	232	1537.4	234
2 comps 51 grids	10,754.3	246	15,915.3	236	14,022.2	236
5 comps 11 grids	1125.1	234	1749.0	234	1344.9	234
5 comps 21 grids	5116.9	212	10,124.5	246	7764.7	248
5 comps 51 grids	64,571.8	260	115,865.0	244	95,196.2	246

WENO, weighted essentially nonoscillatory.

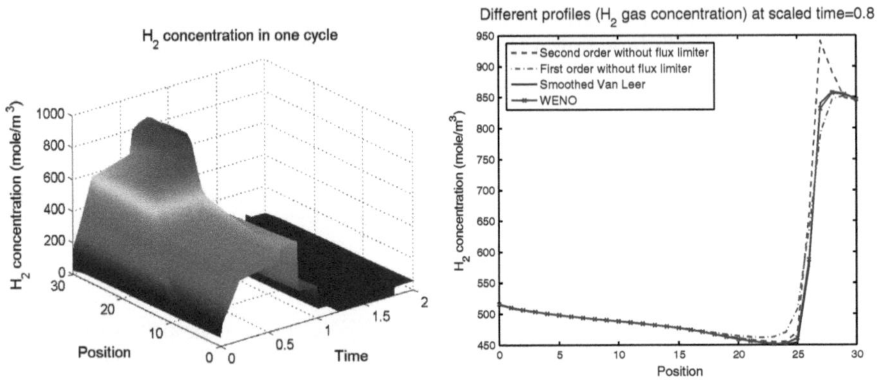

FIGURE 8.4

Difference in bed profiles with van Leer, WENO, and without flux limiter.

than the smoothed van Leer flux limiter. We suspect this lower computation time is partly because the WENO method does not need to be smoothed. Of course, the simulations without flux limiters require the least amount of time, but they are less accurate, as seen in Figure 8.4. Without flux limiters, the first-order method smears the steep front, while the second-order method introduces conspicuous oscillations. In contrast, van Leer and WENO simulate the sharp front without smearing or oscillations.

Thus, with similar accuracy and less computational burden, the WENO flux limiter is used in the remainder of this work. This improves the methods in Dowling et al. (2012), which used the smoothed van Leer flux limiter.

8.4 UTILIZING SPARSE LINEAR ALGEBRA TO IMPROVE COMPUTATIONAL EFFICIENCY

As discussed by Dowling et al. (2012), the sensitivity equations are computationally expensive to integrate, especially when many components or finite volume cells are considered. In some cases, integration of the sensitivity equations requires several CPU hours. As already demonstrated, careful selection of the flux limiter can help manage these computational loads. Another option is to improve the linear algebra routines in the numerical integrator.

In this work, the BDF method implemented in the CVODES software suite is used to integrate the PSA bed model and accompanying sensitivity equations. As part of the BDF method, a linear system of equations, shown below, is repeatedly solved to obtain the states at the next iteration.

$$M\left[y^{n(m+1)} - y^{n(m)}\right] = -G\left(y^{n(m)}\right) \tag{8.29}$$

$$M \approx I - \gamma J, \quad J = \frac{\partial f^T}{\partial y}, \quad \gamma = h_n \beta_{n,0}. \tag{8.30}$$

Here J is the Jacobian of the system, which should be factorized infrequently due to the high cost of matrix operations (Hindmarsh and Serban, 2002). As seen in Figure 8.5, J is sparse and less than 6% of the elements are nonzero. This sparsity is exploited by coupling HSL MA48 (Duff and Reid, 1993) with CVODES, with the goal of dramatically reducing computation time for integrating the PSA model.

Figure 8.6 and Table 8.2 compare the cumulative number of Jacobian updates in CVODES using the built-in dense solver as well as HSL MA48. The dense solver requires 208 Jacobian recalculations and 1972 Jacobian factorizations, whereas the sparse solver requires only 109 recalculations and 882 refactorizations. Interestingly, the spikes in the number of Jacobian recalculations shown in Figure 8.6 correspond to the opening and/or closing of control valves in the superstructure, as seen in Figure 8.7. These valve actions also correspond to rapid changes in pressure in the beds (see Figure 8.8), explaining the need to reevaluate the Jacobian.

The time savings due to use of sparse linear algebra routines become especially pronounced as the size of the PSA bed model grows (see Figure 8.9). On the other hand, these savings are limited, as the linear algebra routines only account for about 60% of the overall computational time. The remainder of the time is spent on matrix preprocessing, differentiation, and error tests. Further speedups may be possible by fine-tuning the automatic differentiation software settings and optimizing the C/C++ PSA simulation code for performance.

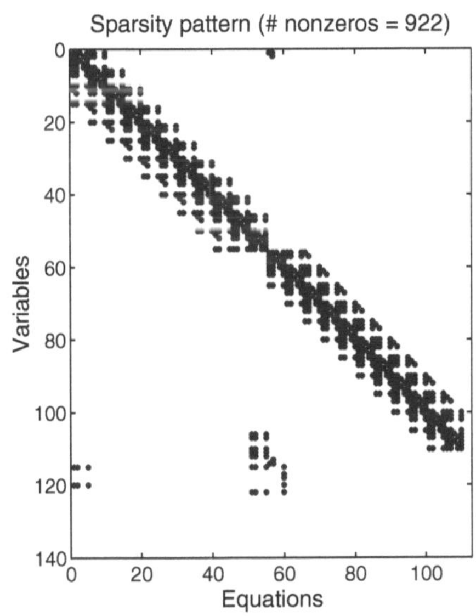

FIGURE 8.5

Sparsity pattern of Jacobian with two components and 11 finite volume cells (5.92% nonzero elements).

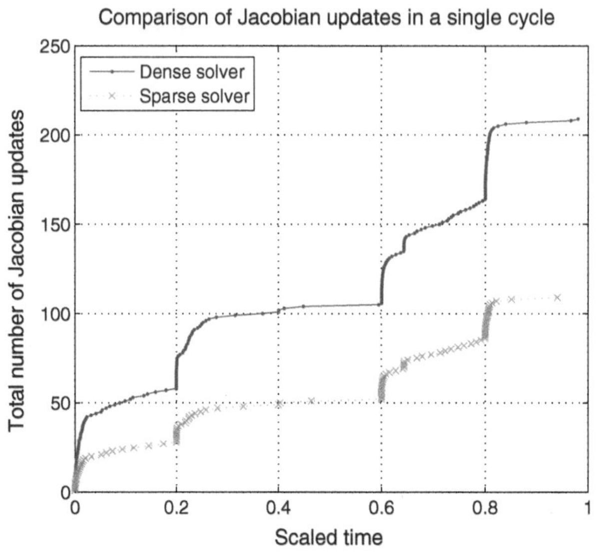

FIGURE 8.6

A comparison of Jacobian updates with dense and sparse linear solvers.

Table 8.2 A Comparison of Jacobian Updates with Dense and Sparse Linear Solvers

Slot		Dense	Sparse
1	#*J* factorizations	611	243
	#*J* calculations	58	27
2	#*J* factorizations	459	195
	#*J* calculations	44	21
3	#*J* factorizations	33	23
	#*J* calculations	5	3
4	#*J* factorizations	520	241
	#*J* calculations	57	34
5	#*J* factorizations	349	180
	#*J* calculations	44	24

8.5 OPTIMIZATION DIAGNOSTICS

To assess the behavior of the updated PSA optimization strategy, we considered a CO_2/H_2 separation case study with a feed composition of 42 mol% CO_2 and 58 mol% H_2, and with CO_2 purity and recovery requirements of 92% and 90%,

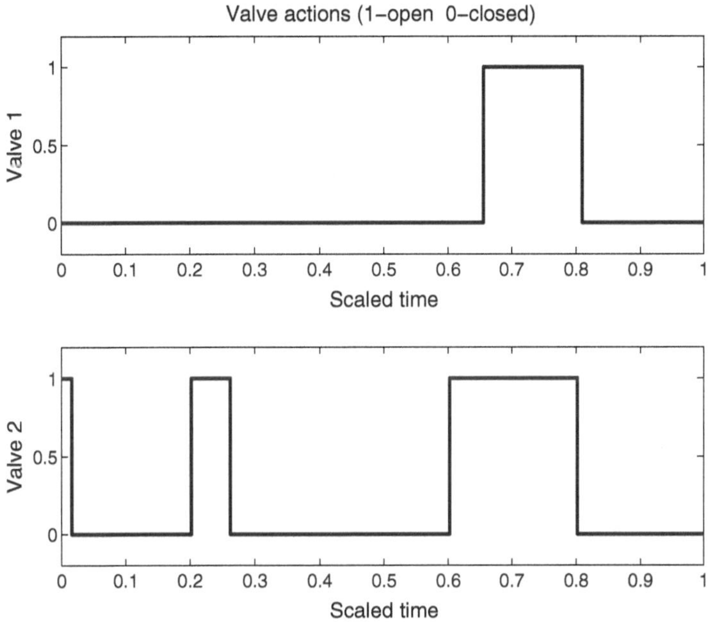

FIGURE 8.7

Valve actions.

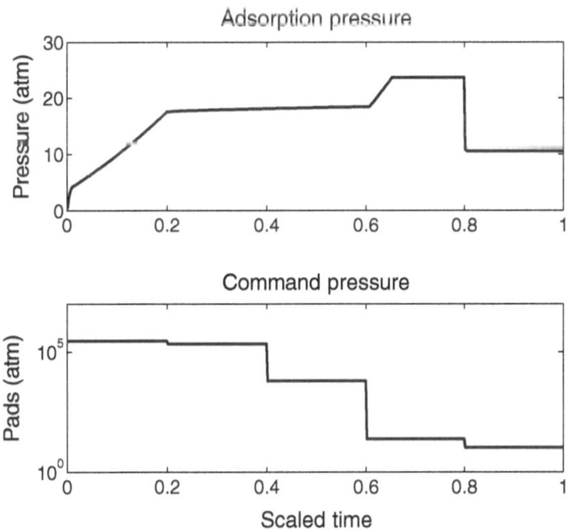

FIGURE 8.8

Adsorption/desorption pressures and command pressures change in one cycle.

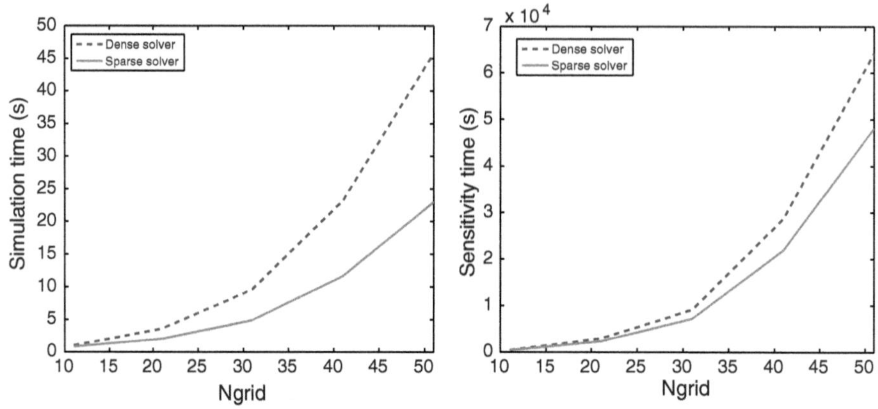

FIGURE 8.9

Performance gap between sparse and dense solvers grow as the number of finite volume cells increase in the two-component case.

respectively. Details of this case study are described by Dowling et al. (2012). In this study, the MATLAB optimization function `fmincon` with SQP algorithm is used. C code for PSA modeling and derivative calculation with CVODES, ADOL-C, and MA48 in HSL are linked with MATLAB. Tests and optimizations are done on an Intel Xeon E5-2440 CPU (2.40 GHz) with 16 GB of RAM.

Despite the computational speedups from switching to the WENO flux limiter and using sparse linear algebra, optimization of the CO_2/H_2 PSA separation system still experiences convergence difficulties. More specifically, the three optimization methods (`active-set`, `sqp`, and `interior-point`) available through MATLAB's `fmincon` function rarely terminate at the point with karush -kuhn -tucker condition (KKT) errors less than 20. This behavior is not surprising given the highly nonconvex nature and steep profiles of the PSA bed model, and the potential for inaccurate derivatives due to inevitable noise in the CVODES calculations. (This is despite relative and absolute error tolerances of 10^{-10} in CVODES.)

8.5.1 IMPORTANCE OF SCALING FEED FRACTION

We applied the `sqp` algorithm in MATLAB for the binary separation case, with each column discretized into 11 spatial grids, and imposed periodic boundary conditions to ensure CSS, we applied the `sqp` algorithm in MATLAB. The initial formulation terminated at an infeasible point, with results shown in the first row of Table 8.3.

To analyze the optimizer performance, we checked the quasi-Newton Hessian (2nd derivative) approximation at the final point, and found it was ill-conditioned. There were several extremely large elements, many of which corresponded to a single control variable: feed fraction (ϕ). We fixed the feed fraction ϕ and reran the optimization with the remaining control/decision variables. As shown in the second

Table 8.3 Convergence Comparisons for Pressure Swing Adsorption Optimization

	Objective Φ (kW h/ tonne CO_2)	Iterations	Constraint Violation	KKT Violation	Termination Condition
Unscaled ϕ	85.08	92	0.99	29.69	Not feasible[a]
Set ϕ constant	84.71	232	7.7×10^{-7}	4.4×10^{-3}	Min possible[b]
Scaled ϕ	84.71	178	2.7×10^{-7}	3.7×10^{-2}	Min possible[b]
High purity case	99.84	327/46[c]	1.79×10^{-7}	1.03×10^{-1}	Min possible[b]

[a]*Not feasible: "no feasible solution found. The size of the current step is less than the tolerance (10^{-6}) but constraints are not satisfied."*
[b]*Min possible: "local minimum found that satisfies the constraints. The objective function is nondecreasing and constraints are satisfied (tolerance 10^{-6})."*
[c]*Iterations for the 5-slot case, followed by the 4-slot case, as described in Section 8.6.*

row of Table 8.3, the optimizer terminated at a primal feasible point (purity, recovery, and CSS constraints are satisfied within a tolerance) with a smaller objective function value and lower KKT error. To further investigate the behavior of this variable, we adjusted the scaling of ϕ. The third row of Table 8.3 shows the results with ϕ scaled by a factor of 1000 and the optimization problem solved from the initial starting point, with ϕ included as a decision variable.

Results listed in the third row took 37 h to converge with dense solver, equivalent to 12.6 min per iteration. 1.8 min per iteration (15%) on average were saved using the spare solver.

8.5.2 HESSIAN APPROXIMATION, FLAT LAGRANGIAN, AND INEXACT FIRST DERIVATIVES

Second derivative information is not directly available from the sensitivity equations and the Hessian information is approximated in SQP using a quasi-Newton update. This may explain the convergence behavior seen in Table 8.3. The SQP algorithm terminated at a point with moderate KKT errors because the calculated quasi-Newton step was smaller than a termination tolerance. An ill-conditioned approximate Hessian would explain the small step size.

We calculated the reduced Hessian at the termination point (row three from Table 8.3) by perturbation to characterize the response surface of the Lagrange function. Some very small eigenvalues of the reduced Hessian matrix illustrate the flatness of this surface. These small eigenvalues correspond to control variables in the 3rd and 5th slots (P_{ads}, P_{des}, α, and β). In these two slots, valves are closed, making the objective function and constraints insensitive to these variables.

We then calculated the steepest descent steps in the reduced space of the active constraints to further investigate the stationarity at this point. Figure 8.10 shows

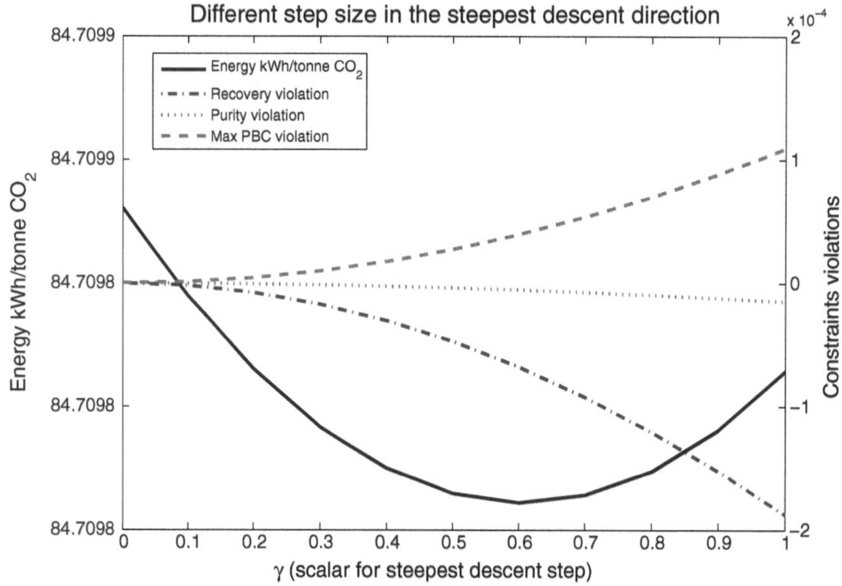

FIGURE 8.10

Objective value and constraint violations with different step size in the steepest descent direction.

plots the objective function and constraints in the steepest descent direction, scaled by the step size γ. When $\gamma = 0.6$, the largest decrease in the objective function is observed. As step size (γ) increases, violations of the equality constraints (CSS) increase, while inequality constraints (CO_2 purity and recovery) are improved. When $\gamma = 0.09$, the objective value reduces by 2.2×10^{-5} and the inequality and equality constraints remain within the tolerance (10^{-6}). This indicates that a descent direction in the reduced space exists, although the objective function improvement is small. The 2-norm of this step is 1.8×10^{-3}, much larger than the termination tolerance (10^{-6}) in `fmincon`. However, due to the inexact Hessian approximation, the 2-norm of the quasi-Newton step was only 2.7×10^{-7}. As the latter step was within the tolerance, `fmincon` terminated. Figure 8.10 also indicates a flat objective around the optimum; hence the optimizer terminates at a near KKT-point.

8.6 CASE STUDY: OPTIMIZATION OF PSA SYSTEMS FOR CO_2 CAPTURE AND HIGH H_2 RECOVERY

From the above case study, we obtained a better understanding of the convergence difficulties. In this section, we consider a higher CO_2 purity and H_2 recovery case and present optimal design results.

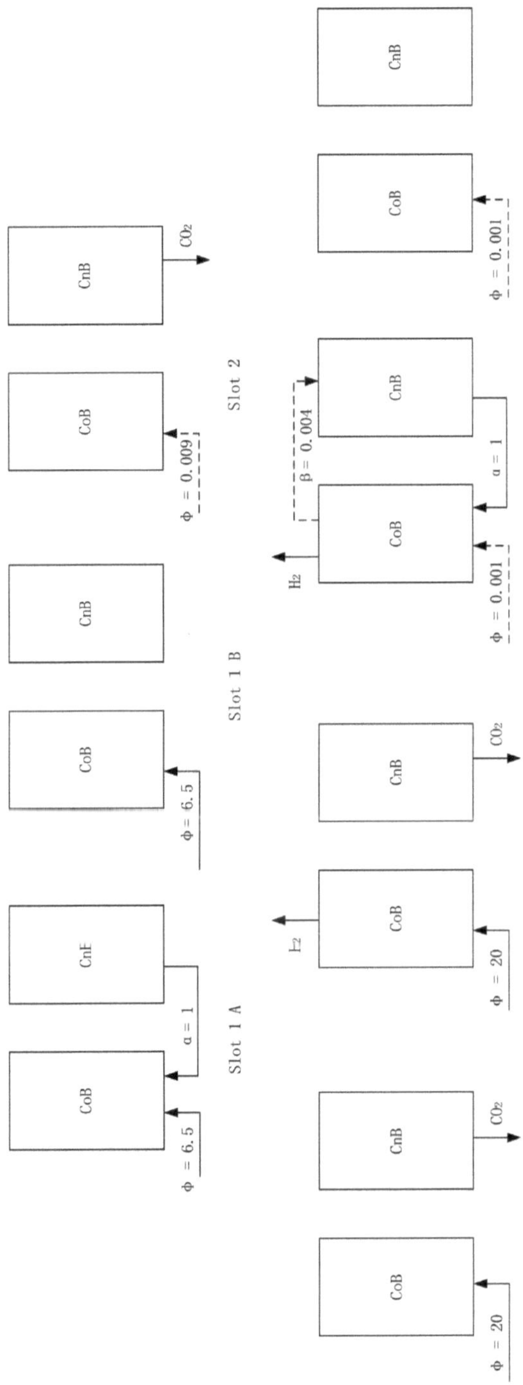

FIGURE 8.11

Optimal cycle for the CO_2/H_2 separation case with high H_2 recovery requirement.

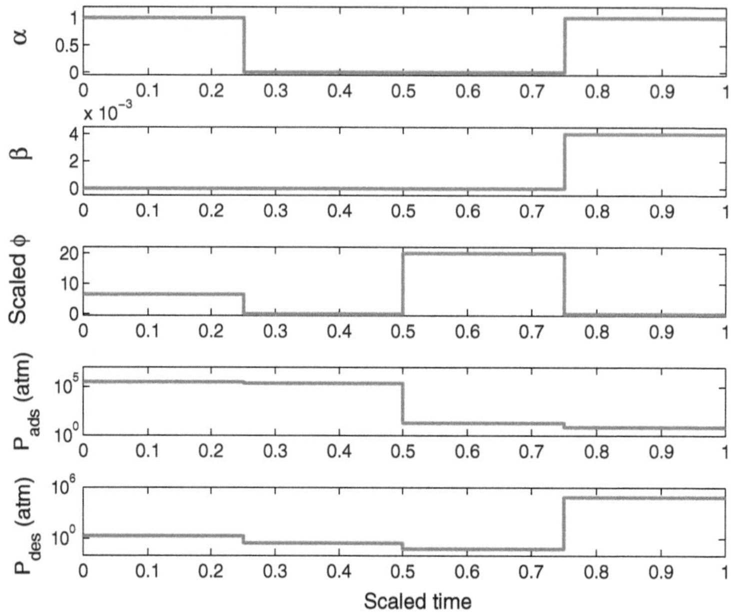

FIGURE 8.12

Optimal control variables for the CO_2/H_2 separation case with high H_2 recovery requirement.

High H_2 recovery is essential because H_2 is combusted to generate power. For a binary separation case, H_2 recovery is related directly to CO_2 recovery and purity. In the previous section, an optimal cycle for 92% CO_2 purity and 90% CO_2 recovery is discussed, with an H_2 recovery of 94.33%. Here, a higher CO_2 purity of 98.49% is considered, which corresponds to an H_2 recovery of 99% when the CO_2 recovery is 90%. As seen in Table 8.3, the optimal specific energy is 99.84 kW h/tonne CO_2. Compared with the previous low H_2 recovery case, an additional 15.13 kW h/tonne CO_2 is required.

The optimal cycle for the high H_2 recovery case is shown in Figure 8.11. Due to valve actions, some of the slots are subdivided. For example, in slot 1, the valve at the bottom of CnB closes after opening for a short time, thus dividing the slot into two subslots (Slot 1 A and Slot 1 B). Feeds keep the same flow rate during the whole slot, while heavy product reflux only exists in Slot 1 A. Furthermore, when considering a 5-slot process, the optimizer selected no activity for slot 3. Thus this slot was not needed. The process was resimulated to verify the results, ultimately recreating the 4-slot process divided into 7 steps (subslots). Optimal control profiles are shown in Figure 8.12.

In the first slot, CoB is pressurized by feed and heavy product recycling, while pressure drops rapidly in the CnB. Pressure in CoB increases and adsorption occurs. Then the heavy product purge stream flow is set to zero. In the next two slots, CoB continues to be pressurized by feed, and the gases (especially CO_2) are adsorbed at high pressure. CnB is depressurized along with CO_2 desorption and release. At the end of slot 3, the top valve is open, producing high-purity H_2 product from the top of CoB. In the last slot, heavy product recycling and (slight) light product recycling occur. The small light recycle maintains high H_2 recovery. After completing this half cycle, with 4 slots and 7 steps, the two beds switch their CoB and CnB roles and the other half cycle continues.

8.7 CONCLUSION AND FUTURE DIRECTIONS

Despite their wide-scale application in industry, systematic optimization of PSA systems remains a very challenging task. PSA processes are difficult to optimize due to their inherent nonlinearities and dynamic nature. Development of sustainable processes that utilize PSA for efficient, low-cost separations would benefit from advancement of PSA optimization methods. In this work, the methods of Dowling et al. (2012) are updated to improve computational performance, both by using the WENO flux limiter and sparse linear algebra routes with the numeric integrator. These advances decrease computation times for integrating the PSA bed model and sensitivity equations by approximately 15% for 2 components and 11 finite volume cells, and it is even more promising with more components and cells. Convergence issues are diagnosed, revealing the importance of scaling the feed fraction. It is hypothesized that poor Hessian approximations lead to premature termination with large KKT violations.

These methods are demonstrated in a PSA case study for the CO_2/H_2 separation in IGCC power plants. The results of Dowling et al. (2012) are also updated to emphasize the importance of a high H_2 recovery rate in this application. This case study establishes the utility of advanced optimization methods to develop efficient and economic separation systems that either directly improve process sustainability or enable sustainable modifications, such as the addition of carbon capture technology for fossil fuel power plants.

There are several future directions of PSA optimization research. As a consequence of the steepest descent direction findings in this work, it may be promising to restart the SQP methods with tighter tolerances and converge to a slightly better solutions. Another promising approach is the application of reduced order models to the PSA optimization problem (Agarwal and Biegler, 2013; Hasan et al., 2012; Khalighi et al., 2014). The central goal is to construct surrogate models that approximate the PDAE bed models, but are much easier to optimize. For example, exact derivative information of the algebraic surrogate bed models can be obtained directly with automatic differentiation. On the other hand, optimization of surrogate models requires additional development to prove convergence to the optimum of the fully detailed model (Biegler et al., 2014).

NOTATIONS

b_{mi}	Langmuir isotherm coefficients for species i site m (1/Pa)
b_L	Lower bound
b_U	Upper bound
BR_i	Flux in the bottom reflux stream for species i (g mol/(m^2 s))
$C_{a,i}$	Gas concentration for species i exiting adsorption bed (mol/m^3)
$C_{d,i}$	Gas concentration for species i exiting desorption bed (mol/m^3)
$C_{feed,i}$	Feed gas concentration for species i (mol/m^3)
C_i	Gas concentration for species i (mol/m^3)
C^i_{pg}	Ideal gas specific heat for species i (J/(mol K))
C_{ps}	Sorbent specific heat (J/(kg K))
d_p	Sorbent particle diameter (m)
E	Net energy (kW h/tonne CO_2)
E^k	Energy consumed/produced by compressor/expander k (kW h/(m^2))
f	Flux, a function of u
F_i	Feed flux for species i (mol/(m^2 s))
h	Enthalpy (J/m^3)
HP_i	Flux in the heavy product stream for species i (g mol/(m^2 s))
ΔH^{ads}_i	Enthalpy of adsorption for species i (J/mol)
J	Jacobian
k_i	Effective mass transfer coefficient for species i (1/s)
K^j_{mi}	Langmuir isotherm coefficients for species i site m (mol/kg, mol/(kg K), mol/kg, K)
L_{bed}	Length of the bed
LP_i	Flux in the light product stream for species i (g mol/(m^2 s))
M^i_w	Molecular weight for species i (g/mol)
N_{aux}	Number of axillary equations
N_{cntrl}	Number of control variables
N_{comp}	Number of gas components
N_{grid}	Number of bed discretization grids
N_{opt}	Number of optimal variables
N_{sens}	Number of sensitivity equations
N_{slot}	Number of time slots
N_{state}	Number of state equations
p	Control variable
P	Bed pressure (atm)
P_{ads}	Adsorption pressure (atm)
P_{des}	Desorption pressure (atm)
q_i	Sorbent loading for species i (mol/kg)
$q^\star_i$	Equilibrium sorbent loading for species i (mol/kg)
q^s_{mi}	Langmuir isotherm coefficients for species i site m (mol/kg)
$Q_{feed,i}$	Flux in the feed stream for species i (g mol/(m^2 s))
R	Ideal gas constant (J/(mol K))
S	Sensitivity
t	Time (s)
t_s	Length of one time slot (s)

T	Temperature (K)
T_w	Wall temperature (K)
TR_i	Flux in the top reflux stream for species i (g mol/(m^2 s))
u	Conserved variable
U_A	Effective heat transfer coefficient (J/(m^3 s K))
v	Gas velocity (m/s)
v_a	Gas velocity exiting adsorption bed (m/s)
v_d	Gas velocity exiting desorption bed (m/s)
v_{feed}	Feed gas velocity (m/s)
x	Bed length (m)
z	Differential variable
α	Bottom reflux fraction
β	Top reflux fraction
ε	Smoothing constant
ε_b	Internal bed void fraction
ε_t	External void fraction
μ	Gas viscosity (Pa s)
ϕ	Feed fraction
Φ	Objective function for specific energy consuming
ψ	Objective function, equality, and inequality constraints
ρ_s	Sorbent density (kg/m^2)

REFERENCES

Agarwal, A., 2010. Advanced Strategies for Optimal Design and Operation of Pressure Swing Adsorption Processes (Ph.D. thesis). Carnegie Mellon University.

Agarwal, A., Biegler, L.T., 2013. A trust-region framework for constrained optimization using reduced order modeling. Optim. Eng. 14, 3–35.

Agarwal, A., Biegler, L.T., Zitney, S.E., 2010. Superstructure-based optimal synthesis of pressure swing adsorption cycles for precombustion CO_2 capture. Ind. Eng. Chem. Res. 49, 5066–5079.

Alpay, E., Kenney, C., Scott, D., 1993. Simulation of rapid pressure swing adsorption and reaction processes. Chem. Eng. Sci. 48, 3173–3186.

Ayoub, S., Tondeur, D., Latifi, M.A., 2011. New formulation of optimization-based simulation of a pressure swing adsorption process: hybrid dynamic optimization. AIChE J. 57, 1367–1373.

Balakrishna, S., Biegler, L.T., 1992. Targeting strategies for the synthesis and energy integration of nonisothermal reactor networks. Ind. Eng. Chem. Res. 31, 2152–2164.

Biegler, L.T., 2010. Nonlinear Programming: Concepts, Algorithms, and Applications to Chemical Processes. SIAM.

Biegler, L.T., Jiang, L., Fox, V.G., 2005. Recent advances in simulation and optimal design of pressure swing adsorption systems. Sep. Purif. Rev. 33, 1–39.

Biegler, L.T., Lang, Y., Lin, W., 2014. Multi-scale optimization for process systems engineering. Comput. Chem. Eng. 60, 17–30.

Chou, C.T., Chen, C.Y., 2004. Carbon dioxide recovery by vacuum swing adsorption. Sep. Purif. Technol. 39, 51–65.

Cruz, P., Santos, J., Magalhes, F., Mendes, A., 2005. Simulation of separation processes using finite volume method. Comput. Chem. Eng. 30, 83–98.

Dowling, A.W., Vetukuri, S.R.R., Biegler, L.T., 2012. Large-scale optimization strategies for pressure swing adsorption cycle synthesis. AIChE J. 58, 3777–3791.

Duff, I.S., Reid, J.K., 1993. MA48, a Fortran Code for Direct Solution of Sparse Unsymmetric Linear Systems of Equations. Science and Engineering Research Council, Rutherford Appleton Laboratory.

First, E.L., Hasan, M., Floudas, C.A., 2014. Discovery of novel zeolites for natural gas purification through combined material screening and process optimization. AIChE J. 60, 1767–1785.

Gomes, V.G., Yee, K.W.K., 2002. Pressure swing adsorption for carbon dioxide sequestration from exhaust gases. Sep. Purif. Technol. 28, 161–171.

Griewank, A., Juedes, D., Utke, J., 1996. Algorithm 755: ADOL-C: a package for the automatic differentiation of algorithms written in C/C++. ACM Trans. Math. Software (TOMS) 22, 131–167.

Haghpanah, R., Majumder, A., Nilam, R., Rajendran, A., Farooq, S., Karimi, I.A., Amanullah, M., 2013. Multiobjective optimization of a four-step adsorption process for postcombustion CO_2 capture via finite volume simulation. Ind. Eng. Chem. Res. 52, 4249–4265.

Hasan, M.F., First, E.L., Floudas, C.A., 2013. Cost-effective CO_2 capture based on in silico screening of zeolites and process optimization. Phys. Chem. Chem. Phys. 15, 17601–17618.

Hasan, M.M.F., Baliban, R.C., Elia, J.A., Floudas, C.A., 2012. Modeling, simulation, and optimization of postcombustion CO_2 capture for variable feed concentration and flow rate. 2. Pressure swing adsorption and vacuum swing adsorption processes. Ind. Eng. Chem. Res. 51, 15665–15682.

Hindmarsh, A.C., Serban, R., 2002. User Documentation for Cvodes, an Ode Solver with Sensitivity Analysis Capabilities. Lawrence Livermore National Laboratory, Livermore (California), p. 189.

IEA, 2012. World Energy Outlook 2012.

Jiang, L., Biegler, L.T., Fox, V.G., 2003. Simulation and optimization of pressure-swing adsorption systems for air separation. AIChE J. 49, 1140–1157.

Jiang, L., Biegler, L.T., Fox, V.G., 2005. Design and optimization of pressure swing adsorption systems with parallel implementation. Comput. Chem. Eng. 29, 393–399.

Jiang, L., Fox, V.G., Biegler, L.T., 2004. Simulation and optimal design of multiple-bed pressure swing adsorption systems. AIChE J. 50, 2904–2917.

Khajuria, H., Pistikopoulos, E.N., 2013. Optimization and control of pressure swing adsorption processes under uncertainty. AIChE J. 59, 120–131.

Khalighi, M., Karimi, I.A., Farooq, S., 2014. Comparing SiCHA and 4A zeolite for propylene/propane separation using a surrogate-based simulation/optimization approach. Ind. Eng. Chem. Res. 53, 16973–16983.

Ko, D., Siriwardane, R., Biegler, L.T., 2003. Optimization of a pressure-swing adsorption process using zeolite 13X for CO_2 sequestration. Ind. Eng. Chem. Res. 42, 339–348.

Ko, D., Siriwardane, R., Biegler, L.T., 2005. Optimization of pressure swing adsorption and fractionated vacuum pressure swing adsorption processes for CO_2 capture. Ind. Eng. Chem. Res. 44, 8084–8094.

Lee, C.H., Yang, J., Ahn, H., 1999. Effects of carbon-to-zeolite ratio on layered bed H_2 PSA for coke oven gas. AIChE J. 45, 535–545.

LeVeque, R.J., 2002. Finite Volume Methods for Hyperbolic Problems, vol. 31. Cambridge University Press.

Liu, Z., Grande, C.A., Li, P., Yu, J., Rodrigues, A.E., 2011. Multi-bed vacuum pressure swing adsorption for carbon dioxide capture from flue gas. Sep. Purif. Technol. 81, 307–317.

Lu, Z., Loureiro, J., Rodrigues, A., LeVan, M., 1993. Pressurization and blowdown of adsorption beds—II. Effect of the momentum and equilibrium relations on isothermal operation. Chem. Eng. Sci. 48, 1699–1707.

Mohamadinejad, H., Knox, J., Smith, J.E., 2000. Experimental and numerical investigation of adsorption/desorption in packed sorption beds under ideal and nonideal flows. Sep. Purif. Technol. 35, 1–22.

Nikolic, D., Kikkinides, E.S., Georgiadis, M.C., 2009. Optimization of multibed pressure swing adsorption processes. Ind. Eng. Chem. Res. 48, 5388–5398.

Nilchan, S., Pantelides, C., 1998. On the optimisation of periodic adsorption processes. Adsorption 4, 113–147.

Ruthven, D., Farooq, S., Knaebel, K., 1994. Pressure Swing Adsorption. VCH Publishers.

Smith, O.J., Westerberg, A.W., 1990. Mixed-integer programming for pressure swing adsorption cycle schedule. Comput. Chem. Eng. 45, 2833–2842.

Smith, O.J., Westerberg, A.W., 1991. The optimal design of pressure swing adsorption systems. Chem. Eng. Sci. 46, 2967–2976.

Smith, O.J., Westerberg, A.W., 1992. Acceleration of cyclic steady state convergence for pressure swing adsorption models. Ind. Eng. Chem. Res. 31, 1569–1573.

Subramanian, D., Ritter, J.A., 1997. Equilibrium theory for solvent vapor recovery by pressure swing adsorption: analytical solution for process performance. Chem. Eng. Sci. 52, 3147–3160.

Takamura, Y., Narita, S., Aoki, J., Hironaka, S., Uchida, S., 2001. Evaluation of dual-bed pressure swing adsorption for CO_2 recovery from boiler exhaust gas. Sep. Purif. Technol. 24.

Tlili, N., Grévillot, G., Vallières, C., 2009. Carbon dioxide capture and recovery by means of TSA and/or VSA. Int. J. Greenhouse Gas Control 3, 519–527.

Walther, A., 2007. Automatic differentiation of explicit Runge-Kutta methods for optimal control. Comput. Optim. Appl. 36, 83–108.

CHAPTER

9

Superstructure-Based Optimal Design of Pipeline Network for CO_2 Transport in Large-Scale Carbon Capture and Sequestration

Chengchuan Zhou*, Pei Liu§,[1], Efstratios N. Pistikopoulos¶

**Department of Earth and Environmental Engineering, Columbia University, New York, NY, USA*

§Department of Thermal Engineering, Tsinghua University, Beijing, China

¶Artie McFerrin Department of Chemical Engineering & Texas A&M Energy Institute, Texas A&M University, College Station, TX, USA

[1]Corresponding author: E-mail: liu_pei@tsinghua.edu.cn

9.1 INTRODUCTION

With the rapid increase in anthropogenic greenhouse gas (GHG) emissions, especially CO_2, many technologies have been developed to reduce GHG emissions in order to prevent subsequent climate change. Carbon capture and sequestration (CCS), which is a technology mainly capturing CO_2 emissions from large point sources of CO_2 and sequestrating the captured CO_2 into geologic formations underground to reduce the amount of CO_2 emitted to the atmosphere, has attracted increasing attention recently, and it has become one of the most promising technologies to mitigate CO_2 emissions in the future. However, it is often the case that CO_2 emission sources and geological formations suitable for CO_2 sequestration are not located in the same places, which means the captured CO_2 need to be transported from the sources to the geologic sequestration sites. It has been widely recognized that pipeline transport is an ideal means for large-scale CO_2 transport over a long distance (Svensson et al., 2004). However, the dispersed distribution of sources and sinks has become a big challenge to achieving satisfactory deployments of infrastructures for CO_2 transport in terms of system reliability and system economy. The problem will become more complicated when considering the heterogeneity of different sources and sinks, i.e., different sources generally have different capture costs and CO_2 purities, while different sinks generally have different sequestration costs and different storage capabilities. Moreover, there are many nonlinear factors embedded in this problem that

Computer Aided Chemical Engineering, Volume 36. ISSN 1570-7946. http://dx.doi.org/10.1016/B978-0-444-63472-6.00009-4

should not be neglected, including the thermophysical and hydraulic properties of CO_2 flows. Accordingly, it is a rather challenging problem to achieve the optimal designs of pipeline networks for CO_2 transport in large-scale CCS projects.

Previous studies have provided many valuable results and insightful conclusions, which are helpful for further delving into this problem. As a powerful tool to solve the problem of optimal design of process configurations and operation conditions, the methodology of superstructure-based modeling has also played an important role in solving the problem of optimal designs of pipeline networks for CO_2 transport in large-scale CCS projects. In this chapter, contributions of previous studies to solving this problem are reviewed and summarized, including the methodology of superstructure-based modeling and its applications to this problem. This chapter will be organized as follows: first, the problem of optimal design of pipeline network will be introduced in detail; next, we present an overview of previous studies on thermophysical and hydraulic properties of CO_2; third, different methods of estimating project costs in previous researches are studied; then, previous models for CCS pipeline network designs are reviewed and compared; a discussion of future research areas about this problem follows before conclusions are drawn.

9.2 OPTIMIZATION PROBLEM DESCRIPTION

Ideally, the optimal design of pipeline network for CO_2 transport should be capable of capturing a target amount of CO_2, while fulfilling the requirements of safe operation and minimizing the total cost of the project. To be specific, the model should integrally solve the following problems (Middleton and Bielicki, 2009; Zhou et al., 2014):

$$\textit{CO}_2\ \textit{emission sources}\begin{cases}\textit{which sources should be retrofitted to capture } CO_2\\ \textit{how much } CO_2 \textit{ should be captured at each proposed source}\\ \textit{the pressure of } CO_2 \textit{ captured at each proposed source}\\ \textit{the temperature of } CO_2 \textit{ captured at each proposed source}\\ \textit{how to connect to the pipeline network and export } CO_2\end{cases}$$

$$\textit{Potential } CO_2 \textit{ storage sites}\begin{cases}\textit{which site should be developed for } CO_2 \textit{ storage}\\ \textit{how much } CO_2 \textit{ should be stored at each proposed site}\\ \textit{the pressure of } CO_2 \textit{ injected at each proposed sites}\\ \textit{the temperature of } CO_2 \textit{ injected at each proposed sites}\\ \textit{how to connect to the pipeline network and import } CO_2\end{cases}$$

$$\textit{Booster pump stations}\begin{cases}\textit{where should stations be deployed}\\ \textit{how to connect to the pipeline network and import } CO_2 \textit{ flows}\\ \textit{the pressure and temparature of each inlet } CO_2 \textit{ flow}\\ \textit{the pressure and temparature of each outlet } CO_2 \textit{ flow}\\ \textit{how to connect to the pipeline network and export } CO_2 \textit{ flows}\end{cases}$$

$$\textit{Pipelines}\begin{cases}\textit{where should pipelines be deployed}\\ \textit{what are the lengths and diameters of these pipelines}\\ \textit{how much } CO_2 \textit{ should be transported in each pipeline and in which direction}\\ \textit{what is the pressure distribution of the } CO_2 \textit{ flow in each pipeline}\\ \textit{what is the temperature distribution of the } CO_2 \textit{ flow in each pipeline}\end{cases}$$

As shown above, mass flow configuration and CO_2 thermophysical property control are two major aspects in the optimization model. The mass flow configuration is considered in the model because the scale and capability of the CCS project have to be guaranteed. The philosophy of studying the thermophysical properties of CO_2 flows in the pipeline networks lies in the fact that CO_2 is transported in liquid phase or as supercritical fluid, and high transport efficiency can thus be achieved (Wilcox, 2012; McCollum and Ogden, 2006). However, the pressure of CO_2 will drop when transported in pipelines, which means CO_2 flows can vaporize if their pressures are not well controlled. In this case, two-phase flows will induce vibration and thus cause fatigue damage to the pipeline network system (Chica et al., 2012). They will also lead to pump cavitation and thus negatively affect the lifespan of the booster pumps (Poullikkas, 1992). Moreover, the volume of CO_2 will explode with rapid vaporization of CO_2 flows and accidents may thus happen (Zhou et al., 2014). Thus, the thermophysical properties of CO_2 flows should be well controlled based on the concerns about the stability and reliability of CCS projects. Accordingly, previous research on CO_2 thermophysical and hydraulic properties will be studied in the third section of this chapter.

Theoretically, an optimization model should simultaneously consider the key decisions mentioned above to achieve the pipeline network design with the minimum project cost. Thus, the cost estimations of different components of CCS projects should also be thoroughly studied, including CO_2 capture costs, CO_2 sequestration costs, and CO_2 transport costs. Thus, an overview of previous studies on cost estimation is provided in the fourth section of this chapter.

Although research on the optimal design of pipeline network for CO_2 transport is still limited, more trials have been conducted to tackle the problem of designing pipeline network for transporting other fluids with similar constraints, including natural gas and water (Andre, 2009). The methodologies of these pipeline network studies can be divided into two categories: one pertains to metaheuristics, like genetic algorithm (Andre, 2009; Fimbres Weihs and Wiley, 2012; Boyd et al., 1994; Abebe and Solomatine, 1998); the other pertains to superstructure-based modeling and continuous relaxation (Zheng et al., 2009; Middleton and Bielicki, 2009; Zhou et al., 2014). Metaheuristics methods can usually provide good results with relatively less computational effort, especially in large-scale problems where a global or local optimum would otherwise take rather long computational time to obtain. However, these methods usually do not guarantee a local optimum and are thus not applicable in cases where exact solutions are required. In this chapter, we will mainly focus on the superstructure-based methods, and they are introduced in the fifth section of this chapter.

9.3 PREVIOUS STUDIES ON CO_2 THERMOPHYSICAL AND HYDRAULIC PROPERTIES

As mentioned above, the thermophysical and hydraulic properties of CO_2 are important here because CO_2 flows in pipelines must be kept in liquid phase or as supercritical fluid so that the CCS project can operate efficiently. Generally, the temperature and the pressure of CO_2 transported in pipelines range between 12 °C and 44 °C and 85 bar and 150 bar, respectively (Serpa et al., 2011; Mohitpour et al., 2000; Forbes, 2008).

According to the phase diagram of CO_2 shown in Figure 9.1, the critical pressure and critical temperature of pure CO_2 are 7.38 MPa and 30.98 °C, respectively (Wilcox, 2012). Thus, to avoid vaporization the pressure of CO_2 flows must be maintained above 7.38 MPa.

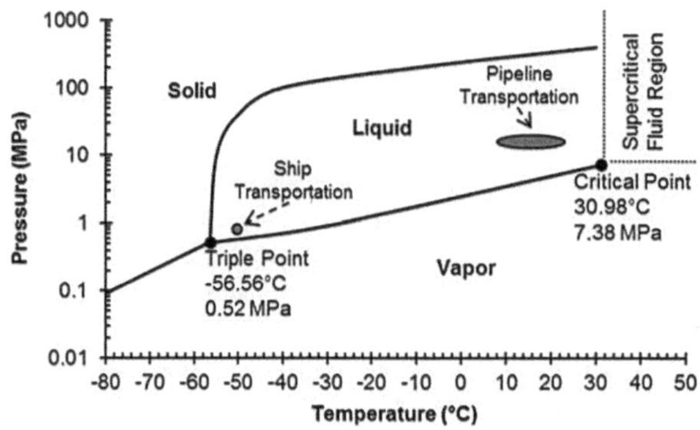

FIGURE 9.1

Phase diagram of CO_2.

Wilcox (2012).

In order to control the pressure of CO_2 transported in pipelines, its thermophysical properties should be thoroughly studied first. According to the CO_2 density diagram, as shown in Figure 9.2, the density of CO_2 transported in

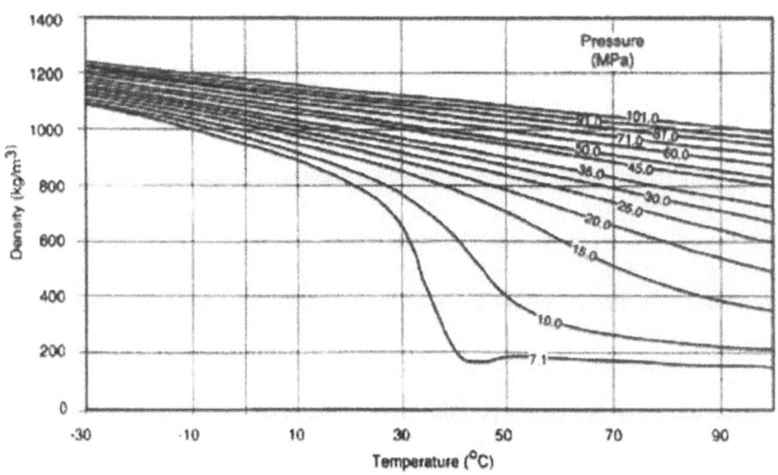

FIGURE 9.2

Density diagram of CO_2.

APEC Energy Working Group (2012).

pipelines ranges approximately between 800 and 1000 kg/m^3 (Wilcox, 2012). Based on experimental data, many equations of state (EOSs) have been developed to describe the thermodynamic behaviors of different fluids, including the van der Waals EOS, the Soave-Redlich-Kwong EOS, and the Peng-Robinson EOS (Medeiros et al., 2008). Moreover, polynomial regression equations of CO_2 density against pressure at different temperatures have also been generated to best fit experimental data in previous studies (McCollum and Ogden, 2006).

The CO_2 viscosity diagram is shown in Figure 9.3, indicating that the viscosity of CO_2 transported in pipelines ranges approximately between 0.05 and 0.1 MPa·s. Similarly, polynomial regression equations of CO_2 viscosity against pressure at different temperatures have also been prepared in previous studies (McCollum and Ogden, 2006). Different theoretical models have also been developed to predict fluid viscosity, like Chung model (Medeiros et al., 2008; Chung et al., 1988).

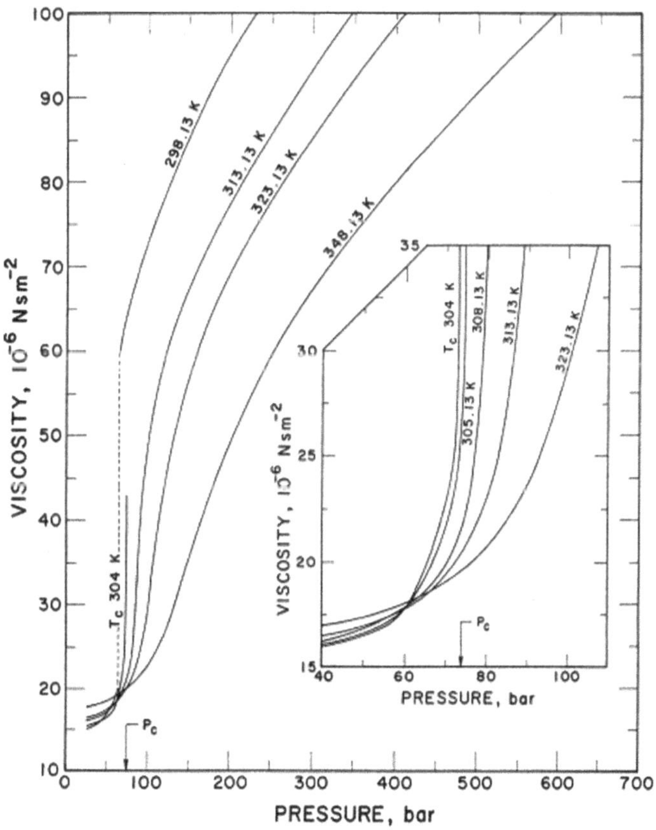

FIGURE 9.3

Viscosity diagram of CO_2.

Stephan and Lucas (1979).

The CO_2 specific heat diagram (calculation results of Refprop 7.0) is shown in Figure 9.4, indicating that the specific heat of CO_2 transported in pipelines ranges approximately between 2 and 30 kJ/kg·K. Generally, the isobaric specific heat of the fluid is calculated by combining the respective ideal gas value and the residue heat capacity. The latter one can be numerically calculated from the EOSs as a function of temperature, pressure, and species molar fractions.

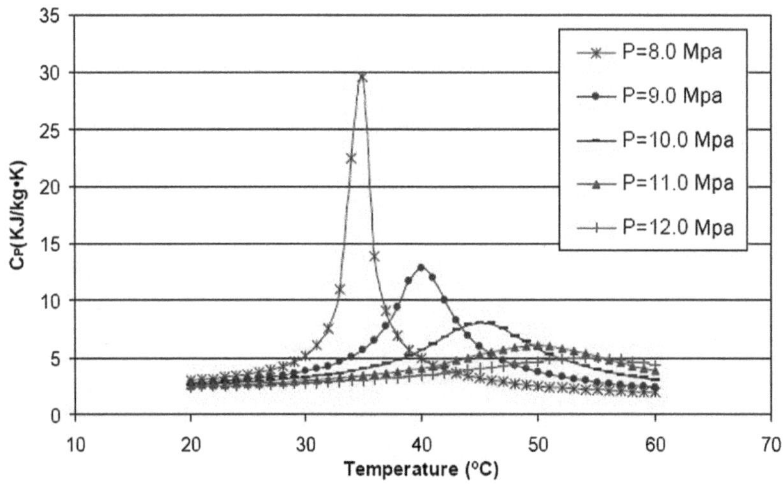

FIGURE 9.4

Specific heat diagram of CO_2.

Chen and Lundqvist (2006).

By applying the balances of momentum, energy, and mass to an infinitesimal length of flow, and simultaneously considering EOS and thermophysical property functions, including viscosity function and specific heat function, we can construct mathematical descriptions about the transport process of CO_2 transported in pipelines (Medeiros et al., 2008). The pressure distribution of the pipeline can thus be numerically calculated by dividing the pipeline into a finite number of segments, and solving the equations mentioned above sequentially from the first segment to the last segment following the flow direction (Medeiros et al., 2008). This methodology was applied in previous studies on an isothermal transmission process with a pipeline of 1000 km in length and 14 inch in internal diameter, carrying 300 t/h of fluid containing CO_2, CH_4, and N_2 (Medeiros et al., 2008). The pipeline was designed to have changing elevations, and there is also an adiabatic booster compressor and an adiabatic recovery turbine assigned along the pipeline, as shown in Figure 9.5 (Medeiros et al., 2008). In this figure, the little orange circles label the system vertices, while the colored diamonds label the system tubes (Medeiros et al., 2008). The pipeline system is thus divided into a series segment, which makes

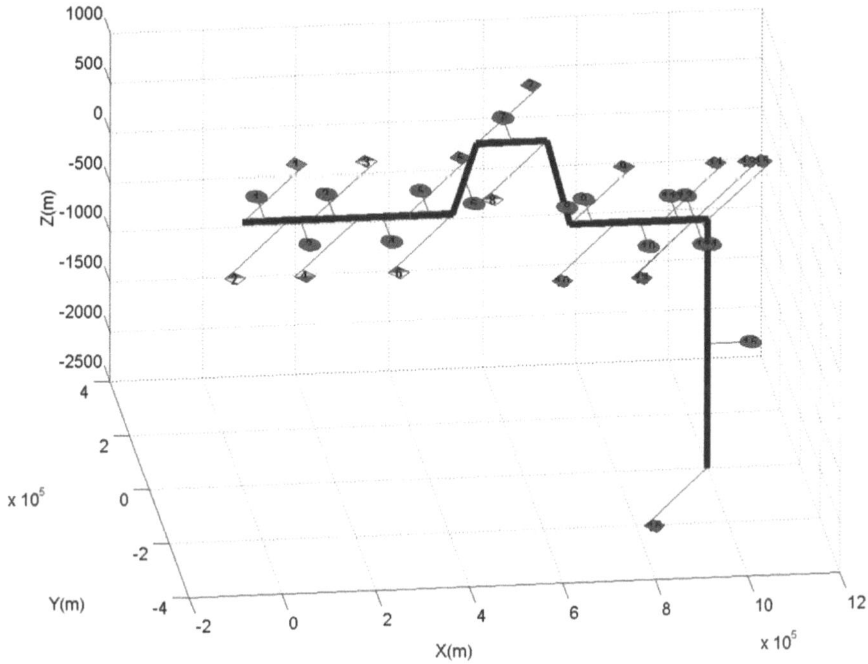

FIGURE 9.5

Spatial representation of the pipeline under consideration.

Medeiros et al. (2008).

the consequent numerical calculation possible (Medeiros et al., 2008). Detailed information of the feed data, tube data, and vertex data can be found in Tables 9.1, 9.2, and 9.3, respectively (Medeiros et al., 2008). The distribution of the fluid pressure is shown in Figure 9.6 (Medeiros et al., 2008).

Table 9.1 Feed Data

Composition	Flow Rate	Temperature	Pressure (abs)	Molar Mass
CO_2 95%mol CH_4 2%mol N_2 3%mol	300 t/h (2.6 Mt/y)	10 °C (283.15 K)	200 bar	0.042971 kg/mol
Calculated Thermodynamic and Transport Properties				
ρ (density) 952.74 kg/m^3	Z 0.3834	c (sonic velocity) 545.35 m/s	$\overline{C}_P$ 0.0917 kJ/mol.K	$\overline{H}$ −387.736 kJ/mol
Fluid state liquid	μ 9.5E- 5 Pa.s	$\gamma = \overline{C}_P/\overline{C}_V$ 2.29404	$\overline{U}$ −388.638 kJ/mol	$\overline{S}$ −0.07306 kJ/mol.K

Medeiros et al. (2008).

Table 9.2 Tube Data

Tube	Length (m)	Diameter (ID) (m)	Roughness (m)	Ω (kW/ m^2K)	Fittings	L_E (m)
1	100E3	0.35	4.572E-5	5E-3	1 *GateVlv*	3.2725
2	50E3	0.35	4.572E-5	5E-3	1 *GateVlv*	3.2725
3	100E3	0.35	4.572E-5	5E-3	1 *GateVlv*	3.2725
4	100E3	0.35	4.572E-5	5E-3	1 *GateVlv*	3.2725
5	100E3	0.35	4.572E-5	5E-3	1 *GateVlv*	3.2725
6	50,005	0.35	4.572E-5	5E-3	1 *GateVlv*	3.2725
7	150E3	0.35	4.572E-5	5E-3	1 *GateVlv*	3.2725
8	50,007	0.35	4.572E-5	5E-3	1 *GateVlv*	3.2725
9	100E3	0.35	4.572E-5	5E-3	1 *GateVlv*	3.2725
10	100E3	0.35	4.572E-5	5E-3	1 *GateVlv*	3.2725
11	70E3	0.35	4.572E-5	5E-3	1 *GateVlv*	3.2725
12	10	0.35	4.572E-5	5E-3	1 *GateVlv*	3.2725
13	990	0.35	4.572E-5	5E-3	1 *GateVlv* + 1T	36.38
14	29E3	0.35	4.572E-5	5E-3	1 *GateVlv*	3.2725
15	2500	0.20	4.572E-5	5E-3	~	~

GateVlv: *gate valve 100% open;* T: *standard tee.*
Medeiros et al. (2008).

However, although the method to accurately calculate the pressure distribution of the CO_2 flow along the pipeline has been well developed, it still seems too complicated for us to apply it for designing pipeline systems or estimating CO_2 transport cost, let alone integrating this process into the models for optimal design of pipeline network for CO_2 transport. In practice, many more assumptions have been made on the transport process and models (Mohitpour et al., 2000; McCoy and Rubin, 2008; McCollum and Ogden, 2006; Ogden, 2004; Heddle et al., 2003; Hendriks et al., 2003; IEA Greenhouse Gas R&D Program, 2002; IEA Greenhouse Gas R&D Program, 2005a,b; Parker, 2004) have been developed to approximately calculate the distribution of the pipelines, i.e., by assuming constant CO_2 flowing velocity, and uniform CO_2 compressibility and temperature along the pipeline, pressure distribution of CO_2 transported in pipelines can be calculated with the following equation, which is derived from energy balance differential equation (Mohitpour et al., 2000; McCoy and Rubin, 2008; Zigrang and Sylvester, 1982):

$$D_i = \left\{ \frac{-64Z_{ave}^2 R^2 T_{ave}^2 f_F \dot{m}^2 L}{\pi^2 \left[M Z_{ave} R T_{ave} \left(p_2^2 - p_1^2 \right) + 2g P_{ave}^2 M^2 (h_2 - h_1) \right]} \right\}^{1/5} \tag{9.1}$$

$$P_{ave} = \frac{2}{3} \left(p_2 + p_1 - \frac{p_2 p_1}{p_2 + p_1} \right) \tag{9.2}$$

Table 9.3 Vertex Data

Vertex	X (m)	Y (m)	Z (m) Elevation	Ground Temp. θ_E	Vertex Specification	Extra Device	Device Specification
1	0	0	150	20 °C	P = 200 bar	~	~
2	100E3	0	150	20 °C	W = 0 *kg/s*	~	~
3	150E3	0	150	20 °C	W = 0 *kg/s*	~	~
4	250E3	0	150	20 °C	W = 0 *kg/s*	~	~
5	350E3	0	150	20 °C	W = 0 *kg/s*	~	~
6	450E3	0	150	20 °C	W = 0 *kg/s*	Comprs + Cooler	P^{OUT} 200 bar T^{OUT} 37 °C *Effic* 60%
7	500E3	0	850	20 °C	W = 0 *kg/s*	~	~
8	650E3	0	850	20 °C	W = 0 *kg/s*	~	~
9	700E3	0	0	20 °C	W = 0 *kg/s*	~	~
10	800E3	0	0	20 °C	W = 0 *kg/s*	~	~
11	900E3	0	0	20 °C	W = 0 *kg/s*	~	~
12	970E3	0	0	20 °C	W = 0 *kg/s*	Heater	T^{OUT} 105 °C
13	970,010	0	0	20 °C	W = 0 *kg/s*	Turbine + Cooler	P^{OUT} 70 bar T^{OUT} 37 °C *Effic* 60%
14	971E3	0	0	20 °C	W = −41.7 *kg/s*	~	~
15	1000E3	0	0	20 °C	W = 0 *kg/s*	~	~
16	1000E3	0	−2500	67 °C	W = −41.7 *kg/s*	~	~

Comprs:1-stage adiabatic compressor; Turbine: 1-stage adiabatic turbine; Effic: efficiency.
Medeiros et al. (2008).

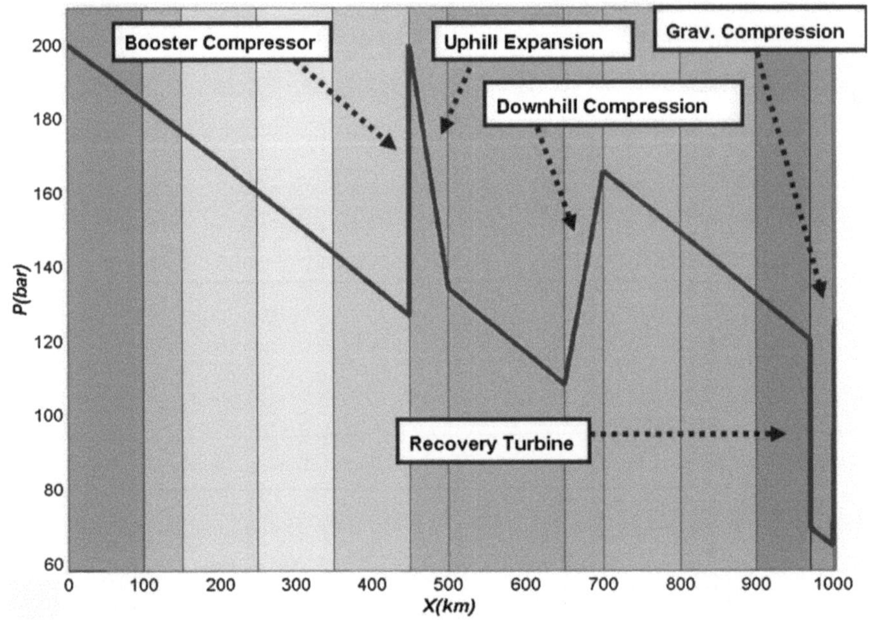

FIGURE 9.6

Pressure profile of the pipeline under consideration.

Medeiros et al. (2008).

$$\frac{1}{2\sqrt{f_F}} = -2\log\left\{\frac{\varepsilon/D_i}{3.7} - \frac{5.02}{Re}\log\left[\frac{\varepsilon/D_i}{3.7} - \frac{5.02}{Re}\log\left(\frac{\varepsilon/D_i}{3.7} + \frac{13}{Re}\right)\right]\right\} \tag{9.3}$$

$$Re = \frac{4\dot{m}}{\mu\pi D_i} \tag{9.4}$$

where D_i is the internal pipeline diameter (meters); Z_{ave} is the average fluid compressibility; R is the universal gas constant ($Pa\cdot m^3\cdot mol^{-1}\cdot K^{-1}$); T_{ave} is the average fluid temperature (kelvin); $\dot{m}$ is the mass flow rate (kg/s); M is the molecular weight of the fluid (kg/kmol); p_1 is the inlet pressure; p_2 is the pressure of any point downstream; h_1 is the inlet elevation; h_2 is the elevation of any point downstream; f_F is the Fanning friction factor (in some other models, the Darcy friction factor is applied instead of the Fanning friction factor); ε is the roughness of the pipeline (meters); R_e is the Reynolds number; μ is the viscosity of the fluid ($Pa\cdot s$). It is worth mentioning that there has not been an analytical expression for the Fanning friction factor or the Darcy friction factor according to the theories of viscous fluid mechanics (McCoy and Rubin, 2008), and different models developed based on experimental data to approximately calculate the Fanning friction factor or the Darcy friction factor have also been provided in previous studies (Heddle et al., 2003; Hendriks et al., 2003; IEA Greenhouse Gas R&D Program, 2002).

Actually, supercritical CO_2 has liquidlike density and gaslike viscosity (Zhang et al., 2006), while its compressibility changes sharply along with the changes of temperature and pressure (McCoy and Rubin, 2008), and models to calculate CO_2 pressure distribution along the pipeline can be quite different based on different assumptions. Equations should be appropriately selected to calculate CO_2 pressure distributions in corresponding cases.

9.4 PREVIOUS STUDIES ON CCS PROJECT COST ESTIMATION

In order to achieve the optimal design of pipeline network with the minimum project cost, methods of estimating costs of different components of CCS projects should be prepared first so that costs of different deployments of CCS infrastructures can be calculated and compared with each other.

Generally, CCS project costs include three parts: CO_2 capture cost, CO_2 transport cost, and CO_2 sequestration cost.

As to CO_2 capture cost, its accurate estimation depends on results of detailed simulations of CO_2 capture processes, and different industrial plants are quite different in terms of emission properties, CO_2 capture technology, and CO_2 capture cost. Detailed simulations of CO_2 capture processes are generally been conducted by software running on computers, i.e., a module developed for simulating amine scrubbing plant performance has been incorporated in the IECM-CS software, with which different components of CO_2 capture equipment can be integrally analyzed, including flue gas blower, absorber, regenerator, solvent processing area, monoethanolamine (MEA) reclaimer, steam extractor, heat exchanger, pumps, and CO_2 compressor (CMU CEES, 2004; Klemeš et al., 2007). CO_2 capture cost can also be estimated with this module, accounting for capital cost, fixed O&M cost, and variable O&M cost (CMU CEES, 2004; Klemeš et al., 2007).

Again, the simulation processes seem too complicated to be implemented in the optimization model for designing pipeline network for CO_2 transport, and simplified methods have been developed in previous studies. Klemeš et al. have developed methods to estimate CO_2 capture cost of coal-fired power plants with linear expressions of plant size, based on IECM-CS software (Klemeš et al., 2007). Apart from estimating total annual cost, this method can also be used to estimate different cost components, including annual sorbent cost and annual electricity cost (Klemeš et al., 2007). Some other studies also provide numerical cost estimations of different industrial plants for reference, as shown in Table 9.4 (Hendriks et al., 2004).

As to CO_2 transport cost, facility parameters are also required before estimating its value. Compressors, pumps, and pipelines are regarded as the major components of pipeline networks for CO_2 transport. D. McCollum and J. Ogden have developed methods to determine parameters of compressors, pumps, and pipelines, while estimating their costs at the same time (McCollum and Ogden, 2006).

Table 9.4 Typical CO_2 Capture Costs of Different Industrial Plants

Facility	€/tCO_2	Facility	€/tCO_2
NGCC	37~43	IGCC	26
Coal fired power plant	29	Nature gas fired power plant	37
Cement	28	Ammonia (flue gas)	36
Iron and steel plant	29	Ammonia (pure CO_2)	3
Refinery	29~42	Hydrogen (flue gas)	36
Petrochemical	32~36	Hydrogen (pure CO_2)	3

Hendriks et al. (2004).

In estimating compressor and pump parameter design and cost, the authors assumed that CO_2 is separated from the flue gases at a pressure of 0.1 MPa, and has to be pressurized to 15 MPa before being transported in pipelines (McCollum and Ogden, 2006). As the critical pressure of CO_2 is 7.38 MPa, compressors should be used to increase CO_2 pressure from 0.1 to 7.38 MPa, while pumps should be used to increase CO_2 pressure from 7.38 to 15 MPa (McCollum and Ogden, 2006). The authors applied the optimal compression ratio equation in previous studies to calculate the number of compressor stages (Mohitpour et al., 2000; McCollum and Ogden, 2006) and compressor power needed for each stage, given CO_2 mass flow rate and inlet temperature (Mohitpour et al., 2000; IEA Greenhouse Gas R&D Program, 2002; McCollum and Ogden, 2006). Based on the maximum size of one compressor train calculated in previous studies (IEA Greenhouse Gas R&D Program, 2002), the number of compressor trains needed can thus be worked out (McCollum and Ogden, 2006). On the other hand, power needed for increasing CO_2 pressure from 7.38 to 15 MPa with pumps can also be calculated with equations provided in the international energy agency (IEA) GHG PH4/6 report (IEA Greenhouse Gas R&D Program, 2002; McCollum and Ogden, 2006). With the results of compressor train numbers, pump sizes, and CO_2 mass flow rates, capital cost and O&M cost can thus be calculated (McCollum and Ogden, 2006; Hendriks et al., 2004; IEA Greenhouse Gas R&D Program, 2002; Kreutz et al., 2005). It is worth mentioning that CO_2 pressure will drop when transported in pipelines, and pump stations are sometimes needed along the pipelines to boost the pressure of CO_2 to avoid vaporization. The processes of parameter design and cost estimation of these booster pumps are the same as mentioned above.

Regarding pipeline parameter design and cost estimation, since pipeline length, CO_2 mass flow rate, CO_2 outlet pressure, CO_2 inlet pressure, and CO_2 inlet temperature are generally regarded as exogenous variables, the only parameter to be determined is the pipeline diameter. The authors applied a pipeline diameter design equation put forward in previous studies (Heddle et al., 2003; McCollum and Ogden, 2006), which was also developed based on the consideration of controlling the CO_2 pressure in pipelines mentioned in the last section, employing the equation form similar to Eqn (9.1). However, for the sake of convenience, the authors still provided equations to estimate pipeline capital cost and O&M cost based on CO_2 mass flow rate and pipeline length (McCollum and Ogden, 2006).

Nevertheless, many more pipeline cost estimation models are constructed based on pipeline length and pipeline diameter, such as the national energy technology laboratory (NETL) model (Grant et al., 2013), the IEA model (IEA Greenhouse Gas R&D Program, 2002), and the Carnegie Mellon University model (McCoy and Rubin, 2008). And the pipeline diameters are also designed with different pipeline diameter design equations, which are derived from different models used for calculating CO_2 pressure distribution in pipelines mentioned in the last section based on different assumptions. Some studies also provide numerical pipeline cost estimations for reference, as shown in Table 9.5 (MIT CCSTP, 2006; McCoy and Rubin, 2008). Some of the previous studies on CO_2 pipeline transport cost estimations are summarized in Table 9.6 (Serpa et al., 2011).

As to CO_2 sequestration cost, facility parameters should be designed first before conducting cost estimations, as well. The major facility parameter that needs to be determined is the number of injection wells (Heddle et al., 2003; McCollum and Ogden, 2006). There are different categories of geological formations suitable for CO_2 storage, including enhanced oil recovery (EOR) site, enhanced coal bed methane recovery (ECBMR) site, depleted gas/oil reservoirs, and saline aquifers. The processes to design the number of injection wells needed for different categories of potential geological CO_2 storage sites are a little bit different from each other. As to EOR or ECBMR sites, the amount of enhanced oil/CBM produced per day should be estimated first, given a certain CO_2 mass flow rate, based on CO_2 effectiveness factors of specific sites (Heddle et al., 2003; McCollum and Ogden, 2006). Then, the number of production wells can be worked out given the oil/CBM production rate of each production well, and the number of injection wells can thus be calculated given the ratio of producers to injectors (Heddle et al., 2003; McCollum and

Table 9.5 CO_2 Pipeline Cost of Different Region in the United States ($/In/mile)

	Central		Southwest		West	
	100 mile	1000 mile	100 mile	1000 mile	100 mile	1000 mile
8 inch	$23,210	$16,560	$28,962	$20,664	$31,117	$22,202
16 inch	$23,777	$16,965	$29,671	$21,170	$31,879	$22,745
24 inch	$24,116	$17,207	$30,093	$21,471	$32,332	$23,069
	Midwest		**Southeast**		**Northeast**	
	100 mile	**1000 mile**	**100 mile**	**1000 mile**	**100 mile**	**1000 mile**
8 inch	$35,194	$25,111	$39,151	$27,934	$41,376	$29,522
16 inch	$36,055	$25,725	$40,108	$28,617	$42,388	$30,244
24 inch	$36,568	$26,091	$40,679	$29,024	$42,992	$30,674

MIT (2006) and McCoy and Rubin (2008).

Table 9.6 Some CO_2 Pipeline Transport Cost-Related Studies

Source	Full Name	Authors	Description	Date
International Energy Agency Greenhouse Gas program (IEA GHG)	Pipeline transmission of CO_2 and energy	Woodhill engineering consultants	Model to estimate the cost and performance of CO_2 transport. Sizing module, in which diameter is calculated, and cost model that calculates capital, fixed and variable operating cost, based on in-house cost equations. Equations based on location and terrain factors, length and diameter	2002
Carnegie Mellon University (CMU)	An engineering-economic model of pipeline transport of CO_2 with application to carbon capture and storage	McCoy, S. T., Rubin, E.S.	Model to estimate the cost per ton of transporting CO_2 for a range of CO_2 flow rates over a range of distances that takes into account regional cost differences within the continental US. Cost equation based on regression analyses of 263 natural gas pipeline project costs published between 1995 and 2005	2008
Massachusetts Institute of Technology	The economics of CO_2 storage	Heddle, G., Herzog, H., Klett, M	Presents pipeline cost calculations based on natural gas pipelines project cost estimates. Determines cost of subsea pipeline based on the literature	2003
Massachusetts Institute of Technology	CO_2 pipeline transport and cost model	Zhang et al.	Presents methodology for the calculation of the optimal pipeline route and estimation of transport costs. Construction costs based on natural gas pipeline cost estimates between 1989 and 1998	2007
Intergovernmental panel on climate change	Special report on carbon dioxide capture and storage	Coleman et al.	Presents onshore and offshore transport costs based on cost information from various sources	2005
Institute of Transportation Studies, University of California	Using natural gas transmission pipeline costs to estimate hydrogen pipeline costs	Parker, N.	Provides equation for the estimation of the cost of construction of a pipeline for a given diameter and length based on construction cost projections of natural gas, oil, and petroleum product pipelines in 893 projects in the United States over 13 years	2004
Institute of Transportation Studies, University of California	Techno-economic models for carbon dioxide compression, transport, and storage & correlations for estimating carbon dioxide density and viscosity	McCollum, D, Ogden, J.	Analytical formulations for CO_2 capture, transport, and storage. Analyses of several studies and their approaches for CO_2 transport. Provides a new equation for the calculation of capital costs based on average output from the various studies. Equation based on flow rate and length. Diameter not included	2006

Serpa et al. (2011).

Table 9.7 CO_2 Sequestration Cost Estimations

Site Category	$/tCO_2$ Avoided	Site Category	$/tCO_2$ Avoided
EOR	−91.26~73.84	Depleted oil reservoir	1.21~11.16
ECBMR	−25.72~18.88	Depleted gas reservoir	1.20~19.43
Saline aquifer	1.14~11.71	Ocean	2.90~14.23

Heddle et al. (2003).

Ogden, 2006). As to depleted gas/oil reservoirs or saline aquifers, given a set of reservoir parameters, including reservoir pressure, thickness, depth, and horizontal permeability, the values of CO_2 downhole injection pressure, CO_2 thermophysical properties, and CO_2 injection rate can be determined together with the method of iterative calculation (Heddle et al., 2003; McCollum and Ogden, 2006). Then, the number of injection wells can be calculated given a certain CO_2 flow rate (Heddle et al., 2003; McCollum and Ogden, 2006). With the design result of the number of injection wells, capital costs of CO_2 sequestration, including site screening and evaluation cost, injection equipment cost, and drilling cost, as well as O&M costs of CO_2 sequestration, including normal daily expenses, consumables, surface maintenance, and subsurface maintenance, can both be calculated with the equations put forward in previous studies (Smith, 2001; Heddle et al., 2003; McCollum and Ogden, 2006). There are also numerical CO_2 sequestration costs provided in some previous studies for reference, as shown in Table 9.7 (Heddle et al., 2003). The reason why negative sequestration costs appear at EOR sites and ECBMR sites is that enhanced oil/CBM is produced during CO_2, which will thus bring about revenues.

9.5 PREVIOUS STUDIES ON MODELS FOR CCS PIPELINE NETWORK DESIGN

In the previous two sections, knowledge about CO_2 pressure control and cost estimations of different components of CCS projects have been introduced. In this section, superstructure-based optimization models for optimal design of pipeline network for CO_2 transport provided in previous studies are reviewed and analyzed.

It has been widely recognized that the methodologies of superstructure-based modeling can play an important role in the field of process system engineering, especially in optimal design of process configurations and operating conditions (Yeomans and Grossmann, 1999). In terms of the problem of optimal design of pipeline network for CO_2 transport, superstructure-based models will employ continuous variables and discrete variables to capture all possible arrangements of sources, sinks, booster pump stations, pipelines, CO_2 flow sequences, and interactions among them, and optimization problems will be constructed via mathematical programming

(Liu et al., 2007). Optimization algorithms will then be applied to these optimization models to work out the optimal design results accordingly.

Many superstructure-based optimization models for designing pipeline networks for CO_2 transport have been developed in previous studies. The models to date can be divided into three categories in terms of network type: source–sink match model (also known as point-to-point model), hub–spoke model, and mature network design model (van den Broek et al., 2010). The complexity and maturity of these three types of models have the same order as listed above, with source–sink match models to be the lowest and mature network design model to be the highest (van den Broek et al., 2010; McKinsey & Company, 2008).

As to the source–sink match models (Dooley et al., 2006), it is often the case that the models match sources and sinks, and connect them directly, aiming to minimize total project costs, including capture cost, sequestration cost, and pipeline cost, while fulfilling the requirement to capture a certain amount of CO_2. In these models, merges or splits of carbon dioxide flows are generally not allowed, and all captured CO_2 must be transported from sources to sinks directly. The problem of pressure drop is generally not considered, or only superficially, and booster pump stations are never employed in these models. A typical source–sink match model can be outlined as follows (Zhou et al., 2014):

$$Minimize\ U\ obj = capture + storage + pipe.$$

$$s.t.$$

$$h(x,y) = 0, g(x,y) \leq 0 \begin{cases} CCS\ target\ (optional) \\ Mass\ balance\ constraint \\ Pipeline\ construction\ decision\ equation \\ Pipeline\ transport\ capability\ constraint\ (optional) \\ Capabilities\ of\ sources\ and\ sinks \end{cases}$$

$$\boldsymbol{x} \in R^n,\ y \in \boldsymbol{Y} = \{0,1\}^m$$

where U is the objective function, h and g are constraints, $\boldsymbol{x}$ is the vector of continuous real variables, and $\boldsymbol{Y}$ is the vector of binary variables.

Source–sink match models have been applied in many early trials proposed to tackle the optimization problem of pipeline network for CO_2, transport. B. Poiencot and C. Brown applied source–sink match models to study the optimal pipeline network design in Florida, USA, while only CO_2 transport costs were taken into consideration (Poiencot and Brown, 2011). Akimoto et al. applied source–sink match models to integrally study the optimal development paths of Japanese energy sector, which included CCS projects, under different scenarios of future energy demands and emission reduction requirements (Akimoto et al., 2004, 2007). The models simultaneously considered transporting CO_2 via pipeline and tanker (Akimoto et al., 2004, 2007). Klokk et al. applied source–sink match models to optimally design pipeline networks in Norway (Klokk, 2010). Some combinations of sources and sinks, which were completely dominated by other combinations, were eliminated by the authors before conducting the optimization process (Klokk,

2010). Meanwhile, some sources and sinks are assigned as intermediate sites, where merges and splits of CO_2 flow sequences are allowed (Klokk, 2010). The model also made decisions on pipeline diameters, and CO_2 transport capacity of pipelines of different diameters were also modeled in the optimization problem, which means the problem of pressure drop was partially considered in this study (Klokk, 2010). R. Tan et al. incorporated time factor into the source−sink match model and constructed a dynamic model for optimal design of CO_2 pipeline network (Tan et al., 2012). Zheng et al. applied the source−sink match model to optimally design the CO_2 pipeline network in the Jing-Jin-Ji region of China (Zheng et al., 2009).

Although the number of studies concentrating on the source−sink match model is relatively large, the drawbacks of this type of model should not be neglected. Apart from its limited capability to consider the problem of CO_2 pressure drop, which pertains to stability and reliability of CCS projects, it can only provide guidance for CCS projects at early stages (van den Broek et al., 2010; Zhou et al., 2014), when the amount of CO_2 to be captured is relatively low, and the numbers of sources and sinks involved in the projects are relatively small, as shown in Figure 9.7 (Zheng et al., 2009). However, when the amount of captured CO_2 becomes large, and more sources and sinks are involved, the design results of this type of model will show a severe problem of redundant pipeline construction, as shown in Figure 9.8 (Zheng et al., 2009), which means the economic performance of source−sink match models under this kind of circumstance is poor.

As a more developed type of optimization model, hub−spoke models can significantly mitigate the problem of redundant pipeline construction. Similar to source−sink match models, the objective function of a hub−spoke model is generally to minimize total project cost, including CO_2 capture cost, CO_2 transport cost, and CO_2 sequestrating cost. However, different from source−sink match models, this type of model does not directly connect sources to sinks. Instead, hub−spoke models assign hubs at places where sources or sinks are clustered. Sources or sinks are directly connected to corresponding hubs nearby, generally designed by source−sink match models, with satellite pipelines. On the other hand, trunk pipelines are employed to connect hubs of different regions, which are also designed by the source−sink match model. CO_2 captured at sources is gathered at corresponding hubs, waiting to be distributed via satellite pipelines or transported to other hubs via trunk pipelines, and finally delivered to sinks for storage. Accordingly, hub−spoke models can be regarded as models repeatedly applying source−sink match models at different regions and different levels. Thus, the outline of a typical hub−spoke model is the same as that of a typical source−sink match model.

Hub−spoke models have also been widely applied to study CO_2 pipeline networks in previous studies. van den Broek et al. once applied hub−spoke models to study the optimal design of pipeline network for CO_2 transport in the Netherlands (van den Broek et al., 2010). Han and I. Lee once applied hub−spoke models to optimally design CO_2 pipeline networks in eastern South Korea (Han and Lee, 2011). Different means for CO_2 transport were considered

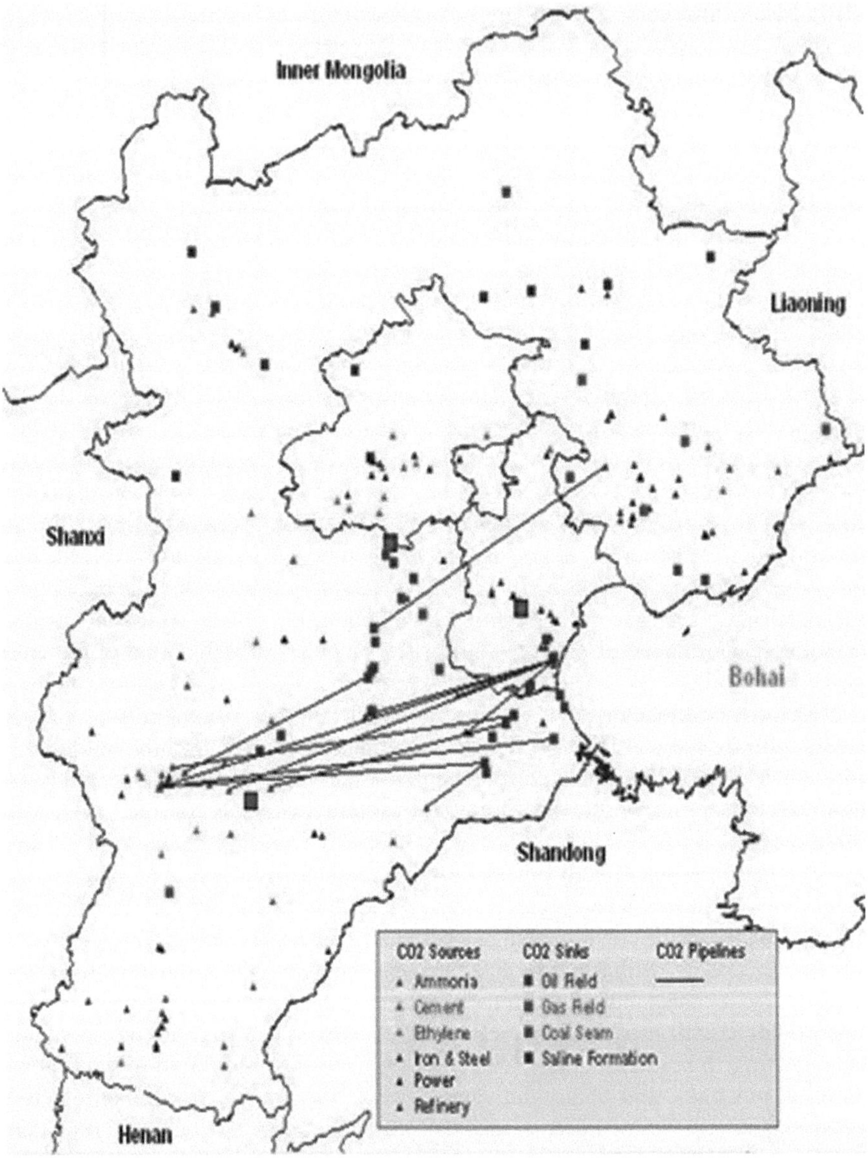

FIGURE 9.7

Source–sink match design at early stage.

Zheng et al. (2009).

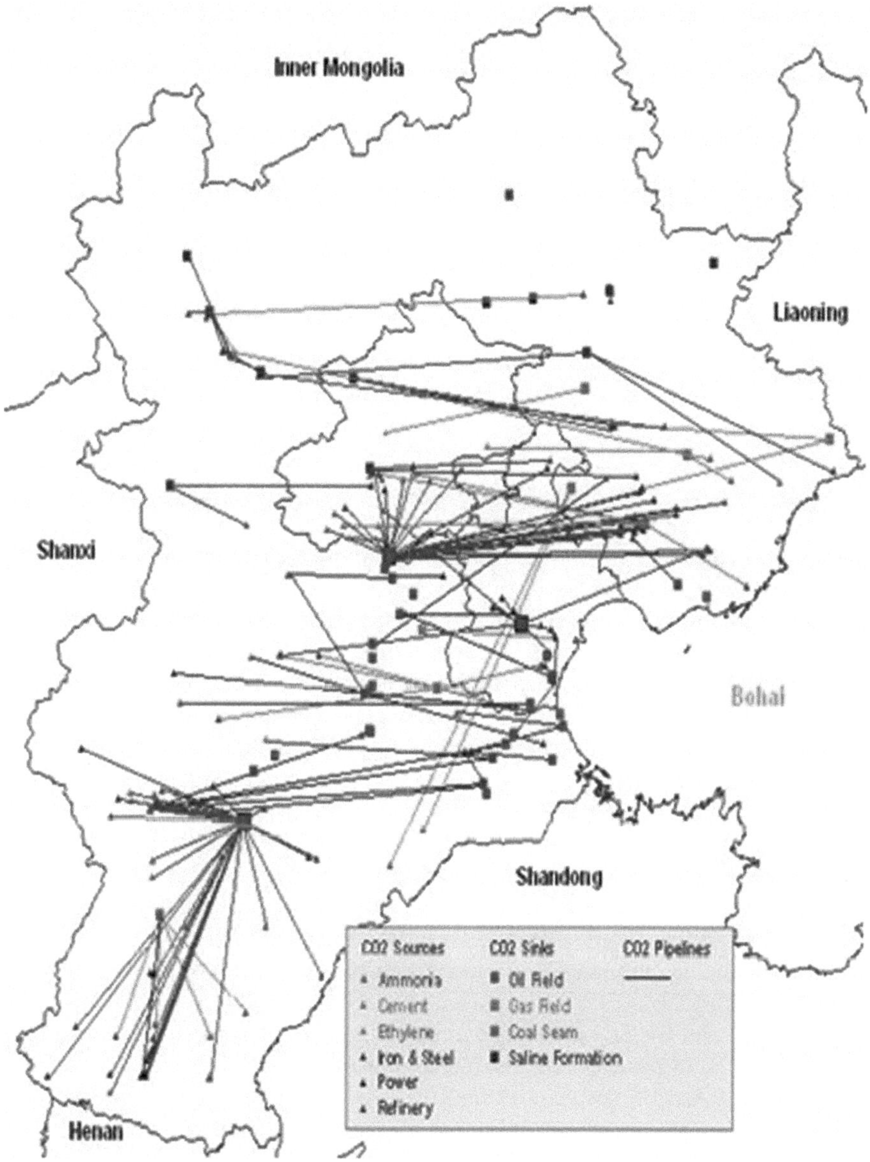

FIGURE 9.8

Large-scale source–sink match design.

Zheng et al. (2009).

simultaneously, including pipeline, tanker truck, tanker ship, and tanker rail car (Han and Lee, 2011). Apart from sequestrating CO_2 into geological formations underground, captured CO_2 is also utilized to produce biobutanol and green polymers (Han and Lee, 2011). Pöyry Energy Consulting once applied hub–spoke models to study CO_2 pipeline networks in the United Kingdom, and the design result is shown in Figure 9.9 (Hare and Ladbrook, 2007). Zhou et al. once applied the hub–spoke model to achieve the optimal design of CO_2 pipeline network in Hebei province of China (Zhou et al., 2013). The problem of CO_2 pressure drop was well implemented in their study, and booster pump stations were employed, with their parameters integrally designed, in the optimization model (Zhou et al., 2013).

Although hub–spoke models have much better model performances in terms of pipeline redundant construction when the scale of CCS project becomes large, it cannot be neglected that there has not been a convincing algorithm to make decisions on the locations of hubs. Put differently, the locations of the hubs are generally selected intuitively by the researchers. Thus, there are still gaps between the design results of this type of optimization models and the global optimal ones, and the resulting CO_2 pipeline networks could be further developed to achieve lower project costs.

As to mature network design models, more reasonable algorithms to design intermediate sites, including the so-called hubs, have been put forward to make the optimization models perform better. This type of model is also developed to minimize total project cost, and intermediate sites are integrally considered and installed in the region under consideration by the model, during the process of optimally designing CO_2 pipeline networks connecting these sources, sinks, and intermediate sites. This type of model is more comprehensive than the previous two models. Apart from fulfilling the requirements of capturing a certain amount of CO_2, many more factors in real-life cases are incorporated in these models, including the selection of pipeline diameters and the installation of booster pump stations. Accordingly, the design results of this type of model are more practical, more reliable, and more cost-effective than those of the previous two types of models.

Middleton and Bielicki developed a mature network design model, which was called SimCCS, in previous studies (Middleton and Bielicki, 2009). The authors first employed ArcGIS software to find the least cost pipeline path of each pair of sites, including sources and sinks, and all these pipeline paths constituted the candidate network (Middleton and Bielicki, 2009). Then, the authors trimmed the candidate network by removing overlapping and superfluous paths, and assigning nodes at junctions of these paths (Middleton and Bielicki, 2009). All the nodes were regarded as potential positions of intermediate sites (Middleton and Bielicki, 2009). The optimization model was developed to construct a pipeline network with the minimum total project cost, based on the trimmed candidate network (Middleton and Bielicki, 2009). It was subject to the requirements of mass balance and capturing a certain amount of CO_2, and it also made decisions on pipeline

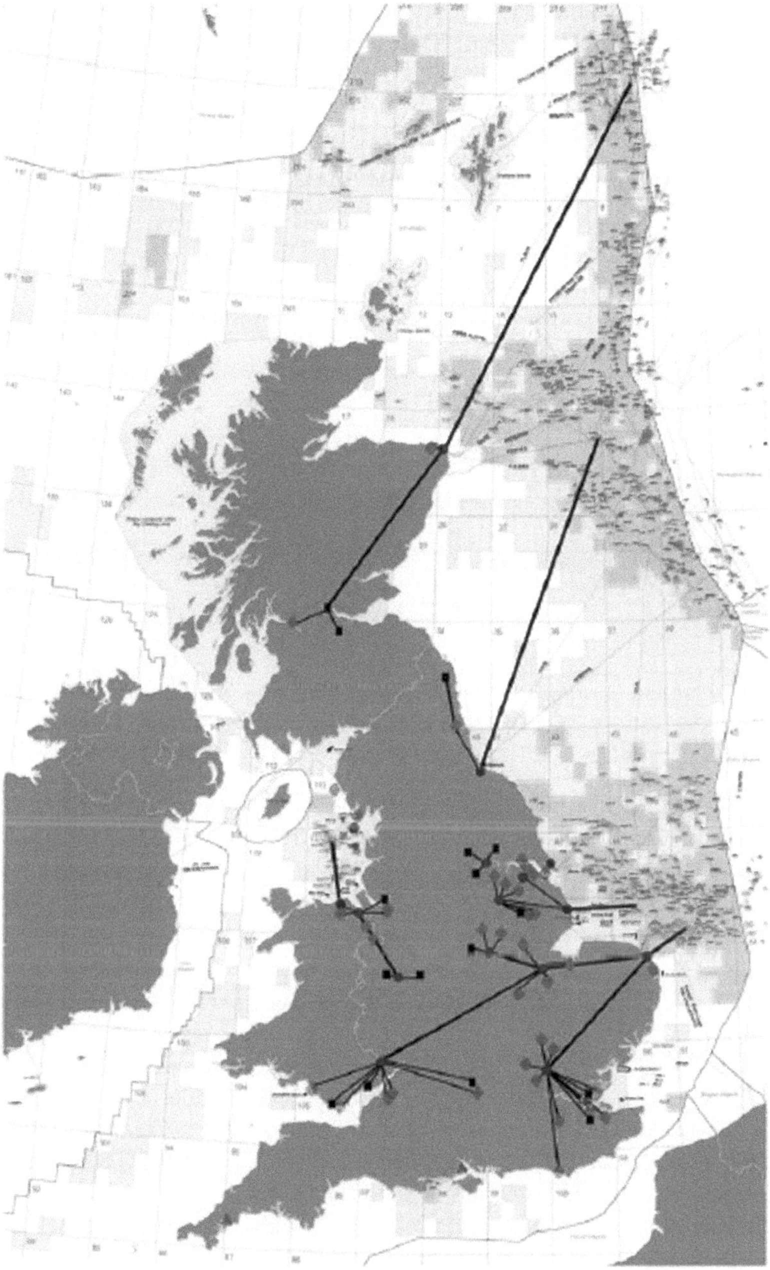

FIGURE 9.9

Hub–spoke design.

Hare and Ladbrook (2007).

diameters (Middleton and Bielicki, 2009). The model was applied to design CO_2 pipeline networks in California, USA, and the design results are shown in Figure 9.10 (Middleton and Bielicki, 2009).

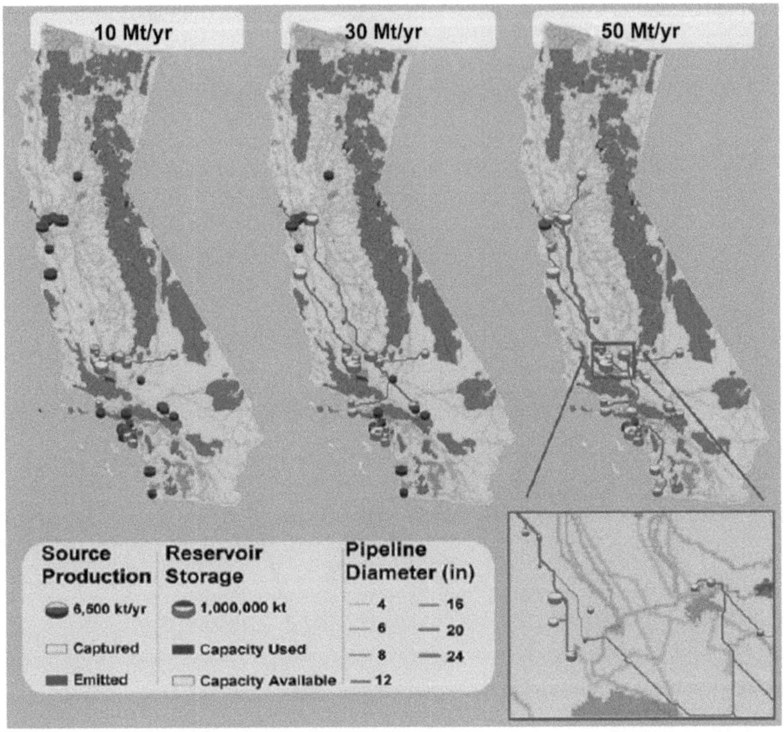

FIGURE 9.10

Design result of CO_2 pipeline networks in California with SimCCS.

Middleton and Bielicki (2009).

Zhou et al. put forward a mature network design model in their previous studies, which is developed to minimize total project cost, while being subject to the constraints of mass balance and pressure requirements (Zhou et al., 2014). Specifically, the model first meshed the region under consideration into grids and regarded the grid nodes as potential places for intermediate sites where booster pump stations could be installed (Zhou et al., 2014). The optimization model then constructed pipelines among these sources, sinks, and potential intermediate sites to minimize total project cost, while fulfilling the requirements of mass balance, capturing a certain amount of CO_2, and maintaining the pressure of CO_2 throughout the pipeline network above the critical point (Zhou et al., 2014). Pipeline diameters were also determined by the model in the optimization process. The model can be outlined as follows (Zhou et al., 2014):

Minimize U obj = capture + storage + pipe + pump + carbon.

s.t.

$$h^{mc}(x,y)=0, g^{mc}(x,y)\le 0 \begin{cases} \text{CCS target} \\ \text{Mass balance constraint} \\ \text{Capabilities of sources and sinks} \\ \text{Pipeline construction decision equation} \end{cases}$$

$$h^{sc}(x,y)=0, g^{sc}(x,y)\le 0 \begin{cases} \text{Pressure drop of carbon dioxide flow} \\ \text{Pressure bounds of carbon dioxide flow} \\ \text{Pressure rise at booster pump stations} \end{cases}$$

$$x\in R^n,\ y\in Y=\{0,1\}^m$$

The authors also provided decomposition algorithms and linear relaxation methods to make the model solvable within a reasonable timescale (Zhou et al., 2014). This model was then applied to a real-life case in Hebei province in China, and some of the design results are shown in Figures 9.11 and 9.12 (Zhou et al., 2014).

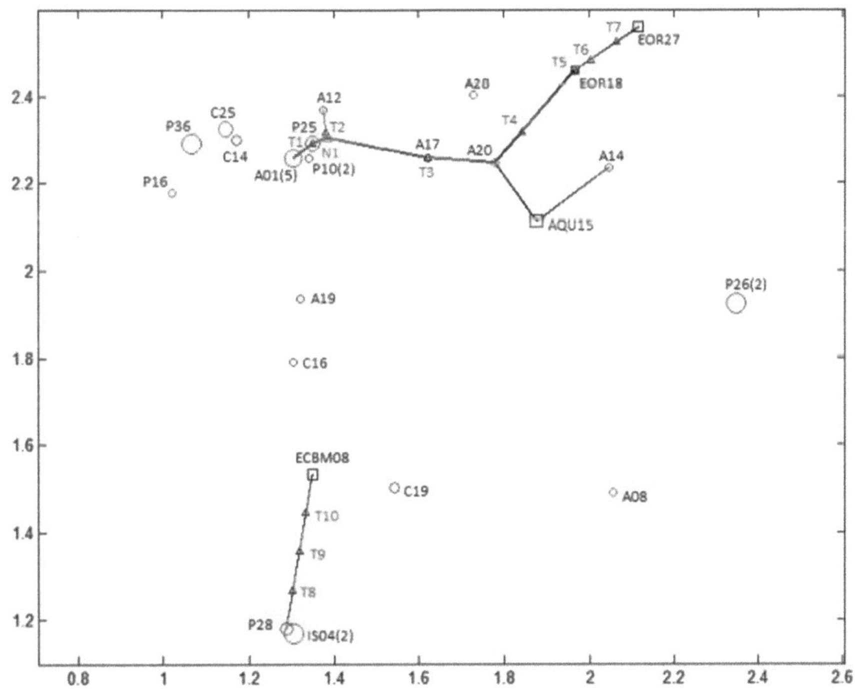

FIGURE 9.11

Pipeline network of 10% capture ratio.

Zhou et al. (2014).

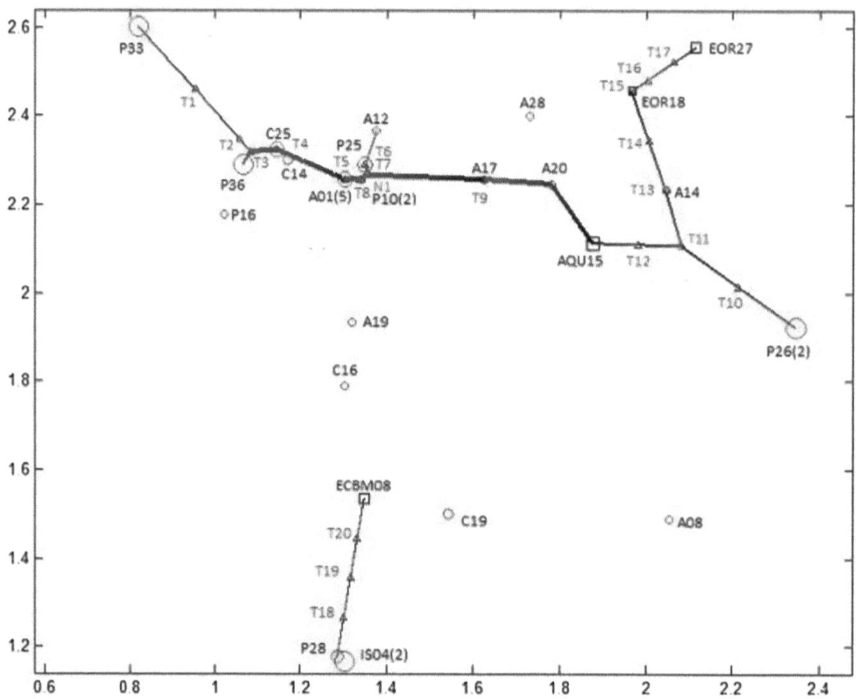

FIGURE 9.12

Pipeline network of 40% capture ratio.

Zhou et al. (2014).

9.6 FUTURE RESEARCH AREAS FOR CO_2 PIPELINE NETWORK DESIGN

Although previous studies have obtained some valuable achievements and provided some insightful results, there are still many problems unsolved in the field of CO_2 pipeline network design.

Constructing more developed and more realistic optimization models for designing CO_2 pipeline networks is always a major concern in this area. However, as mentioned above, nonlinear factors and large model scales are two big challenges to making the optimization models solvable in reasonable timescales. Accordingly, optimization algorithms can be further studied in future researches, and the model performance will be much better if the techniques to solve large-scale optimization problems with nonlinear factors can be well developed.

Moreover, as the problem complexity increases, it becomes increasingly difficult to construct the optimization models based on a single software package. Specifically, it is quite easy to construct superstructure-based models on the platform of

GAMS; however, it is extremely difficult to accurately model the pressure drop of CO_2 transported in pipelines with GAMS, which is nevertheless the advantage of ASPEN PLUS. Similarly, IECM-CS is good at simulating the process of capturing CO_2 at industrial plants, and ArcGIS is good at optimizing the pipeline paths while considering geographical factors. Ideally, if a combined platform of different software packages, including those mentioned above, can be developed, a more sophisticated optimization model can be constructed, and better design results of pipeline networks can thus be achieved.

In order to construct a more comprehensive optimization model, knowledge in relative fields should also be obtained first as preparation, i.e., the uncertainty embedded in constructing CCS pipeline networks needs to be further studied before risk assessments, including leakage risk, ecology risk, market risk, policy risk, and technology development, can be implemented into a more realistic optimization model. Without necessary development of relative research areas, it will be difficult to propose an optimization model with better performance.

9.7 CONCLUSIONS

In this chapter, we review and analyze previous studies on the optimal design of pipeline network for CO_2 transport. Two of the key components of this problem, CO_2 thermophysical and hydraulic properties and project cost estimations, are discussed in detail in separate sections.

As to CO_2 thermophysical and hydraulic properties, the viscous fluid mechanics theoretical models to calculate CO_2 pressure drop in pipelines is provided, followed by an introduction of simplified models for engineering use proposed in previous studies. The thermophysical properties and their value ranges of supercritical CO_2 are also mentioned.

For CCS project cost estimations, previous methods to design facility parameters of different parts of CCS projects are reviewed, followed by a discussion about the corresponding cost estimation models. Numerical cost estimations of different components of CCS projects in previous studies are also listed.

As to superstructure-based models for CCS pipeline network design, a brief introduction of the methodologies of superstructure-based modeling is provided, followed by an overview of applications of this modeling method on previous studies related to CCS pipeline network design. Previous optimization models for designing CO_2 pipeline networks can be divided into three categories: source–sink match model (also known as point-to-point model), hub–spoke model, and mature network design model. Assumptions and configurations of different types of models are summarized, and their applications in real-life cases are also shown.

As to future research areas for CCS pipeline network design, different future research directions are indicated, including developing more advanced optimization algorithms and techniques, constructing combined platforms based on different professional software packages, and studying relative research areas.

REFERENCES

Abebe, A.J., Solomatine, D.P., 1998. Application of global optimization to the design of pipe networks. In: Proc. Int. Conf. Hydroinformatics, vol. 98, pp. 989–995.

Akimoto, K., Kotsubo, H., Asami, T., Li, X., Uno, M., Tomoda, T., Ohsumi, T., 2004. Evaluation of carbon dioxide sequestration in Japan with a mathematical model. Energy 29 (9), 1537–1549.

Akimoto, K., Takagi, M., Tomoda, T., 2007. Economic evaluation of the geological storage of CO_2 considering the scale of economy. Int. J. Greenhouse Gas Control 1 (2), 271–279.

Andre, J., Bonnans, F., Cornibert, L., 2009. Optimization of capacity expansion planning for gas transportation networks. Eur. J. Oper. Res. 197 (3), 1019–1027.

APEC Energy Working Group, 2012. Building Capacity for CO_2 Capture and Storage in the APEC Region: A Training Manual for Policy Makers and Practitioners. Asia-Pacific Economic Cooperation (APEC).

Boyd, I.D., Surry, P.D., Radcliffe, N.J., 1994. Constrained Gas Network Pipe Sizing with Genetic Algorithms. Edinburgh Parallel Computing Centre, University of Edinburgh. Technical Report 11.

van den Broek, M., Brederode, E., Ramírez, A., Kramers, L., van der Kuip, M., Wildenborg, T., Faaij, A., 2010. Designing a cost-effective CO_2 storage infrastructure using a GIS based linear optimization energy model. Environ. Modell. Software 25 (12), 1754–1768.

Center for Energy and Environmental Studies. Carnegie Mellon University, Pittsburgh, PA. http://www.iecm-online.com (accessed 11.08.14.).

Chen, Y., Lundqvist, P., 2006. Analysis of supercritical carbon dioxide heat exchangers in cooling process. In: International Refrigeration and Air Conditioning Conference (Paper 765).

Chica, L., Pascali, R., Jukes, P., Ozturk, B., Gamino, M., Smith, K., 2012. Jumper Analysis with Interacting Internal Two-phase Flow. In: Proceedings of the 22nd International Offshore and Polar Engineering Conference. University of Houston (Doctoral dissertation), pp. 389–395.

Chung, T.H., Ajlan, M., Lee, L.L., Starling, K.E., 1988. Generalized multiparameter correlation for nonpolar and polar fluid transport properties. Ind. Eng. Chem. Res. 27 (4), 671–679.

Dooley, J., Dahowksi, R., Davidson, C., Wise, M., Gupta, N., Kim, S., Malone, E., 2006. Carbon Dioxide Capture Geologic Storage: A Core Element of a Global Energy Technology Strategy to Address Climate Change. The Global Energy Technology Strategy Program, Battelle, Joint Global Change Research Institute.

Fimbres Weihs, G.A., Wiley, D.E., 2012. Steady-state design of CO_2 pipeline networks for minimal cost per tonne of CO_2 avoided. Int. J. Greenhouse Gas Control 8, 150–168.

Forbes, S.M., Verma, P., Curry, T.E., Friedmann, S.J., Wade, S.M., 2008. Guidelines for Carbon Dioxide Capture, Transport and Storage. World Resources Institute.

Grant, T., Morgan, D., Gerdes, K., 2013. Quality Guidelines for Energy System Studies: Carbon Dioxide Transport and Storage Costs in NETL Studies. The National Energy Technology Laboratory, U.S. Department of Energy.

Han, J.H., Lee, I.B., 2011. Development of a scalable and comprehensive infrastructure model for carbon dioxide utilization and disposal. Ind. Eng. Chem. Res. 50 (10), 6297–6315.

Hare, P., Ladbrook, B., 2007. Analysis of Carbon Capture and Storage Cost-supply Curves for the UK. Poyry Energy Consulting Ltd., Oxford (United Kingdom).
Heddle, G., Herzog, H., Klett, M., 2003. The Economics of CO_2 Storage. Laboratory for Energy and the Environment, Massachusetts Institute of Technology.
Hendriks, C., Graus, W., Van Bergen, F., 2004. Global Carbon Dioxide Storage Potential and Costs. Ecofys. Report no. EEP-02001.
Hendriks, C., Wildenborg, T., Feron, P., Graus, W., Brandsma, R., 2003. Ec-case Carbon Dioxide Sequestration. Ecofys. TNO M, 70066.
IEA Greenhouse Gas R&D Programme, 2005a. Building the Cost Curves for CO_2 Storage: European Sector. Report no. 2005/2.
IEA Greenhouse Gas R&D Programme, 2005b. Building the Cost Curves for CO_2 Storage: North America. Report no. 2005/2.
IEA Greenhouse Gas R&D Programme, 2002. Transmission of CO_2 and Energy. Report no. PH4/6.
Klemeš, J., Bulatov, I., Cockerill, T., 2007. Techno-economic modelling and cost functions of CO_2 capture processes. Comput. Chem. Eng. 31 (5), 445–455.
Klokk, Ø., Schreiner, P.F., Pagès-Bernaus, A., Tomasgard, A., 2010. Optimizing a CO_2 value chain for the Norwegian Continental Shelf. Energy Policy 38 (11), 6604–6614.
Kreutz, T., Williams, R., Consonni, S., Chiesa, P., 2005. Co-production of hydrogen, electricity and CO_2 from coal with commercially ready technology. Part B: economic analysis. Int. J. Hydrogen Energy 30 (7), 769–784.
Liu, P., Gerogiorgis, D.I., Pistikopoulos, E.N., 2007. Modeling and optimization of polygeneration energy systems. Catal. Today 127 (1), 347–359.
McCollum, D.L., Ogden, J.M., 2006. Techno-economic Models for Carbon Dioxide Compression, Transport, and Storage & Correlations for Estimating Carbon Dioxide Density and Viscosity. Institute of Transportation Studies, UC Davis.
McCoy, S.T., Rubin, E.S., 2008. An engineering-economic model of pipeline transport of CO_2 with application to carbon capture and storage. Int. J. Greenhouse Gas Control 2 (2), 219–229.
McKinsey & Company, 2008. Carbon Capture and Storage: Assessing the Economics.
Middleton, R.S., Bielicki, J.M., 2009. A scalable infrastructure model for carbon capture and storage: SimCCS. Energy Policy 37 (3), 1052–1060.
Medeiros, J.L.d., Versiani, B.M., Araujo, O.Q.F., 2008. A model for pipeline transportation of supercritical CO_2 for geological storage. J. Pipeline Eng. 7, 253–280.
MIT Carbon Capture and Sequestration Technologies Program, 2006. Carbon Management GIS: CO_2 Pipeline Transport Cost Estimation.
Mohitpour, M., Golshan, H., Murray, A., 2000. Pipeline Design & Construction: A Practical Approach. American Society of Mechanical Engineers Press, New York.
Ogden, J.M., 2004. Conceptual Design of Optimized Fossil Energy Systems with Capture and Sequestration of Carbon Dioxide. Institute of Transportation Studies, UC Davis.
Parker, N., 2004. Using Natural Gas Transmission Pipeline Costs to Estimate Hydrogen Pipeline Costs. Institute of Transportation Studies, UC Davis.
Poiencot, B., Brown, C., 2011. An optimal centralized carbon dioxide repository for Florida, USA. Int. J. Environ. Res. Public Health 8 (4), 955–975.
Poullikkas, A., 1992. Two Phase Flow and Cavitation in Centrifugal Pump: A Theoretical and Experimental Investigation, Loughborough University (Doctoral dissertation).

Serpa, J., Morbee, J., Tzimas, E., 2011. Technical and Economic Characteristics of a CO_2 Transmission Pipeline Infrastructure. Institute for Energy, Joint Research Centre, European Commission.

Smith, L.A., Gupta, N., Sass, B.M., Bubenik, T.A., Byrer, C., Bergman, P., 2001. Engineering and Economic Assessment of Carbon Dioxide Sequestration in Saline Formations. Battelle Memorial Institute.

Stephan, K., Lucas, K., 1979. Viscosity of Dense Fluids. Springer.

Svensson, R., Odenberger, M., Johnsson, F., StrÖmberg, L., 2004. Transportation systems for CO_2—application to carbon capture and storage. Energy Convers. Manage. 45 (15), 2343−2353.

Tan, R.R., Aviso, K.B., Bandyopadhyay, S., Ng, D.K., 2012. Continuous-time optimization model for source−sink matching in carbon capture and storage systems. Ind. Eng. Chem. Res. 51 (30), 10015−10020.

Wilcox, J., 2012. Compression and transport of CO_2. In: Carbon Capture. Springer, New York, pp. 35−51.

Yeomans, H., Grossmann, I.E., 1999. A systematic modeling framework of superstructure optimization in process synthesis. Comput. Chem. Eng. 23 (6), 709−731.

Zhang, Z.X., Wang, G.X., Massarotto, P., Rudolph, V., 2006. Optimization of pipeline transport for CO_2 sequestration. Energy Convers. Manage. 47 (6), 702−715.

Zheng, Z., Gao, D., Ma, L., Li, Z., Ni, W., 2009. CO_2 capture and sequestration source-sink match optimization in Jing-Jin-Ji region of China. Front. Energy Power Eng. China 3 (3), 359−368.

Zhou, C., Liu, P., Li, Z., 2013. A superstructure based mixed-integer programming approach to the optimal design of a pipelines network for CO_2 transport. In: Proceedings of the 23rd European Symposium on Computer Aided Process Engineering, pp. 193−198.

Zhou, C., Liu, P., Li, Z., 2014. A superstructure-based mixed-integer programming approach to optimal design of pipeline network for large-scale CO_2 transport. AIChE J. 60 (7), 2442−2461.

Zigrang, D.J., Sylvester, N.D., 1982. Explicit approximations to the solution of Colebrook's friction factor equation. AIChE J. 28 (3), 514−515.

CHAPTER

10 Optimization and Technoeconomic Analysis of Rapid Temperature Swing Adsorption Process for Carbon Capture from Coal-Fired Power Plant

Subramanian Swernath, Katherine Searcy, Fateme Rezaei, Ying Labreche, Ryan P. Lively, Matthew J. Realff[1], Yoshiaki Kawajiri

School of Chemical & Biomolecular Engineering, Georgia Institute of Technology, Atlanta, GA, USA

[1]Corresponding author: E-mail: matthew.realff@chbe.gatech.edu

10.1 INTRODUCTION

Due to increasing energy consumption and global CO_2 emissions, development of sustainable, low-cost postcombustion CO_2 capture methods are warranted. In order for the CO_2 capture techniques to be sustainable and viable, the amount of external energy (i.e., heating and cooling utilities) needs to be minimized. A significant amount of research has focused on enhancing the efficiency of different postcombustion CO_2 capture methods for coal-fired power plants in the 2000s. The current state-of-the-art technologies include CO_2 absorption by amine solutions (Rochelle, 2009) and pressure and vacuum swing adsorption (Xiao et al., 2008). Hybrid technologies have also been investigated to improve the efficiency of CO_2 capture (Ritter et al., 2012).

The use of a hollow fiber sorbent system operating in a rapid temperature swing adsorption (RTSA) mode for CO_2 capture has been demonstrated experimentally (Lively et al., 2012). In this demonstration, the adsorbent bed consists of hollow fibers of roughly 1000 μm outside diameter that are impregnated with sorbent particles of zeolite 13X. The hollow fibers have an internal bore, of roughly 350 μm, to allow heating and cooling by water. Lively et al. demonstrated the advantages of the RTSA process with hollow fiber sorbents for the capture of CO_2 from flue gas. Preliminary studies indicated that the hollow-fiber RTSA process has low flue gas pressure drop (Lively et al., 2009), due to the structured nature of the bed, and fast heat and mass transfer, due to the thin fiber walls and bore water flow.

As shown in Figure 10.1, the hollow fiber carbon capture system consists of three different components: an inner, impermeable bore; an amine sorbent; and a porous

Computer Aided Chemical Engineering, Volume 36. ISSN 1570-7946. http://dx.doi.org/10.1016/B978-0-444-63472-6.00010-0

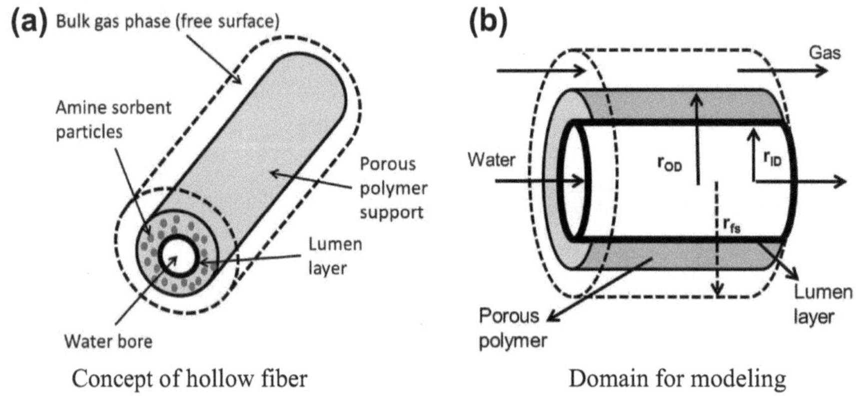

FIGURE 10.1

Hollow fiber sorbent.

polymer. Within the inner impermeable bore, hot and cold water flow. A porous polymeric layer is impregnated with sorbent, which allows for the adsorption of CO_2 from flue gas. The adsorbent particles could be a zeolite (Lively et al., 2012) or an aminosilica material (Labreche et al., 2013; Rezaei et al., 2013). The CO_2-laden flue gas flows across this polymer layer, which can be considered as a third component of a hollow fiber configuration. The presence of the impermeable lumen layer allows for rapid heat transfer between the impregnated solid sorbent particles and the heat transfer fluid in the bore (water) without any seepage of CO_2. Lively et al. (2012) analyzed the effect of cooling water velocity on the sorption performance of a hollow fiber RTSA process for cellulose acetate/13X fiber sorbents. In addition to the heat recovery, the authors (Lively et al., 2012) found that the sorption capacity increases significantly because water flowing through the inner bore removes the heat of adsorption and creates nearly isothermal conditions. However, due to the competitive adsorption behavior shown by zeolite 13X between CO_2 and water, this technology may require a pretreatment process for dehydration. This drawback can be overcome by the use of amine materials as the sorbents in the hollow fiber postcombustion CO_2 capture system (Labreche et al., 2013; Rezaei et al., 2013).

The concept of a hollow-fiber RTSA cycle is shown in Figure 10.2, where multiple hollow fibers are stacked within a module. The hollow-fiber RTSA system for carbon capture offers significant advantages over the traditional packed bed systems. One such advantage is the fast heat transfer between the fiber and the water that flows in the internal bore. During adsorption, cooling water in the bore removes heat of adsorption effectively, thus maintaining the adsorption capacity. The water that recovers the heat from the fiber can be recycled and stored in a tank. Similarly, during desorption, hot water flows in the bore and provides the heat of desorption. Heat integration can be performed between adsorption and desorption steps by cross-exchanging the spent cold and hot water streams. Such a dynamic heat redistribution between the fibers in different steps of an RTSA cycle may achieve a

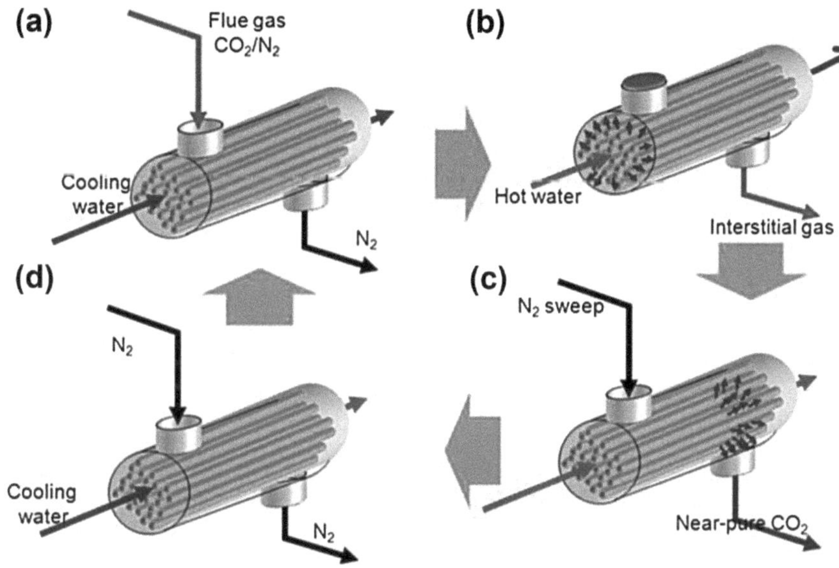

FIGURE 10.2

Schematic of a full-scale RTSA process operated over the four steps of a cycle.

significant reduction in external hot and cold utility usage (i.e., parasitic load). The fast heat transfer also allows quick heating and cooling of the fiber, which reduces the cycle time.

The hollow-fiber RTSA system requires careful design and operation to maximize throughput and minimize energy consumption while meeting constraints. In our previous work, it was found that the velocity of cold and hot water must be selected carefully to achieve efficient heat integration while maintaining the CO_2 purity and recovery specifications at the targeted flue gas throughput (Rezaei et al., 2014). In order to design and operate an RTSA with low capital and utility cost, we employ a dynamic optimization approach. There is a unique dynamic heat integration problem for the hollow fiber RTSA, i.e., recycling the heat and cold streams between different steps in a cycle that has not been encountered in the various studies on the optimization of cyclic adsorption processes such as pressure swing adsorption (PSA) and vacuum swing adsorption (VSA). For this process, multiple objectives must be considered to analyze the trade-off between the minimum external energy utility and minimum capital cost. To the best of our knowledge, there has not been any work to date for optimization of a hollow fiber RTSA process except our preliminary work reported recently (Swernath et al., 2014).

Numerical optimization of cyclic adsorption processes is a significant challenge. Optimization of cyclic adsorption processes have been addressed in many past studies. To avoid expensive computation to find the cyclic steady state (CSS), direct determination of CSS profiles is often employed where a Newton solver determines the CSS concentration and temperature profiles without calculating the transient profiles during startup. The direct determination of CSS can be performed by two different approaches: (1) simultaneous (total discretization) approach and (2) sequential approach (single discretization). In the simultaneous approach, the entire governing partial differential algebraic equations (PDAEs) are discretized both in the temporal and spatial domains into a set of nonlinear algebraic equations. The resulting problem is solved by a nonlinear programming solver (Nilchan and Pantelides 1998). On the other hand, in the sequential approach, the PDAEs are discretized only in the spatial domain and the resulting problem is constrained by ordinary differential equations (Ko et al., 2003). In this work, we use the sequential approach to find the optimal operation of the RTSA process. Furthermore, as mentioned earlier, our process has a unique problem of dynamic heat integration, where the spent hot and cold water from the RTSA process, which gained or provided the heat of adsorption, can be recycled in a time-varying manner for heat integration. Such a dynamic heat integration problem for adsorption processes has not been addressed in past studies.

A primary contribution of this work is to illustrate the use of dynamic optimization to distribute the energy between different steps of a cycle for minimizing the parasitic load while maintaining sufficient CO_2 throughput. Furthermore, in the last part of this chapter we analyze the optimal RTSA operation using a rigorous technoeconomic analysis (TEA) for the CO_2 capture system of a full-scale 550 MW_e (megawatt electricity) coal-fired power plant according to the specifications given by the US Department of Energy (DOE).

10.2 MODELING APPROACH

10.2.1 RAPID TEMPERATURE SWING ADSORPTION CONCEPT

Figure 10.2 shows the schematic of the four steps a single fiber undergoes within one cycle. In a large-scale commercial process, numerous fibers are housed within a single module. In the model we develop in this study, the dynamics of only a single fiber are considered. This assumes that the fiber-to-fiber variation of all variables and properties across a module is negligible.

Figure 10.2(a) shows the adsorption step where the CO_2-laden flue gas is adsorbed on the sorbent impregnated within the fiber. Cold water flows through the bore and removes the heat of adsorption to keep the fiber in a nearly isothermal condition. This helps to ensure a low fiber temperature even in the presence of the heat of adsorption, thus increasing the sorption capacity. The results of a parametric investigation on an adsorption step revealed the importance of water velocity on the trade-off between adsorption heat recovery and sorption capacity (Rezaei et al., 2014).

The second step of the cycle pertains to desorption (Figure 10.2(b)). The desorption step is performed through two substeps: heating and N_2 sweeping. Here, hot water flows through the bore of the fiber and increases fiber temperature from the water inlet side of the module; this process leads to CO_2 desorption. As a result, a wave of CO_2 gas is formed that moves along the fiber in the direction of hot water flow. Due to the efficient heat and mass transfer in the hollow fiber, it is possible to create a relatively sharp CO_2 wave, which allows for high product purity and recovery. In the second substep (Figure 10.2(c)), a continuous N_2 purge stream flows over the fiber at a relatively low velocity. This helps to sweep out the interstitial CO_2.

The final step is the cooling step (Figure 10.2(d)). In this step, cold water flows inside the bore to cool the fiber back to the adsorption temperature. During this step either N_2 or flue gas at an extremely low velocity can be passed across the fiber. In the optimization study presented in this chapter an N_2 stream is used for this sweeping operation. A flue gas that contains both N_2 and CO_2 can also be used as a sweep; this approach may increase the throughput further but reduces the CO_2 recovery. The water velocity in this step needs to be determined carefully with consideration of the trade-off between a short cooling step and high sensible heat recovery.

By carefully designing the above four steps, dynamic heat integration can be achieved by exploiting the unique features of the RTSA process, as discussed in the introduction. The hot water tank supplies hot water for the desorption and N_2 sweeping steps, while the cold water tank supplies the cooling water for adsorption and cooling steps. A detailed schematic is provided later in the optimization problem formulation section.

In this study, we also carry out a TEA based on a full-scale 550 MW_e coal fired power plant to evaluate the scope of a hollow fiber RTSA as a viable process for CO_2 capture. A trade off exists between efficient heat usage by dynamic heat integration, as discussed above, and the capacity of the RTSA process. A high throughput of conventional RTSA systems can be achieved by shortening the cycle time, which is often limited by the relatively slow heating and cooling steps. Nevertheless, an earlier study reported that reducing the cycle time does not achieve efficient heat integration because a high water velocity diminishes high and low temperature peaks in the recycled water, thus reducing the quality of heat (Rezaei et al., 2014). In the initial part of the work, a dynamic optimization study of the cyclic RTSA process is carried out, which provides the optimal operating variables and conditions for the latter TEA. In the final part of the work, a detailed TEA is performed at the optimal operating conditions according to the specifications given by the DOE (US Department of Energy, January 2011).

10.2.2 GOVERNING EQUATIONS

The governing model equations that describe the RTSA process have been developed and are presented in this section. The radial gradients of temperature and concentration within the fiber have been neglected. Mass and heat balances are modeled for the three different phases of a fiber, namely, internal bore, fiber phase, and the free surface around the fiber (bulk gas phase). Figure 10.3 shows the schematic of the different regions of a single fiber. The detailed assumptions considered to develop the model have been presented in an earlier paper (Rezaei et al., 2014; Kalyanaraman et al., 2015).

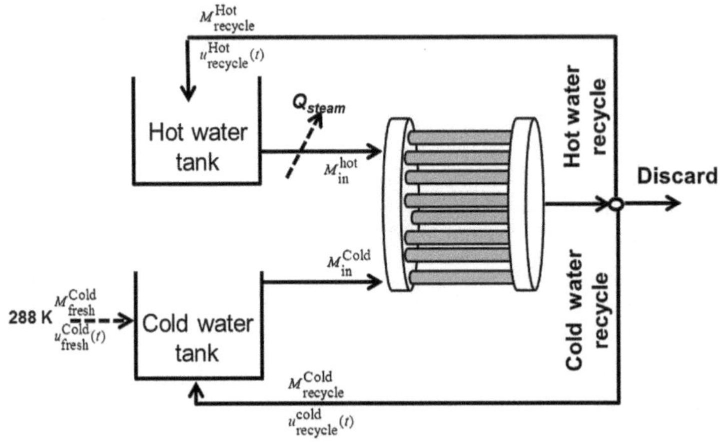

FIGURE 10.3

Heat integration concept for fiber-sorbent RTSA.

The energy balance across the water flowing in the bore neglecting the radial gradients can be written as:

$$\rho_W C_{pw} \frac{\partial T_w(Z,t)}{\partial t} + \rho_W C_{pw} u_w \frac{\partial T_w(z,t)}{\partial z} = A_I h_w (T_w(z,t) - T_f) \tag{10.1}$$

where ρ_w, C_{pw}, $T_w(t,z)$, u_w, and h_w represent the density, specific heat capacity, temperature, velocity, and heat transfer coefficient of water. The subscripts w, f, g represent the water, fiber, and the gas phase, respectively. In addition, $c_{i,g}$ represents the concentration of CO_2 and N_2 in the bulk gas phase while q_i corresponds to the concentration in the solid phase, and ε_f is the void fraction in the fiber phase.

Similarly, the temperature distribution in the fiber across the radial coordinate has been assumed to be a constant. Under such an assumption, the heat balance of the fiber is given by

$$(1-\varepsilon_f)\rho_f C_{pf} \frac{\partial T_f(z,t)}{\partial t} + A_I h_w (T_f(z,t) - T_w(z,t) + A_O h_g (T_f(z,t) - T_g(z,t)) \\ = (1-\varepsilon_f)\rho_f \Delta H_{ads} \frac{\partial_{q_i}(z,t)}{\partial t} \tag{10.2}$$

Here, ρ_f, C_{pf}, T_f, and ε_f represent the density, specific heat capacity, temperature, and porosity of the fiber, respectively; ΔH_{ads} corresponds to the heat of adsorption; $A_o = \frac{4d_o}{d_{fs}^2 - d_o^2}$ is an external specific surface area of the hollow fiber that is defined as the external surface area per free space volume. Similarly, $A_I = \frac{4}{d_I}$ is the internal specific surface area of the bore; h_g represents the convective heat transfer coefficient of the bulk gas phase, and h_w the heat transfer coefficient of water.

The heat balance of the gas phase surrounding the fiber is given as:

$$\rho_g C_{pg}\frac{\partial T_g(z,t)}{\partial t} + \rho_g C_{pg} u_g(z,t)\frac{\partial T_g(z,t)}{\partial z} = A_O h_g\left(T_f(z,t) - T_g(z,t)\right) \tag{10.3}$$

The mass balance for the two components CO_2 and N_2 across the external bulk gas phase is given by:

$$\frac{\partial c_{i,g}(z,t)}{\partial t} + u_g(z,t)\frac{\partial c_{i,g}(z,t)}{\partial z} + c_{i,g}(z,t)\frac{\partial u_g(z,t)}{\partial z} + (1-\varepsilon_f)\rho_f\frac{A_{fiber}}{A_{fs}}\frac{\partial q_i(z,t)}{\partial t} = 0 \tag{10.4}$$

Here, $c_{i,g}(z,t)$ represents the concentration of CO_2 and N_2 in the bulk gas phase, $u_g(z,t)$ is the velocity of the bulk gas phase, and $q_i(z,t)$ is the concentration of CO_2 and N_2 in the solid fiber. N_2 is assumed nonadsorbing, and hence its concentration in the solid phase is zero. $A_{fiber}\,(=\pi(r_{OD}^2 - r_{ID}^2))$ and $A_f(=\pi(r_{fs}^2 - r_{OD}^2))s$ are the cross-sectional areas of the fiber and free surface, respectively.

Furthermore, the total mass balance across the phase under the assumption of negligible pressure drop can be written as:

$$u_g\frac{\partial c_g^T}{\partial z} + c_g^T\frac{\partial u_g}{\partial z} + (1-\varepsilon_f)\rho_f\frac{A_{fiber}}{A_{fs}}\frac{\partial q_{CO_2}}{\partial t} = 0 \tag{10.5}$$

For the mass transfer between the bulk gas phase and the sorbent, we use the linear driving force model:

$$\frac{\partial q_i(z,t)}{\partial t} = k_s\left(q_i^{eq}(z,t) - q_i(z,t)\right) \tag{10.6}$$

where $q_i^{eq}(z,t)$ is the equilibrium concentration in the fiber phase. The thermodynamic adsorption equilibrium between the adsorbed CO_2 on the amine sorbent and the concentration of CO_2 in the bulk gas phase is given by the Langmuir isotherm:

$$q_i^{eq} = \frac{q_{mo}b_{mo}\exp\left(\eta\left(1-\frac{T_f}{T_o}\right)\right)\exp\left(\frac{\Delta H_{ads}}{RT_f}\right)p_i(z,t)}{1 + b_{mo}\exp\left(\frac{\Delta H_{ads}}{RT_f}\right)p_i(z,t)} \tag{10.7}$$

The above set of PDAEs is solved with boundary conditions shown in Table A.1 in the Appendix is simulated/optimized within the modeling environment gPROMS (Process Systems Enterprise L 2011). The isotherm parameters used in this study are from Fauth et al. (2012) and are summarized in Table 10.1. The fiber dimensions were not included in the decision variables but fixed a priori (Table 10.2).

Table 10.1 Isotherm and Mass Transfer Parameters (Fauth et al., 2012)

q_{mo} (mol/kg fiber)	η	b_{mo} (kpa^{-1})	ΔH_{ads} (J/mol)	k_s (s^{-1})
1.75	1.36	2.05×10^{-5}	27,200	0.07

Table 10.2 Fiber Parameters for the Dimension Considered

A_{bore} (m²)	A_{fiber} (m²)	A_{fs} (m²)	L (m)
8.03×10^{-8}	1.05×10^{-6}	7.53×10^{-7}	3.0

10.3 OPTIMIZATION APPROACH

10.3.1 METHODOLOGY

Multiple objectives can be considered for the RTSA process, such as CO_2 throughput, energy usage, CO_2 purity, and CO_2 recovery. In this study, we treat the CO_2 purity and recovery demand as constraints. A higher CO_2 throughput reduces the amount of fiber required to process a given amount of flue gas, thus reducing the capital cost. It should be noted that high throughput is in a trade-off relationship with energy consumption.

We consider a flowsheet of the RTSA system with spent water recycle shown in Figure 10.3. In this process scheme, the temperatures of hot water and cold water that enter the RTSA module are kept constant either at $T^{in}_{w,heat}$ or at $T^{in}_{w,cold}$ depending on the step that the RTSA module is in. The hot and cold water are maintained constant at these temperatures by heating with steam downstream of the hot tank and addition of cold water to the cold tank, respectively. The use of this steam causes the plant electrical generation capacity to be derated. Therefore, an escalation step is required to increase the plant size to reach the original generation capacity. The numerical values for the inlet temperature and concentration conditions are provided in Table A.2 in the Appendix. The spent water from the RTSA module can be recycled back either to the hot or cold tank. To maintain the water volume balance, the spent water can be partially discarded. Note that these decisions, recycling or discarding the spent water, must be done in a time-varying manner. Thus, the problem is an optimal control problem.

In Figure 10.3, the water balance around the water discard and recycle switching point is given by:

$$u_w(t, L) = u^{cold}_{\text{Recycle}}(t) + u^{Hot}_{\text{Recycle}}(t) + u_{discard}(t) \quad (10.8)$$

where $u_{discard}(t)$ is the water velocity in the discard stream, and $u^{cold}_{\text{Recycle}}(t)$ and $u^{Hot}_{\text{Recycle}}(t)$ are cold and hot water recycled to the cold and hot tanks, respectively. In this study, we assume that the water tank volumes must be maintained only over a cycle, allowing temporary changes within a cycle. Under this assumption, the water balances around the hot and cold tanks over a cycle is given by:

$$\int_0^{t_{cycle}} u^{Hot}_{\text{Recycle}}(t)dt - \int_0^{t_{cycle}} u^{in}_{\text{w,hot}}(y_{heat} + y_{sweep})dt = 0 \quad (10.9)$$

$$\int_0^{t_{cycle}} u_{Recycle}^{cold}(t)dt + M_{\text{Fresh}}^{\text{Cold}} - \int_0^{t_{cycle}} u_{w,\text{cold}}^{in}(y_{\text{ads}} + y_{\text{cool}})dt = 0 \tag{10.10}$$

where y_{heat}, y_{sweep}, y_{ads} and y_{cool} are prespecified binary parameters that are one during the heating, sweeping, adsorption, and cooling step and zero otherwise, and $M_{\text{Fresh}}^{\text{Cold}}$ is the mass of the fresh water supplied to the cold tank. Note that the above two equations are enforced over a cycle (end-point constraints) but are not path constraints. Similarly, the heat balances around the hot and cold water tanks are enforced over a cycle, and given by:

$$\begin{aligned} &Q_{steam} + A_{bore}\rho_w C_{pw} \int_0^{t_{cycle}} u_{\text{Recycle}}^{Hot}(t)(T_w(t,L) - T_{ref})dt \\ &-A_{bore}\rho_w C_{pw} \int_0^{t_{cycle}} u_{w,hot}^{in}(y_{heat} + y_{sweep})(T_{w,cold}^{in} - T_{ref})dt = 0 \end{aligned} \tag{10.11a}$$

$$\begin{aligned} &\int_0^{t_{cycle}} u_{\text{Recycle}}^{\text{cold}}(t)(T_w(t,L) - T_{ref})dt + M_{\text{Fresh}}^{\text{Cold}}(T_{\text{Fresh}}^{\text{Cold}} - T_{ref}) \\ &- \int_0^{t_{cycle}} u_{w,\text{cold}}^{in}(y_{\text{ads}} + y_{\text{cool}})(T_{w,cold}^{in} - T_{ref})dt = 0 \end{aligned} \tag{10.11b}$$

where ρ_w is the water density, Q_{steam} is the amount of steam and C_{pw} is the heat capacity of water.

The purity and recovery requirements can be satisfied by fractionating the gas outlet from the RTSA module between two destinations: CO_2 storage and atmospheric release. To represent this decision of gas fractionation, we use a dynamic variable $I_1(t)$, where $I_1(t) \in [0,1]$. This concept is shown in Figure 10.4. When this value is zero, we release the outlet gas to the atmosphere. On the other hand,

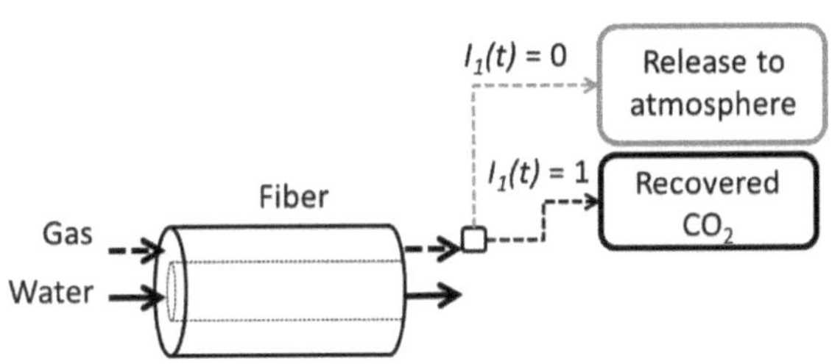

FIGURE 10.4

Binary variable depicting the switching of outlet gas stream between recovered fraction and waste stream ($I_1(t)$).

when $I_1(t) = 1$, the gas is sent to CO_2 storage (Figure 10.4). Using this variable, the purity and recovery constraints can be written as:

$$\text{Recovery}_{CO_2} = \frac{\int_0^{t_{cycle}} I_1(t)u_g(L)c_{CO_2,g}(L)dt}{\int_0^{t_{cycle}} u_g(0)c_{in}^{CO_2}dt} \geq \text{Recovery}_{\text{min}} \tag{10.12a}$$

$$\text{Purity}_{CO_2} = \frac{\int_0^{t_{cycle}} I_1(t)u_g(L)c_{CO_2,g}(L)dt}{\int_0^{t_{cycle}} I_1(t)u_g(L)c_{total,g}(L)dt} \geq \text{Purity}_{\text{min}} \tag{10.12b}$$

It should be noted that we treat this variable $I_1(t)$ as a dynamic integer variable. In order to reduce the system complexity, we assume that the binary variable is fixed to 0 in the adsorption step and fixed to 1 in the N_2 sweeping step. For the other two steps, the optimization algorithm decides the time instant at which the switch occurs. The mathematical definitions of purity and recovery (Eqn (10.12)) have been written using the former idea.

As discussed above, there is a trade-off between the CO_2 throughput and energy usage. The two objectives can then be written as:

$$\text{Throughput} = \frac{\int_0^{t_{ads}} u_g(0)A_{fs}c_{i,g}(0)dt}{V_{fiber}t_{cycle}} \tag{10.13}$$

$$\text{External heat and cold utility cost} = W_1 Q_{\text{steam}} + W_1 M_{\text{Fresh}}^{\text{Cold}} \tag{10.14}$$

where Q_{steam} is the external steam duty and $M_{\text{Fresh}}^{\text{Cold}}$ is the amount of cold water, and $W_1$and W_2 are the prices of steam and cold water per unit amount, respectively. A high throughput reduces the amount of fiber sorbent to process the given amount of CO_2, and a low external utility cost reduces the total utility cost. In this work, we employ the relatively simple functions Eqns (10.13) and (10.14) to represent the economics of the CO_2 capture system. Evaluating rigorous economic metrics, such as the total capital and utility cost, would involve nonlinear discrete functions, and thus the optimization problem would become a discrete optimization problem. The values of W_1 and W_2 in Eqn (10.14) were obtained from Seider et al. (2008) as follows: $W_1 = \$14.5/1000$ kg and $W_2 = \$0.02/1000\ m^3$.

In this study, an RTSA operation that achieves a high CO_2 throughput is pursued as the primary objective, while minimizing the utility cost is considered as the secondary objective. This is because we later find that minimizing the fiber (adsorbent) cost is crucial, which is dominant in the overall capital cost. Furthermore, the sensible heat expended per cycle is a very large component of the utility cost, and this scales with the amount of fiber required, so there is a reasonable degree of correlation between minimizing fiber cost and a large component of the overall utility cost. More specifically, we solve the following sequence of the two optimization problems:

1. Solve the throughput maximization problem

$$\text{Maximize } \phi = \text{Throughput} \tag{10.15}$$

2. Solve the utility minimization problem formulated as follows

$$\text{Minimize } W_1 Q_{\text{steam}} + W_2 M_{Fresh}^{Cold} \quad (10.16a)$$

$$\text{Subject to: Throughput} \geq \phi \quad (10.16b)$$

It was found that there are many nonunique solutions to the problem (10.15) that achieves the highest throughput ϕ. By solving the above sequence of the two problems, (I) and (II), we identify the optimal solution that achieves the lowest utility cost while achieving the highest throughput.

To solve the above optimization problem, we employ the simultaneous approach that finds the optimal operating parameters and internal concentration and temperature profiles at the CSS simultaneously. The PDAEs are discretized in the axial domain only, and the resulting set of ordinary differential equations is solved by an integrator. The computational cost associated with the process transients during startup is eliminated by the direct determination approach, which finds the CSS profiles that satisfy the following constraints: (Jiang et al., 2004; Nichan and Pantelides, 1998)

$$T_w(z,0) = T_w(z, t_{cycle}) \quad (10.17a)$$

$$T_f(z,0) = T_f(z, t_{cycle}) \quad (10.17b)$$

$$T_g(z,0) = T_g(z, t_{cycle}) \quad (10.17c)$$

$$C_{i,g}(z,0) = C_{i,g}(z, t_{cycle}) \quad (10.17d)$$

$$q_i(z,0) = q_i(z, t_{cycle}) \quad (10.17e)$$

Optimization was initialized with a feasible solution. The optimization was performed using the successive reduced quadratic programming solver within gPROMS. All control variables are parameterized as piecewise constants, and the resulting problem was solved by the control vector parameterization approach (Process Systems Enterprise, 2012).

10.4 RESULTS AND DISCUSSIONS

10.4.1 OPTIMIZATION OF RTSA PROCESS

The maximized throughput, found as the solution to Problem (I), is $\phi = 2.46$ mol/(m^3-fiber·s). Problem (II) was solved by using this value of the throughput. The optimal operating conditions are shown in Table A.3 in the Appendix. Figure 10.5 shows the temperature profile at the outlet. It can be seen from this figure that the water temperature is almost isothermal until the hot water fed to the desorption step starts exiting the module. It is also worth noting that the cooling step is short in order to maximize the throughput, and hence the water temperature at the exit does not reach the adsorption feed temperature of 308.15 K.

It can also be seen from Figure 10.5 that the optimal solution does not recover heat of adsorption during the adsorption step but recycles the whole amount of

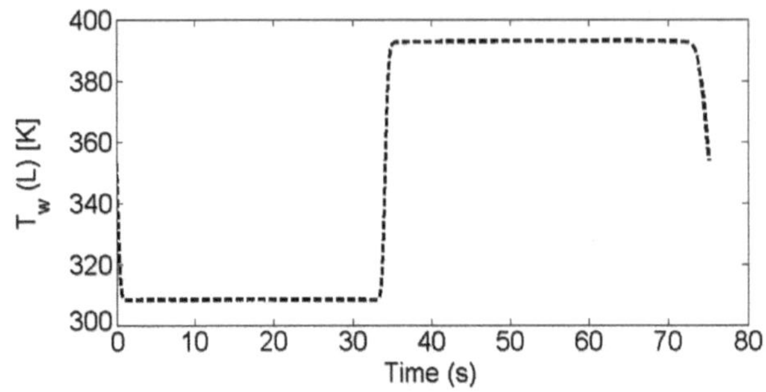

FIGURE 10.5

Water temperature profile at the outlet of fiber.

the spent cooling water back to the cold water tank. This is partly because the temperature increase due to the heat of adsorption was too small, probably due to the relatively low mass transfer coefficient. To maximize the throughput while achieving the high CO_2 recovery, the fiber must be cooled as quickly as possible at the expense of higher utility usage to remove the heat of adsorption so that the capacity would not be decreased. In the optimal operations, the spent cooling water from the adsorption step was recooled by the heat exchanger. Our sensitivity analysis with respect to the mass transfer coefficient revealed that recovering the heat of adsorption is feasible if the mass transfer is enhanced by a potential reduction in the resistance of CO_2 diffusion in the silica particles and polymer sorbent. Further investigation is beyond the scope of the current work.

Figure 10.6 shows the water temperature distribution profile at the outlet. As discussed in the optimization formulation, the water at the outlet could be split into three possible streams, namely, a hot stream (u_{hot}), cold stream (u_{cold}), or a discard stream. Since the objective is to minimize the amount of external energy supplied, the water from the heating and N_2 sweeping steps, which has relatively high temperature, is recycled to the hot tank. The outlet water stream from

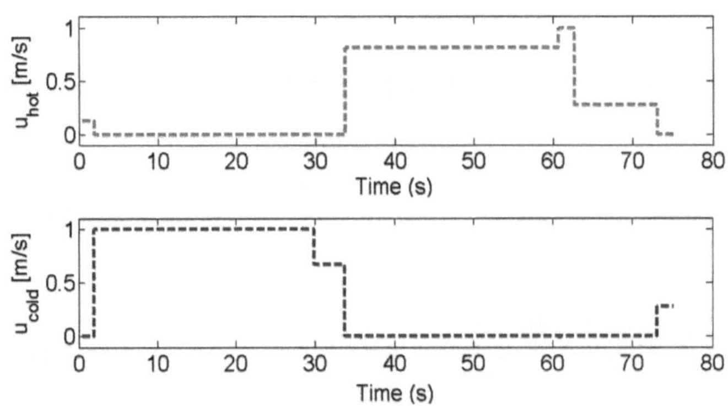

FIGURE 10.6

Water splitting at the outlet of fiber.

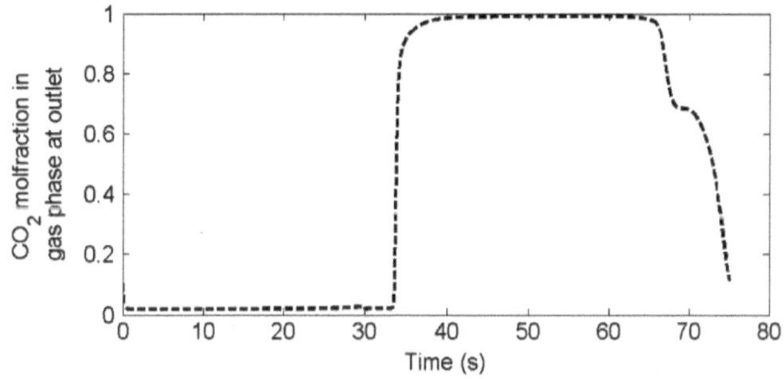

FIGURE 10.7

CO_2 mole fraction profiles in the bulk gas phase at the RTSA fiber outlet.

the adsorption is sufficiently cold and is recycled into the cold tank; thus, recovery of the heat of adsorption is not performed. On the other hand, most of the water from the cooling step is transferred to the hot tank as it captures the sensible heat. Some of the water at the intermediate temperature is discarded. It is possible that the intermediate-temperature water could be used elsewhere in the power plant, but integration of this type has been ruled out of scope. We also observed that the presence of the discard stream in the optimization model facilitates convergence.

In Figure 10.7, the CO_2 mole fraction at the outlet in the bulk gas phase is plotted. This provides the purity and recovery estimates. The purity and recovery of the collected stream is at 90% and 85%, respectively, which was at the minimal values set as constraints in Eqn (10.12).

10.4.2 TECHNOECONOMIC ANALYSIS

10.4.2.1 Basis for TEA

Technoeconomic analysis (TEA) enables prioritization and optimization of design inputs for the fibers, the modules, and the process configuration not only from a technical perspective but also from an economic perspective. The TEA further develops the early-phase RTSA process concept into a practicable, efficient, and economical process design. Development of the TEA begins at the module level and then proceeds to process configuration, heat and material balances, equipment selection and sizing, and capital and operating costs. From these outputs, metrics such as the levelized cost of electricity (LCOE) can be estimated. An overview of the TEA design basis and technical approach is provided below.

The DOE funds significant research and development in carbon capture and has established design bases and specifications that are typically used to conduct TEAs for these systems (US Department of Energy, November 2010, April 2011, August 2012, January 2013). A simplified version of the DOE methodology (US Department of Energy, November 2010, April 2011, August 2012, January

2013) is employed in this work. The design basis comprises technical process inputs and economic inputs, as well as the system boundaries. For this evaluation, the system boundaries are established around the carbon capture system; the inlet flue gas conditions and flow rates, the conditions of utilities (e.g., steam, cooling water), and other technical process inputs are fixed and shown in Table 10.5. Stream data listed in Table 10.3 were given by the DOE for the given system

Table 10.3 Technical Design Basis for TEA (US Department of Energy, January 2011)

Description	Unit	Value
General		
Target net capacity	MW_e	550
Capacity factor	%	85
CO_2 removal	%	90
Stream Data		
Inlet Flue Gas		
Temperature	°C	57.2
Pressure	kPa	101.3
Mass flow rate	kg/h	2,320,000
Composition		
CO_2	Vol%	13.17
H_2O	Vol%	17.25
N_2	Vol%	66.44
O_2	Vol%	2.34
Ar	Vol%	0.80
SO_x	ppmv	42
NO_x	ppmv	74
CO_2 in inlet gas	tonne/h	476
CO_2 captured	tonne/h	428
Outlet CO_2 Specification		
Temperature	°C	51.1
Pressure	bar	152.7
	kPa	15,272
CO_2	mol%	$\geq$95
Cooling Water		
Supply temperature	°C	15.6
Return temperature	°C	26.7
Minimum approach	°C	16.7
Minimum water-cooled process temperature	°C	32.2

boundaries (US Department of Energy, January 2011). According to the specifications, the target CO_2 capture is 90%, and the target product CO_2 purity is 95%. In this work, we reduced these targets to 90% purity and 85% recovery due to the difficulty of achieving both specifications with the system as currently configured. The purity constraint is important for the reuse of the CO_2 for enhanced oil recovery, and will be the target of future system design. The outlet from the RTSA process is product CO_2 delivered at pipeline pressure (152.7 bar).

The targeted application is a greenfield facility that produces 550 MW_e-net electricity. The process design and economic evaluation are based on a 550 MW_e-net capacity after the "parasitic" energy requirements for the CO_2 capture system and the base plant are deducted. Therefore, the plant size for each technology and process design is "escalated" to produce the capacity of 550 MW_e in addition to the parasitic energy consumption for the CO_2 capture process and the base plant. It should be noted that the inlet flue gas flow rate presented in Table 10.3 is based on a plant size of 550 MW_e-net capacity without CO_2 capture.

The technoeconomic evaluation was carried out on a steady state basis for the optimal operating conditions discussed in the previous section. The demands for the hot and cold utilities were obtained by averaging the flow rates over a cycle. Furthermore, the costs associated with the auxiliary units such as flue gas pretreatment and compression of the product stream have also been accounted for in the cost calculations.

10.4.2.2 Analysis on capital cost and energy recovery

Purchased equipment costs for a benchmark plant size were estimated using a combination of vendor quotes, literature values, and Aspen In-plant Cost Estimator. Using these estimated costs, equipment cost curves and scaling factors were selected or developed for all major equipment, which includes compressors and blowers, turbines, heat exchangers, pumps, filters, tanks, and pressure vessels. The cost curve format is the following:

$$(\text{Total cost}) = (\text{Base cost}) \times \left(\frac{\text{Unit size}}{\text{Base size}}\right)^n \times \frac{I}{I_{base}} \tag{10.18}$$

where n is the sizing exponent for a given equipment type, I_{base} is the cost index (Chemical Engineering Plant Cost Index) of the base year, and I is the current cost index. In this study, we employed the Chemical Engineering Plant Cost Index in June 2007 as the current cost index ($I = 532.7$).

To assess the economic feasibility of the proposed process, we define two metrics. First, we define parasitic energy demand of the capture system as follows:

$$\kappa = \left(\sum_i E_i\right) + Q_{steam}(p - p_{ILP}) \tag{10.19}$$

In Eqn (10.13), κ is the parasitic energy demand defined in MW_e, E is the direct electrical energy requirement (MW_e), Q_{steam} is the heat duty in MW_{th} (megawatt thermal), p is the factor to convert thermal energy of steam to electrical power,

p_{ILP} is the electrical energy per unit of thermal energy from a power recovery turbine that lets the low pressure (LP) steam down to 10 °C above the required desorption temperature. The DOE specified an electrical energy penalty per unit mass of LP stream, which is supplied at 395 °C and 1156 kPa; the mass-basis penalty is 0.0911 kW_e/(lb/h steam) (US Department of Energy, January 2011). For a desorption temperature of 120 °C, saturated steam is delivered at 130 °C, p is 0.3025 MW_e/MW_{th}, and p_{ILP} is 0.1313 MW_e/MW_{th}.

Thus, the parasitic energy demand κ is a function not only of the direct electrical requirements, E_i, of major process equipment such as pumps and compressors but also a function of the amount of heat required, Q_{steam}, and the temperature at which that heat is required. The steam delivery temperature impacts the steam energy penalty, which is the opportunity cost of decreased electrical production because steam that would have otherwise driven generating turbines is diverted to the capture system to supply heat. Using the estimated parasitic energy demand, we define the escalation factor, which quantifies the increase in the base plant size required to produce 550 MW_e-net in addition to the energy required for the base plant and for CO_2 capture.

$$\text{Escalation factor} = \frac{\text{Plant capacity}(550\,\text{MW}_e)}{\text{Plant capacity}(550\,\text{MW}_e) - \kappa} \tag{10.20}$$

Table 10.4 lists the number of modules in each cycle step. The number of modules in each step is proportional to the ratio of the step time t^i_{step} to the cycle time (t^i_{step}/t_{cycle}, i = sorption, self sweeping, N_2 sweep, and cooling). The number of modules in the adsorption step is determined by the feed flow rate of CO_2.

Design variables obtained from the optimization were transferred as inputs into the TEA spreadsheet. All calculations for capital and operating costs, parasitic energy demand, and escalation factor were conducted internally within the spreadsheet, which is based on the pseudo-steady state approximation of the original dynamic optimization result. The primary technoeconomic results are presented in Table 10.5 for the unescalated basis.

The energy measures used in Table 10.5 are defined as follows:

The total enthalpy change of the fiber due to the heat of adsorption:

$$H_{\text{sorption}} = \frac{A_{fiber}\Delta H_{ads}\int_0^L\int_0^{t_{ads}} q(z)(1-\varepsilon_f)\rho_f\,dz\,dt}{t_{cycle}} \times \left(\frac{\text{Number of fibers}}{\text{module}}\right) \times (\text{Number of modules}) \tag{10.21}$$

Table 10.4 Number of Modules in Each Step on Pseudo-Steady State Basis

Adsorption	Self Sweeping (Heating Step 1)	N_2 Sweep (Heating Step 2)	Cooling	Total
361	373	25	150	909

Table 10.5 Unescalated Energy Values (Unescalated)

	Values
Sorption enthalpy (MW_{th}) $H_{sorption}$	179
Sensible heat (MW_{th}) $H_{sensible}$	800
Heat recovered (HR)	58%
Cooling recovered (CR)	57%

The total enthalpy change of the fiber by the heating and cooling water:

$$H_{\text{sensible}} = \frac{V_f \rho_f C_f (1 - \varepsilon_f)(T_{f,avg}(t_{sweep}) - T_{f,avg}(t_{ads}))}{(t_{cycle})} \tag{10.22}$$

where $T_{f,avg}(t)$ is the average temperature of the fiber given as:

$$T_{f,avg}(t) = \frac{\int_0^L T_f(z,t)dz}{L} \tag{10.23}$$

Using H_{sorption} and H_{sensible} defined above, the recovered heat can be obtained as:

$$\text{Heat recovered (HR)} = \frac{H_{\text{sorption}} + H_{\text{sensible}} - (\text{Main heater duty})}{H_{\text{sorption}} + H_{\text{sensible}}} \times 100(\%) \tag{10.24}$$

$$\text{Cold recovered (CR)} = \frac{H_{\text{sorption}} + H_{\text{sensible}} - (\text{Main cooler duty})}{H_{\text{sorption}} + H_{\text{sensible}}} \times 100(\%) \tag{10.25}$$

where heater and cooler duties are given by:

$$\text{Main heater duty (MWth)} = m_{hot} C_{p,w} \left(T^{in}_{w,hot} - T^{tank}_{hot}\right) \tag{10.26}$$

$$\text{Main cooler duty (MWth)} = m_{cold} C_{p,w} \left(T^{in}_{w,cold} - T^{tank}_{cold}\right) \tag{10.27}$$

It can be seen in Table 10.5 that over 50% of the heat in hot and cold water is recycled.

Figure 10.8 shows the different cost contributions to the process capital. The major cost contributors are the fiber modules, CO_2 compression, filters, and heat exchangers. Costs for other individual equipment types were less than 3% of the total purchased equipment costs and were aggregated together; the aggregate is reflected by the "Others" category shown in Figure 10.8. These minor expenses include fans, blowers, turbines, CO_2 dehydration, and SO_2 removal. As can be seen in this figure, the fiber cost is the most dominant contribution for the hollow fiber RTSA process. The details of the capital cost calculations can be found in Tables A.4 and A.5 in the Appendix.

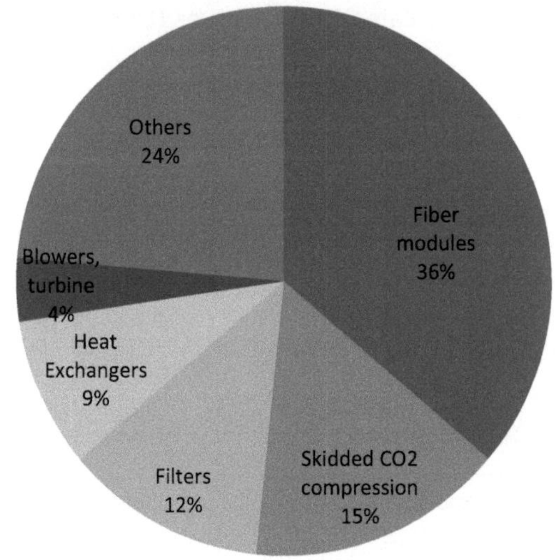

FIGURE 10.8

Distribution of different capital cost contributions.

10.4.2.3 Analysis on operating and maintenance cost

The operating and maintenance (O&M) cost of the RTSA process is analyzed using the capital and energy estimation given above. O&M contributions can be divided into two types: fixed O&M and variable O&M. Fixed O&M costs comprise maintenance and labor. Maintenance costs are calculated as 2% of the total purchased equipment costs while labor costs are fixed at \$995 per day for 365 days a year. On the other hand, variable O&M costs comprise the costs associated with consumables such as makeup chemicals, process water, and waste disposal. Table 10.6 shows the DOE-specified technoeconomic inputs used to calculate the O&M costs. The fiber life was assumed to be 3 years. Costs for transportation, storage, and monitoring, which occur downstream of the capture and compression system, were based on values given by the DOE (\$4.05 per ton CO_2) (US Department of Energy, January 2011).

Using the cost factors provided in Table 10.6, the O&M cost contributions were evaluated. The percentage cost contribution from each of these is shown in Figure 10.9. The primary cost contributors are the steam, fiber replacement, cooling water, and transport, storage, and monitoring of CO_2. The cost is calculated on the escalated basis where the power plant produces 550 MW_e in addition to the energy for CO_2 capture. To further improve the economics of the CO_2 capture process, the fiber replacement cost should be reduced. Typically, the cost of steam is not considered an operating cost when conducting analyses for capture systems that are part of an integrated, coal-fired power plant. The parasitic steam usage of the capture system is accounted for via the required increase in size of the base plant to produce a unit of electricity and the corresponding increase in the cost of electricity. However,

Table 10.6 Inputs for Operating and Maintenance Cost Estimates (US Department of Energy, January 2011).

	Units	Value
Fixed O&M costs		
Maintenance	% of purchased equipment	2.0
Labor	\$/day	995
Variable O&M unit costs (levelized over 20 years)		
Cooling water	\$/1000 gal	0.12
Process makeup water	\$/1000 lb	0.07
Wastewater treatment	\$/lb	0.21
Solid waste disposal	\$/ton	17.87
Toxic waste disposal	\$/ton	89.36
CO_2 TS&M[a]	\$/ton CO_2	4.05
Steam	\$/1000 lb	$0.028\,\kappa + 5.83$

[a]*TS&M indicates transportation, storage and monitoring of the CO_2.*

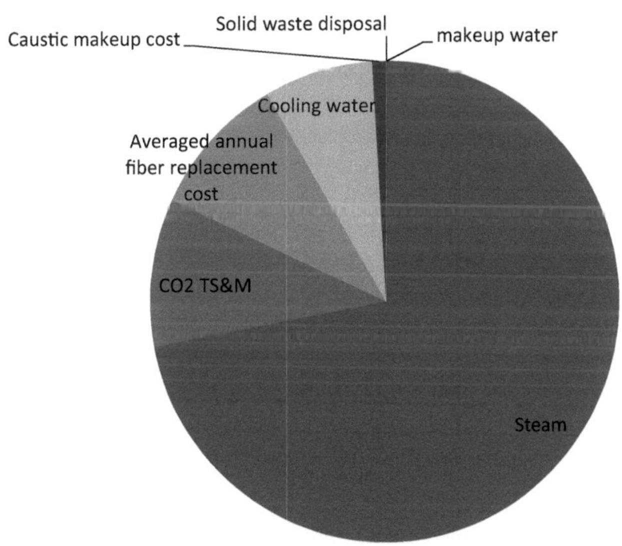

FIGURE 10.9

Variable operating and maintenance cost contribution.

to emphasize the importance of the heat used by this and any capture system, we present steam as an operating cost in this study (Figure 10.9). The "cost" of steam to the isolated capture system was estimated using the equation in Table 10.6 as recommend by the DOE.

In the last step of the TEA, the primary metrics that are often used to compare different CO_2 processes were calculated. LCOE is used as a measure of the overall competitiveness to compare different generating technologies on the same perspective. It represents the electricity cost of building and operating a generating plant over an assumed financial life and duty cycle. For evaluations in which the system included only the capture system, rather than the entire power generation facility, the DOE proposed a simplified equation for estimating the cost of electricity and the cost of CO_2 capture given below (US Department of Energy, January 2011):

$$\text{Levelized cost of electricity (mills/kWh)} = 0.3073\kappa + 64 \tag{10.28}$$

$$\begin{aligned}\text{Cost of } CO_2 \text{ capture (\$/tonne)} = &(\text{Variable O\&M Costs} + \text{Fixed O\&M Costs} \\ &+ \text{Annual Capital Charge}) \\ &\times \text{Captured } CO_2 \text{ (tonne } CO_2\text{/year)}\end{aligned} \tag{10.29}$$

Table 10.7 shows the escalated results for the primary metrics used in CO_2 capture. The LCOE for the Econamine process, the current state-of-the-art capture process, was calculated using the parasitic energy demand report by the DOE (2010) in Eqn (10.28). The cost of electricity estimated for the RTSA process is less than the cost for Econamine, as is the escalation factor. However, it should be noted that this result is based on the reduced values of CO_2 purity and recovery to 90% and 85%, while the target purity and recovery are 95% and 90%, respectively.

It is worthwhile to note that numerous differences may exist in the economic assumptions or bases used to estimate the cost of CO_2 capture and the cost of electricity in this study when compared to other studies. For example, some studies present the cost of electricity on a first-year basis, a 20-year levelized basis, or a 30-year levelized basis. The annual capital charge factor may differ, as it does between the value recommend by the DOE for this study and the DOE 2010 Baseline document, and as may be appropriate for projects with different risk levels. The equations supplied in the FOA to calculate the levelized cost of electricity and steam are based solely on the parasitic energy demands of the CO_2 capture process, and the equations do not account for variability in other operating costs or in capital costs. For a preliminary evaluation of technology under development at the bench scale,

Table 10.7 Technoeconomic Analysis Metrics

	Econamine	Fiber-Sorbent RTSA
Levelized cost of electricity (mills/kWh)	129	121
Cost of CO_2 capture (\$/ton)		39.4
Cost of CO_2 capture (\$/ton)		35.7
Escalation factor	1.382	1.338

the DOE has determined that this is an appropriate approach. If readers plan to compare the results presented in this chapter with results from other studies, the authors strongly caution readers to consider the economic basis of any technoeconomic metric prior to making direct comparisons between evaluations of CO_2 capture technologies.

Given these caveats, a preliminary comparison indicates that the RTSA process with hollow fiber supported amines is competitive with solvent-based amine processes, assuming that further development of the technology can achieve the targeted purity and recovery specifications. Concurrent works available at the time of publication indicates that this success is possible.

10.5 CONCLUSIONS AND FUTURE WORK

In this work, dynamic optimization was performed to determine optimal heat integration for an RTSA process. This heat redistribution approach can pave the way for a reduction in parasitic energy demand. External energy usage was minimized for the maximum throughput. The results of the optimization strategy were later incorporated into a TEA based on a 550 MW_e-net coal-fired power plant and additional specifications given by the DOE. The results of the TEA showed that the RTSA is a viable concept for CO_2 capture.

We expect further improvements in future work. In this study, we reduced the minimum purity and recovery of CO_2 from the target values, which must be increased by improving the fiber adsorbent performance or employing a more advanced operating strategy. Furthermore, we formulated a single objective optimization problem that achieves the highest CO_2 throughput while minimizing the utility usage. A more detailed analysis may be possible by considering these two objectives in a multiobjective optimization framework. Finally, the rigorous economic calculations may be incorporated into the optimization framework directly, if discrete functions can be handled efficiently.

APPENDIX

Table A.1 Boundary Conditions in the Model Equations

Adsorption	Heating	N_2 Sweeping	Cooling
$T_w(t,0) = T^{in}_{w,Cold}$	$T_w(0) = T^{in}_{w,hot}$	$T_w(0) = T^{in}_{w,hot}$	$T_w(0) = T^{in}_{w,cold}$
$T_g(t,0) = T^{in}_{g,ads}$	$T_g(t,0) = T^{in}_{g,heat}$	$T_g(t,0) = T^{in}_{g,sweep}$	$T_g(t,0) = T^{in}_{g,cool}$
$C_{CO_2,g}(t,0) = C^{in}_{CO_2,g,ads}$	$C_{CO_2,g}(t,0) = C^{in}_{CO_2,g,heat}$	$C_{CO_2,g}(t,0) = C^{in}_{CO_2,g,sweep}$	$C_{CO_2,g}(t,0) = C^{in}_{CO_2,g,cool}$
$C_{N_2,g}(t,0) = C^{in}_{N_2,g,ads}$	$C_{N_2,g}(t,0) = C^{in}_{N_2,g,heat}$	$C_{N_2,g}(t,0) = C^{in}_{N_2,g,sweeep}$	$C_{N_2,g}(t,0) = C^{in}_{N_2,g,cool}$
$u_g(t,0) = u^{in}_{g,ads}$	$u_g(t,0)=0.0$	$u_g(t,0) = u^{in}_{g,N_2sweep}$	$u_g(t,0) = u^{in}_{g,Cool}$

Table A.2 Parameters for RTSA Model

Variable	Adsorption	Heating	Sweeping	Cooling
$C^{in}_{CO_2,g,step}\ \frac{mol}{m^3}$	5.97	0.0	0.0	0.0
$C^{in}_{N_2,g,step}\ \frac{mol}{m^3}$	33.58	39.55	39.55	39.55
$T^{in}_{w,step}\ K$	308.15	393.15	393.15	308.15
$r_{ID}\ m$	0.00016			
$r_{OD} m$	0.0006			
$r_{fs}\ m$	0.0007495			

Table A.3 Optimal Operating Variables for RTSA

Operating Variables	Optimal Value	Lower Bound	Upper Bound
$u_{w,ads}$ (m/s)	1.0	0.2	1.0
$u_{w,heat}$ (m/s)	0.81	0.20	1.0
$u_{w,sweep}$ (m/s)	1.0	0.20	1.0
$u_{w,cool}$ (m/s)	0.28	0.20	1.0
$u_{g,sweep}$ (m/s)	0.25	0.13	0.27
$u_{g,cool}$ (m/s)	0.30	0.20	0.35
t_{ads} (s)	29.9	27	60
t_{heat} (s)	30.8	29	60
t_{sweep} (s)	2.0	2.0	15
t_{cool} (s)	12.4	12	35

Table A.4 Parameter Values for TEA

	Value	Units
Fiber Characteristics		
Heat of sorption (ΔH_{ads})	56	kJ/gmol
Fiber module diameter (D)	1	m
Fiber module length (L)	3	m
Module packing fraction (ε_b)	60	vol%
Diameter of fiber + interstitial space (r_{fs})	1549	μm
Fiber OD (r_{OD})	1200	μm
Fiber ID (r_{ID})	320	μm
Fiber porosity (ε_f)	44	%
RTSA Conditions and Cycle Inputs		
Sorption temperature	35	C
Desorption temperature	120	C
Overall CO_2 recovery	85	%

Table A.5 Major Capital Cost Contributors

Equipment	Base Cost ($)	Base Size	Base Cost Index I_{base}	Unit size	Sizing Exponent *n*	Number of Units	Total Cost (June 2007$)
Fiber module	10	1 m^2	564.8	1 m^2	1	5,734,977	54,090,000
Heat exchanger (hot)	28,102	1000 ft^2	564.8	17,304 ft^2	0.73	16	3,392,000
Heat exchanger (cold)	28,102	1000 ft^2	564.8	17,813 ft^2	0.73	20	4,340,000
CO_2 compression	15,525,000	32,619 kW	596	57,901 kW	0.9	1	23,258,000
Purge gas blowers	3,178,500	7860 kW	596	412 kW	0.61	1	470,000
Inlet flue gas blower	3,178,500	7860 kW	596	6510 kW	0.61	2	5,064,000
Warm tank	98,600	100,000 gal	564.8	582,905 gal	0.65	2	584,000
Cold tank	98,600	100,000 gal	564.8	37,438 gal	0.65	1	49,000
Caustic tank	38,300	10,000 gal	564.8	30,042 gal	0.65	1	74,000
Trim SO_2 recirculation pump	22,593	44 kW	560.3	47 kW	0.6	2	44,000
Caustic pump	1410	0.2 kW	560.3	0 kW	0.6	2	2000
Warm water pump	46,467	298 kW	560.3	0 kW	0.86	0	0
Cold water pump	46,467	482 kW	560.3	840 kW	0.86	13	923,000
Hot water pump	46,467	273 kW	560.3	777 kW	0.86	11	1,199,000
Steam condensate pump	10,922	32 kW	560.3	93 kW	0.6	2	40,000
Compression condensate pump	5076	2 kW	560.3	0.06 kW	0.6	2	2000
Dense phase CO_2 pump	494,475	1641 kW	560.3	751 kW	0.37	3	1,056,000
Pressure vessels	216,449	85 m^3	560.3	215 m^3	0.78	1	425,000
CO_2 dehydration system	2,200,000	200 MMSCFD[a]	596	331 MMSCFD[a]	0.6	1	2,660,000
Trim SO_2 removal	475,442	151,100 acfm	560.3	144301 acfm	0.73	12	5,244,000
Cool water filter	36,769	500 gpm	564.8	2950 gpm	0.61	97	9,894,000
Hot water filter	36,769	500 gpm	564.8	2951 gpm	0.61	79	8,058,000
Trim SO_2 filter	40,868	500 gpm	564.8	1671 gpm	0.61	2	160,000

[a]*MMSCFD: Million standard cubic feet/day.*
Georgia Tech Research Corporation (November 2012).

NOMENCLATURE

A_b	Area of bore m^2
A_{fiber}	Area of fiber m^2
A_{fs}	Area of free surface m^2
A_O	External specific surface area, Area of fiber m^2/m^3
A_I	Internal specific surface area, Area of fiber m^2/m^3
b_{mo}	Langmuir adsorption parameter Kpa^{-1}
c	Concentration, mol/m^{-3}
CR	Cold energy recovered
C_p	Specific heat capacity, J/kg K
D	Diameter of module m
k_s	Mass transfer coefficient, s^{-1}
h	Heat transfer coefficient, W/m K
HR	Heat energy recovered
$H_{sorption}$	Total enthalpy of sorption over cycle (MWth)
$H_{sensible}$	Total enthalpy of sensible over cycle (MWth)
L	Fiber length, m
q	Average sorbent loading, mol/kg
t	Time, s
T	Temperature, K
u	Velocity, m/s
z	Axial position, m

Greek Symbols

$-\Delta H$	Heat of adsorption, J/mol
ε	Total fiber/sorbent void fraction
κ	Parasitic energy demand, MW_e
ρ	Density, kg/m

Subscripts

0	Initial state
eq	Equilibrium
f	Fiber
fs	Free space
g	Gas
i	Gas components (CO_2, N_2)
in	Inlet
w	Water
ID	Internal diameter
OD	Outer diameter

ACKNOWLEDGMENTS

This work has been financially supported by the US Department of Energy via grant number DE-FE0007804. However, any opinions, findings, conclusions, or recommendations expressed herein are those of the authors and do not necessarily reflect the views of the DOE.

REFERENCES

Fauth, D.J., Gray, M.L., Pennline, H.W., Krutka, H.M., Sjostrom, S., Ault, A.M., 2012. Investigation of porous silica supported mixed-amine sorbents for post-combustion CO_2 capture. Energy Fuels 26, 2483–2496.

Georgia Tech Research Corporation, November 2012. Rapid temperature swing adsorption using polymeric/supported amine hollow fibers, DOE Award #: DE-FE0007804 Annual Report, Year 1.

Jiang, L., Fox, G.V., Biegler, L.T., 2004. Simulation and optimal design of multiple bed pressure swing adsorption systems. AIChE J. 50, 2904–2917.

Kalyanaraman, J., Fan, Y., Lively, R.P., Koros, W.J., Jones, C.W., Realff, M.J., Kawajiri, Y., 2015. Modeling and experimental validation of carbon dioxide sorption on hollow fibers loaded with silica-supported poly(ethylenimine). Chem. Eng. J. 259 (1), 737–751. http://dx.doi.org/10.1016/j.cej.2014.08.023.

Ko, D., Siriwardane, R., Biegler, L.T., 2003. Optimization of a pressure-swing adsorption process using zeolite 13X for CO_2 sequestration. Ind. Eng. Chem. Res. 42, 339–348.

Labreche, Y., Lively, R.P., Rezaei, F., Chen, G., Jones, C.W., Koros, W.J., 2013. Post-spinning infusion of poly(ethyleneimine) into polymer/silica hollow fiber sorbents for carbon dioxide capture. Chem. Eng. J. 221, 166–175.

Lively, R.P., Chance, R.R., Kelley, B.T., Deckman, H.W., Drese, J.H., Jones, C.W., Koros, W.J., 2009. Hollow fiber adsorbents for CO_2 removal from flue gas. Ind. Eng. Chem. Res. 48, 7314–7324.

Lively, R.P., Chance, R.R., Mysona, J.A., Babu, V.P., Deckman, H.W., Leta, D.P., Thomann, H., Koros, W.J., 2012. CO_2 sorption and desorption performance of thermally cycled hollow fiber sorbents. Int. J. Greenhouse Gas Control 10, 285–294.

Nilchan, S., Pantelides, C.C., 1998. On the optimisation of periodic adsorption processes. Adsorption 4, 113–147.

Process Systems Enterprise, L, 2011. gPROMS Model Developer Guide, v3.4.0 Ed. Process Systems Enterprise Limited, London.

Process Systems Enterprise, L, 2012. gPROMS Optimization Guide, v3.5.0 Ed. Process Systems Enterprise Limited, London.

Rezaei, F., Swernath, S., Kalyanaraman, J., Lively, R.P., Kawajiri, Y., Realff, M.J., 2014. Modeling of rapid temperature swing adsorption using polymer-supported amine hollow fibers. Chem. Eng. Sci. 113, 62–76.

Rezaei, F., Lively, R.P., Labreche, Y., Chen, G., Fan, Y., Koros, W.J., Jones, C.W., 2013. Aminosilane-grafted polymer/silica hollow fiber adsorbents for CO_2 capture from flue Gas. ACS Appl. Mater. Interfaces 5, 3921–3931.

Ritter, J.A., Wu, F., Ebner, A.D., 2012. New approach for modeling hybrid pressure swing adsorption–distillation processes. Ind. Eng. Chem. Res. 51, 9343–9355.

Rochelle, G.T., 2009. Amine scrubbing for CO_2 capture. Science 235, 1652–1654.

Seider, W.D., Seader, J.D., Lewin, D.R., Widagdo, S., 2008. Product and Process Design Principles: Synthesis, Analysis and Design. Wiley Publishers.

Swernath, S., Rezaei, F., Kalyanaraman, J., Lively, R.P., Realff, M.J., Kawajiri, Y., 2014. Optimization and heat integration of hollow fiber based thermal swing adsorption process for CO_2 capture from flue gas. In: FOCAPD Proceedings, pp. 633–638.

US Department of Energy, November 2010. Cost and Performance Baseline for Fossil Energy Plants. Rev 2. In: Bituminous Coal and Natural Gas to Electricity, vol. 1. DOE/NETL-2010/1397.

US Department of Energy, April 2011. Quality Guideline for Energy System Studies (QGESS)–Cost Estimation Methodology for NETL Assessments of Power Plant Performance. DOE/NETL-2010/1455.

US Department of Energy, January 2011. Bench-scale and Slipstream Development and Testing of Post-Combustion Carbon Dioxide Capture and Separation Technology for Application to Existing Coal-fired Power Plants. DOE/NETL DE-FOA-0000403.

US Department of Energy, August 2012. Updated Costs (June 2011 Basis) for Selected Bituminous Baseline Cases. DOE/NETL-341/082312.

US Department of Energy, January 2013. Quality Guideline for Energy System Studies (QGESS), Capital Cost Scaling Methodology. DOE/NETL-341/013113.

Xiao, P., Zhang, J., Webley, P., Li, G., Singh, R., Todd, R., 2008. Capture of CO_2 from flue gas streams with zeolite 13X by vacuum-pressure swing adsorption. Adsorption 14, 575–582.

CHAPTER

11

Toward Sustainable Solvent-Based Postcombustion CO_2 Capture: From Molecules to Conceptual Flowsheet Design

Stavros Papadokonstantakis*[,1], Sara Badr*, Konrad Hungerbühler*, Athanasios I. Papadopoulos[§], Theodoros Damartzis[§,¶], Panos Seferlis[§,¶], Esther Forte[||], Alexandros Chremos[||], Amparo Galindo[||], George Jackson[||], Claire S. Adjiman[||]

**Institute for Chemical- and Bioengineering, Swiss Federal Institute of Technology, Zurich, Switzerland*

[§]Chemical Process and Energy Resources Institute, Centre for Research and Technology-Hellas, Thermi, Greece

[¶]Department of Mechanical Engineering, Aristotle University of Thessaloniki, Thessaloniki, Greece

[||]Department of Chemical Engineering, Centre for Process Systems Engineering, Imperial College London, London, UK

[1]Corresponding author: E-mail: stavros.papadokonstantakis@chem.ethz.ch

11.1 INTRODUCTION

It is well known that carbon dioxide (CO_2) belongs to the group of greenhouse gases (GHGs) acting as an insulation layer within the Earth's atmosphere and stabilizing the global temperature levels within an acceptable range for humanity. However, there is strong evidence that the changes in the natural environment during Holocene, which include the positive effects of GHGs, have been severely countered by a rapidly growing reliance of anthropogenic activities on fossil fuels (worldwide CO_2 emissions of 32 Gt per year in 2009 (Boden et al., 2012)). This has led to higher atmospheric CO_2 concentrations of 387 ppmv, compared to 280 ppmv in the preindustrial time. A further increase (e.g., doubling) of atmospheric CO_2 concentrations may lead to a rise in the global temperature levels of 2–8 °C, threatening the stability of the large polar ice sheets, further intensifying existing adverse impacts on the

Computer Aided Chemical Engineering, Volume 36. ISSN 1570-7946. http://dx.doi.org/10.1016/B978-0-444-63472-6.00011-2

Earth's subsystems (e.g., the loss of mountain glaciers, and the rates of sea level rise) and ultimately severely affect the viability of contemporary human societies (Rockström et al., 2009). Therefore, from sustainability perspective, the incentive for CO_2 capture and storage (CCS) lies in avoiding an irreversible anthropogenic climate change by reducing GHG emissions, while CO_2 capture and utilization (CCU) can contribute by slowing down fossil resource depletion. It is, however, well argued that both CCS and CCU might need a considerable amount of direct and indirect energy input, requiring systematic methodologies (such as life-cycle assessment (LCA) (2009)) to consistently quantify the environmental benefits based on a cradle-to-grave approach (von der Assen et al., 2013).

This chapter focuses on one part of such a cradle-to-grave approach for CCS, namely the carbon capture part, and on one type of technology in this area, namely solvent-based CO_2 capture via chemisorption. In this type of technology, the main idea is that CO_2 is separated from the flue gas or any other process stream leaving a CO_2-emitting processing facility. It is thus inherently a posttreatment approach which requires minimal modifications on the existing CO_2-emitting process. Other types of technology require upstream modifications of the CO_2-emitting processes, either by changing the fuel in a hydrogen/CO_2 mixture and removing the carbon before combustion or by changing the oxidizing agent from air to oxygen with or without chemical looping (KAPSARC, 2012).

The adoption in industry of chemical absorption/desorption systems for postcombustion CO_2 capture is currently challenging due to the high energy penalty in solvent regeneration and the environmental impacts associated with solvent degradation. Research efforts reported in recent years are predominantly based on lab and pilot-scale experiments to select solvents and develop process systems which may potentially improve the overall performance of absorption/desorption CO_2 capture (Ma'mun et al., 2007; Aronu et al., 2009; Puxty et al., 2009). This is very challenging due to the highly nonideal nature of the solvent–CO_2–water chemical interactions and the countless combinations of potential capture solvent candidates, including solvent blends, and process configurations.

The first important task in the design of a solvent-based CO_2 capture system is the choice of a solvent or a solvent blend with optimal thermodynamic properties. The development of tools able to describe the phase behavior, chemical equilibrium, heat capacities, and enthalpies of absorption, among other properties, of potential solvents and their mixtures with CO_2 and water is therefore essential. In particular, predictive molecular models that permit the extent of CO_2 capture by new potential solvents to be explored constitute a very attractive approach to support design activities. To help speed up this effort, computer-aided methods can be used to integrate existing experimental results to predictive models which can in turn be used to guide the search for improved solvents and process schemes and to provide further focus for the experimental effort. Most reported research in CO_2 capture employs rigorous models which are indispensable for the accurate prediction of solvent and process characteristics prior to the selection of a limited set of realistically useful options. Recent progress in physical property prediction has led to the development of increasingly transferable and predictive models. For example, continuum solvation models (Sumon et al., 2012) and density functional theory (Gupta et al., 2012) have been used among other computational chemistry methods for the calculation of the amine basicity to measure and compare few solvents with respect to molecular reactivity. Of particular relevance to this chapter, the statistical associating fluid theory for potentials of variable range (SAFT-VR) equation of state (EoS) (Gil-Villegas et al., 1997; Galindo et al., 1998) has been used to model accurately the fluid phase behavior of aqueous solutions of multifunctional alkanolamines and CO_2 (Rodriguez et al., 2012). Moreover, SAFT-VR has been used in the integrated design of CO_2 capture solvent and process using rigorous rate-based models to identify optimum solvent blends and process characteristics for physical (Pereira et al., 2011) and chemical (Mac Dowell et al., 2010a) absorption systems, while it has also been used in the rigorous dynamic modeling and analysis of chemical absorption columns (Mac Dowell et al., 2013). Perturbed chain polar SAFT has also been used for the design of CO_2 capture solvents using physical absorption (Bardow et al., 2010).

Despite the usefulness of rigorous predictive models, their use in design presents several challenges. They usually require the specification of multiple model parameters which are often unknown and are either selected based on assumptions that introduce uncertainty or determined by costly experiments to obtain validated results. Moreover, the computational effort required is often intense due to the mathematical complexity and diversity of the models required to address issues such as economic, operating, and sustainability performance prior to identifying realistic solvent and process solutions. This effort also prohibits the consideration of a large number of potential solvent and process alternatives required to identify truly optimum design solutions.

These challenges may be addressed through a design approach where the application of conceptual methods precedes the use of rigorous models to enable fast and efficient decision-making at an early stage. Conceptual design methods employ process or molecular models of higher abstraction that are computationally efficient, capture major reactive separation driving forces, and often incorporate engineering know-how, heuristics, and empirical correlations. Then, these models of higher abstraction are combined with optimization algorithms to investigate systematically an enormously wide range of options and to identify the major chemical characteristics and process configurations likely to result in increased economic, operating, and sustainability performance. The solutions obtained require, of course, further refinement using experiments and rigorous models prior to utilization in practical applications. However, they incorporate rich design insights which may be analyzed by engineers prior to transferring a few meaningful conclusions as inputs to experimental validation and rigorous models. Conceptual design methods have been widely utilized in the development of non-CO_2 separations with successful results. Computer-aided molecular design (CAMD) methods (Giovanoglou et al., 2003; Papadopoulos and Linke, 2006a) enable the fast investigation of a vast number of molecular solvent structures that can significantly increase important separation drivers (e.g., selectivity) compared to the use of conventional solvents. The

CAMD framework can be used to identify solvents that present beneficial properties, such as reduced toxicity (Papadopoulos and Linke, 2006b), increased safety (Papadopoulos et al., 2010), inertness to reactions (Papadopoulos and Linke, 2009), and enhancement of reaction rates, (Folic et al., 2007, 2008; Struebing et al., 2013), to name a few. The combined utilization of novel solvents with systematically identified modifications in their host separation systems (Kenig and Seferlis, 2009) results in significant reductions in the associated costs, compared to solvent–process systems used in industry.

On the other hand, early stage CAMD and process design tools have found limited application in postcombustion CO_2 capture. For example, solvents have been proposed to replace methyldiethanolamine in an absorption/desorption process (Chemmangattuvalappil and Eden, 2013) to illustrate the latest algorithmic developments in an approach integrating CAMD and process design through case studies associated with CO_2 capture. The properties utilized as solvent performance criteria in CAMD were calculated using group contribution (GC) methods and included vapor pressure, heat of vaporization, molar volume, heat of fusion, melting and boiling points, soil sorption coefficient, toxic limit concentration, and octanol/water partition coefficient. The resulting solvents were not elaborated or evaluated further in terms of their capture potential (e.g., through rigorous property prediction or process models). Recently, a two-stage approach has been proposed for the selection of CO_2 capture solvents (Papadopoulos et al., 2014). Initially a large solvent set of commercially available amines was evaluated using GC and other early stage property prediction models. The performance criteria included thermodynamic (e.g., vapor pressure, CO_2 solubility), kinetic (e.g., solvent basicity, steric hindrance), and sustainability (e.g., health and safety hazard, life cycle assessment (LCA)) solvent characteristics. A few selected highly performing solvents were then rigorously evaluated using the SAFT-VR and the SAFT-γ SW (square well) approach (Chremos et al., 2013).

The second important task in the design of solvent-based CO_2 capture systems is the choice of the process layout. Early stage design of postcombustion CO_2 capture systems has been mainly addressed through the ad hoc evaluation of flowsheet configurations exploiting empirical knowledge regarding process characteristics which increase the capture performance. Most available research works employ equilibrium models that are generally easy to handle, converge rather easily, and constitute an important design basis for reactive separation processes. Oyenekan and Rochelle (2007) used such a model for the simulation of alternative multifeed and multipressure configurations for the stripper column. Le Moullec and Kanniche (2011) used miscellaneous flowsheet structures and investigated their effect on the overall process efficiency. A more extensive review of similar methods is presented by Damartzis et al. (2014) who proposed a systematic approach for the optimum synthesis of absorption/desorption CO_2 capture flowsheets. The approach combines a superstructure representation with rigorous but flexible modeling techniques for reactive separation processes based on the method of orthogonal collocation on fine elements (OCFE) both to investigate numerous structural and operating flowsheet characteristics and to simulate rigorously the behavior of the equilibrium column.

The third important task in the design of solvent-based CO_2 capture systems is the choice of the sustainability metrics and frameworks to be used for the assessment of the diverse alternatives. Considering the environmental issues that motivate the development of CO_2 capture technologies, sustainable solutions for solvent selection and process design should combine technical and economic assessment as well as environmental and social impact criteria, going beyond the scope of an environmental LCA. To this end, taking into account the diversity of the sustainability objectives and the fact that the hierarchy in decision-making usually implies different levels of prior knowledge utilization, data availability and, therefore, modeling detail, it is inevitable that the structure of the available tools and metrics varies significantly, ranging from simple indicators to holistic frameworks for an overall process assessment. In this context, Ruiz-Mercado et al. (2012a) have recently provided a taxonomy of sustainability indicators and have further discussed the data needs for the calculation of the respective indicators (Ruiz-Mercado et al., 2012b). Other studies have focused on integration of assessment indicators in the decision-making procedure during different stages of process design (Chen and Shonnard, 2004; Sugiyama et al., 2008; Torres et al., 2011; Tugnoli et al., 2011), including also process retrofitting (Carvalho et al., 2008; Bumann et al., 2011).

Besides standard economic indices (i.e., operating and capital expenditure), this study focuses mainly on two types of sustainability indices. The first represents the holistic environmental impact for normal process operation (i.e., LCA type of metrics, such as the cumulative energy demand (Verein-Deutcher-Ingenieure, 1997), the global warming potential (IPCC, 2007), and the eco-indicator 99 (Goedkoop and Spriensma, 2000)). The second refers to the safety, health, and environmental (SHE) hazard identification for evaluation of the harm potential in accidental scenarios. Unlike the cradle-to-gate LCA approach, the SHE hazard identification is performed locally within the boundaries of the CO_2 capture plant, that is, the hazards for the production of the solvent and any other chemical or energy utilities used in the CO_2 capture plant are not considered. Since for the screening of solvents and the assessment of the respective conceptual flowsheets, rigorous process risk analysis methods (e.g., HAZOP by Imperial-Chemical-Industries (ICI) (1974)), fault tree analysis (Parmar and Lees, 1987), and failure mode and effect analysis (Lees, 1996) require detailed plant specifications and process layout information beyond the scope of conceptual process design, this study focuses on quantitative index-based techniques including overall hazard indices suitable for a first screening of process alternatives during early design stages. In this context, Adu et al. (2008) presented a review and a detailed comparison of the most commonly used index-based methods for hazard assessment concluding that the only merit that distinguishes different methods is their level of simplicity, which dictates the early design phase to which they can be applied.

In the rest of this chapter, we will demonstrate how thermodynamic performance, environmental impact, hazard assessment, and economic evaluation metrics can be combined with predictive molecular-based methods for the prediction of thermodynamic properties, CAMD, and process modeling; we will demonstrate how this provides multicriteria insights for the design of solvent-based postcombustion CO_2 capture systems that can be used for benchmarking process alternatives and to initiate rigorous modeling and experimental studies in a more condensed, near-optimum design space.

11.2 THE SOLVENT-BASED CO_2 CAPTURE SYSTEM

11.2.1 SYSTEM BOUNDARIES

The basic layout of an absorption/desorption flowsheet for solvent-based CO_2 capture is presented in Figure 11.1. The main principle is the separation of CO_2 from a gas stream (e.g., a flue gas stream from a power plant) through reactive absorption in an appropriate solvent at low temperatures and the subsequent desorption of CO_2 from the solvent at high temperatures. Both absorption and desorption are typically performed in packed columns with solvent recirculation allowing for heat integration. Additional elements of this basic flowsheet are typically an amine scrubber at the top of the absorber and a postcondenser at the top of the desorption column to remove water and reach a CO_2 mole purity higher than 98%. Oxidative

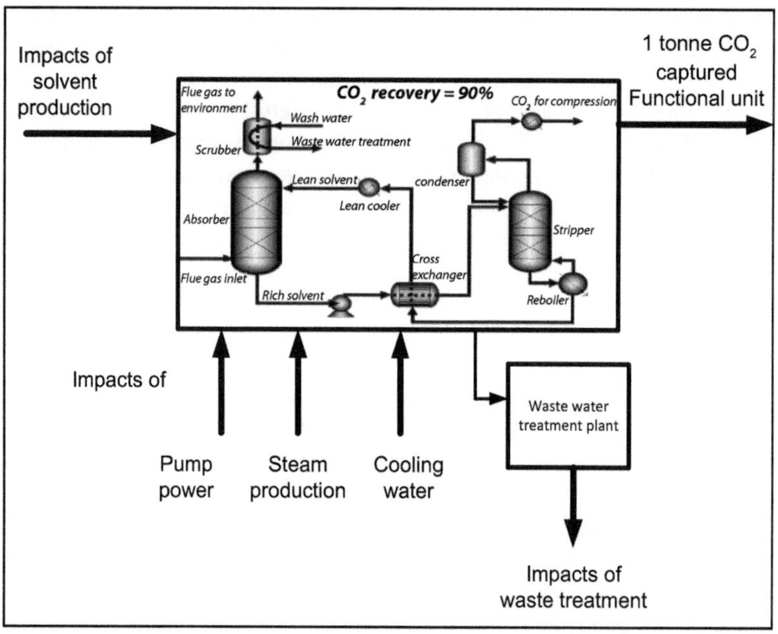

FIGURE 11.1

Scope of the analysis of the solvent-based CO_2 capture system (functional unit 1 tonne of captured CO_2) for a benchmark flowsheet layout with 90% CO_2 recovery and 98 mol% CO_2 purity.

and thermal degradation (Lepaumier et al., 2009a,b) as well as fugitive emissions are two main reasons for a makeup stream of fresh solvent, while a purge stream is required for removing the degradation products generated during steady-state operation of the capture system. This purge stream also comprises nondegraded solvent, which, in principle, can be partially reclaimed (i.e., the reclaiming option is presented with the generic term "wastewater treatment" in Figure 11.1, since in some scenarios of this study the whole stream is considered as waste and in others the reclaiming process has not been modeled in detail).

On the basis of this flowsheet the system boundaries of a solvent-based CO_2 capture system can be defined from an economic, LCA, and SHE hazard perspective. More specifically, here we focus only on a cradle-to-gate sustainability analysis of the capture part (i.e., not integrated with a power plant or any other plant it may be connected to and not including storage or utilization scenarios of the captured CO_2). The functional unit of analysis is 1 tonne of CO_2 captured and the analysis includes both global (i.e., impacts of solvent production) and local effects (process energy consumption, generated liquid waste and air emissions) from an economic and environmental life cycle perspective, but only local effects with respect to the hazard analysis (i.e., only the capture process hazards are considered).

11.2.2 SOLVENT DESIGN AND SELECTION CRITERIA

There are numerous chemical and physical molecular characteristics that need to be considered as criteria in the selection of CO_2 capture solvents. The kinetic constants directly affect the absorber height and the chemical equilibrium affects the required solvent flowrate. The basicity of the solvent is an indicator of its ability to react with CO_2. The miscibility of the solvent in water and its volatility are associated with solvent emissions, while the solvent viscosity is associated with mass and heat transfer characteristics. The slope of the equilibrium curve (i.e., the derivatives of CO_2 partial pressure with respect to loading and temperature) and the heat of absorption are associated with vapor–liquid equilibrium solvent behavior and affect the amount of CO_2 that can be captured as well as the energy required in the desorption column. Moreover, oxidative and thermal solvent degradation are associated with corrosion and solvent losses, the nature of the degradation products is linked to plant emissions, while hazard issues are associated with solvent properties such as toxicity. The feasibility of using these and other similar properties as solvent selection criteria depends on the availability of models which are appropriate for their calculation, and of sufficient data so that these models may be parameterized and made applicable for a wide range of molecular structures. Papadopoulos et al. (2014) have proposed several pure component and mixture properties reflecting the set of solvent thermodynamic, reactivity, and sustainability features mentioned here and which may be calculated using standard GC or other approximate models. These properties can therefore be utilized as solvent selection criteria within an early stage CAMD approach.

The solubility parameter (δ_t) is generally used as a measure of miscibility between materials. It can be represented by the cohesive energy density or the energy required to break up the solvent–solute interactions. Although the solubility parameter provides an overall value for this energy, Hansen proposed to divide it into three individual components, namely the dispersive cohesive energy, the polar cohesive energy, and the hydrogen bonding cohesive energy. This partitioning enables the consideration of both polar and nonpolar interactions and has resulted in the three Hansen or partial solubility parameters (HSPs), namely the dispersion HSP (δ_d), the polar HSP (δ_b), and the hydrogen bonding HSP (δ_h). The difference between each one of these three parameters and the corresponding CO_2 parameters may be used to obtain a distance measure called the relative energy difference (RED) index (Stefanis and Panayiotou, 2008), which provides a means to evaluate the ability of a solvent to dissolve CO_2. Solvents with RED<1 are generally considered as favorable for the dissolution of CO_2. Increasing differences between the solvent–CO_2 solubility are associated with increased energy requirements to enable greater CO_2 solubility in a solvent.

The vapor pressure (P_{vp}) is an important parameter for absorption/desorption CO_2 capture systems because it is associated with solvent losses. Several amines used as solvents exhibit high vapor pressure hence water sprinklers are often used in the absorption process to reduce losses. Clearly, the target during solvent design should be to minimize vapor pressure.

The liquid heat capacity (C_p) plays an important role in determining the heat required to regenerate the amine-based solution in the desorption step. Based on Oexmann and Kather (2010) this heat consists of three parts: the sensible heat, the heat of vaporization of water needed for the steam in the reboiler, and the heat of absorption necessary to desorb the CO_2 from the solution. The sensible heat represents the heat required to raise the temperature of the solution from the absorption to the desorption column. This is directly proportional to the liquid heat capacity of this solution. Hence it is a reasonable assumption that a solution containing an amine with a low heat capacity is expected to contribute toward reducing the sensible heat requirements.

The density (ρ), surface tension (σ), and viscosity (μ) are three properties tightly associated with the design and operating features of the absorption column. In particular, the solvent density should be high either in the vapor or liquid phase because this generally facilitates separation using reduced solvent flow rates, equipment of reduced size, and also reduced pumping power. Surface tension and viscosity should be low because this improves the mass transfer in the packing material.

The basicity (pK_a) is an important property for solvent design and selection as it provides information mainly regarding the reactivity features of amine-based solvents. According to Aronu et al. (2009) reaction rates increase with an increase of the amine basicity, while Versteeg et al. (1996) showed that basicity has a linear relationship with the logarithm of the reaction rate constant for several amines.

The steric hindrance (S) is a property affecting both the absorption rate and capacity of the solvent. It is associated with the ability of amines to form stable carbamates and affects the amine-CO_2 reaction stoichiometry (Sartori et al., 1987). In unhindered amines that form stable carbamates the reaction will stoichiometrically allow for half a mole of CO_2 to be absorbed per mole of amine, while in hindered amines 1 mol of CO_2 may be absorbed per mole of amine. Steric hindrance is associated with the bulkiness of the substituents attached to the amine group in the molecule. Bulky substituents prohibit CO_2 from reaching the reaction center hence reducing the reaction rate (Cao and Liu, 2004).

The melting point temperature (T_m) should be lower than the lowest absorption/desorption temperature to avoid solvent solidification under processing conditions. The boiling point temperature (T_{bp}) should be higher than the highest absorption/desorption temperature to avoid vaporization of the pure solvent under processing conditions.

11.2.3 ABSORPTION/DESORPTION FLOWSHEET DESIGN OPTIONS

Similarly to the chemical solvent characteristics and their indirect effects on process performance, there are several structural and operating modifications in absorption/desorption flowsheets which favor the reactive separation driving forces. Several such structural characteristics were recently analyzed (Damartzis et al., 2014) and are summarized below.

The double-section stripper (DSS) configuration originally proposed by Oyenekan and Rochelle (2007) features the same absorber configuration as in conventional absorption/desorption systems used in CO_2 capture, but the rich stream exiting the heat exchanger is split into two substreams that feed the stripper at two points. Vapor from the part of the stripper below the mid-feed point enters the top part along with the rest of the rich stream from the heat exchanger resulting in a "medium stripping" and thus producing a "semi-lean" liquid stream. This stream is side drawn, mixed with the bottoms product of the column, and recycled to the absorber through the cross-section heat exchanger. This layout combines the favorable effect of material and heat distribution in the stripper through the appropriate feeding of the rich stream with the simultaneous complete removal of the liquid phase from the top part of the column, thus intensifying the role of the stripper.

The multifeed stripper (MFS) configuration proposed by Oyenekan and Rochelle (2007) features two separate columns for the solvent recovery. The absorption process is similar to that of the conventional flowsheet, however, the rich stream exiting the heat exchanger is split into two parts that are directed to the two stripper columns. The bottom product of the first column is subsequently introduced as a side-feed to the second column, whereas a side-draw stream is removed from the second column. The side-draw stream is mixed with the bottom liquid to produce the lean solvent stream recycled to the absorber. The flowsheet configuration attempts to redistribute the driving forces in the stripper through both side material and heat streams to the column. Furthermore, the two different columns enable the operation under different pressure conditions to enhance separation accordingly. In stripping systems operating at high temperatures, elevated pressure ensures higher boiling points for the components and thus, smaller vapor flows. The latter can be translated into smaller column size.

The multifeed absorber (MFA) configuration has an identical stripper configuration with the conventional flowsheet but the lean stream coming from the heat exchanger is fed at two points in the absorption column. The first feeding point is at the top of the absorber, whereas the second feeding point along the absorption column is decided in the optimization stage. The effect of feeding the absorber column at two different locations enhances the driving force distribution since the concentration difference between the liquid and vapor phases is kept at high average values. Therefore, the system improves the overall efficiency of the process.

The intercooled absorber (ICA) configuration targets at the enhancement of the absorption rate of CO_2 during its contact with the liquid solvent in the absorption column. This is achieved through the cooling of the liquid phase using a number of coolers at selected points in the column. In this way, a fraction of the liquid phase is bypassed through a heat exchanger, cooled to a target temperature, and returned to the column. The twofold contribution of the liquid phase cooling in the absorber can be explained by first considering that the lower temperature is expected to improve the chemical equilibrium of the exothermic reactions, allowing for more efficient capture of CO_2. In addition, the solubility of CO_2 in the aqueous solution is favored by lower temperatures, thus enhancing the physical absorption in the aqueous solution as well.

Moving to the operating parameters of the diverse flowsheets, a cross heat exchanger minimum temperature difference of 5 °C instead of 10 °C results in a twofold reduction of the heat required to raise the solvent from the temperature downstream the cross heat exchanger to the reboiler temperature (sensible heat). This may result in approximately 15% reduction of energy requirements in the desorber in a conventional process using a 30 wt% aqueous solution of monoethanolamine (MEA) (Oyenekan and Rochelle, 2007). The aim is to design flowsheets that enable reduction in the sensible heat, while maintaining high CO_2/water ratio and reduced heat exchange area, considering the impact of the temperature differential in the overall solvent and flowsheet design decisions.

Moreover, high pressure desorption results in less water vapor at the desorber head, hence less heat is required for the reboiler. Furthermore, the CO_2 is recovered at greater concentrations and pressures, hence reducing the required compression work (Oyenekan and Rochelle, 2007). It is, therefore, worth investigating cases of uniform high pressure applied to the entire column or cases of varying pressure in different compartments of the column. Designs should aim to recover the latent heat from the water vapor in the overhead stream, which is lost in lower pressure columns. Vacuum pressure desorption should also be considered, as lower temperature steam is used in the reboiler, hence even waste heat from other streams can be utilized. At the same time, the lower desorber temperature reduces the degradation of the solvents and the corrosion. Optimum designs in this case will have to compensate for additional CO_2 compression and slower mass transfer due to low temperature.

11.2.4 SUSTAINABILITY METRICS

The generic flowsheet of Figure 11.1 can also be used to exemplify the type of data required for the sustainability metrics. For the LCA metrics, these are the energy utility consumption in the stripper reboiler (typically steam representing a major LCA impact), the pump, and the condenser (typical electricity and cooling water representing a minor LCA impact), the process water consumption (e.g., for the solution of the solvent and washing of gases), the material and energy utility consumption for the purge stream treatment (either for reclaiming the solvent and/or treating the residues) as well as the process emissions (e.g., CO_2 not absorbed, wastewater treatment or waste incineration referring to the purge stream), and the solvent make up rates due to solvent degradation and fugitive emissions. The optimal flowsheets, although more complicated (e.g., including multiple feed in the columns, division of columns, heat integration, and more complicated recycle structures), require the same type of data for the LCA metrics.

Table 11.1 presents the data required by two hazard assessment methods, the multivariate hazard identification and ranking system by Khan and Abbasi (1998), and the EHS method by Koller et al. (2000). Details about the background calculations for these methods can be found in the original literature, but already from Table 11.1 it becomes clear what type of property and process data are important for carrying out these calculations. Here, it should be noted that these hazard assessment methods are designed to be used for a wide variety of chemicals and to highlight the most common industrial hazards. To highlight specific hazards, for instance, the potential of solvents to form nitrosamines, case-specific modifications may be required. As an example, it is important to consider the dependency of these methods on mass inventories. For the CO_2 capture, this means that less hazardous solvents in larger inventories will be proportionally penalized compared to more hazardous solvents in smaller inventories, and at the same time the hazard of nitrosamines in traces will appear to be negligible. On the other hand, if only the existence of a hazardous chemical is considered, as it is proposed in the EHS method for the health category, the methods would fail to differentiate between different solvents having different potential to form the carcinogenic nitrosamines.

11.3 PROPOSED MULTISCALE DESIGN APPROACH

The properties proposed as performance criteria in solvent and process design need to be utilized within a systematic decision-making framework. The framework proposed in our work (Figure 11.2) consists of three main decision levels aiming to support computational efficiency while identifying optimum capture systems based on solvent–process interactions.

At the molecular level we propose the use of an optimization-based CAMD method (Giovanoglou et al., 2003; Papadopoulos and Linke, 2006a) CAMD is utilized to generate and search a vast number of conventional and novel molecular

Table 11.1 Data Requirements for the HIRA and EHS Hazard Assessment Methods

Hazard	Category	Parameter	Source
Hazard Identification and Ranking System (HIRA) (Khan and Abbasi (1998))			
Safety	Fire and explosion damage index (FEDI)	Heat of combustion	Estimations from molecular descriptors (Benson (1976), Cohen and Benson (1993))
		Mass of chemical in equipment	Process data
		Volume of vessels	Process data
		Equipment pressure	Process data
		Vapor pressure	MSDS/EPI-Suite estimations
		Flash point	MSDS/Estimations from molecular descriptors (Catoire and Naudet (2004), Catoire et al. (2006), Gharagheizi et al. (2008))
		NFPA-R	MSDS
		NFPA-F	MSDS
		Heat of absorption	Process data
Health	Toxic damage index (TDI)	NFPA-H	MSDS
		Vapor density	MSDS
EHS (Koller et al. (2000))			
Safety	Mobility	Difference between boiling point of the pure substance and stripper temperature	Process data MSDS/EPI-Suite estimations
	Fire and explosion	Difference between flash point of the pure substance and stripper temperature	Process data MSDS/Estimations from molecular descriptors (Catoire and Naudet (2004), Catoire et al. (2006), Gharagheizi et al. (2008))
		EC classification	MSDS
		NFPA-F	MSDS
	Acute toxicity	Immediately dangerous to life or health concentrations (IDLH)	MSDS/Statistical model

Table 11.1 Data Requirements for the HIRA and EHS Hazard Assessment Methods—cont'd

Hazard	Category	Parameter	Source
		EC classification	MSDS
		R-codes	MSDS
		$LD50_{oral}$	MSDS/TEST estimations
		$LD50_{dermal}$	MSDS
	Reaction/ decomposition	R-codes	MSDS
		Heat of decomposition	Estimations from molecular descriptors (Benson (1976), Cohen and Benson (1993))
		NFPA-R	MSDS
Health	Irritation	R-codes	MSDS
		EC classification	MSDS
		$LD50_{dermal}$	MSDS
	Chronic toxicity	R-codes	MSDS
		Mutagenicity	TEST estimations
Environment	Water-mediated effects	$LC50_{aquatic}$	EPI-Suite estimations
	Degradation	Half-life in water	EPI-Suite estimations
	Air-mediated effects		Chronic toxicity index
	Solid waste	Amount	Process data
	Accumulation	Bioconcentration factor (BCF)	EPI-Suite estimations
		Octanol/water partitioning coefficient (KOW)	EPI-Suite estimations

EHS, Environmental, Health and Safety method; MSDS, Material Safety Data Sheet; EPI, Estimation Program Interface; TEST, Toxicity Estimation Software Tool.

structures and to identify those that offer the best performance with respect to properties calculated through GC or other approximate pure component models. The main advantages of the proposed approach are the following:

- For the first time in postcombustion solvent-based CO_2 capture numerous properties are considered as performance criteria reflecting solvent thermodynamics, reactivity, and sustainability characteristics.
- The selected properties capture the molecular chemistry effects on important absorption/desorption characteristics.
- They are simultaneously considered within a multiobjective optimization formulation which results in a pool of optimum solvent candidates covering a wide range of trade-offs with respect to the multicriteria CO_2 capture performance.

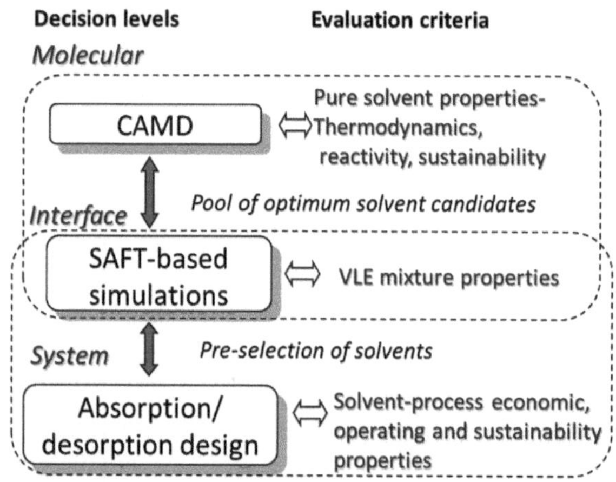

FIGURE 11.2

Outline of the multiscale design approach. CAMD, Computer-aided molecular design; VLE, Vapour-liquid equilibrium.

These characteristics of the proposed technology compensate for the initial utilization of pure component prediction models and enable the identification of a few effective solvents. The reduced accuracy is traded off for high computational efficiency, requiring a few minutes to screen even a very large number (e.g., hundreds of thousands) of molecular structures.

The few solvents selected at the molecular level are introduced in simulations to predict accurately the very nonideal solvent—water—CO_2 mixture vapor—liquid and chemical equilibrium behavior. Predictions are performed through the SAFT-VR and SAFT-γ EoS which enable the calculation of macroscopic process-related properties such as:

- Helmholtz free energies and volumes for given conditions of temperature, pressure, and composition.
- Partial pressure of CO_2 with respect to amine loading, temperature, and concentration.
- Heat of absorption with respect to temperature, pressure, and concentration.

In SAFT-VR and SAFT-γ, molecular and group parameters are used to derive process-related properties, hence SAFT acts as an interface between the CAMD algorithm and the process system level, enabling a further rigorous screening of the designed solvents with respect to their suitability as CO_2 capture options. To this end, SAFT-γ SW (Lymperiadis et al., 2007, 2008) is particularly useful because it is built as a GC framework where molecules are formed from fused heteronuclear segments representing distinct functional groups. In this respect, SAFT-γ SW can be used as a rigorous predictive model for any molecule that can be generated as a combination of the available functional groups. An example of the performance of SAFT-γ SW in the CO_2 capture system is provided in the Section 11.4.2.

The selected solvents are introduced as discrete options in process design to determine the optimum structural and operating characteristics of the absorption/desorption system with respect to economic, operating and sustainability criteria. This level allows process models of higher abstraction to be utilized in order to consider a very large number of structural and operating interactions toward optimum absorption/desorption flowsheets. Prediction of process-related solvent properties such as CO_2 partial pressure and amine loading is directly derived by SAFT, while enthalpy flows are derived based on Helmholtz energy calculated with SAFT. Considering only a few solvents allows an extensive screening for optimum flowsheet characteristics at reasonable computational efficiency. The characteristics may be retained and exploited in a more rigorous process design stage using models of lower abstraction.

11.4 IMPLEMENTATION

11.4.1 CAMD PROBLEM FORMULATION

The CAMD approach employed here is used both to design novel molecular structures and to select commercially available molecules for CO_2 capture from a database of 126 molecules. With respect to the properties utilized as performance criteria, the design and selection goals are represented as follows:

$$\begin{aligned} \max \quad & \rho \\ \min \quad & C_p, P_{vp}, \sigma, n \\ \text{s.t.} \quad & T_m < T_{Abs} \\ & T_{bp} > T_{Des} \\ & RED < RED^{up} \end{aligned} \tag{11.1}$$

$$\begin{aligned} \max \quad & \rho, pK_a \\ \min \quad & C_p, P_{vp}, \sigma, n, RED, (EHS, CED, GWP, EI99) \\ \text{s.t.} \quad & T_m < T_{Abs} \\ & T_{bp} > T_{Des} \\ & \left(S^{lo} \le S \le S^{up}\right) \end{aligned} \tag{11.2}$$

In the above set of equations T_{Abs} is the temperature of the absorption column, T_{Des} the temperature of the desorption column, RED^{up} is an upper limit imposed in the RED index. Set (11.1) is implemented for the design of novel solvents, while set (11.2) is implemented for further refinement of the optimum solvent set obtained from Eqn (11.1) as well as for the selection of few solvents from the database of commercial options. This is because the GC models required for the solvent design stage exist only for the reported properties. Set (11.2) is also implemented in two stages. First, the two solvent sets (database and optimum from CAMD) are refined using all indices except for *EHS*, *CED*, *GWP*, *EI*99, and *S* (i.e., indices associated with sustainability and steric hindrance). The selected solvents are then investigated with respect to sustainability performance and potential for steric hindrance. The brackets are therefore used to indicate the two-stage evaluation approach. The liquid molar density ρ is calculated as the inverse of liquid molar volume V_m (Constantinou et al., 1995), σ, and n from Conte et al. (2008), T_m and T_{bp} from Marrero and Gani (2001), P_{vp} from the Riedel correlation (Poling et al., 2001), C_p from Rayer et al. (2012), pK_a from Chemaxon, and RED from Stefanis and Panayiotou (2008). The steric hindrance S is evaluated qualitatively using the rules proposed by Sartori et al. (1987). The Finechem tool (Wernet et al., 2009) is used to provide estimates (with an average relative error of 30–40%) for LCA-based metrics using molecular descriptors as inputs. The Estimation Program Interface (EPI) Suite and the Toxicity Estimation Software Tool (TEST) (EPA) are used for the calculation of EHS metrics. An average of the dangerous properties calculated by variations of the EHS method (Banimostafa, 2013; Sugiyama et al., 2008) is used to avoid missing scores in case of data gaps. The EHS score is taken as the sum of the safety, health, and environment scores.

The database of molecules considered as candidate CO_2 capture solvents consists of 126 acyclic amines, aliphatic amines, and hydroxylamines. They are obtained from an in-house database available from ETH Zurich, publicly available databases (EPA; NIST) and the commercial catalog of Sigma–Aldrich. The database is not exhaustive but provides an inclusive set of amines with different molecular structures assigned a CAS Registry Number. All the molecules may be obtained from combinations of the following functional groups: CH_3-, $-CH_2-$, $>CH-$, $>C<$, $-OH$, $-CH_2-NH_2$, $-CH_2-NH-$, $-CH_2-N<$, $>CH-NH_2$, $>CH-NH-$, CH_3-NH-, $CH_3-N<$, $\geq C-NH_2$. These groups are selected due the public availability of data required for calculation of pure component molecular properties through GC or other models and they are also used for the design of novel solvents using CAMD.

Overall a database of 283 molecular structures is generated from the implementation of CAMD including the 126 amines found in publicly available databases or previously considered as CO_2 capture solvents in absorption/desorption systems. All the molecules are rank-ordered using the performance index J_i (Papadopoulos et al., 2014) which represents the sum of the scaled values of each property reported in set (11.2) and needs to be minimized. The obtained results are broken down into four classes of molecules as follows:

- The reference (R) class contains only 25 molecules (see also Appendix) specifically selected from a literature search targeting molecules previously employed as CO_2 capture solvents (Papadopoulos et al., 2014).
- The class of designed molecules (D) contains the top 10 molecules resulting from rank-ordering the 157 molecules obtained from CAMD.

- The class of commercially available molecules (C) contains only the 126 solvents of the database.
- The class of commercially available alkanolamines (L) contains a subset of 58 alkanolamines within class C.

11.4.2 THERMODYNAMIC MODELING USING SAFT

To illustrate how thermodynamic behavior can be used in solvent design, in order to obtain optimum solvents for application in carbon capture absorption processes, we use the following example. We consider one of the benchmark amine-based solvents, in particular an aqueous mixture of MEA and compare it with an aqueous mixture of diethanolamine (DEA). Within the SAFT-γ SW framework the thermodynamic behavior of these mixtures can be described based on a number of functional groups that represent the relevant chemical moieties: pure water, CO_2, and those that compose the amines of interest. The amine-related groups needed are CH_2NH_2 and CH_2OH for MEA, and CH_2NH, CH_2, and CH_2OH for DEA. Each of these groups is characterized by a number of parameters that in principle is independent of the molecular environment. Thus, in this case the CH_2OH group is represented by the same parameters in both MEA and DEA. The parameters for some of the groups used have been determined in previous studies of: pure water (Clark et al., 2006), CO_2 and its mixtures with water (Galindo and Blas, 2002), and alkanes (i.e., for the CH_2 and CH_3 groups) (Lymperiadis et al., 2007). Models of amines are currently under development (Chremos et al., 2013). SAFT-γ SW calculations of the solubility of CO_2 in aqueous mixtures of MEA and DEA show very good agreement with experimental data (Lee et al., 1972; Jou et al., 1995), as seen in Figure 11.3(a). Since the group parameters are independent of temperature and composition, a wide range of thermodynamic states can be explored easily. The list of functional groups that have been developed or currently are under development for application to carbon capture processes includes H_2O, CO_2, CH_3, CH_2, $CHCH_3$, CH_2OH, $CHOH$, CH_2NH_2, CH_2NH, and $CHNH_2$. Examples of amines that can be described by the developed groups include, but are not limited to, MEA, 3-amino-1-propanol (MPA), ethylenediamine (EDA), ethylaminoethanol (EMEA), N-methyl-1,3-propanediamine (MAPA), diethylenetriamine (DETA).

The reactions that take place in aqueous solutions of amines and CO_2 are often modeled using quasi-chemical approaches (Chen et al., 1982; Chen and Evans, 1986), which are effective but require a large number of model parameters and hence extensive experimental data. Another modeling route is to adopt a physical approach, such as in the SAFT framework which has been shown to provide results equivalent to chemical and quasi-chemical approaches (Economou and Donohue, 1991) under certain assumptions. This makes it possible to use an implicit-product model where the product species are described as aggregates of the reactants, whose formation is driven by large short-range intermolecular interactions (Mac Dowell et al., 2010b; Rodriguez et al., 2012; Chremos et al., 2013). The use of a physical approach avoids the need for previous knowledge of the equilibrium constants of

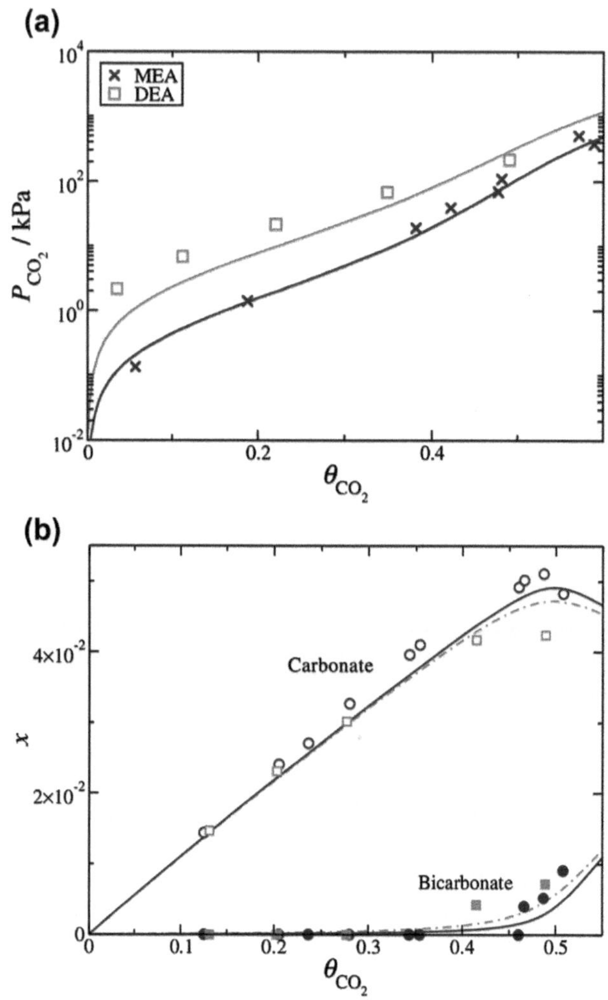

FIGURE 11.3

(a) Solubility of CO_2 in aqueous solutions with a 30 wt% (in mass) of monoethanolamine (MEA) (crosses) and with a 35.4 wt% (in mass) of diethanolamine (DEA) (squares) at $T = 373$ K as a function of the partial pressure of CO_2 along the vapour–liquid equilibrium of the ternary mixture amine + H_2O + CO_2. The solubility is represented as CO_2 loading, θ_{CO_2}, defined as the number of moles of CO_2 absorbed in the liquid phase per mole of amine in the liquid. The symbols correspond to the experimental data of Jou et al. (1995) for MEA and of Lee et al. (1972) for DEA. The solid curves correspond to statistical associating fluid theory based on the square well potential (SAFT-γ SW) calculations. (b) Predicted mole fraction of carbamate and bicarbonate in the liquid phase of a 30 wt% (in mass) MEA aqueous solution at $T = 313$ (red (light gray in print versions)) and 333.15 K (blue (dark gray in print versions)) for the ternary mixture of MEA + H_2O + CO_2 as a function of the CO_2 loading, θ_{CO_2}. The symbols correspond to the experimental data: open symbols for carbamate and solid symbols for bicarbonate; circles correspond to $T = 313.15$ K; and squares to $T = 333.15$ K (Boettinger et al., 2008). The curves correspond to SAFT-γ SW predictions; continuous curves correspond to $T = 313.15$ K; and dot-dashed curves to $T = 333.15$ K.

the chemical reactions, thus reducing the complexity of the problem. The concentration of carbamate and bicarbonate in aqueous mixtures of MEA and CO_2 can thus be predicted based on the concentration of aggregates, as seen through comparison with experimental data (Boettinger et al., 2008) in Figure 11.3(b). The SAFT-γ SW equation of state is shown to provide an accurate prediction of the formation of these components, without the use of speciation data in developing the model. All phase equilibrium calculations performed in this work are carried out using the HELD algorithm (Pereira et al., 2012).

We note that it is not necessary to adopt a physical approach in modeling the reaction products. The SAFT-γ SW framework can be used to develop models of the mixtures in which the reaction products are treated as independent species. This can be achieved by explicitly accounting for the ionic nature of the products, and the chemical equilibrium of each relevant reaction occurring among the components of the mixture, by building on the SAFT framework previously developed to model electrolytes (Galindo et al., 1999; Gil-Villegas et al., 2001; Patel et al., 2003; Schreckenberg et al., 2014). The key challenge in any such model is that it requires knowledge of the types of ionic species formed, as well as the relevant equilibrium constants. The suitability of using speciation data generated from implicit-product models to help reduce the complexity of the problem, especially in mixtures for which there are few or no experimental data, is currently under investigation. One can envisage a two-step approach to solvent selection, in which implicit-product models are first used to identify promising solvent candidates, and the best options are then investigated using explicit-product models that provide a more detailed description of mixture behavior, and of the performance of the capture process.

11.4.3 ABSORPTION/DESORPTION FLOWSHEET MODELING AND DESIGN FRAMEWORK

The optimum design of absorption/desorption CO_2 capture flowsheets is supported by a flexible and inclusive synthesis model incorporated within a generic process modeling framework. The framework serves as a mathematical tool that can be used to reproduce any potentially favorable representation of solvent-based CO_2 capture processes based on an underlying superstructure (Figure 11.4). The proposed superstructure consists of modules representing generic process tasks (e.g., reaction, separation, heat transfer) and interconnecting streams emulating material flows. Each module may be independently assigned a process model representing a particular task, the type of equipment utilized, the desired operating conditions, and so forth. Many different modules may be connected in the same flowsheet using a diverse set of streams (e.g., recycle, bypasses, and so forth). The proposed generic tasks account for (1) the possibility of reaction, mass, and heat exchange between different phases within each module, and (2) the possibility of stream mixing and splitting to enable the distribution of materials among different modules. Modules assigned task (1) may represent a column section, a heater, and a heat exchanger, while modules assigned task (2) may represent a splitter or a mixer. These are used as the building blocks to generate any desired absorption/desorption flowsheet configuration.

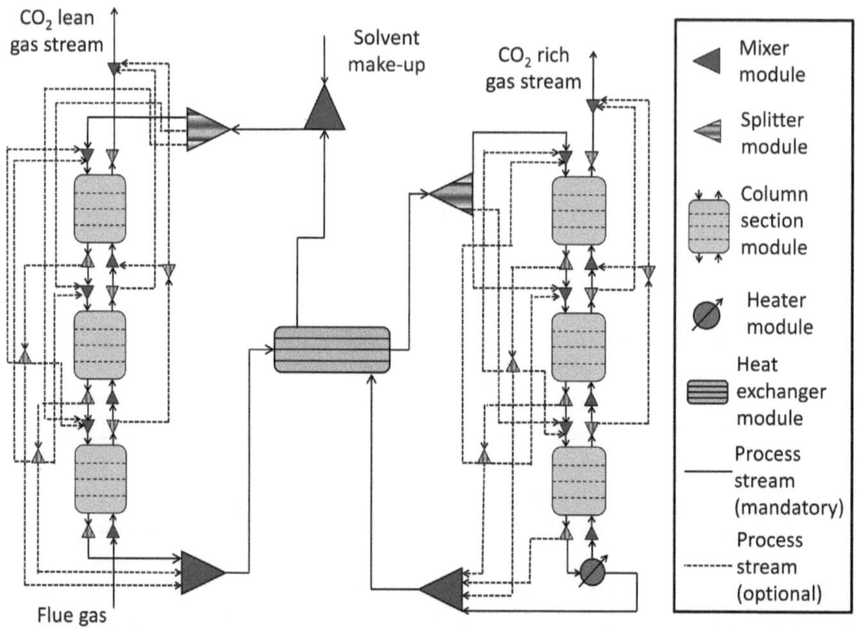

FIGURE 11.4

Superstructure used for optimum absorption/desorption design (Damartzis et al., 2014).

Reactive column sections are modeled using an equilibrium-based set of equations that represent theoretical separation stages. OCFE approximation technique (Seferlis and Hrymak, 1994) is used to subdivide the column section into finite elements of variable size directly associated with column height. The concentration and temperature profiles are approximated as continuous function of the column vertical coordinate. Material and energy balances are assumed to be exactly satisfied at a limited number of points within each element; namely the collocation points. The OCFE approximating technique provides a significant reduction in terms of total number of modeling equations without decreasing the resolution of the model employed for the balance equations. Moreover, the transformation of the key structural variables of the separation column into continuous variables eliminates the need for the use of integer variables to represent discrete separation stages, while providing an equivalent description.

11.5 RESULTS OF MULTI SCALE DESIGN

In Figure 11.5, it can be seen that the average of the top 10 molecules in class D outperforms the average of the top 10 in all other classes, indicating that the novel structures are best when thermodynamic and reactivity properties are considered

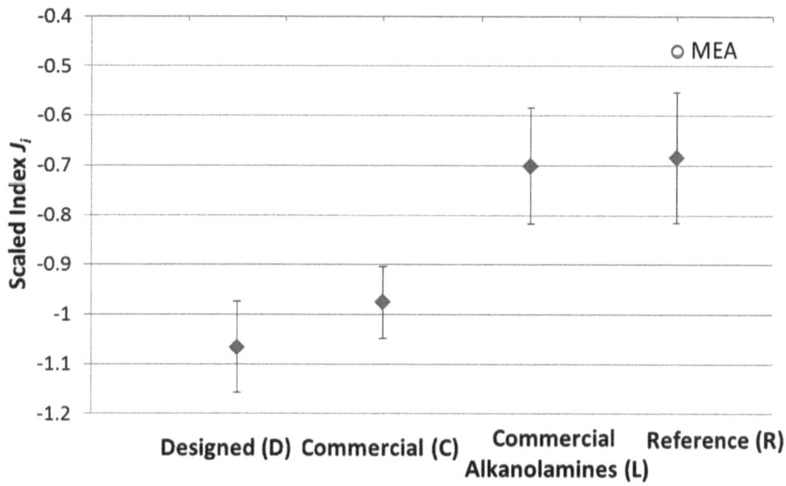

FIGURE 11.5

Comparison of monoethanolamine (MEA) with the top 10 molecules identified in each class based on the average performance index values with 95% confidence intervals. The lower the performance index J_i (Papadopoulos et al., 2014), the better the performance. The performance of MEA in terms of J_i is represented by the circle. Solvents of considerably higher performance than MEA are included in each class.

but sustainability is not taken into account. All classes contain molecules that also considerably outperform MEA. From a comparison of the C and L classes it appears that alkanolamines have a lower performance when all properties are considered. However, alkanolamines like 2-amino-2-methyl-1-propanol (AMP) are well known as CO_2 capture solvents for enabling high CO_2 absorption capacity. The close performance of the L and R classes is due to the fact that most molecules in R are alkanolamines. A more detailed analysis of the results indicates that the molar volume (V_m) and the liquid heat capacity (C_p) are on average worse than MEA in all classes, while all other properties outperform those of MEA considerably. This shows that the proposed approach captures trade-offs among different properties, although molecules exist in these classes with V_m and C_p values relatively close to or better than those of MEA, and which also exhibit better performance in all other properties.

Several highly performing solvents are listed in Table 11.2; they were also briefly reported in (Papadopoulos et al., 2014) but are elaborated for the first time in this work. AMP was originally considered in the R class, but it also ranks among the top options in the C and L classes. AMP was proposed in a mixture with piperazine and water as the solvent that resulted from the CESAR project funded from the European Commission to find efficient postcombustion CO_2 capture solvents (von Harbou et al., 2013). AMP is a well-known moderately hindered amine with a higher CO_2 capacity than unhindered amines but with slower kinetics. Hence, piperazine is used because of the very high reaction rate constant (Bishnoi and Rochelle, 2000).

Table 11.2 Highly Performing Solvents from R, C and L Classes (Papadopoulos et al. (2014))

Solvent	CAS Number	V_m (cm^3/mol)	P_{vp} (Pa)	σ (dyn/cm)	C_p (J/mol K)	n (cp)	*RED*	pk_a
AMP	124-68-5	98.7	315	47.9	176.0	0.66	2.27	9.7
MMEA	109-83-1	80.4	456	34.6	206.1	0.96	3.05	9.85
MAPA	6291-84-5	104.4	1384	32.6	238.4	0.78	1.77	10.5
2AP	6168-72-5	78.5	196	47.0	239.1	0.85	2.60	9.37
EMEA	110-73-6	98.1	255	47.6	237.5	0.69	2.91	10.1
MEA	141-43-5	60.26	150	50.25	214.8	22.6	3.94	9.5
HEXA	111-26-2	132.9	2707	25.8	258.9	0.85	0.74	9.7

RED, relative energy difference; AMP, 2-amino-2-methyl-1-propanol; MMEA, 2-(methylamino)-ethanol; MAPA, n-methyl-1,3-propanediamine; 2AP, 2-(amino)-propanol; EMEA, ethylaminoethanol; MEA, monoethanolamine; HEXA, hexylamine.

MAPA is also an interesting option because it combines a primary and a secondary amine group, which enables increased reaction rates. In this context, it has been combined with N,N-dimethylaminoethanol (DMMEA) (Bruder et al., 2012) and N,N-diethyl-2-aminoethanol (DEEA) (Hartono et al., 2013). MAPA-DEEA mixtures form two liquid phases upon CO_2 loading, one rich and the other lean in CO_2. This behavior enables a reduction in the sensible heat and reduces the stripping steam requirements. Hexylamine (HEXA) is also a phase change solvent (Zhang et al., 2012) which enables the generation of a water-rich phase after the absorption. This is very useful because the water-rich phase may be removed using a decanter prior to stripping, and therefore the stripping may take place at 80–90 °C, reducing significantly the energy requirements compared to conventional stripping at 120 °C. HEXA exhibits high absorption capacity and fast kinetics compared to very similar molecules such as heptylamine and octylamine. However, it also exhibits low solubility in water at low concentrations (Singh et al., 2007). The appearance of this solvent within the above highly performing set, despite the existence in the overall database of heptylamine and octylamine, is a confirmation that the obtained results lead to valid conclusions. This is despite the use of relatively simple models to perform pure component property predictions. Regarding the other amines, 2-(methylamino)-ethanol (MMEA) has been reported to exhibit a CO_2 absorption rate higher than that of DEA and MEA (Kumar and Kundu, 2012), 2-(amino)-propanol (2AP) has been reported to lead to a carbamate stability close to that of MEA (Da Silva, 2011), while EMEA has been reported to have an absorption rate slightly higher than that of MEA and approximately equal to that of MMEA and 2-(butylamino)ethanol (BEA) (Ma'mun et al., 2007). These solvents appear to be promising and are worth further investigation.

The impact of process layout for the MEA case is presented in Figure 11.6. All proposed designs lead to a reduction of the energetic demands of the process. More specifically, in the case of intercooling between the absorber stages (ICA) a drastic reduction of 23.6% was achieved in the required absorption heat load. This is also reflected in the annualized cost for this particular flowsheet, which exhibits a reduction by almost 17%. Furthermore, the ICA configuration showed the lowest optimal operating pressures, almost 150 and 170 kPa for the absorber and stripper columns respectively. Contrary to the case with temperature control in the absorber, splitting of the process streams and redistribution in multiple feed points in the absorber and stripper leads to favorable but less efficient designs in terms of total processing costs. Stream redistribution in the stripper (DSS flowsheet) resulted in a larger column size and capital cost, even in low splitting ratios in the range of 5%. In addition, the DSS configuration showed the lowest cyclic capacity among all cases. The latter is represented by the difference between the rich and lean CO_2 loadings and serves as a measure of the process efficiency, expressing the amount of CO_2 that can be loaded and unloaded into the amine solution through a cycle in the process. Higher cyclic capacity is desirable because it is possible to capture and release a higher amount of CO_2 per mole of amine. Similarly, the redistribution of material in the absorber column (MFA flowsheet) for the optimal split ratio of 50% resulted in a comparatively

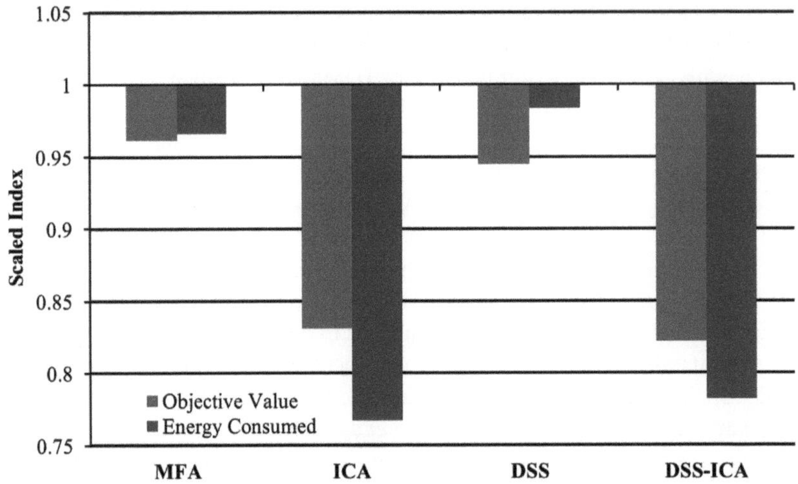

FIGURE 11.6

Optimum design results for flowsheet structures multifeed absorber (MFA), intercooled absorber (ICA) and DSS and DSS-ICA for solvent monoethanolamine. The reported scaled index reflects the improvement achieved in terms of the required heat for solvent regeneration (GJ/tonne CO_2 captured) and the annualized process cost (€/yr, denoted by "objective value") with respect to the conventional absorption/desorption configuration. The annualized process cost comprises the sum of capital and operating process expenses. All cases correspond to a CO_2 capture efficiency of 90% from a flue gas stream containing 14.1% v/v CO_2.

larger column size and an increase in the capital cost of the MFA process, despite the reduction in the energy consumed. Finally, the DSS-ICA flowsheet, which is a combination of the two individual configurations, shows a performance close to that of the ICA configuration, with a slight improvement in the objective function value. The energy consumed in the conventional, MFA, ICA, DSS, and DSS-ICA flowsheets is 4.11, 3.97, 3.16, 4.05, 3.22 GJ/ton CO_2 captured, respectively.

The LCA and hazard metrics for standard (MEA, DEA) and selected solvents from Table 11.2 (AMP, MMEA) in different process configurations are shown in Figure 11.7. Thermal and oxidative degradation rates of MEA are given by Davis (2009) and Goff (2005), respectively. Relative degradation rates for other solvents are calculated in proportion to the percentage solvent loss reported by Eide-Haugmo (2011) and Lepaumier et al. (2009b) for thermal and oxidative degradation, respectively. Degradation products are kept at a steady-state concentration of 1.5% to avoid negative effects on process operation. Solvent recovery through thermal reclaiming is compared to a bleed and feed scenario. The waste generated from the solvent reclaimer is assumed to be landfilled as hazardous waste, while the purge stream from the bleed and feed scenario is assumed to be treated in a wastewater treatment plant. AMP exhibits the lowest process energy demand and as such has the lowest

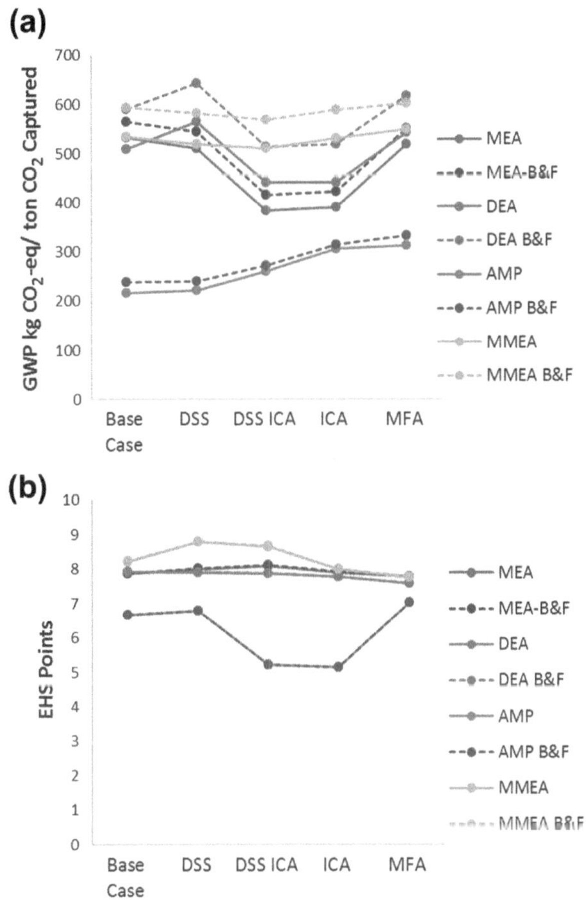

FIGURE 11.7

Sustainability metrics given as (a) global warming potential (GWP) in kg CO_2-eq/tonne CO_2 captured (b) EHS points for 30 wt% monoethanolamine, 35.4 wt% diethanolamine, 30 wt% 2-Amino-2-methyl-1-propanol, and 30 wt% 2-(Methylamino)-ethanol. Two waste treatment scenarios are compared for the purge stream: solvent thermal reclaiming and the bleed and feed (B&F) scenario with wastewater treatment. Multifeed absorber (MFA), intercooled absorber (ICA) and double-section stripper (DSS) and DSS-ICA.

life cycle impact of the different solvents considered. AMP is a more stable solvent with a lower toxicity and it also features a lower solvent circulating rate in the process. This leads to a better performance from an EHS point of view as well. DSS-ICA and ICA process configurations generally display the best performance for all solvents. The addition of a reclaimer highly reduces the solvent makeup rate, leading to a more sustainable operation. The difference between the two scenarios is more pronounced for solvents with higher degradation rates.

11.6 CONCLUSIONS AND PERSPECTIVES

This study demonstrated an efficient computational approach for screening a vast number of commercial and novel solvents and process configurations for postcombustion CO_2 capture. The approach combines CAMD, GC methods, superstructure based process synthesis, and multicriteria sustainability assessment. Moreover, this study provides details for the respective information required by the aforementioned methods, starting from pure substance properties and their respective estimation methods, to solvent–CO_2–water mixture properties and finally to those related to multicriteria sustainability assessment.

The results of the multiscale screening highlighted several high-performance solvents and process configurations and also provided a quantitative analysis of the respective benefits from economic, life cycle, and hazard assessment perspectives. It has been shown that there is significant improvement potential for solvent-based CO_2 capture systems that can be achieved by designing new molecules or considering commercial ones that have not been widely implemented in this system. It has also been demonstrated that further improvement is expected by combining these high-performance solvents with their respective optimal process configurations (e.g., the combination of a DDS with ICA).

The results have also demonstrated that the proposed computational approach, although based on a high level of abstraction with respect to the physicochemical phenomena taking place in an actual CO_2 capture system, was able to identify solvents that have been proposed as promising alternatives by more rigorous studies. This provides a useful validation of the reliability of the approach and of the breadth of designs it generates. Building on this, the results of this study motivate further experimental efforts at the process level for those high-performance molecules that have not yet been rigorously studied as well as at the property estimation level, highlighting the potential benefits of developing additional GC parameters to broaden the solvent design space.

The proposed approach could be further enhanced by expanding its scope in several directions. One promising direction is to extend the CAMD method to consider solvent mixtures in the initial screening phase. The computational effort will certainly be increased (i.e., considering the number of components in the mixture and also their respective concentrations), but one should keep in mind that this remains less demanding than an equivalent experimental study. An additional challenge would be the calculation of the respective performance criteria for these mixtures, especially in some properties needed for hazard assessment (e.g., toxicity). Another promising avenue of development is the use of more accurate shortcut models for predicting solvent degradation rates and the resulting degradation products; this could significantly enhance the first screening of the molecular structures. Moreover, the development of a group contribution method for the prediction of the environmental impact of solvent production could be coupled with information about solvent losses to provide a cradle-to-gate metric in a computationally efficient way for the CAMD approach. The same concept is valid for those hazard assessment related properties, for which easy to use shortcut models are not available (e.g., for threshold limit values).

Finally, it is important to note that the multiscale screening proposed in this study for optimum solvents and process configurations should be validated at further scales of the CO_2 capture design, including the equipment design and the integration potential of the CO_2 capture plant with the CO_2-emitting plant.

APPENDIX

The 25 reference molecules are the following: 2-amino-2-methyl-1-propanol (AMP), 2-(methylamino)ethanol (MMEA), N-methyl-1,3-propanediamine (MAPA), 2-(amino)-propanol (2AP), ethylaminoethanol (EMEA), 1,2-propanediamine (MEDA), N,N-dimethylaminoethanol (DMMEA), 3-amino-1-propanol (MPA), monoethanol-amine (MEA), N,N-diethyl-2-aminoethanol (DEEA), 2-amino-2-methyl-1,3-propane-diol (AMPD), 2-(butylamino)ethanol (BEA), 4-diethylamino-2-butanol (DEAB), diethylenetriamine (DETA), N-(2-aminoethyl)ethanolamine (AEEA), ethylenedi-amine (EDA), 2-amino-2-ethyl-1,3-propanediol (AEPD), 3-(dimethylamino)-1, 2-propanediol (DMAPD), diethanolamine (DEA), methyldiethanolamine (MDEA), diisoprpylamine (DIPA), tetraethylenepentamine (TEPA), tri(hydroxyethyl)amine (TEA), triisopropanolamine (TIPA), tromethamine (TRIS).

ACKNOWLEDGMENTS

The authors are grateful to the Commission of the European Union (project FP7-ENERGY-2011-282789) and the Engineering and Physical Sciences Research Council (EPSRC) of the UK (grants EP/E016340, EP/J014958/1 and EP/J003840/1) for financial support.

REFERENCES

Adu, I.K., Sugiyama, H., Fischer, U., Hungerbühler, K., 2008. Comparison of methods for assessing environmental, health and safety (EHS) hazards in early phases of chemical process design. Process Saf. Environ. Prot. 86, 77–93.

Aronu, U.E., Svendsen, H.F., Hoff, K.A., Juliussen, O., 2009. Solvent selection for carbon dioxide absorption. Greenhouse Gas Control Technol. 9 (1), 1051–1057.

von der Assen, N., Jung, J., Bardow, A., 2013. Life-cycle assessment of carbon dioxide capture and utilization: avoiding the pitfalls. Energy Environ. Sci. 6, 2721–2734.

Banimostafa, A., 2013. Indicator-Based Enhancement of "Green Chemistry and Engineering Principles" in Chemical Process Design. ETH, Zurich. PhD.

Bardow, A., Steur, K., Gross, J., 2010. Continuous-molecular targeting for integrated solvent and process design. Ind. Eng. Chem. Res. 49, 2834–2840.

Benson, S.W., 1976. Thermochemical Kinetics: Methods for the Estimation of Thermochemical Data and Rate Parameters. Wiley.

Bishnoi, S., Rochelle, G.T., 2000. Absorption of carbon dioxide into aqueous piperazine: reaction kinetics, mass transfer and solubility. Chem. Eng. Sci. 55, 5531–5543.

Boden, T., Marland, G., Andres, R., 2012. Global, Regional and National Fossil Fuel CO_2 Emissions. US Department of Energy, Oak Ridge Tennessee, USA.

Boettinger, W., Maiwald, M., Hasse, H., 2008. Online NMR spectroscopic study of species distribution in MEA-H_2O-CO_2 and DEA-H_2O-CO_2. Fluid Phase Equilib. 263, 131–143.

Bruder, P., Owrang, F., Svendsen, H.F., 2012. Pilot study-CO2 capture into aqueous solutions of 3-methylaminopropylamine (MAPA) activated dimethyl-monoethanolamine (DMMEA). Int. J. Greenhouse Gas Control 11, 98–109.

Bumann, A.A., Papadokonstantakis, S., Fischer, U., Hungerbühler, K., 2011. Investigating the use of path flow indicators as optimization drivers in batch process retrofitting. Comput. Chem. Eng. 35, 2767–2785.

Cao, C.Z., Liu, L., 2004. Topological steric effect index and its application. J. Chem. Inf. Comput. Sci. 44, 678–687.

Carvalho, A., Gani, R., Matos, H., 2008. Design of sustainable chemical processes: systematic retrofit analysis generation and evaluation of alternatives. Process Saf. Environ. Prot. 86, 328–346.

Catoire, L., Naudet, V., 2004. A unique equation to estimate flash points of selected pure liquids application to the correction of probably erroneous flash point values. J. Phys. Chem. Ref. Data 33, 1083–1111.

Catoire, L., Paulmier, S., Naudet, V., 2006. Estimation of closed cup flash points of combustible solvent blends. J. Phys. Chem. Ref. Data 35, 9–14.

ChemAxon. Available: http://www.chemaxon.com/marvin/sketch/index.php (accessed 26.03.15).

Chemmangattuvalappil, N.G., Eden, M.R., 2013. A novel methodology for property-based molecular design using multiple topological indices. Ind. Eng. Chem. Res. 52, 7090–7103.

Chen, C.C., Britt, H.I., Boston, J.F., Evans, L.B., 1982. Local composition model for excess Gibbs energy of electrolyte systems. 1. Single solvent, single completely dissociated electrolyte systems. AIChE J. 28, 588–596.

Chen, C.C., Evans, L.B., 1986. A local composition model for the excess Gibbs energy of aqueous-electrolyte systems. AIChE J. 32, 444–454.

Chen, H., Shonnard, D.R., 2004. Systematic framework for environmentally conscious chemical process design: early and detailed design stages. Ind. Eng. Chem. Res. 43, 535–552.

Chremos, A., Forte, E., Papaioannou, V., Galindo, A., Jackson, G., Adjiman, C.S., 2013. Modelling the fluid phase behaviour of multifunctional alkanolamines and carbon dioxide using the SAFT-gamma approach. In: 16th International Conference on Process Integration, Modelling and Optimisation for Energy Saving and Pollution Reduction (Pres'13), 35, pp. 427–432.

Clark, S.N.I., Haslam, A.J., Galindo, A., Jackson, G., 2006. Developing optimal Wertheim-like models of water for use in statistical associating fluid theory (SAFT) and related approaches. Mol. Phys. 104, 3561–3581.

Cohen, N., Benson, S.W., 1993. Estimation of heats of formation of organic compounds by additivity methods. Chem. Rev. 93, 2419–2438.

Constantinou, L., Gani, R., Oconnell, J.P., 1995. Estimation of the acentric factor and the liquid molar volume at 298-K using a new group-contribution method. Fluid Phase Equilib. 103, 11–22.

Conte, E., Martinho, A., Matos, H.A., Gani, R., 2008. Combined group-contribution and atom connectivity index-based methods for estimation of surface tension and viscosity. Ind. Eng. Chem. Res. 47, 7940–7954.

Damartzis, T., Papadopoulos, A.I., Seferlis, P., 2014. Optimum synthesis of solvent-based post-combustion CO_2 capture flowsheets through a generalized modeling framework. Clean Technol. Environ. Policy 16, 1363–1380.

Davis, J., 2009. Thermal degradation of Aqueous Amines Used for Carbon Dioxide Capture.

Economou, I.G., Donohue, M.D., 1991. Chemical, quasi-chemical and perturbation theories for associating fluids. AIChE J. 37, 1875–1894.

Eide-Haugmo, I., 2011. Environmental Impacts and Aspects of Absorbents Used for CO_2 Capture.

EPA. Available: www.epa.gov (accessed 26.03.15).

Folic, M., Adjiman, C.S., Pistikopoulos, E.N., 2007. Design of solvents for optimal reaction rate constants. AIChE J. 53, 1240–1256.

Folic, M., Adjiman, C.S., Pistikopoulos, E.N., 2008. Computer-aided solvent design for reactions: maximizing product formation. Ind. Eng. Chem. Res. 47, 5190–5202.

Galindo, A., Blas, F.J., 2002. Theoretical examination of the global fluid phase behavior and critical phenomena in carbon dioxide plus n-alkane binary mixtures. J. Phys. Chem. B 106, 4503–4515.

Galindo, A., Davies, L.A., Gil-Villegas, A., Jackson, G., 1998. The thermodynamics of mixtures and the corresponding mixing rules in the SAFT-VR approach for potentials of variable range. Mol. Phys. 93, 241–252.

Galindo, A., Gil Villegas, A., Jackson, G., Burgess, A.N., 1999. SAFT-VRE: phase behavior of electrolyte solutions with the statistical associating fluid theory for potentials of variable range. J. Phys. Chem. B 103, 10272–10281.

Gharagheizi, F., Alamdari, R.F., Angaji, M.T., 2008. A new neural network-group contribution method for estimation of flash point temperature of pure components. Energy Fuels 22, 1628–1635.

Gil-Villegas, A., Galindo, A., Jackson, G., 2001. A statistical associating fluid theory for electrolyte solutions (SAFT-VRE). Mol. Phys. 99, 531–546.

Gil-Villegas, A., Galindo, A., Whitehead, P.J., Mills, S.J., Jackson, G., Burgess, A.N., 1997. Statistical associating fluid theory for chain molecules with attractive potentials of variable range. J. Chem. Phys. 106, 4168–4186.

Giovanoglou, A., Barlatier, J., Adjiman, C.S., Pistikopoulos, E.N., Cordiner, J.L., 2003. Optimal solvent design for batch separation based on economic performance. AIChE J. 49, 3095–3109.

Goedkoop, M., Spriensma, R., 2000. The eco-indicator 99: a damage orientated method for life-cycle impact assessment (Methodology report 2000a, Pre-Consultants).

Goff, G., 2005. Oxidative Degradation of Aqueous Monoethanolamine in CO_2 Capture - Processes: Iron and Copper Catalysis, Inhibition, and O_2 Mass Transfer.

Gupta, M., Da Silva, H.F., Svendsen, E.F., 2012. Computational study of thermodynamics of polyamines with regard to CO_2 capture. In: 6th Trondheim Conference on CO_2 Capture, Transport and Storage, 23, pp. 140–150.

Hartono, A., Saleem, F., Arshad, M.W., Usman, M., Svendsen, H.F., 2013. Binary and ternary VLE of the 2-(diethylamino)-ethanol (DEEA)/3-(methylamino)-propylamine (MAPA)/ water system. Chem. Eng. Sci. 101, 401–411.

von Harbou, I., Mangalapally, H.P., Hasse, H., 2013. Pilot plant experiments for two new amine solvents for post-combustion carbon dioxide capture. Int. J. Greenhouse Gas Control 18, 305–314.

2009. ISO14040: Environmental Management - Life Cycle Assessment - Principles and Framework. (Brussels).

Imperial-Chemical-Industries(ICI), 1974. Hazard and Operability Studies: Process Safety Report 2 (London).

IPCC, 2007. Fourth Report.

Jou, F.Y., Mather, A.E., Otto, F.D., 1995. The solubility of CO_2 in a 30-mass-percent monoethanolamine solution. Can. J. Chem. Eng. 73, 140–147.

KAPSARC, 2012. Carbon Capture and Storage: Technologies, Policies, Economics, and Implementation Strategies. CRC Press/Balkema, Leiden, Netherlands.

Kenig, E., Seferlis, P., 2009. Modeling reactive absorption. Chem. Eng. Prog. 105, 65–73.

Khan, F.I., Abbasi, S.A., 1998. Multivariate hazard identification and ranking system. Process Saf. Prog. 17, 157–170.

Koller, G., Fischer, U., Hungerbühler, K., 2000. Assessing safety, health, and environmental impact early during process development. Ind. Eng. Chem. Res. 39, 960–972.

Kumar, G., Kundu, M., 2012. Vapour-liquid equilibrium of CO_2 in aqueous solutions of N-methyl-2-ethanolamine. Can. J. Chem. Eng. 90, 627–630.

Lee, J.I.L., Mather, A.E., Otto, F.D., 1972. Solubility of carbon dioxide in aqueous diethanolamine solutions at high-pressures. J. Chem. Eng. Data 17, 465.

Lees, F.P., 1996. Loss Prevention in Process Industries. Butterworth-Heinemann, Oxford.

Lepaumier, H., Picq, D., Carrette, P.-L., 2009a. New amines for CO_2 capture. I. Mechanisms of amine degradation in the presence of CO_2. Ind. Eng. Chem. Res. 48, 9061–9067.

Lepaumier, H., Picq, D., Carrette, P.-L., 2009b. New amines for CO_2 capture. II. Oxidative degradation mechanisms. Ind. Eng. Chem. Res. 48, 9068–9075.

Lymperiadis, A., Adjiman, C.S., Galindo, A., Jackson, G., 2007. A group contribution method for associating chain molecules based on the statistical associating fluid theory (SAFT-gamma). J. Chem. Phys. 127.

Lymperiadis, A., Adjiman, C.S., Jackson, G., Galindo, A., 2008. A generalisation of the SAFT-gamma group contribution method for groups comprising multiple spherical segments. Fluid Phase Equilib. 274, 85–104.

Le Moullec, Y., Kanniche, M., 2011. Screening of flowsheet modifications for an efficient monoethanolamine (MEA) based post-combustion CO_2 capture. Int. J. Greenhouse Gas Control 5, 727–740.

Ma'mun, S., Svendsen, H.F., Hoff, K.A., Juliussen, O., 2007. Selection of new absorbents for carbon dioxide capture. Energy Convers. Manage. 48, 251–258.

Mac Dowell, N., Galindo, A., Jackson, G., Adjiman, C.S., 2010a. Integrated solvent and process design for the reactive separation of CO_2 from flue gas. In: 20th European Symposium on Computer Aided Process Engineering, 28, pp. 1231–1236.

Mac Dowell, N., Llovell, F., Adjiman, C.S., Jackson, G., Galindo, A., 2010b. Modeling the fluid phase behavior of carbon dioxide in aqueous solutions of monoethanolamine using transferable parameters with the SAFT-VR approach. Ind. Eng. Chem. Res. 49, 1883–1899.

Mac Dowell, N., Samsatli, N.J., Shah, N., 2013. Dynamic modelling and analysis of an amine-based post-combustion CO_2 capture absorption column. Int. J. Greenhouse Gas Control 12, 247–258.

Marrero, J., Gani, R., 2001. Group-contribution based estimation of pure component properties. Fluid Phase Equilib. 183, 183–208.

NIST. Available: http://www.nist.gov/ (accessed 26.03.15).

Oexmann, J., Kather, A., 2010. Minimising the regeneration heat duty of post-combustion CO_2 capture by wet chemical absorption: the misguided focus on low heat of absorption solvents. Int. J. Greenhouse Gas Control 4, 36–43.

Oyenekan, B.A., Rochelle, G.T., 2007. Alternative stripper configurations for CO_2 capture by aqueous amines. AIChE J. 53, 3144–3154.

Papadopoulos, A.I., Badr, S., Chremos, A., Forte, E., Zarogiannis, T., Seferlis, P., Papadokonstantakis, S., Adjiman, C.S., Galindo, A., Jackson, G., 2014. Efficient screening and selection of post-combustion CO_2 capture solvents. Chem. Eng. Trans. 39.

Papadopoulos, A.I., Linke, P., 2006a. Multiobjective molecular design for integrated process-solvent systems synthesis. AIChE J. 52, 1057–1070.

Papadopoulos, A.I., Linke, P., 2006b. Efficient integration of optimal solvent and process design using molecular clustering. Chem. Eng. Sci. 61, 6316–6336.

Papadopoulos, A.I., Linke, P., 2009. Integrated solvent and process selection for separation and reactive separation systems. Chem. Eng. Process. 48, 1047–1060.

Papadopoulos, A.I., Stijepovic, M., Linke, P., 2010. On the systematic design and selection of optimal working fluids for organic rankine cycles. Appl. Therm. Eng. 30, 760–769.

Parmar, J.C., Lees, F.P., 1987. The propagation of faults in process plants - hazard identification. Reliab. Eng. Sys. Saf. 17, 277–302.

Patel, B.H., Paricaud, P., Galindo, A., Maitland, G.C., 2003. Prediction of the salting-out effect of strong electrolytes on water plus alkane solutions. Ind. Eng. Chem. Res. 42, 3809–3823.

Pereira, F.E., Jackson, G., Galindo, A., Adjiman, C.S., 2012. The HELD algorithm for multicomponent, multiphase equilibrium calculations with generic equations of state. Comput. Chem. Eng. 36, 99–118.

Pereira, F.E., Keskes, E., Galindo, A., Jackson, G., Adjiman, C.S., 2011. Integrated solvent and process design using a SAFT-VR thermodynamic description: high-pressure separation of carbon dioxide and methane. Comput. Chem. Eng. 35, 474–491.

Poling, B.E., Prausnitz, J.M., O'Connel, J.P., 2001. The properties of gases and liquids, Vol. 5. McGraw-Hill, New York, USA. http://dx.doi.org/10.1036/0070116822.

Puxty, G., Rowland, R., Allport, A., Yang, Q., Brown, M., Burns, R., Maeder, M., Attalla, M., 2009. Carbon dioxide postcombustion capture: a novel screening study of the carbon dioxide absorption performance of 76 amines. Environ. Sci. Technol. 43, 6427–6433.

Rayer, A.V., Henni, A., Tontiwachwuthikul, P., 2012. Molar heat capacities of solvents used in CO2 capture: a group additivity and molecular connectivity analysis. Can. J. Chem. Eng. 90, 367–376.

Rockström, J., Steffen, W., Noone, K., 2009. A safe operating space for humanity. Nature 461, 472–475.

Rodriguez, J., Mac Dowell, N., Llovell, F., Adjiman, C.S., Jackson, G., Galindo, A., 2012. Modelling the fluid phase behaviour of aqueous mixtures of multifunctional alkanolamines and carbon dioxide using transferable parameters with the SAFT-VR approach. Mol. Phys. 110, 1325–1348.

Ruiz-Mercado, G.J., Smith, R.L., Gonzalez, M.A., 2012a. Sustainability indicators for chemical processes: I. Taxonomy. Ind. Eng. Chem. Res. 51, 2309–2328.

Ruiz-Mercado, G.J., Smith, R.L., Gonzalez, M.A., 2012b. Sustainability indicators for chemical processes: II. Data needs. Ind. Eng. Chem. Res. 51, 2329–2353.

Da Silva, E.F., 2011. Theoretical study of the equilibrium constants for solvents for CO_2 capture. In: 10th International Conference on Greenhouse Gas Control Technologies, 4, pp. 164–170.

Sartori, G., Ho, W.S., Savage, D.W., Chludzinski, G.R., Wiechert, S., 1987. Sterically-hindered amines for acid-gas absorption. Sep. Purif. Methods 16, 171–200.

Schreckenberg, J.M.A., Dufal, S., Haslam, A.J., Adjiman, C.S., Jackson, G., Galindo, A., 2014. Modelling of the thermodynamic and solvation properties of electrolyte solutions with the statistical associating fluid theory for potentials of variable range. Mol. Phys. 112, 2339–2364.

Seferlis, P., Hrymak, A.N., 1994. Optimization of distillation units using collocation models. AIChE J. 40, 813–825.

Sigma-Aldrich. Available: www.sigmaaldrich.com/catalog (accessed 26.03.15).

Singh, P., Niederer, J.P.M., Versteeg, G.F., 2007. Structure and activity relationships for amine based CO_2 absorbents - I. Int. J. Greenhouse Gas Control 1, 5–10.

Stefanis, E., Panayiotou, C., 2008. Prediction of Hansen solubility parameters with a new group-contribution method. Int. J. Thermophys. 29, 568–585.

Struebing, H., Ganase, Z., Karamertzanis, P.G., Siougkrou, E., Haycock, P., Piccione, P.M., Armstrong, A., Galindo, A., Adjiman, C.S., 2013. Computer-aided molecular design of solvents for accelerated reaction kinetics. Nat. Chem. 5, 952–957.

Sugiyama, H., Fischer, U., Hungerbühler, K., Hirao, M., 2008. Decision framework for chemical process design including different stages environmental, health, and safety assessment. AIChE J. 54, 1037–1053.

Sumon, K.Z., Henni, A., East, A.L.L., 2012. Predicting pK(a) of amines for CO_2 capture: computer versus pencil-and-paper. Ind. Eng. Chem. Res. 51, 11924–11930.

Torres, C.M., Gadalla, M.A., Mateo-Sanz, J.M., Esteller, L.J., 2011. Evaluation tool for the environmental design of chemical processes. Ind. Eng. Chem. Res. 50, 13466–13474.

Tugnoli, A., Santarelli, F., Cozzani, V., 2011. Implementation of sustainability drivers in the design of industrial chemical processes. AIChE J. 57, 3063–3084.

Verein-Deutcher-Ingenieure, 1997. Vdi-richtlinie 4600: Cumulative Energy Demand, Terms, Definitions, Methods of Calculation (Düsseldorf).

Versteeg, G.F., Van Dijck, L.A.J., van Swaaij, W.P.M., 1996. On the kinetics between CO_2 and alkanolamines both in aqueous and non-aqueous solutions. An overview. Chem. Eng. Commun. 144, 113–158.

Wernet, G., Papadokonstantakis, S., Hellweg, S., Hungerbühler, K., 2009. Bridging data gaps in environmental assessments: modeling impacts of fine and basic chemical production. Green Chem. 11, 1826–1831.

Zhang, J., Qiao, Y., Agar, D.W., 2012. Improvement of lipophilic-amine-based thermomorphic biphasic solvent for energy-efficient carbon capture. In: 6th Trondheim Conference on CO_2 Capture, Transport and Storage, 23, pp. 92–101.

CHAPTER

12 Selection of Sustainable Processes Using Sustainability Footprint Method: A Case Study of Methanol Production from Carbon Dioxide

Rajib Mukherjee*, **Debalina Sengupta**[§], **Subhas K. Sikdar**[¶,1]

**Vishwamitra Research Institute, Chicago, IL, USA*

[§]*Texas A&M University, TX, USA*

[¶]*NRMRL, US EPA, Cincinnati, OH, USA*

[1]*Corresponding author: E-mail: sikdar.subhas@epa.gov*

12.1 INTRODUCTION

A concept of sustainability was introduced in the Brundtland report known as Our Common Future (Brundtland, 1987). Since the advent of the industrial revolution, economic growth has created environmental degradation and adverse human health impacts, which, at least in the developed nations, had been largely reversed by regulating industrial and municipal activities. Developing countries on the other hand had remained mired in conditions of environment and health that are unacceptable by arguments articulated in the Bruntland report. Since the publication of this report and subsequent acceptance of proposed sustainable development, concern for global warming has been elevated to a point that it appears to supplant the original Bruntland concerns of resource depletion, environmental pollution, and wealth inequity, both intra- and intergenerational. Still, the need of the hour is a model for sustainable development where human health and economic growth should be ensured without compromising the environment while also paying attention to global warming issues, principally through the accounting of emissions of greenhouse gases. Brundtland report talks about measures that can lead to economic growth with simultaneous progress in the condition of the environment as well as progress toward societal equity. Sustainability of such systems as agricultural, industrial, urban infrastructure etc. has been studied, and considerable amount of research has given directions to achieve stability of such systems (Conway and

Computer Aided Chemical Engineering, Volume 36. ISSN 1570-7946. http://dx.doi.org/10.1016/B978-0-444-63472-6.00012-4

Barbier, 2013; Čuček et al., 2012; Graedel and Allenby, 2010). In other cases, countries have been compared on their sustainability status based on their performance in economic, environmental, and societal improvement efforts (Esty et al., 2008; Freudenberg, 2003; Sengupta et al., 2014).

Among various systems differing in types and scales, sustainability of industrial systems is of utmost importance. The dominant way of evaluating a process or a product for sustainability is by making quantitative estimates of values of appropriate indicators or metrics (these terms are used interchangeably here) chosen to measure its impacts on the three domains of sustainability, i.e., the environment, society, and the economy. For example, indicators for process sustainability evaluation make use of process flow and cost data, chemical property data, land use data etc. Comparing these indicator values of competing processes gives an estimate of how good a process is, in terms of sustainability, in comparison to its competitors. Information of this type should be an important aid in making decisions for commercialization.

The use of sustainability indicators and metrics seems promising in principle, but the vast number of applicable indicators already available in the literature can be confounding to a user. Even if the users are able to identify an initial set of indicators pertaining to the process system they are considering, coming to a decision on relative sustainability can be cumbersome. For example, a particular process may be sustainable in the economic sense, while its alternative may be promising from environmental perspectives. To overcome this challenge, the application of a composite index for sustainability evaluation has been proposed by various researchers and methods of aggregation have been applied in different industries. Krajnc and Glavič (2005) have proposed a composite sustainable development index that shows performance of a company in all three areas of sustainability. They have used their composite sustainable development index to study two oil and gas industries. Tokos et al. (2012) have used the composite sustainable development index for sustainability assessment of breweries. Zhou et al. (2012) have extended the use of the composite sustainable development index introduced by Krajnc and Glavič (2005) to a breweries case study by integrating geometric aggregation along with linear aggregation. Singh et al. (2007) also developed a sustainable performance index for comparing steel industries. They have used an analytic hierarchy process to calculate weighting factors for the indicators. Brandi et al. (2014) have suggested the use of Canberra index to aggregate sustainability metrics . They have shown the application of their method in different chemical processes. Olinto on the other hand proposed using a vector space method for aggregating sustainability indicators (Olinto, 2014). In our previous work, we used an aggregated index, called sustainability footprint (D_e), for sustainability evaluation (Sikdar, 2009; Sikdar et al., 2012)

Chemical process flowsheet synthesis has been traditionally done by hierarchical decomposition or by mathematical programming (Grossmann et al., 1999). These methods have helped in choosing the most economic process under optimal condition among competing processes. The primary criterion for process selection in these methods thus is cost (or profit) consideration. In the case of mixed integer superstructure optimization problems, the objective function again is typically maximization of profit. These methods, though robust obviously do not examine sustainability in a comprehensive way and highlight the need for the use of multiple objectives that reflect all relevant sustainability needs for decision making. Sustainability needs are usually expressed with indicators or metrics. An aggregated index such as the sustainability footprint (D_e) provides a convenient surrogate for all included indicators for a multiobjective evaluation for process sustainability.

In this article, sustainability footprint and superstructure optimization method of process selection is applied to six competing processes for methanol production from carbon dioxide. Five of these processes are novel, which involve utilization of carbon dioxide as raw material, and a uniform industrial scale capacity is used for design (Indala, 2004). The processes use different types of catalyst and differ in the amount of raw material required, byproducts formed, and utilities used. These five processes are compared with the traditional process of methanol production from natural gas and carbon dioxide. The results of the process design are used for superstructure formulation and solved using mixed integer linear programming optimization methods to identify a process that maximizes profit under the operating constraints. The results of the process design are also used to compute process sustainability indicators for applying the sustainability footprint method to select the best process in terms of sustainability. The sustainability evaluation includes the use of traditional economic indicators to represent the economic dimension, environmental indicators to represent the environmental dimension, and human health indicators to represent the social dimension. The environmental and human health indicators were computed using the tool for the reduction and assessment of chemical and other environmental impacts (TRACI) impact characterization method (Bare, 2002, 2013). Eight different sustainability indicators were chosen for analysis.

Comparison of industrial processes that produce the same products is central to sustainability evaluation. Adoption of the optimal process among alternatives is an exercise in a compromise among the three objectives of sustainability: profit maximization (or cost minimization), minimization of adverse environmental impacts, and maximization of societal benefits. Oftentimes processes that are found to be economically viable might not be more sustainable from environmental or social perspectives. Data for such a sustainability analysis typically are available as an $m \times n$ matrix, where m is the number of process options for comparison and n is the number of indicators or metrics used for the analysis. A common practice is to plot these data on a spider diagram (Aiche, 2014; Saling et al., 2002). Spider diagrams are convenient as they show on one plot the performance attributes of all the competing process options. When the connected points for one process option is totally subsumed into the space described by the similar points for another process, the former is judged to be clearly superior to the latter, subject to the convention that larger values of indicators signify worse performance for those indicators. But most often this scenario will not be obtained, making an overall visual decision of superiority of one process over another difficult because while some indicators would be superior, others would be inferior. This problem is exceedingly more complex when the number of indictors is large or the process alternatives are too many. To overcome this difficulty, aggregated index in the form of sustainability footprint (D_e) of each process is appropriate. In the present work, results from sustainability footprint are compared with those from mixed integer programming where net profit is used as the objective function. It is to be noted that the former considers multiple indicators spanning all three domains of sustainability while the latter considers only the economic domain. In any case, lower the sustainability footprint, more sustainable is the process.

Beyond making inference on relative superiority of processes from sustainability viewpoint, it may be important to know the relative importance of the indicators in their contributions toward the footprint index. The sustainability footprint D_e is the Euclidean Distance in n dimensional metrics space. It is constructed by measuring a statistical distance between a multidimensional system and another similar reference system. In the present work, the contribution of each metric to the aggregate index is analyzed using a multivariate statistical analysis known as partial least squares-variable importance in projection method (PLS-VIP). Thus, the mixed integer programming is used to choose the most economic process. The aggregate index is used to rank the processes in order of their sustainability, and the PLS-VIP method is used to rank the indicators in order of their importance. Together, they complete an analysis and reinforce decisions made on relative sustainability.

12.2 BACKGROUND LITERATURE

12.2.1 CHEMICAL PROCESS FLOWSHEET SYNTHESIS AND DESIGN: METHANOL PROCESSES

Traditional chemical process flowsheet synthesis and design starts with conceptual design of a process followed by a hierarchical decomposition into five decision levels (Douglas, 1985; Turton et al., 2012). Once a conceptual design is finalized, the detailed design of the process is created, which involves economic analysis, piping and instrumentation, site selection etc. The implementation of the design then enables the commercialization of the process. Thus, chemical process synthesis serves as the first and a very crucial step for the implementation of any new process.

The hierarchical approach used for the synthesis of a process is translated into chemical process flowsheet design. The advent of computing tools in chemical engineering such as Aspen Plus® (2014), Aspen HYSYS® (2014), CHEMCAD® (2014) etc. has facilitated efficient design of chemical processes. These tools have libraries of chemical and thermodynamic properties, process equipment, and cost estimation. One can also compute material and energy use, design for mass and heat exchanges, and recycle networks.

In this section, we provide a brief overview of the conventional process for the production of methanol, followed by five new processes for methanol from carbon dioxide. Indala (2004) designed these new processes in HYSYS, and we include the assumptions used in these designs in this section. The methanol production capacity of the new processes was based on an Ashland Chemical Inc. methanol plant, located in Plaquemine, LA, and the production capacity of this plant was 160 million gallons per year (480,846 metric tons/year) (LC&PPL, 1998).

1. Conventional Methanol Process (Hertwig, 2004)

   The capacity of 200,000 tons/year (548 tons/day) was used for the conventional methanol process (Hertwig, 2004). The two-step process is described below. *Synthesis gas preparation*: The steam reforming of methane is the first step in the preparation of synthesis gas. The reaction is given in Eqn (12.1):

$$CH_4 + H_2O = CO + 3H_2 \tag{12.1}$$

   The reaction is very endothermic, favored by high temperature and low pressure. This reaction produces 1:3 ratio of CO to H_2 instead of the 1:2 needed for methanol synthesis. This ratio can be adjusted either by partial oxidation of methane or by using another source of CO. In case of conventional methanol process, CO_2 is used in water-gas shift reaction to produce CO. The process uses some of the H_2 produced from CH_4 as shown in Eqn (12.2). The overall reaction based on Eqn (12.1) and Eqn (12.2) is Eqn (12.3):

$$CO_2 + H_2 = CO + H_2O \tag{12.2}$$

$$CH_4 + 2H_2O + CO_2 = 4CO + 8H_2 \tag{12.3}$$

   Methanol synthesis in catalytic converter: Methanol synthesis occurs in a converter using Lurgi technology according to Eqn (12.4) (Hertwig, 2004). It comprises a cooled multitubular reactor running at 5–8 MPa at 250–260 °C. Liquid-entrained micron-sized copper-based catalyst can convert as much as 25% per pass.

$$CO + 2H_2 = CH_3OH \quad \Delta H = -103 \text{ MJ/Kgmol} \tag{12.4}$$

   The heat of reaction is directly used to generate high-pressure steam. Methanol is condensed by both heat exchange and pressure reduction and is collected and purified. Steam demand within the methanol plant usually is a close match to the methanol plant's steam production. Extra steam can be produced within the methanol plant for users inside or outside the methanol plant by firing the auxiliary burners in the heat-recovery section of the exhaust gas from the primary reformer.

2. Methanol from CO_2 hydrogenation over Cu(100) catalyst (Nerlov and Chorkendorff, 1999)

   Nerlov and Chorkendorff (1999) described an experimental study for the production of methanol through CO_2 hydrogenation over a Cu (100) catalyst. The following reaction occurs in the reactor:

$$CO_2 + 3H_2 \rightarrow CH_3OH + H_2O \quad \Delta H^\circ = -49 \text{ kJ/mol}, \quad \Delta G^\circ = 3 \text{ kJ/mol} \tag{12.5}$$

The methanol production capacity of the simulated process in HYSYS was 480,000 metric tons per year (54,760 kg/h) (Indala, 2004). The yield of methanol was assumed to be 61%, according to the commercial process from synthesis gas (Wells, 1999) as Nerlov and Chorkendorff (1999) did not report the conversion of CO_2. The unreacted CO_2 was recycled in the design, and 100% conversion of CO_2 was achieved. The produced methanol was 99% pure.

3. Methanol from CO_2 hydrogenation over Cu-Zr Catalyst (Toyir et al., 1998)
Toyir et al. (1998) described the experimental study for the production of methanol by CO_2 hydrogenation. Raney Cu-Zr catalyst leached with aqueous solution of zincate (NaOH + ZnO) was used in this study. The authors reported the formation of carbon monoxide along with methanol by CO_2 hydrogenation. The reactions involved in this process are:

$$CO_2 + 3H_2 \rightarrow CH_3OH + H_2O \quad \Delta H^\circ = -49\ \text{kJ/mol},\ \Delta G^\circ = 3\ \text{kJ/mol} \tag{12.6}$$

$$CO_2 + H_2 \rightarrow CO + H_2O \quad \Delta H^\circ = 41\ \text{kJ/mol},\ \Delta G^\circ = 29\ \text{kJ/mol} \tag{12.7}$$

The methanol production capacity of this simulated process in HYSYS was 481,000 metric tons per year (54,870 kg/h) (Indala, 2004). Toyir and colleagues did not report the conversion of CO_2, and 61% yield similar to commercial process of methanol was assumed (Wells, 1999). The unreacted hydrogen was recycled, and a 100% conversion of CO_2 was achieved. The produced methanol was 99% pure. CO was obtained as a byproduct.

4. Methanol from CO_2 hydrogenation over $Cu/ZnO/ZrO_2/Al_2O_3/Ga_2O_3$ catalyst (Ushikoshi et al., 1998)
This process was designed based on the experimental study described by Ushikoshi et al. (1998), for the production of methanol by CO_2 hydrogenation over a multicomponent catalyst ($Cu/ZnO/ZrO_2/Al_2O_3/Ga_2O_3$). The CO_2 conversion per pass reported in the study was 17%. The reactions involved in the reactor are:

$$CO_2 + 3H_2 \rightarrow CH_3OH + H_2O \quad \Delta H^\circ = -49\ \text{kJ/mol},\ \Delta G^\circ = 3\ \text{kJ/mol} \tag{12.8}$$

$$CO_2 + H_2 \rightarrow CO + H_2O \quad \Delta H^\circ = 41\ \text{kJ/mol},\ \Delta G^\circ = 29\ \text{kJ/mol} \tag{12.9}$$

$$CO + 2H_2 \rightarrow CH_3OH \quad \Delta H^\circ = -90.5\ \text{kJ/mol},\ \Delta G^\circ = -25\ \text{kJ/mol} \tag{12.10}$$

The methanol production capacity of this simulated process was 479,800 metric tons per year (54,730 kg/h) (Indala, 2004). The purity of methanol produced was 99%, and carbon monoxide was obtained as byproduct.

5. Methanol from hydrogenation over $Cu/ZnO/Cr_2O_3$ and CuNaY zeolite catalyst (Jun et al., 1998)
Jun et al. (1998) described the experimental study for production of methanol by CO_2 hydrogenation using a hybrid catalyst of $Cu/ZnO/Cr_2O_3$ and CuNaY zeolite. A small amount of dimethyl ether (DME) was co-produced along with methanol. In this study, the conversion of CO_2 to CO was 10.2% and to oxygenates was 9.4% and the selectivity of dimethyl ether in oxygenates was

36.7%. Using the selectivity to DME in oxygenates and the total conversion to oxygenates, the specific conversion to DME was calculated to be 3.4%. Unreacted CO_2 and H_2 were recycled, thus the conversion was 100%. The following reactions occur in the reactor:

$$CO_2 + H_2 \rightarrow CO + H_2O \quad \Delta H^\circ = 41\ kJ/mol,\ \ \Delta G^\circ = 29\ kJ/mol \tag{12.11}$$

$$CO + 2H_2 \rightarrow CH_3OH \quad \Delta H^\circ = -90.5\ kJ/mol,\ \ \Delta G^\circ = -25\ kJ/mol \tag{12.12}$$

$$CH_3OH \rightarrow CH_3OCH_3 + H_2O \quad \Delta H^\circ = -24\ kJ/mol,\ \ \Delta G^\circ = -17\ kJ/mol \tag{12.13}$$

The methanol production capacity of this simulated process in HYSYS was 479,800 metric tons per year (54,700 kg/h) (Indala, 2004).

6. Methanol from hydrogenation over Pd/SiO2 catalyst (Bonivardi et al., 1998)
   Bonivardi et al. (1998) described the experimental study for the production of methanol by CO_2 hydrogenation over calcium-promoted Pd/SiO2 catalyst. The yield of methanol of 61% was assumed as before (Wells, 1999) as Bonivardi et al. (1998) did not report the conversion of carbon dioxide. Complete conversions of the raw materials were achieved, as the unreacted feed was recycled.

$$CO_2 + H_2 \rightarrow CO + H_2O \quad \Delta H^\circ = 41\ kJ/mol,\ \ \Delta G^\circ = 29\ kJ/mol \tag{12.14}$$

$$CO + 2H_2 \rightarrow CH_3OH \quad \Delta H^\circ = -90.5\ kJ/mol,\ \ \Delta G^\circ = -25\ kJ/mol \tag{12.15}$$

The production capacity of this simulated process was 480,370 metric tons of methanol per year (54,800 kg/h) (Indala, 2004)

12.2.2 MIXED INTEGER PROGRAMMING FOR OPTIMAL PROCESS SELECTION

When competing processes are present, mathematical programming techniques are often used on a superstructure of options to select an optimal solution. Mixed integer programming can be used to select the best option according to the given objective function, which satisfies the constraints, and corresponding variables in the process can be calculated. Grossmann and colleagues discussed the use of mathematical programming methods in chemical process synthesis (Grossmann et al., 1999). The general mathematical model can be represented as given in Eqn (12.16):

$$\begin{aligned} &Min\ Z = f(x,y) \\ &Subject\ to: \\ &h(x,y) = 0 \\ &g(x,y) \le 0 \\ &x \in X;\ y \in \{0,1\} \end{aligned} \tag{12.16}$$

where $f(x,y)$ is the objective function (e.g., cost), $h(x,y) = 0$ are the equations that describe the performance of the system (mass and heat balances, design equations), and $g(x,y) \le 0$ are inequalities that represents the constraints for feasible choices of the variables. The variables x are continuous and represent the state or design variables, while y are the discrete variables, which take 0–1 values corresponding to the selection of an item or an action. If any of the functions are nonlinear, the MIP problem becomes a mixed-integer nonlinear program. If all functions are linear it becomes a mixed-integer linear program. If there are no binary variables, the MIP problem reduces to a nonlinear program or linear program.

Superstructures of process models for the above mathematical programming representation can be classified into three main classes: (1) aggregated models, (2) short cut models, and (3) rigorous models. As the names suggest, the aggregated models are high-level representation of a problem, and the models are greatly simplified representations, which highlight an aspect or objective that tends to dominate the problem. Short cut models involve fairly detailed superstructures that involve cost optimization (investment and operating costs) as the objective function, but the performance of the process units is predicted with simpler nonlinear models in order to reduce the computational costs and/or for exploiting the algebraic structure of the equations, especially for global optimization. Rigorous models also involve detailed superstructures, as well as complex models for underlying process units.

12.2.3 AGGREGATE INDEX D_e AS SUSTAINABILITY FOOTPRINT

The aggregate index (D_e) as sustainability footprint is calculated as the Euclidian distance in the vector space of the indicators. The datasets for the process systems can be represented by an $m \times n$ matrix, where m is the number of process options and n is the number of indicators. Each competing process will have indicator values in economic, societal, and environmental areas, which can be calculated from the process design results using the flow rates, economic data, and TRACI impact characterization factors. The $m \times n$ matrix will have each process represented by a vector X_m and each indicator is an element of the vector. The observed minimum value of each indicator is used to construct a synthetic reference process, X_0. We have used the distance of a process option from the best performing process, which can be synthetically constructed, as a measure of sustainability. Relative to X_0, any of the real process option will always have positive difference of the indicator values. This is essentially shifting the point of reference from one of the processes to an imaginary process, and the exercise is to find out how far a real process is from the reference process. In addition, if the indicator values were defined such that higher values are worse than smaller values, then a smaller difference of an indicator value from that of the same indicator at the imaginary point would be more desirable from a sustainability point of view. To ensure this, every indicator is made unidirectional.

The next step in D_e calculation following the computation of distance from the best performer is to normalize each indicator. This is required as different indicators have different units and D_e represents the Euclidian distance where the vectors are orthogonal to each other. After normalization with respect to x_0, the calculation of the aggregate index D_e for the process $m(D_{e,m})$ is computed by using the Euclidean distance method as shown in Eqn (12.17) (Sengupta et al., 2014; Sikdar et al., 2012).

$$D_{e,m} = \sqrt{\sum_{j}^{n}\left[c_j\frac{(x_j \quad x_{j0})}{(x_j - x_{j0})_{\max}}\right]^2} \tag{12.17}$$

The weighting factor c_j allows use of weighting preference (usually a societal choice) of any of the indicators in comparison to others.

12.2.4 PARTIAL LEAST SQUARE VARIABLE IMPORTANCE IN PROJECTION

PLS-VIP method is used for the ranking of indicators in the order of their importance toward the aggregate index (D_e). The importance of an indicator is measured by comparing the variability of the indicator with the aggregate index. PLS-VIP is a multivariate regression method where information from a data space of a larger number of variables is projected into that with a smaller number of variables. We start with the same data matrix, X, as was used to calculate the aggregate index. The indicators are the variables in the data space. PLS VIP is a supervised model for which an overall data behavior or pattern is required. This overall pattern can be represented by a response vector or a response matrix. In our modeling we have used the aggregate index, D_e as the response vector.

In the PLS-VIP analysis, the number of indicators is reduced in a way such that variations in the set of reduced indicators are most likely to be reflected in the response vector D_e. In other words, the overall data pattern would be unchanged with the reduced number of indicators. The easiest test of this expectation is if one can make the same conclusions about the overall data behavior with this set of fewer number of indicators as was done with the original set. The surplus indicators would thus be understood to be redundant or not important in contributing to the overall performance of the system in question. An application of PLS-VIP for sustainability assessment of manufacturing process was given by Mukherjee et al. (2013) and for OECD and non-OECD countries by Sengupta et al. (2014). In this multivariate method, there are variables that are inferred through a mathematical model of the original (observed) data and are known as latent variables. In contrast to the original variables, the latent variables are not explicit, nor can be tweaked at will. In this paper, the variables are the indicators, and henceforth, we will refer to the original variables as indicators and the latent variables derived from the original indicators as latent indicators.

Partial least squares regression (PLS) model is used to decompose the original data matrix X into two orthogonal matrices, the loadings (L) and scores (T) of

a number of latent indicators, and a residual matrix, E as shown in Eqn (12.18) (Cinar et al., 2004).

$$X = TL^T + E = \sum_{j=1}^{a} t_j l_j^T + E \tag{12.18}$$

The score matrix T is related to the response vector D_e through a regression matrix b as shown in Eqn (12.19) (Chong and Jun, 2005). F is the residual vector of D_e.

$$D_e = Tb^T + F \tag{12.19}$$

The values of the indicators for each process m are represented as a process vector x_m. The process vectors x_m can be related to the score vector through a weight vector w_j as given in Eqn (12.20).

$$t_j = w_j^T x_m \tag{12.20}$$

The variable importance in projection (VIP) for a particular indicator is calculated using the regression coefficient b, weight vector w_j, and score vector t_j as given in Eqn (12.21).

$$VIP_K = \sqrt{n \frac{\sum_{j=1}^{a} b_j^2 t_j^T t_j \left(\frac{w_{kj}}{\|w_j\|} \right)^2}{\sum_{j=1}^{a} b_j^2 t_j^T t_j}} \tag{12.21}$$

where w_{kj} is the kth element of the weight vector w_j.

PLS-VIP is used to identify the importance of each indicator in affecting the aggregate index D_e. In order to avoid the relative variability of the indicators to affect the result, normalized indicators are used for VIP. In contrast to the aggregate index calculation where the normalization is performed with respect to the synthetic reference process, here the normalization is performed with respect to the variance of the indicators. This ensures the use of variability within an indicator but avoid relative variability with respect to those of other indicators. Indicators with lower VIP scores have little influence on D_e and those with the higher VIP scores contribute the most toward D_e. The average of squared VIP scores equals 1. VIP score greater than one is generally used as a criterion for detecting the relative importance of an indicator.

There are different algorithms available to solve PLS regression problems. In the present work, the regression for the PLS is based on PLS1 algorithm (Martens, 1989). In PLS1 algorithm, the correlation coefficient of X and D_e is used to obtain the first extracted score t_1 for the first latent indicator. After obtaining this score for first latent indicator, the regression follows by obtaining that for the second latent indicators from the residuals and so on. This is most appropriate for our problem where response matrix comprises of one column, constituting the aggregate indices, D_e. Regression coefficient and weight vectors from the first three latent indicator score vectors are used for calculating VIP.

12.3 PROBLEM FORMULATION

In this work, we use the design results from five potentially new processes for the production of methanol from carbon dioxide and, for sustainability analysis, compare those results with an existing conventional route. The conventional methanol process parameters including capacity, conversion rates, raw materials, products, byproducts etc. were based on an existing methanol facility in the Lower Mississippi River Corridor and were obtained from consultation with chemical industry associates (Hertwig, 2004). The conventional process involves the steam reforming of methane to syngas, followed by CO:H_2 ratio adjustment using carbon dioxide, and catalytic conversion of carbon monoxide and hydrogen to methanol. Methanol can be produced from hydrogenation of pure carbon dioxide using catalytic conversion over a variety of catalysts. These processes provide an alternative use for carbon dioxide, a greenhouse gas, which is produced from various processes for agricultural chemicals. The new processes were simulated using HYSYS by Indala, 2004. These simulations provide the basis for choosing a process alternative that uses carbon dioxide and is superior to the rest in terms of sustainability. These processes are described in the previous section.

In our problem, we use a shortcut model to generate the superstructure of alternatives, with fairly simple mass and energy balance equations representing the conventional and four of the five new methanol process units. The process by Toyir et al. (1998) is not included in the superstructure since it shows negative profitability. The superstructure is represented in Figure 12.1.

The mathematical formulation of the problem can be represented as below.

Max: $Profit = Gross\ Profit - Raw\ Material\ Costs - Utility\ Costs$

Subject to: $Gross\ Profit = GP(F_i, Y_j)$

$Raw\ Material\ Costs = MC(F_i, Y_j)$

$Utility\ Costs = UC(F_i, Y_j)$

$F_i \geq \text{Demand} \times Y_j$

$F_i \leq \text{Capacity} \times Y_j$

F_i = Flow rate of reactants or products of a particular species in the process.

Y_j = Binary variables denoting the choice of a certain methanol process.

The constraint equations used in the superstructure involved the mass balance, species balance, and energy balance equations, which can be represented in the following general form.

Overall Mass Balance: $F_{in} - F_{out} + F_{generation} - F_{consumption} = 0$

Species Balance: $F_{i,in} - F_{i,out} + F_{i,generation} - F_{i,consumption} = 0$

Overall Energy Balance: $Q_{in} - Q_{out} + \sum H_{i,in}F_{i,in} - \sum H_{i,out}F_{i,out} = 0$

The above process superstructure model formed a mixed integer linear programming problem that is solved using simple branch and bound (SBB) algorithm in General Algebraic Modeling System. The binary variables are constrained such that only one of the Y_j has a value of 1, denoting the choice of an optimal process for methanol.

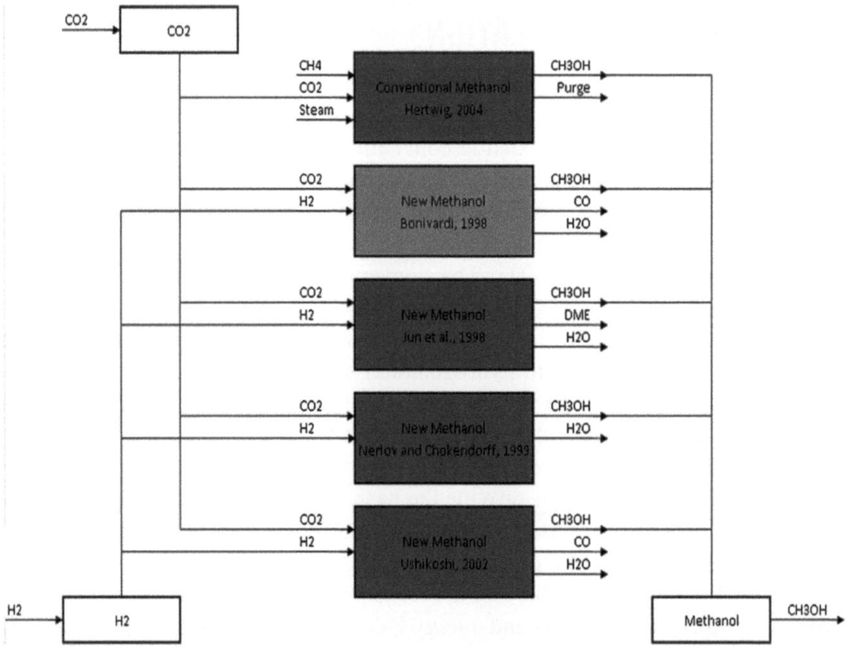

FIGURE 12.1

Superstructure representation and optimal result of methanol process choices.

For the calculation of sustainability footprint, the results from the HYSYS designs (Indala, 2004) were used to compute the values of the chosen indicators. The eight indicators selected for our analysis are given in Table 12.1. Of these eight indicators, Global Warming Air, Smog Air, Energy Intensity, Water Intensity, Mass Intensity, and Carbon Efficiency act as environmental indicators; Net Profit as the economic indicator; and Human Health Particulate Air as the societal indicator. Grouping the chosen indicators into the three sustainability domains is rarely unique, as the indicators are usually representative of more than one dimension (Sikdar, 2003). For example, Smog Air can as well be classified as a societal indicator, or better yet, a socioecological indicator. This classification, however, is not needed for the sustainability analysis performed here. TRACI has other indicators that characterize the social impact through assessment of human health factors of the released chemicals. They were calculated in the present case study. But none of the chemicals involved in these processes has impact in these categories. Because of the difference in the capacity of the design especially for the conventional method, Global Warming Air, Smog Air, Net Profit, and Human Health Particulate Air are represented on per unit of methanol produced basis. To ensure that all the indicators are unidirectional, i.e., a lower value indicating higher sustainability and higher value representing lower sustainability, the following

Table 12.1 Indicators for Sustainability Evaluation and Method to Make Them Unidirectional

Indicator	Desirability of Lower Value	Transformation Applied	Transformed Indicator
Global warming air	Yes	–	Global warming air
HH particulate air	Yes	–	HH particulate air
Smog air	Yes	–	Smog air
Net profit	No	Multiplied with −1	−Net profit
Energy intensity	Yes	–	Energy intensity
Water intensity	Yes	–	Water intensity
Mass intensity	Yes	–	Mass intensity
Carbon efficiency	No	Subtracted from 1 (highest efficiency)	1-Carbon efficiency

transformations are applied to the indicator set. Since a higher value of net profit is always desirable, −1 is multiplied to the net profit indicator to indicate −Net Profit where a lower value is desirable. For carbon efficiency, a higher value is desirable, and transforming the indicator to 1-carbon efficiency will ensure that a lower value is desirable. In this case, we use the transformation of 1-carbon efficiency because efficiency can have a range of values from 0−1, and any other transformation will give erroneous result. The transformed indicator set is used to compute D_e.

12.4 RESULTS AND DISCUSSIONS

All six processes are used to compare their relative sustainability. All of them use CO_2 and H_2 as raw material and produce methanol as the main product along with other byproducts. The processes differ from each other in the type of catalyst used and byproducts formed. Sustainability evaluation is performed in three main steps. In the first step, data are acquired from the processes under consideration. This is followed by the superstructure optimization in the second step and finally sustainability footprint evaluation in the third step. Detail description of each step is given below:

Step 1:

From the process design, we obtain the mass and energy flow rate of each reactant, product and byproduct. These are then used for two different purposes: to decide the upper and lower limits in the constraint equations for superstructure optimization and for the evaluation of the sustainability indicators.

Table 12.2 Unidirectional Values of the Indicators and Sustainability Footprint of the Six Different Processes as Obtained from Eight Different Indicators

Process	Global Warming Air (kg CO_2 eq/kg Methanol)	HH Particulate Air (PM2.5 eq/kg Methanol)	Smog Air (kg O_3 eq/kg Methanol)	−Net Profit ($/kg Methanol)	Energy Intensity (kJ/kg Methanol)	Water Intensity (kg/kg Methanol)	Mass Intensity (kg input/kg Methanol)	1-Carbon Efficiency	D_e
Nerlov and Chokendorff, 1999	−1.38E+00	0.00E+00	6.72E−01	−2.02E−02	2.35E+04	3.32E+02	1.57E+00	3.24E−03	
Toyir et al., 1998	−2.76E+00	3.12E−04	7.21E−01	7.55E−02	3.65E+04	4.87E+02	3.01E+00	5.02E−01	
Ushikoshi et al., 1998	−1.40E+00	3.80E−06	6.73E−01	−3.30E−02	2.10E+04	3.02E+02	1.59E+00	1.56E−02	
Jun et al., 1998	−1.46E+00	0.00E+00	6.72E−01	−5.95E−02	1.83E+04	2.71E+03	1.66E+00	5.68E−02	
Bonivardi et al., 1998	−1.45E+00	1.64E−05	6.75E−01	−5.91E−02	1.59E+04	2.41E+02	1.65E+00	5.33E−02	
Conventional	−9.74E+00	2.33E−07	6.67E−01	−1.52E−01	1.19E+04	1.21E+02	7.23E−01	8.33E−04	
Xo	−9.74E+00	0.00E+00	6.67E−01	−1.52E−01	1.19E+04	1.21E+02	7.23E−01	8.33E−04	
Nerlov and Chokendorff, 1999-Xo	8.36E+00	0.00E+00	5.37E−03	1.32E−01	1.16E+04	2.11E+02	8.46E−01	2.41E−03	
Toyir et al., 1998-Xo	6.98E+00	3.12E−04	5.42E−02	2.27E−01	2.46E+04	3.67E+02	2.29E+00	5.01E−01	
Ushikoshi et al., 1998-Xo	8.35E+00	3.80E−06	5.96E−03	1.19E−01	9.12E+03	1.81E+02	8.65E−01	1.48E−02	
Jun et al., 1998-Xo	8.28E+00	0.00E+00	5.37E−03	9.25E−02	6.37E+03	2.58E+03	9.35E−01	5.59E−02	
Bonivardi et al., 1998-Xo	8.29E+00	1.64E−05	7.93E−03	9.29E−02	3.99E+03	1.20E+02	9.23E−01	5.24E−02	
Conventional-Xo	0.00E+00	2.33E−07	0.00E+00	0.00E+00	0.00E+00	0.00E+00	0.00E+00	0.00E+00	
Max	8.36E+00	3.12E−04	5.42E−02	2.27E−01	2.46E+04	2.58E+03	2.29E+00	5.01E−01	
(Nerlov and Chokendorff, 1999-Xo)/max	1.00E+00	0.00E+00	9.91E−02	5.79E−01	4.73E−01	8.16E−02	3.70E−01	4.81E−03	1.31E+00
(Toyir et al., 1998-Xo)/max	8.35E−01	1.00E+00	1.00E+00	1.00E+00	1.00E+00	1.42E−01	1.00E+00	1.00E+00	2.59E+00
(Ushikoshi et al., 1998-Xo)/max	9.98E−01	1.22E−02	1.10E−01	5.23E−01	3.71E−01	7.00E−02	3.78E−01	2.96E−02	1.25E+00
(Jun et al., 1998-Xo)/max	9.91E−01	0.00E+00	9.91E−02	4.07E−01	2.59E−01	1.00E+00	4.08E−01	1.12E−01	1.55E+00
(Bonivardi et al., 1998-Xo)/max	9.91E−01	5.25E−02	1.46E−01	4.08E−01	1.62E−01	4.63E−02	4.03E−01	1.05E−01	1.17E+00
(Conventional-Xo)/max	0.00E+00	7.47E−04	0.00E+00	0.00E+00	0.00E+00	0.00E+00	0.00E+00	0.00E+00	7.47E−04

Step 2:

The process design parameters as obtained from the HYSYS design are used to construct the mass and energy balance equations for each technology. All of these processes are then combined to form a superstructure as shown in Figure 12.1. The conventional method is limited by the production capacity as provided by Hertwig 2004. A mixed integer linear programming is performed for superstructure optimization with maximizing profit as the objective function. Methanol production from the process by Bonivardi et al. (1998) is found to be most optimum within the constraints of design parameters. This is shown with color green in the superstructure (Figure 12.1).

Step 3:

The flow rate of the reactants, products and byproducts, energy, and cooling water usage as obtained from the process simulation were used for computing eight sustainability indicators. These indicators were then aggregated using the sustainability footprint (D_e) to assess overall sustainability of the processes. The eight indicators chosen for our analysis are computed as shown in Table 12.2.

The sustainability footprints (D_e) from different processes as obtained from the indicators are shown in Figure 12.2. From the results of the sustainability analysis it is clear that the Toyir process is least sustainable. This is obvious from the fact that Toyir has shown a negative profit. Contrary to the optimal result from step 2 where Bonivardi showed the maximum profit under design constrains, the sustainability footprint analysis shows the conventional method as the most sustainable. This is from the fact that though Bonivardi showed more profit per unit of methanol production, the conventional process had lower values for all the eight indicators as shown in Table 12.2.

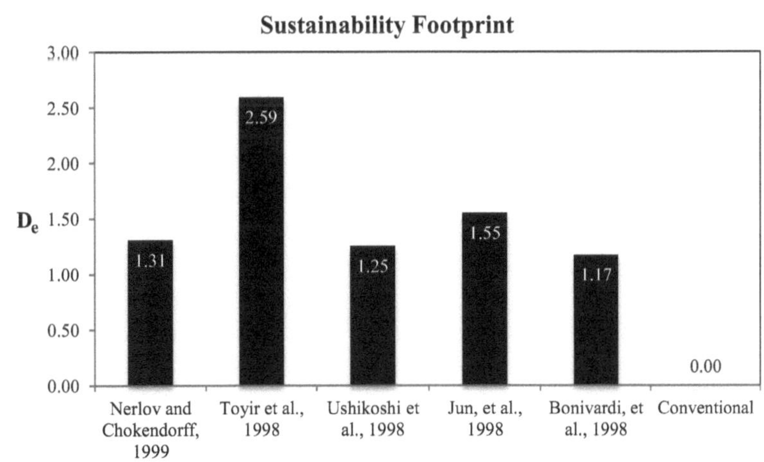

FIGURE 12.2

Sustainability footprint of different process.

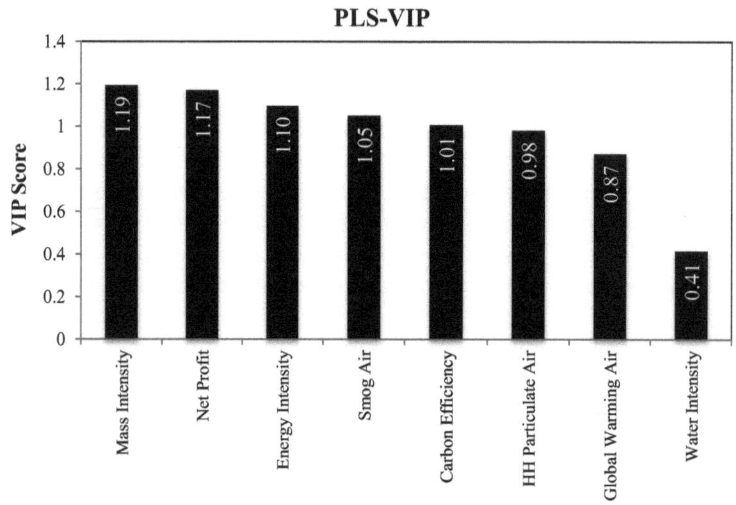

FIGURE 12.3

Importance of different indicators in defining the sustainability footprint (D_e) as obtained from PLS-VIP.

Having found the most sustainable process option and the rank of the processes in terms of sustainability, we proceeded to determining the relative importance of the indicators in influencing the sustainability outcome by the PLS-VIP method. The resultant VIP scores, as shown in Figure 12.3, show that the variance in the mass intensity has maximum influence. This is followed by net profit and so on. From PLS-VIP we can also see that water intensity has the least influence on D_e. If we take 1 as the cutoff then five out of the eight indicators make significant contribution to relative sustainability. Thus Mass Intensity, Net Profit, Energy Intensity, Smog Air, and Carbon Efficiency are the most important indicators for decision making toward sustainable methanol production.

12.5 CONCLUSIONS AND FUTURE WORK

To quantitatively assess the relative sustainability of chemical processes, one has to consider multiple indicators that can be looked upon as sensors of the three domains of sustainability: environmental, societal, and economic. This work uses the sustainability footprint method that is based on the computation of the Euclidean distance D_e to examine six competing processes for synthesis of methanol at the plant scale. This method requires having the indicator data for all the processes for making this comparison possible. The extant technique of process synthesis optimizes processes with respect to objectives, which can be in the simplest case, profit maximization. Integration can also be multiobjective. However, not many multiobjective sustainability-oriented process synthesis methods have been attempted, and they are controversial even when attempted using the triple bottom line theme (Sengupta and Pike, 2012). In the real world, when one includes the relevant indicators from the three domains of sustainability, the number of indicators with respective to which the process synthesis and optimization is to be done, can be large. Developing the ability to do so is the future direction of sustainability analyses.

As a step in this direction, we considered six methanol manufacturing processes and subjected them to both sustainability footprint analysis using multiple indicators and to standard process synthesis with a mixed integer programming with profit as the sole objective function. The results of the analysis showed that process synthesis through superstructure formulation of the competing processes chose the most desirable process different from the choice by the sustainability footprint method. This is understandable as inclusion of the additional indicators changed the overall objective of the exercise from one with profit maximization to one for balancing all interests. In this analysis we used the weighting factor of 1.0 for all indicators. If one were able to choose the indicator weighting differently, it might make a difference in the outcome of the choice of the most sustainable process.

REFERENCES

AIChE, 2014. Sustainability Index (Online). Available: www.aiche.org/ifs/resources/sustainability-index (accessed 14.04.14).

Bare, J.C., 2002. TRACI: the tool for the reduction and assessment of chemical and other environmental impacts. J. Ind. Ecol. 6, 49–78.

Bare, J.C., 2013. Tool for the Reduction and Assessment of Chemical and Other Environmental Impacts (TRACI) (Online). Available: http://www.epa.gov/nrmrl/std/traci/traci.html (accessed 07.02.13).

Bonivardi, A.L., Chiavassa, D.L., Baltanás, M.A., 1998. Promoting effect of calcium addition to Pd/SiO_2 catalysts in CO_2 hydrogenation to methanol. In: Inui, T., M. a. K. I. S. Y., Yamaguchi, T. (Eds.), Studies in Surface Science and Catalysis. Elsevier.

Brandi, H., Daroda, R.J., Olinto, A.C., 2014. The use of the Canberra metrics to aggregate metrics to sustainability. Clean Technol. Environ. Policy 16, 911–920.

Brundtland, G., 1987. Report of the 1987 World Commission on Environment and Development: Our Common Future. Oxford University Press, Oxford.

Chong, I.-G., Jun, C.-H., 2005. Performance of some variable selection methods when multicollinearity is present. Chemom. Intell. Lab. Syst. 78, 103–112.

Cinar, A., Palazoglu, A., Kayihan, F., 2004. Chemical Process Performance Evaluation. CRC press.

Conway, G.R., Barbier, E.B., 2013. After the Green Revolution: Sustainable Agriculture for Development. Routledge.

Čuček, L., Klemeš, J.J., Kravanja, Z., 2012. A review of footprint analysis tools for monitoring impacts on sustainability. J. Cleaner Prod. 34, 9–20.

Douglas, J.M., 1985. A hierarchical decision procedure for process synthesis. AIChE J. 31, 353–362.

Esty, D.C., Levy, M.A., Kim, C.H., De Sherbinin, A., Srebotnjak, T., Mara, V., 2008. Environmental Performance Index. Yale Center for Environmental Law and Policy, 382, New Haven.
Freudenberg, M., 2003. Composite Indicators of Country Performance: A Critical Assessment. OECD Publishing.
Graedel, T.E., Allenby, B.R., 2010. Industrial Ecology and Sustainable Engineering. Prentice Hall.
Grossmann, I., Caballero, J., Yeomans, H., 1999. Mathematical programming approaches to the synthesis of chemical process systems. Korean J. Chem. Eng. 16, 407−426.
Hertwig, T.A., 2004. RE: IMC Phosphate.
Indala, S., 2004. Development and Integration of New Processes Consuming Carbon Dioxide in Multi-Plant Chemical Production Complexes. Citeseer.
Jun, K.-W., Jung, M.-H., Rama Rao, K.S., Choi, M.-J., Lee, K.-W., 1998. Effective conversion of CO2 to methanol and dimethyl ether over hybrid catalysts. In: Inui, T., M. a. K. I. S. Y., Yamaguchi, T. (Eds.), Studies in Surface Science and Catalysis. Elsevier.
Krajnc, D., Glavič, P., 2005. How to compare companies on relevant dimensions of sustainability. Ecol. Econ. 55, 551−563.
Lc&Ppl, 1998. Louisiana Chemical & Petroleum Products List. Office of Policy and Research, Louisiana Department of Economic Development, Baton Rouge, LA.
Martens, H., 1989. Multivariate Calibration. John Wiley & Sons.
Mukherjee, R., Sengupta, D., Sikdar, S.K., 2013. Parsimonious use of indicators for evaluating sustainability systems with multivariate statistical analyses. Clean Technol. Environ. Policy 15, 699−706.
Nerlov, J., Chorkendorff, I., 1999. Methanol synthesis from CO_2, CO, and H_2 over Cu(100) and Ni/Cu(100). Journal of Catal. 181, 271−279.
Olinto, A., 2014. Vector space theory of sustainability assessment of industrial processes. Clean Technol. Environ. 16 (8), 1815−1820.
Saling, P., Kicherer, A., Dittrich-Krämer, B., Wittlinger, R., Zombik, W., Schmidt, I., Schrott, W., Schmidt, S., 2002. Eco-efficiency analysis by basf: the method. Int. J. Life Cycle Assess. 7, 203−218.
Sengupta, D., Pike, R.W., 2012. Chemicals from biomass: Integrating bioprocesses into chemical production complexes for sustainable development. CRC Press.
Sengupta, D., Mukherjee, R., Sikdar, S.K., 2014. Environmental sustainability of countries using the UN MDG indicators by multivariate statistical methods. Environ. Prog. Sustain. Energy 34 (1), 198−206.
Sikdar, S.K., 2003. Sustainable development and sustainability metrics. AIChE J. 49 (8), 1928−1932.
Sikdar, S., 2009. On aggregating multiple indicators into a single metric for sustainability. Clean Technol. Environ. Policy 11, 157−161.
Sikdar, S., Sengupta, D., Harten, P., 2012. More on aggregating multiple indicators into a single index for sustainability analyses. Clean Technol. Environ. Policy 14, 765−773.
Singh, R.K., Murty, H.R., Gupta, S.K., Dikshit, A.K., 2007. Development of composite sustainability performance index for steel industry. Ecol. Indic. 7, 565−588.
Tokos, H., Pintarič, Z., Krajnc, D., 2012. An integrated sustainability performance assessment and benchmarking of breweries. Clean Technol. Environ. Policy 14, 173−193.
Toyir, J., Saito, M., Yamauchi, I., Luo, S., Wu, J., Takahara, I., Takeuchi, M., 1998. Development of high performance Raney copper-based catalysts for methanol synthesis from CO_2 and H_2. In: Inui, T., M. a. K. I. S. Y., Yamaguchi, T. (Eds.), Studies in Surface Science and Catalysis. Elsevier.

Turton, R., Bailie, R.C., Whiting, W.B., Shaeiwitz, J.A., Bhattacharyya, D., 2012. Analysis, Synthesis and Design of Chemical Processes. Pearson Education, Inc., Ann Arbor, MI.

Ushikoshi, K., Moria, K., Watanabe, T., Takeuchi, M., Saito, M., 1998. A 50 kg/day class test plant for methanol synthesis from CO_2 and H_2. In: Inui, T., M. a. K. I. S. Y., Yamaguchi, T. (Eds.), Studies in Surface Science and Catalysis. Elsevier.

Wells, M., 1999. Handbook of Petrochemicals and Processes. Ashgate Brookfield.

Zhou, L., Tokos, H., Krajnc, D., Yang, Y., 2012. Sustainability performance evaluation in industry by composite sustainability index. Clean Technol. Environ. Policy 14, 789–803.

PART

Sustainable supply chains

III

CHAPTER

Design and Planning of Sustainable Supply Chains

13

Bruna Mota*, Ana Carvalho*, Maria I. Gomes§, Ana P. Barbosa-Povoa*,[1]

**Centro de Estudos de Gestao, IST, Universidade de Lisboa, Lisboa, Portugal*

§Centro de Matematica e Aplicacoes, FCT, Universidade Nova de Lisboa, Caparica, Portugal

[1]Corresponding author: E-mail: apovoa@tecnico.ulisboa.pt

13.1 INTRODUCTION

Scarcity of resources and population growth are contributing to an unsustainable situation. Sustainable development, defined by Brundtland (1987) as the "development that meets the needs of the present without compromising the ability of future generations to meet their own needs," gains more importance as we evolve in time. However, since the 2000s the issue is related not only to economic and environmental aspects but also to social aspects (Elkington, 2004).

In this context, supply chains that are nowadays a key system in any organization as they interact globally and involve a large set of resources, need to incorporate sustainability concerns in their design and operations. Supply chain design on its own is a complex problem. A supply chain needs to be efficient, flexible, and resilient and ensure fast delivery at minimum cost in an uncertain environment. Different products need to flow between different entities, using different transportation modes, under different transportation conditions. It involves decisions from the strategic to the operational level, which are related to several stakeholders presenting conflicting needs, and in a global market with different cultures and legislation constraints. Decision support tools for supply chain design have been developed to assist decision makers in this type of decision making (Melo et al., 2009). Sustainability adds even more complexity to the supply chain design. Its complexity and subjectivity and the lack of consensual and clear sustainability measures have hampered research in the field of sustainable supply chain management (Burgess et al., 2006). Sustainable supply chain modeling represents a clear research gap with a lack of models that can simultaneously account for economic, environmental, and social sustainability concerns (Seuring, 2013).

The objective of this chapter is to present an overview of the current state of the art on the topic of sustainable supply chain design and planning, and to present a generic decision support tool developed by Mota et al. (2014) for the design and planning of supply chains that incorporates the three pillars of sustainability, as a

Computer Aided Chemical Engineering, Volume 36. ISSN 1570-7946. http://dx.doi.org/10.1016/B978-0-444-63472-6.00013-6

contribution toward filling the identified research gap. The model evaluates the economic performance of the supply chain, measured through the involved costs, the environmental performance through life cycle assessment (LCA) methodologies, and the social performance through a developed socioeconomic indicator that prefers the location of entities in less developed regions. The augmented ε-constraint method is applied to obtain efficient solutions among the three sustainability objectives. In this chapter two different LCA methodologies are implemented, ReCiPe and Product Environmental Footprint (PEF), and their impact on the final network design is compared. The model is applied to a real case study of a Portuguese food distributor. Important managerial insights are taken from this case study that can be used by both companies and governmental institutions when analyzing potential sustainability measures for sustainable supply chain design and planning.

This chapter is organized as follows. In the next section an overview of the current literature on sustainable supply chain design and planning is presented, with a description of the environmental and social indicators used or possible to use in supply chain design modeling. In Section 13.3 the methodology, based on the generic framework developed by Mota et al. (2014), is presented, with the problem definition and the model formulation detailed in Sections 13.3.1 and 13.3.2, respectively. A real case study is presented in Section 13.4 where the impact of individually considering the three sustainability objectives is analyzed (Section 13.4.1), the efficient frontier considering the three sustainability objectives simultaneously is drawn using the augmented ε-constraint method (Section 13.4.2), and the effect of using different environmental impact assessment methodologies is studied (Section 13.4.3). The results are discussed in Section 13.4.4. Conclusions and future work are presented in Section 13.5.

13.2 LITERATURE REVIEW

Sustainable supply chain design and planning has been gaining more attention since companies are being pressured by both customers and governments to adopt increasingly sustainable strategies. In the literature, sustainability has been identified as one of the driving forces affecting supply chains (Barbosa-Povoa, 2014). However, this field is still considered to be in its infancy since the number of publications has only recently started to grow. This is mainly due to the high complexity involved in this multilevel, multiscale, multiplayer problem with unclear and fuzzy boundaries and measures.

Published works on the topic of sustainable supply chain modeling have addressed the issue in a number of different ways. The environmental pillar of sustainability has been addressed most in the literature. Some authors have studied the impact of existing legislation, such as the emissions trading scheme, on supply chain design and planning (Benjaafar et al., 2013; Chaabane et al., 2012). Optimization models considering closed-loop supply chains (CLSCs) have been gaining momentum, since these models are required to assist company decision makers that are obliged to assume responsibility for the entire life cycle of their products. This aspect is now common practice in several countries. Fleischmann et al. (2001) first introduced this issue into facility location decision models and several other works have followed (Cardoso et al., 2013; Guide and Van Wassenhove, 2002; Mota et al., 2014; Salema et al., 2010). Reviews on the closed-loop literature are given by Guide and Wassenhove (2009) focusing on profitable value recovery, Ilgin and Gupta (2010) where environmentally conscious manufacturing and product recovery (ECMPRO) literature is reviewed, Dekker et al. (2012) where the contribution of operations research to green logistics is highlighted, and Stindt and Sahamie (2014) where CLSC planning in the process industry is analyzed. Building on the previous works, authors started introducing environmental indicators as objective functions or constraints in the CLSC models. LCA methodologies are becoming quite common in this type of work since they allow both retrospective and prospective assessment of the environmental impact of a certain product or process, taking into account its entire life cycle (Commission, 2010). Many different LCA methods are publicly available (Carvalho et al., 2014) and some works have been published comparing their application and results (Pizzol et al., 2011; Renou et al., 2008). Regarding its application to supply chain design models, some authors have attempted to include this aspect as described by Seuring (2013). These works apply several environmental impact methodologies, namely, the environment index (Frota Neto et al., 2008), Eco-indicator 99 (Duque et al., 2010; Guillén-Gosálbez and Grossmann, 2009; Mele et al., 2011; Pinto Varela et al., 2011; Santibañez-Aguilar et al., 2011), IMPACT 2002+ (Bojarski et al., 2009), and ReCiPe (Mota et al., 2014). Eco-indicator 99 has in fact been the one mostly used in the literature. However, ReCiPe (Goedkoop et al., 2009), a follow-up of Eco-indicator 99 combined with CML 2002 (Bare et al., 2006), was according to the European Commission (Commission, 2011) the most accurate and developed among the previously available. However, the European Commission has released a new methodology for environmental impact assessment called PEF, which intends to achieve standardization among the preexistent life cycle approaches (Commission, 2013). Several studies applying different LCA methodologies have been conducted previously to PEF. Therefore, decisions were taken based on the previous LCA results. Now, it is important to verify if the incorporation of PEF will influence the previous decision-making processes.

Evaluating social sustainability is where the most significant research gap is found (Brandenburg et al., 2014; Seuring and Müller, 2008). It is described by the Global Reporting Initiative (GRI) as the dimension of sustainability that "concerns the impacts the organization has on the social systems within which it operates" (GRI, 2013). This is a fairly complex system by itself and developing quantitative measures to evaluate it is quite controversial (Zhao et al., 2012). Several authors have used criteria based on the Sustainability Reporting Guidelines (GRI, 2013) as described by Jørgensen et al. (2008) and Hutchins and Sutherland (2008). However, most measures are identified as being subjective and qualitative.

Moreover, most are based on past occurrences or are designed to evaluate the social performance of the supply chain at the operational level (e.g., Chee Tahir and Darton, 2010; Labuschagne and Brent, 2008; Labuschagne et al., 2005). At a strategic level, none of the measures currently available have been designed to be used in a supply chain context (Hassini et al., 2012). The exceptions are the work of You et al. (2012), which evaluates the full-time equivalent jobs created and, more recently, the work of Mota et al. (2014). In this latter work the authors develop a socioeconomic indicator to be used both by companies and governmental institutions in evaluating the impact of relocalization of facilities on the economic and environmental performance of the supply chain in a multiobjective optimization model.

13.3 METHODOLOGY

13.3.1 PROBLEM DEFINITION

The problem described in this chapter aims to understand how supply chain design decisions are affected by conflicting sustainability objectives as well as different sustainability measures. In short, given

- a possible superstructure for the location of the supply chain entities,
- the investment costs,
- the travel time between each pair of interacting network agents,
- the maximum and minimum flow capacities,
- the costs with human resources,
- the environmental impact factor of each facility (per square meter) for each impact category,
- the environmental impact factor per unit of transportation (per kilometer) for each impact category,
- the regional factor associated to each facility,
- customer's demand volume,
- the unit transportation cost between each pair of interacting network agents for each type of product,

the goal is to determine the network structure and flow amounts that minimize cost, minimize environmental impact, and maximize social benefit in a solution of compromise.

13.3.2 MODEL FORMULATION

The generic modeling framework proposed by Mota et al. (2014) is used but it is tailored to better fit the case study to which it is going to be applied. On one side instead of having a four-echelon structure, which included factories, warehouses, customers, and recovery centers, the model is simplified to account for only two levels: warehouses and customers. On the other side it is further extended to include distribution centers where the warehouses are installed, with the possibility of having different areas and different characteristics according to the type of product.

The supply chain is represented by a graph $G = (V,A)$ with a set of nodes (V) and a set of arcs (A). The model is described below.

13.3.2.1 Sets

Consider the indices:

$i, j, k \in V$ as entities,

$m, \overline{m} \in M$ as products.

The index i is also used interchangeably to describe regions where the entities are located.

13.3.2.1.1 Entities

Each level of the supply chain is defined by just one kind of entity (warehouse, customer), thus node set V is divided into the following subsets:

I_w possible locations for warehouses, $i \in I_w \subseteq V$

I_c locations of customers, $i \in I_c \subseteq V$

13.3.2.1.2 Time

Consider t as index for time units. Let set T be defined as $T = \{t_1, t_2, \ldots, t_h\}$.

13.3.2.2 Parameters

c_i^f	fixed cost of warehouse in location $i \in I_w$
c_i^{uf}	fixed cost of distribution center in location i
c_i^w	cost of warehouse in location i per square meter
c_{mij}^t	transportation cost, per kilometer, per product, from entity i to entity j, $(i,j) \in A$
d_{ij}	distance between entity i and entity j, in kilometers, $(i,j) \in A$
d_{mit}^c	product m demand for entity i for period t, $i \in I_c$
b_{ij}^f	maximum flow capacity of flow $(i,j) \in A$
I_{ac}	environmental impact for activity a (e.g., transport, installation of facilities) in impact category c
s_i	area of entity i, $i \in I_w$
s_m^w	area per unit of product m, $m \in M$
α_i	number of fixed workers in warehouse located in region i, $i \in I_w$
β_i	number of workers per square meter in warehouse located in region i, $i \in I_w$
γ_i	number of workers in distribution center located in region i, $i \in I_w$
c_i^α	cost per fixed worker (e.g., administrative and management positions) in warehouse located in region i, $i \in I_w$
c_i^β	cost per worker (e.g., picking and shipping positions) in warehouse located in region i, $i \in I_w$
c_i^γ	cost per worker in distribution center located in region i, $i \in I_w$
μ_i	regional factor in entity i, $i \in I_w$

13.3.2.3 Variables

13.3.2.3.1 Continuous variables

X_{mijt} amount of product m served by entity i to entity j, at time t
A^M_{mi} area of the warehouse for product m located in i
A_{mit} area of the warehouse for product m located in i necessary at time t

13.3.2.3.2 Binary variables

$Y_{im} = 1$ if warehouse i with product m is opened/served, 0 otherwise
$K_i = 1$ if distribution center i is opened/served, 0 otherwise

13.3.2.4 Constraints

The following set of constraints describes the problem in study.

Distribution:

$$\sum_{i \in I_w} X_{mijt} = d^P_{mit}, \quad m \in M, j \in I_c, t \in T \tag{13.1}$$

Equation (13.1) sets a constraint regarding the demand, where each customer has a demand value defined for each period (d^P_{mit}), which needs to be entirely satisfied within that same period.

Flow constraints:

$$X_{mijt} \le b^f_{ij} Y_{im}, \quad m \in M, i \in I_w, j \in I_c, t \in T, n \in N \tag{13.2}$$

Constraint (2) limits the occurrence of flows between only opened facilities, stating also that the flow between entities i and j need to be small or equal to the maximum flow capacity (b^f_{ij}) allowed between the entities.

Warehouse and distribution center constraints:

$$A_{mit} = \sum_{j \in I_c} s^W_m X_{mijt}, \quad i \in I_w, m \in M, t \in T \tag{13.3}$$

$$A^M_{mi} \ge A_{mit}, i \in I_w, m \in M, t \in T \tag{13.4}$$

$$Y_{im} \le K_i, \quad i \in I_w, m \in M \tag{13.5}$$

Constraint (3) states that the area of each warehouse in each time period needs to equal the flow that passes through it times the necessary area per unit of product (s^W_m). Constraint (4) determines the minimum warehouse area necessary to be installed. Each warehouse is then designed to be large enough to hold all flows that pass through at time t. Inventory is not modeled. Constraint (5) determines that a distribution center is opened at a given location whenever a warehouse is installed at that same location.

$$X_{mijt}, A^M_{mi}, A_{mit} \in \mathbb{R}^+_0; \quad Y_{im}, K_i \in \{0, 1\} \tag{13.6}$$

Finally, constraint (6) presents the variables domains.

13.3.2.5 Objectives

Three types of objectives as mentioned above are considered: economic, environmental, and social.

13.3.2.5.1 Economic assessment

The objective function that takes into account the economic performance of the company is shown in Eqn (13.7). The first term concerns the fixed costs of each warehouse $\left(c_i^f\right)$ controlled by the binary variable Y_i, which equals 1 when a warehouse is opened in location i. The second term concerns the fixed costs of opening a distribution center. The third term relates to the variable costs of the warehouses, according to the area that is installed (A_i) and the cost per square meter $\left(c_i^w\right)$. The fourth term relates to distribution given by the transportation cost per kilometer, per product unit, from entity i to entity j $\left(c_{mij}^t\right)$, times the distance between each pair of entities (d_{ij}) and the flow of each product between entities (X_{mijt}). The fifth and sixth terms concern the costs with human resources that result from opening a given entity. They include the cost of fixed workers per warehouse ($c_i^\alpha \alpha_i$, e.g., administrative and management positions), the cost of workers whose number varies with the installed area ($c_i^\beta \beta_i$, e.g., picking and shipping positions), and the cost of workers per distribution center ($c_i^\gamma \gamma_i$, e.g., positions in day-care and canteen facilities).

$$\begin{aligned} min\ Cost = & \sum_{i\in I_w}\sum_{m\in M} c_i^f Y_{im} + \sum_{i\in I_w} c_i^{df} K_i + \sum_{i\in I_w} c_i^w A_{mi}^M + \sum_{m\in M}\sum_{(i,j)\in A}\sum_{t\in T} c_{mij}^t d_{ij} X_{mijt} \\ & + \sum_{i\in I_w}\sum_{m\in M}\left(c_i^\alpha \alpha_i Y_{im} + c_i^\beta \beta_i A_{mi}^M\right) + \sum_{i\in I_w} c_i^\gamma \gamma_i K_i \end{aligned} \tag{13.7}$$

13.3.2.5.2 Environmental impact assessment

The environmental impact is determined using ReCiPe 2008 (Goedkoop et al., 2009) and PEF (Commission, 2013). The supply chain as a system is used as a functional unit, since this is the system that we wish to compare.

The life cycle inventory of each transportation mode and facility is retrieved from the Ecoinvent database (assessed through the software SimaPro 7.3.2). From this results an inventory list l (e.g., pollutants, resources depleted) and the corresponding quantities (q_l), which are used to determine the environmental impact (I_{ac}) of each activity a (transport and installation of facilities) on impact category c, according to Eqn (13.8). The characterization factors (C_{acl}) are the ones from the ReCiPe 2008 and the PEF methodologies.

$$I_{ac} = \sum_l C_{acl} q_l \tag{13.8}$$

The resulting environmental impacts are used as input data (parameters) to the mathematical model. An overall environmental impact of transport (g_c) and installation of facilities (e_c) is performed for each impact category, according to Eqns (13.9) and (13.10), respectively.

$$g_c = \sum_{m \in M} \sum_{(i,j) \in A} \sum_{t \in T} I_{ac} d_{ij} X_{mijt} \tag{13.9}$$

$$e_c = \sum_{i \in I} I_{ic} A^M_{mi} \tag{13.10}$$

Then these values are aggregated into a single score (*NI*) using the normalization and/or weighting factors (η_c) of the ReCiPe 2008 methodology, as shown in Eqn (13.11). This single score acts as the model's objective function that is to be minimized.

$$min\ NI = \sum_{c} \eta_c (g_c + e_c) \tag{13.11}$$

13.3.2.5.3 Social assessment

The social objective is given by Eqn (13.12),

$$SA = \sum_{i \in I_w} \sum_{m \in M} \mu_i \left(\alpha_i Y_{im} + \beta_i A^M_{mi} \right) + \sum_{i \in I_w} \mu_i \gamma_i K_i \tag{13.12}$$

where α_i is the number of jobs created at warehouse i, β_i is the number of jobs created at warehouse i per square meter, and γ_i is the number of jobs created at distribution center located in region i. μ_i represents a regional factor, which can assume different values according to the intended purpose of the study. Unemployment rate, population density, and income distribution are examples of possible regional factors. The goal is to prefer job creation in regions with given socioeconomic needs. Using this model one can understand how this objective would change the structure of the company and affect its economic and environmental performance. Then proper measures could be introduced, by the company or by governmental institutions, to better contribute to overall supply chain sustainability.

13.3.2.6 Multiobjective approach

Given the conflicting objectives, a multiobjective approach is introduced in the model so that compromise solutions can be found. Among the available methods, we adopted the augmented ε-constraint method (Mavrotas, 2009), since it offers simplicity and enables decision makers to measure the impact of environmental and social constraints on the economic performance of the company. The major drawback of the conventional ε-constraint method relates with weak-efficient solutions. In order to avoid such solutions, augmented ε-constraint method differs from the conventional one in two main points: (1) a lexicographic optimization is performed for every objective function to create the payoff table with only Pareto optimal solutions and (2) the objective function constraints are transformed to equalities by explicitly incorporating slack variables, which are then used as a second term in the objective function, forcing the production of only efficient solutions. For more details on the implementation of the method in a similar problem see Mota et al. (2014).

13.4 CASE STUDY

In this case study the developed model is applied to a Portuguese food retail company. The company receives the products from the suppliers and operates mainly two supply systems: stock and just-in-time (JIT). Nonperishable and frozen products are processed using the stock system, allowing large-quantity orders to optimize price and transport. Perishable products (fruits and vegetables, and fresh products such as fish) and some nonperishable products are handled using the JIT system. Given the different handling and storage conditions of these types of products, five different types of warehouses are necessary: one for nonperishable products, one for frozen products, one for fruits and vegetables, one for fresh products, and one for nonperishable products handled using the JIT system. To centralize product supply the company is organized in distribution centers where the referred warehouses are located, as shown in Figure 13.1. The distribution centers include support facilities and services such as day-care centers, canteens, human resources, maintenance, and cleaning.

The company has committed itself to promote sustainable production, distribution, and consumption practices. In the search to obtain a competitive advantage the company is pursuing a sustainable image and hence has decided to change its strategy to include sustainability concerns, not only environmental but also social. An optimization of the current supply chain has been defined as an important step together with other sustainability practices being implemented at operational levels such as actions that promote energy and water savings, as well as waste management.

The model described in Section 13.4 has been applied to the presented case study and it models the supply chain from the distribution centers to the stores. Regarding

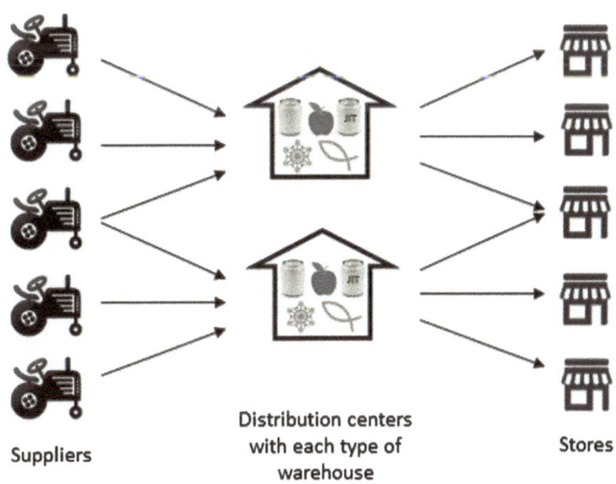

FIGURE 13.1

Supply chain structure.

the social indicator, the regional factor (μ_i) selected is population density to study the impact of potential economic incentives in relocating facilities to less populated regions. A social benefit would arise from the creation of jobs in these less populated regions, not only through direct job creation but also through secondary job creation that would potentially arise from the delocalization of families to these areas, contributing to the development of the region. Therefore, μ_i, see (12), is given by Eqn (13.13).

$$\mu_i = \frac{pd_i}{pd_{Total}} \tag{13.13}$$

where pd_{Total} represents the population density of continental Portugal and pd_i the population density of region i. The higher the regional factor, the higher the population density. Hence, the objective function described in Eqn (13.12) is to be minimized, so as to maximize the obtained social benefit.

To make sure that all the locations chosen have the necessary geographical and physical factors (e.g., infrastructures, appropriate topography) to receive distribution centers, a preliminary selection was made and 18 possible locations were selected. As requested by the company a maximum of four distribution centers is also imposed.

Several analyses will be performed, namely, on the effect of the different sustainability indicators on the obtained network, described in Section 13.4.1, on the solutions obtained when considering the three sustainability indicators simultaneously, in Section 13.4.2, and on the effect of considering different environmental impact assessment methodologies, described in Section 13.4.3.

13.4.1 SUPPLY CHAIN NETWORK RESULTS FOR THE THREE DIFFERENT OBJECTIVES

Figure 13.2 shows the network obtained for each of the three considered scenarios:

- Scenario S1: obtained when minimizing cost;
- Scenario S2: obtained when minimizing environmental impact, measured with ReCiPe;
- Scenario S3: obtained when maximizing social benefit (minimizing *SA* in Eqn (13.12)).

When minimizing cost, scenario S1, the model chooses to open four distribution centers: one in Maia, one in Viseu, one in Vila Franca de Xira, and one in Faro. Only the one in Vila Franca de Xira contains warehouses for all five types of products. The distribution centers in Faro and Viseu are exclusively dedicated to fruits and vegetables. The distribution center in Maia distributes all but frozen products.

When minimizing environmental impact the model chooses to open the exact same four distribution centers as in scenario S1. However, the environmental impact is measured per square meter, which means that there is no difference in terms of environmental impact between installing a large warehouse and installing several small warehouses, as long as the total area is the same. This translates in scenario S2 where all distribution centers have all five types of warehouses, with smaller

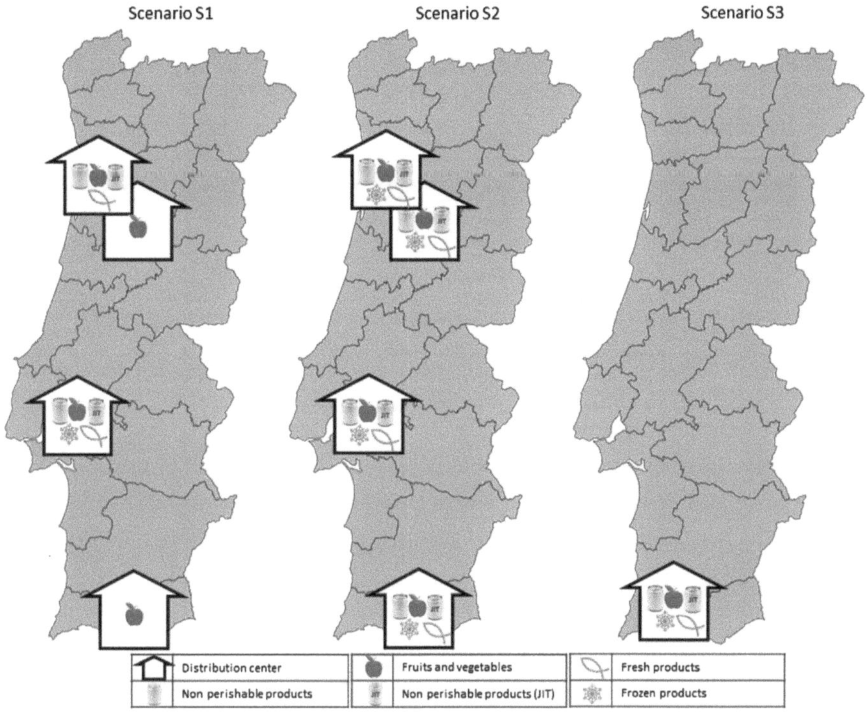

FIGURE 13.2

Network configurations obtained with the three different objective functions.

areas than the minimum cost solution, which reduces the transportation environmental impact.

To achieve the highest possible social benefit, in scenario S3, the model selects only one distribution center, the one located in the less populated region (Silves), where all the workers are employed (a total of more than 14,000). In this way, the company would contribute to solving the socioeconomic problem of low population density in the region that most needs it, from the available ones.

As one can see from Figure 13.3, the main contributor to the total cost is transportation, followed by human resources, warehouses fixed costs, distribution center fixed costs, and warehouse variable costs. A compromise between the transportation and the warehouse fixed cost is obtained to achieve the minimum cost solution, S1. In scenario S2 one can see that the transportation cost slightly decreases, in order to reduce the environmental impact of such activity, as evidenced in Figure 13.4. This is obtained by increasing the number of warehouses in each distribution center, although maintaining the same total area of the minimum cost solution, which leads to a 1.4% cost increase and a 0.5% environmental impact decrease from the minimum cost solution. Regarding scenario S3, for social benefit maximization, the

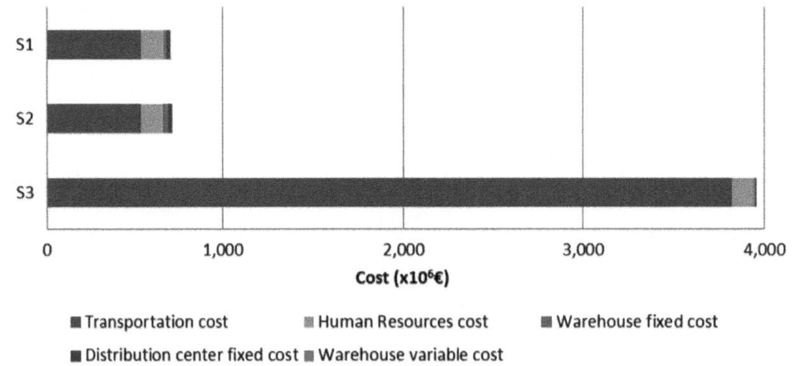

FIGURE 13.3

Cost distribution of each scenario.

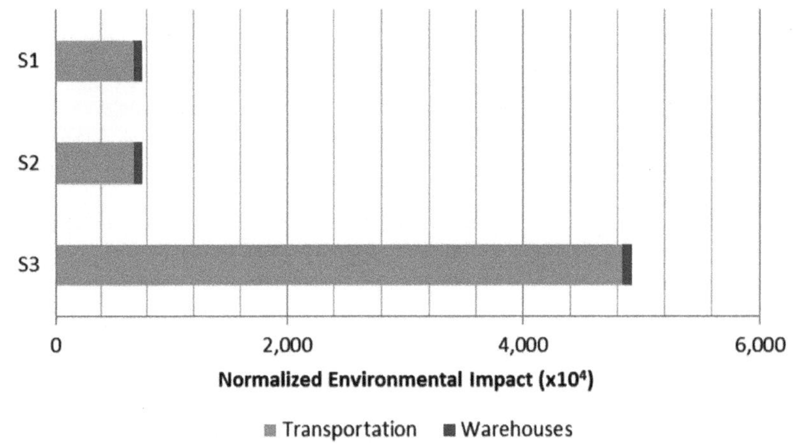

FIGURE 13.4

Environmental impact distribution of each scenario.

cost increases tremendously (by 465% from the minimum cost solution), as expected. The same is reflected in Figure 13.4 with the increase in the environmental impact (by 549% from the minimum environmental impact solution). This is related to the increase in transportation. Since the only distribution center is located in Silves, a strong reinforcement of transportation is necessary to guaranty that all demand is satisfied. As evident this is a nonimplementable solution as it becomes very costly and is not effective in logistic terms.

It is then important to consider the trade-off of the three scenarios to provide the necessary data to the decision makers. This is done in the next section.

13.4.2 MULTIOBJECTIVE APPROACH

A multiobjective approach (the augmented ε-constraint method) was applied to establish the efficient frontier between the economic, environmental (measured through ReCiPe), and social objectives. The results obtained are shown in Figure 13.5.

Solutions A and B represent the solutions with less environmental impact (scenario S2) and lower cost (scenario S1), respectively. Solution Q represents the one with the higher social benefit (lower value obtained with the minimization of the social assessment objective function, equivalent to scenario S3). The ideal solution that would combine the lowest cost (from solution B) with the lowest environmental impact (from solution A) with the highest social benefit (from solution Q) is also represented for comparison (see Figure 13.5 on the left).

One of the first conclusions that can be taken from the two dimensional representations is the fact that the economic and environmental objectives vary linearly, with the exception of solution A where a lower environmental impact is obtained at a higher cost than solution B. This linearity results from the fact that both costs and environmental impact vary linearly with the area of the installed entities and with the distance traveled. However, as mentioned before, the environmental impact methodology used does not take into account the difference between building several smaller warehouses and building only a large one. Hence, these results might in reality be different and more linear than presented. Regarding the social performance, an improvement is only attained with a deterioration of the economic and environmental performances.

Analyzing the three-dimensional representation, it is possible to identify three distinctive groups of solutions. The first group comprehends solutions A through E, which appear to tend to the ideal solution. In fact, a significant increase in the social benefit (decrease in the social assessment objective function) can be obtained with a relatively small compromise of the economic and environmental performances. This is achieved with different combinations of the four distribution centers. By simply replacing one distribution center located in a densely populated region with one that is located in a less populated region the social benefit increases significantly. This is possible without a meaningful increase in the transportation activities, which is the highest contributor to the increase in environmental impact and cost, as shown in Section 13.4.1. In the second group (solutions F, G, and H) the model locates two to three of the four distribution centers in the less populated regions to achieve a higher social benefit. The increase in the transportation is balanced with the increase in the number of warehouses, to achieve the lowest possible cost and environmental impact. In the third group (solutions I through Q), a significant deterioration of the economic and environmental performances is necessary for small improvements in the social performance of the supply chain. This is due to the reduction in the number of distribution centers that occurs from solution I (with three distribution centers) through Q (with only one distribution center), so as to locate the maximum number of workers in the less populated regions. Even though the fixed costs with warehouses and entities decrease, the transportation costs increase significantly.

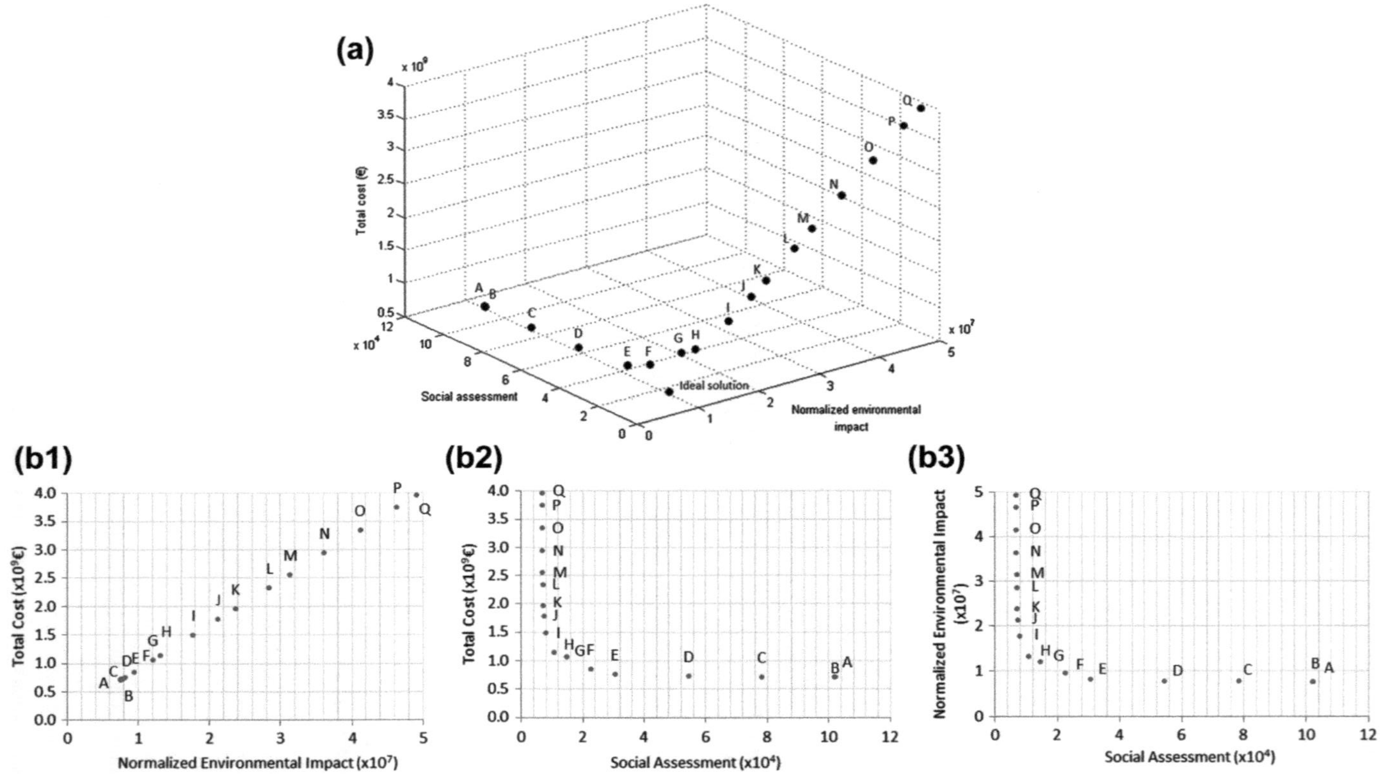

FIGURE 13.5

(a) Multiobjective approach for the minimization of cost with social benefit and environmental impact limitation measured through ReCiPe. (b1) Total cost versus environmental impact. (b2) Total cost versus social assessment. (b3) Environmental impact versus social assessment.

Solutions F, G, and H are the closest to a so-called solution of compromise. Solution F is obtained with a deterioration of the economic and environmental performances of 20% and 25%, respectively. Solution H already requires a decline of 62% and 74%, respectively, for the same objectives, which is nonimplementable at the company level. However, none of the solutions is better than the other. With the conflicting objectives, selecting one of them will always mean attributing a higher weight to one of the objectives over the others.

13.4.3 EFFECT OF DIFFERENT ENVIRONMENTAL INDICATORS ON THE NETWORK—PEF VERSUS RECIPE

For scenario S2, alternatively to ReCiPe, PEF was also used to measure the environmental impact of the supply chain. The goal is to understand if using different environmental measures would affect the supply chain design. Results show that the exact same network is obtained with both methodologies, which minimizes transportation.

In fact, transport is the activity that mostly contributes for the total environmental impact of this supply chain. ReCiPe attributes 90% of the total environmental impact (normalized) to transport, with the remaining 10% coming from the warehouses. PEF attributes 99.9% to transport and the remaining 0.01% to warehouses (results not shown).

Even though the exact same configuration is obtained when minimizing environmental impact measured through ReCiPe and PEF, Figure 13.6 shows that the midpoint categories impact distribution is quite different. In ReCiPe, natural land transformation (NLT), from the Land Use midpoint impact category, is considered to be the main source of concern in this supply chain. However, through PEF, freshwater ecotoxicity (FET) is the highest contributor to the total supply chain environmental impact, followed by water resource depletion (WRD) and human toxicity, cancer effects (HTCE).

13.4.4 DISCUSSION

The work presented in this chapter contributes to fill the identified research gaps and important conclusions can be retrieved from the obtained results.

In the case study presented by Mota et al. (2014) production was the activity that mostly contributed to the environmental impact of the supply chain, followed by transportation and then the installation of entities. In the case study presented in this chapter, since it is a supply chain that does not have production activities, transportation is the one that mostly contributes to its environmental impact. This result helps the decision makers in prioritizing environmental reduction decisions according to the type of supply chain being studied.

Based on the LCA methodologies used, the environmental impact of facilities installation is measured per square meter. This leads to the assumption that the environmental impact of installing several smaller warehouses was the same as installing less but larger ones, if maintaining the same total area. This is most likely not the

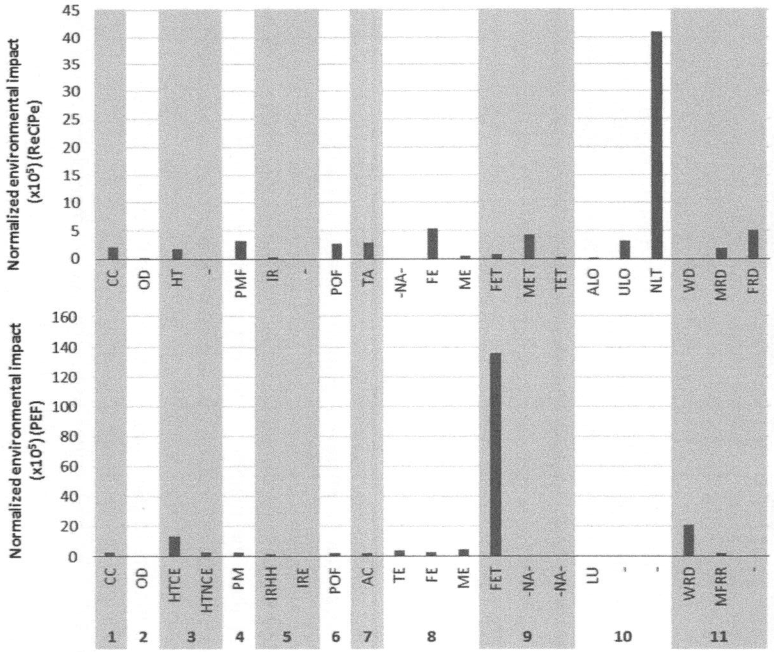

	Midpoint impact category	In ReCiPe	In PEF
1	Climate change	Climate change (CC)	Climate change (CC)
2	Ozone depletion	Ozone depletion (OD)	Ozone depletion (OD)
3	Human toxicity	Human toxicity (HT)	Human toxicity, cancer effects (HTCE) Human toxicity, non-cancer effects (HTNCE)
4	Particulate matter/ Respiratory inorganics	Particulate matter formation (PMF)	Particulate matter (PM)
5	Ionizing radiation	Ionizing radiation (IR)	Ionizing radiation, human health (IRHH) Ionizing radiation, ecosystems (IRE)
6	Photochemical ozone formation	Photochemical ozone formation (POF)	Photochemical ozone formation (POF)
7	Acidification	Terrestrial acidification (TA)	Acidification (AC)
8	Eutrophication	Freshwater eutrophication (FE) Marine water eutrophication (ME)	Terrestrial eutrophication (TE) Freshwater eutrophication (FE) Marine water eutrophication (ME)
9	Ecotoxicity	Freshwater ecotoxicity (FET) Marine water ecotoxicity (MET) Terrestrial ecotoxicity (TET)	Freshwater ecotoxicity (FET)
10	Land use	Agricultural Land Occupation (ALO) Urban Land Occupation (ULO) Natural Land Transformation (NLT)	Land use (LU)
11	Resource depletion	Water depletion (WD) Metal depletion (MRD) Fossil depletion (FRD)	Resource depletion, water (WRD) Resource depletion, mineral, fossil (MFRR)

FIGURE 13.6

Midpoint impact categories comparison measured through ReCiPe and PEF. The categories of each methodology are grouped into 11 categories according to the legend on the right. "-NA-": nonavailable method; "-": nonexistent category.

case in reality and hence some of the results obtained might not be the most real, and this aspect needs further study.

Both LCA methodologies returned the exact same network configuration, which means that PEF and ReCiPe led to the same strategic decision. However, from the midpoint categories analysis it was concluded that the impact categories of main concern are different when measured with the different LCA methodologies. This fact might imply that decisions undertaken in supply chains covering production facilities will be influenced by the method selected. Therefore, it is possible to verify that PEF can bring some changes in the previous assessments conducted in previous methods using ReCiPe. Future work should include studies of different supply chain networks so that stronger conclusions can be taken regarding PEF's influence and exploring the comparison among both methods.

13.5 CONCLUSIONS AND FUTURE WORK

This work explores the design and planning of sustainable supply chains through an optimization approach. A multiobjective mathematical programming model is proposed where the three dimensions of sustainability are incorporated. The model extends the work presented by Mota et al. (2014) to include distribution centers where the warehouses are installed with possible different areas, and of different types depending on the products handled. Cost minimization is used to optimize the supply chain economic performance. On the environmental performance, two different LCA methodologies are implemented, ReCiPe and PEF, and their impact on the supply chains' structure compared. The social performance is evaluated through an indicator that prefers the location of facilities in less developed regions. The model is applied to a real case study of a Portuguese food distributor.

From the work presented it is clear that sustainability in the supply chain is still a topic that is just recently being the focus of researchers and companies. Even though its importance is well established not enough research has been done to develop clear, accurate, and proven guidelines to support the implementation of sustainable practices. Having fuzzy sustainability measures and so many different methodologies available is in fact more confusing for the decision makers who do not want to risk the economic viability of their companies by pursuing strategies based on the uneven grounds of sustainability. This risk is shown in this work when analyzing the results of using two different LCA methodologies. Even though the obtained network is the same, the distribution of its impact is entirely different. Hence, the decisions and investments that could have been made based on a previous analysis using ReCiPe in terms of prioritization of environmental impact reduction measures would now be redundant. Another issue is the inclusion of LCA methodologies in supply chain design and planning models, which must be carefully done. As shown in this work, the fact that the environmental impact is measured per given unit (e.g., square meters), as is the case of the environmental impact of the installation of entities, might lead to inaccurate results. These results then need to be further validated and future work should be performed on accurately modeling this issue.

The social sustainability pillar is the most intricate one to evaluate. One cannot quantitatively and accurately measure the impact of a given relocation decision on the society, given the amount of interconnected factors involved. However, decisions are still made both by governments and municipalities in developing measures that attempt to reduce given socioeconomic problems such as unemployment and desertification. These same measures are analyzed by company decision makers when designing their strategy. The social indicator presented in this work relates to the socioeconomic problem of the uneven distribution of people in the country, which has an impact on the region's economic development. It can be used to study potential economic incentives both by the governments or municipalities that attribute such incentives and by companies that take them into consideration when designing their strategy. Future work is, however, needed in carefully understanding this intricate, complex system so that such type of indicators can be improved and supported by social studies.

Adding to the complexity of the problem studied is the fact that the three conflicting pillars of sustainability need to be considered simultaneously. This justifies the need for a multiobjective approach that facilitates the visualization of the trade-offs involved and the impact of environmental and social concerns on the financial performance of the company. However, no solution is better than the others and the decision maker still needs to select one of the solutions, which means attributing a higher weight to one objective over the other. This will always imply a certain degree of subjectivity. Methodologies exist to deal with such type of decisions, as is the example of multicriteria decisions analysis methodologies, which should be further explored. Even though not returning the exact best solution to the decision maker, since there is not one due to the subjectivity and complexity of the problem, this and other works of the same nature contribute to understanding the impact of sustainability considerations in supply chain design and planning. Additional case studies and models that account for other environmental and social pressures should be performed so as to generate enough understanding of the problem, facilitating the creation of sustainable legislation, so that it does not overcompromise the financial performance of companies.

REFERENCES

Barbosa-Povoa, A.P., 2014. Process supply chains management—where are we? Where to go next? Process Energy. Syst. Eng. 2, 23.

Bare, J., Gloria, T., Norris, G., 2006. Development of the method and US normalization database for life cycle impact assessment and sustainability metrics. Environ. Sci. Technol. 40, 5108–5115.

Benjaafar, S., Li, Y., Daskin, M., 2013. Carbon footprint and the management of supply chains: insights from simple models. In: Automation Science and Engineering, IEEE Transactions on, vol. 10, pp. 99–116.

Bojarski, A.D., Laínez, J.M., Espuña, A., Puigjaner, L., 2009. Incorporating environmental impacts and regulations in a holistic supply chains modeling: an LCA approach. Comput. Chem. Eng. 33, 1747–1759.

Brandenburg, M., Govindan, K., Sarkis, J., Seuring, S., 2014. Quantitative models for sustainable supply chain management: developments and directions. Eur. J. Oper. Res. 233, 299–312.

Brundtland, G.H., 1987. World Commission on Environment and Development. Our Common Future, pp. 8–9.

Burgess, K., Singh, P.J., Koroglu, R., 2006. Supply chain management: a structured literature review and implications for future research. Int. J. Oper. Prod. Manage. 26, 703–729.

Cardoso, S.R., Barbosa-Póvoa, A.P.F., Relvas, S., 2013. Design and planning of supply chains with integration of reverse logistics activities under demand uncertainty. Eur. J. Oper. Res. 226, 436–451.

Carvalho, A., Mimoso, A.F., Mendes, A.N., Matos, H.A., 2014. From a literature review to a framework for environmental process impact assessment index. J. Cleaner Prod. 64, 36–62.

Chaabane, A., Ramudhin, A., Paquet, M., 2012. Design of sustainable supply chains under the emission trading scheme. Int. J. Prod. Econ. 135, 37–49.

Chee Tahir, A., Darton, R.C., 2010. The process analysis method of selecting indicators to quantify the sustainability performance of a business operation. J. Cleaner Prod. 18, 1598–1607.

European Commission, Joint Research Centre - Institute for Environment and Sustainability, 2010. International Reference Life Cycle Data System (ILCD) Handbook. General Guide for Life Cycle Assessment - Detailed Guidance. Luxembourg.

European Commission, Joint Research Centre - Institute for Environment and Sustainability, 2011. International Reference Life Cycle Data System (ILCD) Handbook. Recommendations for Life Cycle Impact Assessment in the European Context.

European Commission, 2013. Commission Recommendation of 9 April 2013 on the Use of Common Methods to Measure and Communicate the Life Cycle Environmental Performance of Products and Organisations. In (vol. L124, 56).

Dekker, R., Bloemhof, J., Mallidis, I., 2012. Operations research for green logistics—an overview of aspects, issues, contributions and challenges. Eur. J. Oper. Res. 219, 671–679.

Duque, J., Barbosa-Póvoa, A.P.F.D., Novais, A.Q., 2010. Design and planning of sustainable industrial networks: application to a recovery network of residual products. Ind. Eng. Chem. Res. 49, 4230–4248.

Elkington, J., 2004. Enter the Triple Bottom Line. The Triple Bottom Line: Does It All Add up, pp. 1–16.

Fleischmann, M., Beullens, P., Bloemhof-Ruwaard, J.M., Wassenhove, L.N., 2001. The impact of product recovery on logistics network design. Prod. Oper. Manage. 10, 156–173.

Frota Neto, J.Q., Bloemhof-Ruwaard, J.M., van Nunen, J.A.E.E., van Heck, E., 2008. Designing and evaluating sustainable logistics networks. Int. J. Prod. Econ. 111, 195–208.

Goedkoop, M., Heijungs, R., Huijbregts, M., Schryver, A.D., Struijs, J., van Zelm, R., 2009. ReCiPe 2008 - A Life Cycle Impact Assessment Method Which Comprises Harmonised Category Indicators at the Midpoint and the Endpoint Level. Ministry of VROM, The Hague.

GRI, 2013. Global Reporting Initiative - G4 Sustainability Reporting Guidelines. Amsterdam.

Guide Jr., V.D.R., Van Wassenhove, L.N., 2002. Closed-loop supply chains. In: Quantitative Approaches to Distribution Logistics and Supply Chain Management. Springer, pp. 47–60.

Guide Jr., V.D.R., Van Wassenhove, L.N., 2009. OR FORUM - the evolution of closed-loop supply chain research. Oper. Res. 57, 10–18.

Guillen-Gosalbez, G., Grossmann, I.E., 2009. Optimal design and planning of sustainable chemical supply chains under uncertainty. AIChE J. 55, 99–121.

Hassini, E., Surti, C., Searcy, C., 2012. A literature review and a case study of sustainable supply chains with a focus on metrics. Int. J. Prod. Econ. 140, 69–82.

Hutchins, M.J., Sutherland, J.W., 2008. An exploration of measures of social sustainability and their application to supply chain decisions. J. Cleaner Prod. 16, 1688–1698.

Ilgin, M.A., Gupta, S.M., 2010. Environmentally conscious manufacturing and product recovery (ECMPRO): a review of the state of the art. J. Environ. Manage. 91, 563–591.

Jørgensen, A., Le Bocq, A., Nazarkina, L., Hauschild, M., 2008. Methodologies for social life cycle assessment. Int. J. Life Cycle Assess. 13, 96–103.

Labuschagne, C., Brent, A.C., 2008. An industry perspective of the completeness and relevance of a social assessment framework for project and technology management in the manufacturing sector. J. Cleaner Prod. 16, 253–262.

Labuschagne, C., Brent, A.C., van Erck, R.P.G., 2005. Assessing the sustainability performances of industries. J. Cleaner Prod. 13, 373–385.

Mavrotas, G., 2009. Effective implementation of the ε-constraint method in multi-objective mathematical programming problems. Appl. Math. Comput. 213, 455–465.

Mele, F.D., Kostin, A.M., Guillén-Gosálbez, G., Jiménez, L., 2011. Multiobjective model for more sustainable fuel Supply chains. A case study of the sugar cane industry in Argentina. Ind. Eng. Chem. Res. 50, 4939–4958.

Melo, M.T., Nickel, S., Saldanha-da-Gama, F., 2009. Facility location and supply chain management – a review. Eur. J. Oper. Res. 196, 401–412.

Mota, B., Gomes, M.I., Carvalho, A., Barbosa-Povoa, A.P., 2014. Towards supply chain sustainability: economic, environmental and social design and planning. J. Cleaner Prod. http://dx.doi.org/10.1016/j.jclepro.2014.07.052.

Pinto-Varela, T., Barbosa-Póvoa, A.P.F., Novais, A.Q., 2011. Bi-objective optimization approach to the design and planning of supply chains: economic versus environmental performances. Comput. Chem. Eng. 35, 1454–1468.

Pizzol, M., Christensen, P., Schmidt, J., Thomsen, M., 2011. Impacts of "metals" on human health: a comparison between nine different methodologies for life cycle impact assessment (LCIA). J. Cleaner Prod. 19, 646–656.

Renou, S., Thomas, J., Aoustin, E., Pons, M., 2008. Influence of impact assessment methods in wastewater treatment LCA. J. Cleaner Prod. 16, 1098–1105.

Salema, M.I.G., Barbosa-Povoa, A.P., Novais, A.Q., 2010. Simultaneous design and planning of supply chains with reverse flows: a generic modelling framework. Eur. J. Oper. Res. 203, 336–349.

Santibañez-Aguilar, J.E., González-Campos, J.B., Ponce-Ortega, J.M.A., Serna-González, M., El-Halwagi, M.M., 2011. Optimal planning of a biomass conversion system considering economic and environmental aspects. Ind. Eng. Chem. Res. 50, 8558–8570.

Seuring, S., 2013. A review of modeling approaches for sustainable supply chain management. Decis. Support Syst. 54, 1513–1520.

Seuring, S., Müller, M., 2008. From a literature review to a conceptual framework for sustainable supply chain management. J. Cleaner Prod. 16, 1699–1710.

Stindt, D., Sahamie, R., 2014. Review of research on closed loop supply chain management in the process industry. Flexible Serv. Manuf. J. 26, 268–293.

You, F., Tao, L., Graziano, D.J., Snyder, S.W., 2012. Optimal design of sustainable cellulosic biofuel supply chains: multiobjective optimization coupled with life cycle assessment and input–output analysis. AIChE J. 58, 1157–1180.

Zhao, Z.-Y., Zhao, X.-J., Davidson, K., Zuo, J., 2012. A corporate social responsibility indicator system for construction enterprises. J. Cleaner Prod. 29, 277–289.

CHAPTER

14 Operability Analysis of Process Supply Chains—Toward the Development of a Sustainable Bioeconomy

Christopher L. E. Swartz[1], Han Wang, Richard Mastragostino
Department of Chemical Engineering, McMaster University, Hamilton, ON, Canada
[1]Corresponding author: E-mail: swartzc@mcmaster.ca

14.1 INTRODUCTION

Enterprises have to operate in an environment of increasing globalization, market competition, raw material and utility costs, tightening environmental constraints and fluctuating customer demands. Enterprises are moreover subjected to a high level of uncertainty, ranging from raw material availability and quality, utility costs, market volatility, and process uncertainty. These factors require appropriate paradigms for the design and operation of process plants and their associated supply chains to ensure that the overall enterprise is profitable and sustainable.

The trend toward a growing bioeconomy is particularly relevant within this context. Biomass feedstocks typically exhibit a high degree of variability; processes are often new and not as well characterized as more traditional technologies, and markets, particularly for specialty products, may be volatile. These factors, for example, underpin the recognition that in order for the forest products industry to succeed in a new environment of integrated forest biorefineries, a paradigm shift from manufacturing-centric to margins-centric operation is required in order to exploit flexibility within the supply chain and process operations to maximize profit (Mansoornejad et al., 2011).

Operability is defined in Grossmann and Morari (1984) as the ability of a process to operate satisfactorily at conditions away from the nominal operating conditions. They furthermore identify flexibility and resiliency as two major components of operability. Flexibility reflects the range of uncertain parameters over which feasible steady-state operation can be maintained, while resiliency refers to the ability of a process to reject disturbances rapidly and smoothly. The development of quantitative techniques for assessing plant operability and for incorporating operability into plant

Computer Aided Chemical Engineering, Volume 36. ISSN 1570-7946. http://dx.doi.org/10.1016/B978-0-444-63472-6.00014-8

design procedures has been the subject of numerous research studies over the past three decades (Yuan et al., 2012). Since resiliency was originally introduced as a property of the plant independent of the control system, we use here instead the term dynamic operability to reflect the dynamic performance of the plant under more general conditions defined by the context in which the analysis is applied (Swartz, 1996; Baker and Swartz, 2004).

Operability is clearly a property that is desirable as well in process supply chains in order to operate satisfactorily in the face of uncertainty and demand variation. Associated properties within the operations research and management science literature are flexibility and agility. There is no consensus on precise definitions of these terms—indeed, Sethi and Sethi (1990) in their review of manufacturing flexibility remark that at least 50 different terms for various types of flexibility had appeared in the manufacturing literature. However, they bear strong similarities to the notions of flexibility and dynamic operability described above in the context of process plants.

The focus of this article is on operability analysis of process supply chains. In the sequel, we provide a more detailed literature review on operability—in the context of both supply chain systems and process plants. We thereafter present a mathematical framework for supply chain operability assessment, drawing on concepts developed in the context of process plants. A multipart case study is presented to illustrate the application of the formulations presented. Finally, conclusions are drawn and avenues for promising research directions identified.

14.2 SUPPLY CHAIN OPERABILITY

14.2.1 QUALITATIVE CONTRIBUTIONS

Notions of operability appear in the supply chain literature largely through the concepts of agility and flexibility. A number of these contributions present conceptual analyses describing elements of supply chain operability. Backx et al. (1998) identify agile, lean, responsive, smooth, and predictive as key objectives of an operable supply chain and point out the need to cast the assessment in terms of a multicriterion framework with ample consideration of the significance (weight) attributed to each objective. Furthermore, the authors examine the volatility surrounding the plant as an opportunity to be exploited, rather than a disturbance to be compensated for, and propose a strategy for intentionally dynamic operating policies at the manufacturing level. Christopher (2000) suggests that the key to survival in today's challenging business environment is through agility, where significant routes to agility are improved market sensitivity (i.e., responding to real demand), leveraging information, collaboration between buyers and suppliers, and more efficient coordination and management of the supply chain. Christopher and Towill (2001) propose an integrated logistics model for enabling agility in a cost-effective manner by combining "lean" and "agile" concepts through hybrid strategies and suggest a three-level framework for summarizing the attributes that embody an agile supply chain. Naylor et al. (1999) discuss the lean and agile manufacturing paradigms within the supply chain context, as well as their integrated application to different segments of a supply chain. Fisher (1997) explains that the nature of the product, whether it is functional (commodity) or innovative (specialty), should dictate the best supply chain strategy: physically efficient if the objective is to reduce cost or market-responsive if the objective is to improve product availability.

Agility is the ability of a complex process to quickly adjust operations to respond/adapt to change, and a clear attribute of an agile supply chain is flexibility. Synonymous with agility is "quick response" (responsiveness), which enables supply chains to rapidly meet customer demands for shorter lead times. Agility can be achieved through a number of avenues, including end product and raw material availability forecasting, buffering against uncertainty through excess inventory (safety stock), manufacturing flexibility, product customization postponement, and lead time reduction. Forecasting through time-series models can be useful for reacting ahead of time; however, uncertainties in demand can be significant in a continuously changing market environment (You and Grossmann, 2008). As pointed out in Fisher (1997), many companies simply need to accept high uncertainty in demand, particularly for noncommodity type products. Excess inventory or safety stock is a method used to hedge against uncertainty, which allows the supply chain to "catch-up" when a change in demand occurs. Holding an overly conservative quantity of safety stock is not cost-effective, therefore effective strategies for computing safety stock levels have been a vital area of research (e.g., Shen et al., 2003; Jung et al., 2004; You and Grossmann, 2010). Matson and McFarlane (1999) propose a set of audit tools to investigate the responsiveness of existing production operations within the steel industry. Holweg (2005) outlines three key dimensions, i.e., product, process, and volume, which influence responsiveness in a supply chain. A conceptual model is proposed to allow enterprises to align their supply chain strategy to the identified factors in order to find a suitable balance between responsiveness and supply chain efficiency. Furthermore, as pointed out in Holweg (2005), responsiveness is strongly influenced by manufacturing flexibility, yet the link has not been rigorously explored. Lin et al. (2006) propose a methodology for quantifying the agility of a supply chain through a multicriteria decision-making strategy and representing uncertain agility-enablers using fuzzy logic.

Manufacturing flexibility is defined as the ability of a manufacturing system to meet objectives under different conditions by the efficient reconfiguration of available manufacturing resources, where a comprehensive review is given in Sethi and Sethi (1990). Manufacturing flexibility, among other considerations, is highly dependent on process plant configuration, i.e., whether process units operate in parallel or series, and whether a process unit is dedicated, or can accommodate multiple production tasks.

Product differentiation is a related concept that involves the practice of delaying the point of differentiation, with inventory carried in a generic form, and final production/customization completed once true demand is known. There are several advantages of postponement, including reduction of finished product inventory that is normally more expensive, improving responsiveness, and enabling the ability for "mass customization" (Fisher, 1997; Christopher, 2000). Delay reduction is an obvious means for which to improve agility in the supply chain. As argued in Fisher (1997), companies should invest aggressively in ways to reduce delays when distributing products with high-demand volatility.

14.2.2 QUANTITATIVE APPROACHES

There are a number of notable contributions that present quantitative metrics and/or mathematical programming formulations and typically include the application of metrics such as on-time deliveries, back orders, manufacturing lead time, and fill rate. A comprehensive review of supply chain performance measures that evaluate notions of operability are given in Beamon (1999). Performance measures are classified into three categories: resources, output, and flexibility. The first is associated with economics and resource usage; the second includes customer responsiveness and the quality and quantity of product produced; while flexibility includes measures that reflect the ability to accommodate volume and schedule fluctuations from suppliers, manufacturers, and customers.

Graves and Tomlin (2003) consider process flexibility in supply chains, where the focus is on the linkage between products and processes in a supply chain configuration. They propose a flexibility measure based on the capacity available to subsets of products that increases with an increase in the number of product-plant links. Garavelli (2003) considers in addition logistics flexibility to account for logistics of procurement and distribution. He does not propose flexibility metrics, but evaluates different supply chain configurations through simulation studies using performance metrics such as work in progress and lead time. Tang and Tomlin (2008) discuss supply chain flexibility strategies in response to a number of risks: supply cost risk, supply commitment risk, process risk and demand risk. Examples using probability models are provided to quantify the impact of flexibility strategies on mitigating the effects of the risks. The authors remark on the need for further analysis to consider combinations of flexibility strategies. Mansoornejad et al. (2013), in their analysis of biorefinery supply chain performance, define flexibility and robustness metrics. The former is designed to capture volume flexibility, while the latter measures economic robustness to market conditions.

A few recent studies have addressed responsiveness of process supply chains within mathematical programming frameworks. A quantitative approach to evaluate responsiveness is through the lead time, defined as the time for the supply chain to respond to customer demands (You and Grossmann, 2007). This definition associates a long lead time with less responsiveness. Lead time is dependent on a number of factors, which can include material procurement, shipping and manufacturing delays, as well as inventory of raw material and product in the supply chain. For instance, if sufficient inventory exists in the distribution site, then demand can be fulfilled immediately and lead time is effectively the delivery delay between the distribution center and the customer. However, if product stock out occurs at the distribution center, the lead time will depend on the delays and the inventory of product and raw material upstream. In the worst case scenario when there is no stock of raw materials or intermediate products, the lead time is effectively the delay between a supplier and customer.

You and Grossmann (2007) propose worst case lead time, i.e., length of the longest time path of chemical flow from a supplier to a customer by way of several manufacturing sites assuming zero inventory as a quantitative measure of supply chain responsiveness. Furthermore, the authors present a framework for strategic decision making, accounting for the selection of suppliers, manufacturing sites and process technology, and operational scheduling, under consideration of both profitability and responsiveness criteria through a bi-criterion optimization approach. You and Grossmann (2008) extend the prior study by considering demand uncertainty and a probabilistic model for stock-out that is dependent on safety stock levels at distribution centers, treated as optimization variables. The expected lead time is the quantitative measure of responsiveness evaluated, linked to the stock-out probability, and a similar definition of worst case lead time as in previous work. The expression for the production delay in multiproduct plants differs depending on a priori specification of large or small changes in a product demand, to account for a different production sequencing policy under high or low demand. A hierarchical algorithm is proposed based on the decoupling of strategic and operational decisions, to efficiently solve the large-scale mixed-integer nonlinear programming problem. You and Grossmann (2011) further extend this work to address the inventory management of a single-product, multi-echelon inventory system, where the maximum (worst case) guaranteed service time (GST) of the last echelon of the supply chain is proposed as a measure of responsiveness. The GST and processing time of the echelon and the GST of the directly preceding echelon contributes to the net lead time, denoting the time span over which safety stock coverage against demand variation is required. The net lead time is related to the safety stock level through a specified safety stock factor. Mastragostino and Swartz (2014) develop a computational framework for dynamic operability analysis of a multiproduct, multi-echelon inventory supply chain using a dynamic optimization formulation, which directly links production decisions, transportation quantities, inventory levels, demand, and lead time. Recourse actions for different demand scenarios are considered through a two-stage stochastic programming formulation, and a bi-criterion optimization problem is formulated for generating the Pareto frontier between an economic and responsiveness criterion. Two case studies within the biorefinery paradigm rigorously explore the applicability of the framework.

14.3 PROCESS PLANT OPERABILITY

Plant operability refers broadly to the ability of a plant to operate satisfactorily. While there are many facets to operability (Marlin, 2010), we focus our attention here on the notions of steady-state flexibility and dynamic operability as defined earlier. The next subsections provide brief reviews of key concepts and advances in these areas.

14.3.1 FLEXIBILITY

Grossmann and co-workers (Halemane and Grossmann, 1983; Swaney and Grossmann, 1985; Grossmann and Floudas, 1987) associated flexibility with the range of parameter variation over which feasible steady-state operation can be maintained. Consider a steady-state process model defined by

$$h(d,x,u,\theta) = 0 \tag{14.1}$$

with design and/or operating constraints

$$g(d,x,u,\theta) \leq 0, \tag{14.2}$$

where h is a vector of steady-state model equations such as material balances, energy balances, and equilibrium relationships, d is a vector of design parameters, x is a vector of model states of the same dimension as h, u is a vector of control variables corresponding to degrees of freedom that can be adjusted during plant operation, and θ is a vector of uncertain parameters.

A scalar-valued flexibility index, F, is defined through the solution of the following problem:

$$\begin{aligned} &F = \max\ \delta \\ &\text{s.t.: } \max_{\theta \in T} \min_u \max_{j \in J} f_j(d,u,\theta) \leq 0 \\ &T(\delta) = \left\{\theta \middle| \theta^N - \delta\Delta\theta^- \leq \theta \leq \theta^N + \delta\Delta\theta^+\right\} \end{aligned} \tag{14.3}$$

In the above, f_j represents the inequality constraint (14.2) in which the state variables, x, have been eliminated via the equality constraints (14.1). Problem (14.3) determines the largest hyperrectangle in the uncertain parameter space centered at θ^N and scaled by δ, such that feasibility can be maintained for all uncertain parameter values within it. $\Delta\theta^-$ and $\Delta\theta^+$ are parameters that define the range of uncertainty over which feasibility is desired. A value of $F = 1$ implies sufficient flexibility for feasible operation within the specified parameter range, while $F < 1$ denotes insufficient flexibility.

The solution of (3) is significantly simplified under conditions where the flexibility is limited by a vertex of the uncertain parameter set T, which is guaranteed if the feasible region is one-dimensional quasi-convex (1-DQC) in the uncertainty space (Swaney and Grossmann, 1985). Problem (14.3) may in such cases be solved by a vertex enumeration scheme. Grossmann and Floudas (1987) present a mixed-integer programming formulation based on identifying the active constraints that may be applied under conditions where the 1-DQC property does not hold.

A number of extensions and applications of this flexibility analysis framework have been proposed. A recent review of developments in this area is given in Grossmann et al. (2014).

14.3.2 DYNAMIC OPERABILITY

The recognition that the design of a plant can significantly affect its dynamic performance has led to the development of systematic methodologies for analyzing these effects and accounting for the effects of dynamics in plant design procedures. Ziegler and Nichols (1943) allude to a control performance limit independent of controller type or structure. Morari and co-workers (Morari, 1983; Holt and Morari, 1985a,b) refer to the quality of this limiting performance as resiliency and develop techniques for quantifying the performance-limiting factors of input constraints, dead time, right-half-plane transmission zeros, and plant-model mismatch through the internal model control framework.

A number of approaches have used an optimization framework for dynamic operability analysis and for incorporating dynamic performance requirements into plant design calculations (Swartz, 1996; Mohideen et al., 1996; Schweiger and Floudas, 1998). A representative optimization-based formulation takes the following form:

$$\begin{aligned}
&\min \quad \phi(x(t_f), z(t_f), u(t_f), d, \theta) \\
&\text{s.t.} \quad h(\dot{x}(t), x(t), z(t), u(t), d, \theta) = 0 \\
&g(x(t), z(t), u(t), d, \theta) \leq 0 \\
&x(t_0) = x_0
\end{aligned} \tag{14.4}$$

where h represents a differential-algebraic equation (DAE) system model, x and z are differential and algebraic state variables respectively, u is a vector of manipulated inputs, and g represents a set of operational and/or design inequality constraints. d and θ, as before, represent design and uncertain parameters.

The above formulation offers considerable flexibility in how the problem is posed. For example, for a fixed design and finite set of disturbances, (14.4) may be used to measure the set-point tracking performance through a suitably defined objective function. Alternatively, by utilizing an economic objective function and introducing plant design parameters as optimization variables, an economically optimal design that satisfies dynamic performance criteria through the specification of appropriate path constraints can be determined. Reviews on the interaction between design and associated dynamic performance can be found in van Schijndel and Pistikopoulos (2000), Sakizlis et al. (2004), and Yuan et al. (2012).

Concepts of the above formulation will be utilized in the sequel, adapted to supply chain systems. In particular, a discrete-time model will be used to capture supply chain dynamics, rather than a DAE model typically used to represent process plant dynamics.

14.4 MATHEMATICAL PROGRAMMING FRAMEWORK FOR SUPPLY CHAIN OPERABILITY ANALYSIS

A biomass products supply chain network containing multiple echelons (biomass suppliers, biorefineries, distribution centers, and customers), multiple production schemes, and multiple products, as shown in Figure 14.1, is considered in this work. A tactical-level supply chain model with modeling details adapted from Bok et al. (2000) and Mastragostino (2012) is developed to represent a discrete-time multiperiod formulation, where the horizon is divided into equal length intervals, Δt, indexed by $t \in T = \{1, \ldots, \mathcal{H}\}$, where $\mathcal{H}$ is the horizon length.

The chemicals $j \in J$ involved in this model can be categorized into three types: raw materials (J_{RM}), which cannot be produced within the supply chain and are only available through purchasing from suppliers; final products (J_{FP}), which are produced to be sold to customers; and intermediate products (J_{IP}), which are produced and then either consumed in other processes or sold to customers as final products. Raw material and intermediate products can be stored on site within the biorefineries; however, final products are stored in distribution centers. A set of multi-scheme continuous processes convert raw materials into products. There are three types of delays considered in this model. The procurement delay accounts for the time between when an order of raw materials is placed and when these raw materials arrive at the biorefinery available for processing. The production delay for each of the processes is captured by the time between the consumption of inputs and the generation of outputs of a process, which can also be interpreted as the residence time of a processing unit. The transportation delay is mainly affected by the distance between the origin and destination. All delays are assumed to be a multiple of the discretization interval, Δt, to avoid a fractional time index in all related variables, as described later.

During the time period t, an order of chemical $j \in J$, $O_{j,h,m,t,}$ is made to suppliers $h \in H_j$ from biorefinery $m \in M_j$, where H_j represents the set of suppliers supplying raw material j and M_j is the set of biorefineries involving chemical j. The raw material j is then converted into either intermediate products for further processing in biorefineries or into final products shipped to distribution centers to be distributed to satisfy demands from the market. The chemical flows of j involved in the kth scheme of the process i within the biorefinery m are denoted as $W_{j,k,i,m,t}$. $F_{j,m,v,t}$

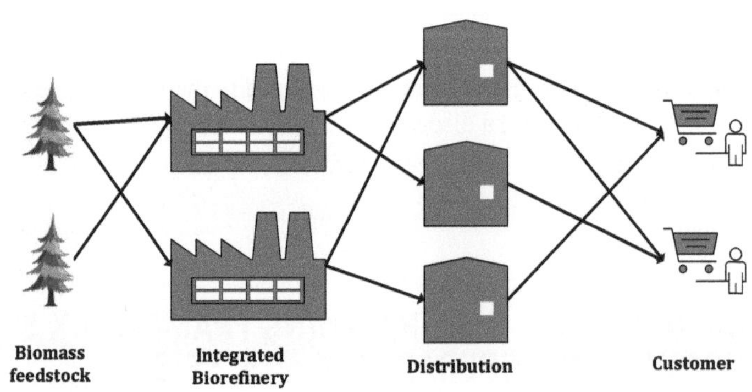

FIGURE 14.1

Biomass supply chain network configuration.

represents the direct withdrawal of final product j from the biorefinery m shipped to distribution center $v \in V_j$, where the final products are stored. A demand of $D_{j,c,t}$ of product j from customer c needs to be fulfilled by the quantity $Q_{j,c,v,t}$ distributed from distribution center v.

In the following subsections, a discrete-time dynamic model of the supply chain is presented, followed by an associated steady-state model.

14.4.1 SUPPLY CHAIN MODEL

14.4.1.1 Discrete-time model

The mass balance of raw materials and intermediate products at a biorefinery is given by Eqn (14.5). The inventory level of chemical j at the current time period at biorefinery m, $I_{j,m,t}$, is the inventory from the last time period, plus the generation by production and the incoming materials purchased from suppliers that arrived at the biorefinery, less the usage during production and the shipment away from biorefinery headed to distribution centers. $\delta^{PRC}_{j,h,m}$ represents the procurement delay between when an order is made to a supplier h and when the material arrives on site and can be processed by a processing unit at biorefinery m.

$$\begin{aligned}&\sum_{h\in H_j} O_{j,h,m,\left(t-\delta^{PRC}_{j,h,m}/\Delta t\right)} + \sum_{i\in I_m}\sum_{k\in K^P_{ji}} W_{j,k,i,m,t} - \sum_{i\in I_m}\sum_{k\in K^C_{ji}} W_{j,k,i,m,t} - \sum_{v\in V_j} F_{j,m,v,t}\\ &+ I^M_{j,m,t-1} = I^M_{j,m,t} \quad \forall j\in J_{RM}\cup J_{IP}, m\in M_j, t\in T\end{aligned} \tag{14.5}$$

The raw material for forest products is often biomass, which is considered perishable. Some studies consider the deterioration of biomass with time by including a deterioration rate in the inventory balances (Eksioğlu et al., 2009; Mastragostino and Swartz, 2014), but we omit this phenomenon in the current model of our study.

Final products are grouped by destination and shipped to various distribution centers, where inventory is kept for distribution to customers, as shown in Eqn (14.6). At distribution centers, incoming products received from biorefineries are stored and distributed to customers $c \in C_j$, who request product j. It is also noted that Eqn (14.6) is consistent with Eqn (14.5) written for final products that are not intermediate products, $j\in J_{FP}\setminus J_{IP}$, where both of the purchase and consumption terms of j are zero.

$$\sum_{v\in V_j} F_{j,m,v,t} = \sum_{i\in I_m}\sum_{k\in K^P_{ji}} W_{j,k,i,m,t} \quad \forall j\in J_{FP}\setminus J_{IP}, m\in M_j, t\in T \tag{14.6}$$

The inventory balance of final products at distribution centers is shown in Eqn (14.7), where $F_{j,m,v,t}$ represents the quantity of final product j shipped from biorefinery m to distribution center v, and $Q_{j,c,v,t}$ stands for the quantity of final products distributed to customer c from distribution center v during time period t. $\delta^{TRP}_{m,v}$ represents the transportation delay of final products shipped from biorefinery m to distribution center v. Delays for products to reach customers from distribution centers are not considered, as they fall outside the system boundary for the responsiveness analysis.

$$\sum_{m\in M_j} F_{j,m,v,\left(t-\delta_{m,v}^{TRP}/\Delta t\right)} - \sum_{c\in C_j} Q_{j,c,v,t} + I_{j,v,t-1}^{V} = I_{j,v,t}^{V} \quad \forall j\in J_{FP}, v\in V_j, t\in T \tag{14.7}$$

Equations (14.8) and (14.9) relate the chemicals involved to the main product defined in each of the schemes.

$$W_{j,k,i,m,\left(t-\delta_{k,i}^{PRD}/\Delta t\right)} = \eta_{j,k,i,m} W_{j',k,i,m,t} \quad \forall j\in J_{ki}^{C}, j'\in J_{ki}^{M}, k\in K_i, i\in I_m, m\in M_j \tag{14.8}$$

$$W_{j,k,i,m,t} = \eta_{j,k,i,m} W_{j',k,i,m,t} \quad \forall j\in J_{ki}^{P}, j'\in J_{ki}^{M}, k\in K_i, i\in I_m, m\in M_j \tag{14.9}$$

Sets J_{ki}^{C} and J_{ki}^{P} represent the chemicals that are consumed and produced respectively in scheme k of process i. J_{ki}^{M}, a subset of J_{ki}^{P}, is the set of main products produced in scheme k of process i. Equation (14.8) captures the relationship between the inputs and the main product, where $\delta_{k,i}^{PRD}$ is the production delay for scheme k of process i. Equation (14.9) relates the output chemicals to the main product in each scheme, and thus the output flows are not affected by the production delays.

The product demands may be only partially fulfilled for some time periods. The unmet demands of product j from customer c, $B_{j,c,t}$, can be either considered as lost sales or logged as back orders to be fulfilled later. Equation (14.10) captures the demand fulfillment, which in particular depicts the scenario of back order accumulation. This equation would be suitable for the scenario counting unmet demands as lost sales, if the term $B_{j,c,t-1}$ is omitted from the left side of the equation, because no previous unmet demands are accumulated throughout the time horizon.

$$D_{j,c,t} - \sum_{v\in V_j} Q_{j,c,v,t} + B_{j,c,t-1} = B_{j,c,t} \quad \forall j\in J_{FP}, c\in C_j, t\in T \tag{14.10}$$

Three types of capacity constraints are included to bound the purchase, production, and inventory quantities. All raw materials ordered for all the biorefineries from any supplier during a time period are bounded by a maximum availability as given in Eqn (14.11). The total output volume of the main product in process i at biorefinery m is constrained by the production capacity $C_{i,m}^{P}$ as in Eqn (14.12).

$$\sum_{m\in M_j} O_{j,h,m,t} \leq O_{j,h}^{max} \quad \forall j\in J_{RM}, h\in H_j, t\in T \tag{14.11}$$

$$\sum_{k\in K_i} W_{j,k,i,m,t} \leq C_{i,m}^{P} \quad \forall j\in J_{ki}^{M}, i\in I_m, m\in M_j, t\in T \tag{14.12}$$

For both raw materials and intermediate products stored at a biorefinery, Eqn (14.13) depicts the inventory capacity constraints. For final products stored at distribution centers, Eqn (14.14) captures such constraints.

$$I_{j,m,t}^{M} \leq C_{j,m}^{M} \quad \forall j\in J_{RM} \cup J_{IP}, m\in M_j, t\in T \tag{14.13}$$

$$I^V_{j,v,t} \le C^V_{j,v} \quad \forall j \in J_{FP}, v \in V_j, t \in T \tag{14.14}$$

The operational profit of successfully converting raw materials into final products in order to meet demands is described in Eqn (14.15) as the difference between revenue from product sales (as computed in Eqn (14.16)) and operating costs.

$$\text{Profit} = \text{Revenue} - \text{Total Cost} \tag{14.15}$$

$$\text{Revenue} = \sum_{t \in T} \sum_{j \in J_{FP}} \sum_{c \in C_j} \sum_{v \in V_j} P^{Sale}_{j,c} Q_{j,c,v,t} \tag{14.16}$$

The total operational costs can be further divided into the following terms:

- Procurement costs for raw materials from suppliers (Eqn (14.17))
- Inventory costs for raw materials, intermediate products, and final products (Eqn (14.18))
- Variable production costs proportional to the main product of the production processes (Eqn (14.19))
- Transportation cost of final products from biorefinery to distribution centers (Eqn (14.20))

$$C_{PRC} = \sum_{t \in T} \sum_{j \in J_{RM}} \sum_{h \in H_j} \sum_{m \in M_j} c^{PRC}_{j,h} O_{j,h,m,t} \tag{14.17}$$

$$C_{INV} = \sum_{t \in T} \sum_{j \in J_{RM} \cup J_{IP}} \sum_{m \in M_j} c^{IM}_{j,m} I^M_{j,m,t} + \sum_{t \in T} \sum_{j \in J_{FP}} \sum_{v \in V_j} c^{IV}_{j,v} I^V_{j,v,t} \tag{14.18}$$

$$C_{PRD} = \sum_{t \in T} \sum_{j \subset J^M_{ki}} \sum_{m \in M_j} \sum_{i \in I_m} \sum_{k \in K_i} c^{PRD}_i W_{j,k,i,m,t} \tag{14.19}$$

$$C_{TRP} = \sum_{t \in T} \sum_{j \in J_{FP}} \sum_{m \in M_j} \sum_{v \in V_j} c^{TRP}_{j,m,v} F_{j,m,v,t} \tag{14.20}$$

To ensure that the inventory level does not drop to unacceptably low values, a penalty term is also included in the economics formulation to penalize the negative deviation from the setpoint (defined by $E^M_{j,m,t}$ and $E^V_{j,v,t}$) as shown in Eqn (14.21). The positive deviation is minimized via the inventory cost term in Eqn (14.18), so the system will avoid storing more materials than necessary in the interest of cost savings.

$$P_{INV} = \sum_{t \in T} \sum_{j \in J_{RM} \cup J_{IP}} \sum_{m \in M_j} c^{EM}_{j,m} E^M_{j,m,t} + \sum_{t \in T} \sum_{j \in J_{FP}} \sum_{v \in V_j} c^{EV}_{j,v} E^V_{j,v,t} \tag{14.21}$$

The capital cost may be computed based on the capacity of the processing units and storage space of inventory using the following equation:

$$C_{CAP} = \sum_{m \in M} \sum_{i \in I_m} c^{CP}_{i,m} C^P_{i,m} + \sum_{j \in J_{RM} \cup J_{IP}} \sum_{m \in M_j} c^{CIM}_{j,m} C^M_{j,m} + \sum_{j \in J_{FP}} \sum_{v \in V_j} c^{CIV}_{j,v} C^V_{j,v} \tag{14.22}$$

14.4.1.2 Steady-state model

The steady-state model equations below follow from the discrete-time dynamic model, with the time index and inventories omitted.

$$\sum_{h\in H_j} O_{j,h,m} + \sum_{i\in I_m}\sum_{k\in K_{ji}^P} W_{j,k,i,m} = \sum_{i\in I_m}\sum_{k\in K_{ji}^C} W_{j,k,i,m} + \sum_{v\in V_j} F_{j,m,v} \quad \forall j\in J, m\in M_j \tag{14.23}$$

$$W_{j,k,i,m} = \eta_{j,k,i,m} W_{j',k,i,m} \quad \forall j\in J_{ki}, j'\in J_{ki}^M, k\in K_i, i\in I_m, m\in M_j \tag{14.24}$$

$$\sum_{m\in M_j} F_{j,m,v} = \sum_{c\in C_j} Q_{j,c,v} \quad \forall j\in J, v\in V_j \tag{14.25}$$

$$D_{j,c} = \sum_{v\in V_j} Q_{j,c,v} \quad \forall j\in J, c\in C_j \tag{14.26}$$

$$\sum_{m\in M_j} O_{j,h,m} \le O_{j,h}^{max} \quad \forall j\in J, h\in H_j \tag{14.27}$$

$$W_{j,k,i,m} \le C_{i,m}^P \quad \forall j\in J_{ki}^M, k\in K_i, i\in I_m, m\in M_j \tag{14.28}$$

The equations for economic criteria are presented in Eqns (14.29)–(14.32) with the difference that no inventory cost is included compared to the discrete-time model.

$$\text{Revenue} = \sum_{j\in J_{FP}}\sum_{c\in C_j}\sum_{v\in V_j} P_{j,c}^{Sale} Q_{j,c,v} \tag{14.29}$$

$$C_{PRC} = \sum_{j\in J_{RM}}\sum_{h\in H_j}\sum_{m\in M_j} c_{j,h}^{PRC} O_{j,h,m} \tag{14.30}$$

$$C_{PRD} = \sum_{j\in J_{ki}^M}\sum_{m\in M_j}\sum_{i\in I_m}\sum_{k\in K_i} c_i^{PRD} W_{j,k,i,m} \tag{14.31}$$

$$C_{TRP} = \sum_{j\in J_{FP}}\sum_{m\in M_j}\sum_{v\in V_j} c_{j,m,v}^{TRP} F_{j,m,v} \tag{14.32}$$

14.4.2 SUPPLY CHAIN FLEXIBILITY FORMULATION

14.4.2.1 Individual flexibility metric

Individual flexibility indices F_k are defined in Eqns (14.33)–(14.36) as the maximum deviation from the nominal conditions of uncertainty θ_k that can be sustained for feasible steady-state supply chain operation. The positive and negative deviation from the nominal value of the uncertain parameter is represented by $\Delta\theta_k^+$ and $\Delta\theta_k^-$, respectively. h_i and g_j are the collection of equality and inequality constraints representing the steady-state supply chain model discussed previously in Eqns (14.23)–(14.28).

$$F_k = \max \delta_k \tag{14.33}$$

subject to:

$$h_i(d, x, u, \theta_k) = 0 \quad \forall i \in C_{eq} \tag{14.34}$$

$$g_j(d, x, u, \theta_k) \leq 0 \quad \forall j \in C_{ineq} \tag{14.35}$$

$$\theta_k^N - \delta_k \Delta\theta_k^- \leq \theta_k \leq \theta_k^N + \delta_k \Delta\theta_k^+ \tag{14.36}$$

In the above, d is a representation of the supply chain design variables and x and u together comprise the supply chain model variables.

Table 14.1 lists a few examples of individual flexibility indices defined using the above formulation that are also commonly discussed within the operations research literature. These indices are defined by the corresponding uncertainty they assess.

One example of uncertainty θ_k is product demand. Under demand volatility, the demand flexibility index, F_d, can be interpreted as the maximum deviation in demand that can be satisfied by a supply chain while still maintaining feasible operation by adjusting the volume of product output. Demand flexibility is also studied quantitatively in operations research and management science literature. Son and Park (1987) quantify product demand flexibility as the ratio of total output of a manufacturing system to the inventory cost of finished products. This approach considers the inventory cost as an opportunity cost and uses demand flexibility as a responsiveness measure of a manufacturing system to match output levels with demands in the effort to reduce cost. Our proposed methodology originates from the feasibility test problem (Swaney and Grossmann, 1985) and therefore recognizes flexibility as a steady-state concept considering the range of parameter variation for which feasible operation can be maintained, rather than the transitional behavior between different steady states. As a result, our supply chain model used to illustrate various types of flexibility does not include inventory as a model variable. One of the flexibility measures often discussed in relation to demand flexibility is volume flexibility, defined as the ability of a supply chain process to operate profitably at different levels of output (Beamon, 1999). Mansoornejad et al. (2013) propose a metric of flexibility to quantify volume flexibility as the deviation from nominal production rates in dimensionless form. Gerwin (1993) views volume flexibility as an adaptive response to uncertainty in demand as it permits changes in the aggregated production level. The possession of sufficient volume flexibility can be considered as a mitigation in the face of demand uncertainty, which is equivalent to the control variables u in the proposed formulation as shown in Eqns (14.34) and (14.35).

Table 14.1 Types of Individual Flexibility Indices

Flexibility (F_k)	Uncertainty (θ_k)
Demand flexibility, $F_d = \max \delta_d$	Product demand
Supply flexibility, $F_s = \max \delta_s$	Raw material supply
Yield flexibility, $F_y = \max \delta_y$	Production yield
Market flexibility, $F_m = \max \delta_m$	Price of final product or costs

Similar to demand, another external source of uncertainty is supply uncertainty. By the definition listed in Table 14.1, the supply flexibility index, F_s, can be evaluated via the proposed formulation to capture the degree of variation in the quantity supplied that can be sustained by the feasible downstream supply chain operations. Recipe flexibility, defined by Mansoornejad et al. (2013) as the ability to utilize a set of adaptable material recipes to achieve different products and/or outputs, can be viewed as a moderation of the impacts caused by supply- or demand-related uncertainties. Variation in raw material quality, reduction of production cost by selecting cheaper recipes, and substitution due to material unavailability are the three main reasons that recipe flexibility is desired during manufacturing processes (Bertrand and Rutten, 1999). In addition, Ferrer-Nadal et al. (2008) also demonstrate that the adoption of flexible recipes has a high benefit in risk management associated with uncertain demands. Therefore, the recipe flexibility shares some connection with both demand and supply uncertainty.

The production yield is the ability of a process to produce, with a simple interpretation that a process with higher yield can generate more output with the same amount of input. Yield is an internal or endogenous source of uncertainty of a supply chain network. The yield flexibility index in Table 14.1 provides a measure of the range of yield variation for a production process while still meeting all constraints. The yield flexibility is usually related to the discussion of production or process flexibility (Sethi and Sethi, 1990).

The final entry in Table 14.1, market flexibility, F_m, is defined using the proposed formulation with an additional constraint to ensure that the profit of the entire supply chain, as defined in Eqn (14.15), does not go below zero, because a losing supply chain is not sustainable in the long run. A more conceptual definition in Sethi and Sethi (1990) refers market flexibility as the adaptability to a changing market environment. This flexibility index can be applied to all market related parameters defined in Eqns (14.29)–(14.32). For instance, the price flexibility index computed as a result of the proposed formulation can be used to further calculate the lowest sale price of final product to remain profitable; the index computed from production cost uncertainty can be used to derive the highest production cost that a supply chain can afford without financial loss. The metric of robustness presented in Mansoornejad et al. (2013) is essentially an assessment of a supply chain's ability to remain profitable. It is computed as the percentage of aggregate deviation from the base case profit for all scenarios with profits less than the base case profit.

14.4.2.2 Overall flexibility metric

The individual metrics for flexibility introduced previously provide insights on how flexible a certain aspect of a process is, while the overall flexibility of a supply chain reflects the potential of the entire network when facing multiple types of uncertainty simultaneously. An overall flexibility index, F, is defined with a similar formulation in Eqns (14.37)–(14.40), but θ here represents the collection of all uncertain parameters (i.e. $\theta = [\theta_1^T, \theta_2^T, \ldots, \theta_k^T]^T$).

$$F = \max \delta \tag{14.37}$$

subject to:

$$h_i(d, x, u, \theta) = 0 \quad \forall i \in C_{eq} \tag{14.38}$$

$$g_j(d, x, u, \theta) \leq 0 \quad \forall j \in C_{ineq} \tag{14.39}$$

$$\theta^N - \delta\Delta\theta^- \leq \theta \leq \theta^N + \delta\Delta\theta^+ \tag{14.40}$$

It is noted that both individual and overall flexibility metrics can be used for analyzing flexibility, but can also be included within a design under uncertainty formulation. Both applications are considered and illustrated in the case studies.

14.4.3 SUPPLY CHAIN RESPONSIVENESS FORMULATION

Building on the formulation presented in Mastragostino and Swartz (2014), the basis of quantitative characterization of dynamic responsiveness is to evaluate the lead time of a supply chain process in response to a step change in demand. The lead time is computed by tracking the time between when a change in demand occurs and when the new demand level is fulfilled and sustained for the rest of the optimization horizon.

Figure 14.2 conceptually illustrates the approach, where in both subfigures (a) and (b) the lead time is the delay between the moment the step change takes place, and the time when demand can be satisfied and sustained. Subfigure (a) considers the scenario when the unmet demand is logged as back orders, which need to be delivered at later times, while subfigure (b) depicts the scenario when the unmet demand is regarded as lost sales, such as products sold to a spot market, which are not needed at later time periods if not fulfilled in the current time period. The demand response has an overshoot in subfigure (a), and the additional delivery is to fulfill the back

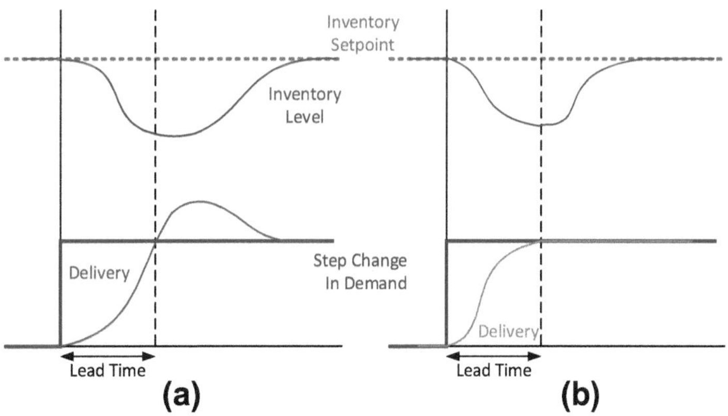

FIGURE 14.2

Illustration of lead time evaluation: (a) unmet demand logged as back order; (b) unmet demand considered as lost sales.

orders accumulated during the lead time. This is also consistent with the response of inventory levels, indicating a faster return to setpoint in subfigure (b) as the inventory starts to replenish right after the new demand is met and sustained.

A binary variable, $Y^{met}_{j,c,t}$, is included in the formulation to track the demand fulfillment.$Y^{met}_{j,c,t} = 1$ if the demand of product j requested by customer c can be served and sustained at time t and $Y^{met}_{j,c,t} = 0$ otherwise.

The inventory setpoint of chemicals, $\bar{I}$, and the inventory offset as the deviation below the setpoint, E, are introduced as a part of the formulation to minimize the inventory offset. The variables are nonnegative, so the inequalities ensure that the offsets do not count when inventory is above the setpoint as shown in the following equations.

$$E^M_{j,m,t} \geq \bar{I}^M_{j,m} - I^M_{j,m,t} \quad \forall\ j \in J_{RM} \cup J_{IP}, m \in M_j, t \in T \tag{14.41}$$

$$E^V_{j,v,t} \geq \bar{I}^V_{j,v} - I^V_{j,v,t} \quad \forall\ j \in J_{FP}, v \in V_j, t \in T \tag{14.42}$$

The following formulation allows demand to be partially satisfied at time period t through the nonnegative variable, $K^1_{j,c,t}$, and restricts the nonnegative variable $K^2_{j,c,t}$, to be zero when $Y^{met}_{j,c,t} = 0$. When $Y^{met}_{j,c,t} = 1$, $K^1_{j,c,t}$, is forced to be 0, and $K^2_{j,c,t}$ serves the demand at time period t in full.

$$\sum_{v \in V_j} Q_{j,c,v,t} = K^1_{j,c,t} + K^2_{j,c,t} \quad \forall\ j \in J_{FP}, c \in C_j, t \in T \tag{14.43}$$

$$K^1_{j,c,t} \leq D_{j,c,t}\left(1 - Y^{met}_{j,c,t}\right) \quad \forall\ j \in J_{FP}, c \in C_j, t \in T \tag{14.44}$$

$$K^2_{j,c,t} = D_{j,c,t} Y^{met}_{j,c,t} \quad \forall\ j \in J_{FP}, c \in C_j, t \in T \tag{14.45}$$

As shown in Eqns (14.44) and (14.45), the system exerts its effort to only fulfill the demand at the current time period, without accounting the accumulated back orders up to the current time period. This is consistent with treating the unmet demand as lost sales (depicted in Figure 14.2(b)) and it is appropriate to model specialty products demand with volatile patterns or spot markets in this manner. On the other hand, the above formulation can be easily adapted to the situations when the unmet demand is treated as back orders and needs to be supplied to customers by the end of the optimization horizon (depicted in Figure 14.2(a)). Equations (14.43) and (14.44) remain unchanged, but the following constraint given by Eqn (14.46) replaces Eqn (14.45).

$$D_{j,c,t} Y^{met}_{j,c,t} \leq K^2_{j,c,t} \leq \alpha Y^{met}_{j,c,t} \quad \forall j \in J_{FP}, c \in C_j, t \in T \tag{14.46}$$

This constraint allows distribution centers to distribute more products to customers than requested in a time period when $Y^{met}_{j,c,t} = 1$, and α is a large number to bound $K^2_{j,c,t}$. The additional amount distributed is to fulfill the back orders accumulated so that all back orders can be cleared by the end of the optimization horizon by including an addition constraint in the model, $B_{j,c,t=\mathcal{H}} = 0$. When $Y^{met}_{j,c,t} = 0$, the demand is not fully served, so $K^2_{j,c,t} = 0$.

Equations (14.47) and (14.48) restrict $Y^{met}_{j,c,t}$ to only change from 0 to 1, since demand fulfillment is required to be sustained over the optimization horizon. The nonnegative continuous variable $A_{j,c,t}$ takes a value of 1 to indicate when $Y^{met}_{j,c,t}$ has switched from 0 to 1.

$$Y^{met}_{j,c,t} - Y^{met}_{j,c,t-1} = A_{j,c,t} \quad \forall j \in J_{FP}, c \in C_j, t = 2\ldots\mathcal{H} \tag{14.47}$$

$$Y^{met}_{j,c,t} = A_{j,c,t} \quad \forall j \in J_{FP}, c \in C_j, t = 1 \tag{14.48}$$

If the worst case lead time of the system is expected to be smaller than the optimization horizon length, a tightening constraint shown in Eqn (14.49) can be applied. The parameter τ indicates the time period by which the demand must be met, and $\tau \le \mathcal{H}$. τ can be determined appropriately to enhance computational performance by exploiting system knowledge. $Y^{met}_{j,c,t}$ is also fixed at 1 from the period τ onward as shown in Eqn (14.50) to indicate that the satisfaction of demand is sustained.

$$\sum_{t=1}^{t=\tau} A_{j,c,t} = 1 \quad \forall j \in J_{FP}, c \in C_j \tag{14.49}$$

$$Y^{met}_{j,c,t} = 1 \quad \forall j \in J_{FP}, c \in C_j, t = \tau\ldots\mathcal{H} \tag{14.50}$$

The following equations describe how the system behaves at the beginning and the end of the time horizon. The inventory setpoints are time independent, equated to the inventory position at $t = 0$ (Eqns (14.51) and (14.52)). In addition, Eqns (14.53) and (14.54) enforce the inventory levels to return to the corresponding setpoints after the response to demand step change, at the end of the optimization horizon.

$$\bar{I}^M_{j,m} = I^M_{j,m,t=0} \quad \forall j \in J_{RM} \cup J_{IP}, m \in M_j, \tag{14.51}$$

$$\bar{I}^V_{j,v} = I^V_{j,v,t=0} \quad \forall j \in J_{FP}, v \in V_j, \tag{14.52}$$

$$\bar{I}^M_{j,m} \le I^M_{j,m,t=\mathcal{H}} \quad \forall j \in J_{RM} \cup J_{IP}, m \in M_j, \tag{14.53}$$

$$\bar{I}^V_{j,v} \le I^V_{j,v,t=\mathcal{H}} \quad \forall j \in J_{FP}, v \in V_j, \tag{14.54}$$

The lead time to completely satisfy the demand of chemical j requested by customer c is defined in Eqn (14.55) as the time elapsed before the demand is fulfilled.

$$L_{j,c} = \sum_t (t-1) A_{j,c,t} \Delta t \quad \forall j \in J_{FP}, c \in C_j \tag{14.55}$$

The system lead time (SLT) given in Eqn (14.56) is depicted as a weighed-average of lead times for all products reaching all customers via different paths, where ω_{jc} is the weight of the lead time assigned for final product j requested by customer c, and $\sum_{j \in J_{FP}} \sum_{c \in C_j} \omega_{jc} = 1$.

$$\text{SLT} = \sum_{j \in J_{FP}} \sum_{c \in C_j} \omega_{jc} L_{j,c} \tag{14.56}$$

14.4.4 METHODOLOGY FOR QUANTITATIVE SUPPLY CHAIN OPERABILITY ANALYSIS

To evaluate the expected lead time under an uncertain demand pattern, a two-stage stochastic programming approach is adopted. The uncertain demand pattern is treated as multiple step changes (denoted by s) with uncertain magnitudes. The system lead time is computed for each of the scenarios considered, and the expected value is taken as the measure of responsiveness.

The responsiveness evaluation for an existing supply chain system given the fixed supply chain design (topology, delays, and capacities) and initial states involves no first-stage decisions, as these are specified. The first-stage decisions are included as optimization variables when considering economic criteria; these include inventory setpoints, which are common for all uncertain scenarios.

The following formulation evaluates the supply chain responsiveness, where the subscript s denoting scenarios is omitted for clarity:

$$\min_{z, Y^{met}_{j,c,t}} \mathbb{E}(\mathrm{SLT}) \tag{14.57}$$

subject to: supply chain model: Eqns (14.5) to (14.14); and responsiveness criteria: Eqns (14.41) to (14.56)

$$z \geq 0, \text{ where } z = \left\{ \begin{array}{ll} A_{j,c,t} & \forall\ j \in J_{FP}, c \in C_j, t \in T \\ B_{j,c,t} & \forall\ j \in J_{FP}, c \in C_j, t \in T \\ C^P_{i,m} & \forall\ m \in M, i \in I_m \\ C^M_{j,m} & \forall\ j \in J_{RM} \cup J_{IP}, m \in M_j \\ C^V_{j,v} & \forall\ j \in J_{FP}, v \in V_j \\ E^M_{j,m,t} & \forall\ j \in J_{RM} \cup J_{IP}, m \in M_j, t \in T \\ E^V_{j,v,t} & \forall\ j \in J_{FP}, v \in V_j, t \in T \\ F_{j,m,v,t} & \forall\ j \in J_{FP}, m \in M_j, v \in V_j, t \in T \\ I^M_{j,m,t} & \forall\ j \in J_{RM} \cup J_{IP}, m \in M_j, t \in T \\ I^V_{j,v,t} & \forall\ j \in J_{FP}, v \in V_j, t \in T \\ K^1_{j,c,t}, K^2_{j,c,t} & \forall\ j \in J_{FP}, c \in C_j, t \in T \\ L_{j,c} & \forall\ j \in J_{FP}, c \in C_j \\ O_{j,h,m,t} & \forall\ j \in J_{RM}, h \in H_j, m \in M_j, t \in T \\ W_{j,k,i,m,t} & \forall\ j \in J_{ki}, m \in M_j, i \in I_m, k \in K_i, t \in T \end{array} \right\} \tag{14.58}$$

Solely assessing the responsiveness of a supply chain system using the expected lead time as described above does not consider the supply chain economics. However, the solutions resulting from such an approach may not be cost-optimal, because there are no cost incentives to prevent overdesigning or to select cheaper options among all the feasible solutions. This suggests inclusion of an economic objective in the responsiveness analysis framework.

Maximizing responsiveness and minimizing costs are two conflicting objectives. Many of the attempts to improve the responsiveness, such as holding more inventory, shortening process and transportation delays, would very likely be costly. One of the remedies is to include the responsiveness measure as a constraint of the cost minimization problem. This can be implemented using the ε-constraint method by specifying that the expected lead time must be less than a parameter (ε) and then solving the problem with a series of ε values to perform Pareto analysis to investigate the trade-offs.

The objective function below is now used to account for the expected cost. Equations (14.17)–(14.20) are also included, as well as a constraint, $\mathbb{E}(\text{SLT}) \leq \varepsilon$, so that the cost-optimal solution can be generated for a range of specified levels of responsiveness, ε.

$$\min_{z, Y^{met}_{j,c,t}} \mathbb{E}(\text{Total Cost} + P_{\text{INV}}) \tag{14.59}$$

The inventory setpoints of all materials ($\bar{I}^{M}_{j,m}\ \forall j \in J_{RM} \cup J_{IP}, m \in M_j$, and $\bar{I}^{V}_{j,v}\ \forall j \in J_{FP}, v \in V_j$) are now treated as first-stage variables rather than model parameters.

14.5 CASE STUDIES

A subset of a forest products supply chain adapted from the case study by Dansereau et al. (2014) is considered. A simplified representation of the supply chain configuration is shown in Figure 14.3, where the blocks represent processes and triangles represent inventory. An integrated biorefinery receives biomass from suppliers as raw materials, which is then converted into a series of products.

The woody biomass is a mixture of three major components: cellulose, lignin, and hemicellulose. The lignin is separated from hemicellulose and can be used further to produce value-added products, but is not considered in this part of the

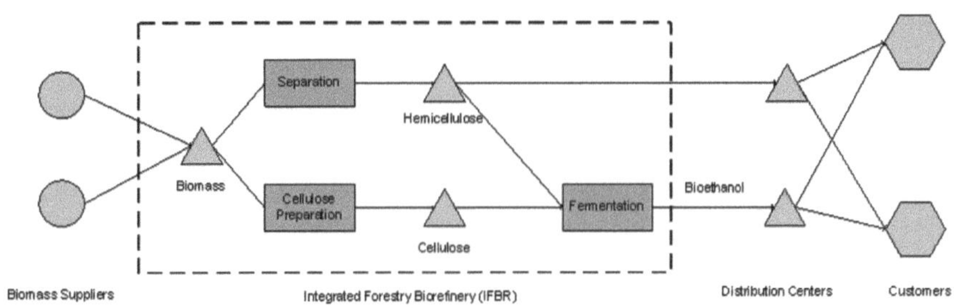

FIGURE 14.3

Biorefinery supply chain configuration for case study.

case study. The hemicellulose is stored as an intermediate product on site in the biorefinery and can be either processed further in the fermentation process or sold as a final product distributed to customers from distribution centers. The cellulose component is also stored on site as an intermediate product and can be consumed in the fermentation process to produce bioethanol.

14.5.1 PROBLEM DESCRIPTION

An integrated biorefinery receiving biomass from two suppliers and serving two customers is considered. The system is assumed to operate at steady state at the beginning of the time horizon, which means all processing units operate at a constant production volume to satisfy customer demands, and there is no net accumulation of materials at any of the inventory facilities. There are three types of delays involved in the supply chain operation for this case, as listed in Tables 14.2–14.4: procurement delay, production delay, and transportation delay.

Table 14.5 summarizes the material balance coefficients for all processes involved in the biorefinery. The coefficients for lignin separation and cellulose preparation processes are based on the information from Sannigrahi and Ragauskas (2013) and the average biomass composition reported by McKendry (2002). The coefficients of fermentation process are in the reference of the theoretical yield by Humbird et al. (2011) and the average yields determined by Lin et al. (2006).

At $t = 0$, the supply chain system is assumed to operate optimally at steady state to satisfy the nominal demand levels as listed in Table 14.6, with the initial inventory levels determined by inventory setpoints.

Three subcases are considered based on this network configuration. First, individual flexibility indices are evaluated. Then the expected lead time, as a measure of supply chain responsiveness, is evaluated by considering different step sizes in demand change using a multiscenario approach. Finally, the trade-offs between

Table 14.2 Procurement Delay ($\delta^{PRC}_{j,h,m}$) in Days

		Supplier (*h*)	
Raw Material (*j*)	**Biorefinery (*m*)**	**S1**	**S2**
Biomass	M1	1	0

Table 14.3 Production Delay (δ^{PRD}_{i}) in Days

Process (*i*)	Production Delay (δ^{PRD}_{i})
Separation	1
Preparation	0
Fermentation	3

Table 14.4 Transportation Delay ($\delta_{m,v}^{TRP}$) in Days

Biorefinery (*m*)	Distribution Center (*v*)	
	V1	V2
M1	2	1

Table 14.5 Material Balance Coefficients ($\eta_{j,k,i,m}$) of Chemical *j* in Process *i* of the Biorefinery

Process (*i*)	Chemical (*j*)			
	Biomass	Cellulose	Hemicellulose	Bioethanol
Separation	2.8	–	1	–
Preparation	2.2	1	–	–
Fermentation	–	2.5	1	2

Table 14.6 Nominal Demands Fulfilled by Steady-State Operation

Customer (c)	Chemical (*j*)	
	Hemicellulose	Bioethanol
C1	120	100
C2	150	120

two conflicting objectives, minimizing expected operating costs and minimizing expected lead time, are investigated.

14.5.2 CASE STUDY 1: FLEXIBILITY ANALYSIS

The individual flexibility indices are evaluated based on the uncertainty information listed in Tables 14.7–14.9. Table 14.7 displays the nominal demand levels of both products by customers and the range of demand deviations. Table 14.8 lists the nominal supply availability levels of biomass and the negative deviations in availability. The nominal selling prices and expected price deviations of bioethanol are listed in Table 14.9 based on the historical data from Trading Ecomomics (2014).

- Demand flexibility index: $F_d = 5.81$, based on the demand data given in Table 14.7. The index is greater than one and thus indicates that a much greater demand deviation than $\Delta D_{j,c}$ listed in the table can be accommodated by this supply chain system. After inspecting the active constraints, the limitations on

Table 14.7 Demand-Related Parameters for Final Product *j* Requested by Customer *c*

Nominal Demand $(\mu D_{j,c})$	**Chemical (*j*)**	
Customer (c)	**Hemicellulose**	**Bioethanol**
C1	120	100
C2	150	120
Demand Deviation $(\Delta D_{j,c})$	**Chemical (*j*)**	
Customer (c)	**Hemicellulose**	**Bioethanol**
C1	15	20
C2	30	10

Table 14.8 Supply-Related Parameters for Raw Material *j* Purchased from Supplier *h*

Nominal Availability $(\mu O_{j,h}^{max})$	**Supplier (*h*)**	
Raw Material (*j*)	**H1**	**H2**
Biomass	4000	5000
Availability Deviation $(\Delta O_{j,h}^{max})$	**Supplier (*h*)**	
Raw material (*j*)	**H1**	**H2**
Biomass	−500	−500

Table 14.9 Sale Price-Related Parameters for Final Product *j* Sold to Customer *c*

Nominal Price $(\mu P_{j,c}^{Sale})$	**Chemical (*j*)**	
Customer (c)	**Hemicellulose**	**Bioethanol**
C1	270	1000
C2	240	1200
Price Deviation $(\Delta P_{j,c}^{Sale})$	**Chemical (*j*)**	
Customer (c)	**Hemicellulose**	**Bioethanol**
C1	–	−200
C2	–	−250

the capacities of both separation and cellulose preparation processes restrict the flexibility.

- Supply flexibility index: $F_s = 7.03$, based on the supply quantity data given in Table 14.8. The minimum quantity of raw material available is 486 and 1486 units from suppliers $H1$ and $H2$, respectively to maintain the production in order to meet the nominal demands given that both quantities available vary at the same proportion indicated by F_s.
- Market flexibility index: $F_m = 4.47$, based on the price data given in Table 14.9. The nonnegativity constraint on the operating profit is active, so the minimum sale price of bioethanol is 105.2 and 81.4 $/ton to each of the customers respectively to reach the break-even point in profit.

14.5.3 CASE STUDY 2: EVALUATION OF EXPECTED LEAD TIME

For this part of the case study, the system is assumed to run at nominal production rates while meeting all demand levels with zero level of inventory at the start for all materials. There are 101 scenarios of uncertain step changes in hemicellulose demand considered, including 50 positive step sizes and 50 negative step sizes from the nominal levels, as well as the step size of zero to represent the nominal scenario. The step sizes range from -100 to 100 with an interval of 2, resulting in the final demands ranging from 20 to 220 units for customer 1. The step sizes for customer 2 range from -50 to 50 with an interval of 1, resulting in the final demands ranging from 100 to 200 units of hemicellulose.

The expected lead time for these 101 scenarios is 0.35 days as an average of satisfying both hemicellulose customers. All the nonzero lead times are generated by the scenarios with a positive step change in the demand, as the excess final products due to reduced demand can be stored as inventory and used to fulfill demands in later time periods without causing further delays in delivering the products.

If only the scenarios with positive step change in demand are counted, the expected lead time is 0.7 days, which is still significantly less than the sum of minimum delays (0 days in procurement from supplier $S2$ + 1 day in separation process + 2 days in transportation = 3 days) incurred to produce and deliver hemicellulose. The sum of minimum delays would have been observed if the system were to start from idle without any nominal production levels or inventory levels, because the system would become a complete "pull" system (made-to-order) and experience every necessary delay during the supply chain activities. The system starting with some levels of production has the option to store the products as inventory for several time periods, so the amount of products available to be drawn from inventory in the next period can satisfy the new demand. This "self-conservation" feature is unique to supply chains starting with material flows in the system and is advantageous in terms of responding to demand changes compared to a complete "pull" system. An analogy can be found in chemical plant operations in that it usually takes a shorter time to transition to another operating point from the current operating point than from start-up.

14.5.4 CASE STUDY 3: ECONOMIC TRADE-OFFS OF RESPONSIVENESS

This case study aims to explore the trade-offs between responsiveness and the total operational cost of the supply chain over 20 operating periods. To simplify, 10 uncertain demand scenarios are considered as listed, with step sizes ranging from -100 to $+80$ with an interval of 20 units for customer 1 and steps ranging from -35 to $+28$ with an interval of 7 units for customer 2. The parameter τ is set to 4 so the expected lead time for hemicellulose to reach the customers cannot exceed 3 days.

The cost parameters used are listed in Tables 14.10–14.12 for procurement, production, and transportation, respectively. The inventory costs are presented in Tables 14.13 and 14.14.

Figure 14.4 displays the expected total operating cost (line plot) and the inventory setpoint of hemicellulose stored in the distribution center (bar plot) for each of the expected lead times specified ranging from 0 to 3 days at an interval of 0.1 days. The inventory setpoint displayed serves as a first-stage design decision and thus is common to all demand scenarios to guarantee feasibility. It is noted that even though the lead times for individual customers must be integer by definition, the expected lead time, as an equally weighted average of all scenarios, can be fractional.

The expected total operating cost exhibits a decreasing trend in response to the greater expected lead time allowed. The decline becomes less dramatic for expected lead time greater than 0.3 days. Such a trend observed in operating cost is consistent with the final product inventory setpoint requirement as shown by the bar plot. The level of initial inventory, captured as the inventory setpoint, increases to accommodate the more stringent requirement of the responsiveness. For instance in the worst case scenario, when demands for both customers step up to the maximum levels,

Table 14.10 Raw Material Purchase Cost ($c_{j,h}^{PRC}$) in \$/ton of Raw Material *j* Purchased from Supplier *h*

	Supplier (*h*)	
Raw Material (*j*)	**S1**	**S2**
Biomass	30	35

Table 14.11 Production Cost (c_i^{PRD}) in \$/day/ton Produced in Process *i*

Process (*i*)	Production Cost (c_i^{PRD})
Separation	10
Preparation	20
Fermentation	50

Table 14.12 Transportation cost $(c^{TRP}_{j,m,v})$ in \$/ton of Final Product *j* Shipped from Biorefinery *m* to Distribution Center *v*

		Chemical (*j*)	
Biorefinery (*m*)	**Distribution Center (*v*)**	**Hemicellulose**	**Bioethanol**
M1	V1	12	–
M1	V2	–	35

Table 14.13 Inventory Cost $(c^{IM}_{j,m})$ in \$/day/ton of Raw Material or Intermediate Product *j* Stored in Biorefinery *m*

	Biorefinery (*m*)
Chemical (*j*)	**M1**
Biomass	1
Hemicellulose	4
Cellulose	3

Table 14.14 Inventory Cost $(c^{IV}_{j,v})$ in \$/day/ton of Final Product *j* Stored in Distribution Center *v*

	Chemical (*j*)	
Distribution Center (*v*)	**Hemicellulose**	**Bioethanol**
V1	5	–
V2	–	8

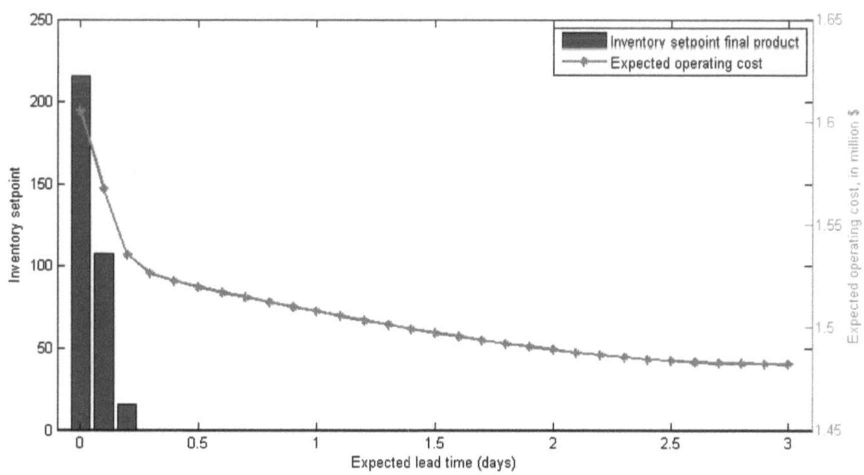

FIGURE 14.4

Optimal operating costs with specified lead time requirement.

$80 + 28 = 108$ extra units of hemicellulose need to be supplied in addition to the nominal demands in each time period. The transportation delay is 2 days, thus the initial inventory of 216 units is exactly sufficient to supply the additional demand during the 2 days of delay, and meanwhile the upstream processes can ramp up production to catch up with the new demand.

The expected total operating cost plotted in Figure 14.4 excludes the penalty cost associated with the inventory setpoint deviations. For the expected lead times less than 0.3 days, the major differences in operating cost are due to maintaining the inventory levels determined by the optimal inventory setpoints during the optimization horizon as the objective function minimizes any negative deviations from the inventory setpoints.

After 0.3 days of expected lead time, the operating cost continues to decline, but at a slower rate. The reason for this steady decline is that the demand is fulfilled at fewer time periods out of the fixed time horizon, when the expected lead time is allowed to be longer. This explanation is confirmed in Figure 14.5, where the cost is normalized as the expected daily operating cost for steady-state operation sustained at the new demand, and computed based on the average of a 5-day period of steady-state operation to fulfill the new demand. However, despite the fact that it seems to have a cost benefit to elongate the expected lead time during supply chain operation, the revenue for satisfying customer demands on time also reduces significantly. Moreover, the consequences for failing to meet the customer demands on time are not limited to the lost revenue in the nowadays competitive business environment. Potential lost orders, lost customers, damaged reputation and credibility, reduced market share, and other liabilities to customers and partners are some

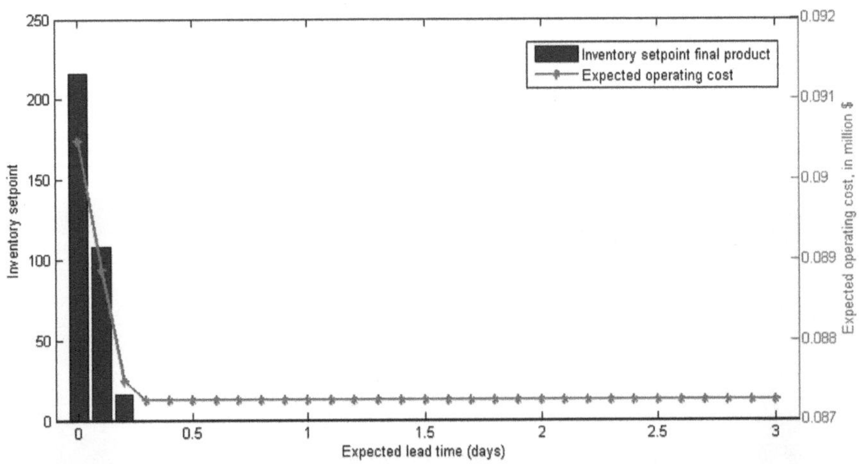

FIGURE 14.5

Optimal normalized daily operating costs to serve the new demand with specified lead time requirement.

examples of the risks that can harm the sustainability and competitiveness of the business with lack of sufficient responsiveness.

14.6 CONCLUSION

Process supply chains are required to operate in an uncertain environment. This is particularly pertinent with the increased shift toward a sustainable bioeconomy, characterized by relatively high levels of uncertainty in raw material availability and quality, market conditions, and process characteristics. Sustainable supply chain systems need to be capable of effectively mitigating and responding to such uncertainty. Key characteristics that reflect this capability are flexibility and responsiveness or agility. Analogous notions have been studied and applied to process plants within the context of plant operability analysis.

This article reviews supply chain and process plant operability, and based upon analogies between them, presents a framework for the quantitative assessment of supply chain operability. Representative dynamic and steady-state supply chain models are presented, followed by optimization-based formulations for the assessment of supply chain flexibility and responsiveness. A multipart case study based on an integrated forest biorefinery is presented to illustrate the application of the approach. Further work is under way to more fully integrate the flexibility and responsiveness formulations.

ACKNOWLEDGMENT

Financial support for this research through the NSERC Strategic Network on Value Chain Optimization, and the McMaster Advanced Control Consortium (MACC) is gratefully acknowledged.

REFERENCES

Backx, T., Bosgra, O., Marquardt, W., 1998. Towards intentional dynamics in supply chain conscious process operations. AIChE Symp. Ser. 320, 5–20.

Baker, R., Swartz, C.L.E., 2004. Rigorous handling of input saturation in the design of dynamically operable plants. Ind. Eng. Chem. Res. 43, 5880–5887.

Beamon, B.M., 1999. Measuring supply chain performance. Int. J. Oper. Prod. Manage. 19, 275–292.

Bertrand, J.W.M., Rutten, W., 1999. Evaluation of three production planning procedures for the use of recipe flexibility. Eur. J. Oper. Res. 115, 179–194.

Bok, J.K., Grossmann, I.E., Park, S., 2000. Supply chain optimization in continuous flexible process networks. Ind. Eng. Chem. Res. 39, 1279–1290.

Christopher, M., 2000. The agile supply chain: competing in volatile markets. Ind. Mark. Manage. 29, 37–44.

Christopher, M., Towill, D., 2001. An integrated model for the design of agile supply chains. Int. J. Phys. Distr. Log. Manage. 31, 235–246.

Dansereau, L.P., El-Halwagi, M., Mansoornejad, B., Stuart, P.R., 2014. Framework for margins-based planning: forest biorefinery case study. Comput. Chem. Eng. 63, 34–50.

Ekşioğlu, S.D., Acharya, A., Leightley, L.E., Arora, S., 2009. Analyzing the design and management of biomass-to-biorefinery supply chain. Comput. Ind. Eng. 57, 1342–1352.

Ferrer-Nadal, S., Puigjaner, L., Guillén-Gosálbez, G., 2008. Managing risk through a flexible recipe framework. AIChE J. 54, 728–740.

Fisher, M., 1997. What is the right supply chain for your product? Harvard Bus. Rev. 75, 105–116.

Garavelli, A.C., 2003. Flexibility configurations for the supply chain management. Int. J. Prod. Econ. 85, 141–153.

Gerwin, D., 1993. Manufacturing flexibility: a strategic perspective. Manage. Sci. 39, 395–410.

Graves, S.C., Tomlin, B.T., 2003. Process flexibility in supply chains. Manage. Sci. 49, 907–919.

Grossmann, I.E., Calfa, B.A., Garcia-Herreros, P., 2014. Evolution of concepts and models for quantifying resiliency and flexibility of chemical processes. Comput. Chem. Eng. 70, 22–34.

Grossmann, I.E., Floudas, C.A., 1987. Active constraint strategy for flexibility analysis of chemical processes. Comput. Chem. Eng. 11, 675–693.

Grossmann, I.E., Morari, M., 1984. Operability, resiliency and flexibility: process design objectives for a changing world. In: Westberg, A.W., Chien, H.H. (Eds.), Proceedings of Int. Conf. on Foundations of Computer-aided Process Design, pp. 931–1030.

Halemane, K.P., Grossmann, I.E., 1983. Optimal process design under uncertainty. AIChE J. 29, 425–433.

Holt, B.R., Morari, M., 1985a. Design of resilient processing plants—V. The effect of deadtime on dynamic resilience. Chem. Eng. Sci. 40, 1229–1237.

Holt, B.R., Morari, M., 1985b. Design of resilient processing plants—VI. The effect of right-half-plane zeros on dynamic resilience. Chem. Eng. Sci. 40, 59–74.

Holweg, M., 2005. The three dimensions of responsiveness. Int. J. Oper. Prod. Manage. 25, 603–622.

Humbird, D., Davis, R., Tao, L., Kinchin, C., Hsu, D., Aden, A., Schoen, P., Lukas, J., Olthof, B., Worley, M., Sexton, D., Dudgeon, D., 2011. Process Design and Economics for Biochemical Conversion of Lignocellulosic Biomass to Ethanol. NREL/TP-5100–47764. Technical Report NERL.

Jung, J.Y., Blau, G., Pekny, J.F., Eversdyk, G.V.R.D., 2004. A simulation based optimization approach to supply chain management under demand uncertainty. Comput. Chem. Eng. 28, 2087–2106.

Lin, C.T., Chiu, H., Chu, P.Y., 2006. Agility index in the supply chain. Int. J. Prod. Econ. 100, 285–299.

Mansoornejad, B., Pistikopoulos, E.N., Stuart, P., 2011. Incorporating flexibility design into supply chain design for forest biorefinery. J. Sci. Technol. For. Prod. Processes 1, 54–66.

Mansoornejad, B., Pistikopoulos, E.N., Stuart, P., 2013. Metrics for evaluating the forest biorefinery supply chain performance. Comput. Chem. Eng. 54, 125–139.

Marlin, T.E., 2010. Teaching "operability" in undergraduate chemical engineering design education. Comput. Chem. Eng. 34, 1421–14311.

Mastragostino, R., 2012. Optimization-based formulations for operability analysis and control of process supply chains (M. A. Sc. thesis). McMaster University, Hamilton, Ontario, Canada.

Mastragostino, R., Swartz, C.L.E., 2014. Dynamic operability analysis of process supply chains for forest industry transformation. Ind. Eng. Chem. Res. 53, 9825–9840.

Matson, J.B., McFarlane, D.C., 1999. Assessing the responsiveness of existing production operations. Int. J. Oper. Prod. Manage. 19, 765–784.

McKendry, P., 2002. Energy production from biomass (Part 1): overview of biomass. Bioresour. Technol. 83, 627–642.

Mohideen, M.J., Perkins, J.D., Pistikopoulos, E.N., 1996. Optimal design of dynamic systems under uncertainty. AIChE J. 42, 2251–2272.

Morari, M., 1983. Design of resilient processing plants—III. A general framework for the assessment of dynamic resilience. Chem. Eng. Sci. 38, 1881–1891.

Naylor, B., Naim, M.M., Berry, D., 1999. Leagility: integrating the lean and agile manufacturing paradigms in the total supply chain. Int. J. Prod. Econ. 62, 107–118.

Sakizlis, V., Perkins, J.D., Pistikopoulos, E.N., 2004. Recent advances in optimization- based simultaneous process and control design. Comput. Chem. Eng. 28, 2069–2086.

Sannigrahi, P., Ragauskas, A.J., 2013. Chapter 10: Fundamentals of biomass pretreatment by fractionation. In: Wyman, C.E. (Ed.), Aqueous Pretreatment of Plant Biomass for Biological and Chemical Conversion to Fuels and Chemicals. John Wiley and Sons, Ltd.

van Schijndel, J.M.G., Pistikopoulos, E.N., 2000. Towards the integration of process design, process control & process operability – current status & future trends. In: Malone, M., Trainham, J., Carnahan, B. (Eds.), Foundations of Computer Aided Process Design Number 323 in AIChE Symposium Series. American Institure of Chemical Engineers, pp. 99–112.

Schweiger, C.A., Floudas, C.A., 1998. Interactions of design and control: optimization with dynamic models. In: Hager, W.W., Pardalos, P.M. (Eds.), Optimal Control: Theory, Algorithms and Applications. Kluwer Academic Publishers, pp. 388–435.

Sethi, A.K., Sethi, S.P., 1990. Flexibility in manufacturing: a survey. Int. J. Flex. Manuf. Sys. 2, 289–328.

Shen, Z., Coullard, C., Daskin, M., 2003. A joint location-inventory model. Transportation Sci. 37, 40–55.

Son, Y.K., Park, C.S., 1987. Economic measure of productivity, quality and flexibility in advanced manufacturing systems. J. Manuf. Sys. 6, 193–207.

Swaney, R.E., Grossmann, I.E., 1985. An index for operational flexibility in chemical process design. Part I: formulation and theory. AIChE J. 31, 621–630.

Swartz, C.L.E., 1996. A computational framework for dynbamic operability assessment. Comput. Chem. Eng. 20, 365–371.

Tang, C., Tomlin, B.T., 2008. The power of flexibility for mitigating supply chain risks. Int. J. Prod. Econ. 116, 12–27.

Trading Ecomomics, 2014. Ethanol. http://www.tradingeconomics.com/commodity/ethanol (accessed 02.09.14.).

You, F., Grossmann, I.E., 2007. Optimal design and operational planning of responsive process supply chains. In: Process Systems Engineering. Wiley-VCH Verlag GmbH & Co. KGaA, pp. 107–134.

You, F., Grossmann, I.E., 2008. Design of responsive supply chains under demand uncertainty. Comput. Chem. Eng. 32, 3090–3111.

You, F., Grossmann, I.E., 2010. Integrated multi-echelon supply chain design with inventories under uncertainty: MINLP models, computational strategies. AIChE J. 56, 419–440.

You, F., Grossmann, I.E., 2011. Balancing responsiveness and economics in process supply chain design with multi-echelon stochastic inventory. AIChE J. 57, 178–192.
Yuan, Z., Chen, B., Sin, G., Gani, R., 2012. State-of-the-art and progress in the optimization-based simultaneous design and control for chemical processes. AIChE J. 58, 1640–1659.
Ziegler, J.G., Nichols, N.B., 1943. Process lags in automatic-control circuits. Trans. ASME 433–444.

CHAPTER

15 Dynamic Simulation-Based Assessment of Supply Chain Sustainability

Arief Adhitya*, Iskandar Halim*, Rajagopalan Srinivasan[§,1]

**Institute of Chemical and Engineering Sciences, Jurong Island, Singapore*

[§]Indian Institute of Technology Gandhinagar, Ahmedabad, Gujarat, India

[1]Corresponding author: E-mail: raj@iitgn.ac.in

15.1 INTRODUCTION

Sustainability has become a key agenda for every industry in the face of mounting environmental challenges, shortages of natural resources, growing awareness of social responsibility, and the need to remain profitable. The chemical process industry, in particular, has a strong emphasis on sustainability due to the nature of its business. The industry involves extraction of raw materials such as crude oil, gas, and minerals, processes which are highly energy intensive, and handling of large volume of toxic, flammable, and hazardous chemicals.

The push for sustainability has pressurized the chemical process industry to consider the sustainability implications of its supply chain operation. The important role of supply chain management for sustainability has been highlighted in surveys conducted by Deloitte (2010) and McKinsey (2010). While sustainability is commonly viewed as having three pillars (also known as the triple bottom line)—economic, environmental, and social aspects (Bonini, 2011), the economic and environmental aspects have received more attention than the social aspect. The integration of environmental thinking and supply chain management is termed as green supply chain management (GSCM) and aims to help companies improve their environmental performance along the supply chain. In the literature, the scope of GSCM is wide and ranges from green purchasing to product design, material sourcing and selection, benign manufacturing, delivery of the final product to the customers as well as end-of-life management of the product after its useful life (Srivasta, 2007).

In the process systems engineering literature, GSCM has received much attention, mostly in the area of supply chain planning and design. A common approach is to formulate a mathematical optimization model for maximum economic benefits and minimum environmental impacts. Decisions include selection of appropriate

Computer Aided Chemical Engineering, Volume 36. ISSN 1570-7946. http://dx.doi.org/10.1016/B978-0-444-63472-6.00015-X

raw materials, suppliers, technologies, or transportation routes (Hugo and Pistikopoulos, 2005; Bojarski et al., 2009; Guillén-Gosálbez and Grossmann, 2010; Mele et al., 2011). Economic performance can be measured through indicators such as profit or customer satisfaction. Environmental indicators include metrics such as waste reduction algorithm (Cabezas et al., 1999) and life cycle assessment (LCA)-based metrics such as CML 2001 (Guinée, 2002). While most of the publications focus on planning and design, less attention is given to operational aspects and dynamics of the supply chain operation.

Supply chain is the network of suppliers, manufacturing plants, warehouses, and distribution channels organized to acquire raw materials, convert them to finished products, and distribute these products to customers. Supply chain decisions such as sourcing of raw materials, intermediates, and packaging; plant sites selection; production, storage, and transportation policies have significant impact on sustainability performance. Supply chain operation is complex as it involves numerous interacting entities with various roles and constraints. This results in complex dynamics, which in turn could lead to unforeseen domino effects. Information delay, limited visibility, and the presence of various uncertainties further complicate supply chain decision-making.

Consequently, a decision taken by an entity may be optimal for its own objective, but its impact on the overall supply chain performance, both economic and environmental, may not be immediately obvious. An integrated analysis of the impacts of decisions on the overall system is required. These motivate the use of supply chain simulation models, which could capture the behavior of the entities, their interactions, the resulting dynamics, and evaluate the overall performance of the supply chain. Dynamic simulation offers a good way to uncover direct and indirect effects of supply chain decisions on sustainability performance. In this work, we present a dynamic supply chain simulation model integrated with indicators from LCA for assessment of supply chain sustainability. We first describe a typical supply chain operation in Section 15.2, followed by dynamic simulation model of the supply chain operation in Section 15.3. Section 15.4 presents sustainability assessment case studies from two different supply chains, i.e., diaper and detergent. Finally, Section 15.5 gives some concluding remarks.

15.2 SUPPLY CHAIN OPERATION AND SUSTAINABILITY

A typical supply chain consists of suppliers, manufacturer, distributors, and retailers/customers (Figure 15.1). The manufacturer and distributor belong to a focal enterprise and have several *departments* performing various supply chain functions while interacting with one another. These departments are sales, operations, procurement, and storage. The sales department deals with customers and accepts their orders. In some cases, the sales department also makes demand forecast. This actual/forecast demand information is communicated to the operations department, who manages production. Based on this demand information and also inventory information

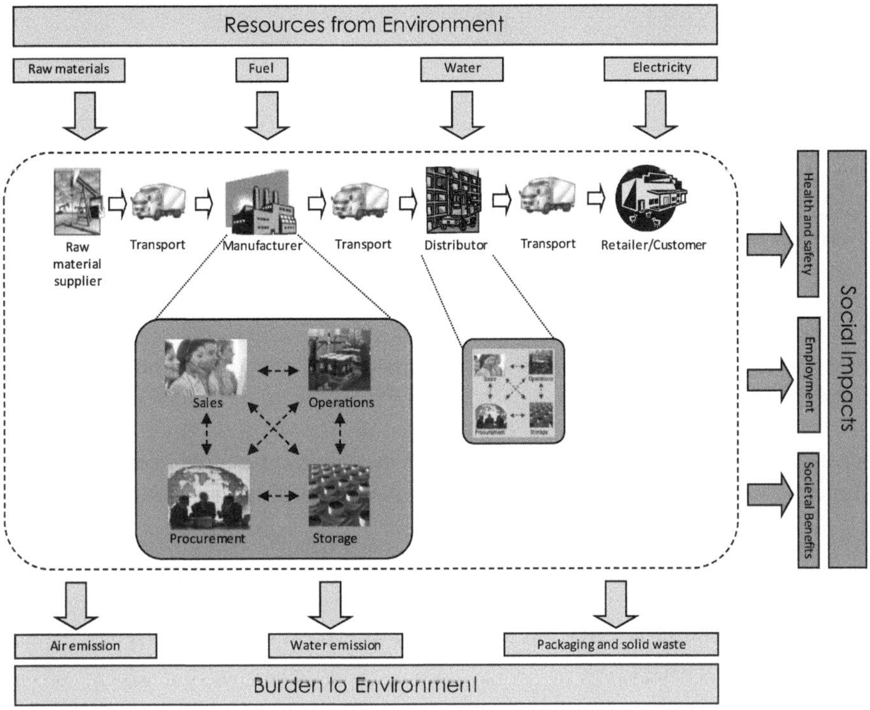

FIGURE 15.1

Typical supply chain with its environmental and social impacts.

from the storage department, the operations department decides how much product to make. The operations department and the storage department communicate with the procurement department for procurement of the required raw materials. The storage department manages both raw materials and product inventory. There may also be a logistics department that manages raw materials and product transportation.

Each of these departments follows certain *policies* in performing its functions. The sales department may have a policy to decide which customer orders to accept, a policy to decide pricing, and a policy for demand forecasting. The operations department may follow a certain scheduling policy to decide production rates and schedule. The procurement department has a procurement policy, e.g., procurement at a regular interval or after reaching a certain inventory level. The storage department may have a storage policy to manage the different raw materials and product inventory. Thus, each of these departments is an independent and autonomous entity making its own decisions in the supply chain operation while interacting with one another.

At the supply chain level, the manufacturer procures raw materials from different suppliers, manufactures the products, and sends them to the distributor. The distributor typically operates in a push-mode; it keeps a certain level of inventory to fulfill retailer/customer orders. Different procurement policies for replenishment of products can be adopted by the distributor. For example, under the fixed interval policy, procurement is done at regular intervals to bring its inventory back to a certain top-up level. The distributor places orders to the manufacturer, which may operate in push-mode (products manufactured based on demand forecast) or pull-mode (products manufactured upon receiving the distributor's orders).

Policies and decisions made by each of the above entities will affect the sustainability performance of the overall supply chain. Sustainability of supply chain operation can be assessed through its economic, environmental, and social impacts, as shown in Figure 15.1. Economic sustainability is related to costs and profitability. The environment is impacted as supply chain operation consumes resources from the environment (raw materials, fuel, water, and electricity) and releases a burden to the environment (air emission, water emission, packaging, and solid waste). Social impacts include employees' health and safety, employment, and other societal benefits. These impacts can be modeled and measured through various indicators. Modeling of the supply chain operation for sustainability assessment is described in the next section.

15.3 DYNAMIC SIMULATION MODEL OF SUPPLY CHAIN OPERATION

A supply chain consists of networks of entities at different levels and can be seen as a sociotechnical system in which the physical and the social networks of actors involved in its operation collectively form an interconnected complex system. The actors determine the operation of the physical network, and the physical network affects the behavior of the actors. Decision-making in the supply chain is distributed across the various actors, each with its own goals and tasks. For example, the operations department makes products and the storage department manages inventory. These actors communicate with one another to get the information necessary to perform their own tasks or to call for certain actions.

Agent-based modeling is a suitable approach to model complex systems comprising of interacting autonomous agents (Macal and North, 2005). In this approach, a system is described by defining the actors (agents) and the interactions between them. The system behavior then emerges from the behavior of the model components and their interactions. Instead of taking a top-down view in modeling, the model is hence constructed from a bottom-up perspective. Because of this, agent-based modeling is considered a natural approach for systems involving distributed and decentralized decision-making. Each actor in the supply chain operation is modeled as an agent with its own activities and interactions with other agents, as summarized in Table 15.1. Policies are implemented to guide the agents' decision-making in performing their activities. In this work, we do not explicitly model the logistics department and transport; material transport is implemented through time delay between dispatch and arrival at destination.

Table 15.1 Agent Activities and Interactions in Supply Chain Operation

Agent		Activity	Interaction with
Retailer/Customer		Place Order	Distributor (Sales)
Distributor	Sales	Receive order Accept order Assign order	Retailer/Customer Retailer/Customer Distributor (operations)
	Operations	Process order	Distributor (sales), distributor (storage)
	Storage	Release product for delivery Manage product inventory	Distributor (operations) Distributor (procurement)
	Procurement	Buy product	Distributor (operations), distributor (storage), manufacturer (sales)
Manufacturer	Sales	Receive order Accept order Assign order	Distributor (procurement) Distributor (procurement) Manufacturer (operations)
	Operations	Request raw materials Make product Store products	Manufacturer (storage) Manufacturer (sales) Manufacturer (storage)
	Storage	Release raw materials for production Manage raw material and product inventory	Manufacturer (operations) Manufacturer (procurement)
	Procurement	Buy raw materials	Manufacturer (operations), manufacturer (storage), raw material supplier
Raw material supplier		Supply raw materials	Manufacturer (procurement)

The agent-based model can be implemented in an agent-based modeling platform such as the Repast simulation toolkit in Java (North et al., 2006), or a mathematical modeling platform such as MATLAB/Simulink (MathWorks, 2010). van Dam et al. (2009) compared the two approaches and identified their strengths in the context of refinery supply chain modeling. Physical elements of the refinery are modeled explicitly in the mathematical platform but not in the agent-based platform. On the other hand, the agent-based platform explicitly has a flexible structure, allowing new agents and connections between agents to be added in extensions to the model. Also, in the agent-based platform it is possible to directly reuse the same ontology and source code for modeling a different supply chain. In general, the efforts required to make changes in the model for various scenarios, ranging from operational to tactical and strategic levels, are similar in the two platforms. In terms of explaining the model and the model results, the agent-based platform offers a natural representation of the decision-making processes and interactions between the entities while the mathematical platform has an edge for explaining the technical process.

The supply chain model has been implemented in both the agent-based platform using Repast (Behdani et al., 2010) and the mathematical platform using MATLAB/Simulink (Adhitya and Srinivasan, 2010). In this work, we use the latter and integrate indicators from LCA into the model to enable sustainability assessment.

The model uses a discrete-time representation; one day is divided into a predefined number of time ticks t. Raw material arrival at the manufacturer is modeled as:

$$RA_r(t) = RS_r(t - LT_r) \tag{15.1}$$

where $RA_r(t)$ is the amount of raw material r arriving at the manufacturer at time t, $RS_r(t)$ is the amount of raw material r ordered from the supplier, and LT_r is the lead time between order and arrival of the raw material r.

Processing in the manufacturer is modeled through the recipe Rcp_r, which specifies the amount of raw material r required to make one unit of product.

$$RU_r(t) = Rcp_r \times CJ^{amt} \tag{15.2}$$

where $RU_r(t)$ is the amount of raw material r used for production at time t and CJ^{amt} is the amount of product to be made in job CJ.

The material balance on raw material inventory at the manufacturer is given by:

$$IR_r(t+1) = IR_r(t) + RA_r(t) - RU_r(t) \tag{15.3}$$

where $IR_r(t)$ is the inventory of raw material r at time t.

All activities and policies of the agents as listed in Table 15.1 have been modeled. For the complete model equations, the reader is referred to Adhitya et al. (2011). Economic performance can be measured through indicators such as manufacturer profit and distributor profit. Environmental and social performance can be measured through indicators from LCA and social/societal LCA, respectively.

LCA is a widely used technique for measuring the environmental (or social for social/societal LCA) consequences assignable to a product or service. Starting from raw material extraction to production process, transportation, point of use, and final disposal, LCA charts the course of all of the consumption of natural resources and release of pollutants into air, water, and soil. However, LCA impacts are derived from a product-centric perspective without considering the effects of supply chain dynamics such as different logistics options, inventories, distribution network, and ordering policy. For example, LCA calculations are typically based on a static, average truckload level and number of trips, whereas in reality these two variables of supply chain operation are dynamic. In this work, we take a supply chain-centric perspective by considering the various supply chain decisions through dynamic simulation. By integrating LCA indicators into the simulation, we can perform sustainability assessment of the supply chain operation, as illustrated through case studies in the next section.

15.4 CASE STUDIES

We present two different case studies—diaper supply chain (Adhitya et al., 2011) and detergent supply chain (Wang et al., 2011). In the former, only economic and environmental aspects are considered while in the latter social impacts are also included.

15.4.1 DIAPER SUPPLY CHAIN CASE STUDY

Modern disposable diapers come in a variety of styles, but their basic raw materials are essentially the same: fluff pulp, superabsorbent polymer (SAP), plastic components such as low density polyethylene and polypropylene (PP), adhesives, and elastics (EDANA, 2005). In this case study, the environmental indicator values are derived from the LCA study of disposable diapers done by the UK Environmental Agency (2005). Eleven environmental indicators are used, namely abiotic resource depletion (ARD), global warming potential (GWP), ozone layer depletion, photochemical oxidation, acidification, eutrophication (EUT), human toxicity (HT), fresh water aquatic ecotoxicity (FWAE), terrestrial ecotoxicity (TE), water usage (WU), and energy consumption (EC). These 11 indicators are evaluated at each stage of the supply chain, namely raw materials, raw material transportation, manufacturing, packaging, product transportation from manufacturer to distributor, distributor operation, and product transportation from distributor to customer. Two economic indicators are used, namely manufacturer's profit and distributor's profit. Table 15.2 summarizes the economic and environmental indicators used in the diaper supply chain simulation model.

Table 15.2 Indicators Used in the Diaper Case Study

Aspect	Indicators	
Economic	PPRO	Manufacturer's profit ($)
	DPRO	Distributor's profit ($)
Environmental	ARD	Abiotic resource depletion (g Sb eq)
	GWP	Global warming potential (g CO_2 eq)
	ODP	Ozone layer depletion (g CFC-11 eq)
	PO	Photochemical oxidation (g C_2H_2)
	ACD	Acidification (g SO_2 eq)
	EUT	Eutrophication (g PO_4 eq)
	HT	Human toxicity (g 1,4-DB eq)
	FWAE	Fresh water aquatic ecotoxicity (g 1,4-DB eq)
	TE	Terrestrial ecotoxicity (g 1,4-DB eq)
	WU	Water use (kg H_2O)
	EC	Energy consumption (MJ)

Two scenarios are considered here. The first involves changes in the product composition; the second looks at the ordering policy in the supply chain. The aim is to assess the sustainability impacts (economic and environmental) of different decisions along the supply chain. The simulation horizon is 60 days. The cumulative profits and environmental impacts over the simulation horizon are compared in the different scenarios

15.4.1.1 Scenario 1: changing the diaper composition

The first scenario analyzes the impact of changes to the diaper composition. The mass composition of a typical diaper from years 1987, 1995, and 2005 are shown in Table 15.3. Supply chain operation with the same demand is simulated for the three different diaper compositions and the results are shown in Figure 15.2.

There is a significant reduction in WU for the 2005 case compared to the 1995 and 1987 cases. This is due to the reduction in the use of fluff pulp from 1987 to 2005 (14.57 g vs 32.83 and 54.94 g). However, the figure highlights an increase in the GWP, ARD, and TE for the 2005 case from the 1995 and 1987 cases. This is caused mainly by the increase in the amount of SAP in the 2005 case (13.65 g vs 4.46 and 0.74 g), and to a lesser extent the increase in the amount of PP (7.22 g vs 3.97 and 4.36 g). The new composition also leads to an increase in the manufacturer's profit since it requires less raw materials (41.95 g vs 49 and 67 g). Although the older compositions use less SAP, they use much more fluff pulp. The results also reveal that the environmental impacts from raw material transportation are smaller in the 2005 case than the 1995 and 1987 cases, because less raw materials are needed to be transported, resulting in fewer trips.

15.4.1.2 Scenario 2: order batching

The second scenario studies the effect of different ordering policies by the distributor. More frequent ordering in smaller batches means more transportation trips and consequently higher transportation impact and cost. Different order batching by the distributor is simulated through three different distributor procurement intervals: 2, 5, and 10 days. The results are shown in Figure 15.3.

Table 15.3 Diaper Compositions

Component	1987 (g)	1995 (g)	2005 (g)
Fluff pulp	54.94	32.83	14.57
Superabsorber (SAP)	0.74	4.46	13.65
Polypropylene (PP)	4.36	3.97	7.22
Polyethylene (PE)	4.29	3.33	2.69
Adhesive	1.34	1.42	1.76
Elastic	0.20	0.20	0.21
Others	1.14	2.79	1.85
Total	67.01	49.00	41.95

Adapted from EDANA (2005)

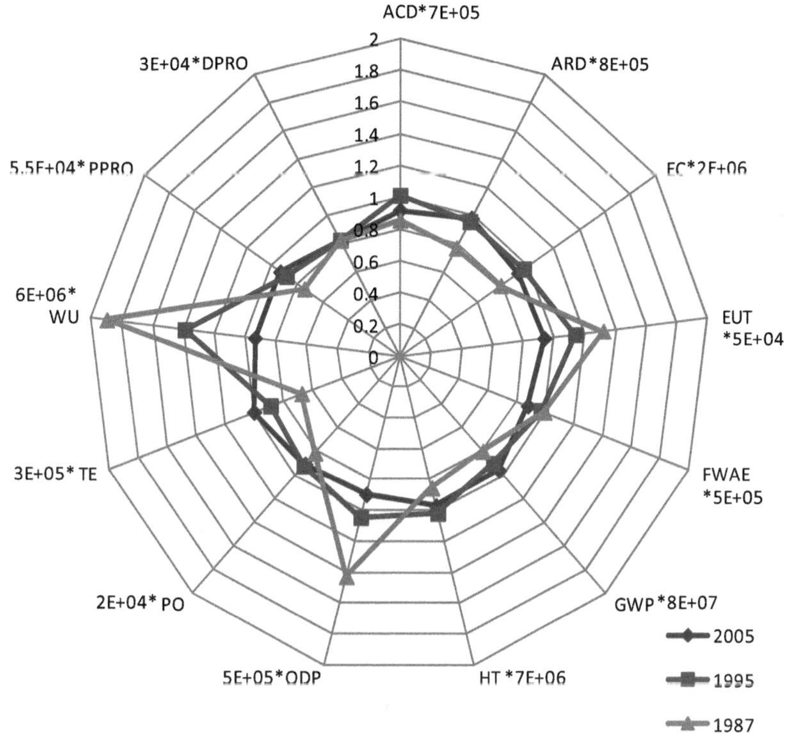

FIGURE 15.2

Economic and environmental indicators for different diaper compositions in the diaper case study (Scenario 1). ACD, acidification; ARD, abiotic resource depletion; EC, energy consumption; EUT, eutrophication; FWAE, fresh water aquatic ecotoxicity; GWP, global warming potential; HT, human toxicity; ODP, ozone layer depletion; PO, photochemical oxidation; TE, terrestrial ecotoxicity; WU, water usage; PPRO, manufacturer's profit; DPRO, distributor's profit.

In general, less frequent ordering (i.e., higher procurement interval) results in lower environmental impacts due to reduced number of transportation trips. The larger batching of order results in longer lead time and the average inventory kept by the distributor is lower (see Figure 15.4). Consequently, the distributor's profit is increased due to lower inventory cost. On the other hand, the average raw material inventory kept by the manufacturer is increased when the distributor order is less frequent, resulting in higher inventory cost and lower manufacturer's profit. Less frequent ordering means reduced inventory and this might have an impact on order fulfillment. This, along with the trade-off between the distributor's and manufacturer's profits, could be studied further to find the optimal supply chain policies for both distributor and manufacturer.

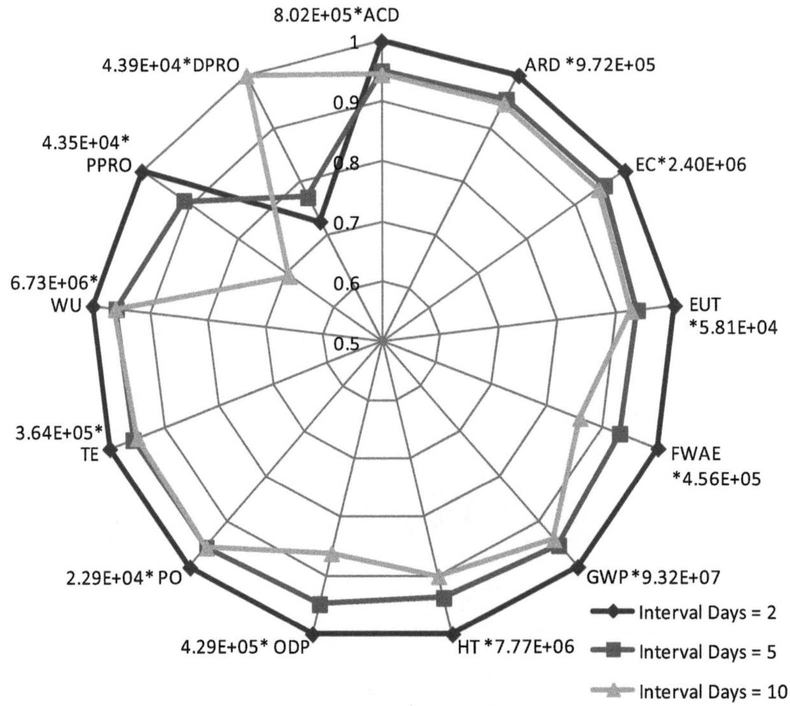

FIGURE 15.3

Economic and environmental indicators for different distributor procurement intervals in the diaper case study (Scenario 2). ACD, acidification; ARD, abiotic resource depletion; EC, energy consumption; EUT, eutrophication; FWAE, fresh water aquatic ecotoxicity; GWP, global warming potential; HT, human toxicity; ODP, ozone layer depletion; PO, photochemical oxidation; TE, terrestrial ecotoxicity; WU, water usage; PPRO, manufacturer's profit; DPRO, distributor's profit.

15.4.2 DETERGENT SUPPLY CHAIN CASE STUDY

In the detergent supply chain case study, we include social indicators in addition to the economic and environmental indicators. Extension of LCA to include the social aspects of products along their life cycle is termed as societal LCA (SLCA). The social implications of a product can be expressed through qualitative, semiquantitative, or quantitative indicators (UNEP, 2009). Qualitative indicators provide information on a particular issue, for example, measures taken by an enterprise to manage employee stress. Semiquantitative indicators categorize qualitative indicators into a yes/no form or a scale-based scoring system, for example, presence of a stress management program (yes/no) or the rating of such stress management program (bad = 0, mediocre = 1, good = 2). Quantitative indicators can be used if the issues at stake are quantifiable.

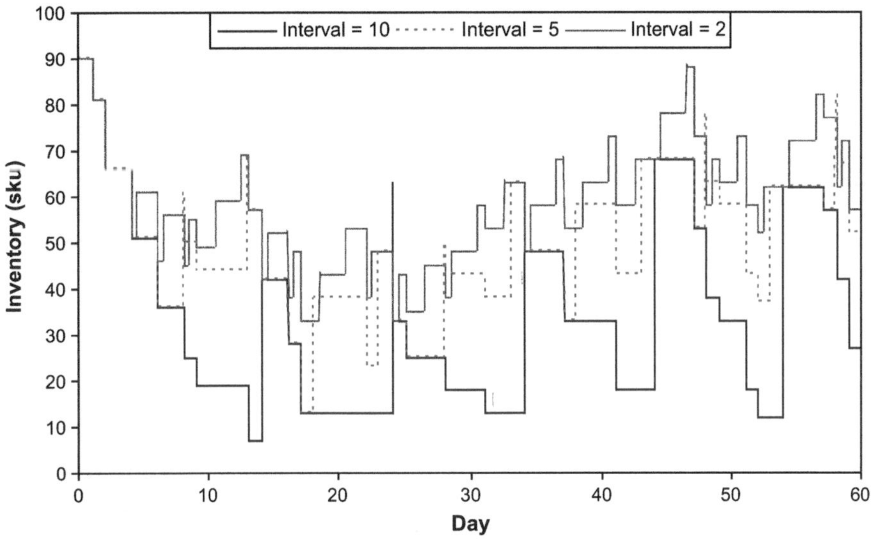

FIGURE 15.4

Distributor inventory for different distributor procurement intervals in the diaper case study (Scenario 2).

Hunkeler (2006) proposed a quantitative SLCA indicator that is based on the number of worker hours to fulfill basic societal needs. The inventories of different unit processes in the product life cycle are associated with the workers' employment incomes and then correlated with the various life necessities they can provide. This is done by dividing the total employment hours by characterization factors that specify the number of labor hours needed to fulfill each unit of societal needs. For example, Table 15.4 shows the societal characterization factors for two countries (France and Germany) and four societal needs (housing, health care, education, and other basic needs). An average worker needs to work for 20,000 h to purchase housing in Germany. Using the life cycle inventory data, we can calculate the total employment hours associated with the supply chain operation. If the total number of hours required for making a detergent ingredient X in Germany is 7 h, then based on Table 15.4, we can infer that this contributes to 0.00035 housing units, 0.0117 health care units, 0.0233 education units, and 0.14 units of other basic needs.

Table 15.4 Wages and Societal Characterization Factors

	Societal Needs (Work hour/unit)			
Country	**Housing**	**Health Care**	**Education**	**Other Basic Needs**
France	25,000	400	375	67.5
Germany	20,000	600	300	50

In this case study, the detergent manufacturer is located in France and the raw materials come from suppliers in seven countries, namely Germany, Russia, France, Canada, South Africa, The Netherlands, and Morocco, by trucks and cargo ships. The societal characterization factors for these countries are taken from Hunkeler (2006). The product is a granular detergent based on the formulation described in Saouter and van Hoof (2002) and sold to local distributors in France. The environmental indicator values are derived from Dewaele et al. (2006). The simulation horizon is 180 days. Two scenarios are presented in this case study: changing detergent formulation and supplier selection policy.

15.4.2.1 Scenario 1: changing the detergent formulation

The first scenario compares the sustainability of traditional detergent production containing a phosphate ingredient (sodium tripolyphosphate or STPP) and modern detergent without STPP. Until the 1980s, STPP was a commonly used ingredient in laundry detergent formulation. Since then, due to environmental concerns, phosphates were banned in Europe and many US states and STPP was phased out and replaced with zeolites, citrates, and other builders. These ingredients are more environmentally friendly but also more costly. Supply chain operation with the same demand is simulated for the two different detergent formulations and the results are shown in Figures 15.5 and 15.6.

Figure 15.5 shows that the phosphate-free detergent has a lower profit because the replacement ingredients for STPP are more costly. It also has a lower social performance than the STPP detergent because fewer employment hours are required for the phosphate-free detergent production. Figure 15.6 shows that the phosphate-free detergent has better environmental performance in all impact categories with significant reductions in EUT, HT, FWAE, solid waste, and GWP.

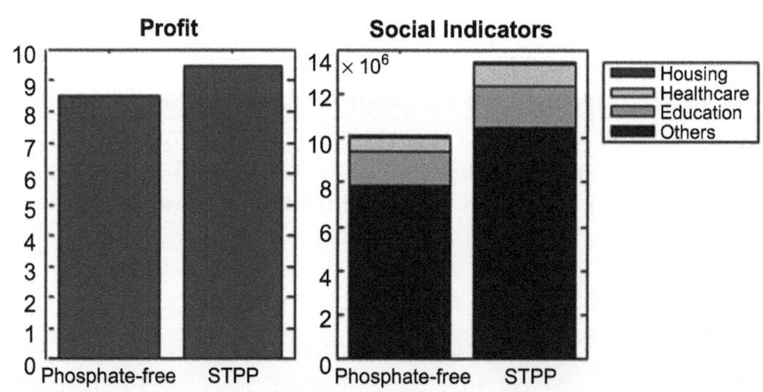

FIGURE 15.5

Profit and social indicators for two detergent formulations (Scenario 1). STPP, sodium tripolyphosphate.

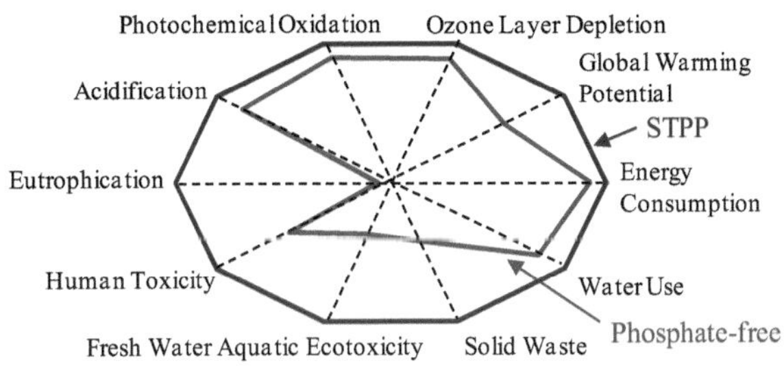

FIGURE 15.6

Environmental indicators for two detergent formulations (Scenario 1). STPP, sodium tripolyphosphate.

15.4.2.2 Scenario 2: supplier selection policy

The second scenario compares two supplier selection policies: favor nearest suppliers versus no preference with regards to supplier distance. Supply chain operation is simulated for the two different policies and the results are shown in Figures 15.7 and 15.8.

Figure 15.7 shows that nearer suppliers favored results in a significantly higher profit because timely arrival of raw materials allows more orders to be fulfilled. More orders completed also means more employment and therefore higher social performance. However, as shown in Figure 15.8, environmental performance of the nearer suppliers favored policy is worse. This can be attributed to the same reason; more orders completed means more raw materials used, more manufacturing, packaging, and transport impacts.

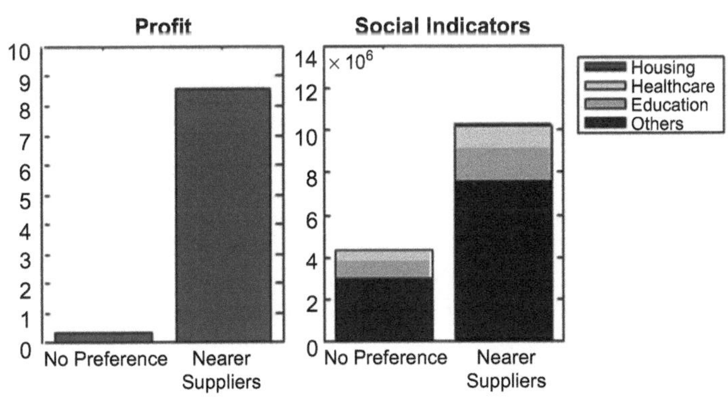

FIGURE 15.7

Profit and social indicators for two supplier selection policies in the detergent supply chain (Scenario 2).

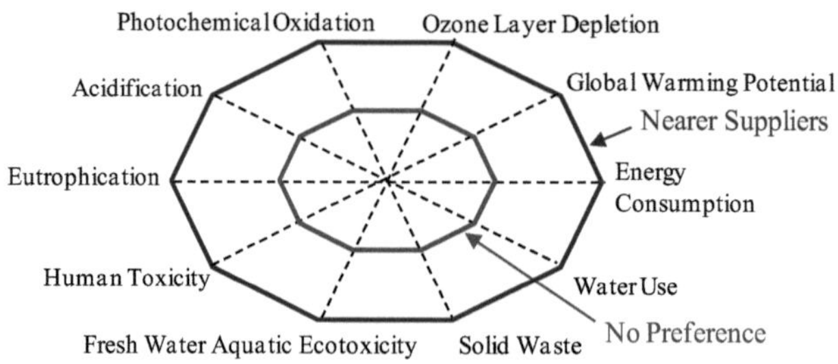

FIGURE 15.8

Environmental indicators for two supplier selection policies in the detergent supply chain (Scenario 2).

15.5 CONCLUDING REMARKS

This paper presents a dynamic simulation-based framework for assessment of supply chain sustainability. Supply chain operation involves numerous interacting entities and thus complex dynamics, which will affect the supply chain sustainability performance. LCA-based assessments are static and do not consider the dynamics in supply chain operation. Thus, there is a need for dynamic simulation integrated with LCA indicators to capture the impacts of these dynamics on the overall sustainability performance. Case studies involving changes in product composition, ordering policy, and supplier selection policy demonstrate how the proposed framework can be used for sustainability assessment while considering the dynamics involved in supply chain operation.

REFERENCES

Adhitya, A., Srinivasan, R., 2010. Dynamic simulation and decision support for multisite specialty chemicals supply chain. Ind. Eng. Chem. Res. 49, 9917–9931.

Adhitya, A., Halim, I., Srinivasan, R., 2011. Decision support for green supply chain operations by integrating dynamic simulation and LCA indicators: diaper case study. Environ. Sci. Technol. 45, 10178–10185.

Behdani, B., Lukszo, Z., Adhitya, A., Srinivasan, R., 2010. Performance analysis of a multiplant specialty chemical manufacturing enterprise using an agent-based model. Comput. Chem. Eng. 34, 793–801.

Bojarski, A.D., Laínez, J.M., Espuña, A., Puigjaner, L., 2009. Incorporating environmental impacts and regulations in a holistic supply chains modeling: an LCA approach. Comput. Chem. Eng. 33, 1747–1759.

Bonini, S., 2011. McKinsey Global Survey Results: The Business of Sustainability. McKinsey & Company.

Cabezas, H., Bare, J.C., Mallick, S.K., 1999. Pollution prevention with chemical process simulators: the generalized waste reduction (WAR) algorithm –full version. Comput. Chem. Eng. 23, 623–634.

van Dam, K.H., Adhitya, A., Srinivasan, R., Lukszo, Z., 2009. Critical evaluation of paradigms for modeling integrated supply chains. Comput. Chem. Eng. 33, 1711–1726.
Deloitte, 2010. Sustainability in Business Today: A Cross-industry View. http://www.deloitte.com (accessed 20 08.14).
Dewaele, J., Pant, R., Schowanek, D., 2006. Comparative Life Cycle Assessment (LCA) of Ariel "Actif à froid" (2006) with previous Ariel laundry detergents (1998, 2001). http://www.avnir.org (accessed 20 08.14).
EDANA (European Disposables and Nonwovens Association), 2005. Sustainability Report: Baby Diapers and Incontinence Products. http://www.edana.org (accessed 20 08.14).
Guillén-Gosálbez, G., Grossmann, I., 2010. A global optimization for environmentally conscious design of supply chains under uncertainty in the damage assessment model. Comput. Chem. Eng. 34, 42–58.
Guinée, J.B. (Ed.), 2002. Handbook on Lifecycle Assessment. Operational Guide to the ISO Standards. Kluwer Academic Publishers, Dordrecht.
Hugo, A., Pistikopoulos, E., 2005. Environmentally conscious long-range planning and design of supply chain networks. J. Cleaner Prod. 13, 1471–1491.
Hunkeler, D., 2006. Societal LCA Methodology and case study. Int. J. Life Cycle Assess. 11 (6), 371–382.
Macal, C.M., North, M.J., 2005. Tutorial on Agent-based Modeling and Simulation. Winter Simulation Conference, Orlando, USA.
MathWorks, 2010. Simulink User's Guide.
McKinsey, 2010. How Companies Manage Sustainability. http://www.mckinseyquarterly.com (accessed 20.08.14).
Mele, F.D., Kostin, A.M., Guillén-Gosálbez, G., Jiménez, L., 2011. Multiobjective model for more sustainable fuel supply chains. A case study of the sugar cane industry in Argentina. Ind. Eng. Chem. Res. 50 (9), 4939–4958.
North, M.J., Collier, N.T., Vos, J.R., 2006. Experiences creating three implementations of the repast agent modeling toolkit. ACM Trans. Model. Comput. Simul. 16 (1), 1–25.
Saouter, E., van Hoof, G., 2002. A database for the life-cycle assessment of procter & gamble laundry detergents. Int. J. Life Cycle Assess. 7 (2), 103–114.
Srivasta, S.K., 2007. Green supply-chain management: a state-of-the-art literature review. Int. J. Manage. Rev. 9 (1), 53–80.
UK Environmental Agency, 2005. Lifecycle Assessment of Disposable and Reusable Nappies in the UK. http://www.environment-agency.gov.uk (accessed 20.08.2014).
UNEP (United Nations Environment Programme), 2009. Guidelines for Social Life Cycle Assessment of Products. http://www.unep.org (accessed 20.08.14).
Wang, P.C., Halim, I., Adhitya, A., Srinivasan, R., 2011. Integrating economic, environmental and social indicators for sustainable supply chains. In: Pistikopoulos, E.N., Georgiadis, M.C., Kokossis, A.C. (Eds.), Computer-aided Chemical Engineering, vol. 29, pp. 1220–1224.

CHAPTER

Transport and Logistics Models for Sustainability

16

H. Giray Resat, Metin Turkay[1]

Department of Industrial Engineering, Koc University, Istanbul, Turkey

[1]*Corresponding author: E-mail: mturkay@ku.edu.tr*

16.1 INTRODUCTION

Global concerns about climate change, energy usage, environmental impacts, and limits to financial resources for transportation infrastructure require new and different approaches for planning, designing, constructing, operating, and maintaining transportation solutions and systems. Today, the importance of sustainability is considered by decision makers as one of the most essential parameter in transportation planning processes. In the literature, there is a widely used definition of sustain ability that was introduced by the United Nations in 1987: "sustainability meets the needs of the present without compromising the ability of future generations to meet their own needs" (WCED, 1987). Although, many definitions of sustainability in various disciplines were developed, there is not a clear and basic description concerning sustainability in transportation operations. The Center for Sustainable Transport (CST, 1998) introduced an approach for sustainable transportation systems which:

- *allows the basic access needs of individuals and societies to be met safely and in a manner consistent with human and ecosystem health, and with equity within and between generations;*
- *are affordable, operate efficiently, offer choice of transport mode, and support a vibrant economy;*
- *limit emissions and waste within the planet's ability to absorb them, minimize consumption of nonrenewable resources to the sustainable yield level, reuse and recycle its components, and minimize the use of land and the production of noise.*

Turkay (2008) examined the environmentally conscious supply chain systems and indicated that transportation-based approaches that should be considered as sustainable transportation are major focal points to include sustainability considerations in supply chain management. He proposed that development of multimodal transportation systems will play an important role in reducing the impact of the

Computer Aided Chemical Engineering, Volume 36. ISSN 1570-7946. http://dx.doi.org/10.1016/B978-0-444-63472-6.00016-1

environment, especially for emission issues. The implementation of this approach into transportation processes cannot be achieved easily; the establishment of a sustainable transportation system requires a well-defined framework to attract all stakeholders from the private sector, public authorities, research, and education. Jeon and Amekudzi (2005) stated this framework as in the Figure 16.1.

One of the major difficulties in analyzing sustainability is the conflict between these four essential factors since each of them imposes some restrictions over the others. Therefore, if a transportation system is improved by focusing on one of these factors, this attempt will deteriorate the other three factors. Therefore, sustainability should not be considered as one of these factors and there should be a new approach which integrates or strengthens all of these factors to develop more complex transportation decision mechanisms. On the other hand, Low and Gleeson (2003) introduced the "paradox of sustainability." In their study, they mentioned the nested box model, which indicates that some serious shifts can take place in some of these sustainability factors. In other words, they explained that environmental factors will be the leading parameters in the decision mechanism and especially social and economic conditions will be integrated with each other and this combined system will be captured by the environmental parameters in the long term. Although the goals of the one factor are often consistent with the others, Low and Gleeson (2003) argue that environmental sustainability should have a priority but not be

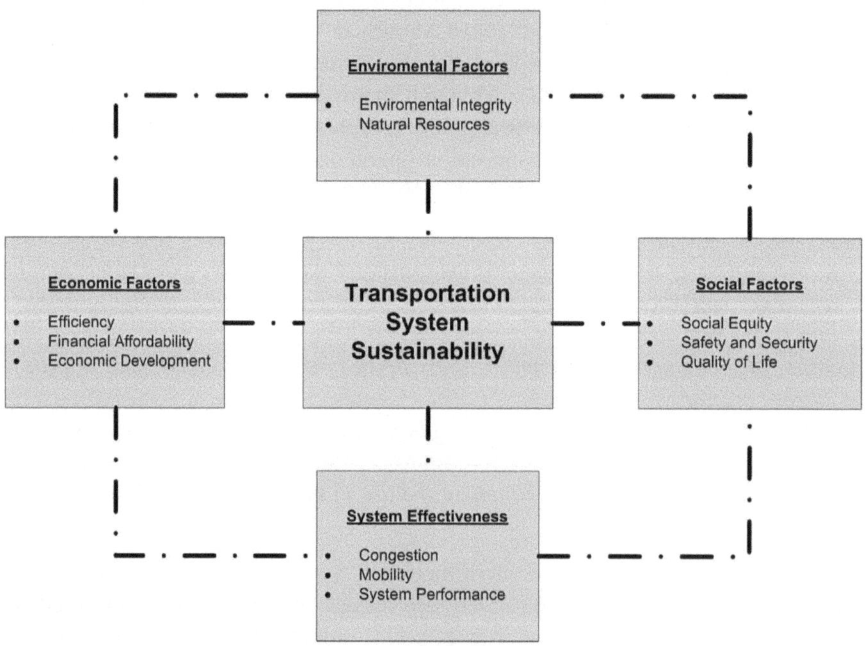

FIGURE 16.1

Four essential factors of transportation system sustainability.

conditional upon economic or social sustainability. Additionally, there are many studies on the conceptualization of sustainability in the literature (e.g., Haughton, 1999; McGranahan and Satterthwaite, 2000; Nichols et al., 2008). All of these studies can help the decision maker and stakeholders in the transportation sector to sustain some measurements.

In addition to the factors, there are some main drivers for the stakeholders of the transportation sector to use and integrate sustainability concepts into their operations and decision mechanisms. A good example is the energy sector that accounts for 29% of the total US energy consumption (EIA, 2009). Also, there is an increasing trend in the environmental issues especially on the green concepts. Therefore, the stakeholders in this sector face very strict regulations and social awareness of greenhouse gas (GHG) emissions and energy usage. Additionally, due to the lack of fuel sources, fuel prices tend to always increase and this trend leads the sector to be more efficient. Also, safety is another important factor for the sector. Passenger and product safety for transportation sector, and worker safety for those involved in manufacturing and operational activities are major triggering topics in the sector.

By considering all of these drivers and factors for transportation sustainability, the stakeholders can easily integrate these concepts into their short- and long-term plans and also transportation activities by defining some performance indexes. Transportation performance indexes predict, evaluate, and monitor what percentage of the adopted objectives of the stakeholders is accomplished by the transportation system. However, such performance indexes should be determined according to the factors given in Figure 16.1 and also assess the regional goals and objectives. They can be easily applied to all stages of the transportation decision-making process. In Figure 16.2, the main decisions steps in which some of the performance measures can be used in the transportation sector are given.

Also, some performance measures to achieve the goals or objectives of sustainability issues in the transportation sector are given in Table 16.1. The classification of these performance measurements is combined into four groups: economic, social, environmental, and effectiveness. The most commonly used indicators directly related to transport sustainability are listed below.

There is a greater range of indicators in the social and environmental categories than in the economic ones. The most frequently applied indicator for measuring social sustainability is the number of transport fatalities; for environmental sustainability, it is land usage of transport infrastructures and emission rates; and for the economic aspect total transportation costs and public expenditures. All of these sustainability indexes play a very important role in tracing and tracking the developments of transportation system.

Yevdokimov (2003) proposed an approach for integration of these factors into the general framework of the transportation problem. The study indicated a system with vertical and horizontal linkages of the transportation network and analysis of sustainability was proposed to be investigated as a whole system of the supply chain, instead of just focusing on a part of the system. In the horizontal case, the transportation network is considered as just a combination of environmental and social factors.

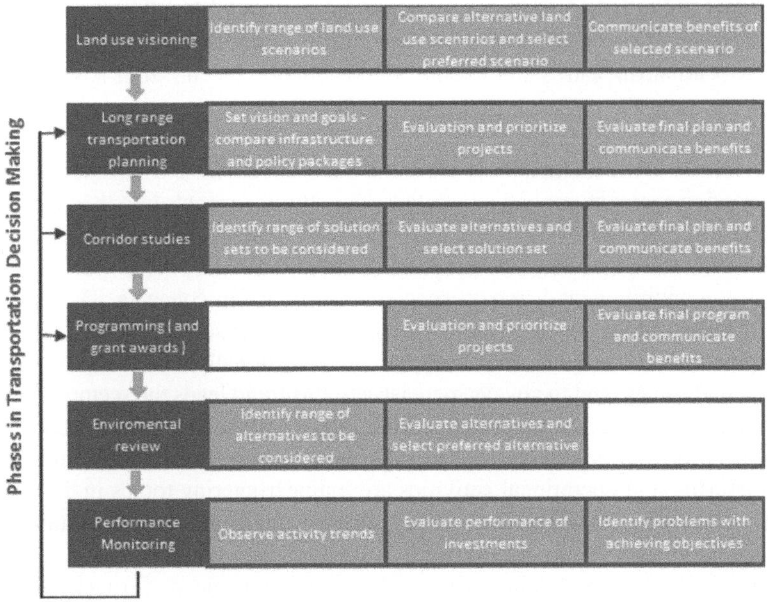

FIGURE 16.2

Main decision steps in transportation sector (EPA, 2011).

Table 16.1 Performance Measures for Different Factors

Sustainability Factors	Objectives	Performance Indexes
Environmental	Minimize GHG emissions	CO_2, NO_x emission rates
	Minimize pollutions (Air, noise etc.)	Noise level CO, VOC emissions
	Minimize resource use	Fuel and land consumption
Economic	Minimize transportation cost	Total travel cost
	Maximize economic efficiency	Total spent time in traffic
	Maximize affordability	Land usage and accessibility
Social	Maximize equity	Equity of welfare Exposure to emission
	Maximize accessibility	Access to the service activity points
	Increase the safety and security	Crash disabilities, fatalities
Effectiveness	Improve mobility	Congestion rate
	Improve system performance	Total carried quantity
		Total distance traveled

Therefore, emissions, noise pollution, and land use considerations are directly combined with the social links such as safety, mobility, and accessibility. In the vertical cases, these horizontal linkages are combined with the economic issues of different parts of the supply chain which means that the approach places the transportation network within the economic perspective of the system and develops its links with environmental and social concepts. These links reflect the idea that the transportation network does not exist on its own. It is an integral part of an economy, and there is a common demand among the transportation and other sectors of the economic system.

In the literature, there are many studies focusing on the main factors of sustainability by considering the performance indexes given in Table 16.1. Most of them adopt the definition of sustainable transportation and propose some concepts or mathematical models based on a single or synthesis of several performance measurements.

16.1.1 ECONOMIC FACTORS

The economic dimension of sustainable transportation is based on the understanding how transportation contributes to or accelerates economic development. Newman and Kenworthy (1999) and Jeon and Amekudzi (2005) indicated some efficiency concepts in the public transportation activities, which directly related with the cost of total system. In addition, there are many studies indicating the importance of the transportation cost and total expenditures in transportation activities. Nicolas et al. (2003), Jeon and Amekudzi (2005), Zhang and Guindon (2006), Zito and Salvo (2011), and Santos and Ribeiro (2013) proposed some methods to balance the competitiveness of the sector and affordability of the services. Moreover, Haghshenas and Vaziri (2012) and Zegras (2006) indicated the time consideration in transportation activities because they mentioned that total spent time in transportation activities directly affects the competitiveness of the sector.

Also, Sharaki and Turkay (2014) indicated that there are a large number of network design studies focusing on the economic factor indexes of sustainable transportation such as Ban et al. (2006), Kim et al. (2008), and Lo and Szeto (2009).

16.1.2 ENVIRONMENTAL FACTORS

In the literature, there are many studies focusing on the total travel distances traveled by the vehicles, which directly have great impact on the environmental issues, such as emissions, use of resources or waste. Newman and Kenworthy (1999), Savelson et al. (2006), Litman (2009), and Santos and Ribeiro (2013) focused on the total land use of the transportation system and total traveled distances to create a linkage over emissions or use of resources. Also, Nicolas et al. (2003), Zhang and Guindon (2006), Zegras (2006), and Santos and Ribeiro (2013) focused on the emission rates such as CO_2, NO_x, noise intensity levels, and energy consumption of the transportation system that reflects on the sustainability aspects according to the environmental dimensions.

The integration of different sustainability measures in transportation sector intends to provide insights into the common issues of the world as discussed in the literature. Vreeker and Nijkamp (2005) suggest that the link between transportation and land use is becoming stronger while transportation is a driver of urban development. However, they also indicate that transportation can sometimes deteriorate the balance of urban development because it is seriously difficult to make any balance between the factors of the transportation sector in real cases. For example, Sharaki and Turkay (2014) indicate that integration of different functionalities such as home, commerce, and service sector requires some efficient infrastructures, links among each function and proper land allocation. Also, Liao et al. (2013) proposed that there is a very strict relationship in the transportation sector between the land usage, urbanization, and GHG emissions. Therefore, land use and environmental factors are closely related and there are many studies on this issue to support the decision authorities in transportation planning. As indicated in the study of Waddell et al. (2007), Paulley and Webster (1991), Southworth (1995), Garret and Wachs (1996), Miller et al. (1999), Wegener (2004), and Dowling et al. (2005) proposed some mathematical models for the land allocation problem, which are directly related with the environmental concepts.

Also, there is a comprehensive literature survey over the models integrating the network design and land usage in the study of Sharaki and Turkay (2014). Lowry (1964), Simmonds and Still (1998), Anderstig and Mattsson (1998), Wegener (1998), Putmann (1983), Martinez (1996), and Timmermans (2003) introduced the mathematical models and solution algorithms of network designs by considering the environmental issues, especially land usage. Studies of Zhou et al. (2008), Yin and Lu (1999), and Huang et al. (2010) focus on environmental sustainability and total traffic emission rates.

16.1.3 SOCIAL FACTORS

Compared to other factors, social impacts are relatively difficult to measure and assess from the mathematical models or algorithms. In order to sustain in accessibility and equity issues, many studies such as Jeon and Amekudzi (2005), Litman (2009), and Santos and Ribeiro (2013) deal with the density of transportation networks, quality and affordability of public transportation by the citizens. Moreover, Nicolas et al. (2003) and Tanguay et al. (2010) indicate the safety consistency with human health by considering the total distance traveled and transport fatalities within the system. Also, Cidell (2012) introduced the linkages between the social factors and land uses. In that study, it is stated that reducing the mobility activities and increase in the sedentary lifestyles may increase the risks of cancer and mortality rates.

16.1.4 EFFECTIVENESS FACTORS

These factors are mostly combined with the other three main factor criteria. In the literature, there are many works covering to measure and assess the performance of the total transportation network (Vuchic, 2007). The important elements of these networks are the speed, density, and frequency in terms of efficiency case. By using these indicators, total movement of the cargos, total distance traveled, total time spent in both transportation activities or mode transfer activities can be easily served for decision makers.

16.1.5 BIOBJECTIVE AND MULTIOBJECTIVE PROBLEMS

Banister (2008), Newman and Kenworthy (1999), and Schiller et al. (2010) studied on the mobility activities and travel time issues. Banister (2008) proposed some balance between the low-carbon solutions for the transportation by reducing the total travel distances of the vehicles. Especially, Schiller et al. (2010) focused on the linkages between land use and mobility activities. In addition to them, Moavenzadeh and Markow (2007) indicated the concept of congestion, which affects the sustainability of the system in terms of economic, social, and environmental factors.

There are some studies that are handling with both economic and environmental conditions. For example, Yin and Lawphongpanich (2006), Ferguson et al. (2010), Yang et al. (2010), Sharma and Mathew (2011), Li et al. (2012), and Zhong et al. (2012) proposed biobjective formulations to focus on the both objective.

16.2 MATHEMATICAL MODELS

Transportation systems are essential components in realizing the transfer of materials among the nodes of supply chains. Therefore, modern supply chain systems continuously seek to improve efficiency of the transportation activities in terms of cost and environmental objectives. As indicated in the study of Turkay (2008), using the multimodal or intermodal transportation systems has strong effects over the economic and environmental performance of the supply chains. The definition of the intermodal or multimodal transportation is proposed by the Jones et al. (2000) as the integration of more than a transportation mode (road, rail, sea, air, and pipeline) in a single transportation chain. In addition to using multitransportation modes in the chain, sustainable transportation problems should include multiobjectives because there are many objectives affecting the efficiency of the system as given in the Figure 16.1. The best way of considering these intermodal transportation activities does not lie in any single factor in reality, so the optimization of the intermodal transportation network can be considered as a multiobjective optimization problem. Therefore, there should be more than a single objective in these problems which can be cost, emission, and utilization of some social indexes. Resat and Turkay (under review) and Fattahi and Turkay (under review a) proposed mixed integer linear (MILP) mathematical models for multiobjective intermodal transportation problems by considering the time, cost, and carbon emission objectives.

16.2.1 OBJECTIVE FUNCTIONS

$$Min\ f_1 = \sum_{t\in T}\sum_{h\in H}\sum_{v\in V}\sum_{i\in I}\sum_{j\in J}\sum_{k\in K}(TC_{ijk}\, X_{ijkvth}) + \sum_{t\in T}\sum_{h\in H}\sum_{i\in I}\sum_{k\in K}\sum_{l\in L}(TrC_{kl}\, w_{iklth}) + \sum_{i\in I}\sum_{t\in T}\sum_{k\in K}(I_{ikt}\, P1) \tag{16.1}$$

$$Min\ f_2 = \sum_{t\in T}\sum_{h\in H}\sum_{v\in V}\sum_{i\in I}\sum_{j\in J}\sum_{k\in K}(J_{ijkvth}) + \sum_{t\in T}\sum_{h\in H}\sum_{i\in I}\sum_{k\in K}\sum_{l\in L}(TrT_{kl}\, w_{iklth}) + \sum_{i\in I}\sum_{t\in T}\sum_{k\in K}\sum_{v\in V}\{((EW_{ikvt} + LW_{ikvt})P2) - (NW_{ikvt}\, P3)\} \tag{16.2}$$

In the model formulation, Eqns (16.1) and (16.2) correspond to the cost and time functions, respectively. In Eqn (16.1) the first part indicates the total transportation cost on the network (multiplication of the total carried quantity, X_{ijkvth}, with the unit transportation cost), the second part gives the total cost due to the transportation mode changes which is again multiplication of binary variable, w_{iklth}, with the unit transfer cost, and the last term shows the cost of keeping the goods in the inventory with a penalty cost. In Eqn (16.2), the first term gives the total transportation duration of the system, the second term indicates the duration of the transport mode change, and the last term corresponds to the penalty costs of early or late arrivals of the vehicles to the nodes. In the last part of the Eqn (16.2), EW_{ikvt}, LW_{ikvt}, and NW_{ikvt} corresponds to the total carried quantity arrived early, late, and within the operational period, respectively.

In addition to these objectives, another contribution along these lines is from Bektas and Laporte (2011) who present the Pollution-Routing Problem as an extension of the classical Vehicle Routing Problem with Time Windows. Figliozzi (2011), Franceschetti et al. (2013), and Demir et al. (2014) developed some models using this pollution routing problems. There is another objective, which combines GHG emission rate with the total distance traveled by vehicles and total carried quantities.

$$\begin{aligned} Min\ f_3 = &\sum_{m\in M}\sum_{i\in I}\sum_{j\in J}\sum_{r\in R}\sum_{k\in K}\Big\{\{(\theta_{ijkmr}g2_{ijkmr} + \omega_{ijkmr}g1_{ijkmr})A \\ &+ (\theta_{ijkmr}g2_{ijkmr} + \omega_{ijkmr}g1_{ijkmr})(S_{kmr})^3B\}\Big\} \\ &+ \sum_{i\in I}\sum_{j\in J}\sum_{k\in K}\sum_{r\in R}\Big\{\{(g2_{ijk2r} + \theta_{ijk2r}g2_{ijk2r} + \omega_{ijk2r}g1_{ijk2r} - \alpha g1_{ijk2r})(S_{k3r})^3B\} \\ &+ \{(\alpha g1_{ijk2r} - g2_{ijk2r})(S_{k2r})^3B\}\Big\} \\ &+ \sum_{t\in T}\sum_{i\in I}\sum_{j\in J}\sum_{k\in K}\sum_{v\in V}\sum_{h\in H} d_{ijk}(y_{ijkvth}\sigma + X_{ijkvth})C \end{aligned} \tag{16.3}$$

Equation (16.3) is for minimizing the total carbon emission rate in the system. First summation indicates the rate caused by the engine module and the speed module, second term shows the total emissions rate generated by the speed module in the all congestion and in all free-flow regions, and the last one represents total rate induced by the weight module.

16.2.2 CONSTRAINTS

16.2.2.1 Traffic congestions

$$TT_{ijkvth} = \frac{d_{ijk}}{S_{mrk}}\left[1 + a\left[\frac{X_{ijkvth}}{C_{ik}}\right]^{b}\right] \quad \forall\ t\in T, i\in I, j\in J, m\in M, r\in R, h\in H, v\in V, k\in K \quad (16.4)$$

To handle the traffic congestion problem, Eqn (16.4), the BPR (Bureau of Public Roads) equation, which was introduced in the Highway Capacity Manual (1965) was used. The BPR equation was modified to obtain transportation times of the modes for every time interval, t. The BPR equation constants were obtained from Horowitz (1985). In general, BPR equation constant b is assumed as "4" but in our case it is taken as "1" to handle the nonlinearity of the Eqn (16.18) because BPR equation depends on carried quantity by road transportation. This lower value of constant b causes speed to be insensitive to volume–capacity ratio until this ratio gets close to 1.0, and then the speed drops suddenly for higher values of constant b but in our case decrease of the speed happens slowly. However, the error gap between these two values is less than 5%. So, it is acceptable and used in our model.

16.2.2.2 Mode transfer

Constraint (16.5) indicates that any material in node i must be transported via at least one of the available transportation modes. This constraint forces the solver to obtain minimum transfer cost and transfer time in node i in terms of transportation modes. Equation (16.6) shows that the arrival transportation mode of the node i must be the same transportation mode of the leaving mode of the node j, otherwise if there is any difference between incoming and outgoing transportation modes, Eqn (16.6) ensures that there should be mode transfer in node i.

$$\sum_{l\in L} w_{iklth} \geq 1 \quad \forall\ t\in T, i\in I, k\in K, h\in H \quad (16.5)$$

$$y_{jikvth} + y_{ijkvth} \geq 2\, w_{iklth} \quad \forall\ t\in T, i\in I, j\in J, k\in K, v\in V, h\in H \quad (16.6)$$

16.2.2.3 Material balance, demand, and supply

Constraint (16.7) shows that incoming and outgoing quantities of carried goods have to be equal to demand of node i for transportation mode k in every time period h, t.

$$\sum_{v\in V}\sum_{j\in J} X_{jikvth} - \sum_{v\in V}\sum_{j\in J} X_{ijkvth} = \sum_{v\in V} D_{ikvth} \quad \forall\ t\in T, i\in I, k\in K, h\in H \quad (16.7)$$

Capacities of vehicle v are satisfied by Eqn (16.8) for transportation mode k in every time period h, t.

$$D_{ikvth}\, y_{ijkvth} \leq X_{ijkvth} \leq \left(Cap_v^{\max} - D_{ikvth}\right) y_{ijkvth} \quad \forall\ t\in T, i\in I, j\in J, k\in K, v\in V, h\in H \tag{16.8}$$

Constraint (16.9) indicates that there is equality between the arriving and leaving quantities in node i for each transportation mode k.

$$I_{ik(t-1)} + \sum_{t'\in T}\sum_{h'\in H}\sum_{v\in V}\sum_{j\in J} X_{jikvh't'} = \sum_{v\in V}\sum_{j\in J} X_{ijkvth} + I_{ikt} \quad \forall\ t\in T, i\in I, k\in K, h\in H \tag{16.9}$$

As the solution methodology, all cost matrices of transportation modes, transfer operations and distances between nodes are initialized as indicated in Eqns (16.10) and (16.11). By using Eqn (16.11), there is assessment of the node i and the node j whether these two nodes are adjacent or not and the model also ensures that there is enough demand of node i by using Eqn (16.12). If they are not connected or there is no demand in origin node i, it is not possible to transport any goods from node i to node j.

$$\sum_{j\in J} y_{ijkvth} \geq 0 \quad \forall\ t\in T, i\in I, k\in K, v\in V, h\in H \tag{16.10}$$

$$y_{ijkvth} = 0 \quad \forall\ t\in T, i\in I, j\in J, k\in K, v\in V, h\in H \ \text{ if}(i,j) \text{ isn't linked} \tag{16.11}$$

$$\sum_{j\in J} y_{ijkt} = 0 \quad \forall\ t\in T, i\in I, k\in K, v\in V, h\in H \quad \text{if } D_{ikvth} = 0 \tag{16.12}$$

Constraint (16.13) indicates that daily keeping inventory of every node i has to fulfill the capacity levels for every transportation mode k.

$$I_{ikt} \leq C_{ik} \quad \forall\ t\in T, i\in I, k\in K \tag{16.13}$$

16.2.2.4 Time windows

In this part of the problem, time-dependent travel durations are the step-wise function for a 24-h time period per day. This material balance in Eqn (16.9) has to be satisfied for every time interval t, h. However, the relationship between t and previous time intervals ($t' < t$) has to be as in Eqn (16.14). Arrival time of vehicle v on link (i, j) for transportation mode k in time period t, h which is equal to summation of departure time of vehicle v on link (i, j) for transportation mode k in time period t, h and transportation duration on link (i, j) for transportation mode k has to be in the range of single time period (between t and $(t-1)$) (Figure 16.3).

$$t-1 \leq t' + TT_{jikvt'h'} \leq t \quad \forall\ t'\in T \tag{16.14}$$

$$EW_{ikvt} \geq \sum_{h\in H}\left(\left[\Delta_{ikt} - t\alpha_{ijkvth}\right] + M3(1 - z1_{ijkvth})\right) \quad \forall\ t\in T, i\in I, j\in J, k\in K, v\in V \tag{16.15}$$

$$LW_{ikvt} \geq \sum_{h\in H}\left(\left[\Omega - t\alpha_{ijkvth} + \Delta_{ik(t+1)}\right] + M3(1 - z2_{ijkvth})\right) \quad \forall\ t\in T, i\in I, j\in J, k\in K, v\in V \tag{16.16}$$

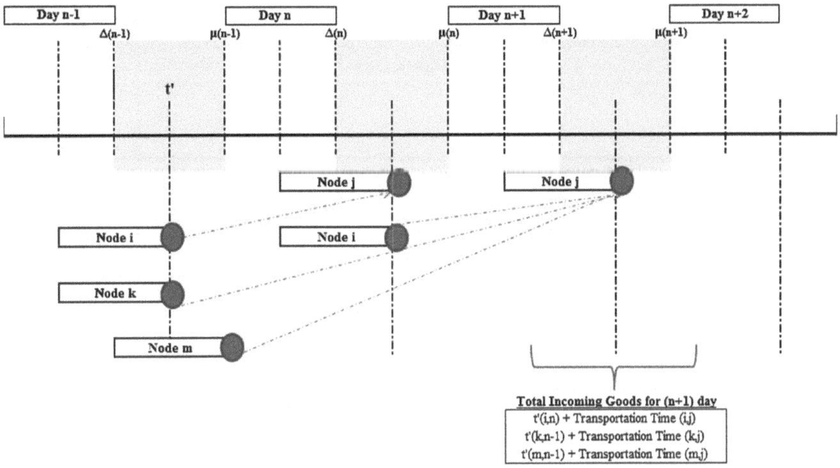

FIGURE 16.3

Sample schema of incoming goods.

$$NW_{ikvt} \geq \sum_{h \in H} \left([td_{ijkvth} - ta_{ijkvth}] + M3(1 - z3_{ijkvth}) \right) \quad \forall\ t \in T, i \in I, j \in J, k \in K, v \in V \quad (16.17)$$

$$z1_{ijkvth} + z2_{ijkvth} + z3_{ijkvth} = y_{ijkvth} \quad \forall\ t \in T, i \in I, j \in J, k \in K, v \in V, h \in H \quad (16.18)$$

Constraints (16.15)–(16.17) show waiting times of vehicle v, which arrives early, late, or in allowed time period, respectively. Constraint (16.18) ensures that vehicle v in node i by transportation mode k in day t has to arrive early, late, or in time, if there is any need to be carried goods on link (i, j). Figure 16.4 shows that waiting times of vehicles according to arrival times. Blue and red arrows indicate

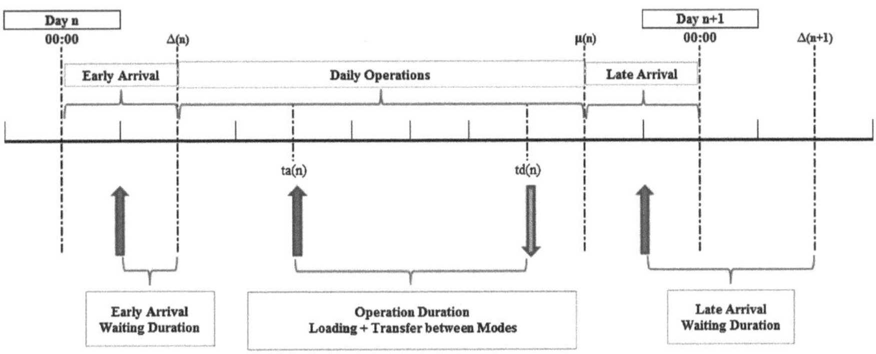

FIGURE 16.4

Timeline for daily operations.

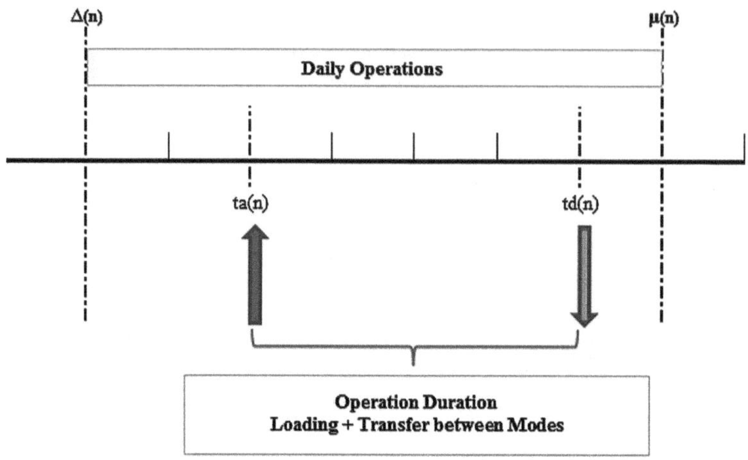

FIGURE 16.5

Schedule for daily operations.

the arrival and departure time of vehicle v in node i by transportation mode k, respectively.

$$td_{ijkvth} \geq ta_{ijkvth} + L_{ikv} + TrT_{kl} - M4(1 - y_{ijkvth}) \quad \forall\ t \in T, i \in I, j \in J, k \in K, v \in V, h \in H \tag{16.19}$$

$$\Delta_{ikt} \leq td_{ijkvth} \leq \mu_{ikt} \quad \forall\ t \in T, i \in I, j \in J, k \in K, v \in V, h \in H \tag{16.20}$$

Constraint (16.19) indicates relationship between departure and arrival times of vehicle v (red and blue flows in Figure 16.5, respectively) in node i for transportation mode k in time period t, h. Time duration between arrival and departure time (indicated as ta and td in Figure 16.5) of vehicle v has to be greater than summation of loading time of vehicle v for mode k and transfer times from mode k to l. Constraint (16.20) ensures that departure time of vehicle v has to be between opening and closing time of node i for transportation mode k in time period t, h.

Constraint (16.21) ensures that time period at which vehicle v leaves the node i by transportation mode k under boundary limits.

$$g1_{ijkvmr} b_{ijkv(m-1)} \leq g2_{ijkvmr} \leq g1_{ijkvmr} b_{ijkvm} \quad \forall\ i \in I, j \in J, m \in M, k \in K, v \in V, r \in R \tag{16.21}$$

Constraint (16.22) indicates that travel duration (of which vehicle v from node i to node j by transportation mode k) used in the carbon dioxide emission calculations is at most equal to transportation duration of speed calculations.

$$\sum_{m \in M} \sum_{r \in R} (\theta_{ijkvmr} g1_{ijkvmr} + \omega_{ijkvmr} g2_{ijkvmr}) \leq \sum_{h \in H} TT_{ijkvth} \quad \forall\ t \in T, i \in I, j \in J, v \in V, k \in K \tag{16.22}$$

16.2.2.5 Supplementary constraints

$$X_{ijkvth} \leq Q_{ijkvth} \quad \forall\ t \in T, i \in I, j \in J, k \in K \tag{16.23}$$

$$X_{ijkvth} \leq M \times y_{ijkvth} \quad \forall\ t \in T, i \in I, j \in J, k \in K \tag{16.24}$$

$$X_{ijkvth} \geq Q_{ijkvth} - M(1 - y_{ijkvth}) \quad \forall t \in T, i \in I, j \in J, k \in K \tag{16.25}$$

$$J_{ijkvth} \leq TT_{ijkvth} \quad \forall t \in T, i \in I, j \in J, k \in K \tag{16.26}$$

$$J_{ijkvth} \leq M \times y_{ijkvth} \quad \forall\ t \in T, i \in I, j \in J, k \in K \tag{16.27}$$

$$J_{ijkvth} \geq TT_{ijkvth} - M(1 - y_{ijkvth}) \quad \forall\ t \in T, i \in I, j \in J, k \in K \tag{16.28}$$

By using Eqns (16.23)–(16.25) for cost functions and Eqns (16.26)–(16.28) for time functions, new linear conditions for the objective functions can be obtained. Constraints (16.29)–(16.34) ensure both integer and nonnegativity conditions.

$$X_{ijkvth}, Q_{ijkvth}, J_{ijkvth}, TT_{ijkvth}, ta_{ijkvth}, td_{ijkvth} \geq 0 \quad \forall\ t \in T, i \in I, j \in J, h \in H, v \in V, k \epsilon K \tag{16.29}$$

$$z1_{ijkvth}, z2_{ijkvth}, z3_{ijkvth} \geq 0 \quad \forall\ t \in T, i \in I, j \in J, h \in H, v \in V, k \in K \tag{16.30}$$

$$EW_{ikvth}, LW_{ikvth}, NW_{ikvth} \geq 0 \quad \forall\ t \in T, i \in I, h \in H, v \in V, k \in K \tag{16.31}$$

$$y_{ijkvth}, w_{iklth}, \in \{0, 1\} \quad \forall\ t \in T, i \in I, j \in J, h \in H, v \in V, k \in K \tag{16.32}$$

$$g1_{ijkvmr}, g2_{ijkvmr} \in \{0, 1\} \quad \forall\ t \in T, i \in I, j \in J, m \in M, r \in R, v \in V, h \in H, k \in K \tag{16.33}$$

$$I_{ikth} \geq 0 \quad \forall\ t \in T, i \in I, k \in K, h \in H \tag{16.34}$$

16.2.3 SOLUTION METHODOLOGIES

For the solution methodology, (AUGMECON) ∊-constraint method (Mavrotas, 2009) should be used to find the Pareto optimal solutions for decision makers. The relationship between objective functions will be determined and applied for different problem cases. However, finding the optimal solutions for this kind of problems is seriously time consuming and difficult for the current solvers because the proposed problems are very complex and NP-hard due to the large number of constraints and indices. Therefore, reducing the complexity of linear integer problems of transportation is the major area of recent studies. In the literature, heuristics, meta-heuristics, and hybrid genetic algorithms form a part of the solution methodologies for linear integer problems (e.g., Limbourg and Jourquin, 2009; Ishfaq and Sox, 2012). However, in the study of Resat and Turkay (under review), an exact method is used to eliminate the lower accuracy of heuristics methods. Especially, preprocessing and logic-cut methods are used to help solvers to obtain more accurate and optimal solutions by decreasing the complexity and division of problems. Although the logic-cuts increase the number of constraints, it would reduce the relaxation gap and would strengthen convex hull approximation.

Therefore, the solver will obtain better, more feasible solutions and finding optimal solution would require shorter CPU times. Also, addition of these extra constraints because of logic-cut and preprocessing methods will not change the integer optimal solutions. As a summary, proposed multiobjective problems can be solved by integer linear optimization models also, as indicated in the study of Fattahi and Turkay (under review a), some logic-cuts and preprocessing methods are proposed and solutions are found in better and faster ways with the help of these logic-cuts and preprocessing methods. Additionally, a new exact algorithm for solving mixed integer multiobjective optimization problems is developed by Fattahi and Turkay (under review b). A penalty parameter is added to the objective function of the sub problems. However, there are some efficiency problems and missing solutions in the epsilon constraint algorithm when the solver search for the efficient solutions below the bounds over function of system parameters The proposed algorithm is tested on various benchmark problems including assignment, knapsack, and set covering. Analysis of the results indicates that the proposed algorithm is able to obtain the entire efficient set for the benchmark problems more effectively.

16.3 CONCLUSIONS

Although there is an increasing amount of studies in the literature, unfortunately there is no commonly agreed definition of sustainability in transportation operations. Therefore, this study tries to foster sustainable transportation issue by giving an extended literature survey. Improvement of the transportation systems by developing more sustainable systems or promoting sustainability within the different functionalities of the sector can be achieved through many performance indexes. In this study, the main indicators for the economic, environmental, and social factors of sustainability of transportation are indicated, and theoretical studies and methodologies are investigated for decision makers and stakeholders from different areas to assess the current situation of the sector. Moavenzadeh and Markow (2007) suggest the following strategies that are based on learning from multiple contexts to improve the sustainability concept in transportation:

- Focusing on demand and supply in order to understand the major requirements and needs of mobility activities;
- Improving existing transport system performance by focusing on specific performance indexes;
- Understanding the needs of competitive market and policies by trade-off analysis; and
- Improving management capability, which directly improves the decisions and flow of information.

APPENDIX

The indices, scalars, parameters, and variables of mixed integer linear programming for multiobjective problem are given below:

INDICES

i	Set of all nodes in the intermodal transportation network ($i = 1,\ldots,I$)
j	Set of all nodes in the intermodal transportation network ($j = 1,\ldots,J$)
k	Set of all transportation modes ($k = 1,\ldots,K$)
l	Set of all transportation modes ($l = 1,\ldots,L$)
r	Set of speed levels of vehicles ($r = 1,\ldots,R$)
v	Set of vehicles used in transportation ($v = 1,\ldots,V$)
t	Set of daily time intervals ($t = 1,\ldots,T$)
h	Set of time intervals in 15 min ($h = 1,\ldots,H$)
m	Set of regions of speed module based on congestion ($m = 0,1,2,3$)

Many constants are used in the both speed and carbon dioxide emission formulations. Also, some big values are used for linearization of this model.

SCALARS

a	BPR equation constant
A	Carbon dioxide emission rate constant
b	BPR equation constant
B	Carbon dioxide emission rate constant
C	Carbon dioxide emission rate constant
$Alpha$	Initial congestion duration (min)
ε	Epsilon value
σ	Curb weight (kg)
Ω	Daily period (h)
$M1, M2, M3, M4$	Big-M parameter values

PARAMETERS

C_{ik}	Capacity of the node i with respect to transportation mode k
Cap_v^{max}	Maximum capacity level of vehicle v
d_{ijk}	Distance between node i to node j by transportation mode k
D_{ikvth}	Demand of the node i with respect to transportation mode k in time interval t
S_k	Speed limit of transportation mode k for free-flow transportation
b_{ijkvm}	Time constant of corresponding travel time module m for vehicles v traveling on link (i, j) by transportation mode k,

$$b_{ijkvm} = \begin{cases} 0 & m = 0 \\ \left(alpha - \dfrac{d_{i,j,k}}{S_k^c} \right) & m = 1 \\ alfa & m = 2 \end{cases}$$

Δ_{ikt} Opening time of node i for transportation mode k in time period t

μ_{ikt} Closing time of node i for transportation mode k in time period t

θ_{ijkvmr} Carbon dioxide emission equation constant,

$$\theta_{ijkvmr} = \begin{cases} 0 & m = 1,3 \\ \dfrac{S_K^{2r} - S_K^{3r}}{S_K^{3r}} & m = 2 \end{cases}$$

ω_{ijkvmr} Carbon dioxide emission equation constant,

$$\omega_{ijkvmr} = \begin{cases} 0 & m = 0 \\ \dfrac{d_{ijk}}{S_k^{1r}} & m = 1 \\ \dfrac{d_{ijk}}{S_k^{1r}} + \left(\dfrac{S_k^{3r} - S_k^{2r}}{S_k^{3r}}\right) & m = 2 \\ \dfrac{d_{ijk}}{S_k^{3r}} & m = 3 \end{cases}$$

$G1$ Any large value greater than upper bound of the secondary objective function

$G2$ Any large value greater than upper bound of the third objective function

$P1$ Penalty cost of waiting time because of early arrival

$P2$ Penalty cost of waiting time because of late arrival

$P3$ Penalty cost of waiting time because of operational process

TC_{ijk} Transportation cost from node i to node j by transportation mode k

TrC_{kl} Transfer cost from transportation mode k to transportation mode l

TrT_{kl} Transfer time from transportation mode k to transportation mode l

L_{ikv} Loading time of vehicle v in node i for transportation mode k

VARIABLES

X_{ijkvth} Quantity of carried goods on link (i, j) by transportation mode k in time t

Q_{ijkvth} Quantity of carried goods on link (i, j) by transportation mode k in time t

J_{ijkvth} Transportation duration of link (i, j) by transportation mode k in time t

TT_{ijkvth} Transportation duration of link (i, j) by transportation mode k in time t

ta_{ijkvth} Arrival time of vehicle v in node i for transportation mode k in time period

td_{ijkvth} Departure time of vehicle v in node i by transportation mode k in period t, h

EW_{ikvt} Waiting time of vehicle v in node i for transportation mode k in time period t because of early arrival

LW_{ikvt} Waiting time of vehicle v in node i for transportation mode k in time period t because of late arrival

NW_{ikvt} Waiting time of vehicle v in node i for transportation mode k in time period t because of in time arrival

$g2_{ijkvmr}$	Time duration for vehicle v traveling from node i to node j by transportation mode k and speed level r in time interval m
I_{ikt}	Inventory of the node i with respect to transportation mode k in time interval t
γ_{vkth}	Occupancy rate of vehicle v for transportation mode k in time period t, h
U	Slack variable of the system

BINARY VARIABLES

$$y_{ijkvth} \quad \begin{cases} 1 & \text{if transportation mode } k \text{ is chosen on link } (i,j) \text{ in time interval } t \\ 0 & \text{else} \end{cases}$$

$$w_{iklth} \quad \begin{cases} 1 & \text{if transfer from transportation mode } k \text{ to } l \text{ in node } i \text{ in time interval } t,\ h \\ 0 & \text{else} \end{cases}$$

$$z1_{ijkvth} \quad \begin{cases} 1 & \text{if vehicle } v \text{ in node } i \text{ for transportation mode } k \text{ in period } t \text{ arrives early} \\ 0 & \text{else} \end{cases}$$

$$z2_{ijkvth} \quad \begin{cases} 1 & \text{if vehicle } v \text{ in node } i \text{ for transportation mode } k \text{ in period } t \text{ arrives late} \\ 0 & \text{else} \end{cases}$$

$$z3_{ijkvth} \quad \begin{cases} 1 & \text{if vehicle } v \text{ in node } i \text{ for transportation mode } k \text{ in period } t \text{ arrives in time} \\ 0 & \text{else} \end{cases}$$

$$g1_{ijkvmr} \quad \begin{cases} 1 & \text{if vehicle } v \text{ departs node } i \text{ in time interval } m \text{ by mode } k \text{ and speed } r \\ 0 & \text{else} \end{cases}$$

REFERENCES

Anderstig, C., Mattsson, L., 1998. Modelling land-use and transport interaction: evaluations and policy analysis. In: Lundqvist, L., Mattsson, L.G., Kim, T.J. (Eds.), Network Infrastructure and the Urban Environment: Recent Advances in Land-Use/Transportation Modelling. Springer, Berlin, pp. 308–328.

Ban, J.X., Liu, H.X., Ferris, M.C., Ran, B., 2006. A general MPCC model and its solution algorithm for continuous network design problem. Math. Comput. Model 43, 493–505.

Banister, D., 2008. Cities, mobility, and climate change. J. Ind. Ecol. 7–10.

Bektas, T., Laporte, G., 2011. The pollution-routing problem. Transp. Res. Part B 45 (8), 1232–1250.

Center for Sustainable Transport (CST), 1998. Sustainable Transportation Monitor. No. 1, Toronto, Ontario, Canada.

Cidell, J., 2012. Sustainable transportation: accessibility, mobility, and derived demand. In: Theis, T., Tomkin, J. (Eds.), Sustainability: A Comprehensive Foundation, pp. 566–576. U of I Open Source Textbook Initiative. Retrieved October 8, 2012.

Demir, E., Bektas, T., Laporte, G., 2014. The bi-objective pollution–routing problem. Eur. J. Oper. Res. 232 (3), 464–478.

Dowling, R., Ireson, R., Skabardonis, A., Gillen, D., Stopher, P., 2005. Predicting Air Quality Effects of Traffic-Flow Improvements: Final Report and User's Guide. Technical Report No. 535. National Cooperative Highway Research Program, Transportation Research Board, National Research Council, Washington, DC.

EPA, 2011. http://www.epa.gov/dced/pdf/Sustainable_Transpo_Performance.pdf.
Fattahi, A., Turkay, M. Sea-Road-Rail Intermodal Network Flow Problem with Preprocessing and Logic Cuts, under review a.
Fattahi, A., Turkay, M. One Direction Search for Exact Nondominated Frontiers of Biobjective Mixed-Integer Linear Programming Problems, under review b.
Ferguson, E.M., Duthie, J., Waller, S.T., 2010. Network Methods for Project Selection Based on Optimizing Environmental Impact. Research Report SWUTC/10/161026-1. University of Texas at Austin, Austin.
Figliozzi, M.A., 2011. The impacts of congestion on time-definitive urban freight distribution networks CO_2 emission levels: results from a case study in Portland, Oregon. Transp. Res. Part C: Emerging Technol. 19, 766–778.
Francheschetti, A., Honhon, D., van Woensel, T., Bektas, T., Laporte, G., 2013. The time–dependent pollution–routing problem. Transp. Res. Part B: Methodol. 56, 265–293.
Garret, M., Wachs, M. (Eds.), 1996. Transportation Planning on Trial: The Clean Air Act and Travel Forecasting. Sage, Thousand Oaks, CA.
Haghshenas, H., Vaziri, M., 2012. Urban sustainable transportation indicators for global comparison. Ecol. Ind. 15, 115–121.
Haughton, G., 1999. Environmental justice and the sustainable city. J. Plan. Educ. Res. 18 (3), 233–243.
Highway Capacity Manual, 1965. Highway Research Board, Special Report 87. National Academy of Sciences, National Research Council, Washington, DC. Publication 1328.
Horowitz, A.J., 1985. Realistic Traffic Delay in Region Wide Traffic Assignment: Implications of the Third Edition of the 1985.
Huang, K., Zhang, J., He, M., Liao, W., 2010. An optimal model and solution algorithm of urban traffic network considering exhaust emission control. In: Proceedings of the 2010 International Conference of Logistics Engineering and Management, pp. 528–534.
Ishfaq, R., Sox, C.R., 2012. Design of intermodal logistics networks with hub delays. Eur. J. Oper. Res. 220 (3), 629–641.
Jeon, C.M., Amekudzi, A., 2005. Addressing sustainability in transportation systems: definitions, indicators, and metrics. J. Infrastruct. Syst. Am. Soc. Civil Eng. (ASCE) 11 (1), 31–50.
Jones, W.B., Cassady, C.R., Bowden, R.O., 2000. Developing a standard definition of intermodal transportation. Transport. Law J. 27 (3), 345–352.
Kim, B.J., Kim, W.K., Song, B.H., 2008. Sequencing and scheduling highway network expansion using a discrete network design model. Ann. Reg. Sci. 42 (3), 621–642.
Li, Z.C., Lam, W.H.K., Wong, S.C., Sumalee, A., 2012. Environmentally sustainable toll design for congested road networks with uncertain demand. Int. J. Sust. Transp. 6 (3), 127–155.
Liao, C.H., Chang, C.L., Su, C.Y., Chiueh, P.T., 2013. Correlation between land-use change and greenhouse gas emissions in urban areas. Int. J. Environ. Sci. Technol. 10 (6), 1275–1286.
Limbourg, S., Jourquin, B., 2009. Optimal rail-road container terminal locations on the European network. Transp. Res. (E) 45 (4), 551–563.
Litman, T., 2009. Sustainable Transportation Indicators: A Recommended Research Program for Developing Sustainable Transportation Indicators and Data. Indicators Subcommittee of the Transportation Research Board.
Lo, H.K., Szeto, W.Y., 2009. Time-dependent transport network design under cost-recovery. Transp. Res. Part B 43 (1), 142–158.

Low, N., Gleeson, B. (Eds.), 2003. Making Urban Transportation Sustainable. Palgrave/Macmillan, Basingstoke, Hampshire (UK).
Lowry, I.S., 1964. A Model of Metropolis. RM-4035-rc. The Rand Corporation, Santa Monica.
Martinez, F.J., 1996. MUSSA: a land use model for Santiago city. Transp. Res. Rec. 1552, 126–134.
Mavrotas, G., 2009. Effective Implementation of the ε Constraint Method in Multi-Objective Mathematical Programming Problems. #213 pp. 455–465.
McGranahan, G., Satterthwaite, D., 2000. Environmental health or ecological sustainability? Reconciling the brown and green agendas in urban development. In: Pugh, C. (Ed.), Sustainable Cities in Developing Countries. Earthscan Publications, London.
Miller, E., Kriger, D., Hunt, J., 1999. Integrated urban models for simulation of transit and land use policies. In: Final Report, Project H-12, TCRP Web Document 9. Transit Cooperative Highway Research Project, National Academy of Sciences, Washington, DC.
Moavenzadeh, F., Markow, M., 2007. Moving Millions: Transport Strategies for Sustainable Development in Megacities. Springer, Dordrecht.
Newman, P., Kenworthy, J., 1999. Sustainability and Cities. Overcoming Automovile Depencence. Island Press, Washington.
Nichols, J.E., Garrick, N.W., Atkinson-Palombo, C., 2008. A framework for developing indicators of sustainability for transportation planning. In: Presented at the 88th Annual Meeting of the Transportation Research Board, Washington DC, January 11–15.
Nicolas, J., Pochet, P., Poimboeuf, H., 2003. Towards sustainable mobility indicators: application to the Lyons conurbation. Transp. Policy 10, 197–208.
Paulley, N., Webster, F., 1991. Overview of an international study to compare models and evaluate land use and transport policies. Transp. Rev. 11, 197–222.
Putman, S.H., 1983. Integrated Urban Models: Policy Analysis of Transportation and Land Use. Pion, London.
Resat, H.G., Turkay, M. Design and Operation of Intermodal Transportation Network in the Marmara Region of Turkey, under review.
Santos, A.S., Ribeiro, S.K., 2013. The use of sustainability indicators in urban passenger transport during the decision-making process: the case of Rio de Janeiro, Brazil. Curr. Opin. Environ. Sustainability 5, 251–260.
Savelson, A., Colman, R., Litman, T., Walker, S., Parmenter, R., 2006. The GPI Transportation Accounts: Sustainable Transportation in Nova Scotia. GPI, Atlantic.
Schiller, P.L., Bruun, E.C., Kenworthy, J.R., 2010. An Introduction to Sustainable Transportation: Policy, Planning, and Implementation. Earthscan, Washington.
Sharaki, N., Turkay, M., 2014. Analysis of interaction among land use, transportation network and air pollution using stochastic nonlinear programming. Int. J. Environ. Sci. Technol. 11 (8), 2201–2216.
Sharma, S., Mathew, T.V., 2011. Multi-objective network design for emission and travel time tradeoff for sustainable large urban transportation network. Environ. Plan B 38 (3), 520–538.
Simmonds, D.C., Still, B.G., 1998. DELTA/START: adding land use analysis to integrated transport models. In: The 8th World Conference on Transport Research, Antwerpen.
Southworth, F., 1995. A Technical Review of Urban Land Use: Transportation Models as Tools for Evaluating Vehicle Reduction Strategies. US Department of Energy, Washington, DC.
Tanguay, G.A., Rajaonson, J., Lefebvre, J., Lanoie, P., 2010. Measuring the sustainability of cities: an analysis of the use of local indicators. Ecol. Ind. 10, 407–418.

Timmermans, H., 2003. The saga of integrated land use-transport modeling: how many more dreams before we wake up? In: 10th International Conference on Travel Behavior Research, Lucerne, International Association for Travel Behavior Research, 10–15 August.
Turkay, M., 2008. Environmentally conscious supply chain management (Chapter 3). In: Papageorgiou, L., Georgiadis, M. (Eds.), Process Systems Engineering, Supply Chain Optimization, vol. 3. WILEY-VCH, Weinheim, pp. 45–86.
US Department of Energy, 2009. US Primary Energy Consumption by Source and Sector, 2008.
Vreeker, R., Nijkamp, P., 2005. Multicriteria evaluation of transport policies. In: Button, K.J., Hensher, D.A. (Eds.), Handbook of Transport Strategy, Policy and Institutions. Elsevier, Oxford, pp. 507–526.
Vuchic, V.R., 2007. Urban Transit Systems and Technology. John WIley & Sons, Hoboken.
Waddell, P., Ulfarsson, G., Franklin, J., Lobb, J., 2007. Incorporating land use in metropolitan transportation planning. Transp. Res. A Policy Pract. 41, 382–410.
Wegener, M., 1998. The IRPUD Model: Overview.
Wegener, M., 2004. Overview of land use transport models. In: Hensher, D.A., Button, K.J., Haynes, K.E., Stopher, P.R. (Eds.), Handbook of Transport Geography and Spatial Systems. Pergamon, Amsterdam, pp. 127–146.
World Commission on Enviroment and Development (WCED), 1987. Our Common Future. Oxford University Press, Oxford, England.
Yang, H., Xu, W., He, B., Meng, Q., 2010. Road pricing for congestion control with unknown demand and cost functions. Transp. Res. Part C 18 (2), 157–175.
Yevdokimov, Y.V., 2003. Vertical linkages of a sustainable transportation system. In: Proceedings of the 38th Annual Conference of Canadian Transportation Research Forum, May 11–14, Ottawa, pp. 136–147.
Yin, Y., Lawphongpanich, S., 2006. Internalizing emission externality on road networks. Transp. Res. Part D 11 (4), 292–301.
Yin, Y., Lu, H., 1999. Traffic equilibrium problems with environmental concerns. J. East Asia Soc. Transp. Stud. 3 (6), 195–206.
Zegras, C., 2006. Sustainable transport indicators and assessment methodologies. In: Biannual Conference and Exhibit of the Clean Air Initiative for Latin American Cities, Sao Paulo, Brazil.
Zhang, Y., Guindon, B., 2006. Using satellite remote sensing to survey transportation related urban sustainability: part 1: methodologies for indicator quantification. Int. J. Appl. Earth Observ. Geoinform. 8, 149–164.
Zhong, R.X., Sumalee, A., Maruyama, T., 2012. Dynamic marginal cost, access control, and pollution charge: a comparison of bottleneck and whole link models. J. Adv. Transp. 46 (3), 191–221.
Zhou, S., Yan, X., Wu, C., 2008. Optimization model for traffic signal control with environmental objectives. In: Proceedings of the 4th International Conference on Natural Computation, Jinan, October 18–20, pp. 530–534.
Zito, P., Salvo, G., 2011. Toward an urban transport sustainability index: an European comparison. Eur. Transp. Res. Rev. 3, 179–195.

CHAPTER

17 Dynamic Optimization for the Optimal Location of New Industrial Facilities Considering the Sustainability of the Watershed

Luis Fernando Lira-Barragán*, Antonio Flores-Tlacuahuac§,1, Martín Rivera-Toledo§, José María Ponce-Ortega*, Medardo Serna-González*

**Chemical Engineering Department, Universidad Michoacana de San Nicolás de Hidalgo, Morelia, Mich, México*

§Departamento de Ingeniería y Ciencias Químicas, Universidad Iberoamericana, México D. F, México

[1]Corresponding author: E-mail: antonio.flores@uia.mx

17.1 INTRODUCTION

Watersheds are the most important natural reservoirs where water streams are transported until their final disposal (i.e., lake, sea, or ocean). Several human activities are established in the surrounding areas of watersheds (such as industrial, domestic, and agricultural) significantly affecting the water availability and quality. These activities are primarily responsible for the pollution of rivers. For this reason, it is very important to study the impact of new industrial discharges prior to determining the location of new industrial facilities.

In this context, in determining the final site to install a new industrial plant usually only economic aspects are considered, without taking into account the environmental impact that the new industrial effluents will have on the watershed (El-Baz et al., 2005a). Once the new plant is installed, the authorities require industry to satisfy environmental regulations; consequently it is often necessary to install treatment networks for the wastewater streams, which represents an additional cost that may yield uneconomical processes. Several methodologies have been reported to address the problem of recycle/reuse for resource conservation inside industrial facilities (see for example, Dunn and El-Halwagi, 2003; El-Halwagi, 1997, 2006; El-Halwagi and Spriggs, 1998). Moreover, resource conservation methodologies

Computer Aided Chemical Engineering, Volume 36. ISSN 1570-7946. http://dx.doi.org/10.1016/B978-0-444-63472-6.00017-3

can be classified as pinch techniques and techniques formulated as mathematical programming models (Pillai and Bandyopadhyay, 2007). The pinch technologies use heuristic rules based on targeting, which conduct to sequential approaches. Foo (2009) published a review of pinch techniques for the water network synthesis (see for example, Wang and Smith, 1994; Dhole et al., 1996; Polley and Polley, 2000; Hallale, 2002; Manan et al., 2004; El-Halwagi et al., 2003; Feng et al., 2007). Other sets of methodologies based on algebraic techniques for the conservation of resources have been reported by Sorin and Bedard (1999), Agrawal and Shenoy (2006), and Gomes et al. (2007). Finally, mathematical programming formulations that automate the synthesis of recycle and reuse networks inside the plant have been reported (see for example, Savelski and Bagajewicz, 2000, 2001, 2003; Alva-Argaez et al., 1998, 1999; Benko et al., 2000; Teles et al., 2008; Kuo and Smith, 1997; Doyle and Smith, 1997; Galan and Grossmann, 1998; Hernandez-Suarez at al. 2004; Gunaratnam et al., 2005; Karuppiah and Grossmann, 2006; Putra and Amminudin, 2008; Ponce-Ortega et al., 2009a; Nápoles-Rivera et al., 2010). However, all previous formulations consider exclusively the processes that happen inside the industrial facilities, ignoring the processes that occur outside the plant, and the impact of wastewater streams when they enter the environment (i.e., watersheds) has not been considered.

On the other hand, watersheds are also impacted by other polluted discharges from different types of sources like municipal zones, agricultural areas, treated and untreated wastewater, precipitation, etc. It is worth noting that all these discharges drastically change the conditions of the streams in the rivers; nonetheless, in the rivers, natural degradation of the pollutants due to the chemical, biochemical, and physical transformations of the flora and fauna (see Brunner and Rechberg, 2004) occurs. For modeling pollutant behavior in a watershed system, Baccini and Brunner (1991) developed a material flow analysis (MFA) model for analyzing ecosystems with human activities where mass, energy, and information are being exchanged with the surroundings. Additionally, Lampert and Brunner (1999) proposed an MFA model to track nutrient discharges into the Danube river basin. El-Baz et al. (2005a) presented an MFA model to track the nitrogenous compounds in the Bahr El-Baqar drainage system, and El-Baz et al. (2005b) presented an MFA model to include mass integration for the Bahr El-Baqar drainage system in Egypt. Lovelady et al. (2009) presented a reverse problem formulation to integrate the discharges of an industrial process with macroscopic watershed systems based on an MFA model, where the key idea consists in fixing the maximum concentration for the pollutants discharged to the environment to avoid hazardous accumulation at the final disposal and ensure sustainable processes. However, the main drawback of this model is that it does not optimize the process simultaneously and it uses a set of heuristic rules.

Lira-Barragán et al. (2011a,b, 2012) have proposed novel approaches to determine the optimal location for new industrial plants, considering the impact of the new polluted discharges on the watersheds and the composition of key pollutants or the properties of interest. However, the previous methodologies have not considered the dynamics related to this type of processes, which yields inadequate solutions owing to the fact that all the above mentioned interactions among the watersheds and their surroundings are not constant and their conditions change with respect to time. Consequently, conditions of the watersheds are modified and the environmental restrictions may not be satisfied during critical periods. In this regard, the reviews by Mendez et al. (2006), Floudas and Lin (2004), Kallrath (2002), and Barbosa-Povoa (2007) highlighted the importance of scheduling optimization methods in the process industry, whereas other authors have successfully applied such techniques to solve various manufacturing problems (see for example, Mouret et al., 2009, 2011; Sharda and Vazquez, 2009; Terrazas-Moreno et al., 2012; Gutiérrez-Limón et al., 2012; Erdirik-Dogan and Grossmann, 2008; Barbosa-Povoa and Macchietto, 1994; Lin and Floudas, 2001; Castro et al., 2005). However, there are no publications on dynamically optimizing the locations of new industrial plants with the watersheds.

To overcome the above drawbacks, this work presents a dynamic model formulation to integrate the water treatment networks for new plants together with the watershed system, where the pollutants from the new plants are discharged and considering the possibility of installing new plants in different available locations. The model considers the fluctuations of the processes related to the watershed to identify the watershed behavior during the year. The optimal quantity and quality for the wastewater discharged from the new plant depends on the location selected to install the plant because the impact on the watershed depends strongly on the location selected (considering the integration with other streams and the chemical and biochemical reactions that occur in the watershed system); and the treatment required is highly influenced by the location of the new plant. The proposed model is a disjunctive programming formulation that is reformulated as a mixed-integer dynamic optimization problem considering the minimization of the total annual cost for the recycle and reuse network and the installation of the new plant to satisfy the process and environmental regulations as well as the sustainability of the watershed system. The model is able to consider multiple pollutants in the waste streams and the selection of several technologies to treat them.

17.2 PROBLEM STATEMENT

The problem addressed in this work can be stated as follows.

- A watershed system with a mouth in a final disposal (i.e., lake, sea, or ocean). The problem considers all the fluctuations and dynamic changes in the subsystem that interact with the watershed such as industrial, residential, agricultural discharges and extractions as well as the seasonal variations for the precipitation, filtration, evaporation etc. The watershed is divided in reaches ($R = \{r|r = 1,2,\ldots,N_r\}$) and tributaries ($T = \{t|t = 1,2,\ldots,N_t\}$) to simulate its behavior.

- A set of alternatives to locate a new industrial facility ($P = \{p|p = 1,2,\ldots,N_p\}$). Each allowed location p has an associated installation cost ($Cland_p$). This term includes the cost associated with the transportation of the raw materials, final products, and services as well as the land cost.
- A set of pollutants ($C = \{c|c = 1,2,\ldots,N_c\}$) discharged from the new industrial plant to the watershed.
- A set of interceptors to treat the first compound (INT^1), the second pollutant (INT^2), and for the N component considered by the system, which is treated in the last set of interceptors (INT^N). In this case, also the operation for the treatment system is carried out under dynamic conditions accounting for the variations in the watershed. Each interceptor has an associated correction factor to simulate its capacity to remove the components.
- Finally, a set of environmental constraints that guarantee the continuity in the typical activities carried out around the watershed and to keep the system under sustainable conditions once the new industrial plant is operating.

Then, the problem consists in determining the optimal location for the new industrial facilities simultaneously with the design and configuration of the interception network under dynamic operation that satisfies environmental regulations and constraints for the pollutants through the watershed system. At the same time it must avoid accumulation of the pollutants at the final disposal. The objective function consists in the minimization of the total annual cost (TAC), including the installation cost for the new plant (which depends strongly of the selected alternative) and the treatment costs for the wastewater stream.

The main nonlinearities included in the mathematical formulation can be found in the component balances for the reaches, specifically in the product of the flow rate and the composition for each reach. In this context, the characteristics of each reach are known prior to the installation of the new industrial facilities; then the consideration of installing the new industrial plant and consequently the possible discharge of this new wastewater to the watershed system (the composition of the waste stream is a variable in the model) transform the flow rate and the compositions of each reach into optimization variables. Moreover, the pollutant concentration (this variable is modified by the selected treatment scheme) represents one of the most important variables to be determined by the optimization process since it must satisfy the environmental regulations imposed for industrial discharges. It is also necessary that the new discharge does not affect the human activities carried out over the watershed (for example, if a water stream is extracted for an agricultural use, the new polluted discharge must not prevent water for agricultural deployment from meeting water quality constraints). The accumulation of pollutants at the final disposal must be avoided, keeping the system under sustainable conditions once the new industrial plant is operating.

The above scenario is illustrated in Figure 17.1. This figure also shows the subdivision of the watershed required by the MFA technique to model the system; thus, the main stream of the river is divided in several reaches (these are represented through ovals) and each reach (a zone where no extensive intermediate effluents

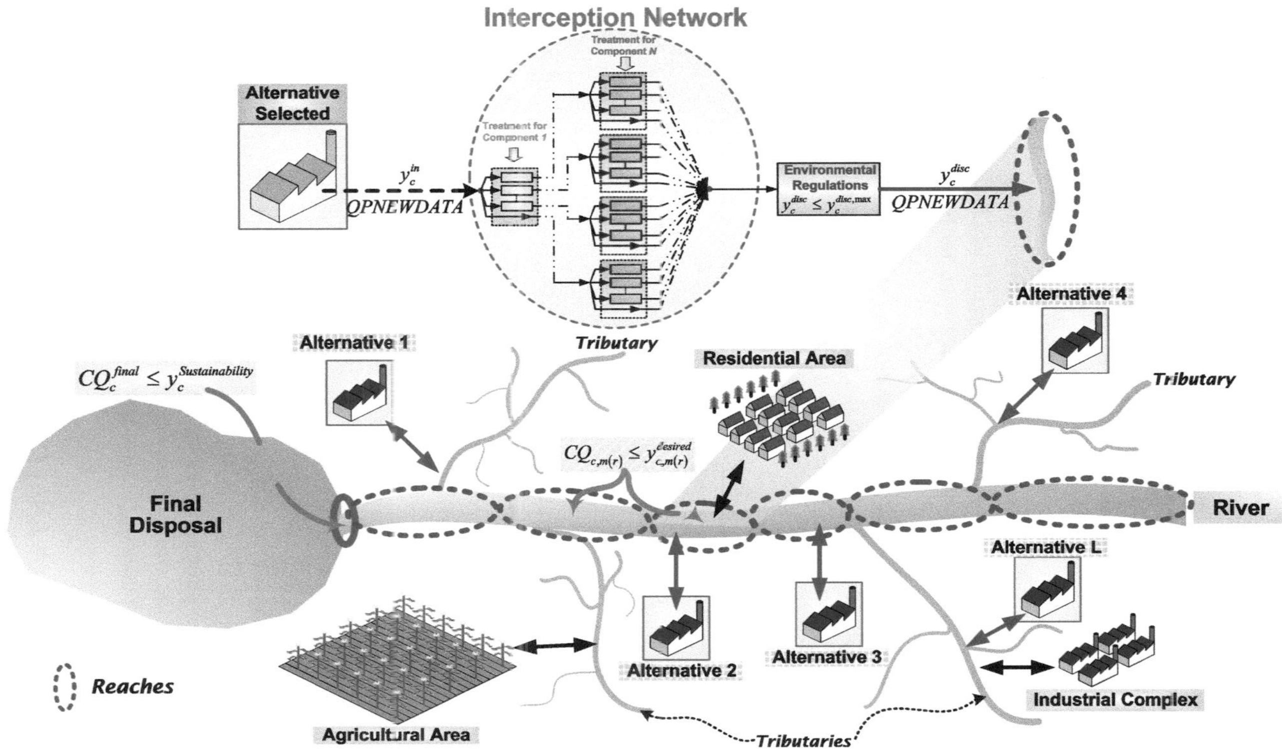

FIGURE 17.1

Alternatives to locate the new industrial plant in a watershed system and the proposed treatment scheme.

are discharged and/or extracted) can receive several tributaries (a tributary can be a channel or a branch of the river). Commonly, watersheds during their trajectories exchange water streams with, e.g., agricultural, industrial, and residential zones. Additionally, some natural phenomena that occur in these systems are precipitation, vaporization, and filtration. All the previous processes significantly modify the characteristics of the river and if a simulation of the river behavior is needed, it is necessary that the model chosen to carry out this task includes the above mentioned items as well as a kinetic term to consider the chemical and biochemical interactions. These processes take place between the pollutants and the aquatic flora and fauna, which can decompose or produce hazardous materials.

Finally, an important aspect accounted for in this work is the optimization of flow rates and compositions included in the mathematical formulation through the dynamic optimization formulation, which generates time profiles for flow rates and pollutant compositions during the year. In this regard, it can be very useful to introduce the seasonal variation of the processes related to the watershed system since they can be drastically different through the year. In particular, the flow rate and components' concentration change strongly during the year.

17.3 MODEL FORMULATION

First, the sets used in the model formulation are defined for a better understanding: the subindex that symbolizes the pollutant or component is c, so N_c corresponds to the total number of pollutants, and C is the set that contains all the components; r is a subindex used to denote a section of the river, whereas N_r corresponds to the total number of sections of the river used for a specific problem, R is the set that contains all r's; j is the subindex for the effluents discharged to the reach r, N_j is the total number of effluents, and J is their set. In addition, the superindex int^N indicates the interceptors' column for the treatment of the pollutant N, whereas $Nint^N$ indicates the total number of interceptors' columns for the treatment, and INT^N is the set adopted to the interceptors' columns; finally, p is the subindex representing the locations for the new plant, while N_p denotes the total number of possible locations, and P is the set associated with the possible locations.

On the other hand, the model is based on the MFA technique to predict the watershed behavior over time. Additionally, the mathematical formulation includes the treatment for the wastewater stream considering multiple pollutants. Finally, the approach takes into account all the water quality restrictions for reaches where some water streams are extracted for human activities. The model generated is described as follows.

17.3.1 MATERIAL FLOW ANALYSIS

The dynamic behavior of the watershed is modeled through the MFA technique to predict the variation of the pollutants along the watershed and over time, considering all the existing interactions with the environment. In this regard, the watershed is segregated in reaches and tributaries to obtain a proper modeling. The proposed methodology only considers the possible location for the new industrial facilities in reaches (generating dynamic balances); then, all the parameters involved in the tributary balances are known and consequently these balances can be determined prior to the optimization process. Thus, the required material and component balances are described as follows.

Overall balance for each reach. This balance establishes that the volume variation in the reach r with respect to time (dV_r/dt) is equal to the inlet flow rate (Q_{r-1}), plus precipitation (P_r), direct industrial discharges (Id_r), agricultural discharges (D_r), residential discharges (Hd_r), the sum of all effluents entering to the reach ($FT_{r,t}$), and the possible discharge for the new plant when the alternative p to install a new industrial plant is near to the reach r ($QPNEW_{r(p)}$), minus the exit flow rate (Q_r) as well as the extractions for filtration and evaporation (L_r) and the used water (U_r) in the r section of the river. This dynamic balance can be stated as follows:

$$\frac{dV_r}{dt} = Q_{r-1}(t) + P_r(t) + Id_r(t) + D_r(t) + Hd_r(t) + \sum_{t=1}^{N_{t(r)}} FT_{r,t}(t) + QPNEW_{r(p)}(t) - Q_r(t) - L_r(t) - U_r(t), \forall r \in R \tag{17.1}$$

where $N_{t(r)}$ refers to the total number of tributaries that are discharged to the reach r.

To solve the differential equations, this work proposes transforming the differential equations (required for the dynamic optimization) into a set of algebraic equations using the orthogonal collocation method on finite elements. In this context, the previous differential equation is discretized as follows:

$$V_{r,i,j} = V_r^o + h_i \sum_{k \in K} A_{k,j} \dot{V}_{r,i,k}, \quad \forall r \in R, i = 1, j \in J \tag{17.2}$$

$$V_{r,i,j} = V_{r,i-1,N_j} + h_{i-1} \sum_{k \in K} A_{k,j} \dot{V}_{r,i-1,k}, \quad \forall r \in R, i \neq 1, j \in J \tag{17.3}$$

$$\dot{V}_{r,i,j} = Q_{r-1,i,j} + P_{r,i,j} + Id_{r,i,j} + D_{r,i,j} + Hd_{r,i,j} + \sum_{t=1}^{N_{t(r)}} FT_{r,t,i,j} + QPNEW_{r(p),i,j} - Q_{r,i,j} - L_{r,i,j} - U_{r,i,j}, \quad \forall r \in R, i \in I, j \in J \tag{17.4}$$

As can be seen in the previous equations, the time is discretized through the finite elements i and the internal collocation points j; additionally h_i is the finite element length, $A_{k,j}$ is the collocation matrix, V_r^o represents the volume of the reach r at the beginning of the time and $\dot{V}_{r,i,j}$ is the first-order derivative for the volume.

Component balances for each reach. This balance is described as follows: the mass accumulation for each hazardous compound c in each reach r is equal to the mass at the inlet; plus the mass contained in the precipitation, industrial discharges, agricultural discharges, residential discharges, the sum of tributaries, and the possible discharge from the new plant; minus the mass leaving the reach, loses, uses, and the mass decomposed by the chemical reactions that are carried out in that section of the river. The reactive term considers the chemical and biochemical reactions that take place in the river by the flora and fauna owing to the interaction between the components and the system, which can produce or remove the pollutants. Therefore, the component balance for each compound c in each section of the river r is stated as follows:

$$\begin{aligned}\frac{d(V_r CQ_{c,r})}{dt} &= Q_{r-1}(t)CQ_{c,r-1}(t) + P_r(t)CP_{c,r} + Id_r(t)CId_{c,r}(t) + D_r(t)CD_{c,r}(t)\\ &+ Hd_r(t)CHd_{c,r}(t) + \sum_{t=1}^{N_{t(r)}} FT_{r,t,i,j}(t)CT_{c,r,t,i,j}(t) - L_r(t)CL_{c,r} - U_r(t)CU_{c,r}(t)\\ &- \int_{V=0}^{V_r} r_{c,r}(t)dV_r(t) - Q_r(t)CQ_{c,r}(t) + QPNEW_{r(p)}(t)CPNEW_{c,r(p)}(t),\\ &\forall c\in C, r\in R\end{aligned} \tag{17.5}$$

Similarly to the overall balance for each reach, the concentration at each one of the discretized points ($CQ_{c,r,i,j}$) is also approximated using a similar monomial basis representation:

$$CQ_{c,r,i,j} = CQ^o_{c,r} + h_i \sum_{k\in K} A_{k,j}C\dot{Q}_{r,i,k}, \quad \forall r\in R, i=1, j\in J \tag{17.6}$$

$$CQ_{c,r,i,j} = CQ_{c,r,i-1,N_j} + h_{i-1} \sum_{k\in K} A_{k,j}C\dot{Q}_{r,i-1,k}, \quad \forall r\in R, i\neq 1, j\in J \tag{17.7}$$

$$\begin{aligned} V_{r,i,j}C\dot{Q}_{c,r,i,j} + \dot{V}_{r,i,j}CQ_{c,r,i,j} &= Q_{r-1,i,j}CQ_{c,r-1,i,j} + P_{r,i,j}CP_{c,r,i,j} + Id_{r,i,j}CId_{c,r,i,j}\\ &+D_{r,i,j}CD_{c,r,i,j} + Hd_{r,i,j}CHd_{c,r,i,j} + \sum_{t=1}^{N_{t(r)}} FT_{r,t,i,j}CT_{c,r,t,i,j}\\ &-L_{r,i,j}CL_{c,r,i,j} - U_{r,i,j}CU_{c,r,i,j} - \int_{V=0}^{V_r} r_{c,r,i,j}dV_{r,i,j}\\ &-Q_{r,i,j}CQ_{c,r,i,j} + QPNEW_{r(p),i,j}CPNEW_{c,r(p),i,j},\\ &\forall c\in C, r\in R, i\in I, j\in J\end{aligned} \tag{17.8}$$

where $CQ^o_{c,r}$ is the concentration of the pollutant c for each reach r at the beginning of the time and $C\dot{Q}_{r,i,k}$ represents the first-order derivative for the pollutants concentration.

Overall balance for each tributary. The total flow rate discharged from the tributary t to the reach r ($FT_{r,t}$) is equal to the sum of the following types of discharges: discharges without ($S^{untreated}_{r,t}$)/with ($S^{treated}_{r,t}$) treatment, industrial discharges ($It_{r,t}$), pluvial discharges ($P_{r,t}$), direct discharges ($D_{r,t}$), minus the losses ($L_{r,t}$), and use or extraction ($U_{r,t}$) of water. Thus, the overall balance for the tributary t that discharges to the reach r is described as follows:

$$FT_{r,t} = S_{r,t}^{untreated}(t) + S_{r,t}^{treated}(t) + It_{r,t}(t) + P_{r,t}(t) + D_{r,t}(t) - L_{r,t}(t) - U_{r,t}(t), \quad \forall r \in R, t \in T \tag{17.9}$$

Component balances for each tributary. The mass for the component c discharged from the tributary t to the reach r is calculated as follows:

$$\begin{aligned} FT_{r,t}(t)CT_{c,r,t}(t) = {} & S_{r,t}^{untreated}(t)CS_{c,r,t}^{untreated}(t) + S_{r,t}^{treated}(t)CS_{c,r,t}^{treated}(t) + It_{r,t}(t)CIt_{c,r,t}(t) \\ & + P_{r,t}(t)CP_{c,r,t}(t) + D_{r,t}(t)CD_{c,r,t}(t) - L_{r,t}(t)CL_{c,r,t}(t) - U_{r,t}(t)CU_{c,r,t}(t) \\ & - \int_{V=0}^{V_{r,t}} r_{c,r,t}(t)dV_{r,t}(t), \quad \forall c \in C, r \in R, t \in T \end{aligned} \tag{17.10}$$

Here, again appears the term to account for the reactions that can be carried out in the tributary.

It should be noted that the overall and component balances for tributaries only contain known flow rates and concentrations and do not include optimization variables; therefore, the flow rates and pollutants concentrations discharged by tributaries can be provided prior to the implementation of the approach.

Additionally, note that most of the terms involved in the above balances have temporal fluctuations, which lead to dynamic equations.

Reactive terms. The reactive terms for the degradation of the pollutants in reaches and tributaries are represented by the following power law kinetic reaction:

$$\int_{V=0}^{V_r} r_{c,r}dV_r = k_c\left[CQ_{c,r}(t)\right]^{\sigma_c}V_r(t), \quad \forall c \in C, r \in R \tag{17.11}$$

$$\int_{V=0}^{V_{r,t}} r_{c,r,t}dV_{r,t} = k_c\left[CT_{c,r,t}(t)\right]^{\sigma_c}V_{r,t}(t), \quad \forall c \in C, r \in R, t \in T \tag{17.12}$$

where k_c represents the kinetic constants for the reactions carried out in the watershed, which is obtained experimentally, σ_c symbolizes the order reaction for the compound c, $CQ_{c,r}$ and V_r are the concentration of each component and the volume of the reach r, respectively; while $CT_{c,r,t}$ and $V_{r,t}$ are the concentration and volume for tributaries, respectively.

Agricultural discharges and uses. Typically, water is extracted from watershed to be supplied in some agricultural areas to achieve an adequate production of the field. The amount of water for this purpose is a function of the area employed for the crops (which also can change during the year), as well as the water which returns to the watershed through filtration can be determined as a function of the cultivated area as follows:

$$D_{r,t}(t) = \alpha_{r,t} A_{r,t}(t), \quad \forall r \in R, t \in T \tag{17.13}$$

$$U_{r,t}(t) = \beta_{r,t} A_{r,t}(t), \quad \forall r \in R, t \in T \tag{17.14}$$

where $\alpha_{r,t}$ is the amount of water per area for agricultural purposes and its unit is m^3/ha s, $\beta_{r,t}$ is the discharged flow rate per area and its unit is m^3/ha s. $A_{r,t}$ is the area used for agricultural purposes near to the reach r.

17.3.2 LOCATION OF THE NEW PLANT

The proposed approach accounts for a set of alternatives to place the new industrial plant p, and the optimization process must determine the optimal location for the industrial facilities considering that the installation cost and the impact of the new wastewater stream in the watershed depends upon the selected site. To carry out this task, a disjunctive optimization formulation has been included, which establishes that if the new industrial plant is located in site 1, then the wastewater and its composition associated with this site must be greater than zero; otherwise, if the new plant is not located in the position 1, the wastewater and its compositions discharged should be zero; the same logic is used for all options. This behavior is modeled through the following disjunction:

$$\begin{bmatrix} Y_1 \\ CPNEW_{c,1}(t) \geq 0, QPNEW_1(t) \geq 0 \\ CPNEW_{c,2}(t) = 0, QPNEW_2(t) = 0 \\ \vdots \\ CPNEW_{c,p}(t) = 0, QPNEW_p(t) = 0 \end{bmatrix} \vee \begin{bmatrix} Y_2 \\ CPNEW_{c,1}(t) = 0, QPNEW_1(t) = 0 \\ CPNEW_{c,2}(t) \geq 0, QPNEW_2(t) \geq 0 \\ \vdots \\ CPNEW_{c,p}(t) = 0, QPNEW_p(t) = 0 \end{bmatrix} \vee \cdots$$

$$\vee \begin{bmatrix} Y_p \\ CPNEW_{c,1}(t) = 0, QPNEW_1(t) = 0 \\ CPNEW_{c,2}(t) = 0, QPNEW_2(t) = 0 \\ \vdots \\ CPNEW_{c,p}(t) \geq 0, QPNEW_p(t) \geq 0 \end{bmatrix}, \quad \forall c \in C$$

In this disjunction, Y_p represents the Boolean variable associated with the location p of the new plant, $QPNEW_p$ and $CPNEW_{c,p}$ are the flow rate and concentration for the compound c of the new wastewater stream generated by the new industrial plant in location p, respectively. Additionally, the disjunctive model states that if the optimal location for the new plant is on site 1, then the Boolean variable Y_1 is set as true and $CPNEW_{c,1}$ and $QPNEW_1$ must be greater than zero, while all the other Boolean variables must be false and the $CPNEW_{c,p}$ and $QPNEW_p$ associated with their locations must be set as zero. Similar situations occur for the case when the selected location is 2, 3, or so on for the p possible locations. It is important to remark that although the installation costs are not considered in the previous disjunction, this value is also affected by the chosen alternative to locate the new industrial facilities, as well as the value of the flow rate and concentration of the wastewater through the MFA model. Some of the constraints to be met are related to environmental regulations and are used to maintain under control the pollutants. These restrictions include the environmental regulations (imposed for the wastewater stream), constraints to maintain the water quality for some specific reaches (restricting the pollutant concentrations for some reaches of interest), and the sustainability constraints (which implies that the compound concentrations at the final disposal are restricted). It should be noted that the main tool to satisfy all the environmental restrictions is the interception network. Finally, the previous disjunction system is reformulated using the convex hull technique (see Raman and Grossmann, 1994; Ponce-Ortega et al., 2009b) as follows.

The Boolean variables (Y_p) are transformed into a set of binary variables (y_p), when the Boolean variables are true, the associated binary variables are one; otherwise, when the Boolean variables are false, the associated binary variables are zero. Therefore, the following relationship allows selecting only one location to place the new industrial plant:

$$\sum_{p \in P} y_p = 1 \qquad (17.15)$$

Moreover, upper limits for the variables $QPNEW_p$ and $CPNEW_{c,p}$ are required:

$$QPNEW_p(t) \leq QPNEW^{Up} y_p, \quad \forall p \in P \qquad (17.16)$$

$$CPNEW_{c,p}(t) \leq CPNEW^{Up} y_p, \quad \forall c \in C, p \in P \qquad (17.17)$$

where $QPNEW^{Up}$ and $CPNEW^{Up}$ represent the upper limits for $QPNEW_p$ and $CPNEW_{c,p}$, respectively. Thus, taking into account Eqn (17.15) (where only one site is activated), previous relationships ensure that $QPNEW_p$ and $CPNEW_{c,p}$ take values greater than zero only for the selected position. In addition, $QPNEW_p$ and $CPNEW_{c,p}$ must be equal to zero when the alternative to place the new plant is not selected.

17.3.3 SUPERSTRUCTURE MODELING FOR IN-PLANT TREATMENT SYSTEM

First, the variables $QPNEW_p$ and $CPNEW_{c,p}$ (used in the MFA model) are related to the variables employed in the treatment system for the wastewater stream through the following relationships:

$$\sum_{p \in P} QPNEW_p(t) = QPNEWDATA(t) \qquad (17.18)$$

$$\sum_{p\in P} CPNEW_{c,p}(t) = y_c^{disc}(t), \quad \forall c \in C \tag{17.19}$$

Then, previous equations establish that the flow rate ($QPNEW_p$) and pollutants composition ($CPNEW_{c,p}$) of the wastewater stream that receives the watershed in the selected position are the same than the flow rate ($QPNEWDATA$) and compositions (y_c^{disc}) discharged by the new industrial plant at the exit of the treatment system. Thus, previous equations connect the MFA model with the treatment system model.

The proposed superstructure for the in-plant treatment is shown in Figure 17.2, which is based on the works reported previously by Gabriel and El-Halwagi (2005) and Ponce-Ortega et al. (2010). The size for this formulation increases exponentially as a function of the number of treated compounds (N_c). However, the model obtained for this superstructure features linear behavior, which is the main advantage of this formulation because it avoids the mixing of different streams and it is useful for the optimization process. As can be seen in this figure, for each pollutant to be treated, there is a set of available treatment units to remove the compound c. Moreover, a bypass has been included to consider that a fraction of the total wastewater stream does not require to be treated (which is modeled as a fictitious interceptor). Therefore, each component must take into account a column of interceptors, generating N columns of treatment units. Note that there is a set of constraints for the wastewater stream discharged to the environment to meet the environmental regulations. Additionally, it is important to remark that the optimization process should indicate the variation of the required treatment with respect to time. It is important to note that the treatment requirements depend strongly on the location of the new industrial plant to satisfy the environmental constraints imposed on the system.

According to the proposed superstructure, the waste stream leaving the new industrial facility is segregated to treat the first component in the set of interceptors (i.e., available technologies for the treatment of the specific pollutant). At the exit of the first treatment, the output streams are segregated again and sent to the following interceptors to treat the second component. The same sequence is used to treat all required components at the interception network, where the mixing of different streams is avoided in all steps of the treatment. Finally, all streams that leave the treatment units are mixed to be sent to the wastewater stream discharged to the environment. Note that the final wastewater streams must satisfy the environmental constraints.

In this sense, a correction factor $\left(\gamma_p^{int^1,\ldots,int^N}\right)$, which can be calculated prior to the optimization process, is used to describe the interceptors' performance. This approach leads to linear models for the interceptors. The correction factor depends on several variables such as the value of the concentration at the inlet conditions of the interceptor, design and operating parameters, as well as the flow rate. The available treatment technologies considered in the interception network can be simulated (i.e., using process simulators such as Aspen Plus) to get their correction factors. It should be noted that the values for the concentrations of the components at the inlet conditions of the waste stream are also known prior to the optimization process

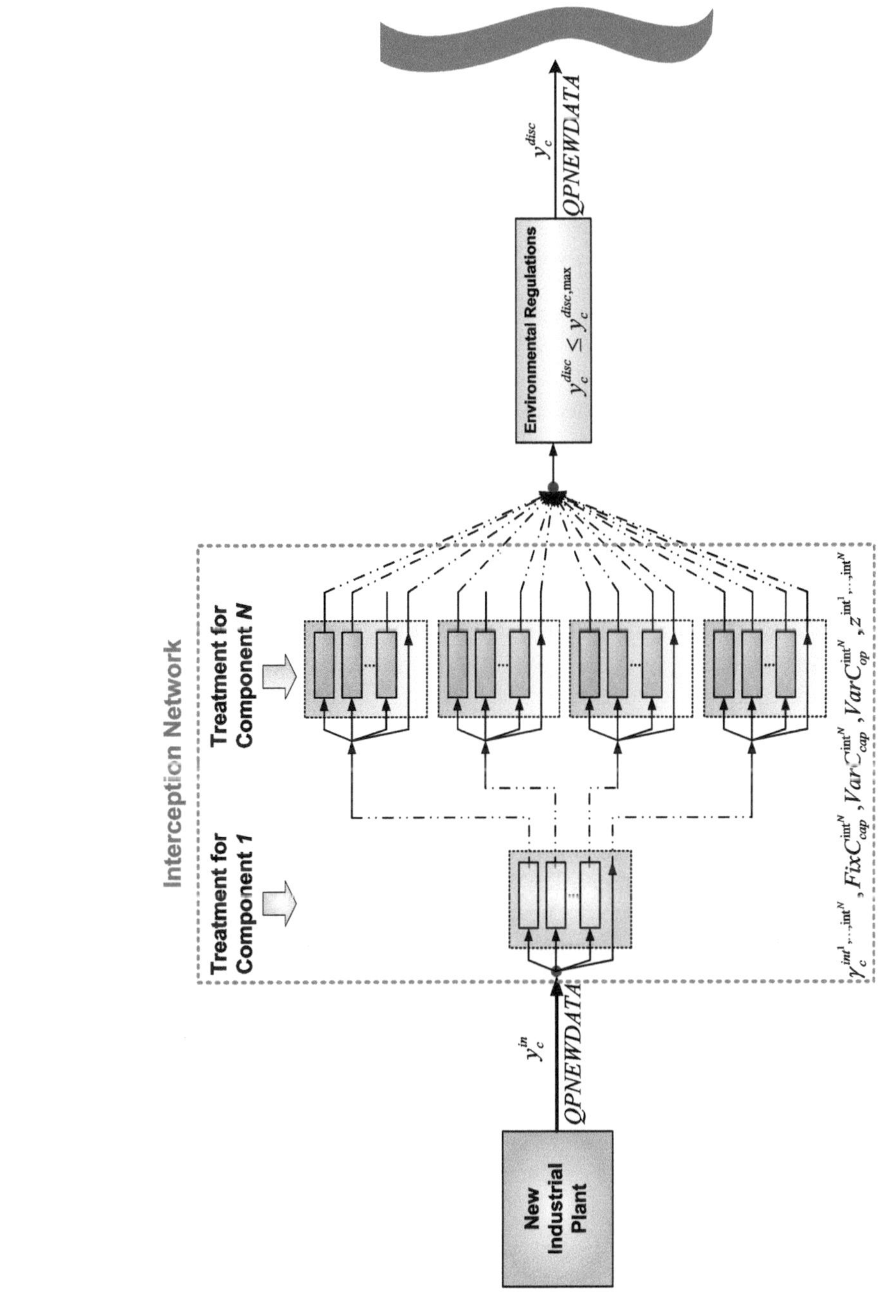

FIGURE 17.2

Interceptors network for wastewater treatment.

(y_c^{in}). Therefore, the values for the concentrations of the components at the exit conditions of the interception network can be calculated before optimization as follows:

$$y_c^{out,\mathrm{int}^1,\ldots,\mathrm{int}^N}(t) = \gamma_c^{\mathrm{int}^1,\ldots,\mathrm{int}^N} y_c^{in}(t), \quad \forall c \in C, \mathrm{int}^1 \in INT^1, \ldots, \mathrm{int}^N \in INT^N \tag{17.20}$$

where $y_c^{out,\mathrm{int}^1,\ldots,\mathrm{int}^N}$ represents the concentrations for the components at the exit of the interception network, according to the path of interceptors followed.

On the other hand, the following relationships model the proposed superstructure for the wastewater treatment in this work.

Splitting of the waste stream. The waste stream leaving the industrial plant is segregated to treat the first component on the first set of available interceptors.

$$QPNEWDATA(t) = \sum_{\mathrm{int}^1 \in INT^1} f_1^{\mathrm{int}^1}(t) \tag{17.21}$$

Then, to treat the second component, the flow rate at the exit of the first line of interceptor is segregated and these are fed to the second set of interceptors.

$$f_1^{\mathrm{int}^1}(t) = \sum_{int^2 \in INT^2} f_2^{\mathrm{int}^1,\mathrm{int}^2}(t), \quad \forall int^1 \in INT^1 \tag{17.22}$$

Notice that it does not allow mixing of different process streams. This approach eliminates the nonlinearities involved in the component balances for the wastewater stream (which is shown below), although the problem size increases exponentially.

Following the same sequence for N components to be intercepted, the next equation models the segregations:

$$f_{N-1}^{int^1,\mathrm{int}^2,\ldots,\mathrm{int}^{N-1}}(t) = \sum_{int^N \in INT^N} f_N^{int^1,\mathrm{int}^2,\ldots,\mathrm{int}^N}(t), \quad \forall int^1 \in INT^1, \ldots, int^{N-1} \in INT^{N-1} \tag{17.23}$$

It should be noted that if the waste stream does not require any treatment, it is sent to a bypass or fictitious unit (located at the end of each set of interceptors) with effectiveness and cost equal to zero.

Overall balance for the waste stream. The proposed superstructure considers that there is no volume variation owing to the treatment. Therefore, the flow rate of the wastewater stream discharged to the environment is equal to the waste stream leaving the industrial plant (*QPNEWDATA*). All the streams leaving from the interception network ($f^{int^1,\ldots,\mathrm{int}^N}$) are mixed to obtain the final wastewater stream.

$$QPNEWDATA(t) = \sum_{\mathrm{int}^1 \in INT^1} \cdots \sum_{\mathrm{int}^N \in INT^N} f^{int^1,\ldots,\mathrm{int}^N}(t) \tag{17.24}$$

Component balances for the wastewater stream. The next relationship is useful to compute the concentrations of the pollutants for the wastewater streams discharged to the environment.

$$QPNEWDATA(t) y_c^{disc}(t) = \sum_{\mathrm{int}^1 \in INT^1} \cdots \sum_{\mathrm{int}^N \in INT^N} f^{int^1,\ldots,\mathrm{int}^N}(t) y_c^{out,\mathrm{int}^1,\ldots,\mathrm{int}^N}(t), \quad \forall c \in C \tag{17.25}$$

where y_c^{disc} is the concentration of the component c discharged to the environment.

Environmental regulations. The environmental regulations impose a set of maximum concentrations which can be discharged to the environment.

$$y_c^{disc}(t) \leq y_c^{disc,\max}, \quad \forall c \in C \tag{17.26}$$

Notice that the flow rates and concentrations contained in the previous equations to model the treatment of the wastewater stream depend strongly on the processing time, since the treatment requirements change according to the dynamic behavior of the processes that occur in the watershed system.

17.3.4 ADDITIONAL ENVIRONMENTAL CONSTRAINTS

In addition to the environmental regulations, the problem considers a set of environmental restrictions imposed over specific reaches and other set of environmental constraints at the final disposal (called sustainability constraints) for the watershed system.

Constraints for the quality of specific reaches. Usually the water of a watershed is used for different purposes (i.e., residential, industrial, or agricultural zones). Therefore, it becomes necessary to impose a set of concentration constraints for the specific reaches to guarantee the quality of the extracted water. The proposed model is able to consider these constraints through the following relationship:

$$CQ_{c,m(r)}(t) \leq y_{c,m(r)}^{desired}, \quad \forall c \in C, m(r) \in M(R) \tag{17.27}$$

where $M(R)$ represents a subset of reaches requiring a specific water quality, while $y_{c,m(r)}^{desired}$ represents the upper limits for the component concentrations to ensure desired conditions in the water quality for specific reaches. It is important to remark that although the concentration in the reach has a significant variation with respect to time ($CQ_{c,m(r)}$), the desired concentration must be kept constant during all the time.

Sustainability constraints. Previous to the implementation of the proposed approach, an analysis to know the maximum concentration of each compound c ($y_c^{sustainable}$) in the final disposal (lake, sea, or ocean) should be carried out. It should be stressed that for some cases the environmental regulations could not be enough to achieve this purpose. Hence, such a constraint can be posed as follows:

$$CQ_c^{final}(t) \leq y_c^{sustainable} \tag{17.28}$$

where CQ_c^{final} is the concentration of the component c at the final disposal after installing the new industrial facility. Once more, it is important to remark that although the concentration in the final reach has a significant variation with respect to time (CQ_c^{final}), the sustainable concentration must be maintained during all time.

17.3.5 OBJECTIVE FUNCTION

Finally, the objective function consists in minimizing the total annual cost (*TAC*), which is constituted by the installation costs for the new plant and the treatment cost (including operational cost and variable capital cost).

$$Min\ TAC = \int_0^t \left[\sum_{int^1 \in INT^1} VarC_{op}^{int^1} f^{int^1}(t) + \cdots \sum_{int^N \in INT^N} VarC_{op}^{int^N} f^{int^1,\ldots,int^N}(t) \right] dt$$
$$+ \int_0^t \left[\sum_{int^1 \in INT^1} VarC_{cap}^{int^1} f^{int^1}(t) + \cdots \sum_{int^N \in INT^N} VarC_{cap}^{int^N} f^{int^1,\ldots,int^N}(t) \right] dt + \sum_{p \in P} Cland_p y_p \quad (17.29)$$

where k_f is the factor used to annualize the inversion. Additionally, a set of available technologies must be selected to treat the waste stream yielding capital $\left(VarC_{cap}^{int^N}\right)$ and operational $\left(VarC_{op}^{int^N}\right)$ costs for each one of the required interceptors. $Cland_p$ represents the installation cost, which takes into account the land cost and the transportation costs for raw materials, products, and services and finally y_p is the binary variable associated with each alternative p to locate the new industrial facilities. It should be noted that the objective function includes integral terms; then an adequate procedure to approximate numerically the integrals must be implemented (i.e., the Radau quadrature method).

17.3.6 REMARKS

- The proposed methodology embeds the dynamic model of the watershed to select the optimal place to locate new industrial facilities considering the impact of the new waste stream in the system.
- In this type of natural systems there are chemical and biochemical processes that gradually degrade the pollutants; the kinetic term included in the component balance for each reach can represent the global effect in the compound's concentration through a simple kinetic model, even when several chemical and biochemical reactions can occur.
- The installation costs ($Cland_p$) include three different costs: transportation cost of raw materials, transportation cost of products, services, and the land cost for alternative p. These costs depend directly on the location chosen for the new plant.

17.4 RESULTS AND DISCUSSION

The following example illustrates the applicability of the proposed methodology.

17.4.1 CASE STUDY

The Balsas system represents one of the most important watersheds in Mexico, it has a hydrologic surface of 117,305 km^2 covering eight states and 420 municipalities. This watershed is affected by several cities, industrial and agricultural zones that discharge their effluents over it. To model adequately the Balsas system, it was necessary to take information from CONAGUA (2010). In this regard, Figure 17.3 shows the schematic representation for the proposed problem in this example. The residential, industrial, and agricultural zones shown in this figure are only representative. However, there are many interactions among these systems and the Balsas watershed. In this case, nitrogen, phosphorous, and arsenic were considered to be pollutants. In addition, 11 possible sites for plant location were also considered. Moreover, to locate the new industrial facilities, the Balsas system has been divided in 23 reaches. In addition, most of the information required by the model must be introduced with fluctuations with respect to time such as the precipitation, evaporation, filtration and all the industrial, residential, and agricultural interactions with the system. In this context, the precipitation data employed in this example have a seasonal variation in accordance with Figure 17.4. It should be noted that the period June–September corresponds to the rainy season and consequently there is a significant variation in the precipitation discharges during the year, which impacts the watershed water quality.

Furthermore, Figures 17.5 and 17.6 show the flow rate and pollutant concentration for the new industrial plant prior to the inlet to the treatment network. In this sense, there are three significant variations in the flow rate and pollutants concentration during the year. The arsenic has concentrations lower than nitrogen and phosphorous. For this reason and in order to visualize the variation in the concentrations over the year, its scale has been separated from the other components as can be seen in Figure 17.6.

On the other hand, Table 17.1 shows the annualized installation cost for each alternative considered to install the new industrial plant, while Table 17.2 lists the

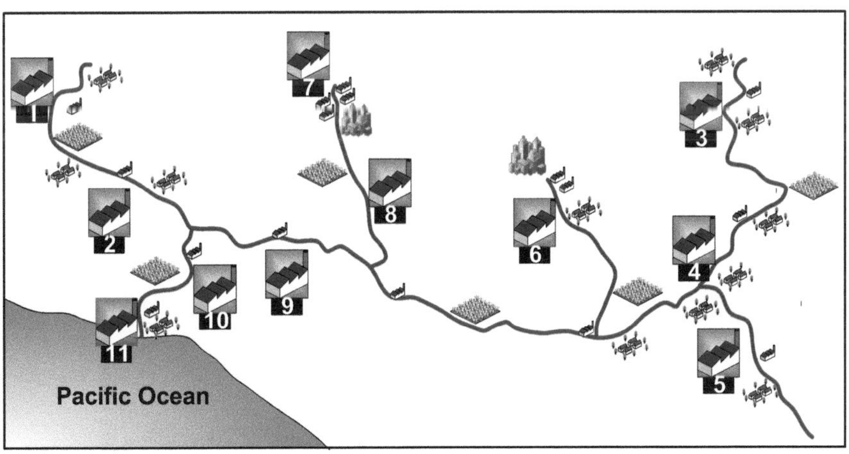

FIGURE 17.3

Schematic representation of the Balsas system, including the alternatives to install the new industrial plant.

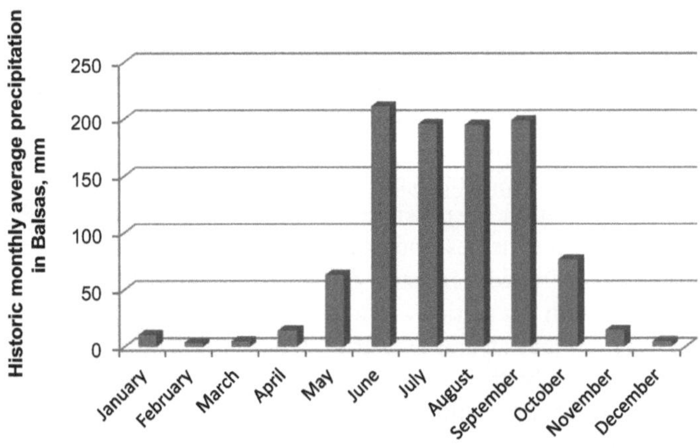

FIGURE 17.4

Historic monthly average precipitation in the Balsas watershed, CONAGUA (2010).

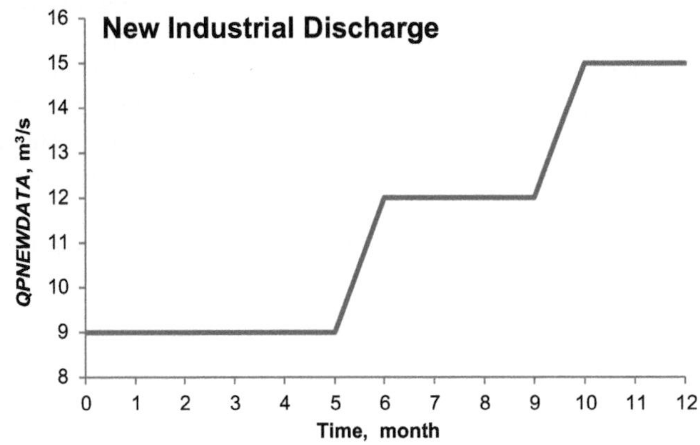

FIGURE 17.5

Flow rate behavior for the new industrial plant.

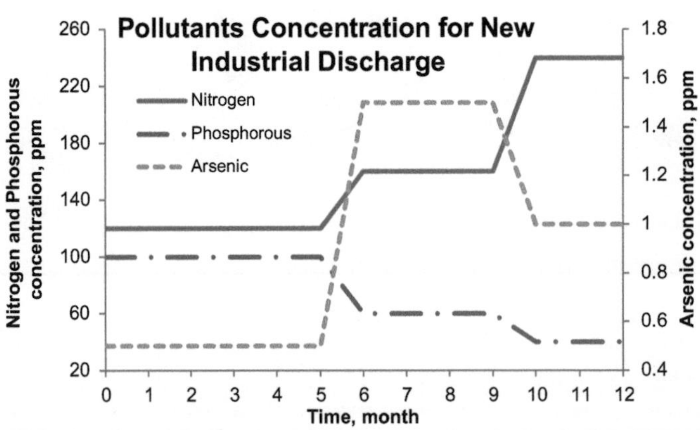

FIGURE 17.6

Pollutant concentrations previous to the interception network for the new industrial plant.

Table 17.1 Installation Costs for Each Site

Site	Annualized Installation Cost, $/Year
1	7.7×10^6
2	7.4×10^6
3	8.5×10^6
4	8.2×10^6
5	8×10^6
6	7.3×10^6
7	7.2×10^6
8	7×10^6
9	8.1×10^6
10	7.6×10^6
11	8×10^6

costs and efficiency factors for each technology available to treat the pollutants considered in this example. It should be noted that for each compound, two interceptors have been considered, where the first option is always the most efficient but is also the most expensive. Additionally, the environmental regulations and sustainability constraints are shown in Table 17.3, which must always be enforced. Note that the environmental regulations restrict the maximum pollutant concentration that can be discharged by the new industrial plant. However, it will be necessary to take into account the impact of this wastewater stream once it enters the environment. In other words, fulfillment of the environmental regulations does not necessarily lead to sustainable systems.

Table 17.2 Characteristics of the Available Interceptors

Interceptor	Efficiency Factor	Cost, $/Month m^3
Nitrogen		
Rec1	0.98	14,515.20
Rec2	0.78	10,627.20
Phosphorous		
Dec1	0.97	20,736
Dec2	0.82	15,811.20
Arsenic		
Aer1	0.99	16,848
Aer2	0.85	5,702.40

Table 17.3 Environmental Constraints for the Considered Example

Pollutant	Environmental Regulation $y_c^{disc,max}$, ppm	Sustainability Constraint $y_c^{sustainable}$, ppm
Nitrogen	40	22
Phosphorous	20	11.1
Arsenic	0.2	0.083

Finally, to implement this case study, the values of the following parameters were set:

- The operation time for the new plant is 8000 h/year.
- The agricultural discharges and uses are: $\alpha_{r,t} = 0.000148$ m^3/ha s and $\beta_{r,t} = 0.000296$ m^3/ha s, respectively.
- The reactive term is represented through first-order kinetic reactions employing the following kinetic constants $k_{nitrogen} = 3.23764 \times 10^{-6}$ s^{-1}, $k_{phosphorous} = 1.492 \times 10^{-6}$ s^{-1}, and $k_{arsenic} = 4.5205 \times 10^{-7}$ s^{-1}.

All the above information is used in the software GAMS deploying the solver DICOPT; in addition, the optimization formulation employs the orthogonal collocation on finite elements to discretize the differential equation into a set of algebraic equations. Once the required discretization of the dynamic model is carried out, the resulting MINLP problem is solved. From the solution, we found that the optimal alternative to locate the new industrial facilities is site 10 (see Figure 17.7) with a *TAC* of \$11.928 MM/year, of which \$7.6 MM/year corresponds to installation cost and the treatment cost is \$4.328 MM/year.

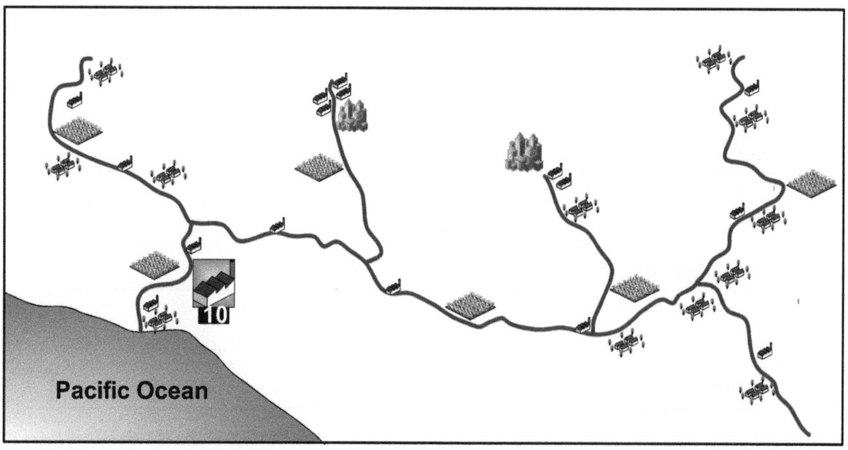

FIGURE 17.7

Optimal solution for the considered example.

Figure 17.8 shows the resulting optimal interception network required to meet the environmental regulations as well as the sustainability constraints. It should be noted that almost all the interceptors considered in the proposed superstructure are required in the optimal treatment configuration. Although this figure does not include the fluctuation in flow rates and concentrations, these have been determined by the optimization process.

Additionally, the variations in nitrogen concentration discharged to the final disposal (CQ^{final}) as well as for the new industrial discharge (y^{disc}) are shown in Figure 17.9. This figure also includes the environmental regulation (regulation) and the sustainability constraint ($y^{sustainable}$) for this compound. From this figure we observe the following behaviors:

- During the rainy season (June–September), the watershed conditions allow discharge of the waste stream from the new industrial plant to satisfy only the environmental regulation. In other words, in this period of the year the precipitation discharges dilute the pollutants discharged to the Balsas system and there are no problems related to the fulfillment of the sustainability constraint.
- When there are no significant precipitation discharges, the wastewater stream must be treated beyond the environmental regulations to meet the sustainability constraint; otherwise, if during this period the new waste stream only satisfies the environmental regulation, the system capacity to degrade the

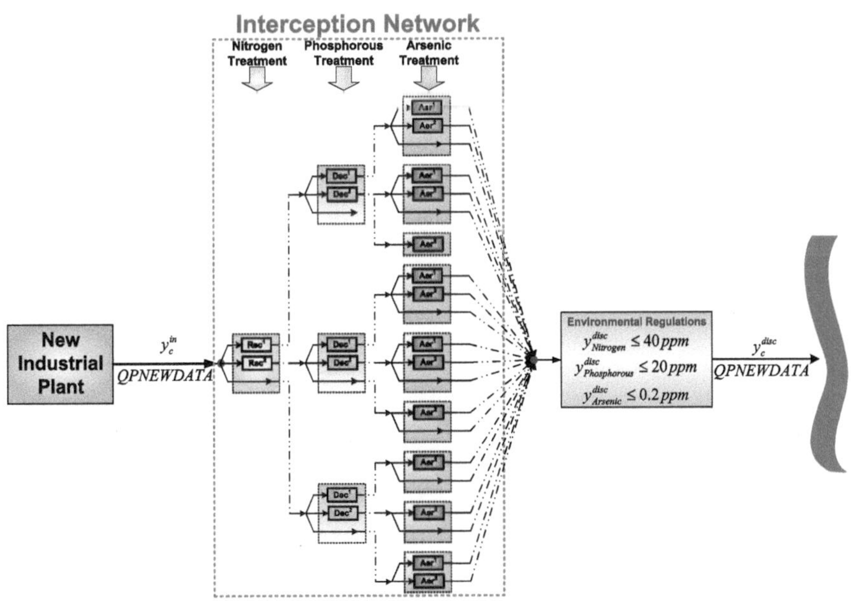

FIGURE 17.8

Optimal network for wastewater treatment obtained in the solution.

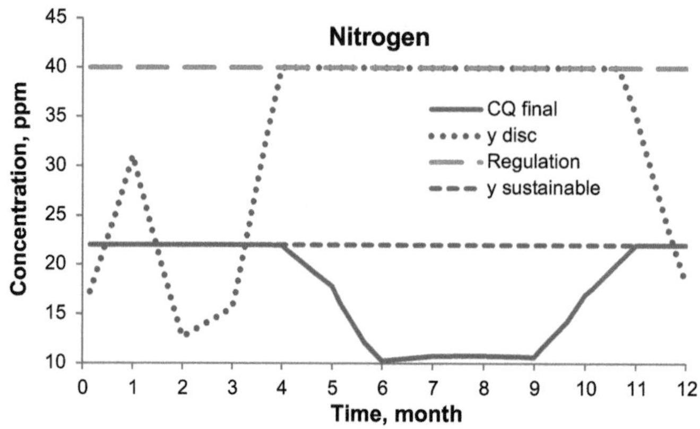

FIGURE 17.9

Nitrogen concentration profiles.

pollutants will be exceeded. Consequently, there will be pollutants accumulation at the final disposal putting the watershed and surrounding areas at risk.

Figures 17.10 and 17.11 demonstrate that the same analysis occurs for the pollutants phosphorous and arsenic, respectively.

Finally, it is important to note the importance of considering the dynamic behavior of the underlying system, which shows the problems that can be generated under critical conditions of the system. Typical models where the dynamic behavior is not considered cannot detect these types of problems and their solutions are not satisfactory. In addition, this example has demonstrated that the environmental

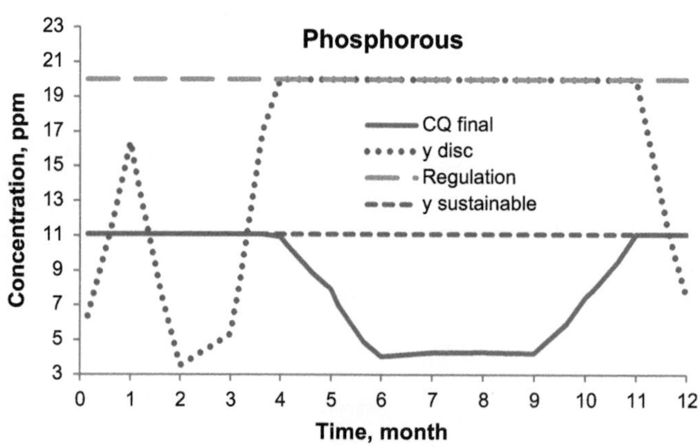

FIGURE 17.10

Phosphorous concentration profiles.

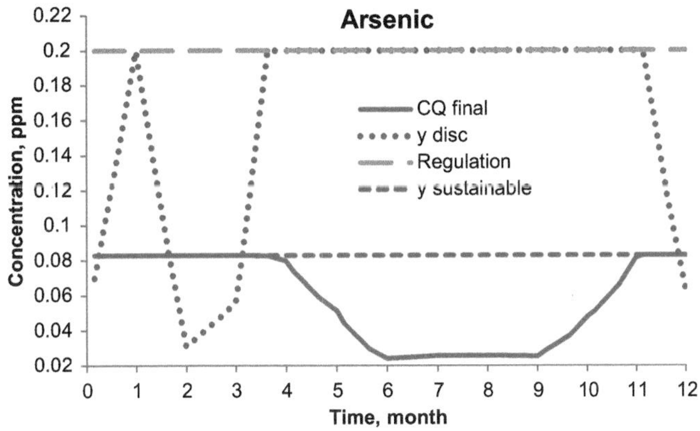

FIGURE 17.11

Arsenic concentration profiles.

regulations for industrial wastewater streams are not adequate for ensuring the sustainability of the watersheds, and it is required to model the watershed to guarantee the sustainability of the system.

17.5 CONCLUSIONS

This work has presented a new dynamic mixed integer nonlinear programming model for the optimal location of new industrial facilities that satisfy environmental regulations, constraints for specific sites of the rivers with water extractions, and sustainability constraints imposed at the final disposal to ensure the sustainability of the watershed and its surroundings. The objective function consists in minimizing the total annual cost, comprising wastewater treatment costs and the installation cost for the new industrial plant. The MFA technique is employed to predict the behavior of the watershed when it is impacted by the new polluted discharges as well as considering all the interactions existing in the watershed, such as industrial, residential, agricultural, and others discharges (i.e., precipitation, treated and untreated discharges) and the natural phenomena like evaporation, filtration, and the chemical and biochemical reactions occurring in the rivers. The dynamic model allows identification of the time periods where the watershed conditions are critical and the treatment required must be readjusted to satisfy all the environmental constraints. To achieve this task, the mass integration inside the industrial plant is considered together with the surrounding watershed and the MFA technique is combined with a disjunctive model for the optimal location of a new industrial plant. The solution of the case study shows that the location of a new industrial plant has a significant impact on the treatment requirements to yield a sustainable process. A further relevant aspect to obtain holistic solutions is the incorporation of dynamic processes to visualize the real behavior of the watershed throughout the year, including the seasonal fluctuations, and identifying the critical periods associated with overall sustainability; if this is not carried out, it could lead to inadequate solutions since critical periods would not be considered. Finally, results show that sometimes the environmental regulations for industrial wastewater streams are not adequate for ensuring the sustainability of the watersheds; to achieve this, more accurate approaches like the proposed methodology are required.

NOMENCLATURE

Indexes

c	Component
i	Finite element
int	Interceptor
In	Inlet
j	Internal collocation point
max	Maximum
min	Minimum
N_c	Number of pollutants to be treated
$m(r)$	Specific reaches that requires a composition constraint
p	Site to locate the new plant
r	Reach
t	Tributary
UP	Upper value

Sets

C	Set for the components, $\{c\|c = 1,\ldots,N_c\}$
I	Set for the finite elements, $\{i\|i = 1,\ldots,N_i\}$
J	Set for the internal collocation points, $\{j\|j = 1,\ldots,N_j\}$
INT	Set for the interceptor, $\{int\|int = 1,\ldots,N_{int}\}$
P	Set for the sites to locate the new plant, $\{p\|p = 1,\ldots,N_p\}$
R	Set for the reaches, $\{r\|r = 1,\ldots,N_r\}$
T	Set for the tributaries, $\{t\|t = 1,\ldots,N_t\}$

Parameters

$A_{r,t}$	Area covered by effluent t in reach r, acre or ha
$A_{k,j}$	Collocation matrix
$CD_{c,r}$	Pollutant concentration of agricultural discharges to the reach r, ppm
$CD_{c,r,t}$	Pollutant concentration of agricultural discharges to the tributary t to the reach r, ppm
$CHd_{c,r}$	Pollutant concentration of residential discharge to the reach r, ppm

$CId_{c,r}$	Pollutant concentration of industrial discharge to the reach r, ppm
$CIt_{c,r,t}$	Pollutant concentration of industrial discharge from the tributary t to the reach r, ppm
CQ_c^{final}	Pollutant concentration discharged to the final disposal, ppm
$CP_{c,r}$	Pollutant concentration of precipitation discharged to the reach r, ppm
$CP_{c,r,t}$	Pollutant concentration of precipitation discharged to the tributary t to the reach r, ppm
$CT_{c,r,t}$	Pollutant concentration of the tributary t to the reach r, ppm
$Cland_p$	Annualized installation cost for the new plant in site p, \$/year
$CL_{c,r}$	Pollutant concentration of total losses (filtration and evaporation) from the reach r, ppm
$CL_{c,r,t}$	Pollutant concentration of total losses (filtration and evaporation) from tributary t to the reach r, ppm
$CS_{c,r,t}^{untreated}$	Pollutant concentration of residual wastewater discharged without treatment to the reach r for tributary t, ppm
$CS_{c,r,t}^{treated}$	Pollutant concentration of residual treated wastewater discharged to the reach r for tributary t, ppm
$CU_{c,r}$	Pollutant concentration of water used from reach r, ppm
$CU_{c,r,t}$	Pollutant concentration of water used from tributary t discharge to reach r, ppm
D_r	Agricultural discharges to the reach r, m^3/s
$D_{r,t}$	Agricultural discharges to the tributary t to the reach r, m^3/s
$FT_{r,t}$	Discharge for the tributary t to the reach r, m^3/s
Hd_r	Residential discharge to the reach r, m^3/s
h_i	Finite element length
Id_r	Industrial discharge from the reach r, m^3/s
$It_{r,t}$	Industrial discharge from the tributary t to the reach r, m^3/s
k_c	Kinetic constant for the degradation for pollutant c
L_r	Total losses (filtration and evaporation) from the reach r, m^3/s
$L_{r,t}$	Total losses (filtration and evaporation) from tributary t of the reach r, m^3/s
$M(R)$	Subset for the specific reaches that require composition constraints
N_c	Total number of components
N_i	Total number of finite elements
N_{int}	Total number of interceptors
N_j	Total number of internal collocation points
N_p	Total number of sites allowable to locate the new plant
N_r	Total number of reaches
N_t	Total number of tributaries
P_r	Precipitation discharged to the reach r, m^3/s
$P_{r,t}$	Precipitation discharged for the tributary t to the reach r, m^3/s
$QPNEWDATA$	Flow rate discharged for the new plant
$S_{r,t}^{untreated}$	Residual wastewater discharged without treatment to the reach r for tributary t, m^3/s

$S_{r,t}^{treated}$	Residual treated wastewater discharged to the reach r for tributary t, m^3/s
$V_{r,t}$	Volume for tributary t from reach r, m^3
$VarC_{cap}^{int^N}$	Capital variable cost for interceptor int^N, $\$/m^3$
$VarC_{op}^{int^N}$	Operational variable cost for interceptor int^N, $\$/m^3$
$y_{c,m(r)}^{desired}$	Desired concentrations in some reaches, ppm
$y_c^{disc,\max}$	Environmental regulation for the pollutant concentration to be discharged by the new plant, ppm
y_c^{in}	Pollutant concentration at the inlet to interception network, ppm
$y_c^{sustainable}$	Sustainability constraints at the final disposal, ppm

Greek symbols

$\alpha_{r,t}$	Agricultural flow rate per area, m^3/ha s
$\beta_{r,t}$	Agricultural use of water from tributary t, m^3/ha s
γ_c	Efficiency factor to remove the pollutant c
σ_c	Order reaction for the compound c

Variables

$CQ_{c,r}$	Pollutants concentration of flow rate exit to the reach r, ppm
$CQ_{c,r,i,j}$	Discretized pollutant concentration of flow rate exit from the reach r, ppm
$CQ_{c,r}^{o}$	Concentration of the pollutant c for each reach r at initial conditions, ppm
$\dot{CQ}_{r,i,k}$	First order derivative for the pollutant concentration, ppm
$CQ_{c,r-1}$	Pollutant concentration of flow rate inlet to the reach r, ppm
$CPNEW_{c,p}$	Concentration for the contaminants discharged for the new plant installed in site p, ppm
dV_r/dt	Volume variation in the reach r with respect to the time, m^3/s
f_N	Segregated flow rate for the treatment network, m^3/s
Q_r	Outflow from the reach r, m^3/s
Q_{r-1}	Inflow to the reach r, m^3/s
$QPNEW_p$	Flow rate discharged for the new plant for the location p
$r_{c,r}$	Reaction carried out in the reach r, g/s
$r_{c,r,t}$	Reaction carried out in the tributary t that discharges to reach r, g/s
TAC	Total annual cost, \$/year
U_r	Water used from reach r, m^3/s
$U_{r,t}$	Water used from tributary t discharged to reach r, m^3/s
V_r^o	Volume of the reach r at the initial conditions, m^3
V_r	Volume for reach r, m^3
$V_{r,i,j}$	Discretized volume for reach r, m^3
$\dot{V}_{r,i,j}$	First order derivative for the volume for reach r, m^3/s
Y_p	Boolean variable for the location of the new plant
y_p	Binary variable for the location of the new plant
y_c^{disc}	Pollutant concentration for the final discharge of the new plant to the river, ppm

REFERENCES

Agrawal, V., Shenoy, U.V., 2006. Unified conceptual approach to targeting and design of water and hydrogen networks. AIChE J. 52 (3), 1071–1082.

Alva-Argaez, A., Kokossis, A.C., Smith, R., 1998. Wastewater minimization of industrial systems using an integrated approach. Comput. Chem. Eng. 22 (Suppl.), S741–S744.

Alva-Argaez, A., Vallianatos, A., Kokossis, A., 1999. A multi-contaminant transshipment model for mass exchange networks and wastewater minimization problems. Comput. Chem. Eng. 23 (10), 1439–1453.

Baccini, P., Brunne, P., 1991. Metabolism of the Anthroposphere. Springer-Verlag, Berlin.

Barbosa-Povoa, A., 2007. A critical review on the design and retrofit of batch plants. Comput. Chem. Eng. 31-7, 833–855.

Barbosa-Povoa, A., Macchietto, S., 1994. Detailed design of multipurpose batch plants. Comput. Chem. Eng. 18 (11–12), 1013–1042.

Benko, N., Rev, E., Fonyo, Z., 2000. The use of nonlinear programming to optimal water allocation. Chem. Eng. Commun. 178, 67–101.

Brunner, P.H., Rechberg, H., 2004. Practical Handbook of Material Flow Analysis. CRC.

Castro, P., Barbosa-Povoa, A., Novais, A., 2005. Simultaneous design and scheduling of multipurpose plants using resource task network based continuous-time formulations. Ind. Eng. Chem. Res. 44-2, 343–357.

CONAGUA, 2010. Water Statistics in the Watershed of Balsas River 2010. Mexican National Water Commission.

Dhole, V.R., Ramchandani, N., Tainsh, R.A., Wasilewski, M., 1996. Make your process water pay for itself. Chem. Eng. 103 (1), 100–103.

Doyle, S.J., Smith, R., 1997. Targeting water reuse with multiple contaminants. Process Saf. Environ. Prot. 75 (B3), 181–189.

Dunn, R.F., El-Halwagi, M.M., 2003. Process integration technology review: background and application in chemical process industry. J. Chem. Technol. Biotechnol. 78 (9), 1011–1021.

El-Baz, A.A., Ewida, K.T., Shouman, M.A., El-Halwagi, M.M., 2005a. Material flow analysis and integration of watersheds and drainage systems: I. Simulation and application to ammonium management in Bahr El-Baqar drainage system. Clean Technol. Environ. Policy 7 (1), 51–61.

El-Baz, A.A., Ewida, K.T., Shouman, M.A., El-Halwagi, M.M., 2005b. Material flow analysis and integration of watersheds and drain systems: II. Integration and solution strategies with application to ammonium management in Bahr El-Baqar drain system. Clean Technol. Environ. Policy 7 (1), 78–86.

El-Halwagi, M.M., 1997. Pollution Prevention through Process Integration: Systematic Design Tools. Academic Press, San Diego.

El-Halwagi, M.M., Spriggs, H.D., 1998. Solve design puzzles with mass integration. Chem. Eng. Prog. 94 (8), 25–44.

El-Halwagi, M.M., Gabriel, F., Harell, D., 2003. Rigorous graphical targeting for resource conservation via material recycle/reuse networks. Ind. Eng. Chem. Res. 42 (19), 4319–4328.

El-Halwagi, M.M., 2006. Process Integration. Academic Press, New York.

Erdirik-Dogan, M., Grossmann, I.E., 2008. Simultaneous planning and scheduling of single-stage multiproduct continuous plants with parallel lines. Comput. Chem. Eng. 32 (11), 2664–2683.
Feng, X., Bai, J., Zheng, X.S., 2007. On the use of graphical method to determine the targets of single-contaminant regeneration recycling water systems. Chem. Eng. Sci. 62 (8), 2127–2138.
Floudas, C.A., Lin, X., 2004. Continuous-time versus discrete-time approaches for scheduling of chemical processes: a review. Comput. Chem. Eng. 28 (11), 2109–2129.
Foo, C.Y., 2009. State-of-the-art review of pinch analysis techniques for water network synthesis. Ind. Eng. Chem. Res. 48, 5125–5159.
Gabriel, F.B., El-Halwagi, M.M., 2005. Simultaneous synthesis of waste interception and material reuse networks: problem reformulation for global optimization. Environ. Prog. 24 (2), 171–180.
Galan, B., Grossmann, I.E., 1998. Optimal design of distributed wastewater treatment networks. Ind. Eng. Chem. Res. 37 (10), 4036–4048.
Gomes, J.F.S., Queiroz, E.M., Pessoa, F.L.P., 2007. Design procedure for water/wastewater minimization: single contaminant. J. Cleaner Prod. 15 (5), 474–485.
Gunaratnam, M., Alva-Argaez, A., Kokossis, A., Kim, J.K., Smith, R., 2005. Automated design of total water systems. Ind. Eng. Chem. Res. 44 (3), 588–599.
Gutiérrez-Limón, M.A., Flores-Tlacuahuac, A., Grossmann, I.E., 2012. A multiobjective optimization approach for simultaneous single line scheduling and control of CSTRs. Ind. Eng. Chem. Res. 51 (17), 5881–5890.
Hallale, N., 2002. A new graphical targeting method for water minimization. Adv. Environ. Res. 6 (3), 377–390.
Hernandez-Suarez, R., Castellanos-Fernandez, J., Zamora, J.M., 2004. Superstructure decomposition and parametric optimization approach for the synthesis of distributed wastewater treatment networks. Ind. Eng. Chem. Res. 43 (9), 2175–2191.
Kallrath, J., 2002. Planning and scheduling in the process Industry. OR Spectrum 24 (3), 219–250.
Karuppiah, R., Grossmann, I.E., 2006. Global optimization for the synthesis of integrated water systems in chemical processes. Comput. Chem. Eng. 30 (4), 650–673.
Kuo, W.C.J., Smith, R., 1997. Effluent treatment system design. Chem. Eng. Sci. 52 (23), 4273–4290.
Lampert, C., Brunner, P.H., 1999. Material accounting as a policy tool for nutrient management in the Danube Basin. Water Sci. Technol. 40 (10), 43–49.
Lin, X., Floudas, C., 2001. Design, synthesis and scheduling of multipurpose batch plants via an effective continuous-time formulation. Comput. Chem. Eng. 25 (4–6), 665–674.
Lira-Barragán, L.F., Ponce-Ortega, J.M., Nápoles-Rivera, F., Serna-González, M., El-Halwagi, M.M., 2012. Incorporating property-based water networks and surrounding watersheds in site selection of industrial facilities. Ind. Eng. Chem. Res. 52 (1), 91–107.
Lira-Barragán, L.F., Ponce-Ortega, J.M., Serna-González, M., El-Halwagi, M.M., 2011a. Synthesis of water networks considering the sustainability of the surrounding watershed. Comput. Chem. Eng. 35 (12), 2837–2852.
Lira-Barragán, L.F., Ponce-Ortega, J.M., Serna-González, M., El-Halwagi, M.M., 2011b. An MINLP model for the optimal location of a new industrial plant with simultaneous consideration of economic and environmental criteria. Ind. Eng. Chem. Res. 50 (2), 953–964.

Lovelady, E.M., El-Baz, A.A., El-Monayeri, D., El-Halwagi, M.M., 2009. Reverse problem formulation for integrated process discharges with wastewater and drainage systems: managing phosphorus in lake Manzala. J. Ind. Ecol. 13 (6), 914–927.

Manan, Z.A., Tan, Y.L., Foo, D.C.Y., 2004. Targeting the minimum water rate using water cascade analysis technique. AIChE J. 50 (12), 3169–3183.

Mendez, C.A., Cerda, J., Grossmann, I.E., Harjunkoski, I., Fahl, M., 2006. State-of-the-Art review of optimization methods for Short-term scheduling of batch processes. Comput. Chem. Eng. 30 (6–7), 913–946.

Mouret, S., Grossmann, I.E., Pestiaux, P., 2011. A new lagrangian decomposition approach applied to the integration of refinery planning and crude-oil scheduling. Comput. Chem. Eng. 35 (12), 2750–2766.

Mouret, S., Grossmann, I.E., Pestiaux, P., 2009. A novel priority-Slot based continuous-time formulation for crude-oil scheduling problems. Ind. Eng. Chem. Res. 48 (18), 8515–8528.

Nápoles-Rivera, F., Ponce-Ortega, J.M., El-Halwagi, M.M., Jiménez-Gutiérrez, A., 2010. Global optimization of mass and property integration networks with in-plant property interceptors. Chem. Eng. Sci. 65 (15), 4363–4377.

Pillai, H.K., Bandyopadhyay, S., 2007. A rigorous targeting algorithm for resource allocation networks. Chem. Eng. Sci. 62, 6212–6221.

Polley, G.T., Polley, H.L., 2000. Design better water networks. Chem. Eng. Prog. 96 (2), 47–52.

Ponce-Ortega, J.M., Hourtua, A.C., El-Halwagi, M.M., Jiménez-Gutiérrez, A., 2009a. A property-based optimization of direct recycle networks and wastewater treatment processes. AIChE J. 55 (9), 2329–2344.

Ponce-Ortega, J.M., Serna-González, M., Jiménez-Gutiérrez, A., 2009b. A disjunctive programming model for simultaneous synthesis and detailed design of cooling networks. Ind. Eng. Chem. Res. 48 (6), 2991–3003.

Ponce-Ortega, J.M., El-Halwagi, M.M., Jiménez-Gutiérrez, A., 2010. Global optimization of property-based recycle and reuse networks including environmental constraints. Comput. Chem. Eng. 34 (3), 318–330.

Putra, Z.A., Amminudin, K.A., 2008. Two-step optimization approach for design a total water system. Ind. Eng. Chem. Res. 47 (16), 6045–6057.

Raman, R., Grossmann, I.E., 1994. Modeling and computational techniques for logic based integer programming. Comput. Chem. Eng. 18 (7), 563–578.

Savelski, M.J., Bagajewicz, M.J., 2000. On the optimality conditions of water utilization systems in process plants with single contaminants. Chem. Eng. Sci. 55 (21), 5035–5048.

Savelski, M.J., Bagajewicz, M.J., 2001. Algorithmic procedure to design water utilization systems featuring a single contaminant in process plants. Chem. Eng. Sci. 56 (5), 1897–1911.

Savelski, M., Bagajewicz, M., 2003. On the necessary conditions of optimality of water utilization systems in process plants with multiple contaminants. Chem. Eng. Sci. 58 (23–24), 5349–5362.

Sharda, B., Vazquez, A., 2009. Evaluating Capacity and Expansion Opportunities at Tank Farm: A Decision Support System Using Discrete Event Simulation. Proceedings of the 2009 Winter Simulation Conference. Institute of Electrical and Electronic Engineers, New York, pp. 2218–2224.

Sorin, M., Bedard, S., 1999. The global pinch point in water reuse networks. Process Saf. Environ. Prot. 77 (B5), 305–308.

Teles, J., Castro, P.M., Novals, A.Q., 2008. LP-based solution strategies for the optimal design of industrial water networks with multiple contaminants. Chem. Eng. Sci. 63 (2), 376–394.

Terrazas-Moreno, S., Grossmann, I.E., Wassick, J.M., 2012. A mixed-integer linear programming model for optimization the scheduling and assignment of tank farm operation. Ind. Eng. Chem. Res. 51 (18), 6441–6454.

Wang, Y.P., Smith, R., 1994. Wastewater minimization. Chem. Eng. Sci. 49 (7), 981–1006.

PART

Bioenergy supply chains

IV

CHAPTER

Strategic Planning of Biomass Supply Chain Networks for Co-combustion Plants

18

José M. Laínez-Aguirre*, Mar Pérez-Fortes§, Luis Puigjaner¶, [1]

**Department of Industrial and Systems Engineering, University at Buffalo, Amherst, NY, USA*

§Joint Research Centre, Institute for Energy and Transport, Petten, The Netherlands

¶Department of Chemical Engineering, Universitat Politècnica de Catalunya, Barcelona, Spain

[1] Corresponding author: E-mail: luis.puigjaner@upc.edu

18.1 INTRODUCTION

The electricity generation sector is seeking to reduce its environmental impact and fossil fuel dependence by decreasing its consumption of coal and natural gas. Due to time for deployment and the investment required, co-combustion of biomass and coal is presented as a transition alternative toward a carbon-neutral energy sector. However, the use of biomass in large centralized energy systems requires establishing supply channels capable of providing the desired feedstock with the required attributes, at the right time and place. Hence, there is a need for the adequate assessment of the effective introduction of co-combustion projects in the current electricity production share, preferably by using biomass waste. In this context, mathematical programs are a promising tool to assist in this quantitative evaluation.

Decision-making tools and tailor-made approaches will allow the appropriate exploitation of the potential of biomass in power generation, heat and cooling applications, and as a transportation fuel. Biomass intrinsic heterogeneity (due to its different origins), high moisture content, low bulk density, lower heating value (LHV), degradation, and seasonality make its direct use as a coal substitute difficult. Biomass may be improved when needed, using a pretreatment: enhanced biomass is more stable with regard to degradation and moisture content variations, and has higher LHV, higher bulk density, and lower moisture content than raw biomass. The replacement of 5−15% of coal by biomass in a power plant (in inlet energy terms) does not affect its performance (Gómez et al., 2010). Therefore, this work is focused on coal substitution by biomass-based power within this interval. The proposed approach considers the moisture content, dry matter, LHV, and bulk density as changeable properties of the biomass feedstock along the supply chain (SC).

Computer Aided Chemical Engineering, Volume 36. ISSN 1570-7946. http://dx.doi.org/10.1016/B978-0-444-63472-6.00018-5

Modeling and optimization of SCs is becoming more popular and any type of industry or process can benefit from this approach. Specifically, the work by Laínez and Puigjaner (2012) reviews the application of SC optimization in the chemical process industry. SC modeling derives from classical approaches that only consider operations, and extends to the consideration of business functionalities, and its dynamics. What is more, it takes into account not only material flows but also energy flows. More concretely, the optimization of biomass SCs using a mathematical formulation is also becoming popular, with different approaches to accommodate the particularities of different case studies. The particularities of each case study depend on the selected context, usually characterized by different raw materials and products to be synthesized. The review by Cambero and Sowlati (2014) remarks that the use of biomass has an important potential to substitute fossil fuel, while all three aspects of sustainability (economic, environmental, and societal) have to be considered in the optimization problem. Uncertainty has been increasingly considered when modeling biomass SCs: in Osmani and Zhang (2014), a stochastic mixed integer linear program (MILP) model is developed, which considers uncertainty in the supply of biomass-to-bioethanol, demand of biofuel, biomass, and biofuel prices. The purpose is to determine the location and the efficiency of the biorefineries on the one hand and the connections and storage sites until the bioethanol is sold, on the other. Gebreslassie et al. (2012) also model a stochastic MILP program to address the optimal design of a biorefinery SC under supply and demand uncertainties. They present a risk management optimization approach. Shastri et al. (2011) point out the seasonal and distributed origin of biomass in its BioFeed MILP model, which is developed to optimize a farm, i.e., the biomass collection step. Highly complex and computational demanding programs are resulting from the development of MILPs for the evaluation of biomass-related SCs, thus calling for decomposition methods that can attenuate the heavy computational load that is needed for the solution of stochastic programs. The work of Osmani and Zhang (2014) employs a decomposition based on the sample average approximation method. Gebreslassie et al. (2012) utilize the Multicut L-shaped method, while the work by Shastri et al. (2011) uses a decomposition scheme together with a distributed computing approach. The works by Liu et al. (2014) and Akgul et al. (2012) also consider the production of liquid biomass under economic, energy, and environmental criteria, and only a cost criterion, respectively. Liu et al. (2014) consider conversion of biomass into bioethanol, biomethanol, and biodiesel as possibilities for biomass raw material, while Akgul et al. (2012) model a bioethanol SC. More recently, Yilmaz Balaman and Selim (2014) designed an anaerobic digestion SC, under cost and environmental criteria optimization taking into account uncertainties. In this case, the method employed is a fuzzy multiobjective (MO)-MILP.

This work is a follow-up to our two previous works on optimization of biomass SCs that select (1) optimum flow rates, sites connectivity, and capacity of units and (2) the best technology to be employed from a superstructure of technological options for biomass processing. Pérez-Fortes et al. (2012) optimized a decentralized biomass problem that uses cassava waste as raw material to power a gasifier—gas engine plant. This problem considers MO optimization along three criteria: economic, environmental, and social. The algorithm developed is the starting point of the work of Pérez-Fortes et al. (2014): the MILP program is adapted to solve an SC problem in Spain that considers biomass waste to fulfill the biomass required to replace a portion of the power in the existing coal power plants. In this case, two objectives are optimized, net present value (NPV) and Impact 2002+, as environmental metric. These two works show (1) the versatility of the mathematical model developed to adapt to various biomass case studies, (2) the need to include biomass properties to take into account their changes along the whole SC, and (3) the increase in computational load that makes the second problem considerably time consuming (up to 13,971 central processing unit (CPU) seconds to reach a solution with an optimality gap of 2.5%). Therefore, the current work starts from the case study described in Pérez-Fortes et al. (2014) and compares the proposed decomposition algorithm with the conventional monolithic approach. The problem statement and case study are briefly discussed since all detailed information can be found in our previous references. The next sections describe the mathematical approach, the results obtained, and the advantages of using a Lagrangian decomposition, specifically, the Optimal Condition Decomposition (OCD), which is a particular case of the Lagrangian relaxation procedure.

18.2 PROBLEM STATEMENT

The biomass SC considered comprises four well-distinguished steps: (1) feedstock collection, (2) storage, (3) pretreatment, and (4) co-combustion (treatment). The need to design efficient networks implies estimating the amount of biomass required at each biomass treatment echelon. It includes the consideration of the different echelons through which biomass is processed, with the consequent changes of properties, the selection of the most appropriate processing/distribution channels from biomass suppliers to energy markets, while dealing with seasonal raw material availability and variable demand. Torrefaction, torrefaction combined with pelletization, pelletization, fast pyrolysis, and fast pyrolysis combined with char grinding compose the superstructure of pretreatment technologies considered in this approach. As for the granularity of the problem, monthly planning periods are used to evaluate feedstock and demand seasonality. The challenges that bioenergy projects and, in particular, the developed model, addresses are (1) integration of sourcing, location, and capacity decisions, (2) biomass properties modeling, (3) seasonal raw matter and demand, and (4) computational load.

Figure 18.1 represents a general SC in which the dashed line outlines the frontier of the modeled SC. Suppliers are modeled as fixed locations where biomass is available. The product can be processed at several sites. The properties of raw biomass may be modified by means of the pretreatment units so as to allow intermediates

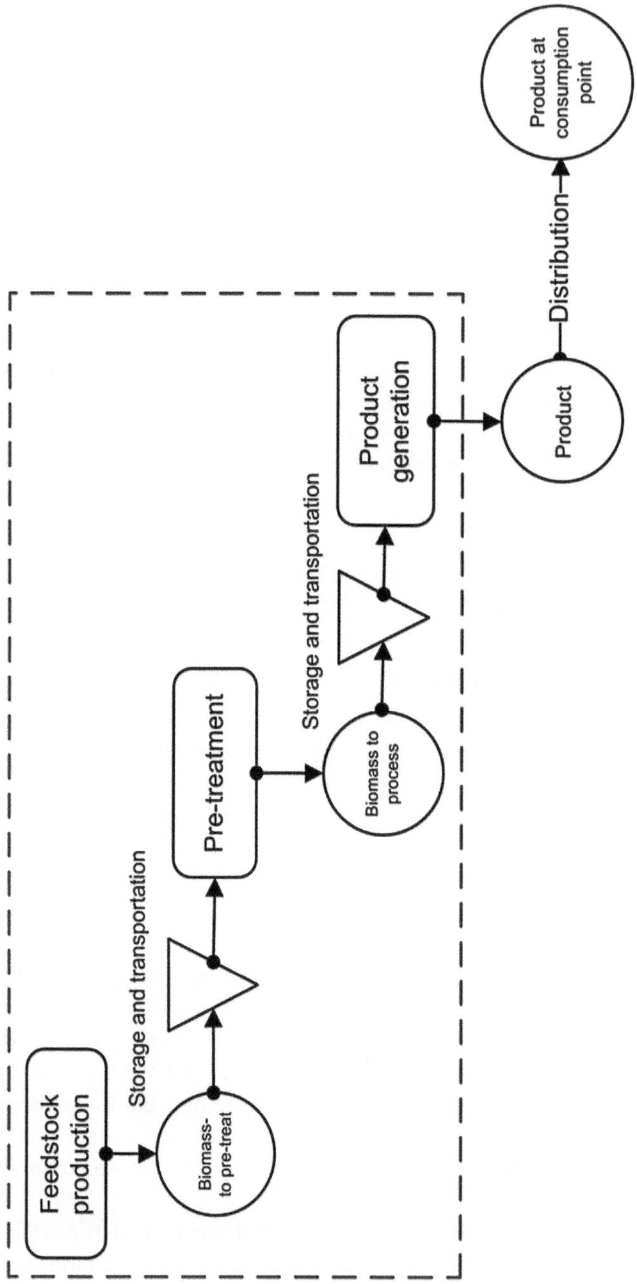

FIGURE 18.1

Biomass-to-energy general SC. The dashed line represents the system limits considered in the study.

to meet the characteristics required by subsequent processes in the SC. Pretreatments may also be convenient to induce savings in transportation costs. The capacity of each processing site is modeled by using a piecewise linear approximation relating the nominal production rate to the availability of the equipment per year (i.e., the number of working hours per year and equipment). Inventory and transportation are modeled by considering upper and lower bounds based on their corresponding biomass handling capacity. Distribution centers can be supplied from more than one pretreatment plant or supplier. Material flows among facilities may appear if selecting such flows improves the overall performance of the SC. The share of energy to be replaced at each co-combustion plant can be met from more than one processing site.

The solution of the problem will suggest size and location for biomass collection sites, storage of raw material, pretreatment units, and storage of pretreated biomass, and will select the most appropriate pretreatment technologies. Moreover, it will designate material flows, allocation of products, and monthly planning of activities, while providing estimates for the resulting costs and environmental impact. As performance indicators, the overall NPV as well as the environmental impact savings associated with the whole SC are modeled.

18.3 MATHEMATICAL MODEL

This chapter describes a tool that can be used to assist in the planning and design of a biomass SC. The mathematical formulation of the SC design-planning problem is briefly described next. The reader is directed to Pérez-Fortes et al. (2014) for further details. The variables and constraints of the model can be roughly classified into three groups. The first one concerns process operations constraints given by the SC topology. The second one deals with the environmental model used, while the third refers to the economic metric applied.

18.3.1 DESIGN-PLANNING MODEL

The design-planning approach presented is a translation of the State Task Network (STN) formulation (Kondili et al., 1993) to SC modeling, which is adapted from the work of Laínez-Aguirre et al. (2009). Such a formulation is suitable to collect all SC node information through a single variable, which eases the environmental and economic metrics formulation. The model's most relevant variable is $P_{ijff't}$, which represents the specific activity of task i performed using technology j receiving input materials from site f and "delivering" output materials to site f' during period t. Indeed, to model a production activity it must receive and deliver material within the same site (P_{ijfft}). In case of a distribution activity, facilities f and f' must be different. The model's equations are briefly described in the next paragraphs.

Materials balances must be satisfied in each of the nodes and time periods. The expression for the balance for each material (stated in the STN formulation) s consumed at each potential facility f in every time period t is given in Eqn (18.1). Parameter α_{sij} is defined as the mass fraction of material s that is produced by task i performed using technology j; T_s set refers to those tasks that have material s as output, while $\overline{\alpha}_{sij}$ and $\overline{T}_s$ set refer to tasks that consume material s.

$$S_{sft} - S_{sft-1} = \sum_{f'} \sum_{i \in T_s} \sum_{j \in \left(J_i \cap \tilde{J}_{f'}\right)} \alpha_{sij} P_{ijf'ft} - \sum_{f'} \sum_{i \in \overline{T}_s} \sum_{j \in \left(J_i \cap \tilde{J}_f\right)} \overline{\alpha}_{sij} P_{ijff't} \quad \forall s, f, t \tag{18.1}$$

The model assumes that process parameters are fixed (such as reaction conversion, separation factors, and temperatures). Notice that the material coefficients (consumption/production factors) for a given activity are fixed and given by the parameters α_{sij} and $\overline{\alpha}_{sij}$; however, there are activities for which it may be desirable to let the model specify the mixture of inputs in order to achieve a given value for a specific biomass property (i.e., moisture content). For such activities the proportion of the different possible feedstocks should be variable. In order to account for those activities the mass balance must be modified as shown in Eqn (18.2).

$$\begin{aligned} S_{sft} - S_{sft-1} = & \sum_{f'} \sum_{i \in T_s} \sum_{j \in \left(J_i \cap \tilde{J}_{f'}\right)} \alpha_{sij} P_{ijf'ft} - \sum_{f'} \sum_{i \in \overline{T}_s} \sum_{j \in \left(J_i \cap \tilde{J}_{f'}\right)} \overline{\alpha}_{sij} P_{ijff't} \\ & + \sum_{i \in T_s} \sum_{j \in \left(J_i \cap \tilde{J}_{f'}\right)} Pv_{sijft} - \sum_{i \in \overline{T}_s} \sum_{j \in \left(J_i \cap \tilde{J}_f\right)} Pv_{sijf't} \end{aligned} \quad \forall s, f, t \tag{18.2}$$

For these flexible activities, it is necessary to ensure that the energy balance is achieved. This is done by introducing Eqn (18.3). Here, HV_{si} is the heating value for material s in activity i. Notice that heating values for feedstock are fixed. An activity changing the output heating value whether (1) it is a pretreatment task that modifies the biomass properties or (2) it is a task that just changes the shape/appearance of biomass but it is receiving different kinds of biomass as input. Notice that the heating value for the feedstock is fixed.

$$\sum_{s \in S_i} HV_{si} Pv_{sijft} = \sum_{s \in \overline{S}_i} HV_{si} Pv_{sijf't} \quad \forall i \in \overline{I}, j, f, t \tag{18.3}$$

Let us consider that a flexible activity must accomplish a total moisture content. In such a case, constraint (4) must be satisfied. Parameters $Water_s$ and $Water_{ij}^{\max}$ express the moisture content for material s and the maximum moisture content permitted for task i performed in equipment j.

$$\sum_{s \in S_i} Water_s Pv_{sijft} \leq Water_{ij}^{\max} \sum_{s \in \overline{S}_i} HV_{si} Pv_{sijf't} \quad \forall i \in \overline{I}, j, f, t \tag{18.4}$$

Equation (18.5) models the temporal changes in facility capacities. The model allows for the simultaneous consideration of design and retrofit of SCs. Equation (18.6) serves for total capacity (F_{jft}) bookkeeping taking into account the amount increased/decreased during planning period t (FE_{jft}).

$$V_{jft}FE^{L}_{jft} < FE_{jft} < V_{jft}FE^{U}_{jft} \quad \forall f, j \in \tilde{J}_{f}, t \tag{18.5}$$

$$F_{jft} = F_{jft-1} + FE_{jft} \quad \forall f, j \in \tilde{J}_{f}, t \tag{18.6}$$

Equation (18.7) ensures the total production rate in each plant is greater than or equal to a minimum desired production rate and lower than or equal to the available capacity. Furthermore, parameter β_{jf} defines a minimum utilization rate of technology j in site f, while $\theta_{ijff'}$ determines the resource utilization factor.

$$\beta_{jf}F_{jft-1} \leq \sum_{f'}\sum_{i\in I_j} \theta_{ijff'}P_{ijff't} \leq F_{jft-1} \quad \forall f, j \in \tilde{J}_{f}, t \tag{18.7}$$

$\theta_{ijff'}$ is the capacity utilization rate of technology j by task i whose origin is location f and destination location f'. This parameter is one of the key factors to be determined when addressing aggregated planning problems that generally consider strategic and tactical decisions. It is important to bear in mind that capacity is expressed as the available time of equipment j during one planning period, then $\theta_{ijff'}$ represents the time required to perform task i in equipment j per unit of produced material. Thus, once operation times are determined this parameter can be easily estimated. Equation (18.8) forces the amount of raw material s purchased from site f at each time period t to be lower than or equal to an upper bound given by physical limitations (A_{sft}). Also, the model assumes that part of the demand can actually be left unsatisfied because of limited production or supplier capacity. Thus, Eqn (18.9) forces the sales of product s carried out in market f during time period t to be less than or equal to maximum demand.

$$\sum_{f'}\sum_{i\in \overline{T}_s}\sum_{j\in J_i} P_{ijff't} \leq A_{sft} \quad \forall s \in RM, f \in Sup, t \tag{18.8}$$

$$\sum_{f'}\sum_{i\in T_s}\sum_{j\in J_i} P_{ijf'ft} \leq Dem_{sft} \quad \forall s \in FP, f \in Mkt, t \tag{18.9}$$

For further model details the reader should refer to Laínez et al. (2009).

18.3.2 ECONOMIC MODEL

Many economic performance indicators have been proposed to assess the economic performance of an SC network design. The most traditional indicators are profit, NPV, and total cost. Other more holistic measures have been proposed (Laínez et al., 2007). Next, expressions to calculate (1) operating revenue, (2) operating cost, and (3) capital investment are presented, which would eventually permit integration with detailed financial models. Here, NPV will be used for the sake of simplicity.

Operating revenue is calculated by means of net sales, which are the income source related to the normal SC activities. Thus, the total revenue incurred in any period t can be easily computed from product sales executed in period t as stated in Eqn (18.10).

$$ESales_{ft} = \sum_{s \in FP} \sum_{f' \in Mkt} Sales_{sff't} Price_{sf't} \quad \forall f \notin (Mkt \cup Sup), t \tag{18.10}$$

In order to calculate overall operating cost an estimation of indirect costs and direct costs are required. The total fixed cost of operating a given SC structure can be computed using Eqn (18.11), where $FCFJ_{jft}$ is the fixed unitary capacity cost of using technology j at site f.

$$FCost_{ft} = \sum_{j \in \tilde{J}_f} FCFJ_{jft} F_{jft} \quad \forall f \notin (Mkt \cup Sup), t \tag{18.11}$$

The cost of purchases from supplier e, which is computed through Eqn (18.12), includes raw material purchases, transportation, and production resources.

$$EPurch_{et} = Purch_{et}^{rm} + Purch_{et}^{tr} + Purch_{et}^{prod} \quad \forall e, t \tag{18.12}$$

The purchases ($Purch_{et}^{rm}$) associated with raw materials made to supplier e can be computed through Eqn (18.13). Variable χ_{est} represents the cost associated with raw material s purchased from supplier e.

$$Purch_{et}^{rm} = \sum_{s \in RM} \sum_{f \in F_e} \sum_{i \in \overline{T}_s} \sum_{j \in J_i} P_{ijfft} \chi_{est} \quad \forall e \in E_{rm}, t \tag{18.13}$$

The costs of production and transportation are determined by Eqns (18.14) and (18.15), respectively. Here, $\rho_{eff't}^{tr}$ denotes the e provider unitary transportation cost associated with material delivery from location f to location f' during period t. τ_{ijfet}^{ut1} represents the unitary production cost associated with performing task i using technology j, whereas τ_{sfet}^{ut2} represents the unitary inventory costs of material s storage at site f.

$$Purch_{et}^{tr} = \sum_{i \in Tr} \sum_{j \in J_i \cap \tilde{J}_e} \sum_{f} \sum_{f'} P_{ijff't} \rho_{eff't}^{tr} \quad \forall e \in \overline{E}_{tr}, t \tag{18.14}$$

$$Purch_{et}^{prod} = \sum_{f} \sum_{i \notin Tr} \sum_{j \in (J_i \cap \hat{J}_f)} P_{ijfft} \tau_{ijfet}^{ut1} + \sum_{s} \sum_{f \notin (Sup \cup Mkt)} S_{sft} \tau_{sfet}^{ut2} \quad \forall e \in \tilde{E}_{prod}, t \tag{18.15}$$

In the case of τ_{ijfet}^{ut1}, it is assumed that the plant operates at the same fixed conditions, meaning that the amount of utilities and labor spent is proportional to the amount of material processed. However, the utilities and labor unitary cost may change in time. Finally, the total investment on fixed assets is computed through Eqn (18.16). This equation includes the investment made to expand the technology's capacity j in facility site f in period t ($Price^{FJ}_{jft} FE_{jft}$).

$$FAsset_{ft} = \sum_{j} Price^{J}_{jft} FE_{jft} + I^{J}_{ft} JB_{ft} \quad \forall f, t \tag{18.16}$$

Equation (18.17) represents the calculation of profit at period t. To conclude, NPV is computed by means of Eqn (18.18), where the discount rate is the rate of return used in a discounted cash flow analysis to determine the present value of future cash flows.

$$Profit_{ft} = ESales_{ft} - (FCost_{ft} + \sum_{e} EPurch_{eft}) \quad \forall t \tag{18.17}$$

$$NPV = \sum_{f} \sum_{t} \left(\frac{Profit_{ft} - FAsset_{ft}}{(1 + rate)^{t}} \right) \tag{18.18}$$

18.3.3 ENVIRONMENTAL MODEL

The application of the LCA methodology to an SC requires four steps, namely, (1) goal setting, (2) life cycle inventory (LCI), (3) life cycle impact assessment (LCIA), and (4) results interpretation.

With regard to goal setting, it is important to define the boundaries of the system under study, and which is the functional unit (FU) that the SC will provide. Boundaries in the case of the chemical industry in general and in the case of biomass SCs are usually drawn from cradle to gate. This is due to the fact that most of these SC products (chemicals and electricity) are used in different ways and their use phase is too difficult to model appropriately. Consequently, raw material extraction, its processing, and shipment to a market are considered as part of the SC system. The purpose of the FU is to allow the comparison of different alternatives in terms of the environmental impact generated. In the case study concerning this work, the FU is given by the amount of inlet coal replaced by the biomass. In that way, the alternatives can be assessed by comparing their associated environmental impact to replace 15% of the inlet coal in the electricity network.

The LCI step involves the estimation of SC environmental interventions (emissions or natural raw material consumptions), which requires assessment of raw material producers, transportation, and product processing impacts. This step is the most time consuming within an LCA, due to the large amount of information that is required to be gathered; however, the usage of LCI databases facilitates this issue. More importantly, the use of a mathematical model helps in calculating the appropriate LCI for the optimal SC configuration.

The results of the LCI step of the LCA can be interpreted by means of different environmental metrics. Environmental interventions are translated into metrics related to environmental impact as end-point or mid-point metrics by the usage of characterization factors (CFs), this translation is the LCIA. These metrics differ in their position along the environmental damage chain (environmental mechanism).

The equations of the environmental model are briefly described next. Equation (18.19) models IC_{aft}, which represents the midpoint a environmental impact associated with site f, arising from activities in period t; $\psi_{ijff'a}$ is the a environmental category impact CF for task i performed using technology j, receiving materials from node f and delivering them to node f'.

$$IC_{aft} = \sum_{j \in \tilde{J}_f} \sum_{i \in I_j} \sum_{f'} \psi_{ijff'a} P_{ijff't} \quad \forall a, f, t \tag{18.19}$$

Similar to the case of α_{sij} and $\overline{\alpha}_{sij}$, the value of $\psi_{ijff'a}$ is fixed and constant, provided that all environmental impacts are directly proportional to the activity performed in that node (P_{ijfft}). This assumption is common practice in LCA, where all direct environmental impacts are considered linearly related with respect to the FU (Heijungs and Suh, 2002). In the case of transportation the FU commonly considered is the kilograms of material transported over a given distance (kilograms per kilometer). Consequently, the value of $\psi_{ijff'a}$ can be calculated by Eqn (18.20) in the case of transportation, considering the distance between sites ($distance_{ff'}$). Here, ψ^{T}_{ija} represents the a environmental category impact CF for the transportation of a mass unit of material over a length unit. It should be noted that in this work environmental impacts associated with materials distribution are assigned to their origin node.

$$\psi_{ijff'a} = \psi^{T}_{ija} distance_{ff'} \quad \forall i \in Tr, j \in J_i, a, f, f' \tag{18.20}$$

Equation (18.21) introduces $DamC_{gft}$, which is a weighted sum of all midpoint environmental interventions combined using g end-point damage factors ζ_{ag} and then further normalized with $NormF_g$ factors.

$$DamC_{gft} = \sum_{a \in A_g} NormF_g \zeta_{ag} IC_{aft} \quad \forall g, f, t \tag{18.21}$$

Equations (18.22) and (18.23) sum up the environmental damage category results for each site f and for the whole SC, respectively.

$$Impact^{2002}_{f} = \sum_{g} \sum_{t} DamC_{gft} \quad \forall f \tag{18.22}$$

$$Impact^{2002}_{overall} = \sum_{f} \sum_{g} \sum_{t} DamC_{gft} \tag{18.23}$$

Finally, the SC network design-planning problem whose objective is to optimize a given set of objective functions (OFs) can be mathematically posed as follows:

$$\underset{X,Y}{Min} \left\{ -NPV, DamC^{SC}_{g}, Impact^{2002}_{overall} \right\}$$

subject to Eqns (18.1)−(18.23)

$$X \in \{0, 1\}; Y \in R^+$$

Here X denotes the binary variables set, while Y corresponds to the continuous variable set.

18.3.4 THE DECOMPOSITION SCHEME

We will carry out a spatial decomposition. For this, the geography of the problem is divided into regions, as depicted in Figure 18.2. Such regions may contain different potential facilities/locations such as suppliers, processing sites, or energy plants (markets). Subsequently, a new set r is introduced for the regions as well as a subset FR_r, which indicates those sites f belonging to region r.

A subproblem is posed for each region, which includes all the material balances, capacity-related equations, and demand constraints for those sites that are confined within it. The variables that are shared among regions are those associated with the flow of materials among sites belonging to different regions, as shown in Figure 18.2. They represent the complicating variables of the problem.

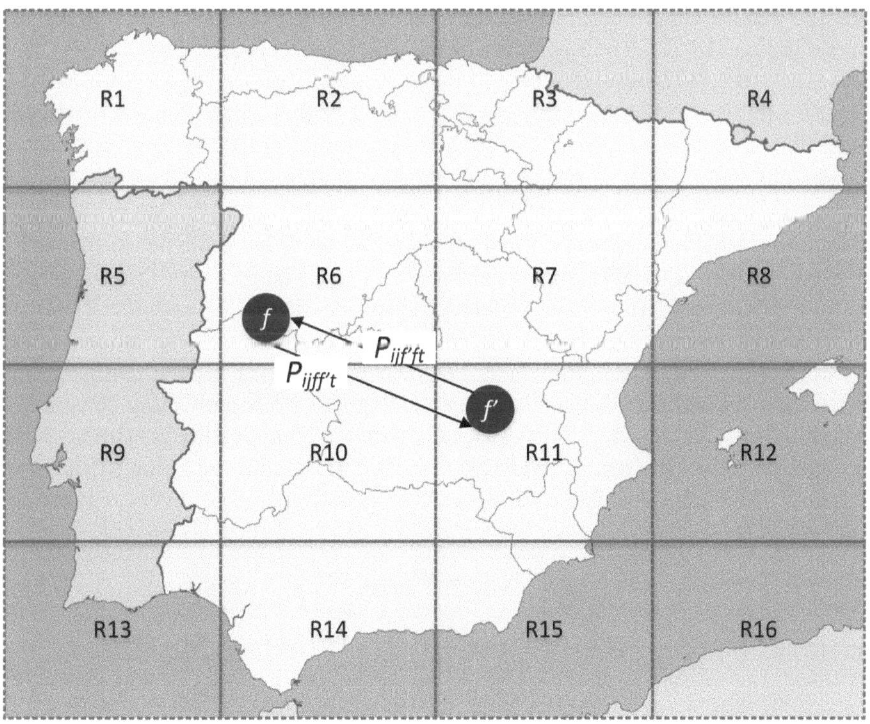

FIGURE 18.2

Geographical division by regions that is employed in the decomposition scheme.

The variables related to flows that are originated from sites confined in a region are considered as part of its corresponding subproblem. Material flows that are originated in other regions are duplicated employing Eqn (18.24) in order to facilitate the convergence of the Lagrangian decomposition scheme. Note that each flow is associated with only one region. This facilitates the definition of the duplicated variables since it is not necessary to define them using the index r. The sufficient condition to enforce the duplication of a flow variable that belongs to other subproblems is that the origin location does not belong to region r.

$$P_{ijf'ft} - \widetilde{P}_{ijf'ft} = 0 \forall i, j \in J_i, f \in FR_r, t, r, f' \notin FR_r \tag{18.24}$$

The OFs are modified to account only for those facilities that are part of the region of the subproblem as described by Eqns (18.25) and (18.26).

$$NPV_r^{region} = \sum_{f \in FR_r} \sum_t \left(\frac{Profit_{ft} - FAsset_{ft}}{(1+rate)^t} \right) \tag{18.25}$$

$$Impact_r^{2002} = \sum_{f \in FR_r} \sum_g \sum_t DamC_{gft} \tag{18.26}$$

Equation (18.24) is then dualized into the OF for the duplicates of the variables belonging to the subproblem, which are included in other regions (subproblems) as shown in Eqn (18.27). The Lagrangian multipliers are represented by λ.

$$OF_r = -NPV_r^{region} + \sum_r \sum_{f' \notin FR_r} \sum_{f \in FR_r} \sum_i \sum_{j \in J_i} \sum_t \lambda_{ijff't} \left(P_{ijff't} - \widetilde{P}_{ijff't} \right) \tag{18.27}$$

We will apply the OCD (Conejo and Nogales, 2002), which is a particular case of the Lagrangian relaxation procedure as depicted in Figure 18.3. One of the OCD advantages is that it provides information to update multiplier estimates in each subproblem iteration; therefore, no master problem exists and the algorithm converges in fewer iterations. The main difference between the OCD and the classical Lagrangian decomposition is that the OCD does not dualize all the complicating constraints. Instead, a subproblem is obtained by dualizing all the complicating constraints of other subproblems and maintaining its own complicating constraints (Eqn (18.24)). The dual variables associated with the complicating constraints kept in each subproblem are readily available from its corresponding solution for the next iteration. Then, the subproblem can be mathematically posed for every region r as follows:

$$\underset{X,Y}{Min} \left\{ -NPV_r^{region}, Impact_r^{2002} \right\}$$

subject to Eqns (18.1)–(18.17), (18.19)–(18.22) $\forall f \in FR_r$; (18.24)–(18.27)

$$X \in \{0,1\}; Y \in R^+$$

It is important to note that the OCD does not need any procedure to update the Lagrange multipliers since this updating process is automatic and results from

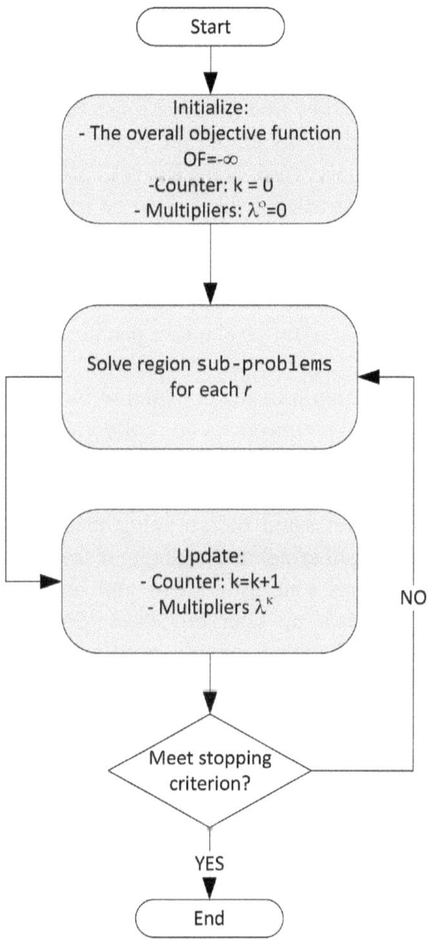

FIGURE 18.3

Schematic representation of the decomposition procedure.

keeping the complicating constraints in the subproblems. A typical stopping criterion is the relative difference between the OF of two consecutive iterations. Details about the convergence properties and the procedure of this decomposition technique can be found in Conejo and Nogales (2002).

18.4 CASE STUDY

The case study presented in this work evaluates co-combustion in the electricity network of Spain; thus, we consider processing existing biomass woody residues (forest wood residues (FWR) and agricultural woody residues (AWR)) to partially feed the existing park of coal power plants. The amount of biomass available is estimated using a "hierarchy of potential" approach, which provides an upper bound for the biomass supply. Only the sites that produce more than 50 kton of biomass per year are considered as candidates to supply the power plants. The total amount of energy contained in the biomass waste is 44 PJ/y (15 PJ/y from FWR and 29 PJ/y from AWR). The total amount of coal that has to be replaced is 22 PJ/y, which is a total of 10% of the whole inlet thermal power (distributed along 19 power plants). Yearly profiles of biomass supply are considered. The model allows biomass storage for upcoming months and considers that its properties, especially humidity, change during storage. AWR has a seasonal production profile, while FWR availability is constant along the year.

In Figure 18.4 a representation of the considered biomass SC is depicted. The SC under study comprises biomass waste collection sites, processing sites where biomass is pretreated and used, distribution centers and energy marketplaces (i.e., current coal power plants).

The biomass has to be pretreated before being finally processed. Figure 18.4 shows that biomass, after gathering and storage if needed, has to achieve the adequate shape and properties (energy content and humidity) through chipping and drying processes to follow up later processes. Specific biomass conditions have been assumed to select the adequate advanced pretreatment process from the superstructure for each feedstock. They basically depend on the moisture content, shape, and LHV of the input required for the treatment plant. Energy densification, matter densification through drying, and dry matter loss in the different processes affecting the heating value have been considered. Co-combustion uses biomass combined with sub-bituminous and bituminous coals, local and imported. It is assumed that the plants work at an efficiency of 36% on LHV basis. Here, we consider the optimization of NPV and the computation of the associated environmental impact reduction derived from the fraction of coal that is replaced by biomass waste.

Three scenarios are analyzed in Pérez-Fortes et al. (2014) to study the tradeoffs among the advanced pretreatments considered in the superstructure: (1) all alternatives are considered, (2) all alternatives are considered, also the no-use of advanced pretreatments, and (3) only the production of liquid biomass is possible. Here, we focus on the most economic-efficient solution for the second scenario to demonstrate the capabilities of the proposed decomposition scheme.

The resulting model is solved for a planning horizon of 10 years, with yearly strategic decisions and monthly planning decisions. General algebraic modeling system (GAMS) is used to implement the MILP and decomposition scheme. The geography of Spain has been divided in 16 regions to apply the proposed decomposition strategy. The monolithic model has in total 59,057 equations, 5,925,346 continuous variables, and 589 discrete variables. Using the monolithic model, it has taken a maximum of 7636 CPU seconds to reach a solution with an integrality gap of 2.5% on a desktop computer dual four core Intel Core i7 920, with 2.66 GHz

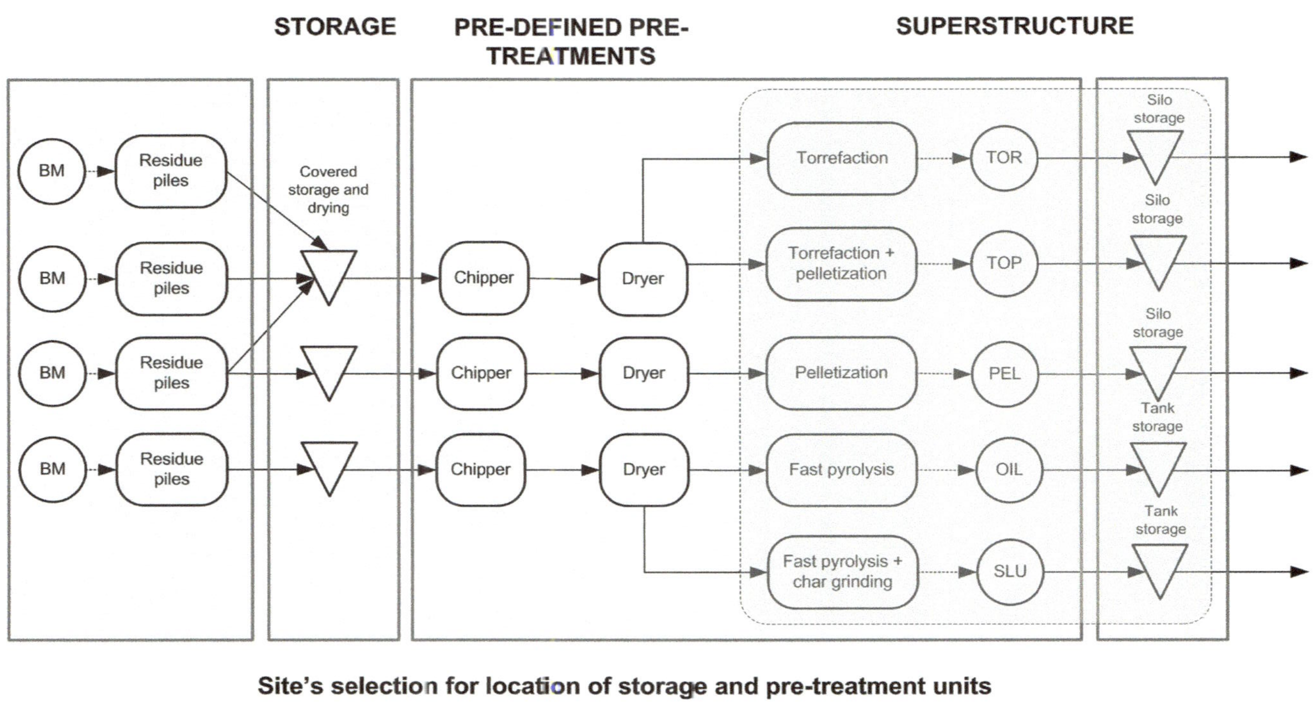

FIGURE 18.4

Schematic representation of the modeled biomass SC (Pérez-Fortes et al., 2014).

CPU and 24 GB random-access memory with eight processors, using MIP solver CPLEX. Preliminary results demonstrate that this computational expense can be significantly reduced by using the proposed decomposition approach.

18.4.1 RESULTS

Figure 18.5 represents the convergence of the proposed decomposition strategy. Fourteen iterations are needed to converge to a solution after 3130 CPU seconds. For instance, a subproblem comprises 36,640 equations, 212,005 continuous variables, and 14 discrete variables. Note that this includes the equations required by the duplication of the complicating variables and that the size of subproblems varies depending on the number of sites within the region. It is noteworthy that a duality gap may exist for the solution obtained from the decomposition. Moreover, the solution resulting from the final iteration may be primal infeasible. Nevertheless, note that all the capacity decisions and their associated binary variables are not causing any conflict in the primal solution since they appear in one single subproblem. Thus, fixing the binary variables to the values obtained from the decomposition algorithm and solving the resulting LP provides a feasible quasi optimal solution. In fact, this results in an NPV value of −0.769 billion €, which differs only by approximately 2% in comparison with the optimal solution obtained by solving the monolithic model (Pérez-Fortes et al., 2014).

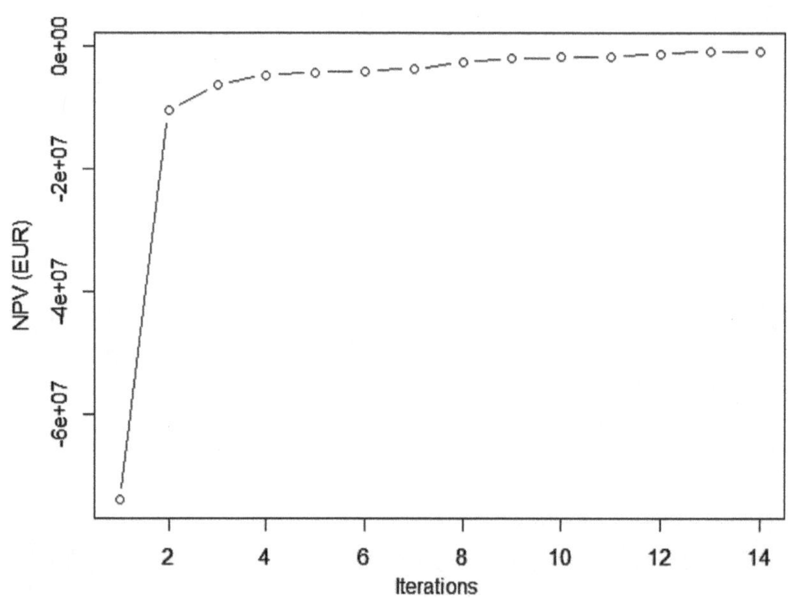

FIGURE 18.5

Convergence of the decomposition scheme for the case study.

Figure 18.6 shows the obtained biomass-based SC that maximizes NPV. The size of biomass providers is relative to the gigajoules per year used from the processing sites to supply the demand. The size of co-combustion plants represents the corresponding replaced thermal input in gigajoules per year. The lines represent the distribution of biomass between sites in terms of kilotons per month. It is distinguished distribution of raw biomass from suppliers to intermediate (pretreatment) sites, distribution of pretreated matter between intermediate sites, and distribution of pretreated matter to the co-combustion plants. For a link to appear in the map, it is sufficient to occur in one single month. This means that the distribution links shown in Figure 18.6 represent shipments during the planning horizon, whether they occur the same month or not. Although the connections among sites are represented by straight lines, the case study is considering a tortuosity factor of 1.4.

The model suggests that 28 locations are opened. Both chipper and driers are installed in 19 of these locations, only a chipper is installed in eight of them, and only dryers are installed in one of these locations. Notice that for this network configuration there are some intersite flows, clearly showing the capabilities of the model to deal with intersite distribution tasks. The investment required to establish the pretreatment process in the biomass SC is provided in Table 18.1. According to the results, the optimized model prefers to ship raw material in short distances (green lines) in comparison to pretreated material (orange and blue lines).

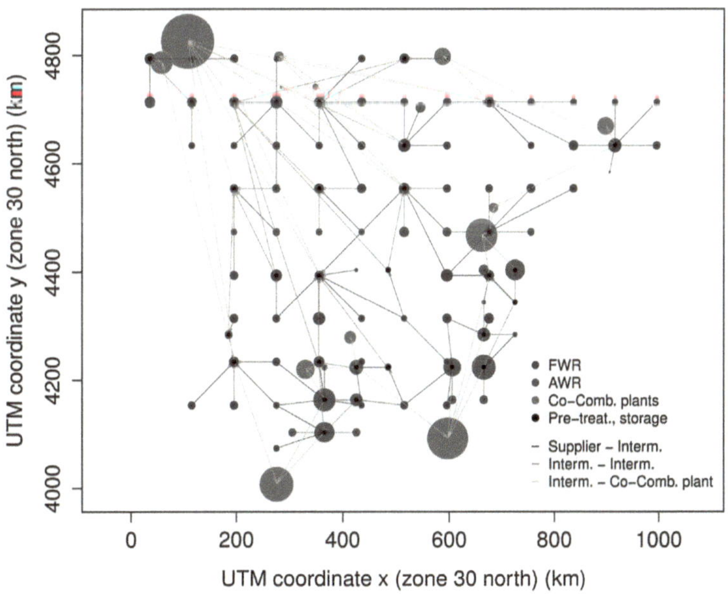

FIGURE 18.6

Optimal NPV network configuration.

Table 18.1 Investments Suggested by the Quasi-Optimal Solution (€ 2010 Million)

Equipment	Investments	Utilities	O&M
Chipper	10.6	0.1	2.1
Dryer	78.9	36.1	2.3
Adaptation for co-combustion plants	461.63	0	18.5

This is due to the higher energy density gained by pretreating the raw biomass. It is also important to mention that there is a trade-off between the savings in transportation costs and the economies of scale obtained by concentrating the pretreatments in fewer locations. Note that virtually only half of the potential pretreatment locations are opened (Table 18.2).

With regard to the environmental impact, the NPV optimal biomass network to replace 10% of the power generated by co-combustion plants provides savings of 410 kpts per year. It is noteworthy that the impact associated with the climate change category (i.e., greenhouse effect gases) is higher when replacing coal by biomass (including all the echelons of the SC). One has to bear in mind that CO_2 is still emitted when using biomass-based generated energy; nevertheless, the other categories are significantly decreased specially for the case of resources and human health. Biomass is regenerated faster than fossil fuels. This is the reason why the impact related to the resources category is larger in the coal-based case. This fact emphasizes the importance of having an overall impact indicator instead of a partial indicator such as CO_2 kg for evaluation purposes. Notice that if the coal- and biomass-based SCs are compared based on CO_2 kg, there would not be a reduction in the emissions to environmentally justify the project.

We would like to highlight that the project to be feasible would require some sort of subsidy. The current solution requires an increase of 9.25% of the revenues to bring the project to an NPV equal to zero.

Table 18.2 Savings in Environmental Impacts from the NPV Optimization Results (Impact 2002+ pts/year)

End-Point Impact Category	Normalized Environmental Impact (Points)
Human health	356,450
Ecosystem quality	15,085
Climate change	−126,932
Resources	165,611
Impact 2002+	410,213

18.5 CONCLUSIONS

This work presents a mathematical model to support the decision-making associated with the strategic and tactical design of biomass SCs. The biomass, which is geographically distributed, is collected and pretreated to feed existing co-combustion plants. The proposed approach is used to deal with a biomass SC located in Spain. The resulting MILPs are medium-size models, which require intensive computational power. In order to tackle more efficiently medium-scale problems, we have used a Lagrangian relaxation-based approach, the OCD. One of the advantages of the OCD is that the procedure to update the multipliers is embedded in the solution of the subproblems, thus avoiding the solution of a master problem. We are exploring an augmented relaxation approach to reduce the duality gap that may appear in the solution of MILPs models. However, it has been demonstrated that the variables associated with capacity decisions are embedded in the subproblems and therefore are not complicating variables, thus given the strategic nature of the problem, fixing the corresponding binary variables to the results from the decomposition scheme and solving the resulting LP provides a feasible quasi-optimal primal solution. Future work is also devoted to devise efficient strategies to add the additional constraints on the OFs to tackle the mono-objective optimizations required in the multiobjective case.

NOTATION

AWR	Agricultural woody residue
FWR	Forest wood residues
LCA	Life cycle assessment
LHV	Lower heating value
MILP	Mixed integer linear program
MO	Multiobjective
NPV	Net present value
OCD	Optimal Condition Decomposition
pts	Points
SC	Supply chain

Indices

e	Suppliers
f, f'	Facility locations
i	Tasks
j	Equipment technology
s	Materials (states)
t, t'	Planning periods
a	Midpoint environmental impact categories
g	End-point environmental impact categories

Sets

A_g	Set of midpoint environmental interventions that are combined into endpoint damage factors g
E_{rm}	Set of suppliers e that provide raw materials
$\widehat{E}_{prod}$	Set of suppliers e that provide production services
$\overline{E}_{tr}$	Set of suppliers e that provide transportation services
F_e	Set of locations f where supplier e is placed
FP	Set of materials s that are final products
I_j	Set of tasks i that can be performed in technology j
$\overline{J}_e$	Technology j that is available at supplier e
$\tilde{J}_f$	Technology j that can be installed at location f
J_i	Technologies that can perform task i
Mkt	Set of market locations
RM	Set of materials s that are raw materials
Sup	Set of supplier locations
T_s	Set of tasks producing material s
$\overline{T}_s$	Set of tasks consuming material s
Tr	Set of distribution tasks

Parameters

A_{sft}	Maximum availability of raw material s in period t in location f
Dem_{sft}	Demand of product s at market f in period t
$distance_{ff'}$	Distance from location f to location f'
$FCFJ_{jft}$	Fixed cost per unit of technology j capacity at location f in period t
I^J_{ft}	Investment required to establish a processing facility in location f in period t
$NormF_g$	Normalizing factor of damage category g
$Price_{sft}$	Price of product s at market f in period t
$Price^J_{jft}$	Investment required per unit of technology j capacity increased at facility f in period t
$rate$	Discount rate
$Water_s$	Moisture for material s
$Water^{max}_{ij}$	Maximum moisture for task i performed in equipment j
α_{sij}	Mass fraction of task i for production of material s in equipment j
$\overline{\alpha}_{sij}$	Mass fraction of task i for consumption of material s in equipment j
β_{jf}	Minimum utilization rate of technology j capacity that is allowed at location f
ζ_{ag}	g end-point damage characterization factor for environmental intervention a
$\theta_{ijff'}$	Capacity utilization rate of technology j by task i whose origin is location f and destination location f'
$\rho^{tr}_{eff't}$	Unitary transportation costs from location f to location f' during period t
τ^{ut1}_{ijfet}	Unitary cost associated with task i performed in equipment j from location f and payable to external supplier e during period t
τ^{ut2}_{sfet}	Unitary cost associated with handling the inventory of material s in location f and payable to external supplier e during period t
χ_{est}	Unitary cost of raw material s offered by external supplier e in period t
$\psi_{ijff'a}$	a environmental category impact CF for task i performed using technology j receiving materials from node f and delivering it at node f'
ψ^T_{ija}	a environmental category impact CF for the transportation of a mass unit of material over a length unit

Binary Variables

V_{jft}	1 if technology j is installed at location f in period t, 0 otherwise

Continuous Variables

$DamC_{gft}$	Normalized end-point damage g for location f in period t
$DamC_g^{SC}$	Normalized end-point damage g along the whole SC
$EPurch_{et}$	Economic value of purchases executed in period t to supplier e
$ESales_{ft}$	Economic value of sales executed in period t from site f
$FAsset_{ft}$	Investment on fixed assets in period t for site f
$FCost_{ft}$	Fixed cost in period t for site f
F_{jft}	Total capacity of technology j during period t at location f
FE_{jft}	Capacity increment of technology j at location f during period t
IC_{aft}	Midpoint a environmental impact associated with site f, which rises from activities in period t
$Impact_f^{2002}$	Total environmental impact for site f
$Impact_{overall}^{2002}$	Total environmental impact for the whole SC
LHV_{si}	Lower heating value for material s in task i
NPV	Net present value
$P_{ijff't}$	Activity magnitude of task i in equipment j in period t whose origin is location f and destination location f'
$Pv_{sijff't}$	Amount of material s for flexible task i in equipment j in period t whose origin is location f and destination location f'
$Profit_{ft}$	Profit achieved in period t for site f
$Purch_{et}^{pr}$	Amount of money payable to supplier e in period t associated with production activities
$Purch_{et}^{rm}$	Amount of money payable to supplier e in period t associated with consumption of raw materials
$Purch_{et}^{tr}$	Amount of money payable to supplier e in period t associated with consumption of transport services
$Sales_{sff't}$	Amount of product s sold from location f in market f' in period t
S_{sft}	Amount of stock of material s at location f in period t

Superscripts

L	Lower bound
U	Upper bound

REFERENCES

Akgul, O., Shah, N., Papageorgiou, L.G., 2012. Economic optimisation of a UK advanced biofuel supply chain. Biomass Bioenergy 41, 57–72.

Cambero, C., Sowlati, T., 2014. Assessment and optimisation of forest biomass supply chains from economic, social and environmental perspectives—a review of literature. Renewable Sustainable Energy Rev. 36, 62–73.

Conejo, A.J., Nogales, F.J., 2002. A decomposition procedure based on approximate Newton directions. Math. Program., Ser. A 93, 495–515.

Gebreslassie, B.H., Yao, Y., You, F., 2012. Design under uncertainty of hydrocarbon biorefinery supply chains: multiobjective stochastic programming models, decomposition algorithm, and a comparison between CVaR and downside risk. AIChE J. 58 (7), 2155–2179.

Gómez, A., Zubizarreta, J., Rodrigues, M., Dopazo, C., Fueyo, N., 2010. An estimation of the energy potential of agro-industrial residues in Spain. Resour., Conserv. Recycl. 54, 972–984.

Heijungs, R., Suh, S., 2002. The computational structure of life cycle assessment. Kluwer Academic Publishers, Dordrecht The Netherlands.

Kondili, E., Pantelides, C., Sargent, R., 1993. A general algorithm for short term scheduling of batch operations. Comput. Chem. Eng. 17, 211–227.

Laínez-Aguirre, J., Kopanos, G., Espuãa, A., Puigjaner, L., 2009. Flexible design-planning of supply chain networks. AIChE 55, 1736–1753.

Laínez, J.M., Guillén, G., Badell, M., Espuña, A., Puigjaner, L., 2007. Enhancing corporate value in the optimal design of chemical supply chains. Ind. Eng. Chem. Res. 46 (23), 7739–7757.

Laínez, J.M., Puigjaner, L., 2012. Prospective and perspective review in integrated supply chain modelling for the chemical process industry. Curren. Opin. Chem. Eng. 1, 430–445.

Liu, Z., Qiu, T., Chen, B., 2014. A study of the LCA based biofuel supply chain multi-objective optimisation model with multi-conversion paths in China. Appl. Energy 126, 221–234.

Osmani, A., Zhang, J., 2014. Economic and environmental optimisation of a large scale sustainable dual feedstock lignocellulosic-based bioethanol supply chain in a stochastic environment. Appl. Energy 114, 572–587.

Pérez-Fortes, M., Laínez-Aguirre, J.M., Arranz-Piera, P., Velo, E., Puigjaner, L., 2012. Design of regional and sustainable bio-based networks for electricity generation using a multi-objective MILP approach. Energy 44, 79–95.

Pérez-Fortes, M., Laínez-Aguirre, J.M., Bojarski, A.D., Puigjaner, L., 2014. Optimisation of pre-treatment selection for the use of woody waste in co-combustion plants. Chem. Eng. Res. Des. 92 (8), 1539–1562.

Shastri, Y., Hansen, A., Rodríguez, L., Ting, K.C., 2011. A novel decomposition and distributed computing approach for the solution of large scale optimisation models. Comput. Electron. Agricu. 76, 69–79.

Yilmaz Balaman, S., Selim, H., 2014. A fuzzy multiobjective linear programming model for design and management of anaerobic digestion based bioenergy supply chains. Energy 74, 928–940. http://dx.doi.org/10.1016/j.energy.2014.07.073.

CHAPTER

19 Supply Chains and Optimization for Biorefineries

José E. Santibañez-Aguilar*, J. Betzabe González-Campos[§], José María Ponce-Ortega*, Medardo Serna-González*, Mahmoud M. El-Halwagi[¶,1]

**Chemical Engineering Department, Universidad Michoacana de San Nicolás de Hidalgo, Morelia, Mich, México*

[§]Institute of Chemical and Biological Research, Universidad Michoacana de San Nicolás de Hidalgo, Morelia, Michoacán, México

[¶]Chemical Engineering Department, Texas A&M University, College Station, TX, USA

[1]Corresponding author: E-mail: el-halwagi@tamu.edu

19.1 INTRODUCTION

A biorefinery is an industrial facility that is capable of producing multiple products (e.g., fuels, specialty chemicals, polymers, etc.) from a wide variety of bioresources (e.g., bagasse, sugar cane, grains, biowaste, etc.). Figure 19.1 is a schematic representation of the concept of a biorefinery and how it may be tied to existing industrial infrastructure (e.g., oil refineries) and CO_2 sequestration cycles. As such, the concept of a biorefinery has gained a lot of attention in recent years because it contributes to enhancing the sustainability of the fossil-based refineries that are restricted by the depletion of oil reserves. Additionally, biorefineries can help reduce the emission of greenhouse gases (GHGs). Biomass production involves the sequestration of CO_2 through photosynthesis. This is a relatively fast phenomenon that can quickly reduce the level of CO_2 concentration in the atmosphere. On the other hand, while fossil fuels involved the capture of carbon sources, the phenomenon took millions of years to yield the fossil fuels. Figure 19.2 is an illustration of the differences.

Figure 19.3 shows a representation for the main steps involved in the operation of biorefineries. These include the cultivation for the bioresources. This step is carried out mainly in the fields yielding important jobs that positively impact the society in these regions. These bioresources need to be harvested and transported to the biorefineries where the bioresources are processed to finally be separated to obtain the desired products and sent to the markets where they are consumed. One important step in the analysis of biorefineries is to account for the consumption of natural resources such as water, other forms of mass, and energy with the possibility of integrating these resources.

Computer Aided Chemical Engineering, Volume 36. ISSN 1570-7946. http://dx.doi.org/10.1016/B978-0-444-63472-6.00019-7

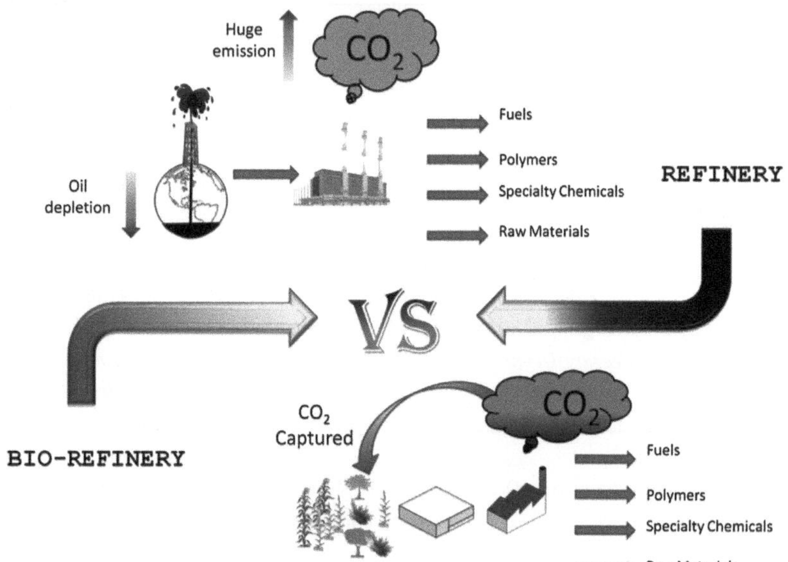

FIGURE 19.1

A fossil-based refinery versus a biorefinery.

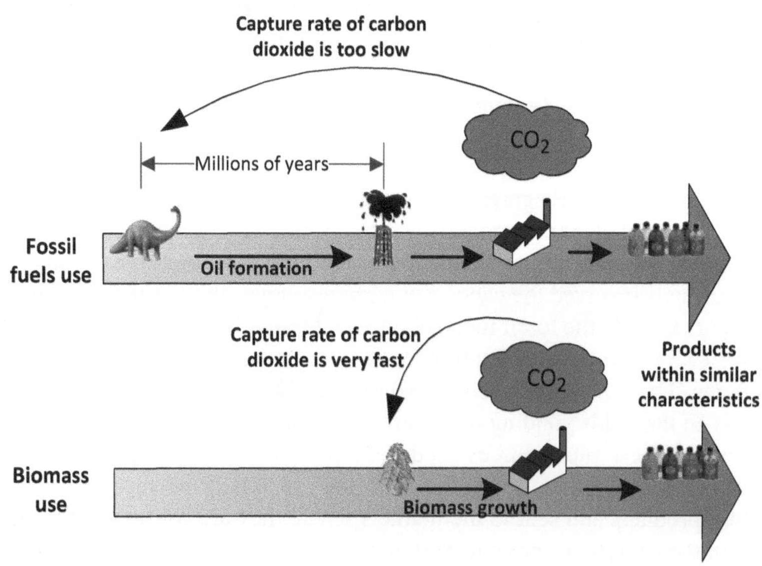

FIGURE 19.2

Comparison for capturing carbon dioxide for a biorefinery and an oil refinery.

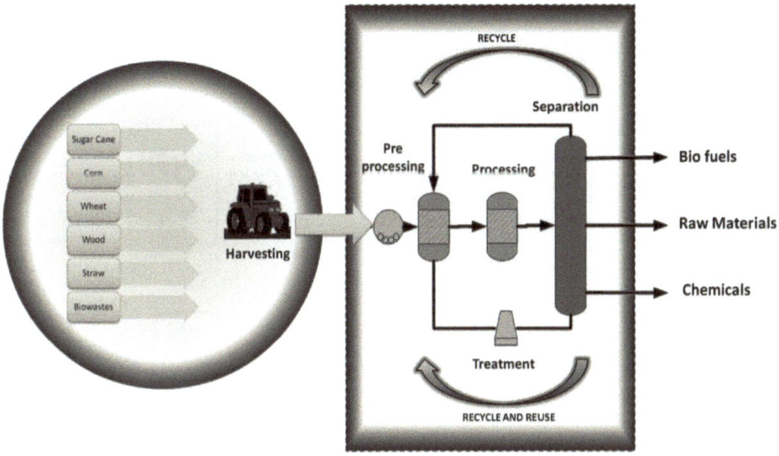

FIGURE 19.3

Representation of key steps in a biorefinery.

It is worth mentioning that there are various types of biomass to produce the portfolio of useful products. There are also several biomass conversion technologies and each one of these alternatives is characterized by various factors such as conversion efficiency, processing cost, processing limits, etc. These technologies may be evaluated and screened by taking into account their technoeconomic viability, environmental impact, and social impact. Furthermore, most of the biomass conversion systems have to be designed under given conditions. Because the suppliers of raw materials are typically located in rural regions, the processing facilities in regions with adequate infrastructure and the main consumers are usually located in urban zones. Consequently, a distributed system like the one shown in Figure 19.4 is necessary to model these types of processes.

In this chapter, several aspects are analyzed for the supply chain of a biorefinery system, which are highlighted as follows:

- Economies of scale: In this case, the supply chain is considered as a distributed system to decrease the transportation costs. Since the processing facilities have different capacities of production, the operating and capital costs depend on the capacity of the plants. It is worth noting that there are two main possibilities for the processing facilities: secondary processing facilities and centralized processing industries (see Figure 19.4). The secondary processing plants are typically small facilities to reduce the costs associated with transportation of bioresources and products. There is also the possibility to install large centralized processing facilities to take advantage of the economies of scale leading to cheaper processing costs. Therefore, the optimization for the

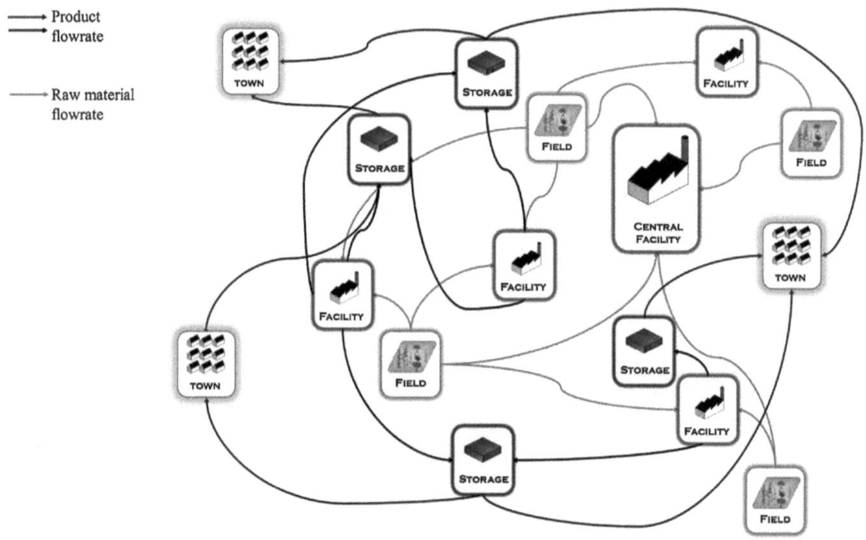

FIGURE 19.4

Schematic representation of a distributed system for a supply chain based on bioresources.

planning of the supply chain for the biorefinery must consider the facility location and capacity in a distributed system.

- Seasonal dependency: One key issue in the supply chain of biorefineries is the seasonal variability of the availability and price of bioresources. This is mainly attributed to the seasonal weather in the different seasons during the year which may affect the availability of the bioresources as well as the demand for bioproducts. Therefore, the storage of bioresources and products in the different locations associated with the supply chain in the biorefineries must be accounted for.
- Sustainability of the supply chain: Sustainability criteria impact drastically the supply chains associated with bioresources. As such, it is necessary to implement a multiobjective methodology to obtain the optimal configuration of the supply chain to enable the consideration of environmental, economic, and social objectives. For the planning of a supply chain of a biorefinery, the economic objectives (the maximization for the net profit or the minimization for the associated costs) must be considered together with the minimization of the overall environmental impact (GHG emissions, ecoindicators, etc.) and the social impact (creation of jobs, competitive uses for the feedstocks, water use, etc.).

The purpose for this chapter is to present several multiobjective optimization approaches for planning supply chains of biorefinery systems while considering the maximization of the net profit, the minimization of the entire environmental impact, and the maximization of the generated jobs. The presented models explicitly take into account the economies of scale through the installation of different preprocessing hubs and central facilities to decrease the transportation costs. Section 19.2 provides a selective literature review of the addressed problems. Section 19.3 presents some case studies and their application for the planning of different supply chains. Finally, Section 19.4 offers the conclusions for this chapter.

19.2 LITERATURE REVIEW

Several studies have focused their attention on the supply chain optimization of biorefineries. Some methodologies for the supply chains of biorefineries have considered a steady state operation (see for example Akgul et al., 2011; Cucek et al., 2010). Other research efforts have included the dynamic behavior (Cruz et al., 2009; Dal-Mas et al., 2011; Santibañez-Aguilar et al., 2014). Recently, research efforts have included environmental criteria for the optimal selection of the biorefinery configuration, raw materials, products and processing paths (Santibañez-Aguilar et al., 2011; You et al., 2011), as well as the optimal site selection involving the transportation costs (Murillo-Alvarado et al., 2014; Santibañez-Aguilar et al., 2014). You and Grossmann (2013) accounted for uncertainties in the design of biorefining supply chains. The generation of products with a high added value from biomass into a distributed system is an important issue and has been studied by several researchers. Aksoy et al. (2008) proposed a method for designing integrated biorefineries, Sammons et al. (2007) presented a flexible superstructure for the optimal planning in the biorefinery production, Sammons at al. (2008) and Van Dyken et al. (2010) proposed mixed integer linear programming (MILP) models for the supply chain in a biorefinery and Elms and El-Halwagi (2009) proposed a scheduling approach for the biodiesel production. Bowling et al. (2011) included the idea of economies of scale in the optimal planning of a biorefinery. Additional works have been undertaken for the optimal planning in the supply chain networks for biofuels production (see for examples Akgul et al., 2011; Aksoy et al., 2008; Corsano et al., 2011; Dal-Mas et al., 2011; Mansoornejad et al., 2010). The aforementioned approaches have considered the economic aspects in the supply chain. Other works have included economic and environmental objectives in the supply chain optimization for the biofuels (see for examples Mele et al., 2009; You and Wang, 2011; Santibañez-Aguilar et al., 2011). Additionally, You et al. (2011) proposed an approach for the optimal planning of supply chains for biofuels taking into account economic, environmental, and social objectives, and also, Parksoy et al. (2013) proposed a MILP model for the optimal planning of supply chain networks in biorefineries.

It is worth noting that in previous works, the environmental objective has been taken into account mainly by accounting for the GHG emissions. Other environmental objectives may be considered. For instance, environmental impact on water bodies may be considered by accounting for the growth of biomass and the discharge of processing pollutants in wastewater streams from the biorefineries. An example of this case is the water hyacinth since this plant grows rapidly and can completely cover water bodies, causing difficulty of navigation and depletion of nutrients and dissolved oxygen which are essential for aquatic lives; in this context, the environment is affected because water hyacinth is not a desired plant in water bodies and the survival of other species is compromised. On the other hand, the water hyacinth consumes a lot of water from water bodies and this amount of water may be used by other organisms. Then the depletion of the water hyacinth is an option to assessment the environmental impact.

Recently, several uses for the water hyacinth have been reported, including its use for wastewater treatment (see Navarro et al., 2011; Chuang et al., 2011; Isarankura-Na-Ayudhya et al., 2007) and as feedstock for biofuels production. For instance, several techniques have been proposed for the production of bioethanol using water hyacinth as feedstock through fermentation (see Nigam, 2002; Kumar et al., 2009; Ganguly et al., 2012). Also, water hyacinth can be used as a biomass feedstock for combustion to provide heat in industrial facilities (Fedler and Duan, 2011) and other chemical products (Gunnardson and Petersen, 2007). However, these approaches did not consider the interaction between the production and processing costs with respect to the economies of scale including the supply chain optimization. To make the processing of water hyacinth economically attractive and sustainable, the economies of scale must be considered along with the distributed supply chain. Consequently, Santibañez-Aguilar et al. (2013) proposed a multiobjective mathematical programming model based on a superstructure of a distributed biorefinery system for the sustainable elimination of the water hyacinth from water bodies in a macroscopic system.

In the context of supply chain optimization, it is crucial to consider an optimization approach that can address the following questions:

- What products should be produced?
- What technologies should be used?
- Where should the processing facilities be located?
- What should be the sizes of these processing facilities?
- What is the optimal size and location for the central processing facilities?
- What markets can be fulfilled with the produced goods?
- What are the associated transportation costs?
- What tax credit (if any) is required to make the biomass processing economically attractive?

It is worth noting that the solution of the above-mentioned problem must address sustainability criteria (economic, environmental, and social issues). Therefore, this chapter presents several multiobjective mathematical programming models based on a superstructure of a distributed biorefinery system for the sustainable biomass conversion in a macroscopic system while accounting for the economies of scale and the production of biofuels and specialty chemicals as well as their optimal distribution satisfying technical and economic constraints.

19.3 CASE STUDIES

This section is divided into three subsections. The first one considers the optimal planning for a biomass conversion system considering economic and environmental aspects. The second one addresses to the optimal planning for a distributed biorefinery system considering the time depending of the parameters and also social, environmental, and economic objectives. Finally, the third one presents the optimal planning of a biomass conversion system focused on the treatment of the water hyacinth from the water bodies; in this case, the environmental objective is the elimination of this plant to get sustainable water bodies.

19.3.1 OPTIMAL PLANNING OF A BIOMASS CONVERSION SYSTEM CONSIDERING ECONOMIC AND ENVIRONMENTAL ASPECTS

This subsection addresses a multiobjective optimization model based on a mathematical programming formulation for the optimal planning of a biomass conversion system. This mathematical formulation considers simultaneously environmental and economic objectives. Furthermore, the optimization approach is able to select the raw materials, products, and processing technologies for the biomass conversion. This way, the economic objective function involves the availability of the bioresources, demand of products, processing limits as well as the costs associated to raw materials, products, and processing. The environmental objective function may be measured through the life cycle analysis methodology by the assessment of the Eco-Indicator 99.

The addressed problem in this subsection can be described as follows: given a set of raw materials is possible to produce a set of products through a portfolio of processing routes. In addition, each alternative can be evaluated technically, environmentally, and economically. The optimal solution represents the best combination of the sets of variables with the maximum economic and environmental benefit. It is worth noting that one of the main objectives for this work is to propose a methodology to consider simultaneously the economic and environmental aspects as is shown in Figure 19.5.

Figure 19.6 shows the superstructure for the mathematical formulation of a biomass conversion system considering simultaneously environmental and economic aspects (see Santibañez-Aguilar et al., 2011).

The mathematical model generated from the superstructure of Figure 19.6 was coded in the software GAMS and its equations have been reported by Santibañez-Aguilar et al. (2011). The model resulting is an LP problem that can be easily solved without numerical complications.

On the other hand, the case study takes into account 21 bioresources available for the system, 3 different products and 10 processing routes. This case study was applied to Mexico but it can be applied to any country. It is important to note that when both objectives contradict each other, then a multiobjective optimization strategy is needed to identify the Pareto curve that describes the set of optimal solutions that simultaneously satisfies the economic and

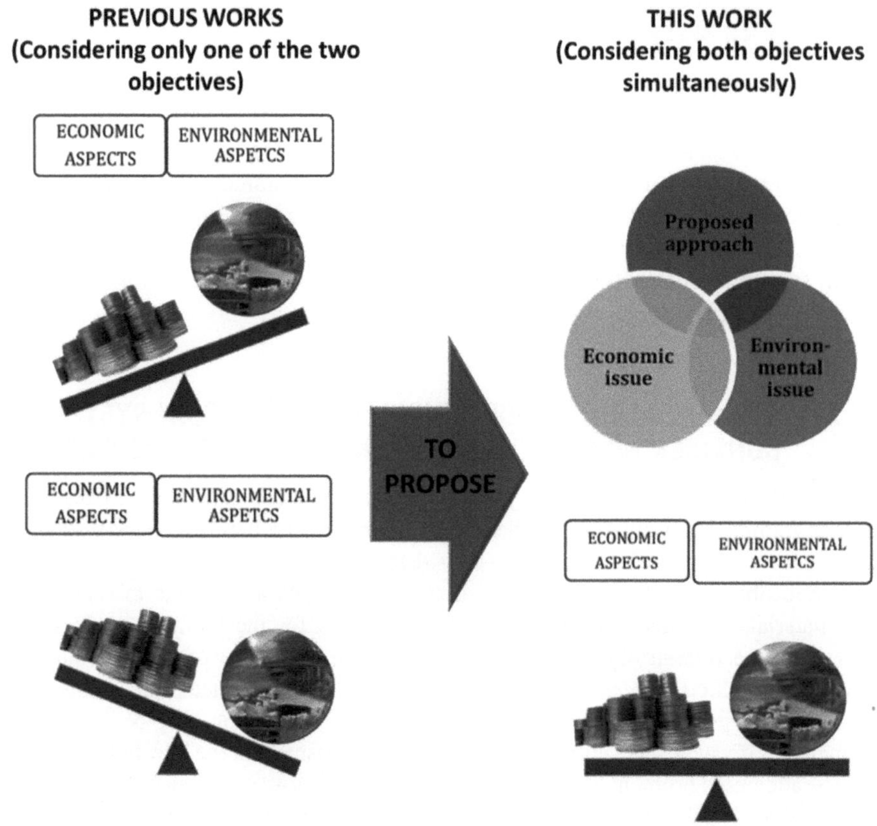

FIGURE 19.5

Main objective for the proposed approach in Section 19.3.1 for the supply chain optimization.

environmental objectives (Figure 19.7) and from which the decision maker can take the one that best fits the specific requirements. Additionally, the required data such as feedstock, product and by-product cost, availability, and demand may be taken from governmental institutions depending on the region where the model is applied. The data and equations for this mathematical model are reported by Santibañez-Aguilar et al. (2011).

In this case, it is possible to identify the solution B (see Figure 19.7) that corresponds to about a quarter of the maximum environmental impact and about 80% of the maximum profit, the configuration of the biomass conversion system is shown in Figure 19.8. It is worth noting that none of the optimal solutions involve the use of food-grade feedstocks.

With respect to ethanol production, the demand is 100% satisfied when the sugar cane is used as the main bioresource, because with 52% of the available sugar cane is possible to satisfy 60% of the bioethanol demand. Regarding the biodiesel and hydrogen production, the fulfilled percentages are very low with respect to the

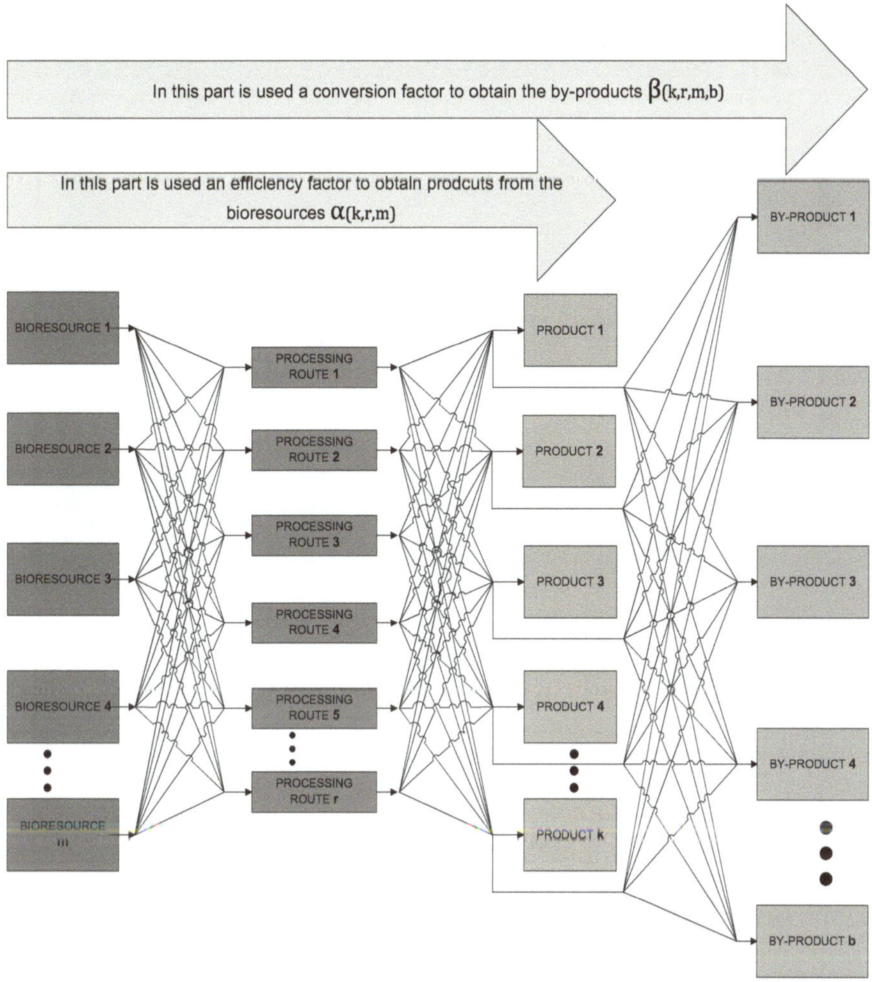

FIGURE 19.6

Superstructure for the optimal planning of a biomass conversion system (Santibañez-Aguilar et al., 2011).

bioethanol production and the consumed bioresource is very high, because the available raw material is not sufficient to satisfy a greater percentage.

19.3.2 OPTIMAL PLANNING OF A DISTRIBUTED BIOREFINERY SYSTEM CONSIDERING ENVIRONMENTAL, ECONOMIC, AND SOCIAL ASPECTS

The problem addressed in this subsection concerns with the long-term planning of a biomass processing supply chain while optimizing three objectives: the profit,

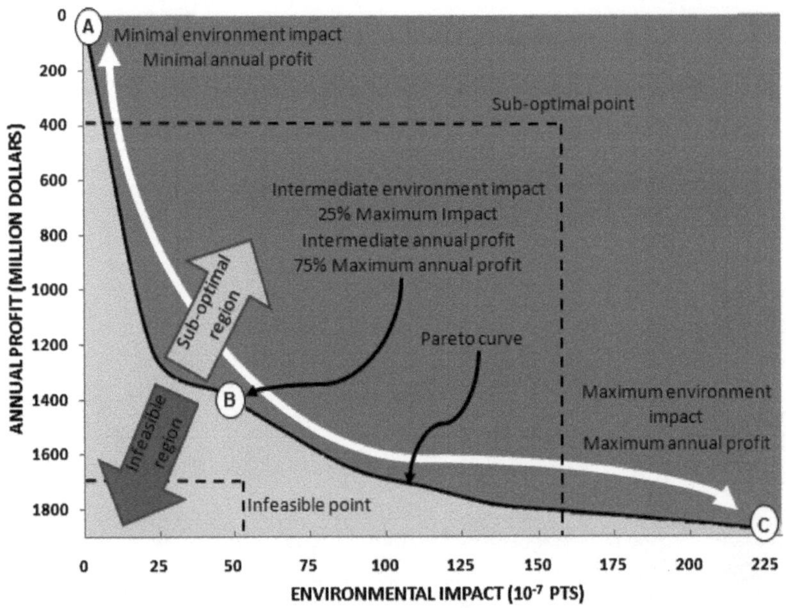

FIGURE 19.7

Pareto solution for the case studied (Santibañez-Aguilar et al., 2011).

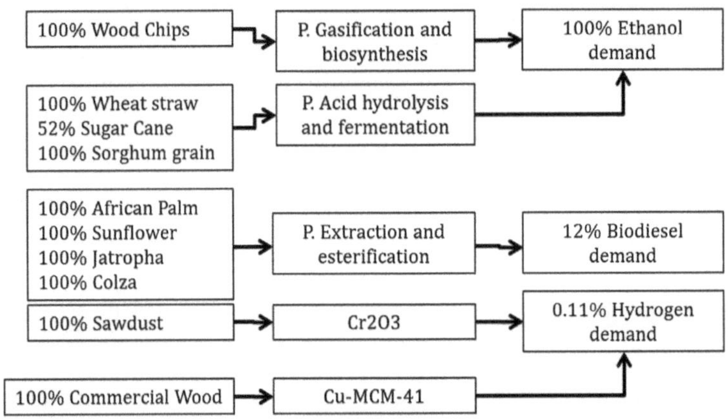

FIGURE 19.8

Configuration of the biomass conversion system for point B of Figure 19.7.

environmental impact, and social impact. This work was based on a state-task network methodology that had been implemented by Maravelias and Grossmann (2003) for the short-term scheduling of multipurpose batch plants. Furthermore, this methodology was used by Castro and Grossmann (2006) for the short-term scheduling of single-stage multiproduct batch plants. Besides, the considered problem includes several harvesting sites (raw material suppliers), a set of multiproduct processing plants and a set of consumers. It is important to note that each considered allocation have storage of material (product or bioresource). Moreover, the supply chain produces biofuels and other chemicals from one or more biomass feedstocks that are available from a given number of raw material suppliers. The harvesting sites have a capacity to produce the bioresources, and also the raw material availability depends on the time. The set of raw materials is purchased and transported from the set of harvesting sites to each processing plant, which transforms them into different products that are either sold in a set of markets or further processed. Thus, each processing facility can distribute the products to different markets or consumers. Figure 19.9 shows the superstructure for this case of modeling for the supply chain based on biomass conversion.

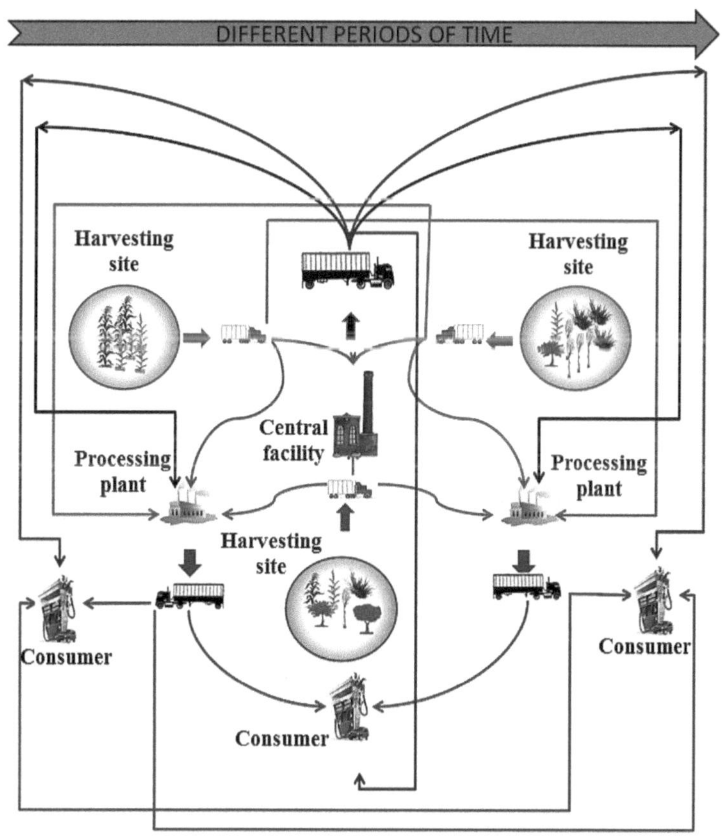

FIGURE 19.9

Schematic representation for the addressed problem in Section 19.3.2.

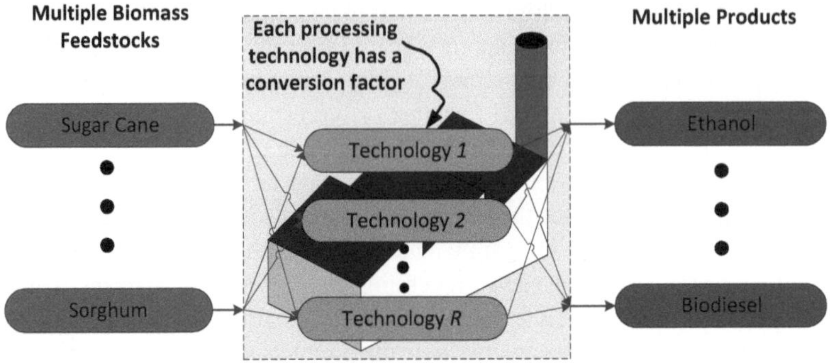

FIGURE 19.10

Representation of the production process for each facility within multiple feedstocks and products.

It is worth noting that each processing facility can use different production technologies to obtain multiple final products from the set of feedstocks. The amount of each product generated by all available manufacturing technologies is assumed to relate to the biomass feedstock consumption through a linear mass balance equation that is expressed in terms of a constant production yield (i.e., the mass of raw materials consumed per unit mass of product manufactured for each technology); this is explained in Figure 19.10. Finally, the goal is to determine the optimal configuration of the distributed biorefining supply chain and the associated planning decisions that simultaneously maximize profit and social impact, minimize the environmental impact, and satisfy the demands of each market over the specified planning horizon.

The proposed model considers that the biomass processing distributed to secondary plants will occur at relatively small scales near the harvesting sites, while the main plant is larger and located in an industrialized zone. This situation arises because feedstocks are obtained from multiple, widely distributed harvesting sites. These distances result in excessive transportation costs of raw materials to a single large processing plant. Therefore, the distribution of facilities in this model takes advantage of the economies of scale.

In order to show the mathematical model for the considered problem, the equations are listed in Table 19.1 and their explanations have been given by Santibañez-Aguilar et al. (2014). Furthermore, some of the more important points for this model are the following:

- The mathematical formulation is a multiobjective, multiperiod MILP problem, where the objective functions are the simultaneous maximization of the net profit, minimization of the environmental impact, and maximization of number of jobs generated in the entire supply chain.
- The proposed approach explicitly accounts for the economies of scale of the production technologies and distributed biomass processing to reduce the

Table 19.1 Equations for the Proposed Mathematical Model from Santibañez-Aguilar et al. (2014)

Equations for the Proposed Mathematical Model	
• Mass balances for harvesting sites	• Transportation constraints (cost and limits)
• Mass balances for processing facilities • For raw materials • For products	• Processing equations (cost, conversion, and limits)
• Mass balances for markets	• Continuity constraint for the beginning and end of the time horizon
• Constraints for the maximum fulfilled demand	• Economic objective function (net profit)
• Constraints for the materials storage	• Environmental objective function (overall environmental impact)
• Availability raw material constraints	• Social objective function (generated jobs)

transportation costs of raw materials. This method also includes the optimal processing plant locations and capacities, as well as the optimal selection of a number of processing technologies to obtain multiple products using a variety of biomass feedstocks.

- The model explicitly considers the seasonal dependence of potential feedstock bioresources for the biorefinery supply chain.
- The time periods used in the considered planning problem is divided into time periods of equal size, where aggregated production is planned by assuming that the tasks must begin and end at a specified time point.

To test the proposed methodology, a case study for planning the supply chain of a biorefinery system in Mexico is considered. This case involves nine bioresources (wood chips, commercial wood, sugar cane, corn, sorghum, sweet sorghum, sunflower, African palm oil, and jatropha) available in Mexico that can be used as feedstock to obtain two products (bioethanol and biodiesel); in addition, there are available four different processing routes (gasification and biosynthesis; hydrolysis and fermentation; gasification and chemical synthesis; and transesterification).

To solve the multiobjective, multiperiod MILP problem and identify the trade-offs between the three objectives considered was proposed the following solution strategy. Limits to clearly visualize the relation between the objectives of two planes must be obtained because it is difficult to illustrate their trade-offs in a Pareto curve with more than two dimensions. A general description of the solution approach is shown in Table 19.2.

Results of the proposed case study are shown through several Pareto curves to determine the trade-offs between the different objectives. Figure 19.11 presents the Pareto curves for the maximization of the net profit and the minimization of the environmental impact while measuring the number of generated jobs under

Table 19.2 Description of the Solution Strategy for the Presented Mathematical Model

Strategy Step	Result
Maximize the net profit without considering the environmental impact or social benefits	Provide information about the maximum net profit, as well as an upper limit for the environmental impact and a lower limit for the social impact
Minimize the environmental impact without considering the net profit or social benefits	Obtain the minimum environmental impact and limits for the lower net profit
Maximize the social benefits without considering the net profit or economic benefits	Provide lower limits for the net profit and upper limits for the environmental impact
Implement the constraint method	Determine the Pareto solutions for the combinations of the three objectives

several constraints for the fulfilled demand of products. In this figure, two different behaviors are identified; in the first section, the profit increases drastically for small increments in the environmental impact; whereas in the second section, the profit increases in small portions with respect to the increment in the environmental impact. In this optimal Pareto curve, it can also be observed that the total number of jobs

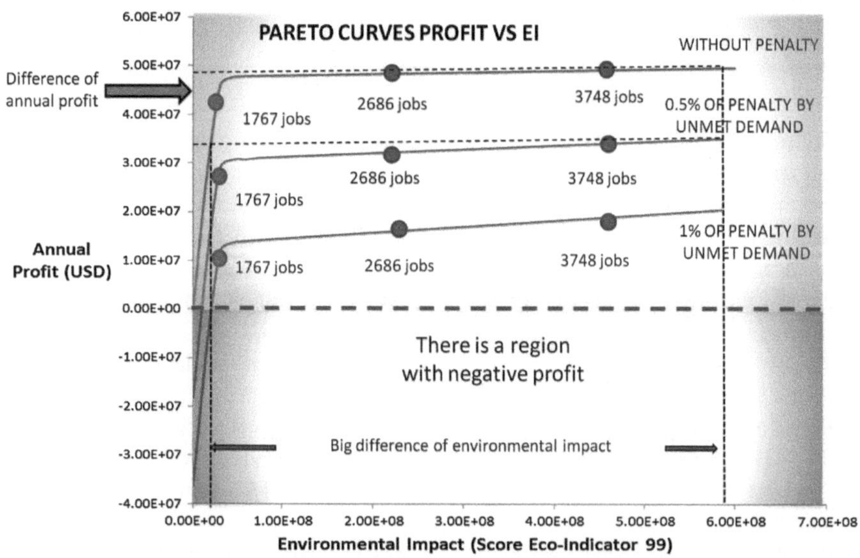

FIGURE 19.11

Pareto curves for the maximization of the annual profit and the minimization of the environmental impact (Santibañez-Aguilar et al., 2014).

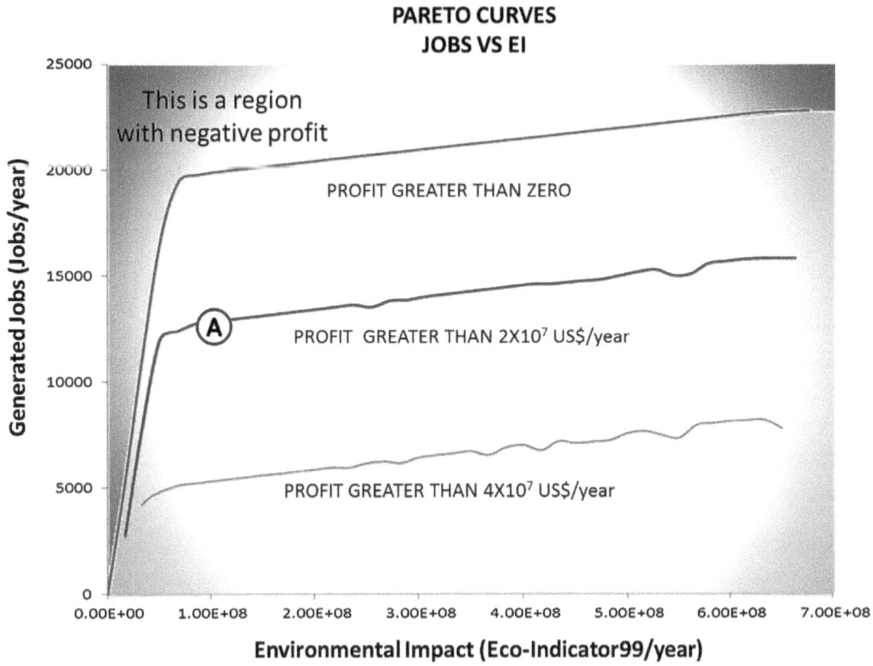

FIGURE 19.12

Pareto curve for the social benefit versus environmental impact for the minimum annual profit (Santibañez-Aguilar et al., 2014).

increases with the increment in the net profit and the increment in the Eco-Indicator 99. For the Pareto curves, the number of created jobs increases until a maximum, where the selected feedstocks are wood chips and sorghum for ethanol and jatropha and African palm for biodiesel.

The following analysis maximizes the social benefit while minimizes the environmental impact for different fixed values of the minimum net profit between zero and $\$4\times10^7$ USD. Figure 19.12 shows the resulting Pareto curves for this scenario. Here, the social benefit is highly significant, generating up to 23,000 jobs. The number of generated jobs strongly depends on the environmental impact. In addition, the number of generated jobs correlates positively with the environmental impact (Figure 19.12). This effect positively impacts the society due to the number of generated jobs, but negatively impacts the environment. Thus, there is a trade-off between these objectives.

Finally, Figure 19.13 shows the Pareto sets of optimal solutions for the number of jobs generated versus the maximum net profit for given values of the environmental impact. These results clearly indicate that both objectives are in conflict: improving the social benefit decreases the net profit and vice versa. This figure indicates that the number of generated jobs is a decreasing near-linear function of the net profit,

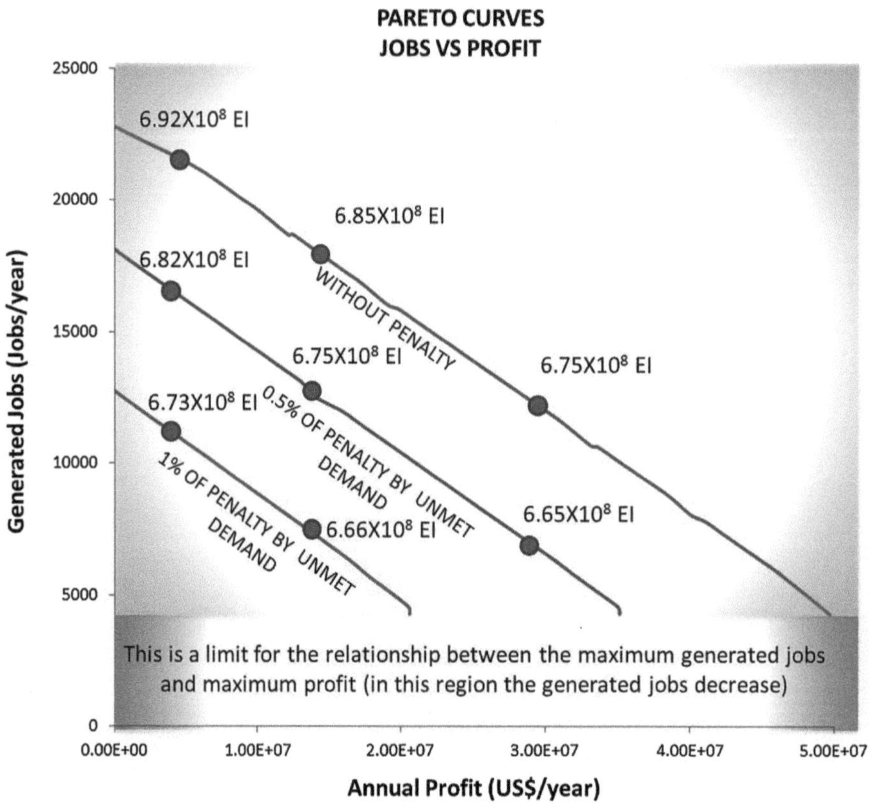

FIGURE 19.13

Pareto curve for the social benefit versus the annual profit (Santibañez-Aguilar et al., 2014).

attaining its lowest values for large net profits and low environmental impacts. Furthermore, the social benefit decreases as the penalty for unmet demand decreases when the environmental impact is constant.

19.3.3 SUSTAINABLE ELIMINATION OF THE WATER HYACINTH THROUGH A DISTRIBUTED BIOREFINERY SYSTEM

In order to define the considered problem in this subsection is important to mention that water hyacinth can be considered as a raw material to produce a set of products because water hyacinth has biomass in its structure; however, this plant has a lot of water too. This way, water hyacinth is a source for clean water and dry biomass, when the water hyacinth is harvested, this plant can be dried to produce water and biomass, also, the produced water has some pollutants but these pollutants may be eliminated, so, the water bodies are cleaned and the biomass is used to obtain a profit.

The proposed model is based on the superstructure presented in Figure 19.14, which was proposed by Santibañez-Aguilar et al. (2013). Based on Figure 19.14 the addressed problem in this subsection can be defined as follows: given are a set of water bodies with specified amounts of water hyacinth; also, given are several markets for the products manufactured with known demands. The problem then consists of determining the location for the processing facilities and their sizes, the technologies used the requirements of water hyacinth in the different places and the distribution of the products.

It should be noted that the optimization model involves maximizing the net annual profit subject to constraints of production and harvesting rate for the water hyacinth, distribution of raw material and products in the different allocations of the supply chain, technical constraints for the processing technologies as well as equations to obtain the capital and operational cost for the supply chain. It is worth mentioning that the model considers simultaneously the supply chain for the bioproducts obtained from the water hyacinth, the remediation of the water bodies, and the recycle and reuse for the recovered water. Furthermore, the optimization model simultaneously optimizes the selection of bioconversion technologies for water hyacinth and water treatment; it also considers the optimization for the size and location for hubs and central treatment facilities to account for the economies of scale; in addition, the model considers what products are selected (bioproducts and water) and the optimal distribution for the products.

To show the applicability of the proposed methodology, a case study from Mexico is used. This case is considered because the central part of Mexico has a lot of water bodies highly infested with water hyacinth; therefore, the sustainability of these systems is in risk.

In order to define the case study, it is necessary to define some of the main allocations. First the potential harvesting sites are identified as the water bodies seriously infested with water hyacinth. Lake of Chapala is one of the proposed harvesting sites since it is a primary lake in Mexico and is infested with water hyacinth. This lake is close to Guadalajara, the second largest city of Mexico. Another site is the Balsas River, which is one of the greatest rivers in the central part of Mexico. This river supplies cooling water to the most important thermoelectric plant in the state of Michoacan in Mexico. In addition, the Lakes of Patzcuaro and Cuitzeo are proposed as harvesting sites. These lakes are close to the cities of Patzcuaro and Morelia, respectively; the Lake of Patzcuaro has an endemic species of fish named white fish, and tourism and fishing are the most important activities in this region. Also, the lake of Yuriria is an important lake of the state of Guanajuato infested with water hyacinth. This way, these harvesting sites are very important water bodies in the considered system. Furthermore, the refinery candidates correspond to the main cities around such water bodies, since the objective of a distributed system is to decrease the costs (manly transportation costs) and, hence, the transportation distances must be short. Besides, the markets are located in cities around the water bodies, because these locations have a given demand of biofuels, water, and chemicals among others products.

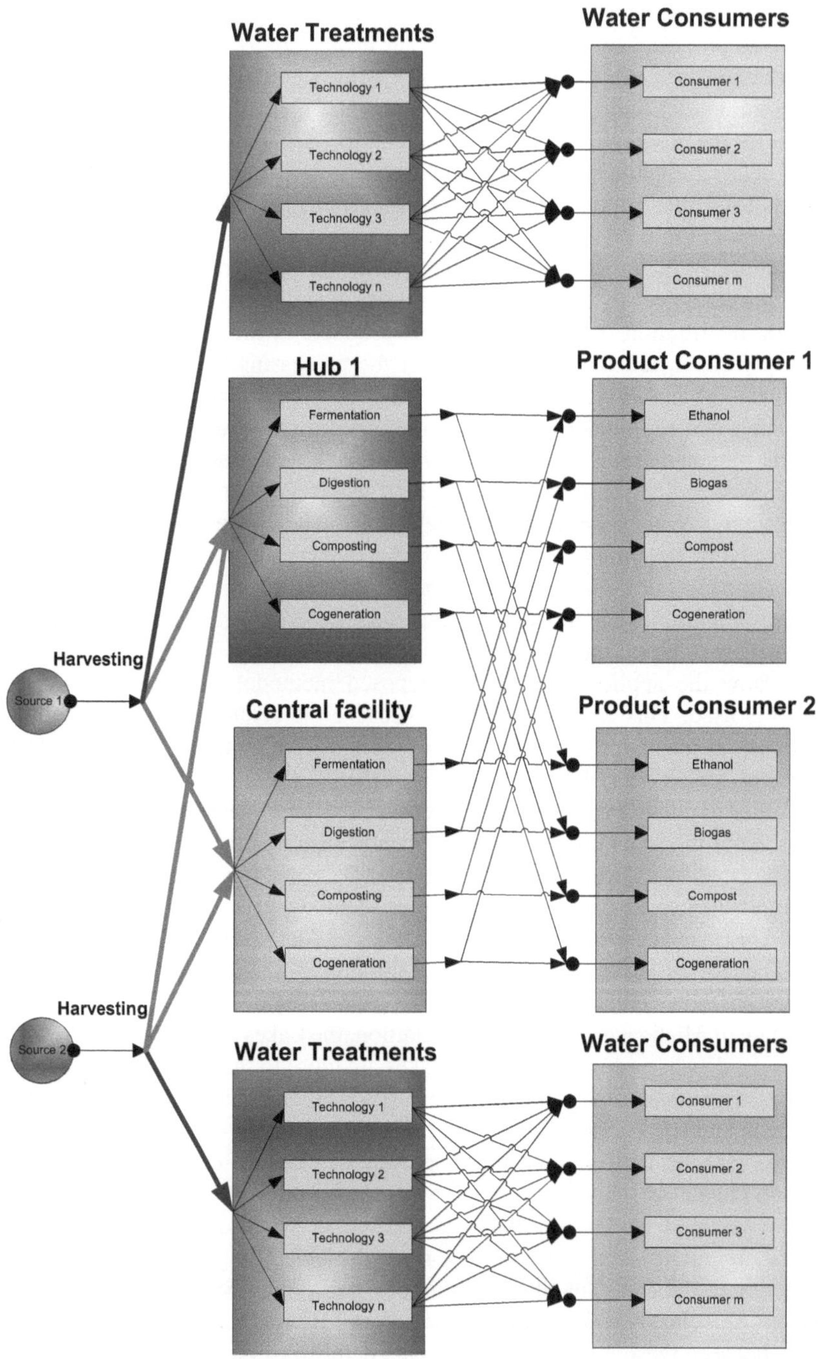

FIGURE 19.14

Superstructure for the distributed supply chain biofuels based on water hyacinth (Santibañez-Aguilar et al., 2013).

The proposed optimization approach was applied to the case study that was previously described. The data for the case study are reported by Santibañez-Aguilar et al. (2013). Results are presented through Pareto curves between the maximization amount of harvested water hyacinth and the minimization of an economic objective, which is defined as the negative of the net annual profit. In this context, a negative value of the economic objective function represents a positive profit. When the value of the economic objective function is positive, it represents that the option is not attractive economically. This economic objective function is taken as the negative of the net profit in order to compare with the harvesting cost of the water hyacinth, which is the cost for the no implementation of the system based on a supply chain. Figure 19.15 shows the trade-off between the maximization of the percentage eliminated of water hyacinth and the maximization of the net annual profit. In this case, it is observed that net annual profit decreases when the eliminated water hyacinth increases. It leads to a decrease in the difference between the harvesting cost and the economic objective function, while the eliminated water hyacinth increases.

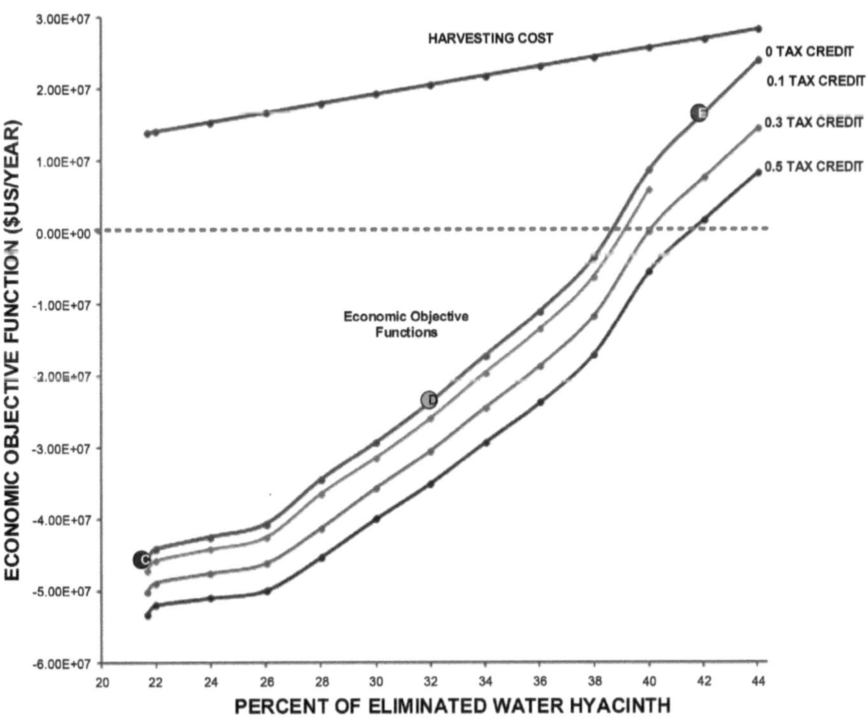

FIGURE 19.15

Pareto curves for the maximization of the net annual profit and the maximization of the percentage of eliminated water hyacinth (Santibañez-Aguilar et al., 2013).

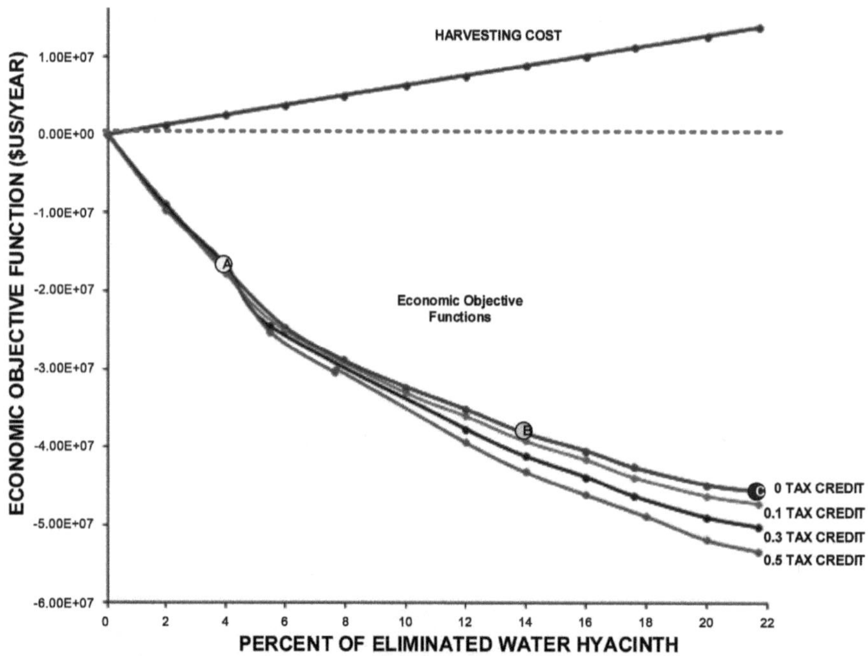

FIGURE 19.16

Pareto curves for a low consume of water hyacinth (Santibañez-Aguilar et al., 2013).

On the other hand, in Figure 19.16, the Pareto curves contradict each other because the economic objective function begins in a value of zero and it decreases, while the harvesting cost increases. The maximum amount of eliminated water hyacinth is about 22% and this point is represented with the letter C, for a tax credit of zero. The point C in Figure 19.16 is the same solution represented as point C in Figure 19.15. As can be observed, the point C represents an inflexion point between both curves. Other analyses can be done when each point is shown based on the percentage eliminated of water hyacinth from each water body.

Figure 19.17 shows the individual consumption of water hyacinth at each source. The main water body for the case study is Lake of Chapala, so it is selected to consider the specific elimination of water hyacinth with different percentages. For example, point A represents an elimination of 30% of the water hyacinth in this lake, whereas in point C almost 100% of the water hyacinth is eliminated from this water body. On the other hand, the second most important water body for this system is the Balsas river because it has a large amount of this plant; however, this river is located close to the west coast and far from other water bodies and cities; therefore, the elimination of water hyacinth from the Balsas river is selected only in point E. Notice in Figure 19.15 that this point represents a high consumption of water hyacinth but it also has a greater *TAC* with respect to the other solutions. In

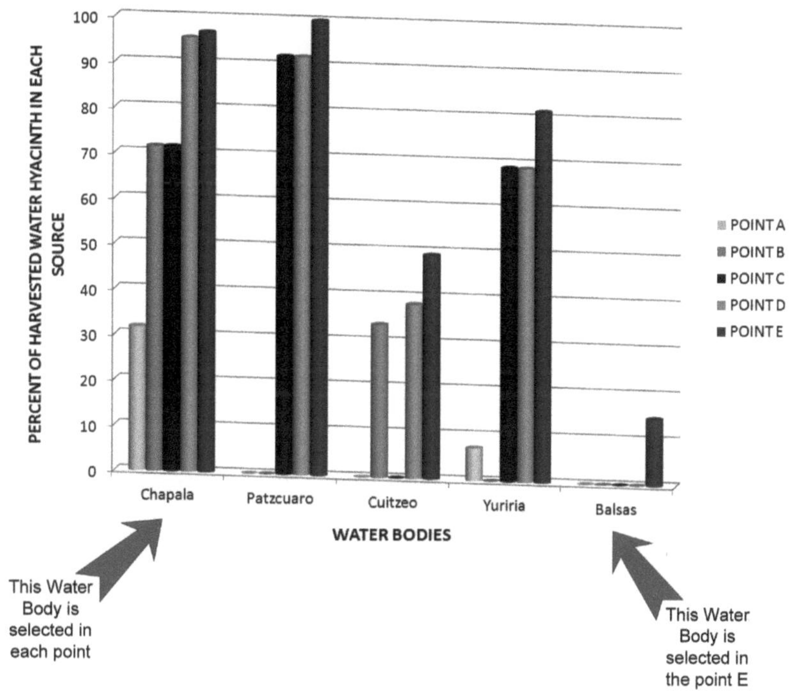

FIGURE 19.17

Representation of eliminated water hyacinth at each source and overall (Santibañez-Aguilar et al., 2013).

addition, in the addressed problem the water bodies corresponding to the lakes of Patzcuaro, Cuitzeo, and Yuriria, which are important because they are the closest ones to the main cities and are related to significant economic activities.

19.4 CONCLUSIONS

Several optimization models have been developed for the optimal planning of the production of distributed biorefinery systems accounting for different key points in the supply chain. The formulations include linear programming, MILP, and mixed integer nonlinear programming models. In addition, all of the methodologies are multiobjective approaches. The consideration of the objectives is different for different cases. The addressed case study shows that it is possible to consider the environmental objective by a global environmental impact and the reduction of pollutants in water bodies. Additionally, the social objective has been included in one of the methodologies by accounting for the generated jobs in the entire supply chain. The results show that these three aspects have a strong impact on the supply chain optimization. The application of the proposed optimization approaches to several case studies in Mexico shows that the economic, environmental, and social objectives may be methodically reconciled and traded off.

REFERENCES

Akgul, O., Zamboni, A., Bezzo, F., Shah, N., Papageorgiou, L.G., 2011. Optimization-based approaches for bioethanol supply chains. Ind. Eng. Chem. Res. 50 (9), 4927–4938.

Aksoy, B., Cullinan, H.T., Sammons Jr., N.E., Eden, M.R., 2008. Identification of optimal poultry litter biorefinery location in Alabama through minimization of feedstock transportation cost. J. Environ. Prog. 27 (4), 515–523.

Bowling, I.M., Ponce-Ortega, J.M., El-Halwagi, M.M., 2011. Facility location and supply chain optimization for a biorefinery. Ind. Eng. Chem. Res. 50 (10), 6276–6286.

Castro, P.M., Grossmann, I.E., 2006. An efficient MILP model for the short-term scheduling of single stage batch plants. Comput. Chem. Eng. 30 (6–7), 1003–1018.

Chuang, Y.S., Lay, C.H., Sen, B., Chen, C.C., Gopalakrishnan, K., Wu, J.H., Lin, C.S., Lin, C.Y., 2011. Biohydrogen and biomethane from water hyacinth (*Eichhornia crassipes*) fermentation: effects of substrate concentration and incubation temperature. Int. J. Hydrogen Energy 36, 14195–14203.

Corsano, G., Vecchietti, A.R., Montagna, J.M., 2011. Optimal design for sustainable bioethanol supply chain considering detailed plant performance model. Comput. Chem. Eng. 35 (8), 1384–1398.

Cruz, J.B., Tan, R.R., Culaba, A.B., Ballacillo, J.A., 2009. A dynamic input-output model for nascent bioenergy supply chains. Appl. Energy 86 (1), 586–594.

Cucek, L., Lam, H.L., Klemes, J.J., Varbanov, P.S., Kravenja, Z., 2010. Synthesis of regional networks for the supply of energy and bioproducts. Clean Technol. Environ. Policy 12 (6), 635–645.

Dal-Mas, M., Giarola, S., Zambini, A., Bezzo, F., 2011. Strategic design and investment capacity planning of the ethanol supply chain under price uncertainty. Biomass Bioenergy 35 (5), 2059–2071.

Elms, R.D., El-Halwagi, M.M., 2009. Optimal scheduling and operation of biodiesel plants con multiple feedstocks. Int. J. Process Syst. Eng. 1 (1), 1–28.

Fedler, C.B., Duan, R., 2011. Biomass production for bioenergy using recycled wastewater in a natural waste treatment system. Resour., Conserv. Recycl. 55 (8), 793–800.

Ganguly, A., Chatterjee, P.K., Dey, A., 2012. Studies on ethanol production from water hyacinth-a review. Renewable Sustainable Energy Rev. 16, 966–972.

Gunnardsson, C.C., Petersen, C.M., 2007. Water hyacinth as a resource in agriculture and energy production: a literature review. Waste Manage. 27 (1), 117–129.

Isarankura-Na-Ayudhya, C., Tantimongcolwat, T., Kongpanpee, T., Prabkate, P., Prachayasittikul, V., 2007. Apropiate technology for the bioconversion of water hyacinth (*Eichhornia crassipes*) to liquid ethanol: future prospects for community strengthening and sustainable development. EXCLI J. 6, 167–176.

Kumar, A., Singh, L.K., Ghosh, S., 2009. Bioconversion of lignocellulosic of water-hyacinth (*Eichhornia crassipes*) hemicelluloses acid hydrolysate to ethanol by *Pichia stipitis*. Bioresour. Technol. 100 (13), 3293–3297.

Mansoornejad, B., Chambost, V., Stuart, P., 2010. Integrating product portfolio design and supply chain design for the forest biorefinery. Comput. Chem. Eng. 34 (9), 1497–1506.

Maravelias, C.T., Grossmann, I.E., 2003. New general continuous-time state-task network formulation for short-term scheduling of multipurpose batch plants. Ind. Eng. Chem. Res. 42 (19), 3056–3074.

Mele, F.D., Guillén-Gosálbez, G., Jiménez, L., 2009. Optimal planning of supply chains for bioethanol and sugar production with economic and environmental concerns. Comput. Aided Chem. Eng. 26, 997–1002.

Murillo-Alvarado, P.E., Santibañez-Aguilar, J.E., Ponce-Ortega, J.M., Castro-Montoya, A.J., Serna-González, M., El-Halwagi, M.M., 2014. Optimization of the Supply chain associated to the production of bioethanol from residues of agave from the tequila process in Mexico. Ind Eng. Chem. Res. 53 (13), 5524–5538.

Navarro, A.R., Rubio, M.C., Maldonado, M.C., 2011. A combined process to treat lemon industry wastewater and produce biogas. Clean Technol. Environ. Policy 14 (1), 41–45.

Nigam, J.N., 2002. Bioconversion of water-hyacinth (*Eichhornia crassipes*) hemicellulose acid hydrolysate to motor fuel ethanol by xylose-fermenting yeast. J. Biotechnol. 97, 107–116.

Paksoy, T., Özceylan, E., Weber, G.W., 2013. Profit oriented supply chain network optimization. Cent. Eur. J. Oper. Res. 1–24.

Sammons Jr., N., Eden, R.M., Yuan, W., Cullinan, H., Aksoy, B., 2007. A flexible framework for optimal biorefinery product allocation. Environ. Prog. Sustainable Energy 26 (4), 349–354.

Sammons Jr., N.E., Yuan, W., Eden, M.R., Aksoy, B., Cullinan, H.T., 2008. Optimal biorefinery product allocation by combining process and economic modeling. Chem. Eng. Res. Des. 86 (7), 800–808.

Santibañez-Aguilar, J.E., González-Campos, J.B., Ponce-Ortega, J.M., Serna-González, M., El-Halwagi, M.M., 2011. Optimal planning of a biomass conversion system considering economic and environmental aspects. Ind. Eng. Chem. Res. 50 (14), 8558–8570.

Santibañez-Aguilar, J.E., Ponce-Ortega, J.M., González-Campos, J.B., Serna-González, M., El-Halwagi, M.M., 2013. Synthesis of distributed biorefining networks for the value-added processing of water hyacinth. ACS Sustainable Chem. Eng. 1 (2), 284–305.

Santibañez-Aguilar, J.E., González-Campos, J.B., Ponce-Ortega, J.M., Serna-González, M., El-Halwagi, M.M., 2014. Optimal planning and site selection for distributed multiproduct biorefineries involving economic, environmental and social objectives. J. Cleaner Prod. 65, 270–294.

Van Dyken, S., Bakken, B.H., Skjelbred, H.I., 2010. Linear mixed-integer models for biomass supply chains con transport, storage and processing. Energy 35 (3), 1338–1350.

You, F., Grossmann, I.E., 2013. Multi-cut Benders' decomposition algorithm for process Supply chain planning under uncertainty. Ann. Oper. Res. 210, 191–211.

You, F., Wang, B., 2011. Life cycle optimization of biomass-to-liquids supply chains with distributed-centralized processing networks. Ind. Eng. Chem. Res. 50 (17), 10102–10127.

You, F., Tao, L., Graciano, D.J., Snyder, S.W., 2011. Optimal design of sustainable cellulosic biofuel supply chains: multiobjective optimization coupled with life cycle assessment and input-output analysis. AIChE J. 58 (4), 1157–1180.

CHAPTER

Optimal Supply Chains for Biofuel Production

20

W. Alex Marvin, Prodromos Daoutidis[1]

Department of Chemical Engineering and Materials Science, University of Minnesota, Minneapolis, MN, USA

[1]*Corresponding author: E-mail: daoutidi@cems.umn.edu*

20.1 INTRODUCTION

The current U.S. biofuel industry is dominated by ethanol (mainly produced from corn grain starch) and biodiesel (mainly produced from soybean oil and recycled cooking oils), but cellulosic biofuels (based on nonfood resources such as agricultural waste, grasses, and woody crops) are predicted to play a large role in the future. Annual ethanol production has quickly expanded from 83 million gallons in 1981 to nearly 15 billion gallons today, and biodiesel production has similarly grown to 700 million gallons. Biofuels became controversial during this rapid growth for influencing the price of food by acting as a competing outlet for agricultural commodities. About a third of US corn production is used to produce ethanol, resulting in ethanol comprising 9.5% of finished gasoline blends in the US (McPhail et al., 2011). Alternatively, cellulosic biofuels are produced from low resource-intensive biomass, and can provide much greater supplies and environmental benefits than food-based biofuels (Hill et al., 2006). The US EPA determined that 0.81 million gallons of cellulosic biofuel were produced in 2013 (U.S. Environmental Protection Agency, 2014), while government mandates require the use of 16 billion gallons of cellulosic biofuel in motor-vehicle fuel by 2022.

Development of the cellulosic biofuel industry to meet these mandates will impose significant logistical challenges at each stage of the biofuel supply chain (biomass production, harvesting, storage, processing and transportation, and biofuel distribution). This industry will require about a billion tons of biomass annually from a variety of sources that each may be seasonally available, with uncertainty in composition, quantity, and price. Furthermore, there are many proposed technologies (e.g., crop residue fermentation and biomass gasification with syngas upgrading) competing for biomass processing, and their adoption will determine the economic and environmental impacts of the industry as a whole.

Biofuel supply chain optimization has emerged as a popular method for understanding the expected expansion of the biofuel industry. These studies seek to determine the optimum supply chain configuration, including all activities from the

Computer Aided Chemical Engineering, Volume 36. ISSN 1570-7946. http://dx.doi.org/10.1016/B978-0-444-63472-6.00020-3

biomass production to the biofuel end-use, which maximizes a desired performance index (e.g., profitability, greenhouse gas emissions, water use or net energy use). Often a real-world study region is chosen with high-resolution spatial information about the type, quantity, and location of biomass availability or production capability, harvest and logistics, and biorefinery plant capabilities to convert biomass to biofuels. A number of thorough reviews have been done of the research in this area, including its overlap with petroleum and chemical operations research (An et al., 2011), with hybrid energy system process and supply chain optimization (Floudas et al., 2012), and with the broader set of multiscale challenges surrounding biomass-to-bioenergy supply chains (Yue et al., 2014; Daoutidis et al., 2013).

In our previous study (Marvin et al., 2013), we proposed a biofuel supply chain optimization problem for determining economical biomass processing facility locations and capacities and applied it to a 9-state region in the Midwestern United States to assess the feasibility of meeting governmental biofuel mandates in 2015. This was formulated as a mixed integer linear program with an objective of maximizing total net present value (TNPV). A detailed cash flow analysis that includes capital depreciation and taxation was embedded into the model formulation to give insights into the minimum biofuel selling price for each facility and for each Renewable Fuel Standard biofuel classification (renewable fuel, advanced biofuel, and cellulosic biofuel). This chapter builds off of that work by adding a second objective of minimizing total greenhouse gas emissions (TGHG) relative to gasoline (i.e., produced biofuel is assumed to offset an energy-equivalent amount of gasoline). Thus the scope of the life cycle analysis is from cradle to grave (or in the case of biofuels, from field to wheels). Recent life cycle analysis literature on biofuels is leveraged to quantify the impact of each activity in the biofuel supply chain. The resulting multiobjective mixed integer linear program is solved using an ε-constraint approach that yields the Pareto Frontier of efficient supply chain designs. This allows for the identification of supply chain decisions that greatly reduce greenhouse gas emissions for minimal economic losses.

The remainder of this article is organized as follows. Multiobjective optimization techniques are reviewed in Section 20.2. The supply chain optimization scope and formulation is described in Section 20.3, and its solution strategy is described in Section 20.4. Life cycle analysis for each activity in the biofuel supply chain is described in Section 20.5. The Pareto Frontier for the multiobjective supply chain optimization is identified and discussed in Section 20.6, along with two Pareto points that have particularly interesting supply chain configurations. Concluding remarks are in Section 20.7.

20.2 MULTIOBJECTIVE OPTIMIZATION

Multiobjective optimization seeks to identify one or a set of Pareto optimal solutions (Pareto points) (Gonzalez, 2007). Pareto points are solutions that are noninferior to all other feasible solutions. This is shown in Figure 20.1 for the two objectives of maximizing TNPV and minimizing TGHG for the entire supply chain. Point (b) is

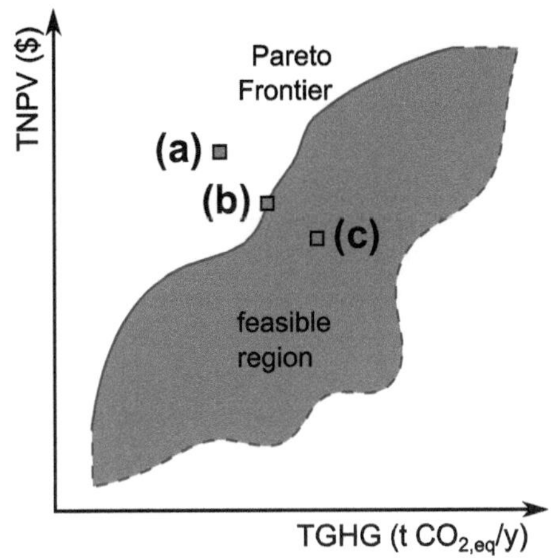

FIGURE 20.1

Example Pareto Frontier for maximization of TNPV and minimization of TGHG. The feasible region for the optimization problem is shown with its efficient boundary being the Pareto Frontier. Point (a) is infeasible, point (b) is a Pareto point, and point (c) is feasible but is dominated by point (b).

a Pareto point, because there is no feasible solution that simultaneously improves both objectives. There are usually many Pareto points, and the whole set of them is referred to as the Pareto Frontier or surface. It is desirable to lower TGHG and raise TNPV, so the upper left boundary in Figure 20.1 is the Pareto Frontier. Each solution along the Pareto Frontier is mathematically equal (i.e., a solution that is better for one objective cannot be found without necessarily worsening at least one other objectives) and thus it depends on the preference of a decision maker to select a "right" solution.

The general form of a multiobjective optimization problem involving K objectives to be minimized is

$$\begin{array}{ll} \min & (f_1(x),\ldots,f_K(x)) \\ \text{s.t.} & x \in X \end{array}$$

where x is the vector of decision variables with feasible set X defined by constraints.

Multiobjective optimization techniques vary in how a decision maker may express their preference, and the techniques fall into three major groups (Grossmann and Guillén-Gosálbez, 2010; Collette and Siarry, 2003; Miettinen, 1999):

- A priori methods
- A posteriori methods
- Interactive methods

A priori methods require preference information to be specified before finding a solution. The solution returned is a Pareto point that best satisfies the decision maker's preference. Examples include scalarization (agglomeration), no-preference methods and utility function methods. Applying a so-called carbon tax to greenhouse gas emissions is a common example of linear scalarization to convert a multiobjective problem involving emissions and economics into a single-objective problem. In general, the weights assigned to the objectives convert them into uniform units (e.g., dollars) and represent the relative importance for each objective. Minimizing the Euclidean norm (L_2) to an ideal state (i.e., desired value for the objectives that is unobtainable) is an example of a no-preference method.

A posteriori methods generate a representative set of Pareto points that the decision makers can then evaluate based on their preferences. These methods require additional computational effort, as determining each Pareto point may require as much time as an a priori method. For example, many well-known a posteriori methods (e.g., Normal-boundary Intersection (Das and Dennis, 1998) and Successive Pareto Optimization (Mueller-Gritschneder et al., 2009)) construct and solve several scalarization problems. Metaheuristics, such as simulated annealing, particle swarm optimization, and evolutionary algorithms, generate a representative set of solutions by searching the feasible region quickly, allowing for an approximation of the Pareto Frontier, but Pareto optimality of the solutions is not guaranteed. The ε-constraint method is the most widely applied a posteriori method in multiobjective optimization. In this method, single-objective problems are solved where one of the original objectives is minimized while the other objectives are constrained:

$$\begin{aligned} \min \quad & f_1(x) \\ \text{s.t.} \quad & x \in X \\ & f_K(x) \le \varepsilon_k, \qquad \forall k \ne 1 \end{aligned}$$

where the parameters ε_k are varied between ε_k^L and ε_k^U. These bounds are determined by solving the K single-objective problems for each objective function separately. Thus,

$$\begin{aligned} \varepsilon_k^L = \min \quad & f_k(x) \\ \text{s.t.} \quad & x \in X \end{aligned}$$

and ε_k^U is the maximum value of $f_k(x)$ among the solutions when minimizing the other objectives.

Solution filtering ensures that the ε-constraint method returns Pareto points, as the solved single-objective problems themselves do not ensure feasibility or that they dominate the solutions with other $\{\varepsilon_k\}$ values. The advantages of the ε-constraint method are that it (1) obtains a representation of the Pareto Frontier that can be quickly improved with a finer grid of $\{\varepsilon_k\}$ values, (2) does not suffer from scaling problems of scalarization (e.g., numerical precision and comparable order of magnitude), and (3) can identify islanded (unsupported) Pareto points that are common to integer and mixed integer problems (Mavrotas, 2009).

Interactive methods require that the decision maker is continuously interacting with the method while searching for the most preferred Pareto point (Geoffrion et al., 1972; Miettinen et al., 2008). Instead of the method converging mathematically, these methods end when the decision maker has psychologically converged on a desirable solution. At each iteration of the method, the decision maker can modify their preferences in a number of ways. Some methods allow for tradeoff information, where the decision maker ranks objective trade-offs for a number of possible search directions. Other methods allow for modifying the target objectives f^{ideal} within a no-preference method search for Pareto points. Classification of objective functions is another approach in which the objectives are ranked at the current Pareto point and then the method seeks to find a better Pareto point. Although intuitive and transparent for the decision maker, iterative methods suffer from the drawbacks of requiring constant decision-maker attention and limiting the decision maker to seeing only a subset of the whole range of possibilities available along the Pareto Frontier.

20.3 SUPPLY CHAIN SCOPE

This chapter builds off of our previous work (Marvin et al., 2013) by adding a second objective of minimizing TGHG and focusing on identifying the Pareto Frontier. Interested readers can find a more complete description of the mathematical formulation and parameter values for the supply chain optimization problem, specifically for the economic objective of maximizing TNPV, in our previous work (Marvin et al., 2013). Those details are not all repeated here, for simplicity.

This article considers the biofuel supply chain as a network of biomass producers, conversion facilities, and markets that performs the functions of biomass production, harvesting, storage, processing and transportation, and biofuel distribution to markets. The optimization problem is to determine how to best utilize the biomass resources in the 12-state region of the Midwestern United States to satisfy the market demands for biofuels in 2015 (driven by government mandates). This model includes

- 3109 biomass producers, each corresponding to a county in the contiguous US
- 159 existing biofuel facilities and 98 candidate sites for new biofuel facilities
- Three markets, corresponding to the three RFS2 biofuel classifications: renewable fuel (RF), advanced biofuel (AB), and cellulosic biofuel (CB)
- Eight biomass types with availability estimates in 2015 from the Billion Ton Study 2011 update (U.S. Department of Energy, 2011): corn, corn stover, five types of forest biomass and wood wastes, and perennial grasses
- Four biofuel types: RF ethanol, AB ethanol, CB ethanol, and CB naphtha and diesel
- Seven biomass processing technologies:
  - Dry-grind starch fermentation of corn
  - Integrated corn and stover process
  - Dilute acid pretreatment and C5 + C6 cofermentation of corn stover
  - Indirect gasification and syngas upgrading to mixed alcohols of wood
  - Direct gasification and syngas upgrading to mixed alcohols of wood
  - Fast pyrolysis and bio-oil upgrading of corn stover
  - AFEX pretreatment and C5 + C6 cofermentation of perennial grasses
- Two transportation methods: road and rail

Note that the seven biomass processing technologies included here can produce four unique biofuel types from the eight biomass types. These technologies were chosen because each is expected to be commercialized in the near-term and the set is intended to be representative of the diversity seen in biomass processing. Furthermore, detailed technoeconomic models are available for each technology from National Laboratory reports and related publications (Wallace et al., 2005; Humbird et al., 2011; Dutta et al., 2011; Dutta and Phillips, 2009; Wright et al., 2010; Tao et al., 2011).

The mathematical formulation from our previous study (Marvin et al., 2013) was expanded to incorporate an environmental objective function. Thus, the multiobjective optimization problem examined in this chapter has the objectives to (1) maximize TNPV and (2) minimize TGHG.

TGHG for the entire supply chain is the sum of the greenhouse gas emissions corresponding to each of the facility sites:

$$\mathrm{TGHG} = \sum_{r} \mathrm{GHG}_r$$

Each facility site r in the study has emissions associated with

- Biomass production, $\mathrm{EF}_r^{\mathrm{Biomass}}$
  - Fertilizer production
  - Fertilizer N_2O
  - Farming
  - Land use change
- Biomass transportation, $\mathrm{EF}_r^{\mathrm{TransportBiomass}}$
- Biorefinery, $\mathrm{EF}_r^{\mathrm{Biorefinery}}$
  - Operations
  - Coproduct credit
- Biofuel transportation, $\mathrm{EF}_r^{\mathrm{TransportProducts}}$
- Market, $\mathrm{EF}_r^{\mathrm{Market}}$
  - Biofuel combustion
  - Fuel credit

The greenhouse gas emission for each facility is then

$$\mathrm{GHG}_r = \mathrm{EF}_r^{\mathrm{Biomass}} + \mathrm{EF}_r^{\mathrm{TransportBiomass}} + \mathrm{EF}_r^{\mathrm{Biorefinery}} + \mathrm{EF}_r^{\mathrm{TransportProducts}} + \mathrm{EF}_r^{\mathrm{Market}}, \quad \forall\, r$$

Emissions from each of these sources are assumed proportional to the corresponding supply chain activity, as described in the following equations:

$$\mathrm{EF}_r^{\mathrm{Biomass}} = \sum_{n,v,b} E_b^N \cdot f_{n,r,v,b}^{NR}, \qquad \forall\, r$$

where E_b^N is the amount of emissions associated with producing a unit of biomass type b (e.g., t $CO_{2,\mathrm{eq}}$/dt):

$$\mathrm{EF}_r^{\mathrm{TransportBiomass}} = \sum_{n,v,b} E_{b,v}^{NR} \cdot D_{n,r}^{NR} \cdot f_{n,r,v,b}^{NR}, \qquad \forall\, r$$

where $D^{NR}_{n,r}$ is the round-trip distance between biomass producer n and facility site r, and $E^{NR}_{b,v}$ is the amount of emissions associated with transporting a unit of biomass type b via method v per unit distance (e.g., t $CO_{2,eq}$/mi/dt):

$$\mathrm{EF}^{\mathrm{Biorefinery}}_{r} = \sum_{b,l} E^{RB}_{b,l} \cdot u^{R}_{r,b,l}, \qquad \forall\ r$$

where $E^{RB}_{b,l}$ is the amount of emissions associated with converting a unit of biomass type b using biorefinery technology l (e.g., t $CO_{2,eq}$/mi/dt):

$$\mathrm{EF}^{\mathrm{TransportProducts}}_{\mathrm{r}} = \sum_{m,v,p} E^{RM}_{p,v} \cdot D^{RM}_{r,m} \cdot f^{RM}_{r,m,v,p}, \qquad \forall\ r$$

where $D^{RM}_{r,m}$ is the round-trip distance between facility site r and market m, and $E^{RM}_{p,v}$ is the amount of emissions associated with transporting a unit of product p via method v per unit distance (e.g., t $CO_{2,eq}$/mi/gal):

$$\mathrm{EF}^{\mathrm{Market}}_{r} = \sum_{m,v,p} E^{M}_{p} \cdot f^{RM}_{r,m,v,p}, \qquad \forall\ r$$

where E^{M}_{p} is the amount of emissions associated with the market/customer use of a unit of product p (e.g., t $CO_{2,eq}$/gal).

The multiobjective mixed integer linear program is then

$$\begin{aligned} \max \quad & (\mathrm{TNPV},\ -\mathrm{TGHG}) \\ \text{s.t.} \quad & \text{TGHG equations} \\ & \text{TNPV equations} \\ & \text{mass balance equations} \end{aligned}$$

where the TGHG equations include all others listed in this section. The TNPV equations and mass balance equations, along with their associated parameter values, are described in detail in our previous study (Marvin et al., 2013). It is important to note a few things about the equations adopted from our previous study: (1) it includes a detailed cash flow analysis for the lifetime of the facilities complete with capital depreciation and taxation, (2) installed capacity for each facility is allowed to vary, and (3) piecewise linear functions for capital investment of each biomass processing technology are used to approximate the effect of economies of scale.

20.4 SOLUTION STRATEGY

The ε-constraint method is used to solve the multiobjective mixed integer linear program, by first transforming it into the single-objective mixed integer linear program (SoMP):

$$\begin{aligned} \max \quad & \varepsilon_1 \cdot (\mathrm{TNPV} - (1 - \varepsilon_1) \cdot \mathrm{TGHG}) \\ \text{s.t.} \quad & \text{TGHG equations} \\ & \text{TNPV equations} \\ & \text{mass balance equations} \\ & \mathrm{TGHG} \le \varepsilon_2 \end{aligned}$$

and then identifying Pareto points iteratively, following the strategy:

Step 1: Maximize TNPV without considering TGHG by solving problem (SoMP) with $\varepsilon_1 = 1$ and $\varepsilon_2 = \text{inf}$. Store the solution value of TGHG as the highest TGHG along the Pareto Frontier $TGHG^{max}$.
Step 2: Minimize TGHG without considering TNPV by solving problem (SoMP) with $\varepsilon_1 = 0$ and $\varepsilon_2 = \text{inf}$. Store the solution value for all decision variables as the initial guess for the next problem. Store the solution value of TGHG as the lowest TGHG along the Pareto Frontier $TGHG^{min}$.
Step 3: Maximize TNPV while constraining TGHG by solving problem (SoMP) with $\varepsilon_1 = 1$ and $\varepsilon_2 = TGHG^{min}$. Store the solution value of all decision variables as the initial guess for the next problem.
Step *K* (*K* ≥ 4): Maximize TNPV while constraining TGHG by solving problem (SoMP) with $\varepsilon_1 = 1$ and ε_2 set to a value within the range ($TGHG^{min}$, $TGHG^{max}$). Store the solution value for all decision variables as the initial guess for the next problem. Note that this initial guess is only a feasible solution for the next problem if the next problem has a higher value for ε_2; therefore, there is an advantage to monotonically increasing ε_2 in the search for Pareto points.
End when the desired number of Pareto points (e.g., resolution of the Pareto Frontier) has been obtained. Solution filtering is then required to confirm that each step returned a feasible solution that is not dominated by any other solution and is thus Pareto optimal.

20.5 LIFE CYCLE ANALYSIS

Biofuels are promoted as alternatives to petroleum-based transportation fuels, in part, for their reduced greenhouse gas emissions, determined by thorough life cycle analysis. This attribute of biofuels has been recognized in policies aimed at reducing the transportation sector's GHG emissions (e.g., the Low-Carbon Fuel Standard in California (CARB (California Air Resources Board), 2010), the US Renewable Fuel Standard program (U.S. Environmental Protection Agency, 2010), and the Renewables Directive in the European Union (European Parliament, 2009)). However, there is still active debate and research over the emissions of biofuels. The emissions of the widespread corn ethanol dry-grind process are even disputed, with some authors concluding that it offers lower emissions when compared with gasoline (Hsu et al., 2010; Wang et al., 2011), some authors concluding the opposite (Searchinger et al., 2008), and others concluding that emissions are nearly equal (Farrell et al., 2006). Most studies agree that cellulosic biofuels offer lower emissions when compared with gasoline (Wang et al., 2012; Tilman et al., 2009; Wu et al., 2006; Han et al., 2011; Schmer et al., 2008). Such life cycle analysis studies differ in their system boundary assumptions for biofuels or petroleum fuels (e.g., including land use change, coproduct credits, petroleum supply defense via military actions) and in their parameter values (e.g., biorefinery technology updates, better estimates).

This chapter utilizes results for the life cycle analysis of select biofuels performed using the GREET (Greenhouse gases, Regulated Emissions, and Energy Use in Transportation) model. The GREET model was developed at Argonne National Laboratory and is a trusted tool for researchers and politicians to examine greenhouse gas emissions from vehicles and transportation fuels on a consistent basis (Argonne National Laboratory—Transportation Technology R&D Center, 2010). Parameters and assumptions are continuously updated as the biofuel technologies develop, small-scale experiments are performed, and large-scale field trials are completed. Furthermore, numerous national laboratory reports and papers have been published on biofuel life cycle analysis using GREET, with some based on technoeconomic models corresponding to the seven biorefinery technologies considered in this chapter.

The system boundary considered for the biofuel life cycle analysis is shown in Figure 20.2. Greenhouse gases considered here are measured in CO_2-equivalent tonnes (t $CO_{2,eq}$). This measure is an aggregation of emissions of CO_2, CH_4, and N_2O, weighted by their 100-year global warming potentials of 1, 25, and 298, respectively. In other words, 1 unit of CH_4 is predicted to have the same global warming potential as 25 units of CO_2 over a 100-year time horizon. These global warming potentials are as specified by the International Panel on Climate Change's 2007 assessment (International Panel on Climate Change (IPCC), 2007).

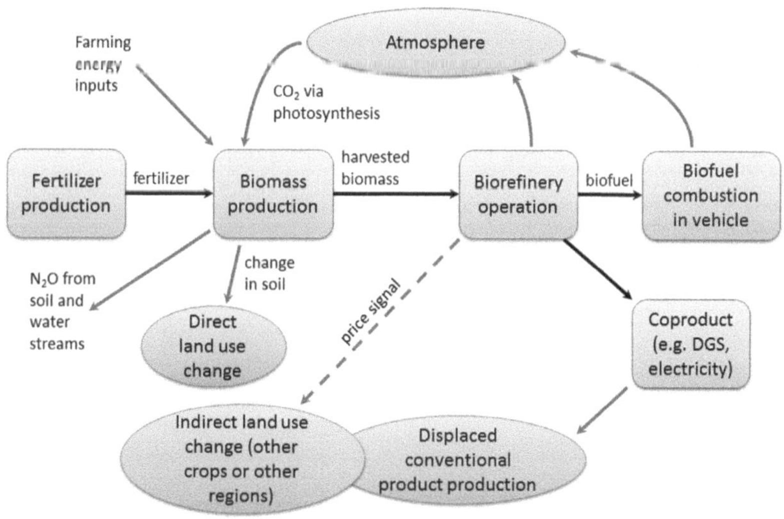

FIGURE 20.2

System boundary of the biofuel life cycle analysis (inspired by Figure 20.2 in (Wang et al., 2012)).

The specific emission parameter values used in this chapter are shown in Tables 20.1–20.5. They are mainly determined from the GREET model and associated literature (Wang et al., 2012; Han et al., 2011; Wu et al., 2006), using process data from the technoeconomic models corresponding to each of the seven considered biorefinery technologies. Many of those values are obtained from a report,

Table 20.1 Emission Parameter Values for Biomass Production

Biomass Type *b*	E_b^N
Corn	0.00916 t $CO_{2,eq}$/Bu
Corn-stvr	0.0636 t $CO_{2,eq}$/dt
All wood	0.111 t $CO_{2,eq}$/dt
Perngrass	0.111 t $CO_{2,eq}$/dt

Table 20.2 Emission Parameter Values for Biomass Transportation

Biomass Type *b*	Method *v*	$E_{b,v}^{NR}$
Corn	Rail	8.58E-7 t $CO_{2,eq}$/mi/bu
all biomass	Truck	1.98E-4 t $CO_{2,eq}$/mi/dt

Table 20.3 Emission Parameter Values for Biorefinery Operation

Biomass Type *b*	Technology *l*	$E_{b,l}^{RB}$
Corn	1	0.00389 t $CO_{2,eq}$/Bu
Corn	2	0.00389 t $CO_{2,eq}$/Bu
Corn-stvr	2	−0.0557 t $CO_{2,eq}$/dt
Corn-stvr	3	−0.0557 t $CO_{2,eq}$/dt
All wood	4	0.145 t $CO_{2,eq}$/dt
All wood	5	0.145 t $CO_{2,eq}$/dt
Corn-stvr	6	0.145 t $CO_{2,eq}$/dt
Perngrass	7	−0.0477 t $CO_{2,eq}$/dt

Table 20.4 Emission Parameter Values for Biofuel Transportation

Product *p*	Method *v*	$E_{p,v}^{RM}$
RF ethanol	Truck	5.91E-7 t $CO_{2,eq}$/mi/gal
AB ethanol	Truck	5.91E-7 t $CO_{2,eq}$/mi/gal
CB ethanol	Truck	5.91E-7 t $CO_{2,eq}$/mi/gal
CB naphtha & diesel	Truck	5.54E-7 t $CO_{2,eq}$/mi/gal

Table 20.5 Emission Parameter Values for Market/Customer Product Use (e.g., Biofuel Combustion)

Product p	E_p^M
RF ethanol	−0.00746 t $CO_{2,eq}$/gal
AB ethanol	−0.00746 t $CO_{2,eq}$/gal
CB ethanol	−0.00746 t $CO_{2,eq}$/gal
CB naphtha & diesel	−0.0112 t $CO_{2,eq}$/gal

which summarizes the recent GREET model updates and performs life cycle analysis on a number of biorefinery technologies along with gasoline (Wang et al., 2012). This is the 2012 update of their earlier study from 2006 (Wu et al., 2006). Transportation emission values for biomass and biofuels are taken from Zamboni et al. (2009).

A comparison of two competing products (e.g., a biofuel and gasoline) requires a choice of functional unit. Functional unit options for fuels include volume or mass basis, energy equivalence, and end-use equivalence. Conclusions depend on the choice of functional unit. For example, comparing ethanol and gasoline using a volume functional unit (i.e., per gallon) leads to the conclusion that ethanol is cheaper, which may not be the case for the consumer considering its lower energy density compared to gasoline. In this article, we will consider the functional unit of gasoline equivalent gallon (i.e., volumes with equal energy content). Thus, the effect of lower energy density in some biofuels is removed. Additionally, the market/customer product use in Table 20.5 accounts for the biofuel offsetting an equal energy volume of gasoline. A value of 94 g $CO_{2,eq}$/MJ, found by Wang et al. (2012), was assumed for gasoline.

It should also be noted that end-use equivalence functional units may be appropriate in some cases. Most notably, miles per gallon of gasoline equivalent gallon is useful for vehicles designed to take advantage of the properties of ethanol (e.g., higher compression ratio and different fuel-air mixture) to extract additional engine performance. However, as the gasoline equivalent gallon functional unit does not depend on particular vehicle design or performance, it is more suitable for the present supply chain optimization.

20.6 RESULTS

The mixed integer linear program (SoMP) from Section 20.4 was coded in IBM ILOG CPLEX Optimization Studio v12.2 (IBM, 2011) and has 156,947 continuous variables, 5488 binary variables, and 40,982 constraints. The Pareto Frontier of supply chains is presented along with some notable Pareto points, with an evaluation of the tradeoff between economic and environmental performance.

20.6.1 PARETO FRONTIERS

The two scenarios of no biofuel mandates and strict mandates were considered, and their corresponding Pareto Frontiers are shown in Figure 20.3. Note that strict mandates are enforced in the mathematical formulation by including a constraint such that the biofuel sales meet or exceed the RFS2 mandates in 2015 for each RFS2 biofuel classification type. The Pareto points are identified using the ε-constraint approach described in Section 20.4. Each of the single-objective problems are solved with a target optimality gap of 0.5% and a time limit of 10 h. Most of the Pareto points are identified to an optimality gap less than 2.0% in 10 h.

The addition of the biofuel mandate constraints reduces the size of the feasible region, and thus the Pareto Frontier for the no biomass mandates scenario dominates the Pareto Frontier of the strict mandates scenario (i.e., has higher TNPV for the same TGHG values). This is largely the result of the strict mandate scenario hitting the upper bound on production of biofuels, whereas the no mandates scenario can improve both economics and emission objectives by producing more biofuel than is truly demanded. For this reason, the remainder of this section will focus on results for the strict mandates scenario.

The effect of constraining TGHG can be seen by plotting the total biofuel production for each Pareto point, as shown in Figure 20.4. Recall that the 2015 RFS2 mandates are for the production of 3 B gal/y cellulosic biofuel, 2.5 B gal/y of other advanced biofuel, and 15 B gal/y of implicit nonadvanced biofuel. CB produced by technologies $l = \{3, 4, 5, 6, 7\}$ may be sold in each of those three categories, whereas AB produced by technology $l = 2$ may be sold in the latter two categories, and RF produced by technology $l = 1$ may only be sold in the last category. Much of the

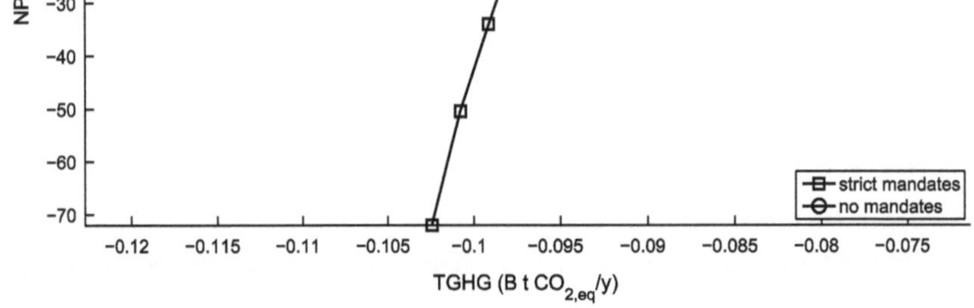

FIGURE 20.3

Pareto Frontiers for the scenarios of no biofuel mandates and strict mandates.

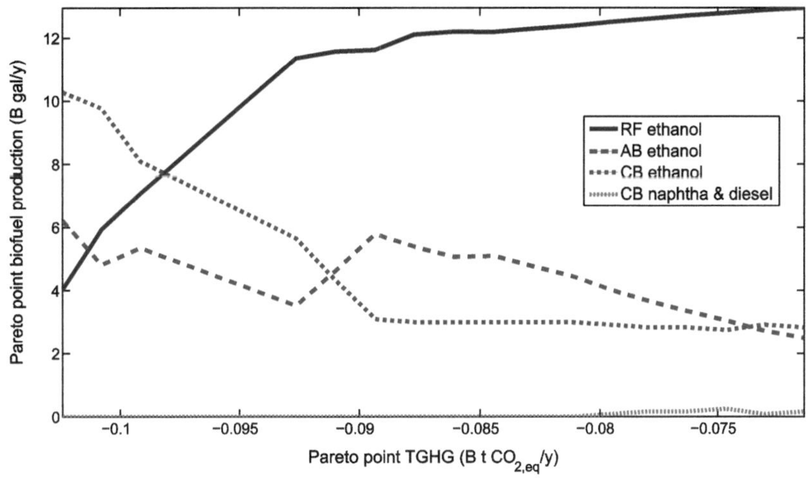

FIGURE 20.4

Total biofuel production for each Pareto point.

biofuel demand is satisfied then by CB for low TGHG Pareto points. Limited cellulosic biomass availability prevents all of the demand to be satisfied this way, and thus AB and RF are also produced. As the ε-constraint is relaxed, an increasing amount of the biofuel demand is satisfied by RF. Also, recall that technology 2 competes for corn (also an RF feedstock) and corn stover (also a CB feedstock) as feedstocks; thus AB production from technology 2 remains relatively constant. CB naphtha and diesel-range is produced in small amounts from technology 6, as it competes for corn stover with technologies that typically have lower greenhouse gas emissions.

20.6.2 KEY PARETO POINTS

This section will provide some analysis of the strict mandates scenario Pareto Frontier, by examining key Pareto points. The goal is to identify modest supply chain changes that may lead to large emission gains for a small loss of profit.

20.6.3.1 Highest TNPV

It turns out that the same supply chain configuration identified in our previous study (Marvin et al., 2013) for the strict mandates scenario is a Pareto point for the multi-objective problem. For simplicity, those results are summarized here. In the solution, five different technologies are installed with a total of 168 facilities producing RF (15,000 M gal/y), 26 producing AB (2500 M gal/y), and 24 producing CB (3000 M gal/y). All produced biofuels are sold in their highest RFS fuel classification. By examining the breakeven prices for each facility, it was found that all facilities are economical for facility-gate prices of 2.39, 2.13, and 1.71 \$/gal for CB, AB, and RF, respectively.

20.6.3.2 Lowest TGHG

The Pareto point with the lowest TGHG is found by first minimizing TGHG without considering TNPV, then maximizing TNPV while constraining TGHG to be less than or equal to the previous solution value. Recall that the previous solution is used as an initial guess to expedite the solution of the second problem. The supply chain configuration for this Pareto point is shown in Figure 20.5. Five of the seven technologies are installed by developing 90 of the 98 candidate sites, with a total of 60 facilities producing RF (4020 M gal/y), 33 producing AB (6211 M gal/y), and 57 producing CB (10,269 M gal/y). CB and AB are produced far in excess of the RFS mandates for their classifications, so they are also sold at lower fuel classifications. Limited supply of cellulosic material in 2015 prohibits CB from dominating the biofuel markets, although that would have resulted in a lower TGHG. Technology 4 (indirect gasification) is favored over technology 5 (direct gasification), which was installed in the highest TNPV Pareto point, for production of CB ethanol from all-wood, because it has higher product yield for nearly identical emissions.

The average distribution of discounted costs for each installed technology is shown in Figure 20.6. By examining the breakeven prices for each facility, it was found that all facilities are economical for facility-gate prices of 4.18, 2.28, and 1.74 $/gal for CB, AB, and RF, respectively. Thus, CB costs are 75% higher for this Pareto point than the highest TNPV Pareto point, while the costs of AB and RF are equivalent. The average distribution of emissions is shown in Figure 20.7. Technologies 2, 3, and 7 have lower net emissions than technology 1 in the solution due to lower emission biomass feedstock(s) and emission-negative biorefinery operation (the result of emission credits for coproducts like electricity). Compressor operation in the high-pressure gasification of technology 4, along with no turbine installation for possible electricity generation, leads it to have emission-positive biorefinery operation. Technology 4 still has lower emissions than technology 1 due to wood being a lower emission biomass feedstock than corn.

20.6.3.3 Balanced

The "balanced" Pareto point described here offers large emissions gains for a small loss on profit, compared to the highest TNPV case. This point was chosen by examining the Pareto Frontier (see Figure 20.3) as a decision maker would. The highest TNPV Pareto point offered TNPV of 19.4 B$ and TGHG of −71.4 M t $CO_{2,eq}$/y, while the lowest TGHG Pareto point offered TNPV of −72.0 B$ and TGHG of −102 M t $CO_{2,eq}$/y. Thus, there is an opportunity to reduce biofuel production emissions by nearly half, from the highest TNPV case. Presented here is the Pareto point that captures more than half of those emission reductions, with a TNPV of 11.3 B$ and a TGHG of −87.7 M t $CO_{2,eq}$/y. This Pareto point was calculated in 7.6 h with a 0.50% gap and is shown in Figure 20.8. Three of the seven technologies are installed by developing 60 of the 98 candidate sites, with a total of 161 facilities producing RF (11,900 M gal/y), 43 producing AB (5600 M gal/y), and 15 producing CB (3000 M gal/y). Just enough CB is produced to satisfy the mandate, but excess

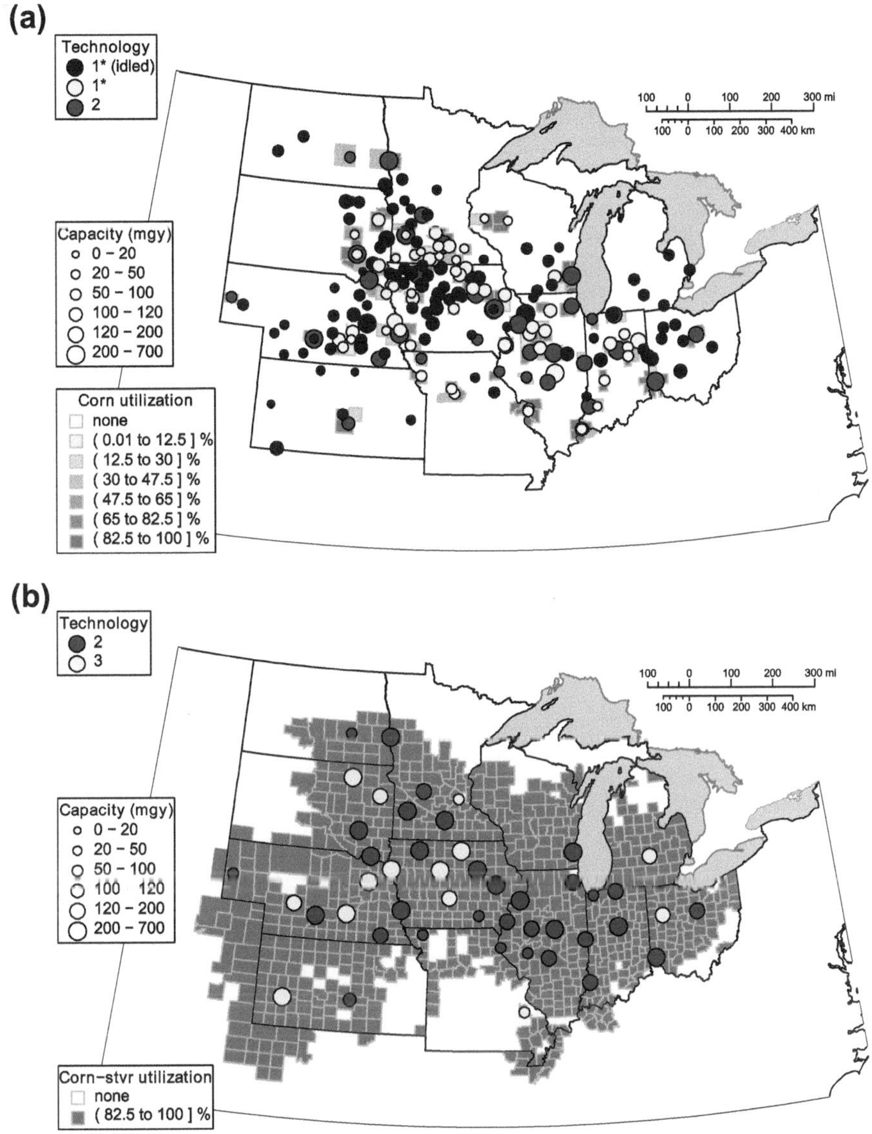

FIGURE 20.5

Optimum supply chain configuration for the lowest TGHG Pareto point in the case of strict biofuel mandates with fuel credits. Installed facility capacity shown in million gallons per year (mgy). RF, AB, and CB are produced. Subfigure (a): Corn grain utilization with corn-utilizing facilities shown. RF and AB ethanol are produced from corn using technology 1 and 2, respectively. Existing facilities are denoted with an asterisk (*). Maximum corn utilization is 34%. In this Pareto point, many existing facilities are idled (zero production) in favor of lower emission biofuel technologies. Subfigure (b): Corn stover utilization with associated facilities shown. AB and CB ethanol and CB naphtha and diesel are produced from corn stover. Subfigure (c): Aggregated forest biomass and wood waste utilization with associated facilities shown. CB ethanol is produced from all wood. Subfigure (d): Perennial grass utilization with associated facilities shown. CB ethanol is produced from perennial grasses.

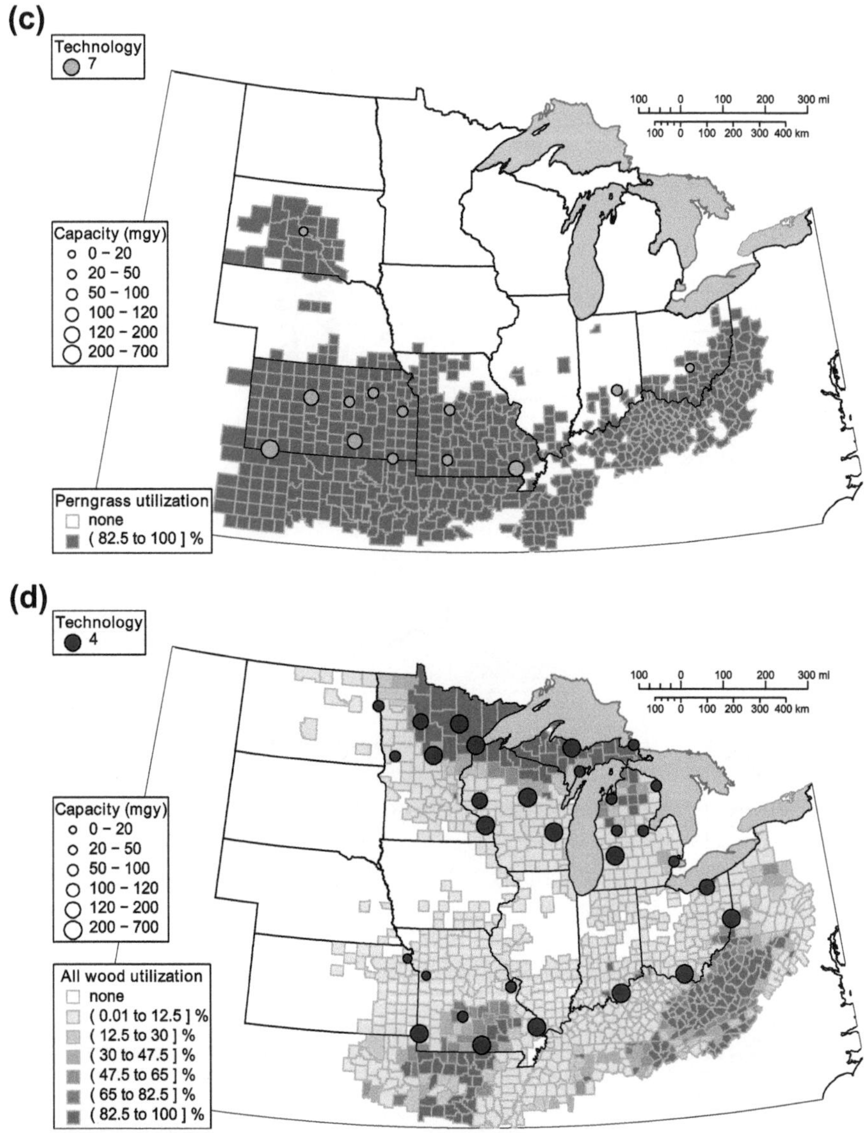

FIGURE 20.5—Cont'd

AB is produced and sold as the lower classification (RF). This offsetting of some RF (produced from corn grain) with AB (produced from in an integrated process that utilizes corn and corn stover) is what gives this Pareto point improved emissions.

The average distribution of discounted costs for each installed technology is shown in Figure 20.9. By examining the breakeven prices for each facility, it was

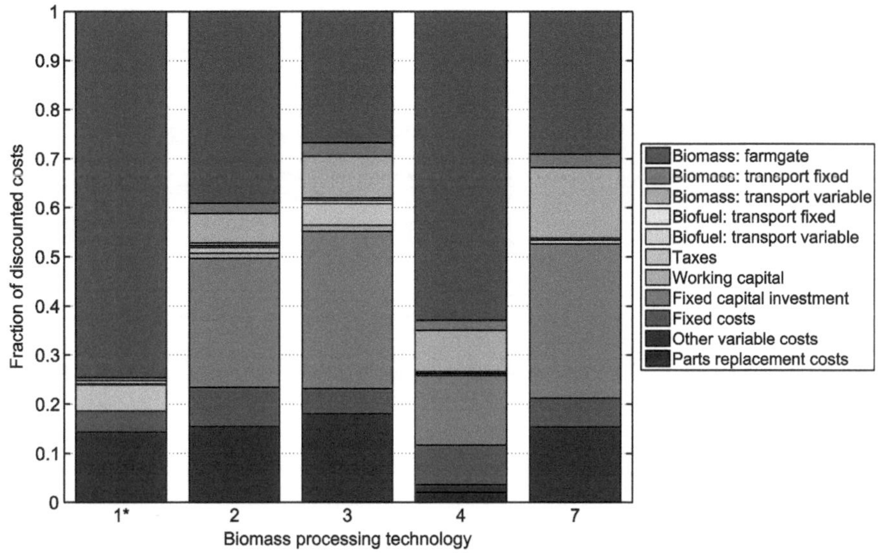

FIGURE 20.6

Breakdown of cost sources for installed technologies for the lowest TGHG Pareto point in the case of strict biofuel mandates with fuel credits. Existing facilities are denoted by an asterisk (*). New facilities of technologies $l = \{5, 6\}$ were not constructed in the optimum supply chain.

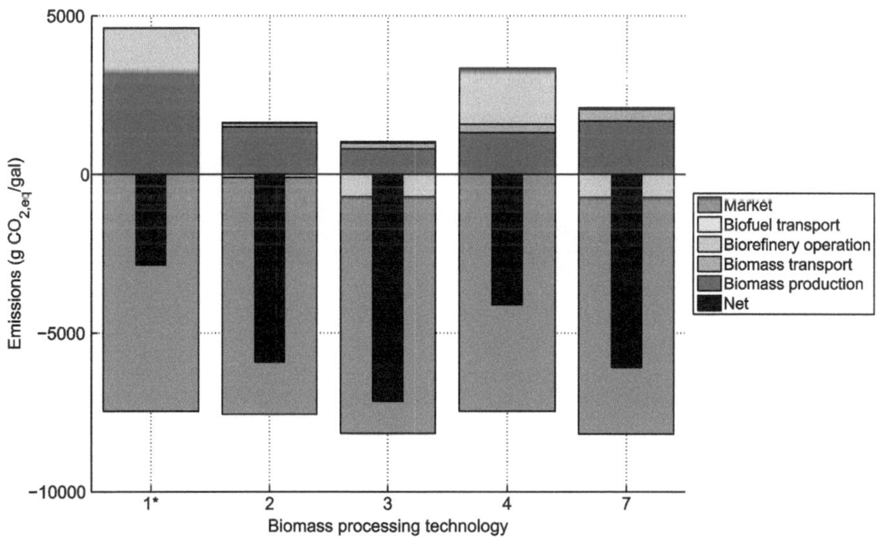

FIGURE 20.7

Breakdown of emission sources for installed technologies for the lowest TGHG Pareto point in the case of strict biofuel mandates with fuel credits. Net emissions per gallon of produced biofuel are also shown. Existing facilities are denoted by an asterisk (*). New facilities of technologies $l = \{5, 6\}$ were not constructed in the optimum supply chain.

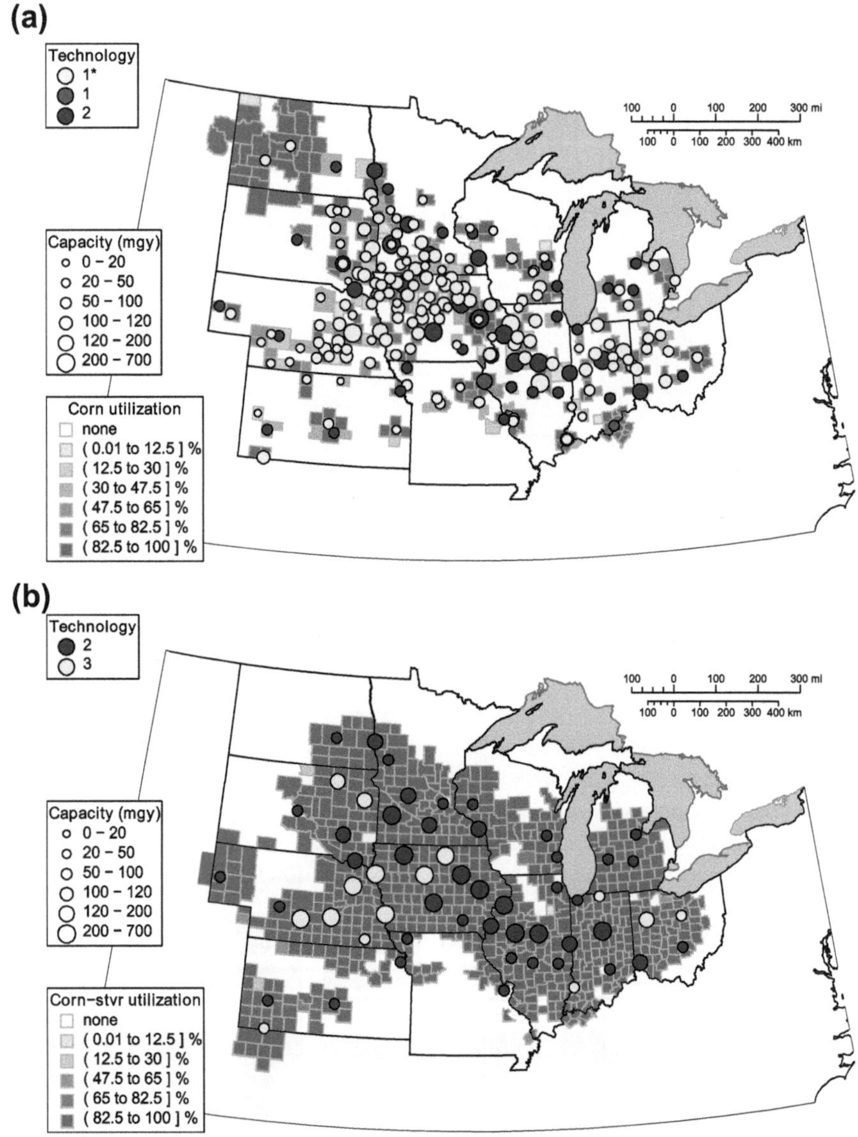

FIGURE 20.8

Optimum supply chain configuration for the "balanced" Pareto point in the case of strict biofuel mandates with fuel credits. Installed facility capacity shown in million gallons per year (mgy). RF, AB, and CB are produced. Forestry resources and perennial grasses were available but were not utilized in the optimum supply chain. Subfigure (a): Corn grain utilization with corn-utilizing facilities shown. RF and AB ethanol are produced from corn using technology 1 and 2, respectively. Existing facilities are denoted with an asterisk (*). Maximum corn utilization is 34%. Subfigure (b): Corn stover utilization with associated facilities shown. AB and CB ethanol and CB naphtha and diesel are produced from corn stover.

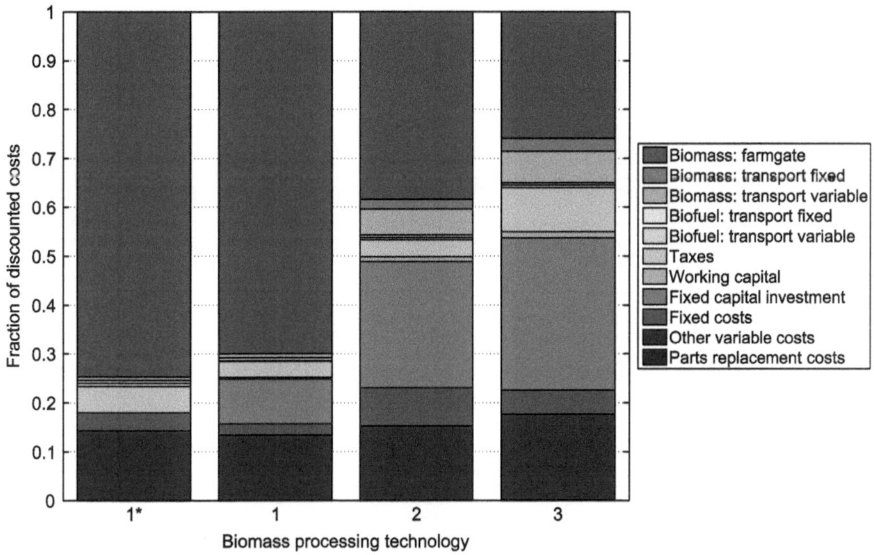

FIGURE 20.9

Breakdown of cost sources for installed technologies for the "balanced" Pareto point in the case of strict biofuel mandates with fuel credits. Existing facilities are denoted by an asterisk (*). New facilities of technologies $l = \{4, 5, 6, 7\}$ were not constructed in the optimum supply chain.

found that all facilities are economical for facility-gate prices of 2.47, 2.23, and 1.71 \$/gal for CB, AB, and RF, respectively. Thus, the biofuel prices are at most only 0.10 \$/gal higher than in the Pareto point with highest TNPV. The average distribution of emissions is shown in Figure 20.10. As expected, the emissions distribution for technologies 1, 2, and 3 are similar to what was found in the other Pareto points; however, there are small differences in the transportation emissions.

20.7 CONCLUSIONS

This chapter proposed a multiobjective mixed integer linear program to determine biofuel supply chain configurations that are Pareto optimal for profit and emissions. It was applied to the Midwestern US 2015 case study introduced in Marvin et al. (2013) to determine facility location, capacity, and technology selection. Pareto Frontiers were found for the two case studies of strict RFS mandates and no mandates. In the strict mandates case study, it was found that the lowest TGHG Pareto point offers an additional 50% reduction in emissions compared to the highest TNPV Pareto point. Unfortunately, the lowest TGHG Pareto point corresponds with a large negative TNPV. One of the Pareto points showed that half of these

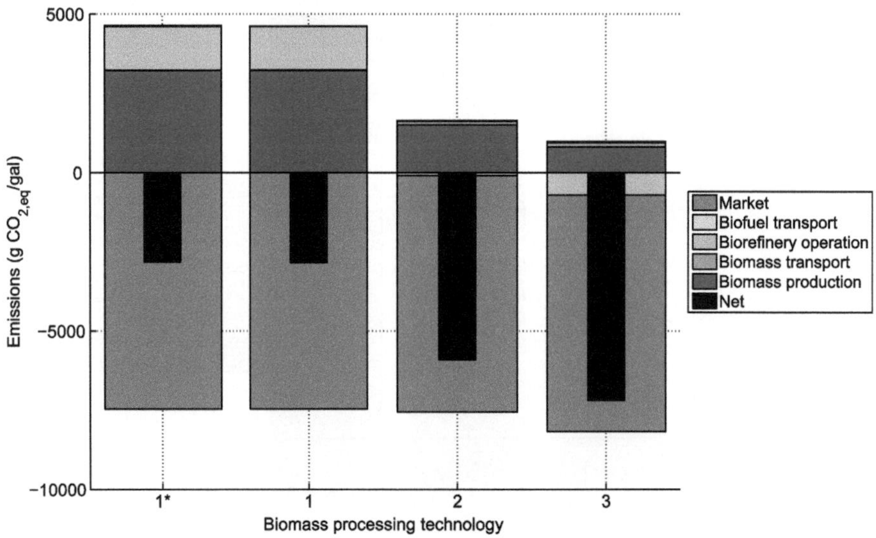

FIGURE 20.10

Breakdown of emission sources for installed technologies for the "balanced" Pareto point in the case of strict biofuel mandates with fuel credits. Net emissions per gallon of produced biofuel are also shown. Existing facilities are denoted by an asterisk (*). New facilities of technologies $l = \{4, 5, 6, 7\}$ were not constructed in the optimum supply chain.

emission reductions are obtainable by shifting some renewable fuel production to advanced biofuel technologies, with biofuel prices only 0.10 $/gal higher than in the Pareto point with highest TNPV.

ACKNOWLEDGMENTS

Partial financial support from the National Science Foundation, CBET award number 1307089, is gratefully acknowledged.

REFERENCES

An, H., Wilhelm, W.E., Searcy, S.W., 2011. Biofuel and petroleum-based fuel supply chain research: a literature review. Biomass Bioenergy 35 (9), 3763–3774.

Argonne National Laboratory–Transportation Technology R&D Center, 2010. GREET Model: The Greenhouse Gases, Regulated Emissions, and Energy Use in Transportation Model.

CARB (California Air Resources Board), 2010. Low Carbon Fuel Standard (LCFS) Regulation. California Environmental Protection Agency, Air Resources Board, Sacramento, CA.

Collette, Y., Siarry, P., 2003. Multiobjective Optimization: Principles and Case Studies. Springer.
Daoutidis, P., et al., 2013. Engineering biomass conversion processes: a systems perspective. AIChE J. 59 (1), 3–18.
Das, I., Dennis, J.E., 1998. Normal-boundary intersection: a new method for generating the pareto Surface in Nonlinear Multicriteria optimization problems. SIAM J. Optim. 8 (3), 631–657.
Dutta, A., et al., 2011. Process Design and Economics for Conversion of Lignocellulosic Biomass to Ethanol: Thermochemical Pathway by Indirect Gasification and Mixed Alcohol Synthesis. National Renewable Energy Laboratory. NREL/TP–5100–51400.
Dutta, A., Phillips, S.D., 2009. Thermochemical Ethanol via Direct Gasification and Mixed Alcohol Synthesis of Lignocellulosic Biomass. National Renewable Energy Laboratory. NREL/TP–510–45913.
European Parliament, 2009. Directive 2009/28/EC: On the Promotion of the Use of Energy from Renewable Sources and Amending and Subsequently Repealing Directives 2001/77/EC and 2003/30/EC.
Farrell, A.E., et al., 2006. Ethanol can contribute to energy and environmental goals. Science 311 (5760), 506–508.
Floudas, C.A., Elia, J.A., Baliban, R.C., 2012. Hybrid and single feedstock energy processes for liquid transportation fuels: a critical review. Comput. Chem. Eng. 41, 24–51.
Geoffrion, A.M., Dyer, J.S., Feinberg, A., 1972. An interactive approach for multi-criterion optimization, with an application to the operation of an academic department. Manage. Sci. 19 (4), 357–368.
Gonzalez, T.F., 2007. Handbook of Approximation Algorithms and Metaheuristics. Chapman & Hall/CRC, Boca Raton.
Grossmann, I.E., Guillén-Gosálbez, G., 2010. Scope for the application of mathematical programming techniques in the synthesis and planning of sustainable processes. Comput. Chem. Eng. 34 (9), 1365–1376.
Han, J., et al., 2011. Well-to-Wheels Analysis of Fast Pyrolysis Pathways with GREET. Argonne National Laboratory. ANL/ESD/11–8.
Hill, J., et al., 2006. Environmental, economic, and energetic costs and benefits of biodiesel and ethanol biofuels. Proc. Natl. Acad. Sci. U.S.A. 103 (30), 11206–11210.
Hsu, D.D., et al., 2010. Life cycle environmental impacts of selected US ethanol production and use pathways in 2022. Environ. Sci. Technol. 44 (13), 5289–5297.
Humbird, D., et al., 2011. Process Design and Economics for Biochemical Conversion of Lignocellulosic Biomass to Ethanol: Dilute-acid Pretreatment and Enzymatic Hydrolysis of Corn Stover. National Renewable Energy Laboratory. NREL/TP–5100–47764.
IBM, 2011. IBM ILOG CPLEX Optimization Studio 12.2.
International Panel on Climate Change (IPCC), 2007. Climate Change 2007: Synthesis Report. Switzerland, Geneva.
Marvin, W.A., Schmidt, L.D., Daoutidis, P., 2013. Biorefinery location and technology selection through supply chain optimization. Ind. Eng. Chem. Res. 52 (9), 3192–3208.
Mavrotas, G., 2009. Effective implementation of the ε-constraint method in Multi-Objective Mathematical Programming problems. Appl. Math. Comput. 213 (2), 455–465.
McPhail, L., Westcott, P., Lutman, H., 2011. The Renewable Identification Number System and U.S. Biofuel Mandates. United States Department of Agriculture.
Miettinen, K., 1999. Nonlinear Multiobjective Optimization. Kluwer Academic Publishers, Boston.

Miettinen, K., Ruiz, F., Wierzbicki, A.P., 2008. Introduction to multiobjective optimization: interactive approaches. In: Multiobjective Optimization. Springer, pp. 27–57.

Mueller-Gritschneder, D., Graeb, H., Schlichtmann, U., 2009. A successive approach to Compute the Bounded pareto Front of Practical multiobjective optimization problems. SIAM J. Optim. 20 (2), 915–934.

Schmer, M.R., et al., 2008. Net energy of cellulosic ethanol from switchgrass. Proc. Natl. Acad. Sci. U.S.A. 105 (2), 464–469.

Searchinger, T., et al., 2008. Use of US croplands for biofuels increases greenhouse gases through emissions from land-use change. Science 319 (5867), 1238–1240.

Tao, L., et al., 2011. Process and technoeconomic analysis of leading pretreatment technologies for lignocellulosic ethanol production using switchgrass. Bioresource Technol. 102 (24), 11105–11114.

Tilman, D., et al., 2009. Beneficial biofuels—the food, energy, and environment trilemma. Science 325 (5938), 270–271.

U.S. Department of Energy, 2011. U.S. Billion-ton Update: Biomass Supply for a Bioenergy and Bioproducts Industry, R.D. Perlack and B.J. Stokes (Leads), ORNL/TM–2011/224. Oak Ridge National Laboratory, Oak Ridge, TN.

U.S. Environmental Protection Agency, 2010. EPA Finalizes Regulations for the National Renewable Fuel Standard Program for 2010 and beyond, EPA–420–f–10–007. United States Environmental Protection Agency.

U.S. Environmental Protection Agency, 2014. Regulation of fuels and fuel additives: 2013 cellulosic biofuel standard. U.S. Environ. Prot. Agency 79 (85), 25025–25031.

Wallace, R., et al., 2005. Feasibility Study for Co-locating and Integrating Ethanol Production Plants from Corn Starch and Lignocellulosic Feedstocks. National Renewable Energy Laboratory. NREL/TP–510–37092.

Wang, M., Han, J., Dunn, J.B., Cai, H., Elgowainy, A., 2012. Well-to-wheels energy use and greenhouse gas emissions of ethanol from corn, sugarcane and cellulosic biomass for US use. Environ. Res. Lett. 7 (4), 045905.

Wang, M.Q., et al., 2011. Energy and greenhouse gas emission effects of corn and cellulosic ethanol with technology improvements and land use changes. Biomass Bioenergy 35 (5), 1885–1896.

Wright, M.M., et al., 2010. Techno-economic Analysis of Biomass Fast Pyrolysis to Transportation Fuels. National Renewable Energy Laboratory. NREL/TP–6A20–46586.

Wu, M., Wang, M., Huo, H., 2006. Fuel-cycle Assessment of Selected Bioethanol Production Pathways in the United States. Argonne National Laboratory. ANL/ESD/06–7.

Yue, D., You, F., Snyder, S.W., 2014. Biomass-to-bioenergy and biofuel supply chain optimization: overview, key issues and challenges. Comput. Chem. Eng. 66, 36–56.

Zamboni, A., Bezzo, F., Shah, N., 2009. Spatially explicit static model for the strategic design of future bioethanol production systems. 2. Multi-objective environmental optimization. Energy Fuels 23 (10), 5134–5143.

CHAPTER

21

Optimal Harvest Management Adaptation for a New Biorefinery Investment in a Timberlands Supply Chain Using a Modified Cyclic Scheduling Model

Kevin Yeh*, Craig Whittaker§, Matthew J. Realff*,[1], Jay H. Lee¶,[1]

**School of Chemical & Biomolecular Engineering, Georgia Institute of Technology, Atlanta, GA, USA*

§Research & Development, Weyerhaeuser NR, Federal Way, WA, USA

¶Department of Chemical and Biomolecular Engineering, Korea Advanced Institute of Science and Technology, Yuseong-gu, Daejeon, Republic of Korea

[1]Corresponding authors: Email: jayhlee@kaist.ac.kr; matthew.realff@chbe.gatech.edu

21.1 INTRODUCTION

The United States has abundant agricultural and forest resources, which makes the adoption of renewable fuels strategies attractive. Its sophisticated agricultural sector has afforded the development of first-generation biofuels from corn ethanol, and residuals or bioenergy crops may support second-generation biofuels. In addition, the forestry sector is equally well developed and faces significant competition in its traditional markets. Therefore, timber managers are looking to expand product portfolios and diversify into new market segments.

This study focuses on the scheduling problem for harvesting and planting cycles in a timberlands system combined with the introduction of new biorefinery infrastructure. The underlying question is how to adjust the balance of timberland age and land use to support the new biorefinery as well as the existing timber industry. Cyclical models are often used to simulate systems where product demand is consistent over

Computer Aided Chemical Engineering, Volume 36. ISSN 1570-7946. http://dx.doi.org/10.1016/B978-0-444-63472-6.00021-5

long time periods (You et al., 2009). In a cyclical model, the values of the final state variables must be within a range of the values of the initial state variables, creating a consistent cycle. Cyclical programming is a common approach used in chemical engineering process scheduling problems (Pinto and Grossmann, 1994; Sahinidis and Grossmann, 1991; Wu and Ierapetritou, 2004; Castro, 2003). It has also been seen in supply chain research (Pundoor and Chen, 2009; You and Grossmann, 2008; Merkuryeva et al., 2010). We believe cyclical scheduling is a reasonable approach for a timberlands system because it simulates sustainable planting/harvesting cycles. Furthermore, the duration of 20 years is chosen because it is the approximate age range of loblolly pine, the tree species used in this model.

The goal of this model is to determine the optimal timberlands planting and harvest rotation to sustain previous manufacturers as well as a new biorefinery. This work utilizes two novel ideas. The first is using a start-up period before reaching the cyclical time rotation. This start-up period gives the harvesters time to adapt their management strategies to accommodate a new biorefinery. The other uses two time indices to track land maturity. Based on the age, different amounts of biomass types are obtained from harvesting, but the complexity of the growth dynamics is buried in coefficients that multiply variables, which appear linearly in the equations.

21.2 PROBLEM STATEMENT

21.2.1 SYSTEM OVERVIEW

In this study, the timberlands supply chain is composed of two major sectors, the timber providers and the manufacturers. The harvesters make decisions about planting and harvesting rotations. This scheduling problem requires the harvesters to maintain a consistent flow of material to manufacturers over time.

To accommodate for a new biorefinery, any changes in planting and harvesting strategies must be made years before the facility begins production. This is due to the long time requirement for timber biomass to mature. Therefore, in our model, we implemented a start-up period to adapt the cyclical planting rotation. Investment decisions are also made in the start-up period.

21.2.2 DEFINITIONS

Definitions of the indices, variables, and parameters are shown in Tables 21.1–21.6

21.2.3 BASIC CONSTRAINTS

Several restrictions on variables must be noted before discussion of the state equations and constraints. These restrictions prevent unrealistic behavior in the system.

Table 21.1 Index Information

Description	Index	Size	Range
Harvesting zone	i	21	1–21
Manufacturer	m	9	Lumber mills [1–8] Pulp mill [9]
Biomass type	k	7	Saw logs Fiber logs Thinnings Prunings Logging residuals Shavings Grass
Land type	h	3	Timber Switchgrass Thinned land
Product type	l	3	Lumber boards Pulp Gasoline
Biorefinery capacity	u	3	Small Medium Large
Biorefinery location	e	3	Loc. 1–Loc. 3
Present time	t	31	[0 30]
Origin time	τ	51	[–20 30]

Table 21.2 State Variables

$S_{iht\tau}$	Occupied land in zone i of type h planted at time τ at present time t
U_{it}	Unplanted land in zone i at present time t

Within these models, it should be noted that two time indices are used: present time (t) and origin time (τ). Origin time tracks the year that the land is planted on. Therefore, land maturity can be determined with $t - \tau$.

$$x_{iht\tau} = 0 \quad \forall i, h \text{ for } t - \tau \notin [X_{min} X_{max}] \tag{21.1}$$

$$x_{iht\tau} \geq 0 \quad \forall i, h \text{ for } t - \tau \in [X_{min} X_{max}] \tag{21.2}$$

$$\hat{x}_{iht\tau} = 0 \quad \forall i \text{ for } h = \text{timber}, \ t - \tau \notin [8 \quad 12] \tag{21.3}$$

$$\hat{x}_{iht\tau} \geq 0 \quad \forall i \text{ for } h = \text{timber}, \ t - \tau \in [8 \quad 12] \tag{21.4}$$

Table 21.3 Decision Variables

P_{iht}	Quantity to plant land type h in zone i at time t
$x_{iht\tau}$	Full harvest amount of zone i of land type h planted at time τ at present time t
$\widehat{x}_{iht\tau}$	Thinning amount of zone i of land type h planted at time τ at present time t
y_{imklt}	Purchase quantity from harvester i by manufacturer m of biomass type k for product l at present time t
$\widehat{y}_{m'mklt}$	Purchase quantity from manufacturer m' by manufacturer m of biomass type k for product l at present time t
π_{ikluet}	Purchase quantity from harvester i of biomass type k for product l at location index e and capacity index u at present time t
$\widehat{\pi}_{mkluet}$	Purchase quantity from manufacturer m of biomass type k for product l at location index e and capacity index u at present time t

Table 21.4 Binary Decision Variables

z_{iht}	Binary variable that controls the harvester i's activity for harvesting land type h at present time t
$\widehat{z}_{iht}$	Binary variable that controls the harvester i's activity for thinning land type h at present time t
$\theta_{ih\tau}$	Binary variable that controls the harvester i's activity for planting land type h at origin time τ
D_{uet}	Binary decision of biorefinery capacity index u of at location e at present time t

Table 21.5 Parameters

α_{kl}	Conversion factor for biomass type k into product l
G_m	Maximum capacity of manufacturer m
A_{ml}	Capacity of product l that must be produced from manufacturer m
$\gamma_{kk'}$	Fraction of resources of biomass type k' obtained from manufacturing biomass type k
κ	Number of biorefineries that can be introduced to the system
B_i	Land area of zone i
ω	Tolerance for cyclical constraint for beginning and end states
H	Minimum fraction of harvested land
β_{ue}	Capital cost of biorefinery of capacity u at location e
$\hat{G}_u$	Maximum capacity of biorefinery for capacity index u
R	Minimum fraction of planting area

Table 21.6 Time-Dependent Parameters

c_{ikt}	Selling price from harvester *i* for biomass type *k* at present time *t*
$\hat{c}_{mkt}$	Selling price from manufacturer *m* for biomass type *k* at present time *t*
p_{mlt}	Selling price from manufacturer *m* of product *l* at present time *t*
δ_t	Discount factor at present time *t*
o_{ikt}	Selling price of harvester *i*'s biomass type *k* to external buyers at present time *t*
$\widehat{p}_{luet}$	Selling price of biorefinery at location *e* of capacity *u* of product *l* at present time *t*
$\rho_{kht-\tau}$	Density from harvesting material type *k* from land type *h* at age $t-\tau$
$\widehat{\rho}_{kht-\tau}$	Density from thinning material type *k* from land type *h* at age $t-\tau$
f_{iht}	Cost per acre of harvesting land type *h* at harvester *i* at present time *t*
$\widehat{f}_{iht}$	Cost per acre of thinning land type *h* at harvester *i* at present time *t*
$\overline{c}_{iht}$	Price of planting in zone *i* of land type *h* at present time *t*
s_{imt}	Shipping cost from harvester *i* to manufacturer *m* at present time *t*
$\overline{s}_{met}$	Shipping cost from manufacturer *m* to biorefinery at location *e* at present time *t*
$\check{s}_{m'mt}$	Shipping cost from manufacturer *m* from manufacturer *m'* at present time *t*
n_{mt}	Processing cost per unit of material for manufacturer *m* at present time *t*
$\widehat{n}_{ut}$	Processing cost per unit of material for biorefinery of capacity index *u* at present time *t*
$\hat{s}_{iet}$	Shipping cost from harvester *i* to biorefinery at location index *e* at present time *t*

Table 21.7 Harvest Age $(t-\tau)$ to Harvest Land Types

Land Type (*h*)	Minimum Harvest Age (χ_{min})	Maximum Harvest Age (χ_{max})
Timber	8	12
	18	22
Switchgrass	2	10
Thinned Land	18	22

This model used a more complex representation for timberlands management behaviors than models typically developed for this application. Land types and age restrictions for harvesting are shown in Table 21.7. Both $x_{iht\tau}$ and $\widehat{x}_{iht\tau}$ can be nonzero within these ranges (Eqns (21.1)–(21.4)). The decision to plant timber ($h =$ timber) yields two time periods of harvest. From 8 to 12 years, young timber can be fully harvested or thinned (changing the land to $h =$ thinned) to provide more space for the maturing trees. The land cleared from thinning cannot be immediately used for new planting decisions. After about 20 years, the trees are large enough to harvest. Furthermore, thinning the timberlands is a decision for the harvesters. Timberlands can remain unthinned; however, due to crowding, the trees cannot reach the size that those in thinned land can, providing a different distribution of biomass types. Unthinned timberlands can provide larger amounts of biomass but may not be able to provide as much specific material that production facilities require.

The decision to plant switchgrass with timber leads to a different behavior. For this planting arrangement, every other row of trees is instead a row of switchgrass. After 2 years, harvestable grass-type biomass is provided yearly until the 10-year point. Afterward, the tree canopy becomes too large, blocking out sunlight and hindering future growth. The remaining layout of trees is the same as post-thinned timberlands.

Further infeasible actions must be prevented by restricting decision and state variables. To prevent a negative and 0 age, $S_{iht\tau}$, $x_{iht\tau}$, $\widehat{x}_{iht\tau}$ are ≤ 0 when $t < \tau$. To prevent incorrect thinning decisions, $\widehat{x}_{iht\tau} = 0$ when h was switchgrass or thinned land. Thinned land could only be generated from other land types, so $P_{iht} = 0$ when $h =$ thinned land. Finally, planting decision P_{iht} can only be nonzero when $t = \tau$ for the constraint in which it is present. Since index τ is the origin time of planting, it makes sense that the planting decision can only occur when the indices are equal.

Also, all variables except for $x_{iht\tau}$, z_{iht}, $\widehat{x}_{iht\tau}$, $\widehat{z}_{iht}$, $S_{iht\tau}$, U_{it}, and D_{uet} are 0 at the final time point. This is due to the setup of the time schedule. The problem ends at the beginning of the final year. Therefore, state variables can have a value, since they describe the beginning of the state. Harvest and thinning variables can contain a value since the decisions with index $t+1$ are made at time t. Conversely, this means variables $x_{iht\tau}$ and $\widehat{x}_{iht\tau}$ are 0 in the first year.

These restrictions will apply to the state equations, constraints, and objective function discussed in the next section.

21.2.4 STATE EQUATIONS

In this model, the state variables (U_{it} and $S_{iht\tau}$) describe the state of the timberlands. U_{it} is the amount of unplanted land for harvest zone i at time t. $S_{iht\tau}$ is the amount of land of type h at harvest zone i at present time t planted at origin time τ. The amount of unplanted land is calculated by Eqn (21.5), where B_i is the total land capacity of harvest zone i. Also, all state variables are constrained to be nonnegative (Eqns (21.6) and (21.7)).

$$U_{it} = B_i - \sum_h \left(P_{iht} + \sum_\tau S_{iht\tau} \right) \forall i, t \tag{21.5}$$

$$S_{iht\tau} \geq 0 \ \ \forall \ i, \ h, \ \text{for} \ \ t \geq \tau \tag{21.6}$$

$$U_{it} \geq 0 \ \ \forall \ i, t \tag{21.7}$$

The different behaviors of switchgrass and timber require separate state equations. Also, a distinction is made for thinned land due to different growth dynamics of thinned versus unthinned lands.

21.2.4.1 Timber state equations

Equation (21.8) is the state equation governing land type h = timber. The value of the next state is determined by the amount of land harvested $x_{iht+1\tau}$ and thinned $\widehat{x}_{iht+1\tau}$ subtracted from the value of the present state. The harvest decisions can only occur at the maturities shown in Table 21.7.

A point regarding the thinning decision $\widehat{x}_{iht+1\tau}$ needs to be clarified. Whenever a certain amount of land is chosen to be thinned ($\widehat{x}_{iht+1\tau}$), the land type index h changes from timber to thinned land. While, physically, trees have been removed from land by this decision, that cleared land is still considered occupied by the remaining trees. What this means is that, in this model, a thinned land state variable has half the number of planted trees that a timber land variable of the same acreage has.

$$S_{iht+1\tau} = S_{iht\tau} - x_{iht+1\tau} - \widehat{x}_{iht+1\tau} \; \forall \; i, \; t, \; \tau \text{ for } h = \text{timber} \tag{21.8}$$

At intervals when $t = \tau$, Eqn (21.8) becomes Eqn (21.9), adding planting decision P_{iht}.

$$S_{iht+1\tau} = S_{iht\tau} - x_{iht+1\tau} - \widehat{x}_{iht+1\tau} + P_{iht} \; \forall \; i, \;\; \text{for } h = \text{timber}, t = \tau \tag{21.9}$$

21.2.4.2 Switchgrass state equations

State behavior for switchgrass-type land is shown in Eqns (21.10) through (21.12). Equation (21.10) defines the amount of land planted. Equation (21.11) represents the behavior of the first 9 years of planted switchgrass. A harvest term $x_{iht\tau}$ does not exist in the state equation. Since a consistent amount of material is generated each year of its lifespan, harvesting does not subtract from the state variable. Equation (21.12) removes the switchgrass-type land after 10 years. This land is converted to thinned land, which is seen in Eqn (21.14).

$$S_{iht+1\tau} = P_{iht} \;\; \forall \; i \text{ for } h = \text{switchgrass}, \;\; t - \tau = 0 \tag{21.10}$$

$$S_{iht+1\tau} = S_{iht\tau} \;\; \forall \; i \text{ for } h = \text{switchgrass}, \;\; 0 < t - \tau < 10 \tag{21.11}$$

$$S_{iht+1\tau} = 0 \;\; \forall \; i \text{ for } h = \text{switchglass}, \;\; t - \tau \geq 10 \tag{21.12}$$

Equation (21.13) must also be included to constrain switchgrass harvesting. Due to the nature of switchgrass, biomass is provided on a yearly basis over a duration. Harvesting has no impact on the amount it provides, so the state variable is not changed by the harvesting decision. Equation (21.13) sets an upper bound that ensures that the amount of switchgrass harvested does not exceed its planted land.

$$x_{iht+1\tau} \leq S_{iht\tau} \quad \forall i, \;\; \text{for } h = \text{switchglass}, \; t \geq \tau \tag{21.13}$$

21.2.4.3 Thinned land state equations

Equation (21.14) is the state equation for thinned land. Thinned land can only be generated from thinning the timber land type or choosing the switchgrass planting decision. The acreage of the thinned land state variable coming from a timber type land origin contains the land of both the trees remaining after thinning (or switchgrass dying off) as well as the land cleared from thinning (land that was originally planted with switchgrass). After 18 years, the thinned land can be fully harvested.

$$S_{iht+1\tau} = S_{iht\tau} - x_{iht+1\tau} + \widehat{x}_{ih't+1\tau} \ \forall i, t, \tau \quad \text{for } h = \text{thinned}, \ h' = \text{timber} \tag{21.14}$$

When the switchgrass land has matured 10 years ($t - \tau = 10$), Eqn (21.14) becomes Eqn (21.15), which adds the switchgrass state term $S_{ih''t\tau}$.

$$S_{iht\tau+1\tau} = S_{iht\tau} - x_{iht+1\tau} + \widehat{x}_{ih't+1\tau} + S_{ih''t\tau} \ \forall i \quad \text{for } h = \text{thinned}, h' = \text{timber}, h'' = \text{switchgrass}, t - \tau = 10 \tag{21.15}$$

21.2.5 DECISION VARIABLE BOUNDS

The following constraints cover the activity of certain decisions. Each of these constraints comes in pairs, creating an upper and a lower bound. These bounds prevent decisions with unrealistically small values. Equation (21.16) is the upper and lower bounds of the planting decision. Equations (21.17) and (21.18) control the harvesting and thinning decisions. In these equations, B_i is the total amount of land (acres) in zone i. R and H are ratios that are the minimum fraction of land that must be planted/harvested. θ_{iht} controls the activity of the planting decision while $z_{iht+1}$and $\widehat{z}_{iht+1}$ control the activity for the harvesting and thinning decisions.

$$RB_i\theta_{iht} \le P_{iht} \le B_i\theta_{iht} \ \forall i, h, t \tag{21.16}$$

$$HB_iz_{iht+1} \le \sum_{\tau} x_{iht+1\tau} \le B_iz_{iht+1} \ \forall i, h, t \tag{21.17}$$

$$HB_i\widehat{z}_{iht+1} \le \sum_{\tau} \widehat{x}_{iht+1\tau} \le B_i\widehat{z}_{iht+1} \ \forall i, h, t \tag{21.18}$$

21.2.6 BIOREFINERY CONSTRAINTS

The biorefinery decision D_{uet} can only be made in the first 10 years. The number of biorefineries that can be built is limited to κ (Eqn (21.19)). Once the investment has been made, binary variable D_{uet} remains active (Eqn (21.20)). After the initial 10 years, the biorefinery decisions are fixed (Eqn (21.21)). Equation (21.22) ensures that only one biorefinery can exist at a location.

$$\sum_{u}\sum_{e} D_{uet} \le \kappa \ \text{for} \ 1 \le t \le 10 \tag{21.19}$$

$$D_{uet+1} \ge D_{uet} \ \forall u, e \ \text{for} \ 1 \le t < 10 \tag{21.20}$$

$$D_{uet+1} = D_{uet} \forall u, e \ \text{for} \ 10 \le t \le 29 \tag{21.21}$$

$$\sum_{u} D_{uet} \le 1 \ \forall e, \text{for} \ 1 \le t \le 10 \tag{21.22}$$

Equation (21.23) determines the manufacturer biomass byproduct availability. This is determined by the amount of material processed by manufacturers ($y_{imk'l't}$). The amount generated is determined by yield term $\gamma_{k'k}$. The pulp mill can also utilize the lumber mill byproduct. The term $\widehat{y}_{mm'klt}$ represents the amount of material that the pulp mill (m') purchases from manufacturers (m).

$$\sum_l \left(\sum_u \sum_e \widehat{\pi}_{mkluet} + \sum_{m'} \widehat{y}_{mm'klt} \right) \le \sum_{k'} \gamma_{k'k} \sum_i \sum_{l'} y_{imk'l't} \quad \forall m, k, t \tag{21.23}$$

Equation (21.24) is the capacity constraint for the biorefineries. The capacity is controlled by biorefinery investment variable D_{uet}.

$$\sum_k \sum_l \left(\sum_i \pi_{ikluet} + \sum_m \widehat{\pi}_{mkluet} \right) \le D_{uet} \widehat{G}_u \quad \forall u, e, t \tag{21.24}$$

21.2.7 MANUFACTURER CONSTRAINTS

The manufacturer constraints are given by Eqns (21.25)–(21.27). Equation (21.25) is an upper bound limited by the facility capacity. Equation (21.26) is the demand from retailers on the manufacturer. Equation (21.27) is the biomass availability constraint. It restricts the amount of biomass that can be purchased by both manufacturers and biorefineries by the amount of material harvested.

$$\sum_k \sum_l \left(\sum_i y_{imklt} + \sum_{m'} \widehat{y}_{m'mklt} \right) \le G_m \quad \forall m, t \tag{21.25}$$

$$\sum_k \alpha_{kl} \left(\sum_i y_{imklt} + \sum_{m'} \widehat{y}_{m'mklt} \right) \ge A_{ml} \quad \forall m, l, t \tag{21.26}$$

$$\sum_l \left(\sum_m y_{imklt} + \sum_u \sum_e \pi_{ikluet} \right) \le \sum_h \sum_t \left(\rho_{kht-\tau} x_{iht+1\tau} + \widehat{\rho}_{kht-\tau} \widehat{x}_{iht+1\tau} \right) \quad \forall i, k, t \tag{21.27}$$

21.2.8 CYCLICAL CONSTRAINTS

Equations (21.28) and (21.29) ensure that the cyclical conditions of the problem are met. T is the time period of the new cyclic schedule initial state, and its value is based off the duration of the transition period (if transition period is 10 years, the new cyclical initial state is given by the state variables at time $t = T = 10$). The final state variables occur at time $t = T + 20$ and can deviate from the initial state variables within tolerance ω. The τ indices for the final state must also be adjusted by 20 years. These cyclical constraints are quite restrictive in that they require age and biomass type distribution among each harvest zone to be constrained within ω of the starting values.

$$S_{ih30\tau+20} \geq (1-\omega)S_{ih10\tau} \quad \forall i, h, \tau \tag{21.28}$$

$$S_{ih30\tau+20} \leq (1+\omega)S_{ih10\tau} \quad \forall i, h, \tau \tag{21.29}$$

21.2.9 OBJECTIVE FUNCTION

The objective function for the model is the overall system profit. This means the distribution of value within the supply chain does not affect the overall value of the objective function. This is further discussed in Section 2.9.3. The revenues and costs described below are specific to each time period t. Each time period is adjusted for inflation and then discounted backward by a factor of 5% to reflect current interest rates for the calculation of net present value. The inflation values were taken from usinflationcalculator.com (2014).

21.2.9.1 Revenue for each time period

Material harvested in excess to the demands of manufacturers and biorefineries within the supply chain system are sold externally at price o_{ikt} (Eqn (21.30)).

$$\text{External Revenue} = \sum_i \sum_k o_{ikt} \left(\sum_\tau (\rho_{kht-\tau} x_{iht+1\tau} + \hat{\rho}_{kht-\tau} \hat{x}_{iht+1\tau}) - \sum_l \left(\sum_m y_{imklt} + \sum_u \sum_e \pi_{ikluet} \right) \right) \tag{21.30}$$

Both the biorefineries and manufacturers generate profits from selling their final products. Their revenues are calculated by Eqns (21.31) and (21.32).

$$\text{Biorefinery Revenue} = \sum_l \sum_u \sum_e \hat{p}_{luet} \sum_k \alpha_{kl} \left(\sum_i \pi_{ikluet} + \sum_m \hat{\pi}_{mkluet} \right) \tag{21.31}$$

$$\text{Manufacturer Revenue} = \sum_m \sum_l p_{mlt} \sum_k \alpha_{kl} \sum_i y_{imklt} \tag{21.32}$$

21.2.9.2 Costs for each time period

Costs within the system include harvesting costs (Eqn (21.33)), capital costs (Eqn (21.34)), planting costs (Eqn (21.35)), transportation costs (Eqns (21.36)–(21.39)), and processing costs (Eqns (21.40) and (21.41)). The capital costs in Eqn (21.34) are a one time fee incurred if a biorefinery is built. Therefore, if variable D_{uet} is 1 when $t = T$ (the beginning of the new cyclic schedule), then the cost is included. This part of the objective function is not reliant on t.

$$\text{Harvesting Costs} = \sum_i \sum_h \sum_\tau \left(f_{iht} x_{iht\tau} + \widehat{f}_{iht} \widehat{x}_{iht\tau} \right) \tag{21.33}$$

$$\text{Annualized Capital Cost} = \sum_u \sum_e \beta_{ue} D_{ueT} \tag{21.34}$$

$$\text{Planting Cost} = \sum_i \sum_h c_{iht} P_{iht} \tag{21.35}$$

$$\text{Manufacturer Transportation Costs from Harvesters} = \sum_i \sum_m s_{imt} \sum_k \sum_l y_{imklt} \tag{21.36}$$

$$\text{Biorefinery Transportation Costs From Harvesters} = \sum_i \sum_e \widehat{s}_{iet} \sum_k \sum_l \sum_u \pi_{ikluet} \tag{21.37}$$

$$\text{Manufacturer Transportation Costs from Manufacturers} = \sum_{m'} \sum_m \breve{s}_{m'mt} \sum_k \sum_l \widehat{y}_{m'mklt} \tag{21.38}$$

$$\text{Biorefinery Transportation Costs from Manufacturers} = \sum_m \sum_e \bar{s}_{met} \sum_k \sum_l \sum_u \widehat{\pi}_{mkluet} \tag{21.39}$$

$$\text{Manufacturer Processing Costs} = \sum_m n_{mt} \sum_i \sum_k \sum_l y_{imklt} \tag{21.40}$$

$$\text{Biorefinery Processing Costs} = \sum_u \widehat{n}_{ut} \sum_k \sum_l \sum_e \left(\sum_i \pi_{ikluet} + \sum_m \widehat{\pi}_{mkluet} \right) \tag{21.41}$$

21.2.9.3 Additional objective values

Equations (21.42) to (21.44) represent the costs of acquiring biomass materials between sectors. These costs are complemented by the revenue gained from sales by a different sector, which yields no effect on the value of the objective function. These equations are included for the calculations of profits specific to the harvesting, manufacturing, and biofuel sectors of the supply chain.

$$\text{Biomass Costs from Harvesters to Manufacturers} = \sum_i \sum_k c_{ikt} \sum_l \sum_m y_{imklt} \tag{21.42}$$

$$\text{Biomass Costs from Harvesters to Biorefineries} = \sum_i \sum_k c_{ikt} \sum_l \sum_u \sum_e \pi_{ikluet} \tag{21.43}$$

$$\text{Biomass Costs from Manufacturers to Manufacturers and Biorefineries}$$

$$= \sum_m \sum_k \widehat{c}_{mkt} \sum_l \left(\sum_{m'} \widehat{y}_{mm'klt} + \sum_u \sum_e \widehat{\pi}_{mkluet} \right) \tag{21.44}$$

21.3 PROBLEM GENERATION AND CASE STUDIES

The two major decision makers in our system are the harvesters and manufacturers. In our case studies, the harvesters provide loblolly pine-type biomass, which is assumed to have a life cycle of about 20 years. After 10 years, the timberlands can be maintained with row thinning, which removes every other row of trees (Dickens et al., 2005). This prevents congestion of the forestlands, allowing the remaining trees more space to grow. Timber from thinned land provides saw logs, fiber logs, and residuals. Unthinned timber also provides these types of biomass, but less saw logs are available due to limited growing space. Instead, a larger total amount of material is harvested, and larger percentage of fiber logs is provided. Materials obtained from thinning decisions are thinnings, residuals, and prunings.

The established manufacturers within the geographic scope of this problem are lumber mills and a pulp mill in the southeast of the United States. Lumber mills process saw logs into timber while pulp mills can use fiber logs, residuals, shavings, thinnings, and prunings to produce pulp. About 90% is the minimum processing capacity for these established facilities. We felt that the established manufacturers should not have to sacrifice productivity for the biorefinery. These facilities could be contractually obligated to provide a consistent flow of product to retailers.

Introducing biofuel production to the supply chain could require alterations to the planting rotations, such as adding an alternative biomass crop of switchgrass. Switchgrass would provide an early consistent source of biomass for biofuel manufacturing. Also, instead of the more crowded timber planting strategy, every other row of trees could be replaced with rows of switchgrass. This means that row thinning is not needed because the switchgrass grows between the rows that are spaced further apart during the first 10 years before the canopy closes over it.

The model developed in these studies contained 21 harvest zones, eight lumber mills, and one pulp mill. The system is limited to one biorefinery investment. The total land area is 724,000 acres. The case studies are assumed to begin at year 1974, and the historical inflation rates from usinflationcalculator.com (2014) are used to accurately inflate the prices through the time horizon.

Two case studies were performed with this timberlands model. First, several different scenarios with varying initial states of the forest and land capacity were generated. These initial states were tested in the modified cyclic model to determine their viability to sustain biofuel production. For the second set of case studies, the duration of the transition period was varied to study if additional time would lead to different outcomes.

Initial states were generated by the model. Without biorefinery investment decisions, the 20-year cycle emulates established timberlands strategies. The problem was run without an initial state and biofuel production, and the initial state of the cycle was saved as an initial state for the case studies. These case studies utilized parameter data received from our industrial partner. The parameter values are shown in Tables (A.1-A.10).

21.4 RESULTS

Optimization was performed in General Algebraic Modeling System (GAMS) using the CPLEX solver on the Georgia Institute of Technology Joe cluster. Information about the specifications of the cluster can be found at Georgia Tech (2014).

21.4.1 INITIAL STATE GENERATION

Figure 21.1 shows a generated established cycle for our timberlands system. This cycle was generated by optimizing a 20-year cyclic model without biorefinery decisions. The established schedules generated in this section and Section 4.2 will be used in the system transition problems.

The three smaller plots on the left of Figure 21.1 are the land amounts utilized for timber and thinned lands. The amount of unplanted land is also shown. The x-axis range of the graphs goes from year 0 to 21 to show the value of the states at year 20; the values of year 20 are shown in the bar between years 20 and 21. The larger figure on the right is a stacked representation displaying the distribution of land at each time point. The labels on the left and bottom of the figure correspond to the y-axis (land usage) and x-axis (time period) for all subplots and stacked graphs. All similar figures in the results section are labeled in the same way.

This initial solution showed inconsistent behavior, particularly at years 1, 5, 9, 13, and 17. These dips occurred through decisions to harvest large amounts of material in the previous year, leaving the next state at a lower land usage. When viewing the subplots, it can be seen that the large dips occurred only in the

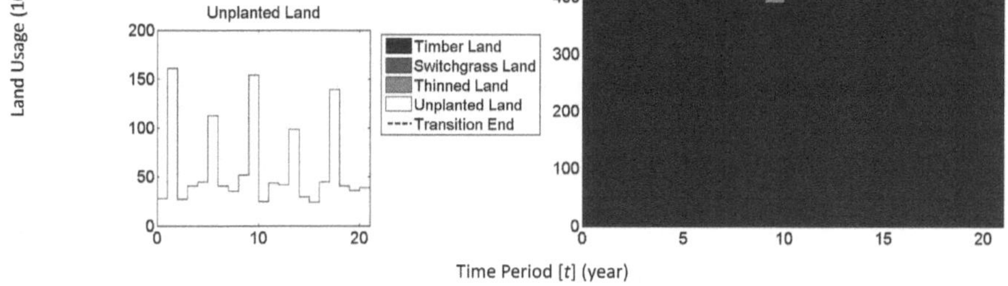

FIGURE 21.1

Established timberlands schedule (unconstrained).

timber-type land while the thinned land stayed relatively consistent. What happened in these scenarios was that the harvested thinned land was replenished by thinning decisions. Another interesting pattern was that these dips occurred at intervals of 4 years, but this may have been due to the specifics of our case study.

Timberlands management may want to avoid these large dips in land usage, so smoothing constraints were introduced to promote consistent harvesting and thinning at each time period (Eqns (21.45) and (21.46)). We felt these constraints were reasonable in that they enforced consistent behavior through the time horizon. The upper bound values of these constraints were determined by averaging the amount of material harvested in each time period in an unconstrained problem.

$$\sum_i \sum_h \sum_\tau x_{iht\tau} \le X \ \ \forall t \tag{21.45}$$

$$\sum_i \sum_h \sum_\tau \widehat{x}_{iht\tau} \le \widehat{X} \ \ \forall t \tag{21.46}$$

$$X = \widehat{X} = 72 \text{ acres} \tag{21.47}$$

Three cases were run around these constraints: "Fully constrained" case where both smoothing constraints of Eqns (21.45)–(21.46) are active, "harvester constrained" case where only the harvesting constraint (Eqn (21.45)) is active, and "unconstrained" case with neither constraint active.

For the constrained problems (Figures 21.2 and 21.3), the stacked representations show a much more consistent behavior for land usage; the drops in planted land are not present in these systems. While no clear difference in consistency can be seen between the harvester constrained and fully constrained cases, on average, the fully

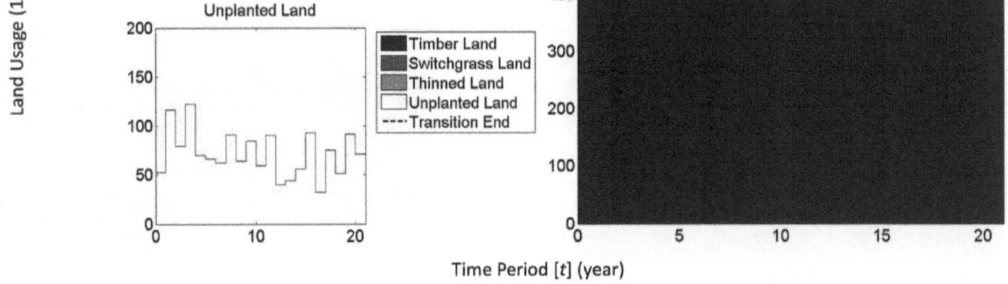

FIGURE 21.2

Established timberlands schedule (harvester constrained).

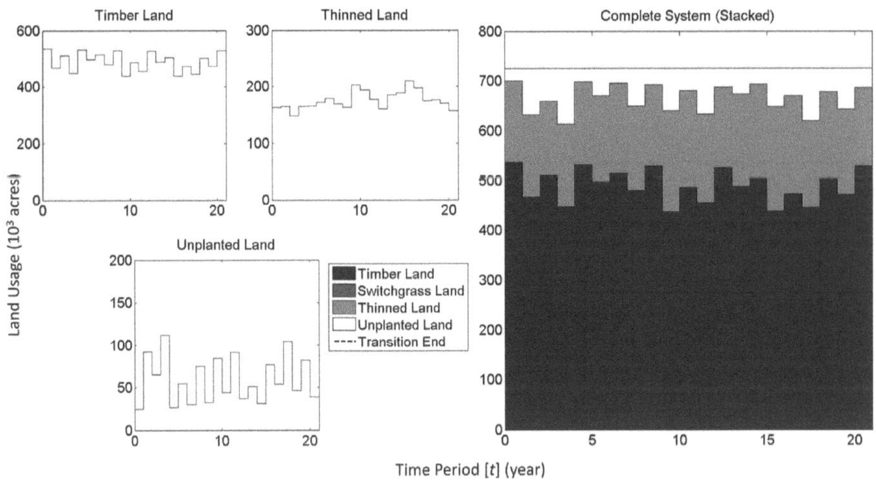

FIGURE 21.3

Established timberlands schedule (fully constrained).

constrained case utilizes more land through the 20-year period. These three case studies were solved within 10% optimality.

21.4.1.1 Variable land study

A study was performed that minimized land usage but still fulfilled the demands of the system throughout the 20-year cycle. This was done by introducing a new set of binary variables that controlled the activity of the harvesters. In the objective function, these variables were heavily penalized for taking on a value of 1.

For the first study, the problem was unconstrained. Not only were the smoothing constraints not present, but the planting, harvesting, and thinning upper and lower bounds were also removed (Eqns (21.16)–(21.18)). Solving this problem revealed that five harvesters was the absolute minimum number required to keep the manufacturers running at 90% capacity. These five harvesters had a total of 0.359 million acres of land, whereas the total system had 0.724 million acres.

Figure 21.4 shows the results of the unconstrained variable land study. The land usage remained consistent with very small increases and decreases except at the beginning and end. The results from this study were used as a lower bound for the constrained problem.

This set of five harvesters was used in two ways to determine the initial state for a system with low timber resources. For the first method, the variable land model was reused and the constraints were reintroduced. For the second method, the initial set of case studies in Section 4.1 was run with this set of five harvesters. After 8-h calculation times for both methods, none of the problems converged to within 40% of optimality. The reintroduction of Eqns (21.16)–(21.18) severely tightened

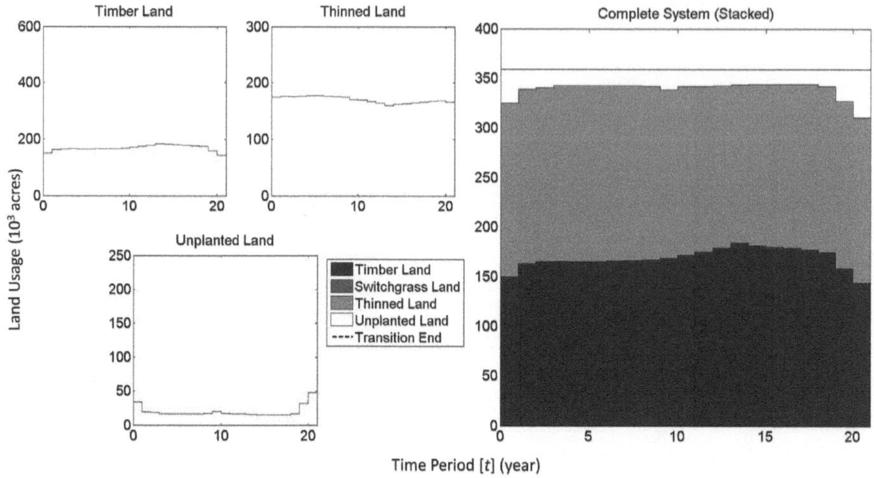

FIGURE 21.4

Schedule for variable land model.

the problem and created a smaller feasible region that was making difficult for GAMS to find a solution.

21.4.2 MODERATE LAND CAPACITY INITIAL STATE GENERATION

Another set of initial conditions was generated by setting up a system with moderate levels of land. This system contained the 13 largest harvesters, including the 5 from the variable land study. This system was run in the three smoothing constraint cases and solved to within 15% optimality. Less land capacity yielded a more difficult problem to solve, which was why the solution tolerance was higher. Figure 21.5 shows an initial state problem under moderate land availability.

The unconstrained case was similar to the scenario with normal land capacity except the drops in year 5 and 13 were not as severe. It seems that the major decreases in land usage occurred at intervals of 8 years with smaller decreases in between. The subplots for thinned and timber land usage were more consistent than their normal land scenario counterparts. With less land available in the system, the managers may require more consistent land management to provide to the manufacturers. The constrained problems were included in the supplementary information due to space constraints. The results (Figures A.1 and A.5) were similar to the normal land cases where the constraints enforced more consistent behavior. The harvester constrained problem seemed to have more periods of level land usage or gradual increases of land usage (years 6–9, 10–13, 13–16) while the fully constrained problem seemed to show larger increases and decreases every year.

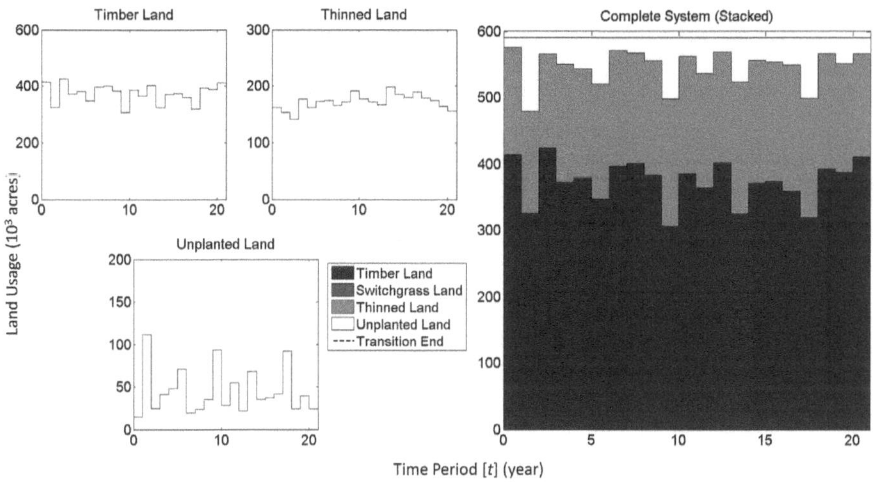

FIGURE 21.5

Established timberlands schedule (moderate land, unconstrained).

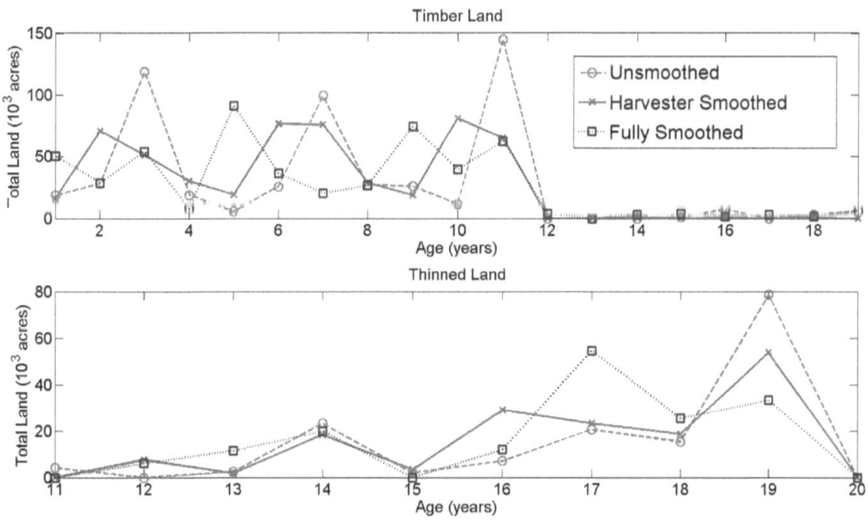

FIGURE 21.6

Initial state age distribution (normal land).

The stacked representations in the previous figures do not give a complete view of the initial state, particularly the age distribution of the land at that time. Figure 21.6 shows the age distribution of the lands at the initial state for normal land availability. Thinned land had an age range of 11–20 years. Timber land had

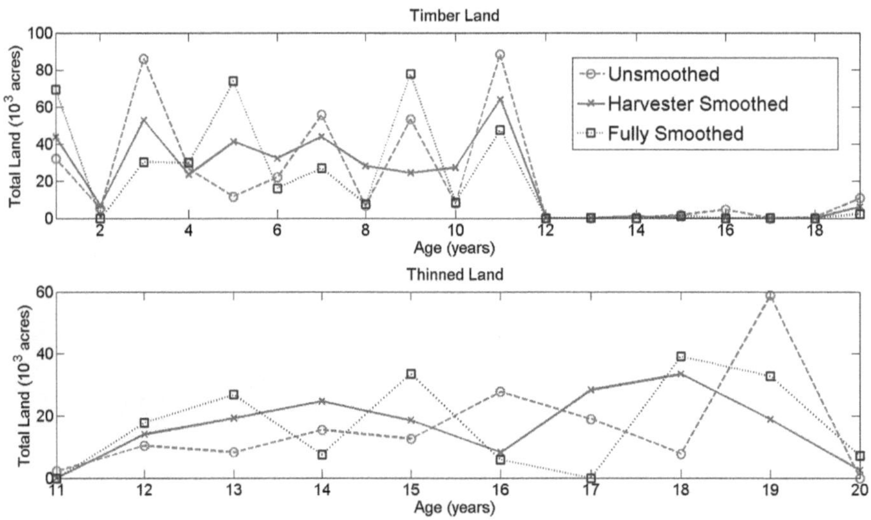

FIGURE 21.7

Initial state age distribution (moderate land).

an age range of 1–20 years, with the majority of land 12 years and younger. Large amounts of 19-year-old thinned land and 11-year-old timber land were seen for the unsmoothed case. The maturity of these lands was within the range of harvesting and thinning, which explains the large drop in land use in year 1. The peaks of the constrained cases never reached the level of the unconstrained case. The restriction from the constraints could limit the height of these peaks.

Figure 21.7 shows the initial state age distribution of the moderate land cases. Again, for the unconstrained case, large amounts of land could be seen at years 11 (timber) and 19 (thinned). When comparing the age distributions of the thinned and timber-type lands, the timberland distribution was much more erratic compared to the thinned land solution for both the normal and moderate land cases. The peaks in the timber land graphs corresponded to drops in land use in the previous figures.

21.4.3 SCHEDULE ADAPTATION RESULTS

The six sets of initial conditions generated above (six different combinations of smoothing constraints and land availability) were used in the schedule adaptation studies. In these studies, a 10-year transition period was used to adapt the harvest schedule to incorporate a new biorefinery. The land availability of the system was the same as the situation from which the initial state was generated, but the smoothing constraints were not included. Because the system was changing to a new harvest schedule, the previous bounds could be invalid. Therefore, initial unsmoothed solutions could be used to set new bounds for the constraints.

For these six sets of initial conditions, five converged on a solution. The normal land cases converged within 10% optimality. The moderate land cases converged within 30% except for the fully constrained problem with an 8-h calculation time. To obtain a solution, this model was split into two cases: one with biorefinery investment enforced and one with no biorefinery. The solutions were compared, and the biorefinery enforced model provided a higher best possible objective value in GAMS. This solution, as well as the harvester constrained normal land and unconstrained moderate land cases, yielded an optimal solution with a biorefinery. These cases are shown in Figures 21.8 and 21.9.

For these cases, timberland usage decreased during the transition period. The thinned land usage at the new cyclic starting year was consistent with the initial state. After the clearing of timber-type land, it was transitioned toward switchgrass. Figure 21.9 started the new cycle with less switchgrass land than Figure 21.8. We believe this could be attributed to the starting behavior of the 30-year period. Figure 21.8 started with a large drop in land usage, thereby freeing more space for a larger initial planting decision for switchgrass. In all three cases, the switchgrass usage dipped around the middle of the new cycle before rising again to satisfy the cyclic constraint.

For the three cases where a biorefinery investment was not optimal, a second case was studied where a biorefinery investment was enforced. Figures 21.10 and 21.11 are the two sets of results from the unconstrained normal land scenario, Figures 21.12 and 21.13 are the results from the fully constrained normal land case. The harvester constrained moderate land case was the last scenario. It is not shown here due to space constraints but is located in this chapter's supplementary information (Figures A.1-A.3). For the biorefinery enforced cases, the system also transitioned land to switchgrass. The unconstrained case had a larger initial drop, which was transitioned to larger amounts of switchgrass land.

Upon examination of the land type subplots, it can be seen that, for all presented cases, both normal and moderate land systems, the initial state of thinned land was at a consistent level, around 150,000−180,000 acres. Furthermore, it seemed that this initial state was a relative lower bound for thinned land usage through the time period. Harvesters may need to maintain this level of thinned land to provide consistent biomass flow to the manufacturers.

When viewing Figures 21.8 to 21.13 alongside Figures 21.6 and 21.7, it was not clear why certain scenarios chose biofuel investments. There was no apparent correlation between the initial state and the choice of biofuel investment. This implies that there is a more complex relationship between initial state and biorefinery decision. Further study is needed to better understand this interaction.

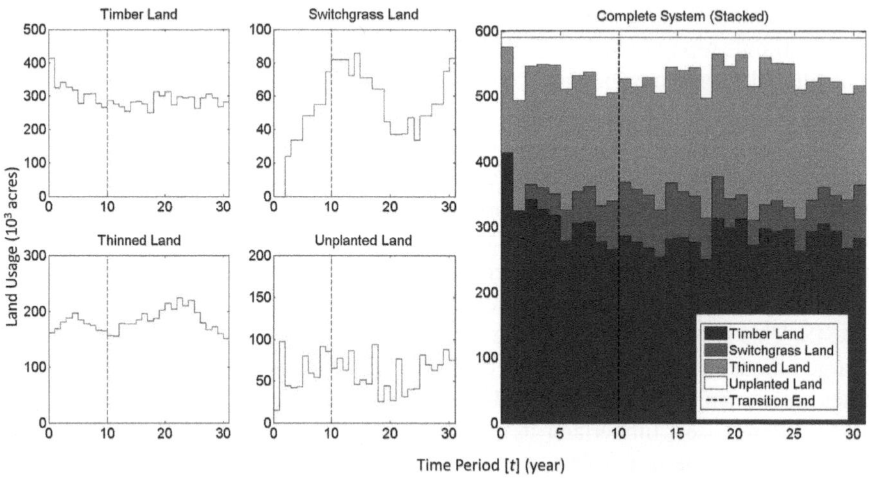

FIGURE 21.8

Ten-year transition timberlands schedule (moderate land, unconstrained initial state).

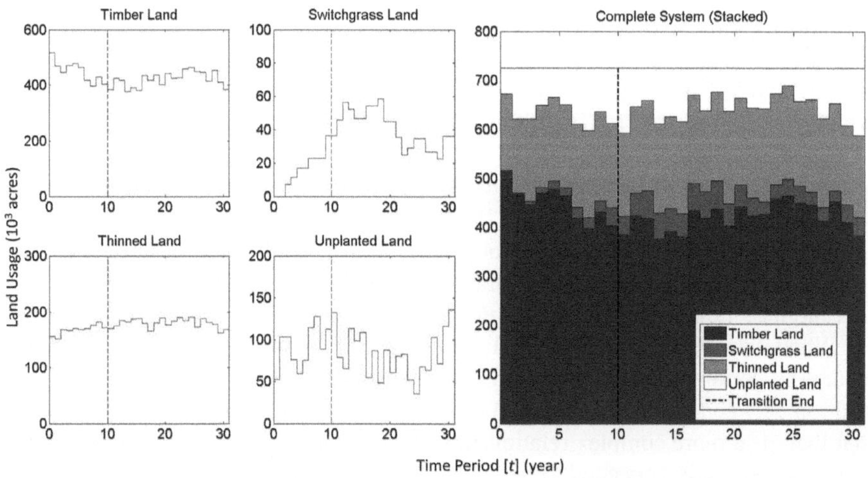

FIGURE 21.9

Ten-year transition timberlands schedule (normal land, harvester constrained initial state).

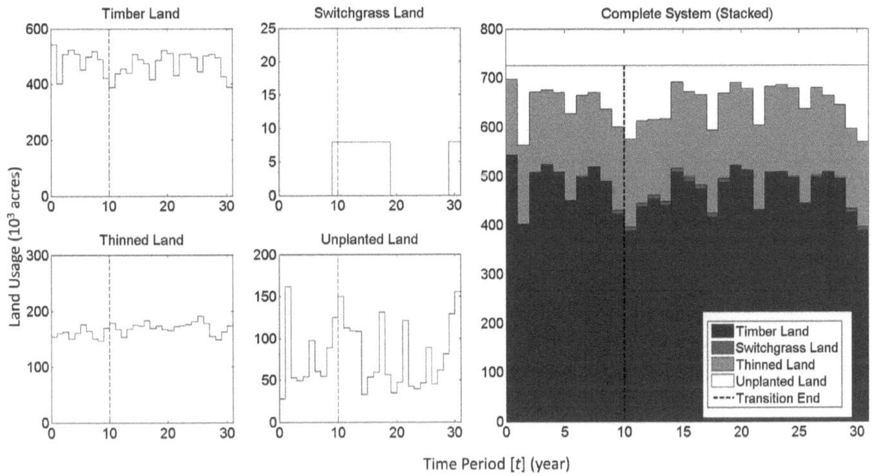

FIGURE 21.10

Ten-year transition timberlands schedule (normal land, unconstrained initial state).

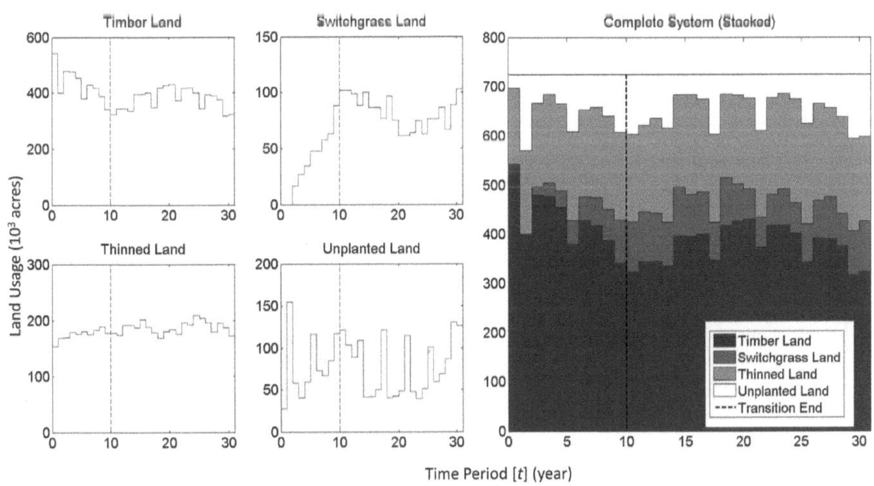

FIGURE 21.11

Ten-year transition timberlands schedule (normal land, unconstrained initial state, enforced biorefinery).

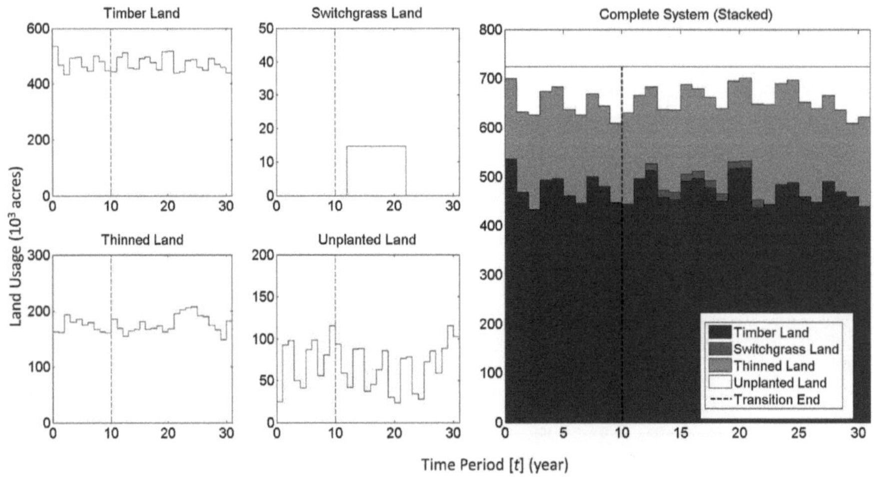

FIGURE 21.12

Ten-year transition timberlands schedule (normal land, fully constrained initial state).

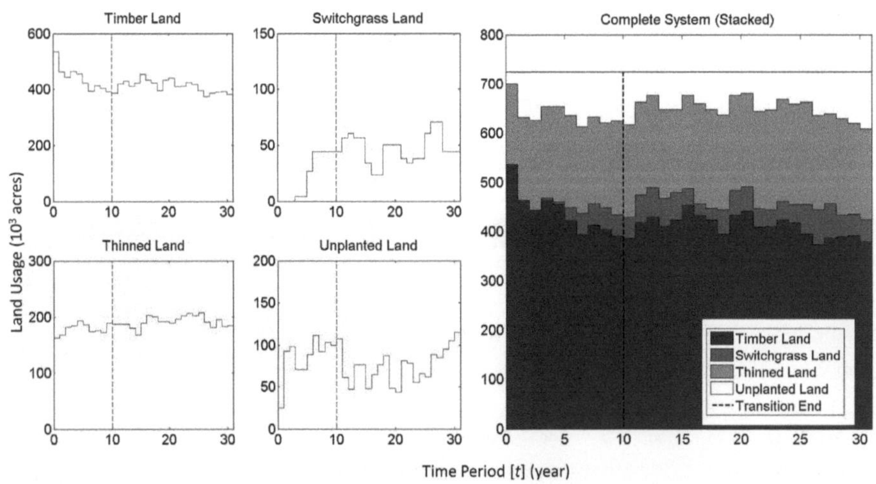

FIGURE 21.13

Ten-year transition timberlands schedule (normal land, fully constrained initial state, enforced biorefinery).

21.4.4 TRANSITION TIME STUDIES

An additional study was conducted where the transition time period length was varied. The initial 10 years was extended to 15- and 20-year intervals. All of the previous initial states were used, yielding 21 different cases. For the 15-year calculation times, only four solutions were obtained through GAMS. The other two did not converge in reasonable time. These results are shown in Figures 21.14–21.16 and Figure A.4.

Figures 21.14 through 21.15 show a 15-year transition period scenario for initial states that did not invest in biofuels with a 10-year transition period. The harvester constrained moderate land initial state was also run in the transition period model, but was not shown due to limited space. The result is shown in the supplementary information (Figure A.4). With five additional years, all three of the originally no biorefinery cases chose to invest. Similar to previous results, a large portion of timber-type land was converted to switchgrass-type land. Also, the subplot of switchgrass-type land shows a large decrease in usage around the middle of the transition period.

Figure 21.16 shows the unconstrained moderate land case with a 15-year transition period. The subplot of switchgrass land for this case did not yield the same single drop in land usage that the other biorefinery cases did (both 10- and 15-year cases). The fact that this case initially invested in a biorefinery led us to believe that adding more time to the initial transition period allowed for the harvesters to change to a more consistent management schedule.

None of the 20-year transition period scenario solutions were obtained. With 8-h-long calculation times, GAMS was unable to converge on solutions within 50%

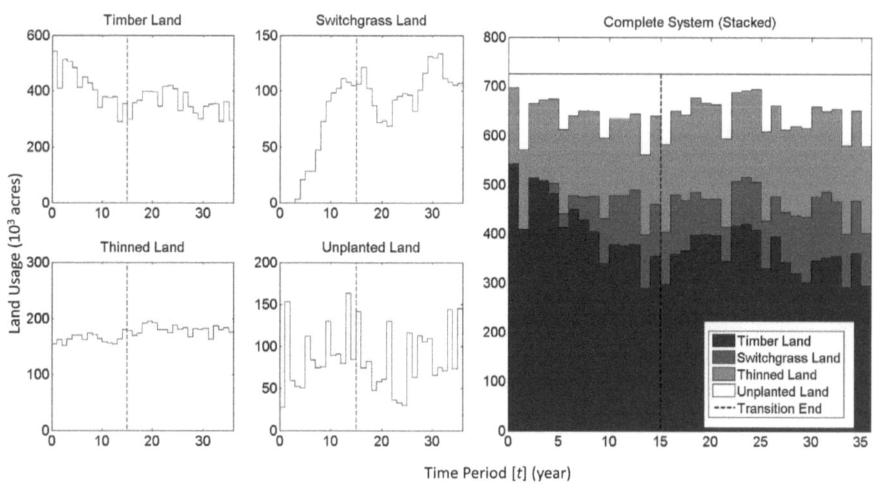

FIGURE 21.14

Fifteen-year transition timberlands schedule (normal land, unconstrained initial state).

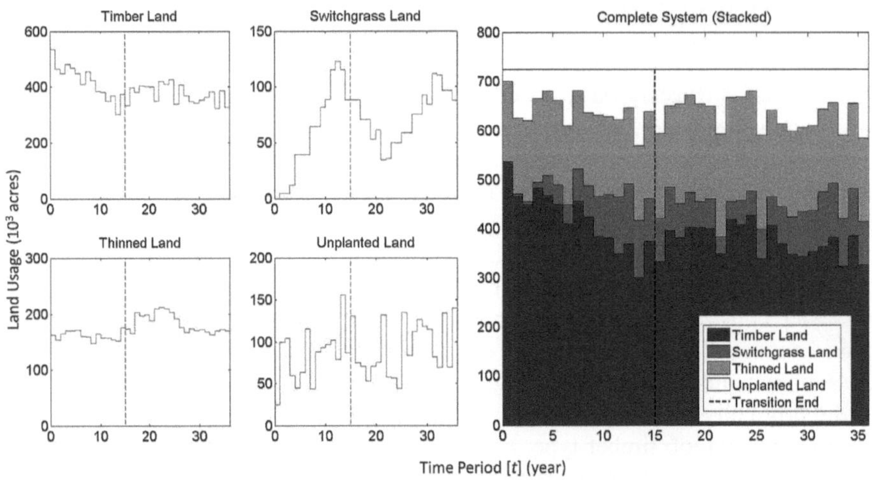

FIGURE 21.15

Fifteen-year transition timberlands schedule (normal land, fully constrained initial state).

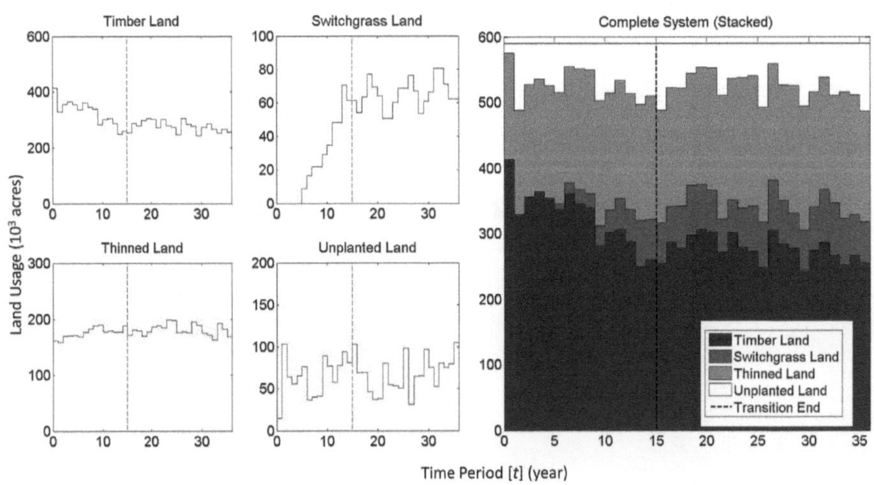

FIGURE 21.16

Fifteen-year transition timberlands schedule (moderate land, unconstrained initial state).

tolerance of the optimal solution. The problem size and complexity could yield a problem too difficult for CPLEX to find a solution. Alternative methods are needed to optimize these problems.

21.5 CONCLUSIONS

This chapter discussed how harvesting and planting schedules for a timberlands supply chain can be adapted to provide material to a new biofuel facility. An initial in-depth study was performed on the generation of established harvesting cycles and initial states. We implemented smoothing constraints to promote consistency throughout the time schedule as well as generate different initial states for the transition state analysis. The behavior of systems with less land capacity was also studied. The 10-year transition period study led us to conclude that the decision to invest in biofuels had a complex relationship with the initial state. We also determined that systems that previously did not invest in biofuels would change their decision with a longer transition period. The studies in this chapter provided an in-depth initial look at the forestry land management problem from a mathematical programming perspective. Additional studies and further development on the model could yield a useful decision-making tool for timberlands managers.

APPENDIX

SIMULATION PARAMETERS

Table A.2 is the amount of material obtained if young timber is harvested. The amount of material gained from thinning $(\widehat{p}_{kht-\tau})$ is half of these values due to row harvesting.

The costs to harvest timber are \$11.00/acre for timber, \$5.00/acre for switchgrass, and \$7.00/acre for thinned land. The cost for thinning is \$6.00/acre. Yearly prices for saw logs and fiber logs were taken from Bardon (2009). The years for this reference range from 1976 to 2009. To fulfill the ranges for the simulation, the inflation numbers from usinflationcalculator.com (2014) were taken from 1974 to 2014. The prices for missing years were generated by utilizing the inflation data. The first column of Table A.5 shows a set of prices over 21 zones that were generated by our industrial source. Using this as a base, the prices from Table A.5 were varied over the 21 harvesters. Prices to external buyers were 80% of the internal prices.

Table A.1 Mature Timber Biomass Densities (ton/acre) ($\rho_{kht-\tau}$)

Age (years) (t-τ)		18	19	20	21	22	23	24	25
Timber	Saw logs	64.97	69.65	79.94	83.20	92.18	95.66	99.13	102.60
	Fiber logs	105.15	112.41	114.06	119.76	119.72	125.20	130.68	136.16
	Residuals	56.71	60.69	64.67	67.65	70.63	73.62	76.60	79.59
Thinned	Saw logs	129.93	139.29	159.87	166.39	184.36	191.31	198.26	205.20
	Fiber logs	27.84	29.84	20.62	21.46	10.84	11.25	11.66	12.07
	Residuals	32.75	34.76	36.76	39.44	42.11	44.79	47.46	50.14

Table A.2 Young Timber Biomass Densities (ton/acre) ($\rho_{kht-\tau}$)

Age (years) (t-τ)	8	9	10	11	12
Residuals	121.48	159.92	198.89	247.75	294.95
Thinnings	32.09	42.79	53.48	64.18	74.88
Prunings	20.25	24.60	28.41	33.03	39.33

Table A.3 Transportation Costs (Year 2010 $'s) ($s_{mt}$ and $\widehat{s}_{iet}$)

Harvest Zone	Manufacturer								Biorefinery Location		
	1	2	3	4	5	6	7	8	Loc. 1	Loc. 2	Loc. 3
1	4.20	4.81	7.23	9.62	–	–	–	–	9.39	10.82	10.11
2	6.84	5.82	9.62	7.68	–	–	–	–	7.47	7.48	10.66
3	8.11	5.82	9.34	6.17	–	–	–	–	6.96	5.67	11.45
4	8.75	6.84	7.47	5.47	8.11	–	–	–	7.31	8.21	6.77
5	7.16	5.13	9.39	4.37	9.62	–	–	–	7.88	8.82	4.75
6	5.82	3.62	5.47	5.47	12.75	–	–	–	3.45	4.32	6.23
7	2.61	4.45	4.45	8.12	–	–	–	–	5.37	8.13	4.64
8	4.60	6.50	3.50	9.00	–	–	–	–	9.55	10.58	12.95
9	4.51	5.89	3.26	7.68	–	–	–	–	8.28	6.27	13.58
10	7.54	7.54	3.90	8.23	14.00	–	–	–	9.71	6.27	8.99
11	12.75	6.38	4.60	4.60	12.75	–	–	–	5.04	2.35	8.74
12	7.79	4.75	4.37	4.37	14.00	–	–	–	8.58	6.88	7.22
13	9.39	6.52	5.82	3.62	9.39	–	–	–	7.48	9.15	12.16
14	14.00	9.07	7.47	5.82	7.79	–	–	–	11.77	13.03	5.37
15	13.50	9.07	7.79	5.47	6.84	–	–	–	7.48	10.19	8.61
16	–	–	–	–	10.11	10.82	10.82	4.14	19.27	19.27	9.52
17	–	–	–	–	10.66	7.48	7.48	5.24	17.06	17.06	7.47
18	–	–	–	–	11.45	5.67	5.67	6.88	15.80	15.80	6.96
19	–	–	–	–	6.77	8.21	8.21	5.66	15.80	15.80	7.31
20	–	–	–	–	4.75	8.82	8.82	7.22	13.27	13.27	7.88
21	–	–	–	–	6.23	4.32	4.32	6.49	11.61	11.61	3.45

Table A.4 Conversion Factors (α_{kl})

Biomass Input	Product Lumber (ton/ton)	Gasoline (gallon/ton)	Pulp (ton/ton)	Wet/Dry tons
Saw logs	0.27	47.75	–	Wet
Residuals	–	92.21	0.20	Dry
Shavings	–	98.80	0.54	Dry
Fiber logs	–	47.75	0.27	Wet
Prunings	–	85.62	0.25	Dry
Grass	–	105.38	–	Dry
Thinnings	–	85.62	0.25	Dry

Table A.5 Timberland Costs

	Timber Prices (year 2010 \$/ton) (c_{ik})	Planting Costs ($\overline{c}_{iht}$) (year 2014 \$/acre)		Land Capacity (acre) (B_i)
		Timber	Switchgrass	
1	31.62	70.00	46.00	125.81
2	33.20	75.00	45.00	36.47
3	33.20	70.00	45.00	23.56
4	29.16	65.00	50.00	18.68
5	29.16	70.00	50.00	21.66
6	29.16	70.00	46.00	38.63
7	25.23	75.00	46.00	89.22
8	25.23	75.00	46.00	29.87
9	25.23	60.00	46.00	58.95
10	19.13	70.00	45.00	34.94
11	19.13	70.00	46.00	22.63
12	24.69	70.00	50.00	16.54
13	24.69	70.00	51.00	54.06
14	24.69	65.00	45.00	16.06
15	24.69	80.00	50.00	12.40
16	28.20	70.00	46.00	16.27
17	28.84	70.00	50.00	18.44
18	28.84	65.00	50.00	24.43
19	29.05	60.00	51.00	30.68
20	30.32	80.00	50.00	18.68
21	30.32	60.00	46.00	16.26

Table A.6 Biomass Costs (year 2006 \$/ton) (c_{ikt})

	Prunings	Grass	Thinnings	Residuals
1	2.17	3.79	2.71	18.08
2	2.27	3.98	2.84	18.99
3	2.27	3.98	2.84	18.99
4	2.00	3.50	2.50	16.68
5	2.00	3.50	2.50	16.68
6	2.00	3.50	2.50	16.68
7	1.73	3.03	2.16	14.43
8	1.73	3.03	2.16	14.43
9	1.73	3.03	2.16	14.43
10	1.31	2.29	1.64	10.94
11	1.31	2.29	1.64	10.94
12	1.69	2.96	2.11	14.12
13	1.69	2.96	2.11	14.12
14	1.69	2.96	2.11	14.12
15	1.69	2.96	2.11	14.12
16	1.93	3.38	2.42	16.13
17	1.98	3.46	2.47	16.49
18	1.98	3.46	2.47	16.49
19	1.99	3.48	2.49	16.61
20	2.08	3.64	2.60	17.34
21	2.08	3.64	2.60	17.34

Table A.7 Harvesting Costs (\$/ton) (f_{iht} and $\widehat{f}_{iht}$)

Timber	11
Switchgrass	5
Thinned land	7
Thinning costs	
Timber	6

Table A.8 Manufacturer Data

Fac.	Product	Capacity (10^3 tons) (G_m)	Demand (10^3 tons) (A_{ml})	Process Cost (year 2014 \$/ton) (n_{mt})	Transport Cost to Pulp Mill (year 2010 \$/ton)	Product Price (year 2010 \$/ton) (Pulp: year 2014 \$/ton) (p_{mlt})	Byproduct Price (k = Shavings) (year 2010 \$/ton) ($\widehat{c}_{mkt}$)
1	Lumber	530	128.79	29.00	14.09	588.00	55.00
2	Lumber	520	126.36	28.00	12.90	504.00	55.00
3	Lumber	460	111.78	32.00	8.41	550.00	55.00
4	Lumber	460	111.78	29.00	6.21	588.00	50.00
5	Lumber	510	123.93	29.00	5.61	450.00	55.00
6	Lumber	320	77.76	33.00	5.65	462.00	60.00
7	Lumber	300	72.90	31.00	5.88	490.00	50.00
8	Pulp	2000	486.00	147.00	–	804.35	–

Table A.9 Biorefinery Parameter Data (year 2014 $)

	Capacity ($\widehat{G}_u$) (tons)	Product Loc. 1	Price (/ton) Loc. 2	(p_{mlt}) Loc. 3	Processing Cost ($\widehat{n}_{ut}$) (/ton)	Capital Cost (millions) (β_{ue})		
						Loc. 1	Loc. 2	Loc. 3
Small	1750	479.00	489.00	490.00	73.20	1.234	1.198	1.284
Medium	2500	480.00	482.00	477.00	67.00	1.548	1.518	1.582
Large	3500	469.00	475.00	462.00	57.77	1.885	1.848	1.923

Table A.10 Other Data

Planting Lower Bound (*R*)	25%
Harvesting Lower Bound (*H*)	10%
Log to Shavings Conversion ($\gamma_{kk'}$)	0.3

ADDITIONAL FIGURES

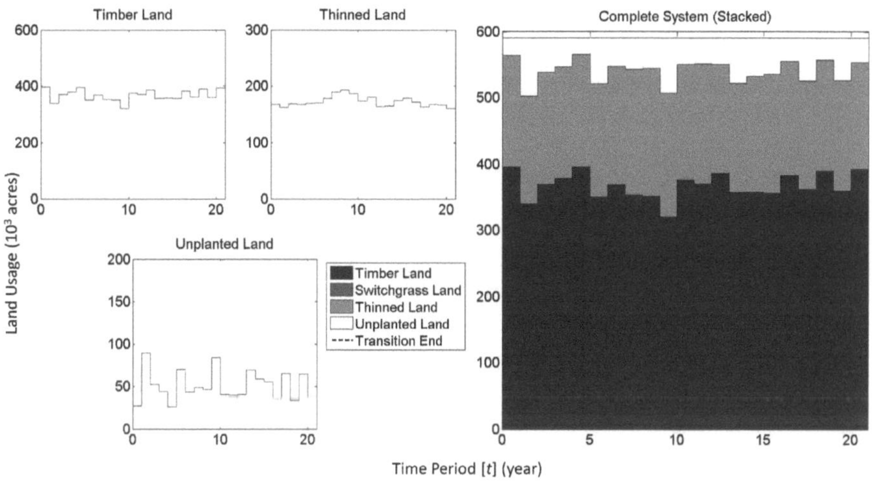

FIGURE A.1.

Established timberlands schedule (moderate land, harvester constrained).

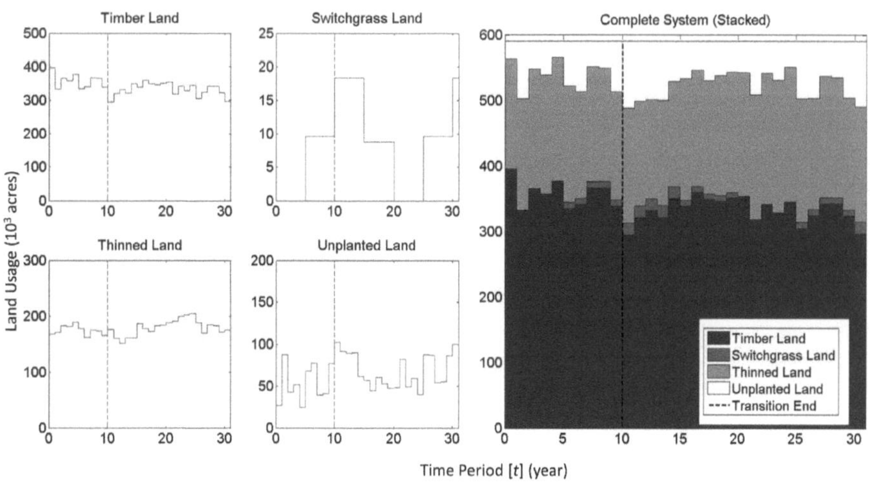

FIGURE A.2.

Ten-year Transition timberlands schedule (moderate land, harvester constrained initial state).

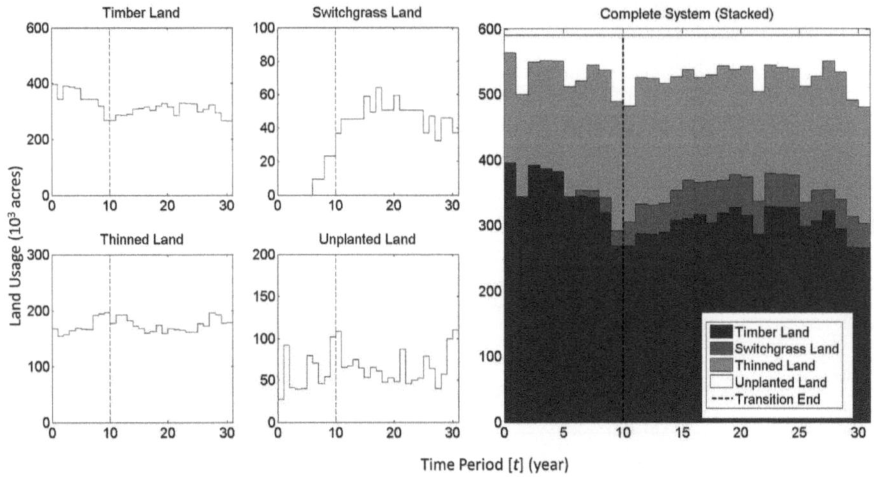

FIGURE A.3.

Ten-year transition timberlands schedule (moderate land, harvester constrained initial state, enforced biorefinery).

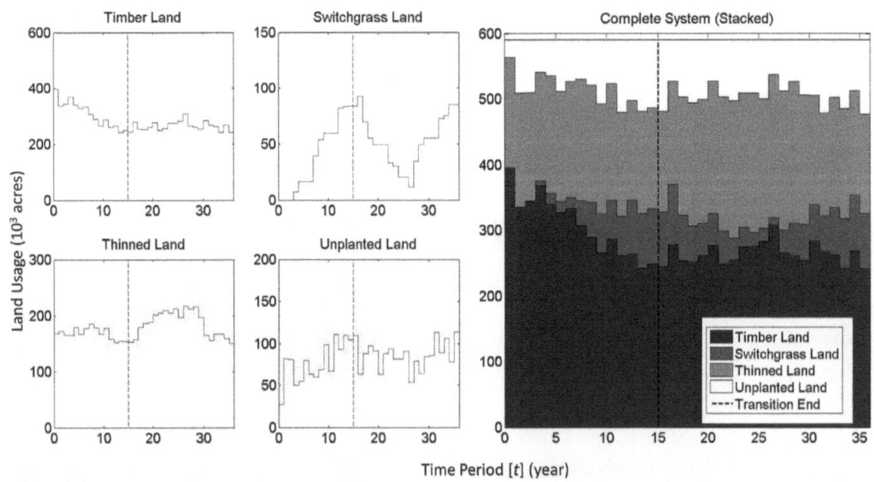

FIGURE A.4.

Fifteen-year transition timberlands schedule (moderate land, harvester constrained initial state).

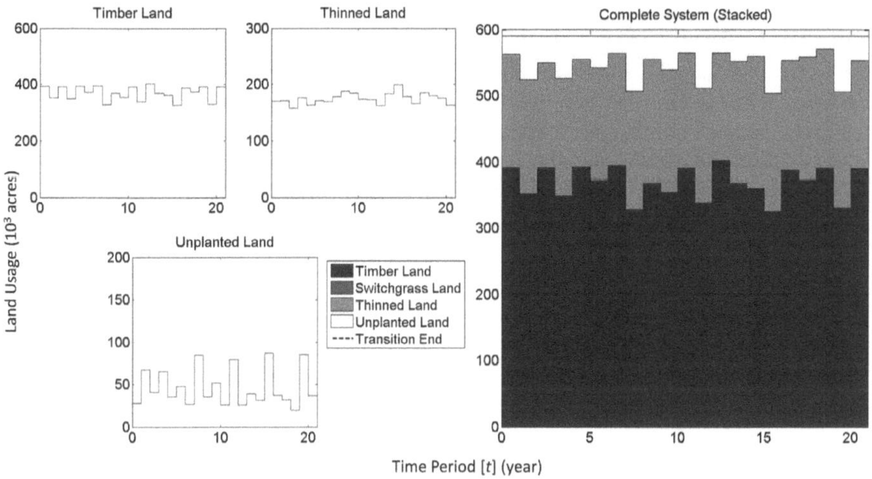

FIGURE A.5.

Established timberlands schedule (moderate land, fully constrained).

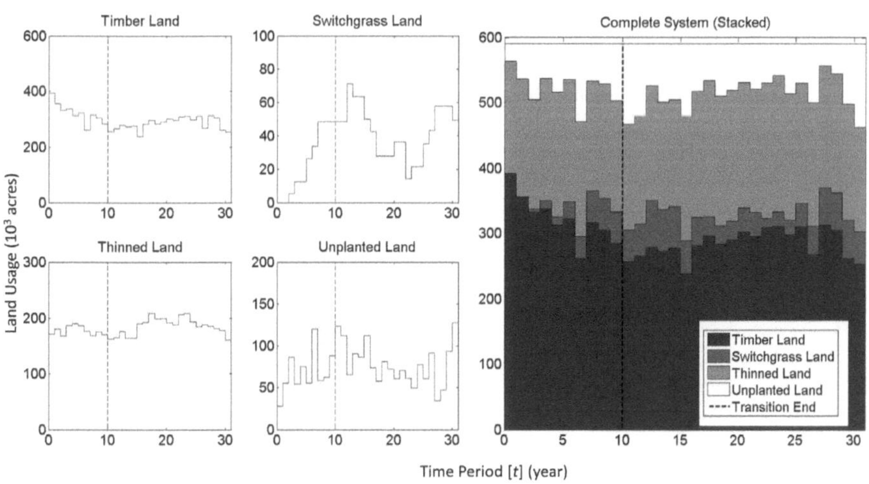

FIGURE A.6.

Ten-year transition timberlands schedule (moderate land, fully constrained initial state).

ACKNOWLEDGMENTS

Partial financial support of the NSF Grant GOALI program, CBET 0933392, is gratefully acknowledged. The research was also supported by the Advanced Biomass R&D Center (ABC) of Global Frontier Project funded by the Ministry of Education, Science, and Technology (ABC-2011-0031354). The authors gratefully acknowledge the assistance of Ludwig Furtner and Todd Wisecarver of Weyerhaeuser NR.

REFERENCES

Bardon, R., 2009. North Carolina State University Extension Forestry. http://www.ces.ncsu.edu/nreos/forest/resources/price_report.php.

Castro, P., 2003. Optimal periodic scheduling of batch plants using RTN-based discrete and continuous-time formulations: a case study approach. Ind. Eng. Chem. Res. 3346–3360.

Dickens, D., Moorhead, D., Dangerfield, C., Chapman, S., 2005. Thinning Pine Plantations. http://www.gfc.state.ga.us/resources/publications/PineThinning.pdf.

Georgia Institute of Technology, 2014. Georgia Tech PACE. http://pace.gatech.edu/joepacegatechedu.

Merkuryeva, G., Merkuryev, Y., Vanmaele, H., May 2010. Simulation-based planning and optimization in multi-echelon supply chains. Simulation 87 (8), 680–695.

Pinto, J., Grossmann, I., 1994. Optimal cyclic scheduling of multistage continuous multiproduct plants. Comput. Chem. Eng. 18 (9).

Pundoor, G., Chen, Z.-L., May 2009. Joint cyclic production and delivery scheduling in a two-stage supply chain. Int. J. Prod. Econ. 119 (1), 55–74.

Sahinidis, N., Grossmann, I., 1991. MINLP model for cyclic multiproduct scheduling on continuous parallel lines. Comput. Chem. Eng. 15 (2).

usinflationcalculator.com, 2014. Historical Inflation Rates, 1914–2014. http://www.usinflationcalculator.com/inflation/historical-inflation-rates/.

Wu, D., Ierapetritou, M., Oct. 2004. Cyclic short-term scheduling of multiproduct batch plants using continuous-time representation. Comput. Chem. Eng. 28 (11), 2271–2286.

You, F., Castro, P.M., Grossmann, I.E., Nov. 2009. Dinkelbach's algorithm as an efficient method to solve a class of MINLP models for large-scale cyclic scheduling problems. Comput. Chem. Eng. 33 (11), 1879–1889.

You, F., Grossmann, I.E., Dec. 2008. Design of responsive supply chains under demand uncertainty. Comput. Chem. Eng. 32 (12), 3090–3111.

CHAPTER

22 Bioethanol Supply Chain Design and Optimization: Some Achievements and Future Challenges for the Development of Sustainable Biorefineries

Sara Giarola*, Fabrizio Bezzo[§,1]

*CPSE, Centre for Process Systems Engineering, Imperial College London, London, UK

[§]CAPE-Lab, Computer-Aided Process Engineering Laboratory, Department of Industrial Engineering, University of Padova, Padova, Italy

[1]Corresponding author: E-mail: fabrizio.bezzo@unipd.it

22.1 INTRODUCTION

The debate on alternative energy sources has been central to policy, economy, and technology research over the past decade. Biofuels have played a key role in this as they have seemed to many the only feasible approach to substitute (at least partially) oil-based fuels in the transport sector. In fact, this has indeed led to the establishment of a large-scale production system, particularly in some countries (notably, the United States; Brazil, where in fact a biofuel industry had been running for quite some time; as well as several European countries). The initial enthusiasm has, however, given place to a more skeptical view, once the multifaceted reality in biofuel production became clear. Production processes were based on first-generation technologies, which rely on "simple" crops, i.e., sugar-, starch-, or oil-based crops. This soon raised a "food-for-fuel" issue with several evident ethical and social implications. In addition, the very economic and environmental sustainabilities of these technologies were questioned as well as their actual capacity to be a substitute for fossil fuels in a significant way (Farrell et al., 2006; Searchinger et al., 2008). One immediate consequence was to increase efforts in searching for alternative production processes to the (alternative themselves) first-generation ones. Second-generation technologies based on lignocellulosic materials were seen as a possible breakthrough to transition the transport fuel sector toward a new oil-independent era. Obviously, the problem was not a trivial one and new issues arose. Several

Computer Aided Chemical Engineering, Volume 36. ISSN 1570-7946. http://dx.doi.org/10.1016/B978-0-444-63472-6.00022-7

researchers and policy makers suggested that a solution should be sought even further, either by looking to microalgae (the so-called third-generation technologies) or moving away from biomass-based fuels altogether and focusing on electric vehicles.

The scope here is not to discuss the general question of "alternative fuels." However, it should be clear by now that the production of biofuels has quickly disclosed a variety of issues requiring both a multiscale and multidisciplinary approach. The multiscale nature of the problem is quickly explained. The development of biofuels ranges from the discovery and development of new technologies (involving nano- or microscale chemical reaction kinetics and/or enzymatic/biological transformations) up to the design and optimization of novel processes and entire biorefineries. Furthermore, the design and assessment of the overall production supply chains (SCs) (in terms of agricultural practice and transport choices) and, eventually, the interactions between different technologies, markets, and policies (at both regional and global scales) need to be tackled, too. Most of the above questions may be analyzed according to a strictly technical (engineering or scientific) standpoint, but often require a complementary (and sometimes even more important) evaluation in terms of economic profitability, financial risk, and environmental impact.

In this chapter, ethanol (or bioethanol) production systems will be considered as a case study. As a matter of fact, bioethanol has been assuming a leading position among biofuels (Carriquiry et al., 2011) and represents the most mature fuel both in terms of first- and second-generation technologies (although the latter are just entering proper industrial deployment). Process systems engineering approaches have been adopted to analyze and tackle several issues discussed above. We will focus on the context of the SC analysis and optimization, which is quite emblematic to expose the complexity and multifaceted nature of biofuel systems. Mathematical programming offers a general modeling framework to address such high-level planning decisions as those involved in the design of future energy systems. This modeling approach also allows the required flexibility to explore diverse alternatives along the entire production network to yield optimal trade-offs between conflicting targets and to guarantee an efficient use of the limited resources (Grossmann and Guillén-Gosàlbez, 2010).

In fact, in recent years a number of contributions have appeared where mathematical programming techniques have been exploited to analyze and optimize several biofuel SCs (see also the review by Yue et al. (2014)). Most of the first contributions focused on the definition of optimal configurations of ethanol SCs according to economic indices, taking into account both first- and second-generation technologies. Dunnett et al. (2008) developed a multiperiod mixed-integer linear programming (MILP) framework to determine the cost-optimal configuration for a lignocellulosic bioethanol SC involving a wide range of technological, system-scale, biomass supply and ethanol demand distribution scenarios in the UK context. Zamboni et al. (2009b) and successively Akgul et al. (2010) presented a spatially explicit MILP model for the integrated management of the key issues affecting corn-based ethanol SCs such as agricultural practice, biomass supplier allocation, production site location and capacity assignment, logistics distribution, and transport system optimization in Italy. Eksioglu et al. (2009) proposed an MILP model for the optimal design of a biomass-to-biorefinery SC and management of logistics by minimizing the total annual cost. Huang et al. (2010) developed an MILP model for the optimal multistage optimization of ethanol SCs based on the minimization of the total system cost. Parker et al. (2010) addressed the problem through an MILP framework involving the biorefinery siting optimization of a multifeedstock and multitechnology biomass-based supply network. Čuček et al. (2010) proposed an MILP model for the economic optimization of a bioenergy network in a regional context; the generation of different types of energy products, heat, electricity, biofuels, and food was taken into account. Akgul et al. (2012a) implemented an MILP cost optimization framework, where the biomass production potential for marginal land was also assessed. Hybrid first- and second-generation ethanol production technologies were analyzed.

Other contributions introduced the issue of uncertainty within the economic optimization framework. Dal-Mas et al. (2011) proposed a multiperiod MILP framework model for the strategic design and investment capacity planning of a first-generation ethanol SC dealing with feedstock and product price uncertainty. The model aimed at maximizing the expected net present value (NPV) and minimizing the financial risk separately. Kim et al. (2011) presented a two-stage mixed integer stochastic programming model for the optimal design of a biofuel SC network including the presence of uncertainties in biomass supply amounts, biofuel market demands, biomass and biofuel market prices, and processing technologies. Kostin et al. (2012a) implemented a multiscenario MILP model for the optimal design and planning of integrated ethanol—sugar SCs considering uncertainty in the product demand. The model sought to optimize the SC economic performance by taking into account different financial risk measures. Marvin et al. (2012) proposed an MILP model to determine the optimal biomass configuration for a biomass-to-ethanol SC situated in the Midwest of the United States. They also performed sensitivity analyses of the result to price uncertainty. Gebreslassie et al. (2012) developed a multiperiod stochastic MILP model to address the design of biorefineries including uncertainty in feedstock supply and biofuel demand. Mazzetto et al. (2013) analyzed a first-generation ethanol SC in terms of its financial performance by incorporating some methodologies for commodity price dynamics forecasting.

Mathematical programming techniques may be coupled to the concept of Green Supply Chain Management (GrSCM). In particular, multiobjective mathematical programming offers the most powerful tool, enabling the exploration of balanced trade-offs between conflicting objectives (Guillén-Gonsálbez and Grossmann, 2009), as GrSCM goals likely are. The GrSCM of a corn-based bioethanol network was first addressed by Zamboni et al. (2009a) through a spatially explicit multiobjective MILP modeling framework minimizing both SC operating costs and greenhouse (GHG) emissions. Giarola et al. (2011) extended the approach to a multiperiod multiobjective MILP model (accounting for NPV and GHG emission) of hybrid first- and second-generation ethanol SCs taking into account corn and corn stover. You and Wang (2011) proposed a spatially explicit multiperiod multiobjective MILP model of a biomass-to-liquid SC. The economic objective was measured by the total annualized cost, and the environmental performance was calculated in terms of GHG emissions. Mele et al. (2011) provided a spatially explicit bicriterion multiobjective MILP framework where environmental and financial criteria were both addressed in ethanol production from sugarcane. They considered the Eco-indicator 99 and the global warming potential (GWP) as the environmental impact metrics to be minimized. Zamboni et al. (2011) developed a whole-systems multiobjective optimization of a first-generation ethanol SC showing how efficient crop management might contribute significantly to mitigate GHG emissions. Giarola et al. (2012) proposed a different approach by incorporating environmental effect within an MILP model where carbon trading effects were included in the economic optimization. The effect of uncertainty on carbon credit and feedstock prices was also taken into account. Akgul et al. (2012b) presented a multiobjective MILP approach for the optimization of hybrid first- and second-generation SCs in the United Kingdom. You et al. (2012) developed a multiobjective MILP model to address the optimal design and planning of cellulosic ethanol SCs under economic, environmental, and social objectives. The model took into account the major characteristics of cellulosic ethanol SCs, including supply seasonality and geographical diversity. Kostin et al. (2012b) presented a multiobjective approach to the design of bioethanol SCs considering several life cycle assessment (LCA) impacts. To overcome the numerical difficulties of dealing with several objective functions, they investigated the application of a rigorous MILP-based dimensionality reduction method that minimized the error of omitting objectives. Several environmental metrics were taken into account by Čuček et al. (2012): different footprints were assessed against the environmental performance within an MILP framework representing a regional biomass SC for energy (and fuel) production. Bernardi et al. (2012, 2013) proposed a general multiobjective MILP framework for the design of first- and second-generation ethanol SCs where the environmental impact was simultaneously assessed in terms of carbon and water footprints (WFs), through a specific analysis in the Italian context. Giarola et al. (2013) discussed a risk management approach where the economic and environmental behavior of ethanol SCs was optimized based on an MILP framework where multiple decision criteria were simultaneously considered in an uncertain market scenario. Osmani and Zhang (2014) proposed a two-stage stochastic optimization model to maximize the expected profit of a lignocellulose-based bioethanol SC simultaneously minimizing the system carbon emissions. They introduced the effect of uncertainties in supply, demand, and prices. They showed how uncertainty might have a significant impact on the capacity of biorefineries to process the biomass and degrade its financial and environmental performance.

In fact, it can be seen that some of the key challenges that have been addressed in the most recent literature are concerned with two main issues. The first one deals with the fact that fuel production systems (or most generally energy systems), when asked to replace an existing infrastructure, should respond to several society demands. These not only have to regard the capabilities of substituting oil and guaranteeing an adequate production capacity, but also need to meet cost/profitability targets as well as other acceptability requirements. In other words, they should be "better" than the fuels they are supposed to substitute, not only by reducing the impact on the environment but also by supporting job creation and improving the social perception of new technologies.

The second issue is about uncertainty. In fact, this may be considered "the issue" in any novel technology. There is uncertainty concerning prices of feedstock and product (How is the supply–demand curve going to behave?); there is uncertainty related to the market of reference (e.g., Will customers accept the new product? Will governments provide incentives for biofuels?) and other markets (e.g., What is the interaction with the electricity market? Are oil refineries allies or competitors?); there is uncertainty concerning environmental legislation (e.g., What kind of environmental targets will be imposed?); there is uncertainty in terms of the technology itself (e.g., What second- or third-generation technologies will prove successful? What are large-scale yields and costs? What are real emissions?); and so on. Analyzing the above is a task of formidable difficulty and yet it is necessary to provide credible scenarios to investors and policy makers.

This chapter will focus on these two points, mainly discussing the work that has been carried out by the authors themselves. First we will describe how the biofuel production system is modeled. Then we will focus on GrSCM issues by taking as reference three of our works discussing the complex interaction between economic and environmental targets in the design of bioethanol SCs. Afterward, the effect of uncertainty on the design of ethanol SCs will be taken into account through two case studies. Note that for the sake of brevity and simplicity, equations and other formal aspects will not be presented here, but they can be found in the original works. At the end we will draw some remarks to comment on what has been achieved, what has just started to be tackled, and what is still unexplored. The perspective clearly will be that of process systems engineers, but we will also try to provide a broader view.

22.2 BIOFUEL PRODUCTION SYSTEM DESIGN: KEY MODELING ASSUMPTIONS

The design of biofuel systems relies on the simultaneous assessment of selected performance indicators for alternative topologies of production networks (or SCs), to steer policy and investments for the promotion of a more sustainable energy sector. The analysis here presented served a dual purpose. The study addressed the profitability of biofuel (i.e., bioethanol) production systems by means of SC analysis techniques. Bioenergy infrastructures were also evaluated from an environmental perspective, i.e., accounting for impacts on global warming and on water resources. Mathematical programming (e.g., MILP) techniques were used to model biofuel systems and support the decision-making process in the energy sector.

SC analysis techniques have traditionally dealt with the integrated management of the activities of the entities (i.e., suppliers, manufacturers, distributors, and retailers) involved in product/service deliveries to consumers (Beamon, 1998). Such techniques also have a remarkable role in supporting national and international policies as well as investments in new (bio)energy infrastructures. The management of biomass flows in these systems involves more challenges compared to the traditional SCs, as it mainly concerned the transportation of low-energy density materials with wide and sparse availability (Mafakheri and Nasiri, 2014). Biofuel system performance metrics here referred to the energy provided by ethanol in a combustion engine. Accordingly, the boundaries of the analysis (or SC nodes) included biomass cultivation, biomass transport, fuel production, and fuel distribution (Figure 22.1).

The environmental assessment of biofuel systems was performed according to the LCA methodology limited to a Well-To-Tank approach (CONCAWE, 2007), thus incorporating feedstock production, processing, fuel production, and delivery. The analysis focused on the estimation of the SC environmental impact on global warming and water resource use. The former was evaluated in GHG emissions produced by the system. Relevant burdens to biofuel production networks (CO_2, CH_4, N_2O) were selected and grouped together in a single indicator, the carbon dioxide equivalent emissions (CO_2-eq), according to the concept of 100-year GWPs (ISO, 2006 series). The impact on water resources was evaluated according to the WF methodology as laid down by Hoekstra et al. (2011) and extended to an LCA approach.

MILP models were developed to address the strategic planning and design of biofuel systems and were solved optimizing design and strategic variables using the CPLEX solver in general algebraic modeling system (GAMS) (Rosenthal, 2006). These techniques are powerful when tackling future infrastructure development (Shah, 2005) as the SC network nodes are simultaneously optimized in the selection among multiple alternatives.

The European Union (EU) directives and targets were taken as references in the analyses discussed in this chapter. The European Commission set minimum levels of biofuels equal to 5.75% in 2010 and to 10% (for renewables) on energy content by

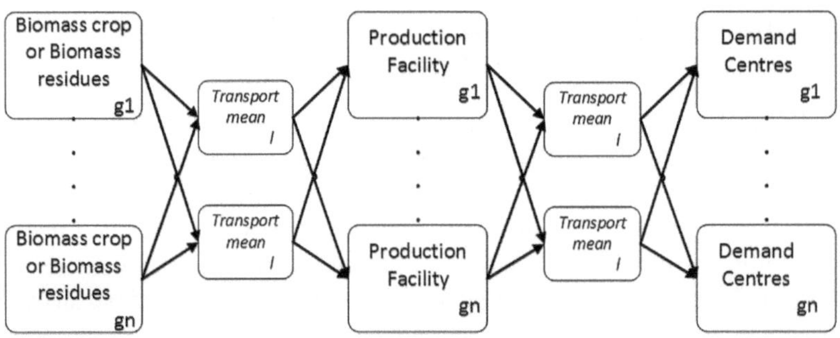

FIGURE 22.1

Biofuels supply chain network.

2020. Also, biofuels were required to meet sustainability criteria. As such, a minimum of 35% (percentage that should increase up to 50% in 2017) of GHG emission savings compared to fossil fuels was targeted (EC, 2009). Here, Northern Italy was used as a reference geographical context to study potentials and drawbacks of biofuel system deployment, in order to suggest political and investment strategies promoting more sustainable energy infrastructures. The territory of Northern Italy was discretized into 59 cells of equal size, also considering biomass importation from an additional hypothetical cell, representing foreign suppliers. Then, the model was populated through the description of the following data sets:

- Each cell was characterized in terms of fraction of arable land and type and amount of biomass available
- The economic and environmental characterization of the biomass pretreatment step was associated to the indirect water consumption and GHG emissions due to the energy consumed in biomass drying and storage.
- A geographically dependent description of economic and environmental burdens related to farming practices was performed (i.e., biomass unitary production costs, GHG emission factors, as well as water needs—including direct and indirect effects due to farm inputs and machinery).
- A grid-dependent description of the transport logistics was performed. Each transport mode available (i.e., trucks, railways, barges, ships, and transships) was characterized in terms of capacities, infrastructure network quality, environmental burdens (accounting for indirect GHG emissions and water needs due to fuel use), and costs.
- A set of technologies of first- and second-generation ethanol (which will be discussed later on) was characterized in terms of technical (yields) and economic parameters (capital and operating costs) as a function of biomass type, production technology, and plant scale. Environmental burdens were associated to the biomass type and production technology, e.g., direct impacts on global warming and water resources as well as indirect GHG emissions and water use due to the production of the chemical reagents, enzymes, and energy (e.g., natural gas, electricity) necessary to run the facilities.
- Bioethanol demand at the blending terminals located in the region was set to vary according to the European mandates.

22.3 GrSCM IN THE STRATEGIC DESIGN OF BIOFUEL SYSTEMS

GrSCM represents a field of SC management where a production system design is realized by coupling environmental and profitability targets. The integration of mathematical programming techniques with environmental principles results in powerful multiobjective mathematical programming tools allowing for the simultaneous optimization of multiple criteria and the exploration of trade-offs between conflicting targets (Srivastava, 2007). Here, we will provide three case studies to show how environmental and economic objectives interact and compete in the design and optimization of bioethanol SCs. All discussion is based on work that we have published (referred to in the text). First, we will discuss the simultaneous environmental (GHG) and economic optimization (NPV) of first- and second-generation technologies for bioethanol production (Giarola et al., 2011). Then, we will observe how the optimization of the cultivation practice, particularly in terms of fertilizer usage, may change the environmental impact radically (Zamboni et al., 2011). Finally, we will analyze how the incorporation of an additional environmental index, namely, the WF, may affect the design of a biofuel SC (Bernardi et al., 2012, 2013).

22.3.1 MULTIOBJECTIVE OPTIMIZATION IN FIRST- AND SECOND-GENERATION TECHNOLOGIES

In this section, the strategic design and planning of a general biofuel SC over a 15-year horizon is proposed (Giarola et al., 2011). In fact, the development of bioethanol SCs requires the incorporation of environmental impact mitigation criteria along with economic aspects in order to accomplish the EU guidelines in terms of emission reduction. The design process was conceived as an optimization problem in which the production system was required to comply with a dual objective: (1) maximization of the financial performance of the business (NPV) and (2) minimization of the impact on global warming (in terms of overall GHG emissions). The modeling framework dealt with agricultural practice, biomass supplier allocation, production site location and capacity assignment, logistics distribution, and transport system optimization.

First- (from corn grain) and second- (from corn stover) generation technologies were included to convert biomass into ethanol. Ten possible solutions (see Table 22.1) were investigated. Among the available technical alternatives, three main processing designs were assessed: (1) the dry-grind process (DGP), i.e., the

Table 22.1 Ethanol Technologies: Identification and Products Description of Each Technology Labeled as *k*

	Process			Input		Output		
k	DGP	LCEP	Hybrid	Grain	Stover	Ethanol	Power	DDGS
1	X			X		X		X
2	X			X		X	X	
3	X			X	X	X	X	X
4		X			X	X	X	
5, 6, 7			X	X	X	X	X	X
8, 9, 10			X	X	X	X	X	

standard corn-based ethanol process (Franceschin et al., 2008); (2) the lignocellulose-to-ethanol process (LCEP), where only stover was converted to ethanol (USDOE, 2002); (3) the integrated grain-stover process (Hybrid), where both corn grain and stover were processed to obtain ethanol.

With regard to DGP, three instances were analyzed depending on the method used to provide energy to the plant: by the grid ($k = 1$), by using the main by-product, i.e., distiller's dried grain with solubles (DDGS) ($k = 2$), or by using stover ($k = 3$) to fuel a combined heat and power (CHP) generation system. In the LCEP technology ($k = 4$), heat and power were provided by thermal valorization of lignin. As regards Hybrid, an Aspen Plus® model was set up to integrate corn- and stover-to-ethanol processes in a configuration where the recovery and utilities sections were shared. Three instances were modeled representing specific (1:1, 1:2, 1:3) corn-to-stover ratios of the biomass fed to the plant. Two alternative DDGS end-uses were considered, i.e., either DDGS was sold to the animal feed market ($k = 5, 6, 7$, where the three indexes stand for the three decreasing corn-to-stover ratios) or converted to energy along with lignin ($k = 8, 9, 10$).

To evaluate the impact on global warming of SC operations, a substitution method was used to account for the end-use of by-products, DDGS, and electricity (ISO, 2006). Credits were modeled as emission savings associated to the replacement of alternative goods by coproducts. Emissions from DDGS end-use were evaluated considering it as a substitute either for soy in the cattle feed market or for fuel in a CHP station supplying energy to ethanol plants. In the latter case, any excess power was sold to the grid.

Table 22.2 collects the economic and environmental performance of the system configurations obtained as a result of the multiobjective optimization (Pareto frontier). As a representative case, the planning strategies involved in configuration C (Table 22.2) are shown in Figure 22.2 and discussed below.

This SC configuration involved the installation of a hybrid production system in which both first ($k = 2$) and second-generation ($k = 4$) technologies were used.

Table 22.2 Pareto Set of Solutions Resulting from the Biobjective Problem

Configuration	Technologies	NPV (€/GJ)	GHG (kg CO_2-eq/GJ)	Emission Reduction[a] (%)
A: economic optimum	1	1.18	76.9	10
B	2	0.88	36.0	58
C	2, 7	0.43	24.9	71
D	4, 7	−0.27	13.7	84
E: environmental optimum	4	−4.26	6	97

[a]*The GHG emission factor for gasoline was assumed to be equal to 85.8 kg CO_2-eq/GJ, according to Brown et al. (2005).*

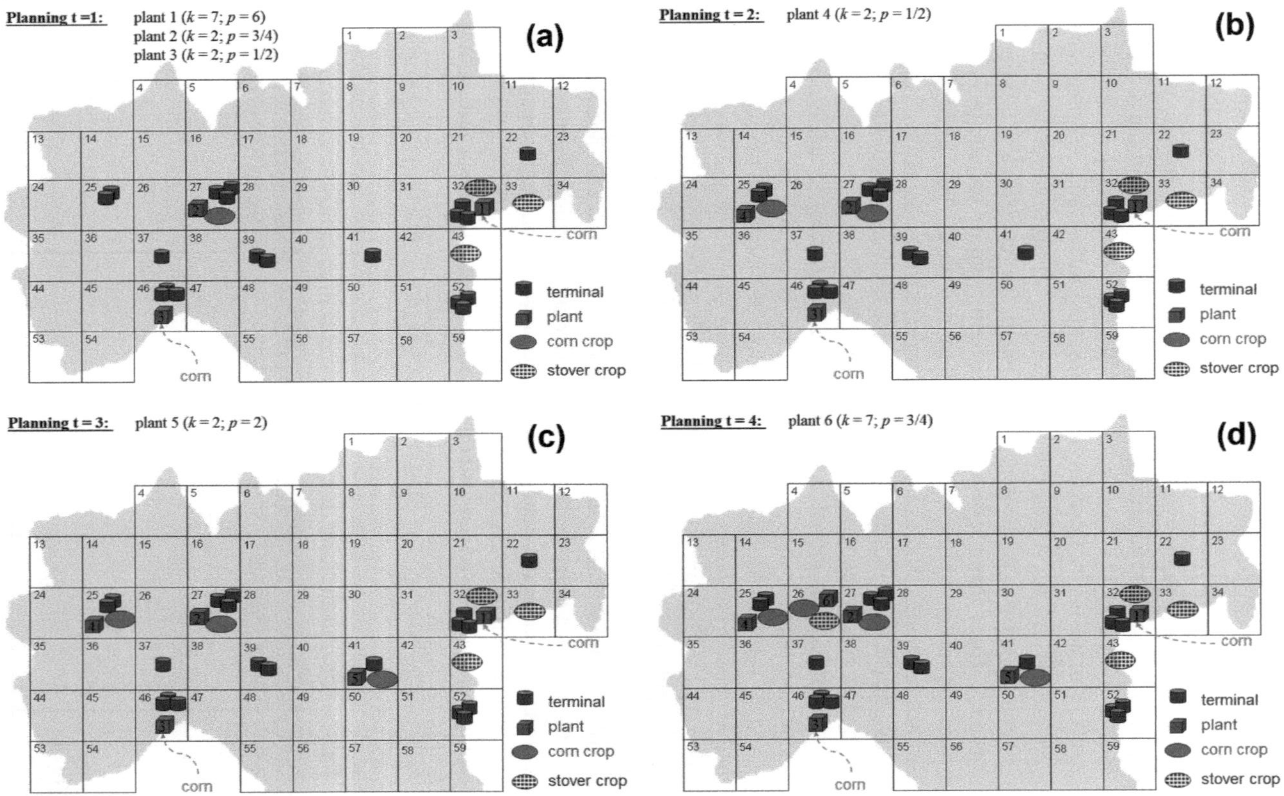

FIGURE 22.2

(a–d) Design and planning strategy along the entire time horizon: case C. t represents the time periods, k the technologies, and p the size of the production facilities, which ranges between 96 and 276 kt/year and includes six intervals.

After Giarola et al. (2011).

- At $t = 1$ (Figure 22.2(a)), three production plants were installed, one exploiting a hybrid technology, while the remaining two, a first-generation-based process. The hybrid generation facility ($k = 7$) was assigned the largest size, about 280 kt/year of ethanol. Residual biomass was available locally, whereas corn was shipped from abroad. A second medium-sized DGP plant ($k = 2$) of about 180 kt/year of ethanol availed of imported corn. A third plant was assigned a small production capacity (about 100 kt/year). A standard DGP technology ($k = 2$) was installed, thus availing of the large corn stocks produced in the neighborhoods.
- At $t = 2$ and $t = 3$ (Figure 22.2(b) and (c)), standard DGP plants of small size ($k = 2$) were settled. Having a relatively small capacity (about 100 kt/year), they were supplied with local biomass (corn).
- With concerns to the planning strategy at $t = 4$ (Figure 22.2(d)), a medium-sized second-generation plant ($k = 7$) of about 190 kt/year was installed. The hybrid production plant would avail of locally produced biomass (both corn and stover).

From the financial perspective, configuration C performed an NPV of 134 M€. The overall investment was paid back in about 10 years. The return on investment (ROI) index was about 16%. This represents quite a lower value compared to the minimum acceptable threshold, usually set at 24% according to Peters et al. (2003).

From a strategic policy standpoint, the modeling outcomes suggested that the EU targets on global warming mitigation could be fulfilled by ethanol fuel systems only through the exploitation of cellulosic biomass. The adoption of integrated corn grain−stover technologies would allow for a significant reduction of GHG emissions (abiding by the EU emission limits), while retaining a good profitability.

22.3.2 WHOLE-SYSTEM OPTIMIZATION: MANAGING NITROGEN BALANCE FOR GLOBAL WARMING MITIGATION

First-generation ethanol represents the only mature technology developed at the industrial scale (EUBIA, 2013). However, its actual contribution to GHG emission reduction compared to gasoline is still debated. This depends not only on the selection of system boundaries but also on variability of local climate, properties of soil, cropping management, and cultivation practice (Kim and Dale, 2008). Indeed, the application of nitrogen in biomass cultivation stage is the primary source of polluting by-products such as nitrous oxide (Robertson et al., 2008).

In this section, MILP programming techniques are applied to optimize nitrogen fertilizer usage in first-generation ethanol SCs ensuring best whole-system performance in terms of both profit and GHG emissions (Zamboni et al., 2011). The problem was studied on a land of fixed surface (40,000 ha), fully cultivated with biomass (wheat) to be supplied to a unique production plant operating with a DGP technology (Franceschin et al., 2008). Referring to Table 22.1, two different technological options for DDGS end-use were addressed: (1) soy meal substitute to be sold in the cattle fodder market ($k = 1$) and (2) fuel fed to a CHP station ($k = 2$). Nitrogen dosage increase was modeled to produce:

- a direct increase in wheat yields;
- a direct increase in grain protein yields (indirectly, improving DDGS yields to the detriment of the ethanol ones);
- a direct increase in costs related to fertilizers;
- an indirect increase in the total global warming impact due to greater GHG emissions coming from both fertilizer production and N_2O release from soil;
- an indirect increase in emission credits coming from product displacements (i.e., DDGS).

Discretized primary response curves, obtained from literature data fitting (Smith et al., 2006), were used to describe the dependence of grain yield, grain protein content, and alcohol yield on the nitrogen dosage used for cropping. Secondary response curves were used to model the protein content in the DDGS, as this affects its market price on a different level depending on the end-use option. Along with the DDGS market price, relevant economic parameters were characterized according to their dependence on nitrogen dosage (i.e., unit production cost for biomass cropping, unit production cost for ethanol, unit cost reduction due to DDGS end-use). The environmental description of the system in terms of impact on global warming was also performed accounting for the dependence on the nitrogen dosage (i.e., global emission factor for biomass cultivation, global emission factor for ethanol production, and emission credits from by-product end-use).

Figure 22.3 reports the Pareto solutions of the optimization problem and the trade-off between the NPV and the total impact (i.e., GHG emissions, kg CO_2-eq/ha/y) of the system. Sensitivity analyses were performed on electricity and DDGS market prices as detailed below.

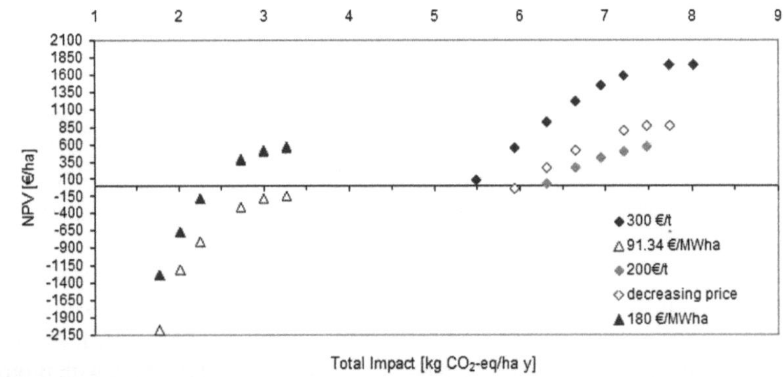

FIGURE 22.3

Pareto set of solutions: NPV versus total impact.

After Zamboni et al. (2011).

Considering an electricity price without incentives, equal to 91.34 €/MWh, best economic performance could be obtained when DDGS was sold as a soy meal substitute. This strategy would encourage a large dosage of mineral fertilizers to increase the yields of subproducts and maximize the grain protein content.

According to the most likely short-term Italian market perspectives, the DDGS price was set equal to 300 €/t (◆ in Figure 22.3). Under these assumptions, the system enabled good profitability across a broad range of cultivation management conditions.
Assuming a medium- to long-term perspective, DDGS selling price would reasonably decrease. Two trends were considered, i.e., a depreciation between 300 and 100 €/t (◇ in Figure 22.3) and a constant selling price of 200 €/t (◆ in Figure 22.3). Both scenarios showed a reasonable profitability, although the second was less sustainable. In the former case, an NPV of about 35 M€ and payback times of 5 years were obtained. This scenario would correspond to a high DDGS productivity due to heavy nitrogen dosages (e.g., of about 225 kg N/ha). In general, large nitrogen dosage was necessary to boost the DDGS productivity dramatically affecting the GHG emissions of the entire production chain. As such, the bioethanol SC GHG emissions were shown to exceed the gasoline emissions by about 14% when the nitrogen application was maximized (250 kg N/ha).

On the other hand, if DDGS was exploited to fuel a power station, significant GHG emissions reductions (ranging between 54% and 63% compared to gasoline) were achieved. This was due to both the emission credits assigned to energy displacement and the lower nitrogen applications on field. This technological solution did not appear to be economically viable, though.

The system profit was confined within the negative region, if subsidies on the electricity price were not included, even when maximizing the DDGS production (and hence the nitrogen dosage) (Δ in Figure 22.3).
Taking into account the current subsidy regime in Italy, the profitability of the business may be increased to attain positive values (▲ in Figure 22.3), while still retaining a GHG emission reduction of about 54%.

22.3.3 INVESTIGATING OPPORTUNITIES FOR CARBON AND WF REDUCTION AT STRATEGIC LEVEL OF BIOENERGY INFRASTRUCTURE DESIGN

The pressure on water resources is forecasted to increase for biofuel production growth and its consequences require appropriate appraisal (Dominguez-Faus et al., 2009). These aspects would need to be included at the design level of future bioenergy systems. The WF index estimates the freshwater consumed or polluted during the production process of a commodity (Hoekstra et al., 2011) and accounts for both direct and indirect uses in the whole SC (Figure 22.4). A general production

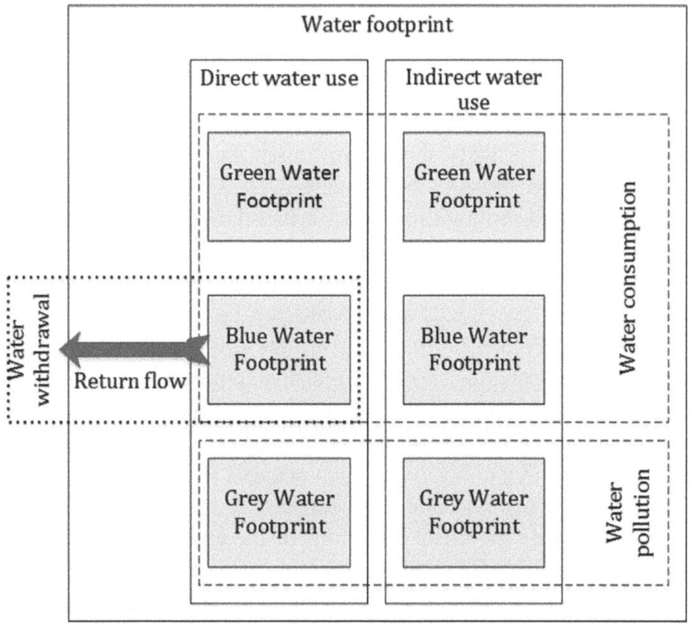

FIGURE 22.4

Representation of the component of a water footprint.

Adapted from Hoekstra et al. (2011).

system WF is assigned three contributions: the green WF (i.e., due to evaporation of rainwater withdrawn by the system), the blue WF (i.e., due to evaporation of surface and groundwater withdrawn by the system), and the grey WF (i.e., accounting for water polluted by the system).

In this section, water use (limited to the blue WF), GHG emission, and profit drivers are integrated in the design and planning of biofuel SCs according to a GrSCM approach (Bernardi et al., 2012, 2013). A multiobjective MILP formulation was proposed where the multiple drivers were optimized at each stage of the production network (Figure 22.1).

The optimization problem used Northern Italy as a case study, according to the hypotheses described in Section 22.2 of this chapter. The geographical dependence of the water directly consumed for corn cropping was determined as the difference between the water loss because of crop evapotranspiration and the amount of water provided by rainfall (Allen et al., 1998). The rainfall contribution, estimated from meteorological data, was diminished by the recharge rate for ground and surface water (about 30% of the irrigation rate). The evapotranspiration was modeled according to Todorovic et al. (2007). This direct water consumption was attributed to corn only, since stover was given a residual value.

The problem was first decomposed in a biobjective optimization between water use and profit. Table 22.3 reports performance indicators (NPV, WF, GHG

Table 22.3 Pareto Set of Solutions Resulting from the Biobjective Problem

Configuration	Technologies	NPV (€/GJ)	WF (m^3/GJ)	GHG (kg CO_2-eq/GJ)	Emission Reduction (%)
A: economic optimum	1	1.17	9.98	77.2	10
B	1, 7	0.23	1.9	39.4	54
C	7	−0.56	0.62	17	80
D: environmental optimum	4	−7.5	−0.06	6	97

emissions—per gigajoule of ethanol—and GHG emission reduction compared to gasoline) for the biofuel SC configurations on the Pareto frontier.

First-generation technologies ($k = 1$) ensure good profitability (configuration A), but poor environmental performance in terms of both impact mitigation on water resources and global warming. In this case, 97% of the overall water consumption is derived from biomass irrigation. The "water-withdrawals-to-availability ratio" (WtAR) index[1] presented a maximum value of 0.43 for underground water, thus showing a high stress on water resources. The deployment of hybrid ethanol facilities, using both corn and stover as feedstocks for fuel production, allowed a general reduction of environmental burdens, but showed a gradual worsening of profitability (cases B and C, Table 22.3). The configuration minimizing the WF involved the installation of exclusively cellulosic facilities, where the local use of residual biomass positively affected the environmental burdens on water resources (i.e., the maximum value of WtAR was 0.012). The system had a negative WF (−0.06 m^3/GJ), for the credits due to excess electricity sold to the grid, but was economically unsustainable (NPV = −7.5 €/GJ).

The most profitable design also complying with EU regulations was studied setting minimum levels of GHG emission reductions in the economic optimization. In Figure 22.5 the final SC design for the biofuel SC is reported. This biofuel SC configuration relied on first-generation technologies ($k = 2$), where DDGS was burned in a power station. The first-generation plants located along the coastline would be supplied by imported corn. One hybrid technology supplied with local biomass would be installed, too. The resulting configuration guaranteed good profitability (the NPV was equal to 0.71 €/GJ). The system WF (equal to 13.1 m^3/GJ or 279 L of water/L of fuel) turned to be higher than in configuration A (Table 22.3).

[1]The WtAR index, evaluated according to van Oel et al. (2009), was determined as the ratio between the direct water withdrawals and the surface or underground water available in each cell. The threshold values for WtAR are 0.2, 0.4, and 0.8, representing a medium, high, and very high stress on water resources, respectively.

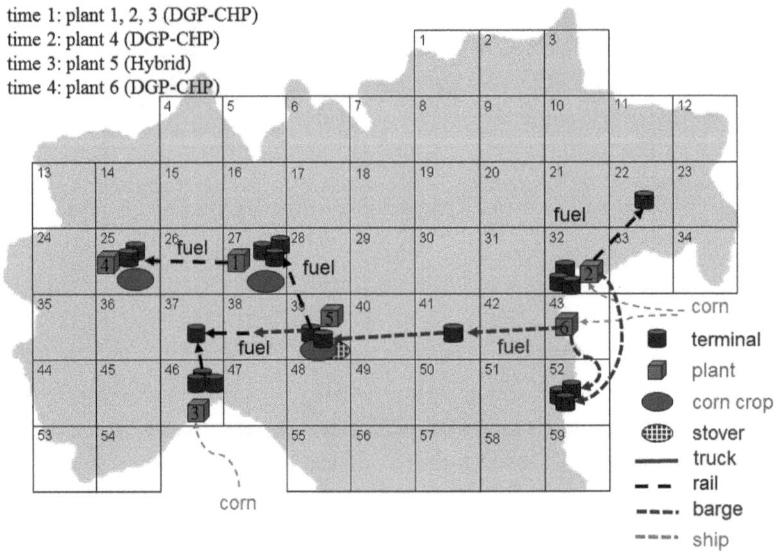

FIGURE 22.5

SC economic optimum with an imposed GHG emission reduction of 60% with respect to gasoline.

This was due to the different end-use of DDGS, which allowed for lower water credits as a fuel than as cattle feed.

It was also verified that technological advancements in farming practices (e.g., by replacing sprinkler with drip irrigation) might lead to a substantial reduction of the WtAR index.

22.4 THE DESIGN OF NEW BIOFUEL SCs IN THE PRESENCE OF UNCERTAINTY FACTORS

As discussed in the introduction, one fundamental issue in the design and optimization of biofuel SCs is represented by a number of uncertainty factors undermining several assumptions. Here we will consider two case studies based on some previous work of ours where first we discuss how feedstock and product price uncertainties may affect the optimization effort (Dal-Mas et al., 2011) and then discuss how carbon trading policies may be introduced in an uncertain scenario (Giarola et al., 2012).

22.4.1 ADDRESSING FINANCIAL RISK DUE TO MARKET PRICE UNCERTAINTY IN FIRST-GENERATION ETHANOL INVESTMENTS

Ethanol production industry expansion has been tempered by changing market conditions and, in particular, by the high variability of biomass costs and revenues from

Table 22.4 SC Configuration for Profit Optimization (Instances A.1, A.2, A.3) and Risk Minimization (Instance B.1)

Period	A.1 (DDGS Price = 300 €/t)	A.2 (DDGS Price = 200 €/t)	A.3[a] (DDGS Price Depreciation)	B.1 (DDGS Price = 300 €/t)
1	[$p = 2$; $g = 46$]; [$p = 3$; $g = 52$]	[$p = 1$; $g = 32$]; [$p = 4$; $g = 46$]	[$p = 1$; $g = 46$]; [$p = 4$; $g = 43$]	[$p = 4$; $g = 43$]
2	[$p = 2$; $g = 27$]	[$p = 3$; $g = 43$]	[$p = 3$; $g = 27$]	[$p = 4$; $g = 46$]
3	[$p = 2$; $g = 32$]	–	–	–
4	–	–	–	–
5	–	–	–	–
Expected NPV, M€	250	135	137	
ROI*	33%	28.1%	28.7%	18.8%

p represents the size of the production facilities, which ranges between 80 and 260 kt/year and includes six intervals, while g represents the square cells in the geographical grid.
[a]*DDGS price is supposed to vary as follows: 300 €/t since 2009 to 2012, 200 €/t since 2013 to 2016, and 100 €/t since 2017 to 2019.*

by-products (Schmit et al., 2009). In order to reduce financial risk on investment, this uncertainty should be addressed at the design and planning stage of systems, where effects of market volatility could be relevant (Papageorgiou, 2009; Awudu and Zhang, 2012).

This section addresses the financial assessment of a bioethanol production system in Northern Italy in a 10-year time framework (i.e., between 2009 and 2019). In order to describe the investors' perspective, the biofuel system, contrary to what was shown in Sections 22.3.1 and 22.3.3, was not required to fulfill ethanol targets. The influence on the production planning of a biofuel SC based on a DGP ($k = 1$, Table 22.1) was addressed via a stochastic approach, embedding uncertainty in the biomass cost and ethanol market price. The analysis of historical data concerning the biomass purchase cost in Italy (Frascarelli, 2008) and the ethanol selling price in Southern Europe (Agra, 2009) led to a compound probability density function, which was discretized into a set of 16 combined scenarios of fuel ethanol selling price and corn cost.

The biofuel system optimized according to the expected NPV (i.e., Bagajewicz, 2008), when high market price values of DDGS were assumed, led to a profitable business as indicated by a greater ROI value (instance A.1 in Table 22.4). In these cases, the planning of ethanol production allowed more distributed SCs where economies of scale were not fully exploited and small-sized facilities were also present.

Figure 22.6 illustrates the optimal SC configuration as planned at the end of the network lifetime for instance A.1. The preferred facility location along the coastline would also make biomass importation generally preferred to domestic production.

When the DDGS price was equal to 200 €/t (instance A.2), three plants characterized by a capacity on average greater than instance A.1 greater average size were planned to be built no later than 2011 (Table 22.4) because of long payback times

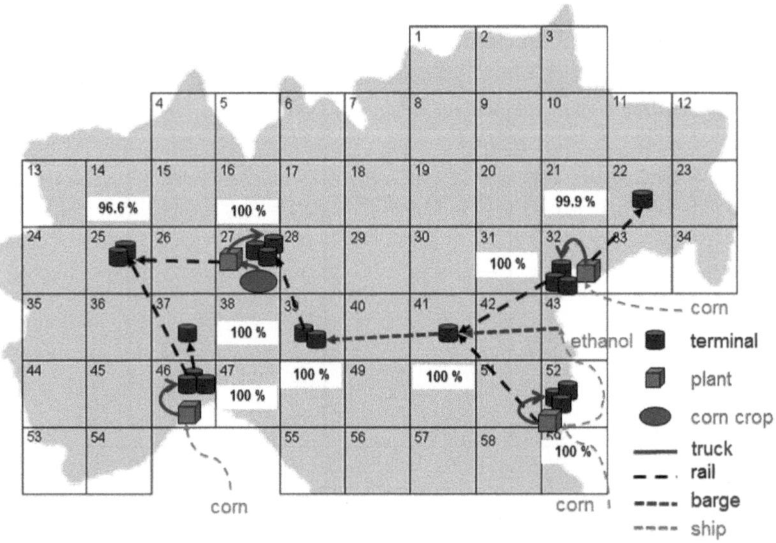

FIGURE 22.6

Medium-scenario probabilities.

After Dal-Mas et al. (2011).

needed to recoup the investment. Economies of scale turned out to be more relevant to improve the business, although this was substantially worsened compared to instance A.1 (the ROI index decreased to 28.1%). Finally, for the case in which DDGS market price progressively decreased (instance A.3), the planning strategy (Table 22.4) resembled instance A.2. Long payback times pushed toward an early realization of the ethanol plant sites. With this configuration, the event probability to have a payback time lower than 5 years was around 69%. On the contrary, the probability not to pay back the investment by the end of the lifetime of the production system (set to 10 years) was about 8%.

Although the expected NPV optimization could steer decision makers toward investments aiming at maximum profit, it could not guarantee that it would prevent economic losses in case of very adverse market conditions. Therefore, in order to model a risk-averse investor, who would be more interested in optimizing the less favorable situations, other indicators, such as the conditional value at risk (CVaR)[2] (Rockafellar and Uryasev, 2000), would be more suitable.

[2]The CVaR represents the NPV evaluated for a number of scenarios of economic losses occurring at a certain confidence level and time period. It directly derives from the value at risk (VaR) index with respect to a specified confidence β-level; the β-VaR of an investment is the lowest amount α such that, with probability β, the loss will not exceed α, whereas the β-CVaR is the expectation that losses are above that amount α. Here, the worst potential market scenarios with a 10% occurrence probability were taken into account.

When biofuel production chain is optimized according to the CVaR index evaluated on the worst-case market scenarios with a 10% confidence interval, if the DDGS price goes below 200 €/t, the discounted cumulative cash position was never positive and the best solution was not to invest in any businesses at all. With reference to instance B.1 and as shown in Table 22.4, a few plants of the bigger size would be planned. All in all, the business would show a low profitability: the average ROI index was equal to 18.8%. The risk prevention attitude would also produce a disinvestment trend in which ethanol production never exceeded the 73.6% of the market requirement. Investment decisions regarding the installation of new plants would occur only at the very beginning of the time horizon, so confirming a less risky investment strategy (Table 22.4). This allowed for a partial mitigation of the economic losses.

22.4.2 CARBON TRADING FOR THE PROMOTION OF ADVANCEMENTS AND TECHNOLOGY DIVERSIFICATION IN BIOFUEL GENERATION SYSTEMS

The acknowledgment of an economic value to GHG emission savings could lead to a cost-effective transition toward more sustainable biofuel systems (Creutzig et al., 2011). This could promote technologies for ethanol production based on nonfood biomass, which have been shown to be less CO_2 emitting, although more expensive than first-generation processes. This section addresses the development of a transport sector wide cap and trade on GHG emissions (Giarola et al., 2012). In assessing the potentials of the trading market to promote low carbon fuels, the effect of the volatility of the CO_2-eq emissions allowances (Chevallier, 2011) and of feedstock cost was addressed via a stochastic MILP model using a scenario-based approach.

The problem focused on the optimization of a single production plant (i.e., upstream SC) with a fixed product demand and with all-year availability of biomass. Bioethanol production in Northern Italy represented the reference system for the economic and environmental analysis of every SC or life cycle stage. Several feedstocks for ethanol production were included (i.e., corn, poplar, willow, miscanthus, corn stover, wheat and barley straw, switchgrass). The environmental impact due to their cultivation was assessed including effects on the amount of organic carbon stored in the soil (soil organic carbon, SOC). Five main processing technologies (k) were studied:

1. the DGP with DDGS sold to the cattle feed market substituting soy (Franceschin et al., 2008);
2. a variation of the DGP where DDGS is burnt to fuel in a CHP system (Franceschin et al., 2008);
3. the dilute acid process (DAP), where cellulosic feedstock is hydrolyzed with dilute sulfuric acid (USDOE, 1999);
4. the steam explosion process (SEP), where the cellulosic biomass is pretreated with high-pressure steam before conversion into ethanol (Sassner et al., 2008)
5. the gasification biosynthesis process (GBP), where biomass-based syngas is fermented to ethanol (Wei et al., 2009).

The long-term capacity planning and technology selection problem for a bioethanol SC aimed at the maximization of the financial performance of the business (expected NPV). The CO_2 emission trading scheme was modeled (Bojarski et al., 2009) constraining the total SC emission to be lower than a maximum value, the so-called cap. This guaranteed that minimum emission reduction compared to fossil fuels was met (EC, 2009). The trading market allowed that extra credits to emit could be bought (sold) by the system, if the emissions produced were above (below) the target. Depending on the case, the system would incur additional costs, or gain additional revenues.

The optimization problem solution showed that the economic optimum of the system in absence of carbon trading would involve the selection of a facility operating with a standard corn-based DGP technology in which DDGS is sold as fodder. This option represented a cost-effective solution (Figure 22.7(b)) and allowed for remarkable revenues coming from the by-product business. Although significant emission credits would come from DDGS, the impact associated with the agricultural phase (instance I in Figure 22.7(a)) was so high that the overall GHG emission savings were very low compared to the gasoline pathway.

If the emissions trading scheme was allowed, a facility using a technology DAP on corn stover would be installed. The higher profitability compared to first-generation bioethanol was partly due to the lower cost and volatility of lignocellulosic residues than corn (Figure 22.7(b)). Also, a relevant share of the annual income came from traded emissions (up to about 3.5%). The system fulfilled the 2017 EU regulations for the low environmental impacts associated with each LCA node and the credits for fossil power displacement (Figure 22.7(a)).

Second best technological options to meet the emission reduction targets were given by miscanthus-based GBP plants. This solution allowed for both favorable economic and environmental performance (Table 22.5 and Figure 22.7). It is important to note that the share of the incomes from trading of permits to emit (between 5.3% and 3.9%) played an important role in determining the profitability of this business, characterized by high capital investments (Figure 22.7(b)). However, if a more cautionary technical performance (Piccolo and Bezzo, 2009) were to be considered, the optimal configuration would only envisage a stover-based SEP. In this case, the expected profitability decreased (e.g., the expected NPV moved from 0.5 down to 0.28 €/GJ), but the impact on global warming was still low (about 12.9 kg CO_2-eq/GJ).

When effects of farming practices on SOC were included in the optimization (instance II in Figure 22.7(a)), the system showed a trend similar to the base case, although energy crops (e.g., miscanthus) were preferred to residual biomass, as straw removal might endanger the overall carbon balance. More precisely, the optimum in the absence of an emissions trading scheme was still given by a DGP plant, but the impact on global warming from the agricultural phase was 25% higher than before. In fact, high carbon emissions occurred from intensive crop practices, as is likely with annual crops. If an emission trading scheme was operating, then GBP

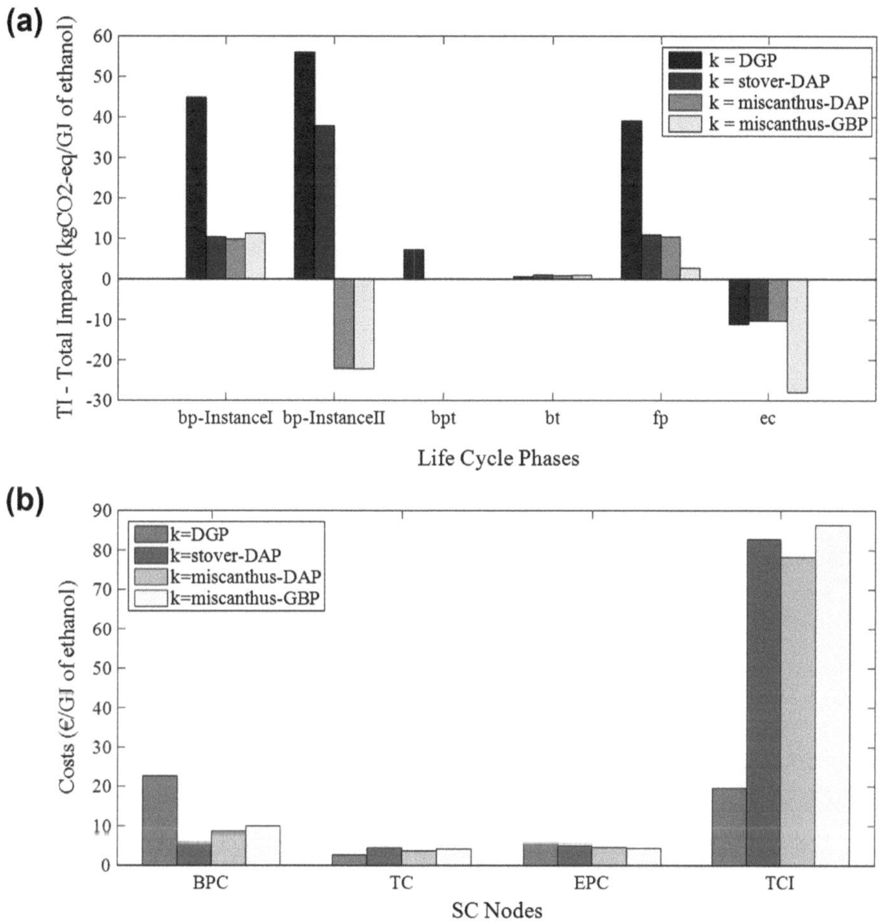

FIGURE 22.7

(a) Impact of the life cycle phases (biomass production, *bp*; biomass pretreatment, *bpt*; biomass transport, *bt*; fuel production, *fp*; and emission credits, *ec*) and (b) costs along the SC nodes (biomass purchase costs, BPC; transport costs, TC; fuel production costs, EPC; and capital costs TCI) for the selected technologies.

After Giarola et al. (2012).

and DAP technologies using miscanthus would be installed (Table 22.5). These solutions appeared promising both on environmental and economic aspects, matching the EU guidelines on emission reductions and gaining revenues from carbon trading (between 7.6% and 6% of the whole income, depending on the emission reduction target).

Table 22.5 SC Configurations for Profit Optimization

Selection Order	Technology	Expected NPV (€/GJ)	TI (kg CO_2-eq/GJ)	GHG Emission Reduction (%)
SOC Not Included				
1	DGP	0.48	80.7	6
1	Stover–DAP	0.69	12	35
1	Stover–DAP	0.52	12	50
1	Stover–DAP	0.57	12	60
2	Miscanthus–GBP	0.53	−13	35
2	Miscanthus–GBP	0.46	−13	50
2	Miscanthus–GBP	0.41	−13	60
SOC Included				
1	DGP	0.48	92	−1
1	Miscanthus–DAP	0.73	−21	35
1	Miscanthus–DAP	0.65	−21	50
1	Miscanthus–GBP	0.62	−50	60
2	Miscanthus–GBP	0.72	−50	35
2	Miscanthus–GBP	0.66	−50	50
2	Miscanthus–DAP	0.59	−21	60

22.5 DISCUSSION AND FINAL REMARKS

It is evident that the analysis of biofuels production is a complex task, particularly when environmental issues should be taken into account. In this chapter, we provided a few examples to highlight some of the challenges that we need to tackle when multifaceted objectives (environmental as well as economic ones) and/or uncertainty sources have to be considered. Although an important research effort has been carried out by the scientific community, we believe that several additional issues are still at stake.

With regard to the environmental analysis, the main difficulty is still about the quantification of the environmental impact. We do not want to review here the long debated and still open question about the application of LCA techniques, the calculation of indirect land usage effects, and the approaches to measure the various effects of human activities on the environment. We simply would like to comment on the fact the choice of the assessment method (e.g., Lucas et al., 2014) may have a significant influence on the results we obtained. Furthermore, in our opinion it is still questionable whether focusing on a comprehensive index (e.g., Eco-Indicator 99, Impact 2002+, and ReCiPe) is a better choice than assessing specific effect singularly (as we did here for the water and carbon footprints). If comprehensive indices allow measuring the overall impact, they may mask specific contributions, which may become critical in the territorial context (water usage is a typical example).

As pointed out in the introduction, sources of uncertainty represent the other key issues when assessing such systems. It is impossible to overstate the effect of uncertainty on the results we obtain in the analysis of production systems that are yet to be established. Relatively small variations in the initial assumptions may determine huge differences in the final outcome (a clear example comes from very pioneering sectors such as algae-based technologies where completely contradictory results can be found in the literature). Although huge modeling and computational challenges may arise, it is the authors' opinion that this represents the key research area in the design of alternative SCs.

To conclude, we believe that more information, analysis tools, and interactions between different types of expertise are still necessary to obtain a full comprehension and representation of such a complex system. Furthermore, most models assume that the SCs being investigated are somewhat isolated instead of interacting with existing infrastructure, and hybrid approaches may also be envisaged (e.g., Baliban et al., 2013).

On the one side, more players should be taken into account and the consumer's attitude needs to be described since alternative fuels and energy options may be available, for instance, through game theory approaches (e.g., Bai et al., 2012; Yue and You, 2014). In fact, even a countrywise (or EU-wise) approach may represent too narrow a system to assess a global effect on the fuel market and related emissions, which would be affected by global oil prices, biofuel price and capacity in tropical countries, indirect land use change, international trading regulation, government subsidies and environmental legislation, and so on. In this case, the best solution is usually to refer to partial equilibrium models or even general equilibrium models such as the Global Trade Analysis Project (www.gtap.org).

It is, nonetheless, important to realize that the more comprehensive and "universal" a model is, the less detailed and precise it becomes. For instance, general equilibrium models may provide extremely important information to assess general trends and cause–effects relationships, but can hardly be helpful to guide investors' decisions or even to steer local policy.

Finally, we are also persuaded that a better assessment of the overall question would be insufficient if we stick to the status quo and do not have the imagination to foresee how things may change and, most importantly, to suggest how things should change. Decision makers need tools which evaluate how the system may react to different options and how it may change if optimized toward specific goals. For instance, agriculture has evolved over tens of thousands of years to produce food crops. However, if the final goal is fuel instead of food, the overall chain might have to be operated in a different way and in fact even the boundary between "first-generation" and "second-generation" biofuels may start blurring. This is always to be kept in mind when moving to new and unexplored technical lands.

REFERENCES

Agra, 2009. European Ethanol Price Report – Period: 2005–2008. Agra Informa Ltd. Available from: www.agra-net.com.

Allen, R.G., Pereira, L.S., Raes, D., Smith, M., 1998. Crop evapotranspiration–guidelines for computing crop water requirements. In: FAO Irrigation and Drainage, Paper 56, FAO, Rome, Italy.

Agkul, O., Shah, N., Papageorgiou, L.G., 2012a. Economic optimisation of a UK advanced biofuel supply chain. Biomass Bioenerg. 41, 57–72.

Agkul, O., Shah, N., Papageorgiou, L.G., 2012b. An optimisation framework for a hybrid first/second-generation bioethanol supply chain. Comput. Chem. Eng. 42, 101–114.

Akgul, O., Zamboni, A., Bezzo, F., Shah, N., Papageorgiou, L.G., 2010. Optimization-based approaches for bioethanol supply chains. Ind. Eng. Chem. Res. 50, 4927–4938.

Awudu, I., Zhang, J., 2012. Uncertainties and sustainability concepts in biofuel supply chain management: a review. Renew. Sust. Energ. Rev. 6, 1359–1368.

Bagajewicz, M., 2008. On the use of net present value in investment capacity planning models. Ind. Eng. Chem. Res. 47, 9413–9416.

Bai, Y., Ouyang, Y., Pang, J.-S., 2012. Biofuel supply chain design under competitive agricultural land use and feedstock market equilibrium. Energ. Econ. 34, 1623–1633.

Baliban, R.C., Elia, J.A., Floudas, C.A., 2013. Biomass and natural gas to liquid transportation fuels: process synthesis, global optimization, and topology analysis. Ind. Eng. Chem. Res. 52, 3381–3406.

Bernardi, A., Giarola, S., Bezzo, F., 2012. Optimizing the economics and the carbon and water footprints of bioethanol supply chains. Biofuels Bioprod. Biorefin. 6, 656–672.

Bernardi, A., Giarola, G., Bezzo, F., 2013. Spatially explicit multi-objective optimization for the strategic design of first- and second-generation biorefineries including carbon and water footprints. Ind. Eng. Chem. Res. 52, 7170–7180.

Beamon, B.M., 1998. Supply chain design and analysis: models and methods. Int. J. Prod. Econ. 55, 281–294.

Bojarski, A.D., Laìnez, J.M., Espuña, A., Puigjaner, L., 2009. Incorporating environmental impacts and regulations in a holistic supply chains modelling: an LCA approach. Comput. Chem. Eng. 33, 1747–1759.

Brown, G., Wood, J., Estrin, A., 2005. Bioethanol Greenhouse Gas Calculator. Users' Guide. Home-Grown Cereals Authority (HGCA), London, U.K.

Carriquiry, M.A., Du, X., Timilsina, G.R., 2011. Second-generation biofuels: economics and policies. Energ. Policy 39, 4222–4234.

Chevallier, J., 2011. Detecting instability in the volatility of carbon prices. Energy Econ. 33, 99–110.

CONCAWE, EUCAR., JRC (CONCAWE), 2007. Well-to-Wheels Analysis of Future Automotive Fuels and Powertrains in the European Context: WELL-to-tank Report. Version 2c. European Commission, Brussels, Belgium.

Creutzig, F., McGlynn, E., Minx, J., Edenhofer, O., 2011. Climate policies for road transport revisited (I): evaluation of the current framework. Energ. Policy 9, 2396–2406.

Čuček, L., Lam, H.L., Klemeš, J.J., Varbanov, P.S., Kravanja, Z., 2010. Synthesis of regional networks for the supply of energy and bioproducts. Clean Technol. Environ. Policy 12, 635–645.

Čuček, L., Klemeš, J.J., Varbanov, P.S., Kravanja, Z., 2012. Total footprints-based multi-criteria optimisation of regional biomass energy supply chains. Energy 44, 135–145.

Dal-Mas, M., Giarola, S., Zamboni, A., Bezzo, F., 2011. Strategic design and investment capacity planning of the ethanol supply chain under price uncertainty. Biomass Bioenerg. 35, 2059–2071.

Dominguez-Faus, R., Powers, S.E., Burken, J.G., Alvarez, P.J., 2009. The water footprint of biofuels: a drink or drive issue? Environ. Sci. Technol. 43, 3005–3010.

Dunnett, A., Adjiman, C.S., Shah, N., 2008. A spatially explicit whole-system model of the lignocellulosic bioethanol supply chain: an assessment of decentralized processing potential. Biotechnol. Biofuels 1, 13.

EUBIA, 2013. Bioethanol Production and Use. EUBIA, Brussels, Belgium.

European Commission (EC), 2009. Directive 2009/28/EC of the European Parliament and of the Council of 23 April 2009 on the promotion of the use of energy from renewable sources and amending and subsequently repealing directives 2001/77/EC and 2003/30/EC. Off. J. Eur. Parliament. Brussels.

Eksioglu, S.D., Acharya, A., Leightley, L.E., Arora, S., 2009. Analyzing the design and management of biomass-to-biorefinery supply chain. Comput. Ind. Eng. 57, 1342–1352.

Farrell, A.E., Plevin, R.J., Turner, T.T., Jones, A.D., O'Hare, M., Kammen, D.M., 2006. Ethanol can contribute to energy and environmental goals. Science 311, 506–508.

Franceschin, G., Zamboni, A., Bezzo, F., Bertucco, A., 2008. Ethanol from corn: a technical and economical assessment based on different scenarios. Chem. Eng. Res. Des. 86, 488–498.

Frascarelli, A., 2008. Available at: www.gruppo2013.it/working-paper/Pagine/default.aspx.

Gebreslassie, B.H., Yao, Y., You, F., 2012. Design under uncertainty of hydrocarbon biorefinery supply chains: multiobjective stochastic programming models, decomposition algorithm, and a comparison between CVaR and downside Risk. AIChE J. 58, 2155–2179.

Giarola, S., Bezzo, F., Shah, N., 2013. A risk management approach to the economic and environmental strategic design of ethanol supply chains. Biomass Bioenerg. 58, 31–51.

Giarola, S., Shah, N., Bezzo, F., 2012. A comprehensive approach to the design of ethanol supply chains including carbon trading effects. Bioresour. Technol. 107, 175–185.

Giarola, S., Zamboni, A., Bezzo, F., 2011. Spatially explicit multi-objective optimisation for design and planning of hybrid first- and second-generation biorefineries. Comput. Chem. Eng. 35, 1782–1797.

Grossmann, I.E., Guillén-Gosàlbez, G., 2010. Scope for the application of mathematical programming techniques in the synthesis and planning of sustainable processes. Comput. Chem. Eng. 34, 1365–1376.

Guillén-Gosálbez, G., Grossmann, I.E., 2009. Optimal design and planning of sustainable chemical supply chains under uncertainty. AIChE J. 55, 99–121.

Hoekstra, A.Y., Chapagain, A.K., Aldaya, M.M., Mekonnen, M.M., 2011. The Water Footprint Assessment Manual: Setting the Global Standard. Earthscan, London and Washington DC.

Huang, Y., Chen, C.W., Fan, Y., 2010. Multistage optimization of the supply chains of biofuels. Transport. Res. E-Log. 46, 820–830.

ISO, 2006. ISO 14040:2006: Environmental Management – Life Cycle Assessment – Principles and Framework. International Standards Organization, Geneva.

Kim, J., Realff, M.J., Lee, J.H., 2011. Optimal design and global sensitivity analysis of biomass supply chain networks for biofuels under uncertainty. Comput. Chem. Eng. 35, 1738–1751.

Kostin, A.M., Guillén-Gosálbez, G., Mele, F.D., Bagajewicz, M.J., Jiménez, L., 2012a. Design and planning of infrastructures for bioethanol and sugar production under demand uncertainty. Chem. Eng. Res. Des. 90, 359–376.

Kostin, A.M., Guillén-Gosálbez, G., Mele, F.D., Jiménez, L., 2012b. Identifying key life cycle assessment metrics in the multiobjective design of bioethanol supply chains using a rigorous mixed-integer linear programming approach. Ind. Eng. Chem. Res. 51, 5282–5291.

Kim, S., Dale, B.E., 2008. Life cycle assessment of fuel ethanol derived from corn grain via dry milling. Bioresour. Technol. 99, 5250–5260.

Lucas, G., Bezzo, F., Carvalho, A., 2014. Optimal design of a bioethanol supply chain considering different environmental impact assessment methods. In: Klemeš, J.J., Varbanov, P.S., Liew, P.Y. (Eds.), Computer-aided Chemical Engineering, 24nd European Symposium on Computer Aided Process Engineering, 33. Elsevier, Amsterdam (The Netherlands), pp. 973–978.

Marvin, W.A., Schmidt, L.D., Benjaafar, S., Tiffany, D.G., Daoutidis, P., 2012. Economic optimization of a lignocellulosic biomass-to-ethanol supply chain. Chem. Eng. Sci. 67, 68–79.

Mazzetto, F., Ortiz-Gutierrez, R.A., Manca, D., Bezzo, F., 2013. Strategic design of bioethanol supply chain including commodity market dynamics. Ind. Eng. Chem. Res. 52, 10305–10316.

Mele, F., Kostin, A., Guillén-Gosàlbez, G., Jimenez, L., 2011. Multiobjective model for more sustainable fuel supply chains. A case study of sugar cane industry in Argentina. Ind. Eng. Chem. Res. 50, 4939–4958.

Mafakheri, F., Nasiri, F., 2014. Modeling of biomass-to-energy supply chain operations: applications, challenges and research directions. Energ. Policy 67, 116–126.

Osmani, A., Zhang, J., 2014. Economic and environmental optimization of a large scale sustainable dual feedstock lignocellulosic-based bioethanol supply chain in a stochastic environment. Appl. Energ. 114, 572–587.

van Oel, P., Mekonnen, M., Hoekstra, A., 2009. The external water footprint of the Netherlands: geographically-explicit quantification and impact assessment. Ecol. Econ. 69, 82–92.

Parker, N., Tittmann, P., Hart, Q., Nelson, R., Skog, K., Schmidts, A., Gray, E., Jenkins, B., 2010. Development of a biorefinery optimized biofuel supply curve for Western United States. Biomass Bioenerg. 34, 1597–1607.

Papageorgiou, L.G., 2009. Supply chain optimization for the process industries: advances and opportunities. Comput. Chem. Eng. 33, 1931–1938.

Peters, M.S., Timmerhaus, K.D., West, R.E., 2003. Plant Design and Economics for Chemical Engineers. McGraw-Hill, New York.

Piccolo, C., Bezzo, F., 2009. A techno-economic comparison between two technologies for bioethanol production from lignocellulose. Biomass Bioenerg. 33, 478–491.

Robertson, G.P., Dale, V.H., Doering, O.C., Hamburg, S.P., Melillo, J.M., Wander, M.M., Parton, W.J., et al., 2008. Sustainable biofuels redux. Science 322, 49–50.

Rockafellar, R.T., Uryasev, S., 2000. Optimization of conditional value-at-risk. J. Risk 2, 21–41.

Rosenthal, R.E., 2006. GAMS—A Users' Guide (Version 22.5). GAMS Development Corporation, Washington, DC, U.S.A.
Sassner, P., Galbe, M., Zacchi, G., 2008. Techno-economic evaluation of bioethanol production of three different lignocellulosic materials. Biomass Bioenerg. 32, 422—430.
Schmit, T.M., Luo, J., Tauer, L.W., 2009. Ethanol plant investment using net present value and real options analyses. Biomass Bioenerg. 33, 1442—1451.
Searchinger, T., Heimlich, R., Houghton, R.A., Dong, F., Elobeid, A., Fabiosa, J., Tokgoz, S., et al., 2008. Use of U.S. croplands for biofuels increases greenhouse gases through emissions from land-use change. Science 319, 1238—1240.
Shah, N., 2005. Process industry supply chains: advances and challenges. Comput. Chem. Eng. 29, 1225—1235.
Smith, T.C., Kindred, D.R., Brosnan, J.M., Weightman, R.M., Shepherd, M., Sylvester-Bradley, R., 2006. Wheat as a Feedstock for Alcohol Production. HGCA, London.
Srivastava, S.K., 2007. Green supply-chain management: a state-of-the-art literature review. Int. J. Manage. Rev. 9, 53—80.
Todorovic, M., Caliandro, A., Albrizio, R., 2007. Irrigated agriculture and water use efficiency in Italy. In: Options Méditerranéennes, Series B, 57, pp. 102—138.
USDOE, 1999. Lignocellulosic Biomass to Ethanol Process Design and Economics Utilising Co-current Dilute Acid Prehydrolysis and Enzymatic Hydrolysis Current and Futuristic Scenarios. National Renewable Energy Laboratory, USDOE, Golden, CO, U.S.A.
USDOE, 2002. Lignocellulosic Biomass to Ethanol Process Design and Economic Utilizing Co-current Dilute Acid Prehydrolysis for Corn Stover. National Renewable Energy Laboratory, USDOE, Golden, CO, U.S.A.
Wei, L., Pordesimo, L.O., Igathinathane, C., Batchelor, W.D., 2009. Process engineering evaluation of ethanol production from wood through bioprocessing and chemical catalysts. Biomass Bioenerg. 34, 255—266.
You, F., Tao, L., Graziano, D.J., Snyder, S.W., 2012. Optimal design of sustainable cellulosic biofuel supply chains: multiobjective optimization couples with life cycle assessment and input output analysis. AIChE J. 58, 1157—1180.
You, F.Q., Wang, B., 2011. Life cycle optimization of biomass-to-liquid supply chains with distributed-centralized processing networks. Ind. Eng. Chem. Res. 50, 10102—10127.
Yue, D., You, F., 2014. Game-theoretic modeling and optimization of multi-echelon supply chain design and operation under Stackelberg game and market equilibrium. Comput. Chem. Eng. 71, 347—361.
Yue, D., You, F., Snyder, S.W., 2014. Biomass-to-bioenergy and biofuel supply chain optimization: overview, key issues and challenges. Comput. Chem. Eng. 66, 36—56.
Zamboni, A., Bezzo, F., Shah, N., 2009a. Spatially explicit static model for the strategic design of future bioethanol production systems. 2. Multi-objective environmental optimization. Energ. Fuel 23, 5134—5143.
Zamboni, A., Murphy, R.J., Woods, J., Bezzo, F., Shah, N., 2011. Biofuels carbon footprints: whole-systems optimisation for GHG emissions reduction. Bioresour. Technol. 102, 7457—7465.
Zamboni, A., Shah, N., Bezzo, F., 2009b. Spatially explicit static model for the strategic design of future bioethanol production systems. 1. Cost minimization. Energ. Fuel 23, 5121—5133.

PART

V

Sustainability analysis methodologies

CHAPTER

An Integrated Multiscale Modeling Framework for Sustainable Process Design Applications

23

Rebecca J. Hanes, Bhavik R. Bakshi[1]

William G. Lowrie Department of Chemical and Biomolecular Engineering, The Ohio State University, OH, USA

[1]Corresponding author: E-mail: bakshi.2@osu.edu

23.1 INTRODUCTION AND CONTEXT

The objective of chemical process optimization has traditionally been to find the most profitable process design within safety and capacity constraints. This single-objective design paradigm yielded profitable, reliable processes but led to growing regulatory pressure to reduce the environmental impacts generated by the chemical process industry. Environmentally conscious design methodologies followed. Early such methodologies include heat integration (Linnhoff and Flower, 1978) and the extension of the integration concept to mass integration (El Halwagi and Manousiouthakis, 1989), which reduced energy and water consumption through retrofitting or improvements on an original design. Integration techniques yielded significant economic benefits through reducing energy and water costs as well as environmental benefits such as reduced fossil fuel and water consumption. However, the analysis boundary of process integration and of waste-reducing methodologies developed around the same time (Douglas, 1992) was limited to a single process or plant. Using these techniques, it was possible to reduce the environmental impact of the plant itself only to increase impacts of upstream or downstream processes such as feedstock production or waste processing (Grossmann et al., 1982; Azapagic and Clift, 1999). Subsequent environmentally conscious design methodologies, particularly the waste reduction algorithm (Hilaly and Sikdar, 1994; Young and Cabezas, 1999), expanded the analysis boundary to include select upstream and downstream processes. The method for environmental impact minimization combined a process optimization problem with life cycle assessment (LCA), thereby expanding the analysis boundary to the life cycle scale (Stefanis et al., 1997; Yang and Shi, 2000). The objective of such sustainable process design (SPD) applications is to design the process of interest such that environmental impacts of all processes in the boundary are balanced with economics of the process of interest.

Computer Aided Chemical Engineering, Volume 36. ISSN 1570-7946. http://dx.doi.org/10.1016/B978-0-444-63472-6.00023-9

The basic SPD problem consists of designing an individual process or supply chain under both economic and environmental objectives, as follows (Cano-Ruiz and McRae, 1998):

$$\begin{gathered}\text{maximize } Z_1; \quad \text{minimize } Z_2 \\ \text{subject to } f(x) \geq 0; \quad Z_1 = Pg(x); \quad Z_2 = Bh(x)\end{gathered} \tag{23.1}$$

Equation (23.1) involves the process model(s) $f(x)$, which constrain the design variables x. The function $g(x)$ represents process inputs and outputs, which are scaled by price data P to calculate the economic objective function Z_1. In a similar manner, the function $h(x)$ represents inputs and outputs scaled by environmental impact data B to calculate the environmental objective function Z_2. Multiobjective optimization (MOO) techniques such as the ε-constraint method are applied to locate the Pareto curve between the economically optimal designs with maximum Z_1 and environmentally optimal designs with minimum Z_2.

Within the past several years such multiobjective SPD techniques have been extensively applied. SPD applications have grown in scale, from individual processes (Kniel et al., 1996) to supply chains at the regional or national scales (Guillén-Gosálbez and Grossmann, 2009; Elia et al., 2014; Yue et al., 2014a,b). In contrast, the sustainability assessment has remained at the same scale used in very early SPD applications.

SPD methods universally use process analysis to build the life cycle inventory. Process analysis is an inventory procedure wherein life cycle processes are added to the inventory individually according to the significance of their contribution to total life cycle impacts. The inventory is complete when the inclusion of additional life cycle processes is determined to have a negligible effect on the total inventory. The analysis boundary for SPD methods is generally limited to a cradle-to-gate life cycle and the life cycle network tends to resemble a supply chain rather than the full life cycle network. In general, downstream processes and contributing processes past second order are neglected. The analysis boundary thus excludes many systems of economic and of environmental interest in the full life cycle (Lenzen, 2000). Moreover, when process analysis is used to compile an inventory, expanding the analysis boundary is computationally intractable for more than a few processes due to the high data requirement and the increasing complexity of the life cycle network.

A process optimized with respect to a partial life cycle is not guaranteed to be optimal when analyzed in the context of the full life cycle. Hybrid life cycle methods offer an efficient way to balance the demands of a comprehensive analysis boundary with the need for accurate and process-specific life cycle data. These methods capture portions of the life cycle ignored by process analysis by combining detailed data for a few key life cycle processes with an environmentally extended input–output (EEIO) model that captures the macroeconomic system within which the processes operate (Moriguchi et al., 1993; Joshi, 1999). The coarser, more aggregated EEIO model captures the full life cycle without requiring data on each distinct process in the life cycle. A hybrid inventory therefore combines the level of detail of a

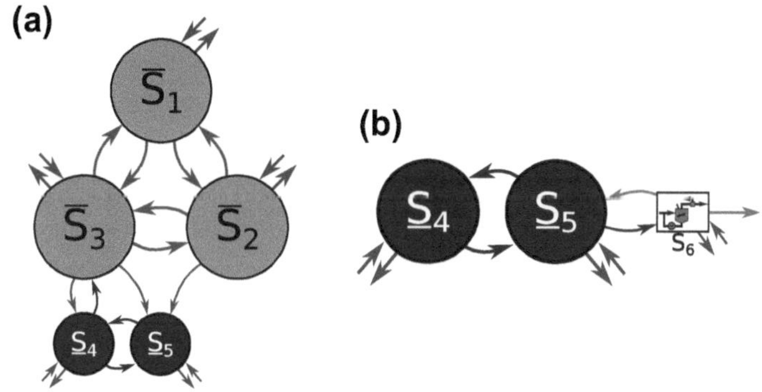

FIGURE 23.1

Comparison between the typical hybrid life cycle model (a) and conventional SPD model (b).

process analysis inventory with a comprehensive analysis boundary that contains the full life cycle (Suh and Huppes, 2005).

Hybrid life cycle models and SPD models are complementary as seen in Figure 23.1 The hybrid model (Figure 23.1(a)) is comprehensive but coarser due to lack of details at the process scale while the SPD model (Figure 23.1(b)) contains details at the process and unit operation scale but has a restricted analysis boundary. The objective of this work is to develop a multiscale modeling framework that combines a comprehensive hybrid life cycle model based on integrated hybrid LCA with fundamental engineering process models and apply it to SPD (Suh, 2004). The proposed framework not only captures an expanded analysis boundary relative to current SPD methods but also models the life cycle as dependent on the process or supply chain design. In contrast, conventional life cycle models are entirely static and empirical, and can only be scaled up or down according to the level of production required by the process of interest. The proposed modeling framework results in more accurate impact accounting as less of the life cycle is neglected and the life cycle models are no longer static. It is expected that the expanded analysis boundary of the proposed framework will also expand the feasible design space, resulting in new design options that offer both environmental and economic benefits.

23.2 MODELING FRAMEWORK

23.2.1 MULTISCALE SYSTEM AND COMPONENTS

The proposed modeling framework captures process systems with components at three scales: process, value chain, and economy. At the process scale, models represent individual chemical plants. Because engineering process models are aggregations of connected unit operation models, unit operations are not considered to be a fourth scale. Processes and unit operations are thus treated as the same scale.

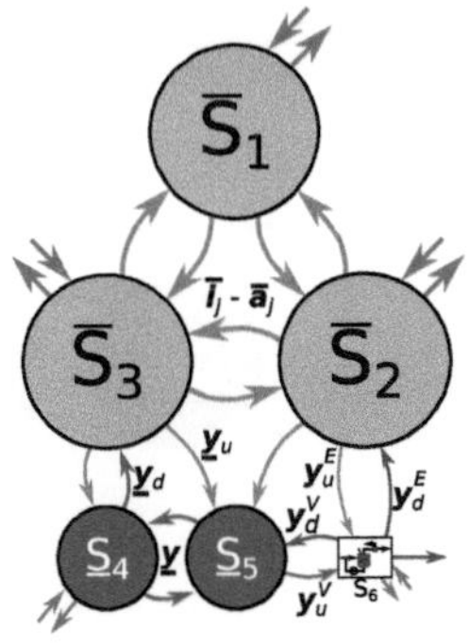

FIGURE 23.2

Multiscale process systems with components at the economy (brown (dark gray in print versions)), value chain (red (gray in print versions)), and process (blue (light gray in print versions)) scales are captured by the proposed modeling framework.

Similarly, supply chains are aggregations of connected process models and are treated at the process scale. The value chain scale takes the place of the process model-based life cycle inventory used in conventional SPD problems. Value chain activities represent average production technology; the value chain models are less specific and, because they are derived from empirical data, potentially less accurate compared to process scale models. Data used to derive activity models is obtained from life cycle inventory databases such as the US LCI database, ecoinvent, or Eco-LCA (National Renewable Energy Laboratory, 2012; Swiss Center for Life Cycle Inventories, 2013; Eco-LCA, 2014).

An EEIO model captures the largest scale in the modeling framework. At this scale system components are industrial sectors, which similar to value chain activities are aggregates of production technologies. Sectors are highly aggregated to keep the EEIO model tractable. A typical value chain activity is coal electricity production while the corresponding industrial sector captures all electricity generation processes.

Figure 23.2 demonstrates the structure of a typical multiscale process system, with the scale of each component indicated by color. Each component is modeled using an input–output (IO) vector, although the derivation, interpretation, and properties of the IO vectors vary between scales.

Chemical process models are typically represented in optimization problems by systems of nonlinear equations. This form makes the chemical process models cumbersome to connect to the larger scale models represented with IO vectors. Section 23.2.3 describes how chemical process models are put in a form suitable for connecting to larger scale models.

23.2.2 ENVIRONMENTALLY EXTENDED INPUT–OUTPUT MODEL

An EEIO model is derived from a transactions matrix $\overline{\mathbf{X}}$ of economic exchanges between sectors, a value added vector $\overline{\mathbf{v}}$, a final demand vector $\overline{\mathbf{f}}$, and an environmental interactions matrix $\overline{\mathbf{B}}$ that contains exchanges between each sector and the environment, normalized to one dollar's worth of sector production (Leontief, 1970; Leontief and Ford, 1972). Value added are sector inputs such as labor that originate outside the economic system. Final demand consists of commodities sold outside the economic system, that is, to purchasers other than sectors (Leontief, 1936; Johansen, 1960). Table 23.1 shows $\overline{\mathbf{X}}, \overline{\mathbf{v}}, \overline{\mathbf{f}}$, and $\overline{\mathbf{B}}$ for a sample two-sector economic system. The transactions matrix, the value added, and the final demand are used to calculate the direct requirements matrix $\overline{\mathbf{A}}$ by normalizing the columns of $\overline{\mathbf{X}}$ by the total output of each consuming sector $\overline{\mathbf{X}}$, as follows (Miller and Blair, 1985):

$$\overline{\mathbf{A}} = \overline{\mathbf{X}}\hat{\overline{\mathbf{x}}}^{-1} \tag{23.2}$$

The elements of $\overline{\mathbf{A}}$ are technical coefficients $\overline{a}_{ij}$, interpreted as the amount of commodity input required from the ith producing sector per dollar output of the jth consuming sector. A fundamental assumption of IO analysis is that the technical coefficients are fixed. Fixed $\overline{a}_{ij}$ imply that (1) the amount of each commodity input consumed by a sector is directly proportional to that sector's level of production and (2) sectors consume commodities in fixed proportions regardless of the level of production.

The total output of sector j, $\overline{\mathbf{x}}_j$, is equal to the output of sector j consumed within the economy plus output sold as final demand. This relationship, an economy-wide monetary balance on commodity exchanges, is shown in Eqn (23.3).

$$\overline{\mathbf{x}} = \overline{\mathbf{A}}\overline{\mathbf{x}} + \overline{\mathbf{f}} \tag{23.3}$$

Equation (23.3) is solved for $\overline{\mathbf{x}}$ by rearranging as follows:

$$\overline{\mathbf{x}} = \left(\overline{\mathbf{I}} - \overline{\mathrm{A}}\right)^{-1}\overline{\mathbf{f}} \tag{23.4}$$

$\left(\overline{\mathbf{I}} - \overline{\mathbf{A}}\right)^{-1}$ is the total requirements matrix or the Leontief inverse. Because the $\overline{a}_{ij}$ are assumed to be fixed, total economic activity $\overline{\mathbf{x}}$ can be calculated for arbitrary values of $\overline{\mathbf{f}}$. Equation (23.4) taken by itself is an economic input–output (EIO) model. The EIO model is environmentally extended with the addition of the environmental interactions matrix $\overline{\mathbf{B}}$, also shown in Table 23.1. An EEIO model, unlike an EIO model, is used to calculate an inventory of environmental interactions generated as a consequence of producing $\overline{\mathbf{f}}$. The inventory vector $\overline{\mathbf{g}}$ is calculated by postmultiplying $\overline{\mathbf{B}}$ with the vector of total sector outputs $\overline{\mathbf{x}}$ as shown in Eqn (23.5).

$$\overline{\mathbf{g}} = \overline{\mathbf{B}}\overline{\mathbf{x}} \tag{23.5}$$

In the derivation of Eqn (23.4), flows from sector i to sector j are implicitly assumed to consist of sector i's primary commodity product. This model, known as the industry–industry approach, does not account for sectors that produce more than one commodity. Sectors often produce multiple commodities, but that information is not captured in the transactions matrix or in the requirements matrices. The make-use approach, also known as the commodity-industry approach, offers greater flexibility when dealing with secondary production. Assuming that there

Table 23.1 Transactions Matrix $\overline{\mathbf{X}}$, Value Added $\bar{\mathbf{v}}$, Final Demand $\bar{\mathbf{f}}$, and Environmental Interactions $\overline{\mathbf{B}}$ for a Two-Sector Economic System

Producing Sectors ($)	Consuming Sectors ($)		Final Demand ($)	Total Output ($)
	Sector 1	Sector 2		
Sector 1	$\bar{x}_{11}$	$\bar{x}_{12}$	$\bar{f}_1$	$\bar{x}_1 = \bar{x}_{11} + \bar{x}_{12} + \bar{f}_1$
Sector 2	$\bar{x}_{21}$	$\bar{x}_{22}$	$\bar{f}_2$	$\bar{x}_2 = \bar{x}_{21} + \bar{x}_{22} + \bar{f}_2$
Value added ($)	$\bar{v}_1$	$\bar{v}_2$		
Total input ($)	$\bar{x}_1 = \bar{v}_1 + \bar{x}_{11} + \bar{x}_{21}$	$\bar{x}_2 = \bar{v}_2 + \bar{x}_{12} + \bar{x}_{22}$		
Environmental interactions	$\bar{b}_{11}$ $\vdots$ $\bar{b}_{r1}$	$\bar{b}_{12}$ $\vdots$ $\bar{b}_{r1}$		

are as many commodities as sectors in the economy, the direct requirements matrix obtained using the commodity-industry approach is as follows:

$$\overline{\mathbf{A}}_c = \overline{\mathbf{U}}\left(\overline{\mathbf{V}}^T\right)^{-1} \tag{23.6}$$

The use matrix $\overline{\mathbf{U}}$ in Eqn (23.6) describes the consumption of commodities by sectors and has dimensions of $I \times J$, where I is the number of commodities and J the number of sectors. The make matrix $\overline{\mathbf{V}}$ describes the production of commodities by sectors and has dimensions of $J \times I$. When $I = J$, the use matrix is equivalent to the transactions matrix; when $I = J$ and there is no secondary production, $\overline{\mathbf{V}}$ is equivalent to the diagonalized total sector output vector (Miller and Blair, 1985). Equation (23.6) also requires the commodity-technology assumption, which states that a commodity is always produced from the same combination of commodity inputs regardless of the sector that produces it. Because the number of sectors and the number of commodities are assumed to be equal, $\overline{\mathbf{A}}_c$ has the same dimensions as the industry–industry direct requirements matrix $\overline{\mathbf{A}}$. However, the elements of $\overline{\mathbf{A}}_c$ are in general not equal to the elements of $\overline{\mathbf{A}}$.

Throughout the rest of the chapter, any reference to direct requirements matrices refers to $\overline{\mathbf{A}}_c$ derived from Eqn (23.6), although the subscript c has been dropped to reduce annotation.

23.2.3 VALUE CHAIN AND PROCESS MODELS IN VECTOR NOTATION

Viewed column-wise, the total requirements and interactions matrices can be interpreted as a set of sector models with each column representing one sector. Each element of the total requirements matrix represents the total direct and indirect consumption in monetary units of commodity i required to produce one dollar's worth of output from sector j, while each element of column j of the interactions matrix represents the amount of environmental interaction r generated as a consequence of producing one dollar's worth of output from sector j (Miller and Blair, 1985).

Vector notation for value chain activities also uses IO vectors as models (Heijungs, 1994). The model for activity l consists of one vector of product flows to and from other activities and one vector of flows to and from the environment or environmental interactions. Different product and interaction flows have different units but all units are physical. Negative vector elements are inputs and positive elements are outputs. Activity models are normalized to some standard level of production, not necessarily one unit of output and not necessarily the same level of production for each activity.

The value chain as a whole is modeled with a technology matrix $\underline{\mathbf{Y}}$, each column of which is one activity model and each row of which captures the production and consumption of one product. A fundamental assumption of life cycle modeling is that process models are linearly scalable (Heijungs and Suh, 2002). This assumption leads to Eqn (23.7):

$$\underline{\mathbf{Ys}} \geq \underline{\mathbf{f}} \tag{23.7}$$

in which $\underline{s}$ is a vector of activity scaling factors and $\underline{\mathbf{f}}$ is the final demand on the value chain. The value of $\underline{\mathbf{s}}$ is determined by $\underline{\mathbf{Y}}$ and $\underline{\mathbf{f}}$ and can be found by solving the following linear program:

$$\begin{aligned} &\text{minimize } \sum_l \underline{s}_l \\ &\text{subject to } \underline{\mathbf{Ys}} \geq \underline{\mathbf{f}} \\ &\underline{s}_l \geq 0, \forall l \end{aligned} \tag{23.8}$$

$\underline{\mathbf{s}}$ represents the minimum level of production required from each value chain activity in order to produce $\underline{\mathbf{f}}$.

Any type of process can be represented with an IO vector. The difficulty lies in determining the relationship between the inputs and outputs, particularly if the relationship is not fixed as in value chain activity and sector models. Chemical processes can be represented in vector notation with the addition of unit operation models *external* to the IO vector that determine the exact values of the IO vector elements. Vector elements are derived from fundamental unit operation models and are determined by the values of unit operation design variables.

Chemical process models consist of two separate portions. The IO vectors shown in Eqns (23.9) and (23.10) contain process n's exchanges of M products with other processes and R interactions with the environment.

$$\mathbf{x}_n = \begin{bmatrix} x_{1n} \\ x_{2n} \\ \vdots \\ x_{Mn} \end{bmatrix} \tag{23.9}$$

$$\mathbf{b}_n = \begin{bmatrix} b_{1n} \\ b_{2n} \\ \vdots \\ b_{Rn} \end{bmatrix} \tag{23.10}$$

The second portion, the unit operation models h_{mn} and h_{rn}, shown in Eqns (23.11) and (23.12) describes the inner workings of the process and relate the process inputs and outputs to each other and to the set of unit operation design variables $\{z_n\}$ upon which process operation is dependent.

$$x_{mn} = h_{mn}(\{z_n\}) \tag{23.11}$$

$$b_{rn} = h_{rn}(\{z_n\}) \tag{23.12}$$

The IO vector is in a form that can easily be combined with and connected to value chain activity and sector models described in Section 23.2.2, which is more difficult when process models are in their usual form as a set of equations that do not distinguish between IO and inner flows.

23.2.4 MULTISCALE SYSTEM MODEL

With system components at all scales represented in the same form, connecting different scales becomes relatively straightforward. The sector models, activity models, and process IO vectors describe exchanges within scales; further IO vectors are required to describe exchanges between scales. A multiscale model of sector j then consists of the jth column of the total requirements matrix $\bar{\mathbf{i}}_j - \bar{\boldsymbol{a}}_j$, one IO vector $\underline{\mathbf{y}}_{dj}$ containing flows from the value chain scale to sector j, and a second IO vector $\mathbf{y}_{dj}^{E}$ containing flows from the process scale to sector j, as follows:

$$\begin{bmatrix} \bar{\mathrm{i}}_j - \bar{a}_j \\ -\underline{\mathrm{y}}_{dj} \\ -\mathrm{y}_{dj}^{E} \end{bmatrix} \tag{23.13}$$

Note that Eqn (23.13) captures only technological and economic exchanges; environmental interactions are still captured in the interactions matrix. The subscript d on $\underline{\mathbf{y}}_{dj}$ indicates that the vector represents flows downstream of the value chain, called downstream cutoff flows. The combined superscript E and subscript d on $\mathbf{y}_{dj}^{E}$ indicates that the vector represents flows downstream of the process scale that are sent to the economy.

Value chain activity models and sectors models are similarly connected to other scales. A multiscale model of activity l consists of an IO vector $\underline{\mathbf{y}}_{ul}$ of input flows from the economy, the standard activity model $\underline{\mathbf{y}}_{l}$, and an IO vector $\mathbf{y}_{dl}^{V}$ of input flows from the process scale as shown in Eqn (23.14):

$$\begin{bmatrix} -\underline{\mathrm{y}}_{ul} \\ \underline{\mathrm{y}}_{l} \\ -\mathrm{y}_{dl}^{V} \end{bmatrix} \tag{23.14}$$

The multiscale model of process n consists of an IO vector of input flows from the economy, an IO vector of input flows from the value chain, and an IO vector of flows to and from other processes.

$$\begin{bmatrix} -\mathrm{y}_{un}^{E} \\ -\mathrm{y}_{un}^{V} \\ \mathrm{y}_{n} \end{bmatrix} \tag{23.15}$$

Figure 23.2 gives the physical interpretation of the different subvectors in Eqns (23.13)–(23.15).

Combining the three scales into one integrated multiscale modeling framework leads to inevitable overlap, particularly between the EEIO model and the smaller scale components. Suppose that, in the multiscale system shown in Figure 23.2, value chain activity $\underline{\mathbf{S}}_4$ was captured in the model for sector $\bar{\mathbf{S}}_3$ and process $\mathbf{S}_6$ was captured in the model for sector $\bar{\mathbf{S}}_2$. Sector $\bar{\mathbf{S}}_3$ is then the parent sector for activity $\underline{\mathbf{S}}_4$ and sector $\bar{\mathbf{S}}_2$ is the parent sector for process $\mathbf{S}_6$. For instance, $\underline{\mathbf{S}}_4$ represents iron ore mining and $\bar{\mathbf{S}}_3$ represents all mining and mining support activities, while $\mathbf{S}_6$ represents an oil refinery and $\bar{\mathbf{S}}_2$ represents oil and natural gas extraction and processing. Then $\underline{\mathbf{S}}_4$ and $\mathbf{S}_6$ are captured twice in the multiscale model, once implicitly at the economy scale and once explicitly at a smaller scale, illustrated in Figure 23.3(a). The production and environmental interactions from $\underline{\mathbf{S}}_4$ and $\mathbf{S}_6$ are thus double counted, leading to the system appearing to have worse economic and environmental performance than it has in reality. To prevent double-counting system components, the EEIO model is disaggregated as shown in Figure 23.3(b). The portions of $\overline{\mathbf{S}}_3$ and $\overline{\mathbf{S}}_2$ corresponding to $\underline{\mathbf{S}}_4$ and $\mathbf{S}_6$ are removed, resulting in disaggregated sectors $\overline{\mathbf{S}}_{3*}$ and $\overline{\mathbf{S}}_{2*}$. Although not shown, the same procedure is applied to any value chain activities that double-count individual processes.

Equations (23.16) and (23.17) define the two primary components of the multiscale modeling framework.

$$\underline{\overline{\mathbf{A}}} = \begin{bmatrix} \overline{\mathbf{I}} - \overline{\mathbf{A}}^*(\{z\}) & -\underline{\mathbf{Y}}_u(\{z\}) & -\underline{\mathbf{Y}}_u^E(\{z\}) \\ -\underline{\mathbf{Y}}_d(\{z\}) & \underline{\mathbf{Y}}^*(\{z\}) & -\mathbf{Y}_u^V(\{z\}) \\ -\mathbf{Y}_d^E(\{z\}) & -\mathbf{Y}_d^V(\{z\}) & \mathbf{Y}(\{z\}) \end{bmatrix} \tag{23.16}$$

$$\underline{\overline{\mathbf{B}}} = \begin{bmatrix} \overline{\mathbf{B}}^*(\{z\}) & \underline{\mathbf{B}}^*(\{z\}) & \mathbf{B}(\{z\}) \end{bmatrix} \tag{23.17}$$

Equation (23.16) is the hybrid transactions matrix in which each row corresponds to a single product and each column is a multiscale model of a single process, activity, or sector. Submatrices not on the diagonal are the upstream and downstream cutoff matrices that model flows between scales. Upstream cutoff matrices are denoted with a subscript u and model flows from larger to smaller scales. Downstream cutoff matrices are denoted with a subscript d and model flows from smaller to larger scales. The three submatrices on the diagonal model flows within scales.

All of the submatrices in Eqn (23.16) are dependent on unit operation design variables z. At the process level, $\mathbf{Y}$ is explicitly dependent on z as the process models

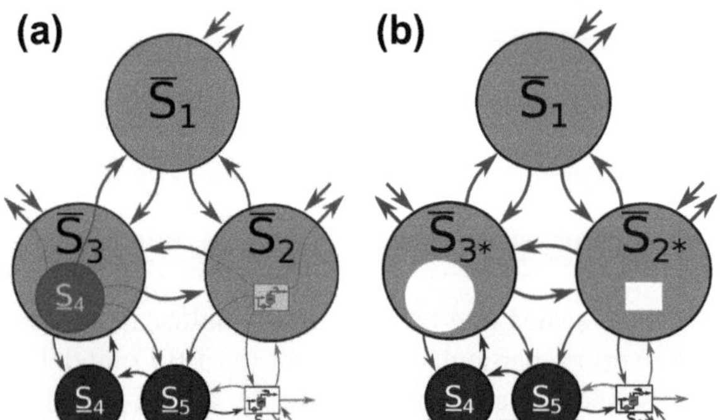

FIGURE 23.3

Value chain activities and individual processes are commonly captured in the EEIO model, causing those components to appear twice in the multiscale model (a). The EEIO model is disaggregated into the smaller scale component(s), modeled separately, and the remainder of the economy, captured in the EEIO model, to avoid double counting such components (b).

Details of the disaggregation procedure are given in Hanes and Bakshi (2014).

in **Y** are derived from unit operation models. Both the activity and sector models in the hybrid transactions matrix and the environmental models in the hybrid interactions matrix of Eqn (23.17) are dependent on z through the disaggregation procedure and through upstream and downstream flows that connect to the process scale.

The modeling framework is intended to represent a large design space, including design alternatives at both the process and value chain scale. At the process scale, alternatives include the choice of z and a superstructure of potential unit operations and connections between unit operations (Yeomans and Grossmann, 1999). The choice of different configurations is represented with the integer variables I that determine which unit operations are operating and how they are connected.

Value chain activities have linear models; the superstructure optimization problem at the value chain scale therefore reduces to the problem of finding the optimal linear combination of value chain activities. The set of alternatives is represented with a rectangular choice-of-technology (RCOT) model. An RCOT model contains one or more alternatives to each activity as additional columns in $\underline{\mathbf{Y}}$. $\underline{\mathbf{Y}}_{\mathrm{R}}$, the RCOT model for $\underline{\mathbf{Y}}$, has $\sum_{l\in L} N_l$ columns and L rows, where N_l are the number of alternatives for value chain activity l (Duchin and Levine, 2011, 2012). An example of $\underline{\mathbf{Y}}_{\mathrm{R}}$ with alternatives for one value chain activity is given in Eqn (23.18).

$$\underline{\mathbf{Y}}_{\mathrm{R}} = \left[\underline{\mathbf{Y}}_{1\ldots l-1} \middle| \underline{\mathbf{y}}_l^1 \quad \underline{\mathbf{y}}_l^2 \quad \cdots \quad \underline{\mathbf{y}}_l^{N_l} \middle| \underline{\mathbf{Y}}_{l+1\ldots L} \right] \tag{23.18}$$

When an RCOT model is used, additional columns representing the environmental interactions of the process alternatives must be added to the interactions matrix. The value chain RCOT interactions matrix is denoted by $\underline{\mathbf{B}}_{\mathrm{R}}$. Value chain design alternatives are included in the modeling framework given in Eqns (23.16) and (23.17) by replacing $\underline{\mathbf{Y}}^*(\{z\})$ and $\underline{\mathbf{B}}^*(\{z\})$ with $\underline{\mathbf{Y}}_{\mathrm{R}}^*(\{z\},\{I\})$ and $\underline{\mathbf{B}}_{\mathrm{R}}^*(\{z\},\{I\})$. The disaggregation process described in Hanes and Bakshi (2015) applies without modification to the RCOT model. The scaling vector and final demand vector in Eqns (23.19) and (23.20) are the last components in the modeling framework.

$$\underline{\overline{\mathbf{s}}} = \begin{bmatrix} \overline{\mathbf{s}} \\ \underline{\mathbf{s}} \\ \underline{\mathbf{s}} \end{bmatrix} \tag{23.19}$$

$$\underline{\overline{\mathbf{f}}} = \begin{bmatrix} \overline{\mathbf{f}} \\ \underline{\mathbf{f}} \\ \underline{\mathbf{f}} \end{bmatrix} \tag{23.20}$$

$\underline{\overline{\mathbf{s}}}$ is a vector that scales each component in the multiscale modeling framework such that the final demand $\underline{\overline{\mathbf{f}}}$ on the entire system is met. In general, the process scale elements of $\underline{\overline{\mathbf{s}}}$, $\mathbf{s}$, will be constants as the scale of the process to be designed is already fixed. The value chain and economy scale elements of $\underline{\overline{\mathbf{s}}}$ are decision variables in the process optimization problem. $\underline{\overline{\mathbf{f}}}$ is a vector of constants. The value chain and economy scale elements in $\underline{\overline{\mathbf{f}}}$ will generally be zero with the exception of the final demand on the process of interest. A product balance on the multiscale system is now written using Eqns (23.16)–(23.20).

$$\underline{\overline{\mathbf{A}}}(\{z\})\underline{\overline{\mathbf{s}}} \geq \underline{\overline{\mathbf{f}}} \tag{23.21}$$

The interpretation of Eqn (23.21) is that the level of production of each process, activity, and sector in the multiscale modeling framework is scaled and the values of the unit operation design variables are chosen such that the final demand on the system is met or exceeded.

The sustainable design problem of Eqn (23.1) can now be restated using the multiscale modeling framework. Equation (23.21) is a set of constraints on the decision variables $\underline{\overline{\mathbf{s}}}$, $\{z\}$, and $\{I\}$. Z_1 and Z_2, the two objective functions, still represent economic and environmental considerations, respectively. Z_1 is effectively unchanged from Eqn (23.1): the expanded analysis boundary of the proposed modeling framework does not affect Z_1, as only the economics of the process scale are of interest. However, Z_2 now involves environmental interactions generated at three distinct scales, modeled as *functions of decision variables* rather than as constant factors. Decision variables affect aspects of the system at all scales, including the macroeconomic structure, value chain structure, and unit operation configurations and operating conditions in each process. Due to the integrated nature of the proposed design framework, these decisions are optimized *simultaneously* and *in the context of the entire system*. There are no sequential decisions, and each decision within the system affects all other decisions regardless of scale. Equation (23.22) states the multiscale SPD problem as a mixed integer non-linear program (MINLP) with two objective functions.

$$\begin{aligned}
&\text{maximize } Z_1 = \sum_{n=1}^{N} Z_{1n}(\{z\},\{I\}) \\
&\text{minimize } \underline{\overline{Z}}_2 = \sum_{j=1}^{J} \overline{Z}_{2j}(\{z\},\{I\}) + \sum_{l=1}^{L} \underline{Z}_{2l}(\{z\},\{I\}) + \sum_{n=1}^{N} Z_{2n}(\{z\},\{I\}) \\
&\text{subject to} \\
&\begin{bmatrix} \overline{\mathbf{I}} - \overline{\mathbf{A}}^*(\{z\},\{I\}) & \underline{\mathbf{Y}}_u^*(\{z\},\{I\}) & \mathbf{Y}_u^E(\{z\},\{I\}) \\ \underline{\mathbf{Y}}_d^*(\{z\},\{I\}) & \underline{\mathbf{Y}}_R^*(\{z\},\{I\}) & \mathbf{Y}_u^V(\{z\},\{I\}) \\ \mathbf{Y}_d^E(\{z\},\{I\}) & \mathbf{Y}_d^V(\{z\},\{I\}) & \mathbf{Y}(\{z\},\{I\}) \end{bmatrix} \begin{bmatrix} \overline{\mathrm{s}} \\ \underline{\mathrm{s}} \\ \mathrm{s} \end{bmatrix} \geq \begin{bmatrix} \overline{\mathrm{f}} \\ \underline{\mathrm{f}} \\ \mathrm{f} \end{bmatrix} \\
&\mathbf{h}_n(\{z\},\{I\}) \geq \mathbf{0}, \quad \forall n \in N \\
&Z_{1n} = \mathbf{p}_n(\{z\},\{I\}) \cdot \mathrm{s}, \quad \forall n \in N \\
&\overline{Z}_{2j} = \overline{s}_j \overline{\mathbf{b}}_j(\{z\},\{I\}), \quad \forall j \in J \\
&\underline{Z}_{2l} = \underline{s}_l \underline{\mathbf{b}}_{l\mathrm{R}}(\{z\},\{I\}), \quad \forall l \in L \\
&Z_{2n} = s_n \boldsymbol{b}_n(\{z\},\{I\}), \quad \forall n \in N
\end{aligned} \tag{23.22}$$

Finding a solution to Eqn (23.22) involves iteration on the disaggregation calculations. Because the process-scale matrices contain elements that depend on decision variables, these matrices will change every time the decision variables are perturbed. The disaggregated direct requirements and technology matrices $\overline{\mathbf{A}}^*$ and $\underline{\mathbf{Y}}_{\mathrm{R}}^*$ will likewise change due to their dependence on the process-scale matrices (Hanes and Bakshi, 2015). This necessitates recalculating the elements of the hybrid technology matrix $\underline{\overline{\mathbf{A}}}$ and of the hybrid interactions matrix $\underline{\overline{\mathbf{B}}}$ at every iteration of the solution algorithm.

23.3 DEMONSTRATION: POLYMER SUPPLY CHAIN DESIGN

A toy process and supply chain design problem demonstrates the proposed modeling framework and its potential application. The supply chain involves the production of two polymer products, only one of which is of interest, from electricity, monomer, and catalyst. Figure 23.4 shows the superstructure of potential supply chain processes, with similar processes grouped by process type. Of the supply chain processes, only the model for expanded and thermoformed polymer production (ETPP), the process of interest, involves design variables; the supply chain processes have fixed, linear process models. The ETPP model depends on two variables, *X* and *C*, which are roughly analogous to reaction conversion and catalyst consumption rate. The desired product of ETPP is expanded polymer; ETPP also produces thermoformed polymer as a by-product, which is sold back to the economy. Due to the simplicity of this system, the value chain scale is neglected, leaving only the supply chain (process) and economy scales.

The polymer production system shown in Figure 23.4 exists within an eight-sector economy with four parent sectors and four nonparent sectors. Three of the parent sectors have two constituent processes each and the fourth has one constituent process; all of the supply chain processes have parent sectors. The eight industrial sectors of the economic system are water and sewage, coal mining, crude oil extraction, natural gas extraction, power generation and supply, organic chemical manufacturing, specialty chemical manufacturing, and plastics manufacturing (PM), with the latter four sectors being the parent sectors. PM is the parent sector of polymer production, and the other three sectors are the parent sectors for the supply chain processes.

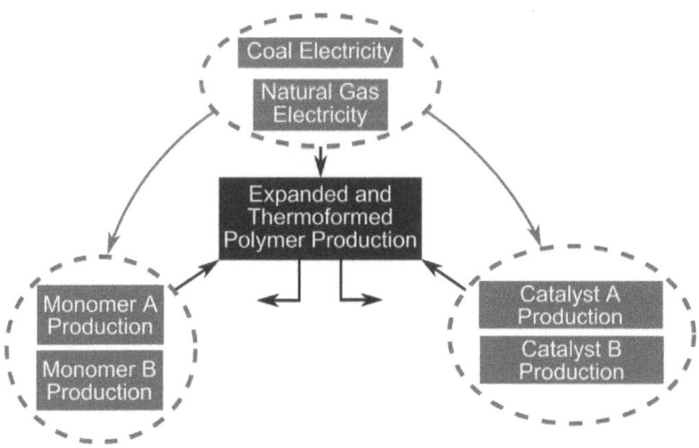

FIGURE 23.4

Superstructure of the polymer supply chain. The bold process and flows indicate dependence on unit operation variables. Processes grouped in the dotted ovals share a parent sector.

Because the ETPP process model involves two unit operation variables *X* and *C*, all inputs and outputs to this process as well as its environmental interactions are dependent on *X* and *C*. Figure 23.5 shows the relationships between parent sectors, constituent processes, and unit operation variables in this production system. PM, the parent sector for ETPP, is disaggregated into two components: ETPP, modeled theoretically at the process scale, and all other plastic production processes that remain in the economy-scale model. Through the disaggregation, the entire multiscale model of the disaggregated PM sector is dependent on *X* and *C*, even though the variables exist at a much smaller scale. The PM sector is connected through its inputs and outputs to all other sectors in the economy, both parent and nonparent. Thus, *the entire economic model and all economy-scale environmental interactions* are dependent, albeit indirectly, on unit operation variables. This interaction between model scales is captured *without* needing to specify the exact relationship between the unit operation variables and each sector, which could prove prohibitively difficult for more realistic systems. The upstream and downstream cutoff matrices $\mathbf{Y}_u(X, C)$ and $\mathbf{Y}_d(X, C)$ contain, respectively, inputs to the supply chain processes produced by economic sectors and by-products of the supply chain processes that are consumed by the economy.

Elements in these two matrices corresponding to ETPP are dependent on the unit operation variables, causing the level of production required from the economy and the level of *avoided* production caused by the sale of thermoformed polymer to vary according to *X* and *C*. In the model, the thermoformed polymer is arbitrarily sold back to the PM sector. However, the framework can also be used to identify the optimal market or combination of markets for by-products with respect to economic and/or environmental criteria of interest.

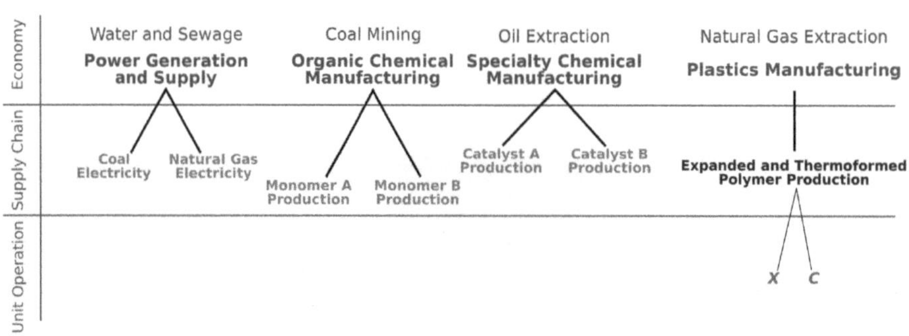

FIGURE 23.5

The toy polymer production system consists of an eight-sector economic system with four parent sectors, six supply chain processes, and one process of interest. All processes are constituent processes, and all save ETPP have empirical Leontief process models. ETPP is dependent on two variables at the unit operation scale, *X* and *C*.

23.3.1 RESULTS

Optimizing the polymer supply chain involves selecting the optimal mix of electricity, monomer and catalyst production processes from the superstructure, and choosing the optimal values of *X* and *C*. In this particular model, it is possible for the supply chain to include duplicate processes, for instance, a 50/50 mix of coal and natural gas electricity. The system was optimized to minimize separately total production cost, supply chain CO_2 emissions and life cycle (supply chain and economy) CO_2 emissions. Designs were obtained using the proposed multiscale modeling framework as well as a standard SPD approach in order to compare results of the two modeling approaches.

The standard SPD approach used the same supply chain model as the multiscale approach and simulates the value chain scale by accounting for only direct contributions from select economic sectors (see Figure 23.5). Minimum CO_2 designs obtained under the standard approach were then evaluated with the multiscale modeling framework in order to compare results of the two approaches under the same system boundary.

The effect of the different approaches and objective functions on the ETPP process design is shown in Figure 23.6. Figure 23.6(a) and (b) shows percentage change in results on the y-axis: all bars represent the percentage by which the result (production cost or process CO_2 emissions) changes compared to the base case, which is the standard SPD approach with the supply chain scale objective function. Figure 23.6(a) compares ETPP process costs in minimum cost supply chain designs, while Figure 23.6(b) compares ETPP emissions in minimum CO_2 supply chain designs. The most significant change from the base case ETPP process design occurred when the multiscale modeling framework was used with the life cycle scale CO_2 objective function: ETPP production cost increased more than 30% while emissions were reduced by about 8%.

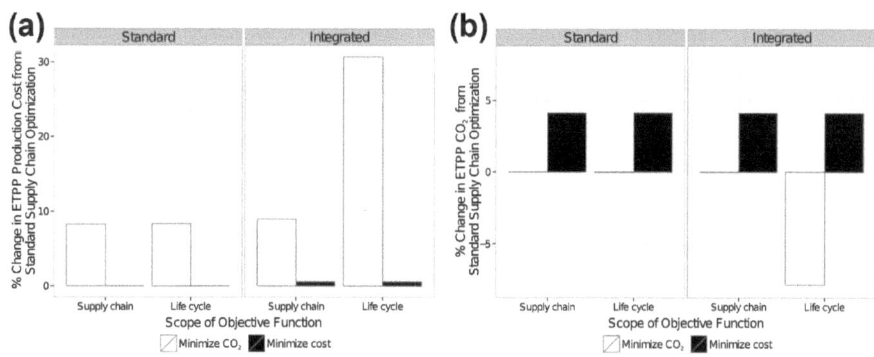

FIGURE 23.6

Percent change in ETPP CO_2 emissions (b) and production costs (a) for minimum CO_2 and minimum cost designs. Results of the standard SPD approach are on the left and results of the multiscale modeling framework are on the right.

If the rest of the supply chain is not considered, the results of Figure 23.6 are quite negative. The process designs obtained using the expanded analysis boundary and objective function scope have a much greater increase in cost than reduction in emissions. However, when emissions from the rest of the supply chain and life cycle are taken into account as in Figure 23.7, the advantages of the multiscale approach's expanded analysis boundary and of the expanded objective function scope are seen clearly. Figure 23.7 indicates that under both the standard and multiscale modeling approaches, the minimum life cycle CO_2 designs have lower system-wide emissions than do the minimum supply chain CO_2 designs. For the standard SPD approach, the emissions savings is about 6% and for the multiscale approach, 17%. By considering a larger analysis boundary and objective function scope, new supply chain designs are found that trade an increase in emissions at the supply chain scale for a much larger decrease in emissions at the economy scale, resulting in an integrated system with lower overall emissions. Moreover, when production cost is considered for the entire supply chain, the increase in cost of the multiscale approach designs is much less significant than when only the ETPP process is considered: 0.4% for minimum supply chain CO_2 designs and 3% for minimum life cycle CO_2 designs. The trade-off between increased production cost and reduced emissions is thus much more

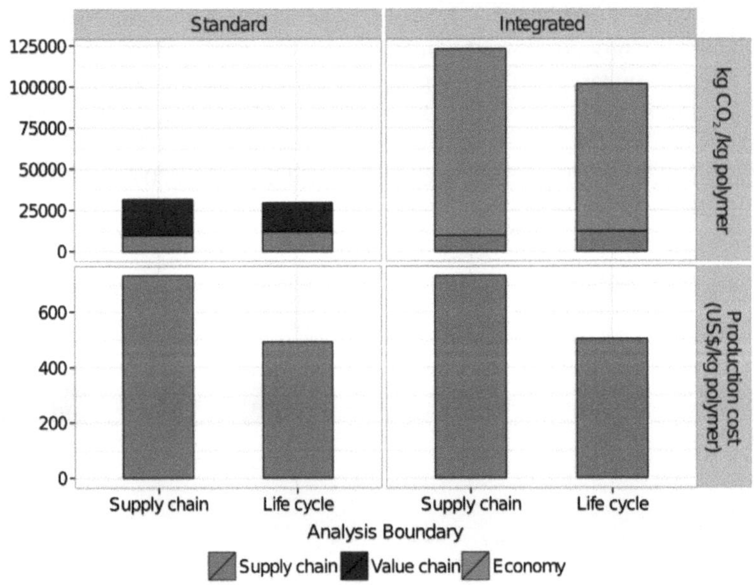

FIGURE 23.7

Production cost and CO_2 emissions for minimum CO_2 designs obtained using a standard SPD approach (left) and the multiscale modeling framework (right). Due to the simplicity of the problem, the multiscale modeling framework was simplified to include only supply chain and economy scale models.

attractive for the supply chain as a whole than for the ETPP process alone. Figure 23.7 also indicates that both approaches located supply chain designs with lower production costs when life cycle CO_2 emissions were minimized compared to when supply chain CO_2 emissions were minimized. This result is due to the differences in the supply chain process models. The supply chain models were set up such that in each pair of processes supplying an input to ETPP, one process had higher process CO_2 emissions and lower cost; the high-CO_2/low-cost processes also generally had fewer inputs drawn from the economy. When CO_2 was minimized at the supply chain scale, the low-CO_2/high-cost processes with more economic inputs were selected, resulting in a design with both higher production costs and higher emissions from the economy scale. On the other hand, minimizing life cycle CO_2 minimized the sum of process and economy emissions, resulting in the high-CO_2/low-cost processes with fewer economic inputs being chosen. Thus, the total emissions as well as the production cost were reduced by the minimum life cycle CO_2 designs, although the process scale emissions increased.

Although Figure 23.7 appears to indicate that the standard SPD approach yielded supply chain designs with lower CO_2 emissions, much of the difference between the two sets of results is due to the different analysis boundaries. The standard SPD approach neglects all indirect contributions from the economy scale while the multiscale approach includes direct and indirect economic contributions. Figure 23.8 shows CO_2 emissions of the four minimum CO_2 supply chain designs in Figure 23.7 with the emissions for the standard SPD approach designs calculated using the multiscale modeling framework.

The minimum CO_2 designs obtained with the multiscale approach have lower emissions than the standard SPD approach designs for both supply chain and life cycle objective functions. At the supply chain scale, the reduction in CO_2 is approximately 5% and at the life cycle scale the reduction is approximately 4%.

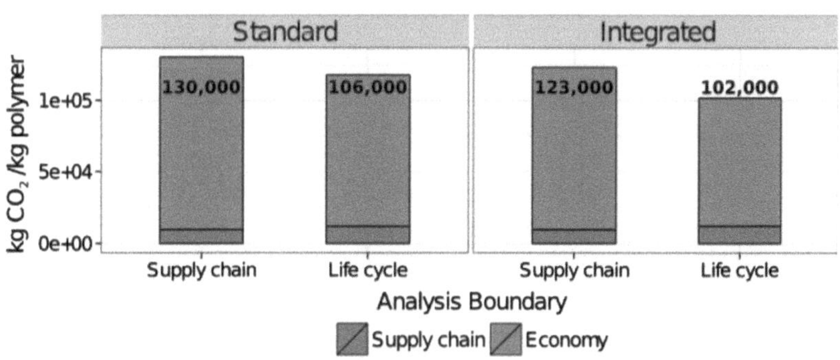

FIGURE 23.8

Minimum CO_2 supply chain designs obtained under the standard SPD approach are evaluated using the multiscale modeling framework (left) and compared to minimum CO_2 designs obtained under the multiscale approach (right).

23.4 CONCLUSIONS

The proposed multiscale modeling framework draws on hybrid LCA methods to integrate models at the process, value chain, and economy scales. Detailed engineering models with unit operation design variables are used at the process scale while coarser empirical models capture the value chain and economy scales. The economy and value chain scale models are disaggregated to avoid implicitly double-counting system components at smaller scales. As a result of the disaggregation, unit operation variables are propagated throughout the model, allowing the full integrated model to be dependent on the process or supply chain design. Applying the multiscale framework to SPD applications expands the typically restricted analysis boundary to the national level, thereby including the full life cycle network in the process or supply chain optimization. Results of a toy supply chain design problem indicate that the proposed framework yields significant environmental benefits. Supply chains designed for minimum emissions using a standard SPD approach were evaluated using the multiscale framework and were found to have higher overall emissions than did minimum emissions designs obtained using the multiscale approach. This indicates that the expanded analysis boundary of the multiscale approach prevents life cycle impacts from being shifted outside the system, resulting in designs that are environmentally superior to those obtained with a more restricted analysis boundary. Additionally, while supply chain production cost increased under the multiscale approach, the increase was proportionally much less than the reduction in emissions.

Although not demonstrated in the toy problem, all MOO techniques currently used in SPD apply without modification to the multiscale modeling framework. Future work will use the multiscale framework to locate the Pareto curve with respect to process or supply chain economics and environmental impacts of the full life cycle.

REFERENCES

Azapagic, A., Clift, R., 1999. Life cycle assessment and multiobjective optimisation. J. Cleaner Prod. 7 (2), 135–143.

Cano-Ruiz, J., McRae, G., 1998. Environmentally conscious chemical process design. Annu. Rev. Energy Environ. 23 (1), 499–536.

Douglas, J.M., 1992. Process synthesis for waste minimization. Ind. Eng. Chem. Res. 31 (1), 238–243.

Duchin, F., Levine, S.H., 2011. Sectors may use multiple technologies simultaneously: the rectangular choice-of-technology model with binding factor constraints. Econ. Syst. Res. 23 (3), 281–302.

Duchin, F., Levine, S.H., 2012. The rectangular sector-by-technology model: not every economy produces every product and some products may rely on several technologies simultaneously. J. Econ. Struct. 1 (1), 1–11.

Eco-LCA, 2014. http://resilience.eng.ohio-state.edu/ecolca-cv/.

El-Halwagi, M.M., Manousiouthakis, V., 1989. Synthesis of mass exchange networks. AIChE J. 35 (8), 1233–1244.

Elia, J.A., Baliban, R.C., Floudas, C.A., 2014. Nationwide, regional, and statewide energy supply chain optimization for natural gas to liquid transportation fuel (GTL) systems. Ind. Eng. Chem. Res. 53 (13), 5366–5397.

Grossmann, I., Drabbant, R., Jain, R.K., 1982. Incorporating toxicology in the synthesis of industrial chemical complexes. Chem. Eng. Commun. 17 (1–6), 151–170.

Guillen-Gosalbez, G., Grossmann, I.E., 2009. Optimal design and planning of sustainable chemical supply chains under uncertainty. AIChE J. 55 (1), 99–121.

Hanes, R.J., Bakshi, B.R., 2015. From Process to Planet: A Multiscale Modeling Framework for Sustainable Engineering. Technical Report. PSE Group, The Ohio State University.

Heijungs, R., 1994. A generic method for the identification of options for cleaner products. Ecol. Econ. 10 (1), 69–81. http://www.sciencedirect.com/science/article/pii/0921800994900388.

Heijungs, R., Suh, S., 2002. The Computational Structure of Life Cycle Assessment. Kluwer Academic Publishers. http://library.ohio-state.edu/record=b5578772_S7.

Hilaly, A.K., Sikdar, S.K., 1994. Pollution balance: a new methodology for minimizing waste production in manufacturing processes. Air Waste 44 (11), 1303–1308.

Johansen, L., 1960. A Multi-Sectoral Study of Economic Growth, vol. 21. North-Holland Publishing Company.

Joshi, S., 1999. Product environmental life-cycle assessment using input-output techniques. J. Ind. Ecol. 3 (2–3), 95–120.

Kniel, G., Delmarco, K., Petrie, J., 1996. Life cycle assessment applied to process design: environmental and economic analysis and optimization of a nitric acid plant. Environ. Prog. 15 (4), 221–228.

Lenzen, M., 2000. Errors in conventional and input-output based life cycle inventories. J. Ind. Ecol. 4 (4), 127–148.

Leontief, W., 1970. Environmental repercussions and the economic structure: an input-output approach. Rev. Econ. Stat. 52 (3), 262–271.

Leontief, W., Ford, D., 1972. Air Pollution and the Economic Structure: Empirical Results of Input-Output Computations. North Holland Publishing, pp. 9–30.

Leontief, W.W., 1936. Quantitative input and output relations in the economic systems of the United States. Rev. Econ. Stat. 18 (3), 105–125.

Linnhoff, B., Flower, J.R., 1978. Synthesis of heat exchanger networks: I. Systematic generation of energy optimal networks. AIChE J. 24 (4), 633–642.

Miller, R.E., Blair, P.D., 1985. Input-Output Analysis. Prentice-Hall, Englewood Cliffs, NJ.

Moriguchi, Y., Kondo, Y., Shimizu, H., 1993. Analysing the life cycle impacts of cars: the case of CO_2. Ind. Environ. 16 (1/2), 42–45.

National Renewable Energy Laboratory, 2012. US Life Cycle Inventory Database (accessed April 2014). https://www.lcacommons.gov/nrel/search.

Stefanis, S., Livingston, A., Pistikopoulos, E., 1997. Environmental impact considerations in the optimal design and scheduling of batch processes. Comp. Chem. Eng. 21 (10), 1073–1094.

Suh, S., 2004. Functions, commodities and environmental impacts in an ecological-economic model. Ecol. Econ. 48 (4), 451–467.

Suh, S., Huppes, G., 2005. Methods for life cycle inventory of a product. J. Cleaner Prod. 13 (7), 687–697.

Swiss Centre for Life Cycle Inventories, 2013. Ecoinvent Life Cycle Inventory Database (accessed September 2013). http://www.ecoinvent.org/.

Yang, Y., Shi, L., 2000. Integrating environmental impact minimization into conceptual chemical process design process systems engineering review. Comp. Chem. Eng. 24 (2), 1409–1419.

Yeomans, H., Grossmann, I.E., 1999. A systematic modeling framework of superstructure optimization in process synthesis. Comp. Chem. Eng. 23 (6), 709–731.

Young, D.M., Cabezas, H., 1999. Designing sustainable processes with simulation: the waste reduction (WAR) algorithm. Comp. Chem. Eng. 23 (10), 1477–1491.

Yue, D., Slivinsky, M.R., Sumpter, J., You, F., 2014a. Sustainable design and operation of cellulosic bio-electricity supply chain networks with life cycle economic, environmental and social optimization. Ind. Eng. Chem. Res. 53 (10), 4008–4029.

Yue, D., You, F., Snyder, S.W., 2014b. Biomass-to-bioenergy and biofuel supply chain optimization: overview, key issues and challenges. Comp. Chem. Eng. 66, 36–56.

CHAPTER

Social Sustainability of Complex Systems

24

Tamara Popovic*,[1], Andrzej Kraslawski*,[§]

**School of Industrial Engineering and Management, Lappeenranta University of Technology, Lappeenranta, Finland*

[§]Faculty of Process and Environmental Engineering, Technical University of Lodz, Lodz, Poland

[1]Corresponding author: E-mail: tamara.popovic@lut.fi

24.1 INTRODUCTION

The great majority of the systems resulting from natural evolution (atmosphere, ocean, earth, etc.), social interactions (group, community, etc.), and engineering activities (city, factory, market, etc.) can be classified as complex systems (Li et al., 2004). Thus, it is not surprising that the interest for better understanding of their functioning is incessant.

Even with the increasing interest in complex systems, their generally accepted definition is still missing. Most commonly they are understood as sets composed of large number of interconnected elements (agents) that create stable structures thanks to their interactions (Mitchell, 2006). That actually means that describing or understanding the function of one part of the system cannot give information about the function of the whole system (Bar-Yam, 1997). Basic characteristics of complex systems are

- Unpredictable behavior,
- Plethora of interconnections,
- Emerging new type of behavior,
- Synergy phenomena (inability to calculate it just by adding new elements), and
- Ability of self-organization (Marczyk and Deshpande, 2008).

The above mentioned characteristics make complex systems closely related to social, technological, and natural processes (Sokolova and Fernández-Caballero, 2012), because there is always a risk that the difficulty of predicting behavior of complex systems will have a negative impact on society and human well-being (Brinsmead and Hooker, 2011). Also, Misra et al. (2010) emphasize the importance of understanding the vulnerabilities of complex systems, as today's society is reliant on their various functions. Therefore, complex systems, as such, can have an impact on various social sustainability aspects such as health, comfort, safety, etc. That motivates the need of research for an assessment of social sustainability of complex systems.

Computer Aided Chemical Engineering, Volume 36. ISSN 1570-7946. http://dx.doi.org/10.1016/B978-0-444-63472-6.00024-0

24.2 SOCIAL SUSTAINABILITY

It is generally accepted that the social pillar is one of the least addressed elements of sustainability (Colantonio, 2011). And as it can be seen in Figure 24.1, social sustainability requires further development in order to be assimilated with environmental and economic sustainability and thus enable a detailed assessment of all sustainability pillars.

The assessment of the sustainability is usually made using sustainability indicators, thus social sustainability indicators should address impact on the social systems (Brent and Labuschagne, 2006). The main role of sustainability indicators is to simplify the evaluation process and to enable easier use of the obtained assessments by stakeholders, project engineers, politicians, etc. (Parkes et al., 2012). The existing methods of social sustainability assessment are diversified and developed by many organizations and researchers. Social sustainability can be assessed by using indicator frameworks, social impact assessment (SIA) guidelines, government actions, socially responsible investment indexes, international standards and guidelines, corporate social responsibility standards, etc. (Brent and Labuschagne, 2006). The existing methodologies and frameworks are usually applying similar social impact categories (Benoît, 2010), for example, social sustainability comprises large set of specific impact categories. Some of them are health, housing, security, empowerment, equity, human rights, etc. (Brent and Labuschagne, 2006). The high diversity of the existing methodologies is a basis for the further presentation of the SIA, social life cycle assessment (S-LCA), and global reporting initiative (GRI).

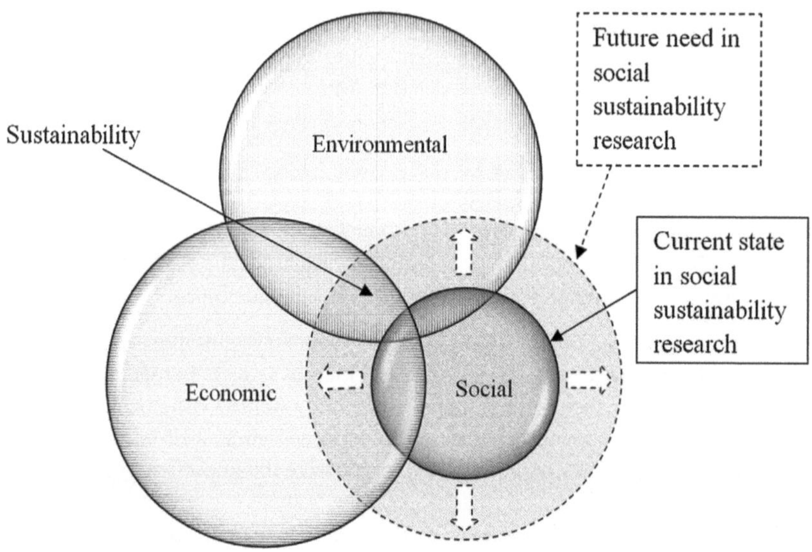

FIGURE 24.1

Three pillars of sustainability.

Table 24.1 Principles of SIA

Principle 1	Evaluate existing human environment, which could be affected by the certain action, program, or policy.
Principle 2	Evaluate social sustainability aspects from human environment that could be affected by certain action, program, or policy.
Principle 3	Identify methods that could be used for evaluation of impacts.
Principle 4	Collect data needed for making evaluation.
Principle 5	Verify that all social sustainability aspects are included in process of evaluation.
Principle 6	Perform evaluation or monitoring of the certain action, program, or policy.

US principles and guidelines, 2003

24.2.1 SOCIAL IMPACT ASSESSMENT

Social impact assessment is developed as addition to the environmental impact assessment (Benoît and Vickery-Niederman, 2010). It is methodology that assesses social sustainability by using qualitative and quantitative indicators useful for decision makers (Benoît and Vickery-Niederman, 2010). Guideline for social impact assessment is comprised of six principles, which are summarized in Table 24.1 (US principles and guidelines, 2003).

24.2.2 SOCIAL LIFE CYCLE ASSESSMENT

S-LCA is a social impact assessment methodology that focuses on life cycle activities that affect people (Dreyer et al., 2006). It is developed as an addition to life cycle assessment (LCA) in order to include the social aspect of sustainability in the LCA methodology (Jørgensen et al., 2008). However, S-LCA can be used individually without combining it with LCA. The phases in the S-LCA are derived from the ISO14044 framework as

- Goal and scope definition,
- Life cycle inventory analysis,
- Life cycle impact assessment, and
- Interpretation (UNEP, 2009).

Stakeholder involvement is highly emphasized, and stakeholders play a crucial role in the S-LCA methodology (UNEP, 2009). Therefore, indicators developed for the evaluation in the S-LCA are phased according by the stakeholders (workers, local community, consumers, society, and value chain actors) affected by the life cycle activities. However, an extensive list of selected indicators does not exist, because their use depends on the goal and scope of the study (UNEP, 2009).

24.2.3 GLOBAL REPORTING INITIATIVE

GRI is one of the sustainability reporting frameworks, which assess social, environmental, and economic sustainability. The scope of GRI as monitoring framework is process industry, which can make reports about their sustainability development. It is the best known voluntary framework for sustainability reporting (Marimon et al., 2012). The framework uses indicators for the assessment of the three pillars of sustainability.

For social sustainability, the subcategories according to which the indicators are sorted could be labor practices and decent work, human rights, society, and product responsibility (GRI, 2013). The main purpose of GRI is to give information to stakeholders so they can make proper decisions about investments or purchase of certain product. Another purpose can be seen in comparing results with other corporations (Marimon et al., 2012).

24.3 SOCIAL SUSTAINABILITY OF COMPLEX SYSTEM

A common characteristic of previously presented assessment methodologies is the inability to make a complete social sustainability assessment of complex systems (UNEP, 2009). The same is stated in Agenda 21 (United Nations Sustainable Development, 1992), that the policies for full assessment of the sustainability of complex systems are still missing and that research actions should be focused on developing assessment tools.

According to some authors, such as Babarenda Gamage et al. (2010), the assessment of sustainability of complex systems can be done by understanding characteristics of systems and their possible impacts on the various aspects of sustainability. If none of the existing methodologies is able to allow for an evaluation, further development of a new model has to be performed (Babarenda Gamage et al., 2010). Based on that approach, the steps in the social sustainability assessment of complex systems can be presented as in Figure 24.2.

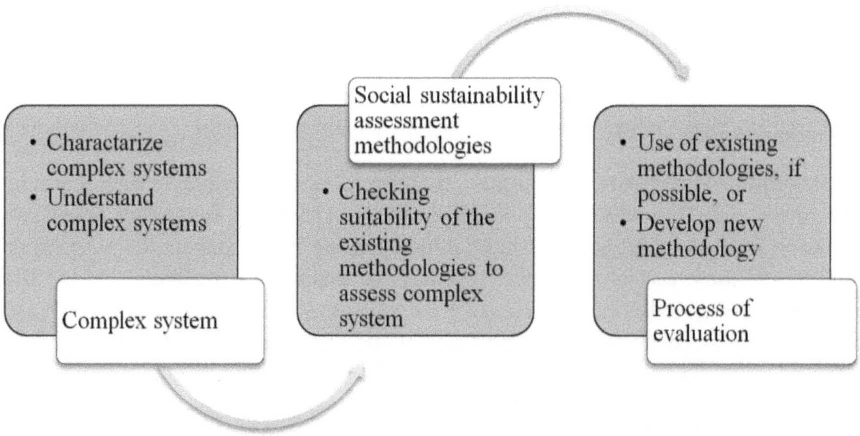

FIGURE 24.2

Steps in assessing social sustainability of complex systems.

The new methodology can be developed by incorporating two or more existing methods in order to encompass all properties and impacts of complex systems (Babarenda Gamage et al., 2010).

24.4 CASE STUDY

Even with the high diversity of assessment methodologies a generally accepted methodology for evaluation of social sustainability of complex systems is still missing. This is because existing methodologies and frameworks are not fully developed and cannot encompass all characteristics of complex systems. Therefore, further research and development on social sustainability is still needed.

Presented methodologies (SIA, S-LCA, GRI) can serve as the foundation for further development of new methodologies for assessing social sustainability of complex systems. Since the main tools for measuring social sustainability are indicators, their development is highly encouraged. The indicators should cover a high range of impacts and be understandable for all stakeholders as crucial elements of assessment of social sustainability.

For instance, social sustainability indicators can be developed and classified according to aspects such as health, comfort, safety and security, social capital, social equity, and social peace (Popovic et al., 2014). Popovic et al. (2014) addressed the social impact in the context of sustainability by proposing a set of indicators. The developed indicators have been used to enable quantitative assessment of social sustainability of wastewater treatment facilities. The proposed indicators are presented in Table 24.2.

The calculation and evaluation of the proposed indicators is performed on the basis of the annual reports of various wastewater treatment plants (Popovic et al., 2014). The detailed calculation is omitted in this chapter and the obtained results are given in Table 24.3.

From the results obtained in Table 24.3 it can be inferred that the social sustainability indicators proposed by Popovic et al. (2014) can be used only for relative evaluation of the process.

The same observations are obtained using methodologies and frameworks presented previously (SIA, S-LCA, GRI). However, the difference between SIA, S-LCA, or GRI and the approach used by Popovic et al. (2014) consists in the fact that SIA, S-LCA, and GRI use general indicators, i.e., the social sustainability indicators are not specifically developed for one type of the system. That kind of approach could cause omission of some relevant impacts on social sustainability. It can provoke the question whether it is better to develop and use generalized indicators or to make them system/process specific. However, analyzing the literature one can easily see that various complex systems have influence on different aspects of social sustainability. Therefore, the approach based on the use of system-specific indicators appears to be more suitable to address the issue of a broader range of impacts.

Table 24.2 Quantitative Social Sustainability Indicators for Wastewater Treatment Facilities

Social Sustainability Aspect	Social Sustainability Indicator	Indicator Definition	Method of Evaluation	Unit of Measure
Health	Violations	Shows concentration of pollutants, which deviate from concentration of pollutants dictated by law (e.g., BOD, TSS)	Concentration at the end of process/concentration dictated by law	–
	Vision	Presents the goal of the wastewater treatment plant to achieve certain cleaning properties, by estimating percentage of annual budget allocated for vision	(Investment for vision/annual plant budget) 100%	%
	Scope of network	Shows percentage of residents connected to sewers	% of households connected to treatment	%
Comfort	Onerousness	Shows occupied area by the wastewater treatment plant, as well as sewer pipe durability by estimating frequency of their change	$km^2$of occupied land; km of sewer pipes changed/year	km^2; km/year
	Scope of sludge treatment	Shows percentage of activated sludge used in agriculture, landfills, or thermal disposal, etc.	% of treated sludge	%
Safety and security	Employee training	Shows hours of training per year and worker satisfaction by the level of training	h of training/year; % of workers satisfied with level of training	h/year; %
	Research and development	Shows number of projects done in the wastewater treatment plant within certain time period	No. of successful projects/total no. of projects	–

Social capital (Popovic et al., 2014)	Customer value	Shows amount of water reuse and for what purpose	% of reused water	%
	Subsidies	Shows percentage of subsidies from public institutions that is used e.g., for improvement of process.	% of subsidies of total plant budget	%
	Efficiency	Shows the number of experts exchange across sectors	Number of experts in a given organization/total number of experts	–
Social equity (Popovic et al., 2014)	Fees	Compares the price of the treatment for customers and industries	Price for residents: price for companies = x:y	–
	Local affairs/autonomy	Shows existence/absence of price variations among different regions	(price at certain distance—minimum price)/price at certain distance	–
	Social compatibleness	Shows financial accessibility to the treatment, by estimating billing method for the treatment	€/m^3 of wastewater or % of income/m^3	€ ($)/$m^3$ or %
Social peace (Popovic et al., 2014)	Responsiveness	Shows ability of plant to react on sudden hazard situations	Number of exercises/year	No./year
	Corporate environmental awareness	Shows approach of different companies/industries to the treatment of its wastewater	Number of companies (industries) treating their own wastewater/total number of companies (industries) in community	–
	Participativeness	Shows possibility for stakeholders to be involved in decision-making	Number of elections/total number of new projects	–

Table 24.3 The Assessment of the Values of Social Sustainability Indicators

Social Sustainability Aspect	Indicator	Obtained Value (result)	Source
Health	Violations	0.11 (BOD); 0.24 (TSS)	Operations report (2012)
	Vision	93%	
	Scope of network	76.3%	
Comfort	Onerousness	630.54 km	
	Scope of sludge treatment	1407 t	
Safety and security	Employee training	91.67% of satisfied workers	
	Research and development	0.71	
Social capital	Customer value	19% for irrigation	Assefaw et al. (2009)
	Subsidies	55.7%	Assefaw et al. (2009)
	Efficiency	Lack of data	–
Social equity	Fees	Price for residents: price for industry = (8.50$ + (3.46$/1000 gallons)): ((8.50$ + (3.46$/1000 gallons)) + surcharge fees)	Rates and Fees (2013)
	Local affairs/autonomy	0—same price regardless of the distance	Rates and Fees (2013)
	Social compatibleness	$/1000 gallons	Rates and Fees (2013)
Social peace	Responsiveness	96 work observations/year	Annual Report (2011)
	Corporate environmental awareness	Lack of data	–
	Participativeness	1—stakeholders included in decision-making	Assefaw et al. (2009)

However, social sustainability as a wide, emerging field requires further development and research as well as the involvement of a wide range of scholars specializing in engineering and social scientists.

REFERENCES

Annual Report, 2011. Water and Wastewater Treatment Branch, City of Saskatoon. saskatoon.ca/DEPARTMENTS/City%20Clerks%20Office/Pages/ReportsAndPublications.aspx (accessed 10.12.13.).

Assefaw, M., Karani, P., Hamza, A., Njuguna, P.E., Askari, K., 2009. In: Gabal El-Asafar Wastewater Treatment Plant – Stage II Phase II Project, Project Appraisal Report. African Development Bank. afdb.org/en/documents/document/egypt-gabal-el-asfar-wastewater-treatment-plant-stage-ii-phase-ii-project-appraisal-report-40978/ (accessed 17.12.13.).

Babarenda Gamage, G., Boyle, C., McDowall, R., 2010. The development of an integrated model for assessing sustainability of complex systems. In: 4th International Conference on Sustainability Engineering and Science, Auckland, NZ.

Bar-Yam, Y., 1997. Dynamics of Complex Systems. Addison-Wesley, Massachusetts.

Benoît, C., Vickery-Niederman, G., 2010. Social sustainability assessment literature review. In: The Sustainability Consortium. Arizona State University and University of Arkansas, USA.

Brent, A., Labuschagne, C., 2006. Social indicators for sustainable project and technology life cycle management in the process industry. Int. J. Life Cycle Assess. 11 (1), 3–15.

Brinsmead, T.S., Hooker, C., 2011. Complex system dynamics and sustainability: conception, method and policy. In: Hooker, C. (Ed.), Handbook of the Philosophy of Science, Philosophy of Complex Systems, vol. 10. Elsevier, BV.

Colantonio, A., 2011. Social sustainability: exploring the linkages between research, policy and practice. In: Jaeger, C.C., Tàbara, J.D., Jaeger, J. (Eds.), European Research on Sustainable Development. Springer, Berlin.

Dreyer, L.C., Hauschild, M.Z., Schierbeck, J., 2006. A framework for social life cycle impact assessment. Int. J. LCA 11 (2), 88–97.

GRI (Global Reporting Initiative), 2013. G4 Sustainability Reporting Guidelines. GRI, Amsterdam, Netherlands.

Jørgensen, A., Le Bocq, A., Nazarkina, L., Hauschild, M., 2008. Methodologies for social life cycle assessment. Int. J. LCA 13 (2), 96–103.

Li, J., Zhang, J., Ge, W., Liu, X., 2004. Multi-scale methodology for complex system. Chem. Eng. Sci. 59, 1687–1700.

Marczyk, J., Deshpande, B., 2008. Measuring and tracing complexity in science. In: Minai, A., Braha, D., Bar-Yam, Y. (Eds.), Unifying Themes in Complex Systems – Proceedings of the Sixth International Conference on Complex Systems. Springer, Berlin.

Marimon, F., del Mar Alonso-Almeida, M., del Pilar Rodríguez, M., Cortez Alejandro, K.A., 2012. The worldwide diffusion of the global reporting initiative: what is the point? J. Clean. Prod. 33, 132–144.

Mishra, V., Dion, H., Bar-Yam, Y., 2010. Vulnerability analysis of high dimensional complex systems. In: Dolev, S., Cobb, J., Fischer, M., Yung, M. (Eds.), Stabilization, Safety, and Security of Distributed Systems. Springer-Verlag, Berlin.

Mitchell, M., 2006. Complex systems: network thinking. Artif. Intell. 170 (18), 1194–1212.

Operations report, 2012. Water and Sewer Department. City of Greeley, CO, USA. http://www.greeleygov.com/Water/documentlibrary.aspx%20 (assessed 01.09.14.).

Parkes, O., Bogle, I.D.L., Littieri, P., 2012. Toward defining quantitative methodology to enhance the sustainability performance of major international events. In: Lockhart Bogle, I.D., Fairweather, M. (Eds.), Computer Aided Chemical Engineering, 30, pp. 46–50.

Popovic, T., Kraslawski, A., Heiduschke, R., Repke, J.-U., 2014. Indicators of social sustainability for wastewater treatment processes. In: Eden, M., Siirola, J.D., Towler, G.P. (Eds.), Computer Aided Chemical Engineering, 34, pp. 723–729. http://dx.doi.org/10.1016/B978-0-444-63433-7.50105-X.

Rates and Fees, 2013. City of Longmonton, CO. Official Government Website. ci.longmont.co.us/pwwu/water/fees.htm#surcharge (accessed 17.12.13.).

Sokolova, M.V., Fernández-Caballero, A., 2012. Decision Making in Complex System. Springer-Verlag, Berlin.

UNEP – United Nations Environmental Programme, 2009. Guidelines for Social Life Cycle Assessment of Products.

United Nations Sustainable Development, 1992. United Nations Conference on Environment & Development. Agenda 21, Rio de Janerio, Brazil.

US principles and guidelines, 2003. Principles and guidelines for social impact assessment in the USA. Impact Assessment and Project Apprasial 21 (3), 231–250.

CHAPTER

Economic Sustainability of Products and Processes 25

Davide Manca
PSE-Lab, Process Systems Engineering Laboratory, Dipartimento di Chimica, Materiali e Ingegneria Chimica "Giulio Natta", Milano, Italy
E-mail: davide.manca@polimi.it

25.1 INTRODUCTION

This chapter focuses on the economic sustainability of both industrial products and processes. Products are produced in industrial plants by specific processes. A process can produce some products. A product can be produced by some processes.

For instance, on the one hand, the SSP (i.e., solid state polycondensation) process can produce different grades of polyethylene terephthalate (PET) by modulating the intrinsic viscosity of that polymer as a function of suitable set points of the process variables. PET products range from food grade packaging to technical yarns (e.g., tire cords, bottle-grades, air bag yarns, cloth fillings, substrates for solar cells). On the other hand, desalinated water can be produced by at least three distinct processes: reverse osmosis, evaporation, freezing.

Products and processes are conceptually dissimilar but at the same time they are interwoven and above all interacting. The industrial horizon covered by a product is wider than that of a process since it requires considering features that operate before and after the production plant where the process is run. Often, the literature refers to that horizon as the "supply chain" (SC) of the product where the process plays the role of transforming the input raw material(s) into some output product(s). The SC comprises further actions such as moving both the input and output goods to and fro the plant while solving the distribution and inventory problems, and meeting the customer(s)' demand. In addition, the SC has to deal with supplementary intangible features (that are compromising, blocking, and mandatory) such as the availability of utilities, transportation means, maintenance support, insurance, and advertising. Here, the concept of sustainability assumes a primary role in running the process to produce the product through a well-balanced SC management (SCM). The extension of the above-mentioned horizon increases further as the sustainability of SC is involved. According to Bakshi and Fiksel (2003), the number of stakeholders implicated in the SC covers new dimensions of the problem. Figure 25.1 shows how the sustainability of SC is a multifaceted question.

Computer Aided Chemical Engineering, Volume 36. ISSN 1570-7946. http://dx.doi.org/10.1016/B978-0-444-63472-6.00025-2

FIGURE 25.1

List of stakeholders involved and contributing to the sustainability of SC.

see also Bakshi and Fiksel (2003).

Usually, the first two stakeholders (i.e., #1 and #2 of Figure 25.1) belong to the standard representation of corporate SC. From #3 to #8, the list of stakeholders widens and starts covering new territories of the SC panorama, which are often unexplored or at least under considered at the corporate level. To better understand the articulation of that list, it is worth presenting three milestones that marked the path toward a progressive familiarization with the sustainability term applied to SC and SCM subjects. These milestones are: WCED (1987), Rio summit, 1992, and World summit, 2002. The first document widely recognized as the minter of the sustainable development (SD) path was published in 1987 with the title "Our common future" from the World Commission on Environment and Development (WCED) of United Nations. Also known as the Brundtland report (from the last name of the coordinator and president of WCED in 1987) that document synthesized the contributions of *senior government representatives, scientists and experts, research institutes, industrialists, representatives of nongovernmental organizations, and the general public* thus displaying a manifold approach to sustainability. WCED (1987) focused on the environmental issues of economic sustainability and recognized that poverty reduction, gender equity, and wealth redistribution play a major role in human resource development and are crucial in the formulation of strategies for environmental conservation. Additionally, the Brundtland report spotted that there are environmental limits to economic growth of industrialized countries. It also laid the foundations for the Earth Summit of 1992, which delivered the Agenda 21 document and the Rio declaration that established the Commission on SD. Ten years later, the 2002 World summit discussed more extensively the SD concept by the United Nations (WSSD 2002) and set a number of targets to be met in the following years.

The WCED (1987) minted the definition of SD as the *development that meets the needs of the present without compromising the ability of future generations to meet their own needs.* This very concise sentence distills the three pillars of sustainability, which are economic, environmental, and social development. For the sake of clarity, some authors (Bakshi and Fiksel, 2003; Sikdar, 2003; Othman et al., 2010; Ouattara et al., 2012) instead of the three-pillar concept prefer the "triple bottom line" to refer to the economic, environmental, and social issues of sustainability.

The social contribution to sustainability allows understanding most of the stakeholders reported by Bakshi and Fiksel (2003) that are numbered from #3 to #8 in Figure 25.1. Some of them are pretty pertinent to the production plant (i.e., they belong to the narrow field) while the remaining ones apply to a background that covers a wider field (i.e., they go beyond the battery limits of the chemical plant and embrace external features such as communities that can be even international respect to the production site). Regarding the regions of influence, Sikdar (2003) identifies four typologies of sustainable systems:

1. systems referring to global concerns or problems, such as global warming, ozone depletion, use of genetically modified crops;
2. systems characterized by geographical boundaries (e.g., cities, villages, defined ecosystems);
3. systems based on businesses, either localized or distributed, which strive to be sustainable by practicing cleaner technologies, eliminating waste products, reducing emissions of greenhouse gases, and reducing energy intensity of processes; and
4. any particular technology that is designed to provide economic value through clean chemistries.

The sequence of sustainable systems proposed by Sikdar (2003) is hierarchical and reduces the dimension of intervention/influence by moving from the first to the last item. Systems 3 and 4 are suitable for chemical engineering problems because they rely more on process and product designs, and manufacturing methods.

25.2 THREE PILLARS OF SUSTAINABILITY

After the WCED (1987) definition of sustainability, a large number of scientific works were published in the following decades either treating single aspects or discussing the complex interactions among economy, environment, and society. As extensively discussed in Azapagic and Perdan (2000), the first step to assess the SD of industry is to identify simple and informative indicators relevant to the economic, environmental, and social aspects. The second step is related to the identification of suitable metrics to track and compare the performance of different options. By focusing first on the identification of indicators, it is possible to introduce two classes. The former includes all-industry indicators, whereas the latter proposes sector-specific indicators. Table 25.1 shows the classification of indicators proposed by Azapagic and Perdan (2000).

Table 25.1 Classification of Indicators for the Three Pillars of SD

Economic Indicators	Environmental Indicators	Social Indicators

Financial indicators
• value added
• contribution to GDP
• expenditure on environmental protection
• environmental liabilities
• ethical investments
	Environmental impacts
	• resource use
	• global warming
	• ozone depletion
	• acidification
	• eutrophication
	• photochemical smog
	• human toxicity
	• eco toxicity
	• solid waste
		Ethics indicators
		• Presentation of cultural values
		  • stakeholder inclusion
		  • involvement in community projects
		• International standards of conduct
		  • business dealings
		  • child labor
		  • fair prices
		  • collaboration with corrupt regimes
		• intergenerational equity

Human-capital Indicators • employment contribution • staff turnover • expenditure on health and safety • investment in staff development	Environmental efficiency • material and energy intensity • material recyclability • product durability • service intensity Voluntary actions • environmental management systems • environmental improvements above compliance levels • assessment of suppliers	Welfare indicators • income distribution • work satisfaction • satisfaction of social needs

Adapted from Azapagic and Perdan (2000).

It is worth observing that each of the three pillars has a distinctive identity but at the same time contains *contaminations* from the other pillars. For instance, the economic indicators show a human-capital subsection that finds its origin in the social indicators. The same happens for the welfare subsection of social indicators that are pretty related to the economic indicators. The environmental indicators are even more connected to the economic issues. Azapagic and Perdan (2000) observe that environmental indicators are often oversimplified and refer to material and energy consumptions that can be transformed easily into economic quantities. For instance, the chemical industry uses often energy consumed, waste production, and by-products recycling as environmental indicators. These indicators are harmonized and normalized to a single measure by means of suitable metrics based on economic criteria.

Even if the classification of indicators in Table 25.1 lists a few qualitative terms, it is often preferable to work with quantitative terms that allow the implementation of a systematic approach (if not a factual methodology) to the SD assessment. The contribution of social terms to the quantitative assessment of SD, which is a superset of SC, is almost missing in most of the methods proposed in the literature (e.g., GRI, ICI/EB, Unilever/OBIA, WRI, AIChE/CWRT). On one hand, this is due to the difficulty of finding a common basis to measure both social and economic/environmental issues. On the other hand, most of the companies involved in SCM have been rather reluctant to recognize the social question as part of their production route. Nonetheless, things are changing due to two separate enablers. The former enabler fines companies that do not respect environment/social constraints/rules. The so-called "name and blame" approach from governmental agencies, with the display of companies to "shame lists," has forced corporations to undertake an improvement path to avoid also the negative publicity that can have an even higher impact on the company business in terms of production loss. The latter enabler consists in improving a priori the company image perceived by the clients, customers, society, and influence groups (e.g., activists) to avoid future risks of image spoilage. The following Starbucks and Foxconn dossiers show the paramount role played by social issues on company image and business.

After crude oil (CO), coffee is the most traded commodity with a market value of about USD 100 billion in 2013 (Business Insider, 2014). Coffee is traded more than natural gas, gold, corn, and wheat. Starbucks is an American global coffee company and coffeehouse chain that in 2003 had revenues of USD 15 billion and employed 160,000 workers. Even if in 2000 Starbucks paid a well-above market price for coffee beans, it had to face the censure of Global Exchange, the leader of "corporate accountability" movement, who criticized the company for profiting at the coffee farmer's expense and not buying "Fair Trade" coffee beans (Argenti, 2004). The business risk for Starbucks was both direct and indirect. The direct risk was that the activists of Global Exchange could block the access to Starbucks' shops and make public sit-ins to awaken public opinion. The indirect risk was that partners could lose faith in the integrity of (social) values and start resigning from contracts, thus compromising the corporate SC of Starbucks. To overcome this risky situation, Starbucks decided to test the reliability of the Fair Trade model by selling that kind of socioeconomically sustainable coffee in its shops even if most of the suppliers did not meet the corporate high-quality standards. This introduced a further risk of alienation from Starbucks by the original suppliers who were already complying with the high-quality standards for coffee beans. Eventually, Starbucks realized that Fair Trade was not the most effective way to help farmers and focused on other initiatives to purchase green coffee at arranged prices independent of the market quotations. In addition, Starbucks signed long-term contracts with suppliers to build seller–buyer loyalty (Argenti, 2004).

Foxconn is the world's largest electronics manufacturing company with USD 128 billion revenues in 2012 and 1.23 million employees. Apple, Dell, Hewlett Packard, Microsoft, Sony, RIM, and Nintendo are some of Foxconn's customers. In the first semester of 2010, 12 employees attempted suicide at Foxconn's biggest factory in Shenzhen, China, where over 420,000 workers are employed.

Ten people died and two reported serious injuries after jumping from the factory buildings (Xu and Li, 2013). The employees, interviewed by an independent committee of Chinese, Hong Kong, and Taiwanese observers, reported that the assembly lines ran too fast and the working conditions resembled those of a prison. Apple, Dell, and Hewlett Packard were severely criticized in most of industrialized countries for their products being manufactured in the "sweatshops" of Foxconn. Consequently, these companies decided to investigate independently on the suicides and the working conditions. However, the investigation reports were never published. To preserve the company image and the robustness of important brands such as iPhone and iPad (assembled by Foxconn factories) Steve Jobs first and Tim Cook after, the CEOs of Apple, pointed out that the Shenzhen plant (also known as "Foxconn City") had its own hospital, fire department, restaurants, entertainment centers, and sport activities. Tim Cook also visited the production line of iPhone in March 2012 just after Apple became the first technology company to join the Fair Labor Association in January of that year by opening up suppliers' factories to inspections (Bloomberg News, 2012). As a matter of fact, it was reported by both the US and Chinese consulting companies that Apple had received negative publicity after human-right and work-quality negligence at Foxconn factories.

Both Starbucks and Apple dossiers allow observing the somehow *forced* attention that these companies had to adopt toward social and environmental issues to preserve their image and brands as prerequisites to their core business. Starbucks and Apple were compelled to interact at different levels with their suppliers to enhance the social and environmental sustainability of their SC.

The illustrations reported above allow introducing the concepts of either self-centered or altruistic management by corporations. The self-centered behavior is focused on the observance of law and the keeping of rules. Often, operator training and safety implementation at the production site are passed off by companies as attention and awareness to both environmental and social issues. On the contrary, an altruistic management would turn the attention to the people involved in the production and delivery of products. These people are not only those working for the corporation but also those interested, at different levels, in the SC all the way to the population living in the surroundings of the production site and warehouses. An altruistic behavior grants better working conditions, flexible working hours, and transparent information to the involved population about critical issues and safety exposure. An environmentally and socially oriented SCM considers, tackles, and solves the issues relevant to the suppliers, workers, population, distributors, and customers of the company by means of a cradle-to-grave approach (see also Section 25.2.2). The altruistic approach to SCM can also produce some synergistic side effects such as customer loyalty and increased company reliability. Indeed, some authors discussed also a possible outflow of altruistic behavior caused by a premeditated opportunistic approach to SCM as far as environmental and social matters are concerned (Dyllick and Hockerts, 2002; Ouattara et al., 2012).

25.2.1 DIMENSIONS OF SUSTAINABILITY

To assess the sustainability of systems and more specifically of SC, there is need for measuring their performance by means of suitable metrics based on indicators. The previous sections introduced both indicators and metrics and showed their interdependency from the economic, environmental, and social categories. Accordingly, it is possible to perform assessments based on one-, two-, and three-dimensional searches as a function of the number of variables taken into account. Indeed, the three pillars of sustainability allow moving in search of the optimal solution on a single axis (1-D) at a time, onto a plane (2-D), or within the whole region (3-D) contained by those attributes. If these axes assume the role of independent objective functions then the problem becomes a multiobjective optimization (MOO) and either a Pareto line (2-D) or a Pareto surface (3-D) describe the set of optimal solutions.

Schwartz et al. (2002) report a study on sustainability metrics performed by the Center of Waste Reduction Technologies (AIChE, USA) to evaluate process alternatives. The indicators used in that study are material, energy, and water uses; toxic and pollutant dispersions; greenhouse gas emissions. IChemE (UK) classified similar indicators and assigned them to the three-pillar classes. At the environmental level, IChemE classified energy, material, water, and land uses. They proposed as environmental impacts: acidification, global warming, human health, ozone depletion, photochemical smog formation, and ecological health. As economic indicators, IChemE specified a number of value-added quantities and the R&D costs. Eventually, the proposed social indicators were based on employee benefits, safety, and how workers are treated by the company. Even if both AIChE and IChemE suggested lists of indicators applied to single pillars of sustainability, Sikdar (2003) advised a new perspective for the classification of indicators. He suggested every indicator should be analyzed in depth to find its possible membership to more than one category of the three-pillar classification. For instance, as far as energy use is concerned, this is a major driver for economic development. If energy is obtained by nonrenewable sources then it has an impact on future generations. In addition, energy has always an impact on the environment, which is played by the emission of pollutants and greenhouse gases. As a consequence, the energy use is intrinsically a 3-D indicator and allows performing a holistic assessment of sustainability as it covers at the same time the economic, environmental, and social pillars. Based on similar considerations, Sikdar (2003) classified also material use and pollutant dispersion as 3-D indicators. It is therefore possible to identify a hierarchy of indicators that from a top-down perspective cover the 3-D, 2-D, and 1-D features of SD. These indicators allow performing a systematic analysis of sustainability applied to products, processes, and enterprises (for instance in terms of SC/SCM). We can conclude that designing for sustainability of either products or processes is an activity that increases the dimensionality of the optimization problem as the number of objectives and constraints to be taken into account is much larger. The interdependency of these objects makes the problem more complex and often drives to a MOO formulation (Ouattara et al., 2012).

25.2.2 CRADLE-TO-GRAVE VERSUS CRADLE-TO-GATE

The crystal clear definition of chemical SC proposed by Sahay and Ierapetritou (2013) as *a network of suppliers, production facilities, warehouses, and markets designed to acquire raw materials, manufacture, and store and distribute products among the markets* depicts perfectly the breathing horizon of SC and the intertwined environments that contribute to its structure. As discussed in previous sections, it is therefore straightforward to adopt a cradle-to-grave approach to the assessment of sustainability. The cradle-to-grave term is inherited from life cycle assessment (LCA) and describes both the spatial and temporal horizon where it is applied. That term requires considering the assessment from the very first seconds when the antecedents of the product are involved (e.g., design procedures, raw materials, provisioning) till the very last seconds of the SC when the product is delivered to the final user. In reality, the LCA goes even beyond the SC endpoint and takes into account the use, life, and disposal/recycle of the product.

Conversely, some authors (Othman et al., 2010; Ouattara et al., 2012; Yue and You, 2013) prefer to reduce the assessment horizon to the production site and adopt the so-called cradle-to-gate approach that covers the production stages that go from manufacturing to the factory gate, i.e., before products are delivered to the customer. By choosing this spatially reduced approach to SC sustainability, the analysis is somehow simpler and tailored to the company's need for knowledge, which often remains within the battery limits of the industrial plant. For instance, Young and Cabezas (1999) proposed the waste reduction algorithm (WAR), which is a typical cradle-to-gate approach to environmental assessment. That algorithm works on the potential environment impact (PEI) of the production plant to assess the role played by a specific process on both energy and mass balances.

Similar conclusions can also be drawn in case of product design. For instance, Gani (2004) proposed a systematic approach to chemical product design and introduced a germinal idea of sustainability by suggesting a compromise among different constraints and specifications that range from quality to physical and chemical properties, from environmental to economic issues, from feasibility to flexibility. According to the horizon width used to design the optimal product, one passes from the cradle-to-gate to the cradle-to-grave perspective.

These last considerations pave the way to the central part of this Chapter that focuses on processes and products viewed from both sides of design and operation under the perspective of economic sustainability.

25.3 ECONOMIC SUSTAINABILITY

Previous sections discussed how the economic attribute is the most ancient and well-established among the three pillars of sustainability as far as the feasibility of products and processes is concerned. On one hand, the economic pillar sustains the environmental and social elements. On the other hand, both environmental and social terms are often traced back to the economic one. As a matter of facts, the economic assessment (EA) has a pretty long history and several indicators and associated metrics were developed to quantify that feature.

In 1988 J.M. Douglas published a book on Conceptual Design (CD) of chemical processes. That book systematized a methodology described in a few papers Douglas published some years before on the EA of chemical processes. His hierarchical approach to the design of chemical processes was aimed at defining a sequence of five steps to design, with an increasing detail, a chemical plant according to a specific process while quantifying its feasibility in terms of economic potentials (EP). Douglas (1988) stated that *The purpose of engineering is to create new material wealth* and suggested that the five decisions to be taken in the design of a chemical plant should be:

1. Batch versus continuous;
2. Input-output structure of the flowsheet;
3. Recycle structure of the flowsheet;
4. General structure of the separation system;
5. Heat exchanger network.

Those decisions are sequential and subject to the economic feasibility of the previous level. The necessary condition to move from level 3 to 4 up to level 5 is that the previous EP is positive. Every level in the hierarchy quantifies in greater detail the equipment and ancillary features involved in the plant. The approach is similar to that of a photographer who moves from a wide angle lens to a telephoto lens to progressively gather more details from the whole vision. Once the first-level binary decision, whether adopting a continuous or batch process, is taken, the second-level EP (EP2) opens the EA to a plethora of quantitative considerations that not only cover the process feasibility, but also are capable of addressing the product viability.

The definition of EP2 is almost trivial:

$$EP2 = (Products\ \ Price + Byproducts\ \ Value) - (Raw\ \ Materials\ \ Cost) \tag{25.1}$$

It is based on drawing up an economic balance between incomes and outcomes produced by selling the products (and/or using the by-products) and buying the raw materials. Neither capital expenses (CAPEX) from equipment nor further operative expenses (OPEX) such as utilities and labor costs are accounted for by EP2. It is, therefore, pretty evident that EP2 must be plentifully positive to make the chemical plant economically attractive and continue with its design. The crucial point is what prices/costs are used in Eqn (25.1) to quantify the EP2. Douglas (1988) used as case study the hydrodealkylation (HDA) process that converts toluene (T) to benzene (B) and provided for both components some fixed economic values that made it appealing. For the sake of completeness, all the EP of CD are evaluated assuming fixed prices/costs over the whole time horizon of the EA. Usually, these prices/costs

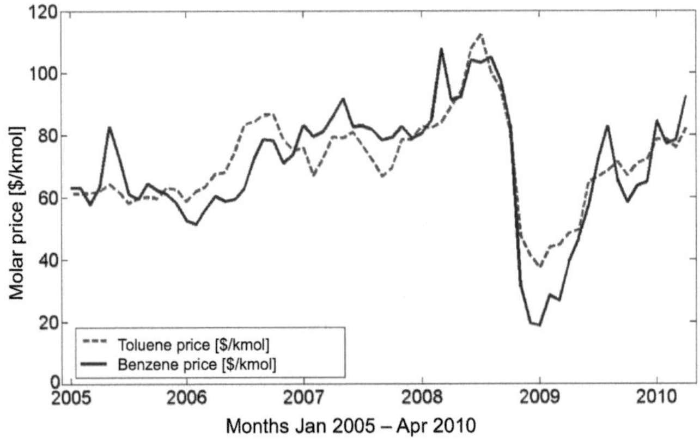

FIGURE 25.2

Monthly quotations of toluene and benzene in the 2005–2010 period.

are assigned when the design activity is carried out. Unfortunately, the real world is quite more complicated as Figure 25.2 shows.

Since the HDA process is based on an equimolar reaction between T and B ($C_7H_8 + H_2 \rightarrow C_6H_6 + CH_4$), if one assumes theoretical conversion and selectivity of 100% (which is a conservative assumption to discuss the forthcoming issues) then it is straightforward deducing that the EP2 changes sign repeatedly and oscillates between positive and negative values in the 5-year interval of Figure 25.2 as soon as B price is respectively either higher or lower than T cost. This is a first critical point about CD that would compromise the whole systematic approach of Douglas to EA of chemical plants and more in general the feasibility studies related to financial investments in industrial processes if one did not take into account the dynamic attribute of market quotations and neglected terms such as volatility, fluctuation, and uncertainty.

25.3.1 VOLATILITY, FLUCTUATIONS, AND UNCERTAINTY

Financial markets are all but stationary and systematic. Indeed, fluctuations and uncertainty are intrinsic to both raw material and commodity markets even if recent years saw unexpected oscillations with volatilities that reached unprecedented peak values.

25.3.1.1 Volatility of raw materials

To better explain the uncertainty of markets, it is worth focusing on the CO quotations that play worldwide a paramount role in conditioning both industrial and financial activities. For instance, in the first months of 2007 CO was quoted at about 50 USD/bbl and its price kept on increasing almost steadily for 16 months until the end of the second quarter of 2008 when it reached a peak value of 140 USD/bbl.

Financial experts forecast that in a few months CO would break the 200 USD/bbl barrier with the worldwide economic expansion being the responsible for that unprecedented increase. Afterward, the US subprime mortgage crisis of 2008 summer was the triggering event that subverted all the bullish predictions and initiated a very tough recession period for most of the industrialized countries. The Lehman Brothers default, the Royal Bank of Scotland bankruptcy, and the Madoff's fraud were just a few sides of the same medal drawing a very intense period of crisis. In the second semester of 2008, the CO price underwent the following losses:

- reduction of 28% from 140 to 100 USD/bbl in 10 weeks;
- reduction of 50% from 80 to 40 USD/bbl in 10 weeks; and
- reduction of 42% from 60 to 35 USD/bbl in 7 weeks.

Eventually, in less than 6 months, the CO quotation witnessed a fourfold contraction touching a minimum of 35 USD/bbl. The repercussions were not only financial but also industrial. For instance, processes that had been kept in standby for decades, due to their economic burden, were planned to be exploited as soon as the CO quotations would overtake some feasibility thresholds. That was the case of tar sands in Canada and Venezuela. Big projects were planned for the years to come at the beginning of 2008 but had to be postponed and finally aborted due to the vast recession of 2008. In the following years, the shale oil and shale gas discovery in the United States played a further role in conditioning the global supply of raw materials. For instance, the sudden availability of raw energetic materials (i.e., natural gas and oil) in the United States, which are at the basis of both energy and petrochemical industries, moved the overproduction of US coal to European countries with a sensible reduction in its prices.

25.3.1.2 Volatility of commodities

Most of commodities derived from CO as either direct distillates or petrochemical derivatives see their prices tightly bound to those of raw materials used for their production. Similar remarks can be applied to derivatives of natural gas and coal. For the sake of brevity, this section focuses again on the HDA process and the assessment of the EP2, which sets a necessary condition for the plant feasibility. Table 25.2 shows some interesting results. Over the 5-year period, the averaged quotations of T and B make the EP2 marginally negative. Conversely, if one uses the spot monthly prices of T and B (for instance March and December 2008) the EP2 assumes respectively quite positive and negative values.

This would mean that the CD of an HDA plant, having the same production capacity, performed just after 9 months would suggest respectively either a profitable or unprofitable solution. This point is absolutely unacceptable as it would subvert any CD deduction and make the methodology weak and unreliable. As a result, there is need for introducing the volatility and uncertainty features in the CD methodology. A dedicated Section 25.3.2 of this Chapter covers the dynamic conceptual design (DCD) of chemical plants (Manca and Grana, 2010). To support the complexity of economic sustainability of chemical processes, it is worth reporting

Table 25.2 Benzene and Toluene Prices and Corresponding EP2 Values for an HDA Plant Based on the Productivity Data of Douglas (1988)

	Benzene Price (USD/kmol)	Toluene Price (USD/kmol)	EP2 (10^6 USD/y)
2005–2010 period	70.16	72.19	−2.0318
March 2008	107.79	84.18	51.122
December 2008	19.41	41.17	−40.589

a short sentence from Ullmann's encyclopedia (2003) *Whether or not hydrodealkylation is used is related directly to the demand and price of benzene relative to toluene. If the relative demand for the two products changes, the dealkylation units may be placed on standby.* This suggests a dynamic allocation of resources for the operation of an HDA plant (Manca, 2013b). In addition, Milmo (2004) reports that *The long-term average utilization rate of HDA plants has been around 40% because of the many occasions when the price of the toluene input was higher than that for the benzene output.*

25.3.1.3 Volatility of utilities

Chemical plants and SC depend significantly from utilities such as electric energy (EE), steam, water, fuel oil, diathermic oil, and cryogenic fluid. For the sake of brevity, this subsection focuses on the volatility of EE prices, which depend also on the unpredictability of its production.

The European Community directive 96/92/CEE forces each European member state to establish a national EE market where its price is negotiated hourly. One of the most important economic indexes is the daily list based on 24 hourly quotations for every day of the week. This is a 24/365 market that is based on demand-and-offer rules.

Figure 25.3 shows the typical trend of the hourly quotations of EE price at the Italian marketplace (GME, 2014). It is possible to observe the typical M-shaped trend of EE prices, which, in that day, covered the 37.35−113.45 €/MWh interval with over a threefold variation in just 5 h (i.e., from 9 AM to 2 PM). On the whole, both maxima and minima of EE prices change every day not only in terms of absolute values, but also of time position. The values of minima and maxima depend on the seasonal, meteorological, and climatic conditions. They are also function of political, economic, and financial circumstances. Similar trends can be observed for both European and non-European countries with each country characterized by peculiar features (Marinovic et al., 2013). To give an idea of these issues, we focus now on a peculiar quarter for the EE European quotations. It will be shown how rather similar EC nations, often neighboring, witnessed pretty different scenarios and trends of EE price. The lesson learned is that specifically tailored econometric models (EM) of EE prices should be developed as a function of both the country and the time horizon to be simulated.

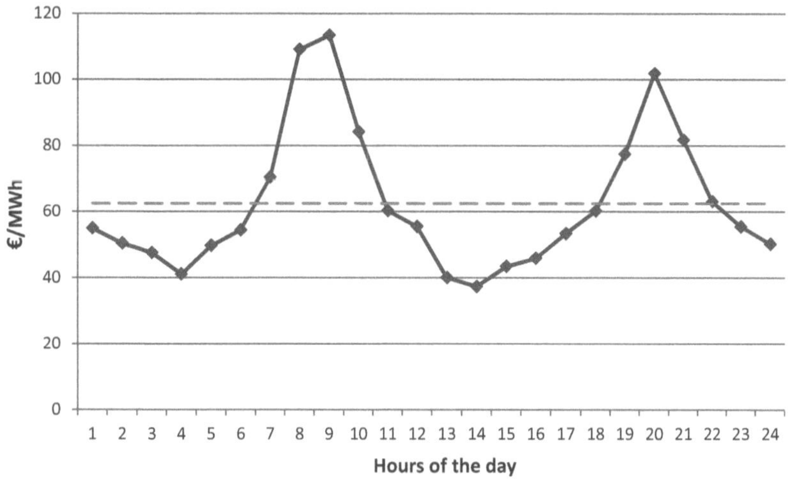

FIGURE 25.3

Italian market electric energy (EE) hourly prices on 31 March, 2014. The dashed line is the average value of EE prices.

Courtesy of Gestore Mercati Energetici at www.mercatoelettrico.org.

The first quarter of 2012 was expected to be an ordinary winter season characterized by rather high EE prices due to few daylight hours and high heating requirements (DG Energy, 2012). Indeed, this is what happened at the beginning (i.e., January) and at the end (i.e., March) of that quarter. In contrast, the first 2 weeks of February 2012 brought an extreme cold weather in most European countries. This 2-week period had a significant impact on power demand. The EE consumption in the EU-27 region was 5.1% higher than that of the same period in February 2011.

Low temperatures caused sudden price increases in the Central Western Europe (CWE) region. This circumstance was amplified by diminishing wind power generation. Daily average power prices reached a peak on February 8–10, 2012, with 85–100 €/MWh values. Prices in France were even higher at 367 €/MWh with a bizarre peak-hour value of 1938 €/MWh. In those days, France became a net power importer, which is an unusual situation. Even if Germany confines with France, things were different. The politically oriented decision of moving toward solar and wind generation supplemented by hydro sources allowed Germany to differentiate from France. Daily average power prices reached the peak value of 78 €/MWh in England. This not-so-significant increase in the EE price was primarily due to the availability of cheap coal sources and less tight-grid margins. It is worth observing in the same days the significant spread of 400%–500% in EE prices of CWE countries.

The cold spell of early February 2012 coupled to a strong increase in gas price, which reached a maximum almost three times higher than conventional quotations, made Italy reach an EE peak price of 134 €/MWh. Conversely, in those days, Spain and Portugal did not experience harsh weather conditions.

In northern Europe countries, the average daily price increased from 37 €/MWh in January to 86 €/MWh in the first days of February 2012. Besides the cold weather conditions, a serious reduction of nuclear generation capacity in Sweden on February 2 produced severe repercussions in the neighboring countries. When the Swedish nuclear reactors were resumed, the system price fell back to 43 €/MWh on February 5.

The increase in EE price of Greece was due to the EU Council restrictive measures against Iran that banned in January 2012 the imports of CO and by-products from that country to the EU. The oil embargo had a considerable impact on the EE quotations in Greece (with a daily average peak at 120 €/MWh) as this country depends heavily on oil for power generation and the share of Iranian oil import was the highest among EU nations.

The role played by political and social events affected also the daily power prices of Balkan countries in that same period. A coal miners' strike in Bulgaria began at the end of January 2012 and the government decided to halt the country's power exports in order to satisfy domestic needs. This had a repercussion on the Hungarian EE daily price that reached its peak at 145 €/MWh and was amplified by the coldest days of February that froze the Danube river and had serious repercussions on the hydro generation in Serbia.

All these issues, as extensively discussed by DG Energy (2012), show how the EE price of European markets is entangled and how meteorological, geographical, political, and social matters play an interconnected and highly dependent role on quotations. Similar considerations can also be applied to countries of other continents.

25.3.2 DYNAMIC EA

Subsections 25.3.1.1–25.3.1.3 showed how the economic sustainability of both products and processes depend significantly on the volatility of demand, supply, and price of all the elements involved in the production, delivery, distribution, and commercialization of chemical products. These elements are: raw materials, final products, and utilities. We showed that the volatility may cover short-, medium, and long-term horizons. In fact, the shortest volatility of prices was that of EE hourly quotations, whereas the longest volatility covered yearly horizons as in case of CO and commodities. The time granularity of volatility goes from hours to days, from weeks to months, up to years as a function of the time horizon investigated by either design or scheduling/planning activities. Therefore, both CD and SC domains are heavily conditioned by economic volatility where uncertainty and fluctuations of prices are elements that cannot be neglected at all.

This section proposes and discusses a dynamic approach to the EA of both CD and SC which is primarily based on EM and dynamic trajectories of prices of raw materials, commodities, and utilities. These EM are implemented and exploited on short-, medium-, long-term horizons with a time granularity (i.e., sampling time, discretization time) that is consistent with the problem to be solved. The following subsections present the forecasting models for the EA of either DCD or Dynamic Supply Chain Management (DSCM). Since DCD and DSCM are based on forecasting models of prices then such methodologies lay their foundations on distributions of economic potentials and probabilistic curves of different economic scenarios.

25.3.2.1 Price forecast methodology

Figure 25.4 shows the sequence of actions needed to build an EM for price forecasting purposes, which can tackle both design and operation tasks.

This procedural methodology allows covering the EA and the first pillar of sustainability for the following activities:

- design of new chemical plants;
- revamping/retrofitting of existing plants;
- process intensification;
- scheduling/planning of production and plants;
- scheduling/planning of SC; and
- product and process design.

25.3.2.1.1 Reference component

The identification of a reference component (RC) is not mandatory but highly recommended. Instead of tracking the price dynamics of every OPEX involved in the process, it is possible to define an RC and quantify the OPEX respect to that component. The RC allows structuring and tuning the following price forecasting models of raw materials, products/by-products, and utilities by using that term as the independent variable of the EM to be identified. As extensively discusses in Manca (2013b), most of chemical products and processes are economically linked to CO as far as OPEX are concerned. Being a raw material, CO is the precursor

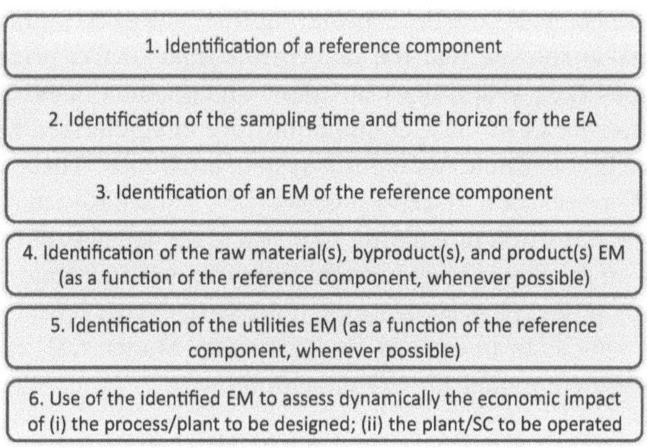

FIGURE 25.4

Sequence of actions to be taken to finalize the econometric models.

in a number of chemical processes. Its cost is well-known, largely available, and periodically updated. CO also plays a significant role respect to a number of industrial utilities (e.g., steam, EE, diathermic oil, refrigeration systems). Hence, CO is a good candidate for both product and utility econometric modeling. By identifying possible future trajectories of CO, it is then possible to assess the corresponding trajectories of both commodities and utilities that play a role in the EA of either DCD or DSCM.

25.3.2.1.2 Sampling time and time horizon

The time horizon of the EA depends on the characteristic time of the phenomenon to be studied. In case of DCD, the time horizon can even be tens of years (i.e., the expected lifetime of the plant to be designed). In case of tactical allocation of resources in both processes and SC, the time horizon is usually few days/weeks. In case of strategic problems in SC planning the time horizon is usually some weeks, few months (generally no more than a quarter). In case of energy-intensive process and EE optimization the time horizon can even be shorter and cover few hours.

As far as the granularity of the EA is concerned, the typical sampling time, to solve the discrete mathematical model of the problem, is a suitable fraction of the time horizon and depends mainly on two factors. The former is the availability of real prices/quotations and depends on the specific economic/financial markets involved and analyzed. The latter depends on the complexity of the numerical simulations involved, which play a significant role mostly on SC problems (rather than on DCD problems).

25.3.2.1.3 EM of the RC

The variation of the quotation of CO from the previous price to the following one is called shock. Since 2000, most of the CO shocks measured on a weekly basis had a maximum $\pm 10\%$ spread with a Gaussian distribution. To determine if the CO shocks have either a stochastic nature or are deterministic and depend from previous quotations, it is worth considering the autocorrelogram of shocks. Manca et al. (2011) showed that the autocorrelogram of CO prices over rather long-term horizons (i.e., years) does not show significant values. Therefore, one can infer that the shocks in the CO quotations are characterized by a stochastic nature and are neither periodic nor exhibit systematic trends. These features allow assuming the CO prices as a Markov process (i.e., a stochastic and time-discrete process where the transition probability to a new state depends only on the previous one and not on the history that brought the system to that state). Based on the fact that a Markov process has no memory of the past, except for the previous state from which the new state is derived stochastically, Manca (2012) proposed the following time-discrete model for CO quotations:

$$P_{CO,i} = P_{CO,i-1} \cdot \left(1 + RAND \cdot \sigma_{CO} + \overline{X}_{CO}\right) \tag{25.2}$$

where $P_{CO,i}$ is the price of CO at time i (usually the sampling time is either a week or a month), $\sigma_{CO}, \overline{X}_{CO}$ are respectively the standard deviation and average value of CO prices over a suitable time interval in the past and *RAND* is a random function that produces a normal distribution of real numbers with average value 0 and standard deviation 1. Equation (25.2) is used step-by-step, over the prediction horizon of the problem to be solved, to forecast a possible price trajectory. As aforementioned, the single trajectory of CO prices is not representative of the whole problem. On the contrary, it is recommended to evaluate a large number of trajectories, subject to the stochastic nature of that EM, so to obtain a probabilistic distribution of possible future scenarios. The higher the number of implemented scenarios, the more robust is the EA. As commented in Rasello and Manca (2014), Eqn (25.2) should be applied to a specific CO market. Indeed, in recent years (i.e., after 2010) the two most important CO markets, i.e., the Brent and WTI ones, saw their quotations taking apart because of shale gas, shale oil, international crises, embargos, and available infrastructures that assumed unequal distributions in the respective geographic domains of influence. Rasello and Manca (2014) suggested neglecting the average value of CO for long-term horizons so as to remove the systematic bullish/bearish trend that would depend only on the time interval chosen to calculate $\overline{X}_{CO}$. For the sake of space, Table 25.3 introduces some additional EM centered on the forecast of CO quotations.

25.3.2.1.4 EMs of raw materials and (by)products

Manca (2013b) in his paper on the commodity fluctuations of OPEX terms, proposed a systematic approach to econometric modeling for forecast purposes in DCD problems. The same modeling approach can also be applied to dynamic SC problems as discussed in Rasello and Manca (2014). The construction of EM for raw materials, by-products, and products of chemical processes that can be used on short-, medium-, and long-term horizons are at the basis of this subsection. The above cited papers from Manca and coworkers propose a sequence of actions to prepare and identify the EM of distillates, derivatives, and commodities that exhibit some economic dependency from an RC. Due to its pivotal role in commodity pricing, we suggest adopting CO as the RC. This is a straightforward assumption for both Oil and Gas and Petrochemical sectors. Nonetheless, Mazzetto et al. (2013, 2015) showed that CO can be taken as the RC also in the strategic design of bioethanol SC where biomaterials are used as raw inlet streams. The candidate EM assumes the following general structure:

$$ADL(q,p) = c_0 + a_0x_0 + a_1x_1 + \ldots + a_qx_q + b_1y_1 + b_2y_2 + \ldots + b_py_p \qquad (25.3)$$

where x_i,y_i are respectively the prices at time t_i of the independent (i.e., CO) and depend (i.e., the specific commodity) variables. The increasing subscript i refers to older terms of the time series. Equation (25.3) is chosen for its formulation that is linear, keeps the single contributions separate, and neglects any direct interactions (i.e., neither rectangular nor quadratic terms are provided). According to the econometric terminology, it depicts an autoregressive distributed lag (ADL) model (Stock and Watson, 2003). According to the identification world of dynamic systems (Ljung, 1998), Eqn (25.3) is also known as autoregressive model with exogenous

Table 25.3 List of Econometric Models Available in the Literature to Forecast the CO Prices

Model	Features	References
AR	The autoregressive (AR) model uses an autoregressive approach to model crude oil (CO) volatility. Five lagged values of past volatility, corresponding to the average number of trading days in a week, are used as drivers.	Sadorsky, 2006
GARCH	Generalized autoregressive conditional heteroskedasticity (GARCH) models estimate jointly a conditional mean and a conditional variance equation. GARCH models are very useful when analyzing data that appear to exhibit volatility clustering (e.g., futures data). The GARCH(1,1) model works well in most applied situations.	Kang et al., 2009
MA	Moving average (MA) methods are widely used in time series forecasting. Daily ex-post volatility is measured by squared daily returns. Here a moving average of m days ($m = 20, 60, 180$ days) is used to generate volatility forecasts.	Sadorsky, 2006
RWM	Use of a random walk model (RWM) to forecast the CO price fluctuations.	Murat and Tokat, 2009
VAR + ANN	Use of value-at-risk (VAR) and artificial neural network (ANN) techniques to forecast ex-post CO price oscillations.	Mirmirani and Li, 2004
WDE	The wavelet decomposed ensemble (WDE) model analyzes the changing nature of CO market and uses the wavelet analysis to determine the dynamic underlying features at a finer timescale. This approach allows analyzing the microstructure of time varying heterogeneous markets.	He et al., 2012
YE model	This model predicts monthly CO prices by using the relationships between the current price (inventories and excess capacities) and the next month quotation.	Ye et al., 2009

inputs (ARX). The definition of time granularity and model detail/complexity in terms of q,p elements are extensively discussed in Manca (2013b). Suffice to say that the optimal number of dependent/independent terms in Eqn (25.3) passes through a correlation and autocorrelation analysis that shows in terms of quantitative diagrams how detailed the EM should be. Frequently, the identified model is rather simple and comprises either $ADL(0,1)$ or $ADL(1,1)$ formulations. The *keep-it-simple* approach to modeling, coupled to avoiding overparameterization risks, reduces the complexity of the model. The evaluation of *a*,*b*,*c* parameters in Eqn (25.3) is carried out by a linear regression procedure (e.g., multidimensional unconstrained optimization) that minimizes the sum of squared errors between real quotations and model values. Once the EM has been identified, over a suitable past time interval, it can be used to forecast future price trajectories (as described in Subsection 25.3.2.1.6). For the sake of completeness, the scientific literature proposes also other EM (e.g., Liu and Wan, 2012; Sung et al., 2012) that are tailored to very specific chemical components over definite time intervals. The generality of Eqn (25.3) makes the ADL model a good candidate to cover a large number of chemical products over different time intervals.

25.3.2.1.5 EMs of utilities

Utilities can be classified roughly into material and energy fluxes. As far as material utilities are concerned, the EM can be traced back to the commodity models described in subsection 25.3.2.1.4. For instance, Manca et al. (2011), and Rasello and Manca (2014) proposed an EM for fuel oils distilled from CO, which are used for heating purposes in chemical plants.

When the modeling activity shifts to energy utilities, such as EE, then it is recommendable to adopt a specific approach. For the sake of space, this subsection presents a methodology to model the EE prices of the Italian energy market (see also Section 25.3.1.3). A similar approach can be applied also to other EE markets provided that one pays attention to the peculiar features of those markets. As mentioned above, the 96/92/CEE directive, concerning common rules for the internal markets of electricity, forced the European member states to establish national EE markets where the EE price is negotiated hourly. These marketing bodies regulate a number of distinct contracts according to the timescale and EE amount that is either bought or sold. Starting from the typical M-shaped profile of Figure 25.3, which covers the 24-h EE quotations of a common day, and passing to the 3D-surface of Figure 25.5, which collects the quotations of 1 year, it is possible to observe how the peaks of the M-shaped curves fluctuate according to some deterministic periodicities and further stochastic contributions. On average, the valleys of the 3D-surface are positioned in the nights and in the afternoons, whereas the peaks are in correspondence of late mornings and evenings. By observing Figure 25.5 on longer time intervals, it is possible to point out differences between working days and holidays, and some periodic trends that can be reasonably ascribed to seasonal influences (e.g., winters and summers). The stochastic contributions can be attributed to those elements discussed in Section 25.3.1.3.

Manca (2013a) showed that the EE markets are significantly conditioned worldwide by CO quotations (see also Figure 25.6). Indeed, in most European countries, a large fraction of EE is produced in power plants where either direct derivatives of CO (e.g., gas oil, diesel, naphtha, kerosene, burning oil, heavy distillates) or components influenced by CO quotations (e.g., natural gas, petroleum gas, coal) are

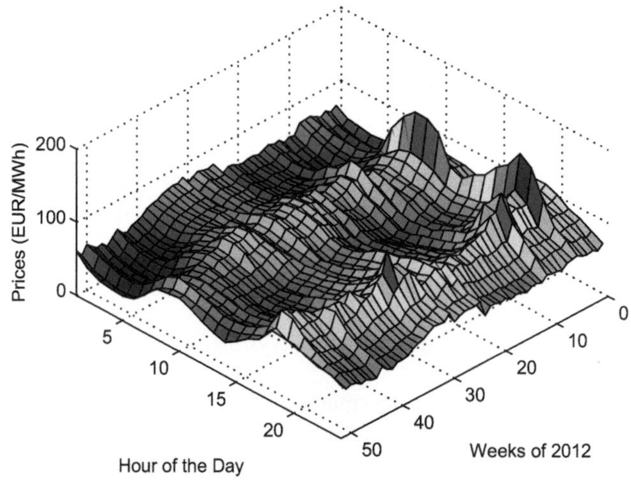

FIGURE 25.5

Italian market electric energy prices in 2012 (hourly quotations averaged on a weekly basis).

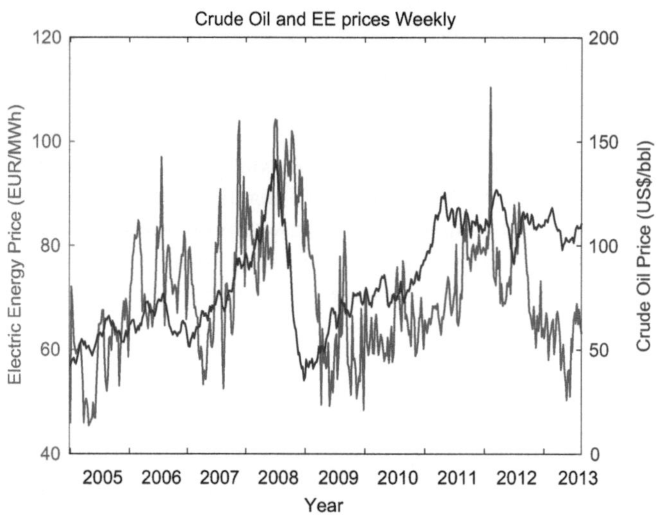

FIGURE 25.6

Italian market electric energy prices and crude oil Brent quotations in the 2005–2013 period.

involved. Even when alternative sources are exploited (e.g., solar, wind, geothermal, nuclear energy), the EE production is still referred to, compared with, and influenced by CO quotations.

Figure 25.6 shows a systematic time delay of about 12–15 weeks between EE and CO quotations. This delay is comparable to the length of a season and enforces the functional dependency of the EM for EE from CO. The recommended features of such a model are as follows:

- time discreteness (as CO quotations are available daily, weekly, monthly);
- simplicity (to avoid overparameterization);
- include:
    - a linear trend with a periodic term due to possible seasonal contributions;
    - a stochastic term to account for unpredictable fluctuations and more complex market behaviors;
- implement the CO quotations of a specific reference market (e.g., Brent market for European nations and WTI market for North American countries).

These elements bring to the following formulation for the EM of EE:

$$P_{EE,i,j} = a_j + b_j \cdot P_{CO,i-t_d} + c_j \cdot \sin\left(\frac{2\pi \cdot i}{T_j} + \varphi_j\right) \tag{25.4}$$

where $P_{EE,i,j}$ is the EE price (€/MWh) at the $i-th$ discrete time for the $j-th$ interval of the day (usually a fraction of the day as discussed in Manca, 2014). a and b are the constants of the linear model. c is the proportional constant of the periodic term based on T period and φ phase. t_d is the time delay of the EE prices respect to P_{CO} that is the CO price [US$/bbl]. T and t_d have the same time units of the discretized quotations of CO and EE. In addition, t_d has been intentionally chosen to be independent of any day intervals as the observed time-delay between EE and CO quotations is supposed to be systematic. Apparently, Eqn (25.4) does not account for any stochastic contribution. Conversely, the stochastic term is embedded in the CO quotations as discussed in Section 25.3.2.1.3 and formulated in Eqn (25.2). A multidimensional unconstrained minimization of the sum of squared errors between the real EE prices and model values allows identi-fying the five sets of parameters and t_d of Eqn (25.2). Manca (2014) determined that:

- the EE price model is primarily affected by the a_j parameter;
- a secondary and often consistent contribution is given by the c_j parameter;
- the 2005–2013 period was rather tumultuous. However, it can be considered altogether as a bullish period as all the b_j parameters are positive and comparable;
- the phase parameter, φ_j, assumes rather constant values and does not produce significant effects on the time shift of the periodic function;
- the T_j parameters are almost all similar and their values, being about 52, are quite comparable to a 1-year interval. This means that the periodic contribution of Eqn (25.4) is systematic and rather reproducible for every hour of the day;
- the $t_d = 15$ value is comparable to the number of weeks in a season (i.e., 13 weeks). Hence, it is possible to maintain that the EE price responds to the CO quotations with a delay of about 3 months.

25.3.2.1.6 Use of EM for the assessment of economic sustainability

Subsections 25.3.2.1.3–25.3.2.1.5 showed a systematic approach to finalize the EM of raw materials, (by)products, and utilities that contribute to OPEX of chemical plants and more in general of SC. These models can then be used to forecast the OPEX terms over quite different time horizons according to the PSE problem to be solved. This subsection focuses on two complementary typologies of PSE problems dealing with the EA of SC and the CD of chemical plants. Both the suggested approaches to EA will introduce the dynamic (D) attribute to SCM and CD giving rise to two new branches of PSE that are known as DSCM and DCD.

If one wants to quantify the economic sustainability of SC over either scheduling or planning horizons subject to market fluctuations, then it is possible to use the EM of OPEX terms to produce a set of scenarios that are distributed according to the modeled fluctuations of dependent variables (i.e., commodities and utilities) and the stochastic contribution of the independent variable (i.e., the RC which is usually CO). Rasello and Manca (2014) discuss an approach to DSCM of oil refineries. Equally, Mazzetto et al. (2013, 2015) use suitable EM for the strategic design of bioethanol DSCM.

As far as DCD is concerned, the EA of OPEX terms in chemical plants can be extended from the original approach proposed by Douglas (1988) and based on the so-called EP by adding the dynamic attribute which results into the DEP concept (i.e., dynamic economic potential). Manca and coauthors (Manca and Grana, 2010; Manca et al., 2011; Manca, 2012, 2013a, 2013b, 2014) proposed the following DEPs to assess the economic sustainability of an HDA plant:

$$DEP2_k[\$/y] = \frac{\sum_{i=1}^{nMonth}\left(\max\left[0, \left\{\sum_{p=1}^{NP} C_{p,i,k}\cdot F_p - \sum_{r=1}^{NR} C_{r,i,k}\cdot F_r\right\}\right]\cdot nHpM\right)}{nMonths/12}$$
$$k = 1, \ldots, nScenarios \tag{25.5}$$

$$Revnues3_{i,k}[\$/h] = \max\left[0, \left\{\sum_{p=1}^{NP} C_{p,i,k}\cdot F_p - \sum_{r=1}^{NR} C_{r,i,k}\cdot F_r - W_{compr}\cdot C_{EE,i,k}\right\}\right]$$
$$DEP3_k[\$/y] = \frac{\sum_{i=1}^{nMonths}(Revenues3_{i,k}\cdot nHpM)}{nMonths/12} - \frac{(IC_{react} + IC_{compr})}{nMonths/12} \tag{25.6}$$
$$k = 1, \ldots, nScenarios$$

$$Revenues4_{i,k}[\$/h] = \max\left[0, \left\{\sum_{p=1}^{NP} C_{p,i,k}\cdot F_p - \sum_{r=1}^{NR} C_{r,i,k}\cdot F_r - \sum C_{EE,i,k}\cdot W_{electr}\right.\right.$$
$$\left.\left. - \sum C_S\cdot F_S - \sum C_{H_2O}\cdot F_{H_2O} - C_{FO,i,k}\cdot F_{FO}\right\}\right] \tag{25.7}$$
$$DEP4_k[\$/y] = \frac{\sum_{i=1}^{nMonths}(Revenues4_{i,k}\cdot nHpM)}{nMonths/12} - \frac{\sum_{e=1}^{nEquip}(IC_e)}{nMonths/12}$$
$$k = 1, \ldots, nScenarios$$

Where N_P,N_R are the number of (by)products and raw materials; *C*,*F* are the price/cost of process streams and their corresponding flowrates; *nHpM*,*nEquip* are the number of production hours in a month and the number of process units involved in the EA; *W*,*IC* are the electric power absorbed by a specific unit and the investment cost of equipment; *S*, H_2O, *FO* are the subscripts that identify respectively the steam used in the reboilers of the distillation columns, the water used in the condensers of those columns, and the fuel oil used in the furnace of the HDA plant (see also http://pselab.chem.polimi.it/ for further details). Equations (25.5)−(25.7) can be extended/adapted to any chemical process. *DEP5* is intentionally not formalized since the optimizations performed on the network of heat exchangers, as suggested by Douglas (1988), would not introduce any additional bits of information to the EA of the DCD framework.

Equations (25.5)−(25.7) consider both CAPEX and OPEX terms. The *max* function in Eqns (25.5) and (25.6) summarizes the concept that the plant should be operated only when the revenues are positive, i.e., the incomes are higher than the expenditures whose role in terms of prices and costs is updated according to the time granularity of the problem (e.g., monthly quotations for the HDA case study). The cornerstone of the DCD methodology is symbolized by the *nScenarios* parameter that calls for quantifying a set of different scenarios depending on the price/cost trajectories obtained by the EM through their intrinsic stochastic contributions. Therefore, the *k* subscript, which connotes every DEP, takes to a probabilistic concept of DCD that is founded on the distribution of possible economic scenarios for a specific process/plant.

Figures 25.7−25.9 summarize the information that can be extracted from the DCD approach to chemical processes and plants (further details are available at

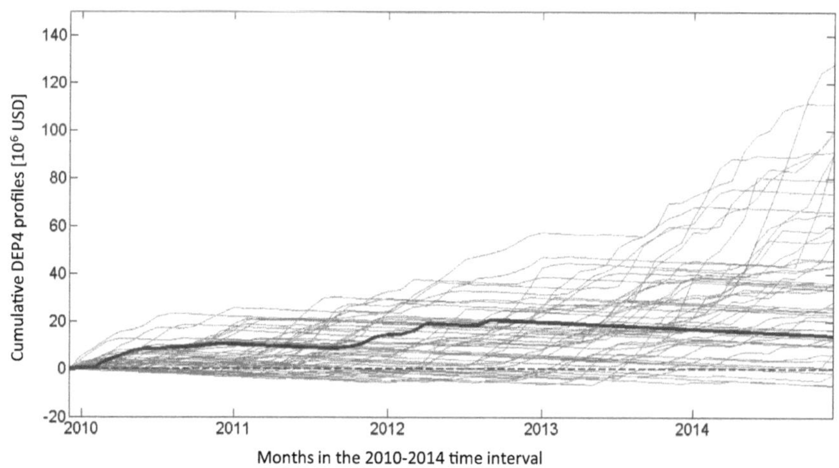

FIGURE 25.7

Trajectories of 50 possible DEP4 as a function of 50 different economic scenarios. The price/cost forecasts start on 2010 and have a 5-year time horizon. The bold line highlights one of the possible economic scenarios that the HDA plant under design could witness during its operation.

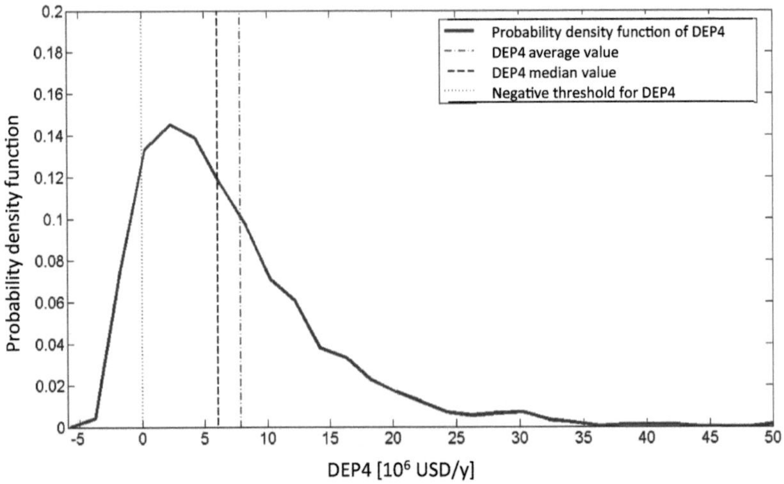

FIGURE 25.8

Probability density function of the DEP4 distribution for a high number of economic scenarios accounting for the uncertainties and fluctuations of OPEX terms.

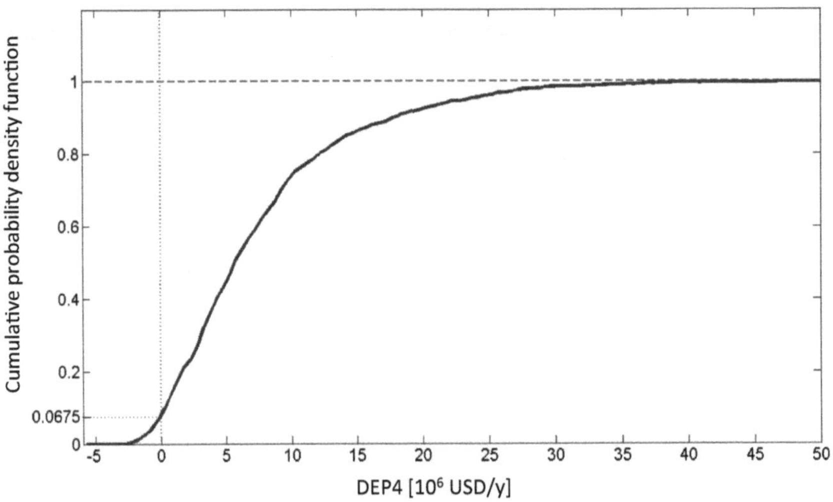

FIGURE 25.9

Cumulative probability density function of the distribution of DEP4. 6.75% of the possible forecast scenarios would produce negative EP of fourth level.

the PSE-Lab Web site reported in the References) and allow quantifying the economic sustainability of both chemical plants and products. A necessary condition for economic sustainability is that DEP are positive. This condition can be raised to higher values to meet a user-defined profitability threshold that makes the chemical investment/operation appealing respect to alternative financial opportunities. Manca et al. (2011) showed that the economic profitability of a conventional HDA plant is in line with the comments of Milmo (2004), reported in Section 25.3.1.2, about the 40% average utilization rate of the process.

25.4 CONCLUSIONS

This chapter defined a systematic approach to the quantification of economic sustainability for both chemical products/processes and SCs under market uncertainty and price/cost/quotation fluctuations. The quantification, which is a characteristic if not intrinsic feature of PSE, was obtained after a long way on defining and classifying the concepts of SD inflected into the three pillars of economy, environment, and society.

The manuscript showed how the more recent environmental and social pillars can be traced back to the assessment of economic sustainability. As soon as market uncertainty and price/cost fluctuations are concerned (which are intrinsic features of modern, international, and intertwined economic marketplaces), the deterministic trends of EM must be supplemented with stochastic elements that allow depicting a palette of possible economic scenarios. This feature brings to a probabilistic approach to EA that is based on distributions of DEP applied to DCD and DSCM.

NOTATIONS

ADL	Autoregressive distributed lag
ARX	Autoregressive model with exogenous inputs
B	Benzene
CAPEX	Capital expenses
CD	Conceptual design
CO	Crude oil
CWE	Central Western Europe
DCD	Dynamic conceptual design
DEP	Dynamic economic potential
DSCM	Dynamic supply chain management
EA	Economic assessment
EC	European Community
EE	Electric energy
EM	Econometric model/modeling
EP	Economic potential
HDA	Hydrodealkylation
LCA	Life cycle assessment
MOO	Multiobjective optimization
OPEX	Operative expenses
PET	Polyethylene terephthalate

PSE	Process systems engineering
R&D	Research and development
RC	Reference component
SC	Supply chain
SCM	Supply chain management
SD	Sustainable development
SSP	Solid state polycondensation
T	Toluene
WCED	World Commission on Environment and Development

ACKNOWLEDGMENTS

I want to thank all the people who contributed, in these years and at different levels, to the development of the DCD framework: Roberto Grana, Andrea Fini, Mirko Oliosi, Rafael Rasello, Roberto Totaro, Clément Boisard, and Salman Nazir. A special thank goes to Piernico Sepiacci for the literature review on Sustainability.

REFERENCES

Argenti, P.A., 2004. Collaborating with activists: how Starbucks works with NGOs. Calif. Manage. Rev. 47 (1), 91–116.

Azapagic, A., Perdan, S., 2000. Indicators of sustainable development for industry: a general framework. Process Saf. Environ. Prot. 78 (4), 243–261.

Bakshi, B.R., Fiksel, J., 2003. The quest for sustainability: challenges for process systems engineering. AIChE J. 49 (6), 1350–1358.

Douglas, J.M., 1988. Conceptual Design of Chemical Processes. McGraw-Hill, New York.

Dyllick, T., Hockerts, K., 2002. Beyond the business case for corporate sustainability. Bus. Strategy Environ. 11 (2), 130–141.

Energy, D.G., 2012. Quarterly report on european electricity markets. Mark. Obs. Energ. 5 (1), 1–37.

Gani, R., 2004. Chemical product design: challenges and opportunities. Comput. Chem. Eng. 28, 2441–2457.

He, K., Yu, L., Lai, K.K., 2012. Crude oil price analysis and forecasting using wavelet decomposed ensemble model. Energy 46, 564–574.

Kang, S.H., Kang, S.M., Yoon, S.M., 2009. Forecasting volatility of crude oil markets. Energ. Econ. 31, 119–125.

Liu, L., Wan, J., 2012. A study of Shanghai fuel oil futures price volatility based on high frequency data: long-range dependence, modeling and forecasting. Econ. Modell. 29, 2245–2253.

Ljung, L., 1998. System Identification: Theory for the User. Prentice Hall, New York.

Manca, D., 2012. A methodology to forecast the price of commodities. Comput. Aided Chem. Eng. 31, 1306–1310.

Manca, D., 2013a. A methodology to forecast the price of electric energy. Comput. Aided Chem. Eng. 32, 679–684.

Manca, D., 2013b. Modeling the commodity fluctuations of OPEX terms. Comput. Aided Chem. Eng. 57, 3–9.
Manca, D., 2014. Price model of electrical energy for PSE applications. Comput. Aided Chem. Eng.
Manca, D., Fini, A., Oliosi, M., 2011. Dynamic conceptual design under market uncertainty and price volatility. Comput. Aided Chem. Eng. 29, 336–340.
Manca, D., Grana, R., 2010. Dynamic conceptual design of industrial processes. Comput. Chem. Eng. 34, 656–667.
Marinovic, M., Makajic-Nikolic, D., Stanojevic, M., 2013. Optimization in day-ahead planning of energy trading. J. Appl. Eng. Sci. 11 (4), 201–208.
Mazzetto, F., Ortiz-Gutiérrez, R.A., Manca, D., Bezzo, F., 2013. Strategic design of bioethanol supply chains including commodity market dynamics. Ind. Eng. Chem. Res. 52, 10305–10316.
Mazzetto, F., Simoes-Lucas, G., Ortiz-Gutiérrez, R.A., Manca, D., Bezzo, F., 2015. Impact on the optimal design of bioethanol supply chains by a new European Commission proposal. Chem. Eng. Res. Design. 93, 457–463.
Milmo, S., 2004. Benzene prices in Europe escalate to tight supply-demand. Chem. Mark. Rep.
Mirmirani, S., Li, H.C., 2004. A comparison of VAR and neural networks with genetic algorithm in forecasting price of oil. Adv. Econom. 19, 203–223.
Murat, A., Tokat, E., 2009. Forecasting oil price movements with crack spread futures. Energ. Econ. 31, 85–90.
Othman, M.R., Repke, J., Wozny, G., Huang, Y., 2010. A modular approach to sustainability assessment and decision support in chemical process design. Ind. Eng. Chem. Res. 49 (17), 7870–7881.
Ouattara, A., Pibouleau, L., Azzaro-Pantel, C., Domenech, S., Baudet, P., Yao, B., 2012. Economic and environmental strategies for process design. Comput. Chem. Eng. 36 (1), 174–188.
Rasello, R., Manca, D., 2014. Stochastic price/cost models for supply chain management of refineries. Comput. Aided Chem. Eng. 33, 433–438.
Sadorsky, P., 2006. Modeling and forecasting petroleum futures volatility. Energ. Econ. 28, 467–488.
Sahay, N., Ierapetritou, M., 2013. Supply chain management using an optimization driven simulation approach. AIChE J. 59 (12), 4612–4626.
Schwartz, J., Beloff, B., Beaver, E., 2002. Use sustainability metrics to guide decision-making. Chem. Eng. Prog. 98 (7), 58–63.
Sikdar, S.K., 2003. Sustainable development and sustainability metrics. AIChE J. 49 (8), 1928–1932.
Stock, J.H., Watson, M.W., 2003. Introduction to Econometrics. Pearson Education, London.
Sung, C., Kwon, H., Lee, J., Yoon, H., Moon, I., 2012. Forecasting naphtha price crack using multiple regression analysis. Comput. Aided Chem. Eng. 31, 145–149.
Ullmann's Encyclopedia of Industrial Chemistry, Sixth ed., 2003. Wiley-VCH, New York.
WCED, 1987. Our Common Future. Oxford University Press, Oxford, U.K.
Xu, K., Li, W., 2013. An ethical stakeholder approach to crisis communication: a case study of Foxconn's 2010 employee suicide crisis. J. Bus. Ethics 117 (2), 371–386.
Ye, M., Zyren, J., Blumberg, C.J., Shore, J., 2009. A short-run crude oil price forecast model with Ratchet effect. Atl. Econ. J. 37, 37–50.

Young, D.M., Cabezas, H., 1999. Designing sustainable processes with simulation: the waste reduction (WAR) algorithm. Comput. Chem. Eng. 23, 1477–1491.

Yue, D., You, F., 2013. Sustainable scheduling of batch processes under economic and environmental criteria with MINLP models and algorithms. Comput. Chem. Eng. 54, 44–59.

INTERNET LINKS

Business Insider, 2014. http://www.businessinsider.com/ last accessed 20.11.14.

Bloomberg News, 2012. http://www.bloomberg.com/news/2012-03-29/apple-says-cook-visited-new-foxconn-plant-in-zhengzhou-china.html last accessed 20.11.14.

GME, Gestore Mercati Energetici, 2014. http://www.mercatoelettrico.org/En/default.aspx last accessed 20.11.14.

PSE-Lab website, 2014. http://pselab.chem.polimi.it/ last accessed 20.11.14.

World Summit on Sustainable Development, 2002. http://www.earthsummit2002.org/.

Additional Material

Additional material on this book chapter is available at the following link: http://pselab.chem.polimi.it/book-chapter-on-sustainability.

Index

Note: Page numbers followed by "f" and "t" indicate figures and tables respectively.

C

H

I

J

K

L

P

Q

R

X

Z